## S-Parameters for Signal Integrity

Master the usage of s-parameters in signal integrity applications and gain full understanding of your simulation and measurement environment with this rigorous and practical guide. Solve specific signal integrity problems, including calculation of the s-parameters of a network, linear simulation of circuits, de-embedding, and virtual probing, all with expert guidance. Learn about the interconnectedness of s-parameters, frequency responses, filters, and waveforms. This invaluable resource for signal integrity engineers is supplemented with the open-source software *SignalIntegrity*, a Python package for scripting solutions to signal integrity problems.

**Peter J. Pupalaikis** is an electrical engineer and inventor who works for Teledyne LeCroy. He is an IEEE Fellow.

"The most modern and up-to-date book on linear network theory with applications. Deep and comprehensive theory is coupled with detailed applications, making this book a must-have not only for signal integrity professionals, but for any microwave engineer."

**Andrea Ferrero, Keysight**

"This book provides unique and consistent description of s-parameters' use for analysis of linear networks, and signal measurement and processing in one volume, supplemented and illustrated with free open-source signal integrity software. The book can be used for learning the subject of emerging microwave signal integrity or as a comprehensive and indispensable reference for every microwave and signal integrity engineer and scientist."

**Yuriy Shlepnev, Simberian Inc.**

"This is an outstanding and refreshing book for the novice and advanced engineer alike. Written by a well-known expert in the field, it provides a rather unique access to the difficult topic of signal integrity, through a systematic learning-by-doing approach. Software which is freely accessible through an open-source Python library, *SignalIntegrity,* allows the user to easily program the numerous examples that accompany the theory. The material ranges from simple to complex problems, using the s-parameter concept for high-speed signal integrity as a unifying theme. The book is appropriate for self-study and as a reference for teaching, and empowers the reader with a very unusual and stimulating blend of competences."

**Peter Wittwer, University of Geneva**

# S-Parameters for Signal Integrity

PETER J. PUPALAIKIS
Teledyne LeCroy, Inc.

CAMBRIDGE
UNIVERSITY PRESS

**CAMBRIDGE**
**UNIVERSITY PRESS**

University Printing House, Cambridge CB2 8BS, United Kingdom

One Liberty Plaza, 20th Floor, New York, NY 10006, USA

477 Williamstown Road, Port Melbourne, VIC 3207, Australia

314–321, 3rd Floor, Plot 3, Splendor Forum, Jasola District Centre, New Delhi – 110025, India

79 Anson Road, #06–04/06, Singapore 079906

Cambridge University Press is part of the University of Cambridge.

It furthers the University's mission by disseminating knowledge in the pursuit of education, learning, and research at the highest international levels of excellence.

www.cambridge.org
Information on this title: www.cambridge.org/9781108489966
DOI: 10.1017/9781108784863



First published 2020

Printed in the United Kingdom by TJ International Ltd, Padstow Cornwall

*A catalogue record for this publication is available from the British Library.*

ISBN 978-1-108-48996-6 Hardback

*This book is dedicated to my father, my favorite engineer, whose passing separated my life into the two parts that are with him and without him...*

# Contents

# Preface

THIS BOOK is the culmination of about ten years of writing that began while I was developing software for signal integrity analysis in oscilloscopes, which my company, Teledyne LeCroy, ended up branding as *Eye Doctor*. During the development of this software, I found many recurring patterns in the solutions of s-parameter systems that allowed me to do a lot of things with a slight change of the recipe. And in the development of the SPARQ and the WavePulser, two time-domain reflectometry instruments used to measure s-parameters, I learned a lot about s-parameters and calibrations.

Why s-parameters and signal integrity? Around the turn of the millennium, the field of signal integrity changed dramatically. I'm not even sure the field had an actual name at that time. As a check, I looked at two books written by a foremost authority on signal integrity and one of my favorite authors, Howard Johnson. His first book [1] published in 1993 has no index entry for signal integrity, while his second book [2], published in 2003, has extensive entries. As signal integrity became more about signal propagation (in the title of Dr. Johnson's second book), the field tended increasingly towards electromagnetic properties and today heavily overlaps with the field of microwaves and radio frequency (RF); with this has come an increasing use of scattering parameters, or s-parameters, and the vector network analyzer (VNA), which is the primary instrument for measuring them.

The fields of s-parameters and microwaves and RF are primarily about the frequency domain, while the business of signal integrity is the time domain. Despite the wide use of frequency-domain electromagnetic analysis, it's in the end mostly about whether a bit that is transmitted as a one or zero gets to the receiver and is interpreted properly.

S-parameters can be a confusing topic in many ways, including how they are measured, what they represent, how they connect to circuit theory, how they are manipulated either to integrate or remove them from a measurement, and especially how they interact with waveforms in the time domain. This book is mostly about these things.

My feeling is that it is always good to understand better what is going on, even if one is one is relying on software simulation tools or measurement instruments. Both of these require their own level of understanding and expertise. Any experienced engineer has been burned by the simulator or measurement instrument, either by doing something incorrectly, or because of the limitations of these tools, which is why we constantly try to correlate the two.

My expertise is mostly in measurement instruments, which are in themselves often quite intricate and expensive, and have an industry of their own. Software and simulation, meanwhile, have grown into a giant industry, and the advances in the technology, especially electromagnetic field solvers, have greatly increased the ability of signal integrity engineers. In some ways, however, they have weakened our minds, especially when overly relied on, and certainly have emptied our wallets. My original goal in the writing of this book was to teach methods that can be employed to solve relatively simple problems. These simple problems include aggregating systems of s-parameter blocks, de-embedding s-parameter blocks, performing linear simulations, and performing time-domain de-embedding and embedding of waveforms measured in systems. Expensive software is often used to perform even these common and not too difficult tasks. But, while writing out the math and explanations of all this, I was struck by a disappointing thought: No one is actually going to use this. I found that even these relatively simple solutions involved huge amounts of math with giant equations. That's when the idea for the *SignalIntegrity* software was born.

The *SignalIntegrity* software is an open-source library that allows you to do everything taught in this book, and it's free. I used this software to test what I had written, and, eventually, a lot of the book became about both the math used to generate the software and the software itself. Finally, somewhat close to the end of this project, I decided to wrap this up in a graphical user interface application which makes it much easier to solve these problems.

I'm a big fan of open-source software, and my desire is to collaborate with others in extending this software and to continue to advance its capability.

To summarize, this book is about:

1. S-parameters – it unifies the analysis of circuits and networks using s-parameters and methods conducive to s-parameters.
2. Four specific signal integrity problems and their solutions, these being the calculation of the s-parameters of a network, the linear simulation of circuits, de-embedding, and virtual probing.
3. A set of software tools called *SignalIntegrity*, a Python package that contains a library of modules used for scripting solutions to various problems in signal integrity; it also contains a graphical user interface (GUI) based application for schematic entry and solution.
4. The interconnectedness of s-parameters, frequency responses, filters, and waveforms, whose underlying theory enables one to deal with many signal integrity problems, especially where the time domain meets the frequency domain.

It is probably equally about these four things.

A colleague of mine who saw my original draft said to me, "This is not a signal integrity book." I was shocked by this. But I think what he meant is, "It's not the standard signal integrity book." It contains none of the things normally found, like board stack-ups, vias, electromagnetic theory, physics, or anything like that. Mostly it is about abstractions and mathematical manipulations of s-parameters and waveforms. It is also about how s-parameters connect with circuit theory.

And despite not being the typical signal integrity book, as another colleague told me, "It's your book." I hope you like it and that you learn something useful.

I would like to take this opportunity to thank and acknowledge several people. I'll start with people who did actual work on this.

First, and foremost, I'd like to thank Tom Reslewic, former CEO of Teledyne LeCroy. The contents of this book are intertwined so much with my company's work and intellectual property that it never could have been published without Tom going to bat for me. Despite the fact that Tom is foremost a businessman and I am foremost an innovator, we've shared a relationship of trust through the years. I cannot understate the value of having a friend, colleague, and, now that I think of it, leader like Tom supporting my efforts.

Next, I'd like to thank my friend and colleague Dr. Kaviyesh Doshi. There was a time at LeCroy when I was *the* mathematics expert. That time ended with his arrival at the company, and we have worked so closely over the years that much of his knowledge was imparted to me. This book would be much less interesting without the things that he taught me, namely in the area of linear algebra, and Kavi contributed significantly to many of the algorithms and solutions presented in this text. Kavi read and checked the math presented here, but, as the saying goes, all errors are my own.

Finally, I'd like to thank my boss for many years, Dave Graef, former CTO of Teledyne LeCroy. Dave negotiated the contract with Cambridge University Press. For some reason, this was a long and drawn out process that I thought would never end. He promised me that the contract would be signed before he retired and he kept his word; it was his last act as CTO prior to leaving the company.

Several people were generous with their time and attention in looking over the first manuscript. Their generosity is a testament to the kinds of people I find myself surrounded with in this field.

Dr. Istvan Novak, a colleague and friend, gave me lots of helpful advice and, most importantly, encouragement to finish this truly enormous task.

My college professor who I've now known for over thirty years, Dr. Peter Wittwer, also gave lots of encouragement and advice on the organization of this book. The multi-port s-parameter calculation came from collaboration with him originally.

During my writing, I reached out to Dr. Gilbert Strang, who is the world's top authority on linear algebra. I asked him some philosophical questions about writing and teaching in an e-mail; he responded in five minutes and, after sending him a chapter that he immediately read and reviewed (and corrected), he allayed my fear of presenting all this math. And his last e-mail comment was "...and back to writing my [his own] book." Thanks a lot Dr. Strang.

Dr. Yuriy Shlepnev provided important feedback on certain areas such as normalization and local port referencing; feedback that I mostly followed. And, despite critical comments, he also gave a lot of encouragement.

Dr. Eric Bogatin, himself a prolific author, supplied the "It's your book" encouragement.

I'd like to thank Julie Lancashire, executive editor at Cambridge University Press, for providing the opportunity to offer this book through such a prestigious publisher, and especially Irene Pizzie, who proofread this book and corrected my numerous grammatical errors.

I'm grateful to everyone involved.

# Abbreviations

| | |
|---|---|
| **AC** | alternating current |
| **ADC** | analog-to-digital converter |
| **AMS** | American Mathematical Society |
| **API** | application programming interface |
| **BERT** | bit error rate tester |
| **CZT** | chirp z-transform |
| **DC** | direct current |
| **DFT** | discrete Fourier transform |
| **DSO** | digital storage oscilloscope |
| **DSP** | digital signal processing |
| **DUT** | device under test |
| **ESD** | electrostatic discharge |
| **ESR** | effective series resistance |
| **FET** | field-effect transistor |
| **FFE** | feed-forward equalizer |
| **FIR** | finite impulse response |
| **GPL** | general public license |
| **GUI** | graphical user interface |
| **IDFT** | inverse discrete Fourier transform |
| **ISI** | inter-symbol interference |
| **KCL** | Kirchhoff's current law |
| **KVL** | Kirchhoff's voltage law |
| **LMSE** | least mean-squared error |
| **NRZ** | non-return-to-zero |
| **PAM** | pulse amplitude modulation |
| **PCB** | printed circuit board |
| **PRBS** | pseudo-random bit sequence |
| **PyPI** | Python Packaging Index |
| **RF** | radio frequency |
| **RLGC** | resistance, inductance, conductance, capacitance |
| **rms** | root mean squared |
| **SNR** | signal-to-noise ratio |
| **SOLT** | short-open-load-thru |
| **SOLR** | short-open-load-reciprocal |
| **SVD** | singular value decomposition |
| **TDR** | time-domain reflectometer |
| **UI** | unit interval |
| **UML** | universal modeling language |
| **VNA** | vector network analyzer |

# Introduction

SIGNAL INTEGRITY is a field of electrical engineering involving the transfer of information from one place to another. The goal is to transfer information efficiently, free of errors, and at high speed. The accomplishment of this goal involves the analysis, measurement, and simulation of waveforms at a transmitter and receiver and the behavior of the channel through which the waveforms pass.

The channel involves interconnected devices like connectors, cables, and printed circuit board (PCB) traces, and the analysis of these devices involves primarily their electromagnetic properties. Most books on signal integrity include teaching on these electromagnetic properties and how the device geometry, materials, and construction affect these properties in order to provide the engineer with useful information such that their designs can be controlled. This book is not about that.

Two integral topics in signal integrity design are measurement and simulation, and the engineer is constantly trying to reconcile their knowledge about how to affect the system, the computer simulation of the system, and the measurements made of the system. The measurements and simulations in signal integrity involve waveforms (either transmitted or received) and the behavior of the channel. Measurements of waveforms are generally made with digital oscilloscopes, and measurements of the behavior of the channel are usually made with vector network analyzers or time-domain reflectometers. The concept of waveforms is not difficult, although their sampled nature might cause some complexity.

The electromagnetic behavior of the channel is generally abstracted using network parameters, called scattering parameters, or s-parameters. S-parameters are the primary output of simulation tools and measurement instruments that simulate and measure the behavior of the channel. The concept of the s-parameters comes from the microwave community and was originally used to describe the behavior of antennas, filters, amplifiers, and other microwave devices. S-parameters are a sampled frequency-domain construct just as waveforms from oscilloscopes are sampled time-domain constructs.

In the microwave community, most often the frequency-domain aspect of a microwave device is used in simulations with other frequency-domain measurements of the signals that interact with them, as measured by a spectrum analyzer, for example. The interactions between frequency-domain signals and microwave devices in the frequency domain are well understood by the microwave engineer, and, often, the narrow-band nature of many designs makes the analysis and measurement relatively straightforward. In signal integrity, the interaction between time-domain waveforms and frequency-domain devices characterized by their s-parameters adds an element of difficulty.

The goals of this book are as follows:

- To address comprehensively the abstract concept of s-parameters and the theory behind them, from both a mathematical and circuit theory standpoint.

- To provide methods and software tools for manipulating s-parameters in ways that are commonly required by engineers.
- To provide solutions for several broad topics encountered involving s-parameters, such as simulation and de-embedding in both the time and frequency domain.
- To provide the theory and mechanics of how time-domain waveforms interact with frequency-domain s-parameters.
- To provide the mathematical concepts of how s-parameters are measured, both by frequency-domain and time-domain instruments and how they can be viewed in both domains.

All of the theory is accompanied by software that is part of an open-source software project, called *SignalIntegrity*. The goal of the software project is to provide free software tools for solving signal integrity problems, and this book describes the construction of this software that would be useful to a project contributor, and connects the mostly mathematical theory with the software implementation that is useful for users of the software.

## Organization

This book is organized in four distinct parts:

- Part I, Scattering Parameters – the mathematical and circuit theory of s-parameters and s-parameter systems.
- Part II, Applications – the theory and programmatic techniques for solving four specific classes of problems: s-parameter determination, de-embedding, virtual probing, and linear simulation.
- Part III, Signal Processing and Measurement – the theory and software for signal processing in signal integrity, notably, how waveforms and filters generated in linear simulations from s-parameters interact. This part also covers the analytical topics of measurement, model extraction, and the impedance profile, an important signal integrity tool that is the time-domain view of return loss.
- Part IV, *SignalIntegrity* – the organization and usage of the *SignalIntegrity* Python software project.

At the end of the book, the following appendices are provided:

- Appendix A, Terminology and Conventions – the mathematics nomenclature used in the book.
- Appendix B, Telegrapher's Equations – some derivations of the telegrapher's equations, used for the derivation of the transmission line in Chapter 7.
- Appendix C - Matrix Algebra – some details of the matrix algebra and nonlinear fitting methods used in this book.
- Appendix D, Symbolic Device Solutions – some details of larger symbolic device solutions for the devices derived in Chapter 6. These provide some insight into how a user might generate symbolic solutions themselves.

Readers and teachers should view Part I and Part III as the theoretical meat of this book, along with the mathematical descriptions of the solutions in Part II.

For a practitioner, perhaps Part II is the most useful for scripted solutions, with Part I forming the theoretical backdrop, especially Chapter 4, S-Parameter System Models, with the remaining chapters used as a reference to be read when required. Certainly, all of the concepts in Part II present useful concepts in the field.

And, of course, software developers are welcome to attempt to alter or even contribute to the development. In this case, Part IV would be where to start, with Part II providing software details of the solutions provided.

# Part I

## Scattering Parameters

### Introduction

SCATTERING PARAMETERS, or s-parameters as they are called, are the primary format for characterizing interconnects in signal integrity. This first part is all about s-parameters.

The first chapter of this part, Chapter 1, begins by explaining network parameters in general, using only voltages and currents. This is a foothold from whence to discuss s-parameters.

In Chapter 2, the critical transformation from voltages and currents to waves is discussed, which is the basis for s-parameters.

With the concept of waves covered, scattering parameters are presented in Chapter 3, along with the mathematical concepts required for using them as a network parameter.

Chapter 4 covers the solution of interconnected networks of devices.

While waves are the basis for s-parameters, they rely on the choice of reference impedance and normalization factor. These are rather confusing, and the details required to understand this topic are presented in Chapter 5.

In order to use s-parameters in circuit simulations, Chapter 6 develops the s-parameter definitions of all possible source and sense elements, which leads further to the development of definitions for dependent sources and amplifiers. These devices enable the use of s-parameters even in circuits where they are not traditionally used.

Finally, in Chapter 7, the topic at the center of signal integrity, transmission lines, is covered. The discussion of transmission lines presented does not cover the physical attributes, but is theoretical and attempts to connect transmission lines with circuit theory.

# 1

# Network Parameter Models

NETWORK PARAMETERS are often taught along with general circuit theory. Whether previously exposed to this topic or not, an electrical engineer would not find the concept of network parameters difficult to understand, in principle. A chapter on network parameters forms a good introduction and bridge between circuit theory and scattering parameters, also a network parameter and the major topic of this book.

Even if the reader is already familiar with network parameters, this chapter provides some necessary distinctions between the style generally taught and a different definition used in this text. What is referred to here is the so-called classic two-port network parameter model used to provide simple methods of circuit combinations. Here, the classic two-port model will be introduced only to distinguish it from an absolute network parameter model that is more closely analogous to s-parameters, which are covered in Chapter 3.

The terminology used here is to distinguish network parameter models from network parameters. This word usage is intended specifically to separate two concepts frequently attributed to network parameters. For now, understand that network parameters are generally lists of numbers at specific frequencies and network parameter models are continuous, analytic, and valid for all frequencies. This is explained in more detail when it comes up, but sometimes the generic term network parameters is used to describe both situations.

This chapter is organized first to inform the reader what, mathematically, network parameter models are and second how they are different from the basic circuit elements with which electrical engineers are most familiar. The calculation of network parameters of simple elements and circuits is explained and their usage in some basic circuit simulations is provided. After some basic calculations and usages of network parameter models for simulation are explained, methods of conversion between different types of network parameters are covered. It is not meant to be a treatise on network parameters, only a preparation for the topic of s-parameters. Therefore, the concepts of combining network devices and simulation with network devices are lightly touched upon, as the remainder of this book will analyze networks with s-parameters only, whose analysis is quite different.

## 1.1 The Concept of Network Parameter Models

This section introduces the concept of network parameters, starting with how they are related to circuits and circuit elements. Then, it will be seen that not all network parameters are the same, necessitating an explanation of the required features of the network parameters, especially with regard to s-parameters, which are introduced later.

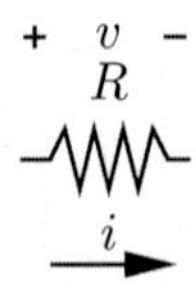


$$v = i \cdot R$$

**Figure 1.1** A resistor

### 1.1.1 Circuits and Circuit Elements

The explanation of network parameter models is best begun using circuit terms. A simple circuit element is a series resistance, as shown in Figure 1.1. A resistor has two *ports*, the locations used to connect it into a circuit. It is labeled with a given resistance $R$. Circuit theory provides everything needed to describe the behavior of this element identified by the resistor symbol and given its resistance value. Specifically, Ohm's law states that the voltage *across* it, $v$, is related to the current *through* it, $i$, as $v = i \cdot R$. A sign convention is used such that the voltage across the element is specified by labeling each terminal either positive or negative, with the voltage across the element specified as the difference between the voltage at the plus terminal and the voltage at the negative terminal. When Ohm's law is applied, the current flows from the positive terminal to the negative terminal. There is a presumption that the positive terminal is higher in potential than the negative terminal and that if it is not, the voltage across the element is negative. Furthermore, there is the presumption that the direction of current flow is from the positively labeled terminal to the negative terminal. If the current flow is in the opposite direction, then the current through the element is negative. In the case of a resistor, the current through the element is related to the voltage across it in a constant manner under all conditions. In circuit theory the current into one port is always equal to the current out of the other port. If the element were a reactive element, such as an inductor or capacitor, the relationship between current and voltage is dependent not only on the values of inductance or capacitance, but also on the currents and voltages at other times. In reactive elements, this relationship is either the time derivative in the time domain or the frequency in the frequency domain. Either way, this time and frequency dependence is encapsulated in a single number of either capacitance or inductance.

A circuit element is the basic building block of a *topological* model, meaning that the behavior of a circuit is dictated by the circuit elements themselves and the interconnections between them. The choice of type and interconnection of circuit elements is referred to as the circuit topology and, when combined with the circuit values, it defines a circuit model. Model, in this context, generally excludes any forcing functions like independent current or voltage sources. These sources of stimuli are variable and do not determine how the circuit model behaves. Instead, the stimuli are combined with a circuit model to form a circuit, and the measurement of the behavior of the circuit with a given set of stimuli is considered to be a simulation.

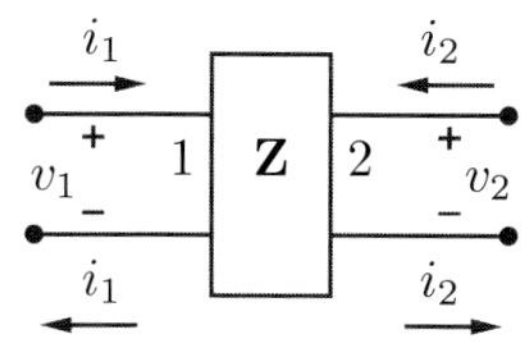


**Figure 1.2** A classic two-port Z-parameter network device

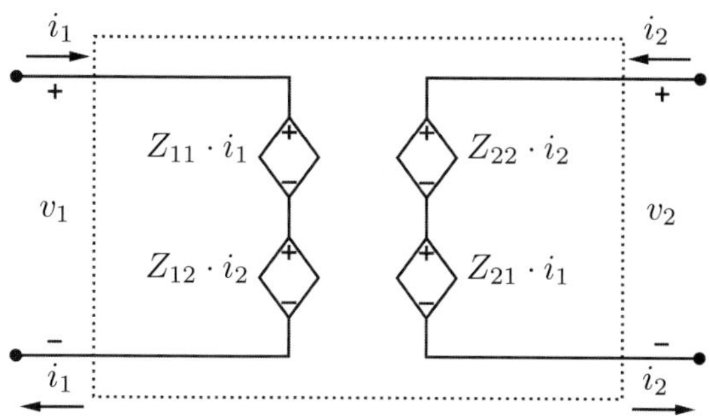


**Figure 1.3** Equivalent circuit for a classic two-port Z-parameter network device

### 1.1.2 Classic Two-Port Network Devices

The idea behind network analysis is that a circuit element or elements configured in a circuit (which is a topological model) can be converted into a *behavioral* model. Behavioral models hide all of the internal workings of a circuit by mimicking the behavior of a circuit at the terminals of the model. A behavioral model has pros and cons specifically as a result of the hiding of internal behavior. The pro is that complicated circuits can be reduced to terminal behavior, thus requiring the complexity of the circuit to be dealt with one time: during the model creation. The con is that the inner workings and topology, and therefore control over the circuit behavior, is lost. There are many problems in circuit and systems design where the behavior of the system is measured and the goal is to convert the behavioral model or set of measurements back into the topological model. This is generally a tough problem, as seen in Chapter 16 – Model Extraction.

In classic two-port network analysis, the standard network element is a two-port network device, as shown in Figure 1.2. Shown in Figure 1.2 is a Z-parameter device that operates by measuring its terminal currents and converting these into terminal voltages internally, as shown in Figure 1.3.[1] For this two-port network there is a matrix of Z-parameters:

$$\mathbf{Z} = \begin{pmatrix} Z_{11} & Z_{12} \\ Z_{21} & Z_{22} \end{pmatrix},$$

[1] Often, the dependent voltage sources that depend on the port current are simply shown as an impedance element (i.e. the voltage source that outputs the voltage $Z_{11} \cdot i_1$ is just an impedance of $Z_{11}$).

where the following relationship is implied:

$$\begin{pmatrix} v_1 \\ v_2 \end{pmatrix} = \begin{pmatrix} Z_{11} & Z_{12} \\ Z_{21} & Z_{22} \end{pmatrix} \cdot \begin{pmatrix} i_1 \\ i_2 \end{pmatrix}.$$

Some things to point out about classic two-port network devices:

1. There are a pair of terminals per port (not a single terminal, as one might expect).
2. There are constraints on the pairs of terminals in that the current leaving the bottom of a pair must equal the current entering the top of the pair. This constraint is called the *port condition.*
3. The voltage at a port is measured across a pair of terminals.

There are many limitations imposed on these networks mostly for the purpose of making generalizations when interconnecting network devices. The limitations that are germane in the context of this text are:

1. The device cannot sink or source current and cannot be connected internally to a voltage source or even to ground.
2. They are generally limited to two ports.
3. Care must be taken when interconnecting these devices with other such devices.

It is preferable to have a device that is not limited to two ports, can form absolute voltage and currents, and can be interconnected in any manner (with the following of some general rules). Furthermore, for the purpose of this book, it is preferable to have a device that maps to the s-parameter device.

### 1.1.3 Absolute Network Devices

The classic two-port network device discussed in the preceding section is not covered in the book and it needs to be distinguished from the style of device that is used in the remainder of this text. There is countless information available in other texts and on the web regarding classic two-port analysis, and the reader will need to be careful in applying that information to the network device used here.

This different style of device is referred to here as an *absolute* network parameter device, for reasons that shall become apparent.

An example absolute network device is shown in Figure 1.4, where Figure 1.4(a) is a box with two ports labeled 1 and 2 and a value of $\mathbf{Z}$ in the box. The device is a Z-parameter device and has Z-parameters of $\mathbf{Z}$, which is in bold because it is a matrix. For network parameters, this matrix is square with a size equal to the number of ports. It defines the relationship between some port variables and others. With a given number of ports $P$, there are $P$ independent port variables and $P$ dependent port variables, and the matrix multiplied by the $P$ independent variables equals the dependent variables. The port variables are listed as $i_1$, $i_2$, $v_1$, and $v_2$, representing the four possible port variables for a two-port network device. Note the convention that port currents always point into the device. Z-parameters are a type of parameter that relates the port currents (the independent variables) to the port voltages (the dependent variables) in the relationship $\mathbf{v} = \mathbf{Z} \cdot \mathbf{i}$, where $\mathbf{v}$ is a list of

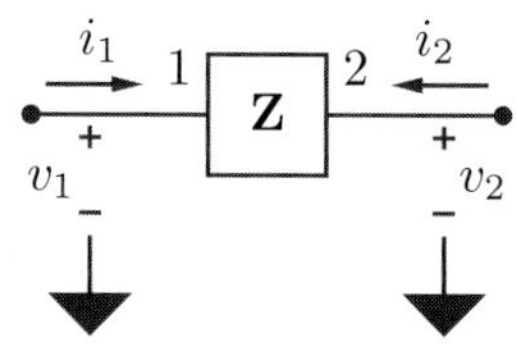


(a) Simple representation

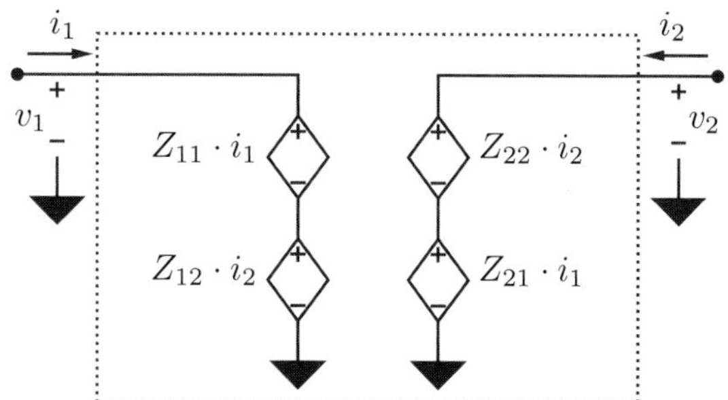


(b) Circuit representation

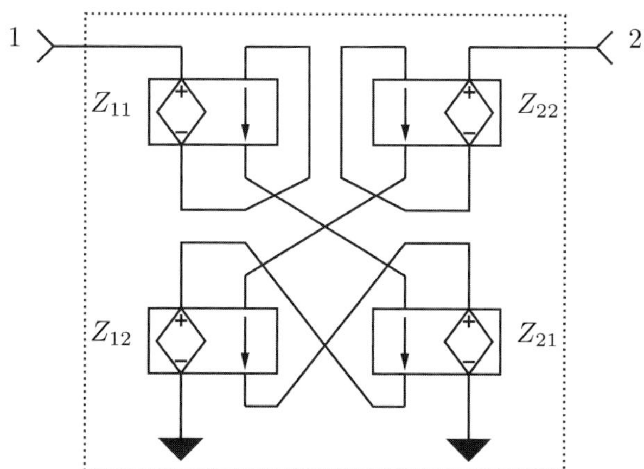


(c) Circuit representation with dependent sources

**Figure 1.4** A Z-parameter network device

port voltages listed in order of port number and **i** is a list of port currents listed in order of port number. Given network parameters **Z**, the convention for a $P$-port device is

$$\mathbf{Z} = \begin{pmatrix} Z_{11} & Z_{12} & \cdots & Z_{1P} \\ Z_{21} & Z_{22} & \cdots & Z_{2P} \\ \vdots & \vdots & \ddots & \vdots \\ Z_{P1} & Z_{P2} & \cdots & Z_{PP} \end{pmatrix},$$

where the following relationship is implied:

$$\begin{pmatrix} v_1 \\ v_2 \\ \vdots \\ v_P \end{pmatrix} = \begin{pmatrix} Z_{11} & Z_{12} & \cdots & Z_{1P} \\ Z_{21} & Z_{22} & \cdots & Z_{2P} \\ \vdots & \vdots & \ddots & \vdots \\ Z_{P1} & Z_{P2} & \cdots & Z_{PP} \end{pmatrix} \cdot \begin{pmatrix} i_1 \\ i_2 \\ \vdots \\ i_P \end{pmatrix}. \tag{1.1}$$

This voltage and current relationship is similar to the classic network parameter model, differing only in what the voltages and currents actually represent. Defined in circuit terms, the implication of Figure 1.4(a) is a circuit model, as shown in Figure 1.4(b). A self-contained model using dependent sources is provided in Figure 1.4(c). From (1.1) and Figure 1.4(b) one can see that a Z-parameter is defined such that

$$Z_{xy} = \left.\frac{v_x}{i_y}\right|_{\text{all other i=0}}$$

The differences between the absolute network device and the classic two-port network device are:

1. The absolute network device has one terminal per port.
2. The absolute network device has its voltage specified as an absolute voltage. This voltage is the absolute voltage to ground, not the voltage across any terminals.
3. The absolute network device has its internal sources connected to ground and is also generating absolute voltage, not a voltage across any terminals.
4. The absolute network device has as many ports as needed.

The main limitation of absolute network parameters is that there are no generalizations to be made about interconnected devices and how the network parameters of interconnected devices combine. There are some benefits regarding this type of device, which is why they are used in this text:

1. They interconnect just like circuit elements, meaning that the restrictions on interconnection are the same as in a circuit. Any port of a device can be connected to any port of another device or multiple devices.
2. There is no limit to the number of ports.
3. They are similar in construction to the s-parameter network device, the main subject of this book.

Regarding the main limitation, this text develops general rules for evaluating interconnected networks of absolute network parameters that are programmatic in nature. In other words, recipes are provided for dealing with interconnected networks that, while imposing more difficulty for simple networks, enable easier analysis of larger networks, especially where a computer or computer program is involved.

It is important to see that in the classic two-port network case, if the terminals where current exits the classic device are tied to ground, the behavior of the two types of network parameters is identical. In classic analysis, this sometimes negates the whole purpose because the terminals are meant to be connected to other devices (albeit in specific ways) and this often changes the behavior of the circuit, but when the behavior is the same, the network parameters are identical and the analysis can be the same. When the terminals at which currents exit cannot be tied to ground, the use of absolute network parameters requires a higher port count device.

On a somewhat tangential note, the classic two-port network parameters can always be obtained from a multi-port absolute network parameter device by simply shorting the unexposed ports to ground. The reverse is never possible.

From now on in this text, reference to the terms *absolute* or *classic* two-port network parameters are no longer used and the term "network parameters" refers only to the absolute type.

There are four classes of network parameters that will appear in this text. These are: impedance parameters, which relate voltage to current; admittance parameters, which relate current to voltage; scattering parameters; and *hybrid* parameters, which relate mixtures of voltages and currents to others. There are five specific types of network parameters that will appear in this text. These are:

1. Z-parameters, which are impedance parameters.
2. Y-parameters, which are admittance parameters.
3. S-parameters, which are synonymous with scattering parameters.
4. ABCD parameters, which are specific hybrid voltage and current parameters defined only for cascading two-port devices.
5. T-parameters, which are transmission parameters used for cascading s-parameter devices.

Only network parameters utilizing voltage and current: Z-parameters, Y-parameters, and ABCD parameters are defined in this chapter. The Z-parameter has already been defined.

A Y-parameter device is shown in Figure 1.5(a). The circuit model is shown in Figure 1.5(b), with an equivalent model utilizing dependent sources provided in Figure 1.5(c). Y-parameters define the relationship of port currents to port voltages as $\mathbf{i} = \mathbf{Y} \cdot \mathbf{v}$. Given network parameters $\mathbf{Y}$, the convention for a device with $P$ ports is

$$\mathbf{Y} = \begin{pmatrix} Y_{11} & Y_{12} & \cdots & Y_{1P} \\ Y_{21} & Y_{22} & \cdots & Y_{2P} \\ \vdots & \vdots & \ddots & \vdots \\ Y_{P1} & Y_{P2} & \cdots & Y_{PP} \end{pmatrix}.$$

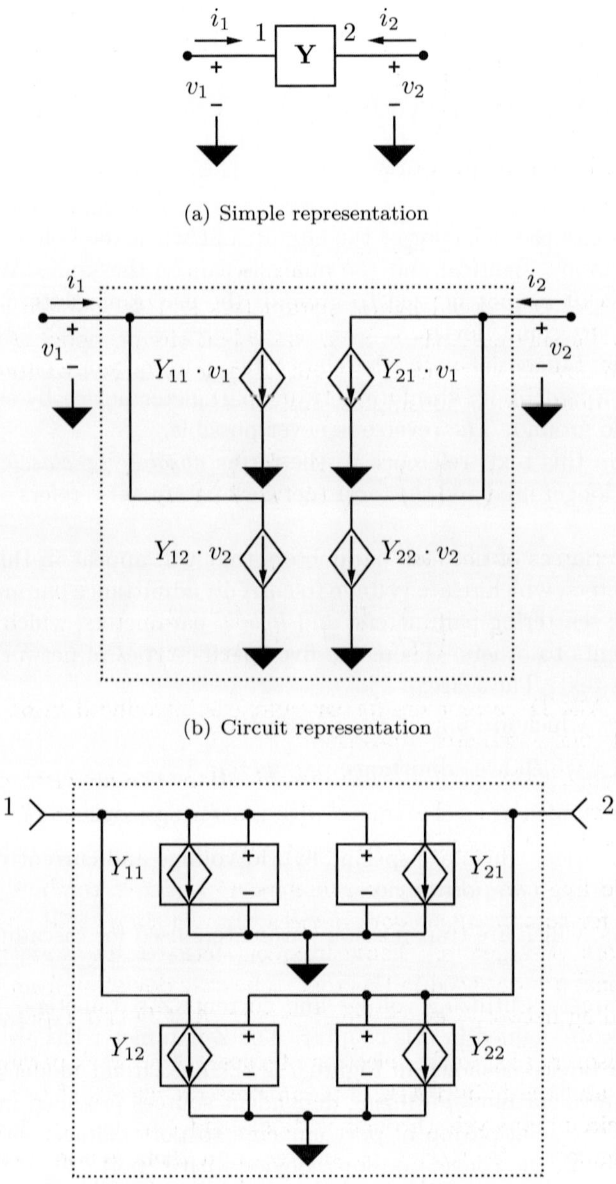


(a) Simple representation

(b) Circuit representation

(c) Circuit representation with dependent sources

**Figure 1.5** A Y-parameter network device

The relationship is implied as follows:

$$\begin{pmatrix} v_1 \\ v_2 \\ \vdots \\ v_P \end{pmatrix} = \begin{pmatrix} Y_{11} & Y_{12} & \cdots & Y_{1P} \\ Y_{21} & Y_{22} & \cdots & Y_{2P} \\ \vdots & \vdots & \ddots & \vdots \\ Y_{P1} & Y_{P2} & \cdots & Y_{PP} \end{pmatrix} \cdot \begin{pmatrix} i_1 \\ i_2 \\ \vdots \\ i_P \end{pmatrix}.$$

A Y-parameter is defined such that

$$Y_{xy} = \left.\frac{i_x}{v_y}\right|_{\text{all other } \mathbf{v}=0}.$$

ABCD parameters are special hybrid parameters and are only formally defined for a two-port network. Furthermore, there is no equivalent circuit model. After understanding why they exist and how they are used, one can imagine variations on the definition, but for now they formally define a relationship as [3, 4]

$$\begin{pmatrix} v_1 \\ -i_1 \end{pmatrix} = \begin{pmatrix} A & B \\ C & D \end{pmatrix} \cdot \begin{pmatrix} v_2 \\ i_2 \end{pmatrix}. \tag{1.2}$$

Here, the convention used is that

$$\mathbf{A} = \begin{pmatrix} A & B \\ C & D \end{pmatrix} = \begin{pmatrix} A_{11} & A_{12} \\ A_{21} & A_{22} \end{pmatrix}.$$

The usefulness of ABCD parameters for cascading is introduced in §1.5. The reason for using different network parameter types is:

1. For certain types of circuit elements or combination of circuit elements, certain types of network parameters cannot be defined. This especially applies to Z- and Y-parameters. S-parameters do not generally have this limitation.

2. When classic two-port network parameters are used to combine devices, there are simple rules for combining network devices through simple addition or multiplication of the network parameters. These interconnection rules (and their accompanying generalizations) are not used in this text. The only deviation from this statement will be for cascading network devices which rely on ABCD and T-parameters.

3. The final reason relates to the selection of classes of network parameters. Impedance, admittance, and the hybrid ABCD parameters are classes of parameters that relate directly to circuit analysis through voltages and current, which are familiar to all electrical engineers. Scattering parameters and their hybrid T-parameters are not directly related to circuit analysis (although they can be related) and are not generally familiar to engineers involved in circuits. These types of parameters are most often encountered in the analysis of microwave systems because they provide better insight into the workings of a key element, the transmission line, which is discussed in Chapter 7. In some sense, this book is a bridge between scattering parameters and their voltage–current equivalents.

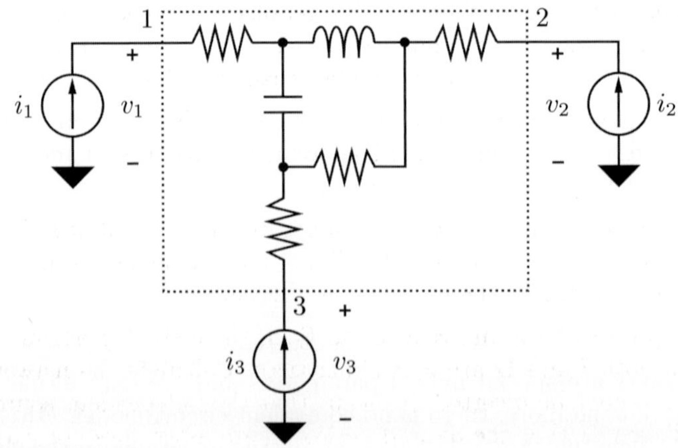


**Figure 1.6** An arbitrary circuit instrumented for Z-parameter measurement

## 1.2 Network Parameter Models of Circuit Elements and Circuit Models

It is useful to understand how to equate network parameter models to basic circuit elements. The methods for analyzing circuit elements apply to all circuit models (i.e. many interconnected circuit elements). The goal here is to reduce a circuit, such as the one shown in Figure 1.6, to a box with terminals, such as the one shown in Figure 1.4(a), with an assumed internal structure like those shown in Figure 1.4(b) or Figure 1.4(c). In other words, to find the network parameter equivalent behavioral model of a circuit or circuit element. As will be shown, the circuit in Figure 1.6 is already instrumented with external current sources for the determination of Z-parameters.

There are a few ways to go about this, but first a simple recipe is provided:

1. Identify the type of network parameter desired along with the independent and dependent variables associated with the network parameter. In the case of Z-parameters, with the relationship $v = Z \cdot i$, the independent variables are the currents and the dependent variables are the voltages. In the case of Y-parameters, with the relationship $i = Y \cdot v$, the independent variables are the voltages and the dependent variables are the currents.

2. Identify and label all ports along with the port voltages and currents at the ports of the new network device.

3. For each independent variable, *instrument* the circuit model or element with sources. For Z-parameters, current sources are applied to each terminal. For Y-parameters, voltage sources are applied to each terminal. Each source is labeled with the port number.

4. Alternately turn on each independent source (leaving all of the others set to zero) and calculate all of the port voltages and currents for each circumstance. Because of

the instrumentation, one of these will be completely dependent on the source applied. Label the measurements of the port voltages as $i_{xy}$ and $v_{xy}$, where $x$ refers to the port number and $y$ is the measurement number (usually, which port is on).

5. Arrange all the independent variables in a matrix (call it $\mathbf{I}$ for now), where an element $I_{xy}$ refers to element $x$ in the list of all independent variables for the network parameter and $y$ refers to the measurement number.
6. Similarly, arrange all the dependent variables in a matrix (call it $\mathbf{D}$ for now), where an element $D_{xy}$ refers to element $x$ in the list of all dependent variables for the network parameter and $y$ refers to the measurement number.
7. These steps produced two matrices $\mathbf{I}$ and $\mathbf{D}$. If there are $P$ ports in the final network device, then both $\mathbf{I}$ and $\mathbf{D}$ are $P \times P$ matrices. Calculate the network parameter as $\mathbf{D} \cdot \mathbf{I}^{-1}$. If $\mathbf{I}$ cannot be inverted, it means that the independent source values chosen are not sufficient or that the desired network parameter does not exist.

These steps are repeated, explicitly stating how this would be done to calculate the Z-parameters of a circuit element or model exposing two terminals:

1. For Z-parameters, there are two independent variables, the currents, which are $i_1$ and $i_2$, the currents into each port. The list, in port order, is $\left(\begin{smallmatrix} i_1 \\ i_2 \end{smallmatrix}\right)$.
2. Since both independent variables are currents, the network is instrumented with two current sources, one on each port. Each current source is labeled with the port number. Since the current source sets the current on a given port, the sources are labeled as $i_1$ and $i_2$.
3. Label all port voltages and currents at the ports of the new network device. These are $i_1$, $i_2$, $v_1$, and $v_2$.
4. Set current source $i_1 = 1$ and $i_2 = 0$ and record all of the port voltages and currents as $i_{11} = 1$, $i_{21} = 0$, $v_{11}$, and $v_{21}$. Then set current source $i_1 = 0$ and $i_2 = 1$ and record all of the port voltages and currents as $i_{12} = 0$, $i_{22} = 1$, $v_{12}$, and $v_{22}$.
5. Since the independent variables are $i_1$ and $i_2$, the independent variable matrix becomes $\mathbf{I} = \left(\begin{smallmatrix} i_{11} & i_{12} \\ i_{21} & i_{22} \end{smallmatrix}\right) = \left(\begin{smallmatrix} 1 & 0 \\ 0 & 1 \end{smallmatrix}\right) = \mathbb{I}$.
6. Since the dependent variables are $v_1$ and $v_2$, the dependent variable matrix becomes $\mathbf{D} = \left(\begin{smallmatrix} v_{11} & v_{12} \\ v_{21} & v_{22} \end{smallmatrix}\right)$.
7. Calculate the Z-parameters as $\mathbf{D} \cdot \mathbf{I}^{-1} = \left(\begin{smallmatrix} v_{11} & v_{12} \\ v_{21} & v_{22} \end{smallmatrix}\right) \cdot \left(\begin{smallmatrix} i_{11} & i_{12} \\ i_{21} & i_{22} \end{smallmatrix}\right)^{-1}$. The analysis is best performed such that $\mathbf{I} = \mathbb{I} = \left(\begin{smallmatrix} 1 & 0 \\ 0 & 1 \end{smallmatrix}\right)$ (though this is not a requirement) so that no matrix inverse is required and the Z-parameters are read directly from $\mathbf{D}$.

### 1.2.1 Network Parameter Model of a Series Resistor

Here, an attempt is made to calculate the Z-parameters of a simple resistor. The goal is to find the values of $\mathbf{Z}$ for a network parameter device such as that shown in Figure 1.4(a), with an internal structure such as that shown in Figure 1.4(b) or Figure 1.4(c), that mimics the behavior of the circuit element shown in Figure 1.7, which has been instrumented with current sources for Z-parameter measurement. Examining Figure 1.7, one can see immediately that there is a problem. It is impossible to control independently the two

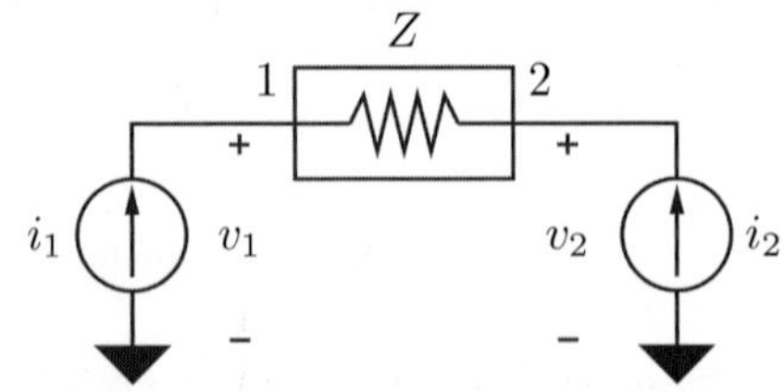


**Figure 1.7** A series impedance instrumented for Z-parameter measurement

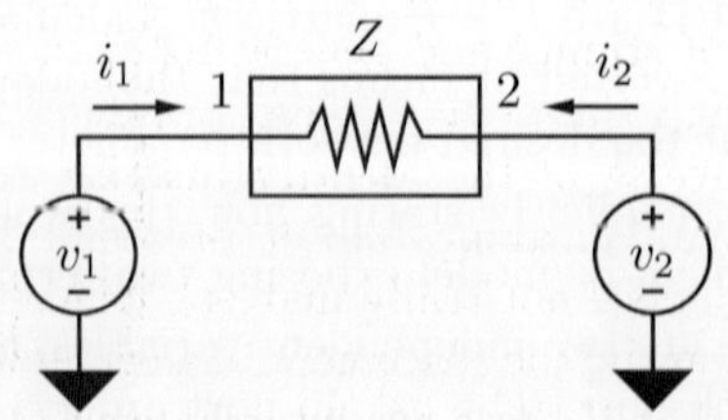


**Figure 1.8** A series impedance instrumented for Y-parameter measurement

current sources because setting either of them is expected to set the current through the resistor. Usually, when the system is instrumented according to the recipe (with the types of sources specified at the designated ports, and the sources cannot be driven independently), this is an indication that the desired network parameter does not exist. Therefore, an attempt is made to measure the Y-parameters for this element by instrumenting the resistor with voltage sources, as shown in Figure 1.8:

1. In calculating Y-parameters, the two independent variables are the voltages, which are $v_1$ and $v_2$, the voltages at each port. The list, in port order, is $\binom{v_1}{v_2}$.
2. Since both independent variables are voltages, the network is instrumented with two voltage sources, one on each port, as shown in Figure 1.8. Since the voltage source sets the voltage on a given port, the sources are labeled as $v_1$ and $v_2$.
3. All port voltages and currents are labeled at the ports of the new network device. These are $i_1$, $i_2$, $v_1$, and $v_2$.
4. The voltage sources are set such that $v_1 = 1$ and $v_2 = 0$, and all of the port voltages and currents are recorded as $i_{11} = 1/R$, $i_{21} = -1/R$, $v_{11} = 1$, and $v_{21} = 0$. Then, the voltage sources are set such that $v_1 = 0$ and $v_2 = 1$, and all of the port voltages and currents are recorded as $i_{12} = -1/R$, $i_{22} = 1/R$, $v_{12} = 0$, and $v_{22} = 1$.
5. Since the independent variables are $v_1$ and $v_2$, the independent variable matrix becomes $\mathbf{I} = \begin{pmatrix} v_{11} & v_{12} \\ v_{21} & v_{22} \end{pmatrix} = \begin{pmatrix} 1 & 0 \\ 0 & 1 \end{pmatrix} = \mathbb{I}$.
6. Since the dependent variables are $i_1$ and $i_2$, the dependent variable matrix becomes $\mathbf{D} = \begin{pmatrix} i_{11} & i_{12} \\ i_{21} & i_{22} \end{pmatrix} = \begin{pmatrix} 1/R & -1/R \\ -1/R & 1/R \end{pmatrix}$.

7. Finally, the Y-parameters are calculated as $\mathbf{D}\cdot\mathbf{I}^{-1} = \left(\begin{smallmatrix} i_{11} & i_{12} \\ i_{21} & i_{22}\end{smallmatrix}\right)\cdot\left(\begin{smallmatrix} v_{11} & v_{12} \\ v_{21} & v_{22}\end{smallmatrix}\right)^{-1} = \left(\begin{smallmatrix} 1/R & -1/R \\ -1/R & 1/R\end{smallmatrix}\right)\cdot\left(\begin{smallmatrix} 1 & 0 \\ 0 & 1\end{smallmatrix}\right)^{-1} = \left(\begin{smallmatrix} 1/R & -1/R \\ -1/R & 1/R\end{smallmatrix}\right)$:

$$\mathbf{Y}_{resistor} = \frac{1}{R}\cdot\begin{pmatrix} 1 & -1 \\ -1 & 1 \end{pmatrix}. \tag{1.3}$$

This example demonstrates the calculation of the Y-parameters of a resistor, although the original goal was to calculate the Z-parameters. Observe that the circuit model shown in Figure 1.5(b) is a completely acceptable replacement for a resistor in a network.

Regarding the Z-parameters of a resistor, there is something that needs mentioning. The $P$-port network parameters for circuit elements and circuit models can be calculated by supplying any of $P$ sets of stimuli and generating $P$ measurements provided that the matrix $\mathbf{I}$ is invertible, although the method recommended is the best in both ease of use (no need to calculate the inverse of the identity matrix) and because the matrix to be inverted is always orthonormal, meaning that it presents the best situation numerically (symbolically, orthonormality does not really matter). Since the difference between Z- and Y-parameters is, most importantly, the placement of the matrix containing the currents and the voltages, it can be seen that if these are interchanged (i.e. $\mathbf{D}$ and $\mathbf{I}$ are interchanged) and $\mathbf{D}\cdot\mathbf{I}^{-1}$ calculated, the other network parameter is obtained. This is the same as saying that $\mathbf{Z} = \mathbf{Y}^{-1}$. Since $\mathbf{Y}^{-1}_{resistor}$ does not exist, it is clear that the series resistor has no Z-parameters. As absurd as it sounds, a series impedance does not have impedance parameters!

To calculate the ABCD parameters, one can try to use the same measurements under the same measurement conditions as for the Y-parameters, recognizing that the matrix might not be invertible. To do this, the measurements are rearranged, resulting in the ABCD parameters of the series resistor:

$$\begin{aligned}\mathbf{A}_{resistor} &= \begin{pmatrix} v_{11} & v_{12} \\ -i_{11} & -i_{12} \end{pmatrix}\cdot\begin{pmatrix} v_{21} & v_{22} \\ i_{21} & i_{22} \end{pmatrix}^{-1} \\ &= \begin{pmatrix} 1 & 0 \\ -1/R & 1/R \end{pmatrix}\cdot\begin{pmatrix} 0 & 1 \\ -1/R & 1/R \end{pmatrix}^{-1} = \begin{pmatrix} 1 & -R \\ 0 & 1 \end{pmatrix}.\end{aligned} \tag{1.4}$$

### 1.2.2 Network Parameter Model of a Shunt Impedance

For Z-parameters of the element in Figure 1.9, the shunt impedance is driven with current sources and the current through the element is measured as the sum of each of the driving current sources, and therefore the voltage across the resistor to ground at $v_1 = v_2 = (i_1 + i_2)\cdot Z$. The Z-parameters are calculated as

$$\begin{aligned}\mathbf{Z}_{shunt} &= \begin{pmatrix} v_{11} & v_{12} \\ v_{21} & v_{22} \end{pmatrix}\cdot\begin{pmatrix} i_{11} & i_{12} \\ i_{21} & i_{22} \end{pmatrix}^{-1} \\ &= \begin{pmatrix} Z & Z \\ Z & Z \end{pmatrix}\cdot\begin{pmatrix} 1 & 0 \\ 0 & 1 \end{pmatrix}^{-1} = \begin{pmatrix} Z & Z \\ Z & Z \end{pmatrix}.\end{aligned} \tag{1.5}$$

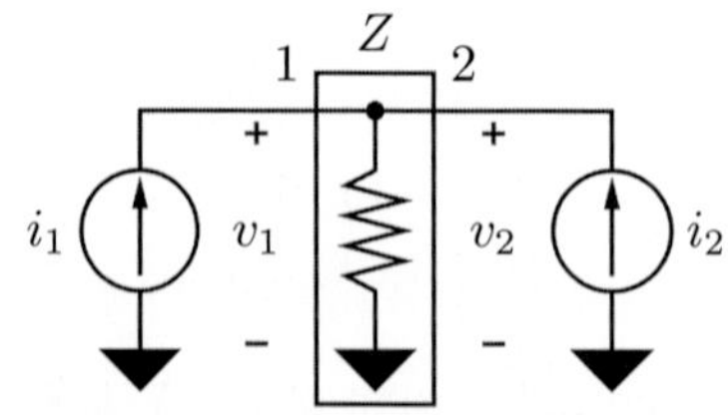


**Figure 1.9** A shunt impedance instrumented for Z-parameter measurement

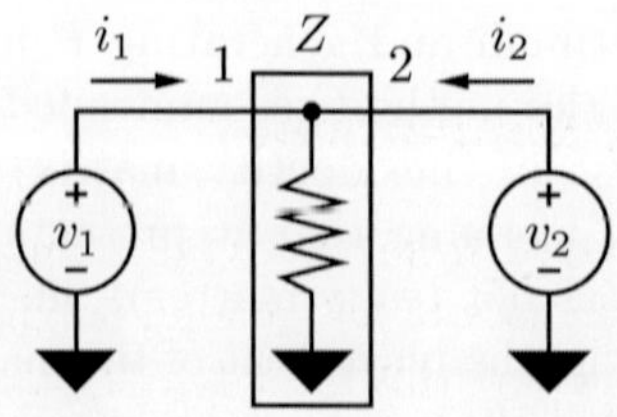


**Figure 1.10** A shunt impedance instrumented for Y-parameter measurement

The Y-parameters can be calculated using Figure 1.10, or as the inverse of the Z-parameters, and they don't exist.

The ABCD parameters are

$$\mathbf{A}_{shunt} = \begin{pmatrix} v_{11} & v_{12} \\ -i_{11} & -i_{12} \end{pmatrix} \cdot \begin{pmatrix} v_{21} & v_{22} \\ i_{21} & i_{22} \end{pmatrix}^{-1}$$
$$= \begin{pmatrix} Z & Z \\ -1 & 0 \end{pmatrix} \cdot \begin{pmatrix} Z & Z \\ 0 & 1 \end{pmatrix}^{-1} = \begin{pmatrix} 1 & 0 \\ -1/Z & 1 \end{pmatrix}. \quad (1.6)$$

### 1.2.3 Network Parameter Model of an Impedance to Ground

For the Z-parameters of the element in Figure 1.11, the impedance is driven with a current source that sets the current through the element as shown in Figure 1.11(a). The voltage across the resistor to ground is $v_1 = i_1 \cdot Z$. Therefore, the Z-parameters are calculated as

$$\mathbf{Z}_{impedance-ground} = v_1 \cdot i_1^{-1} = Z.$$

The Y-parameters can be calculated from the setup in Figure 1.11(b) or calculated as the inverse of the Z-parameters:

$$\mathbf{Y}_{impedance-ground} = i_1 \cdot v_1^{-1} = \frac{1}{Z}.$$

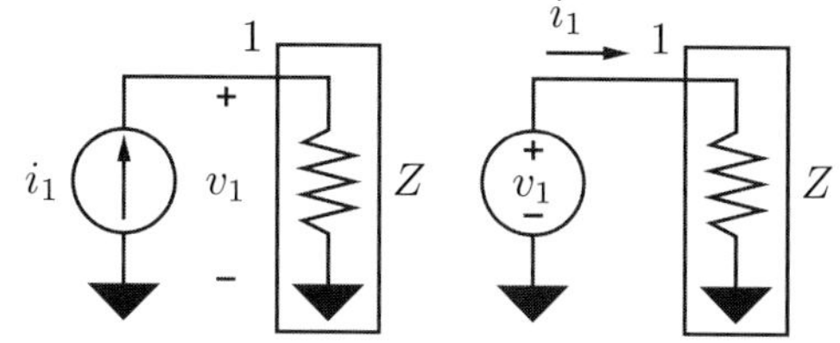


(a) Z-parameters (b) Y-parameters

**Figure 1.11** A terminating impedance instrumented for network parameter determination

## 1.3 Network Parameter Conversions

Network parameters can be converted from one type to another using the circuit methods already discussed for circuit elements and by general solving of linear systems if desired, but there are more elegant ways to to this. First, recognize that a given set of network parameters converts a set of independent variable measurements into a set of dependent variable measurements. If the network parameters are designated as $\mathbf{P}$, it is understood that $\mathbf{P}$ converts a set of independent variable measurements $\mathbf{I}$ into a set of dependent variable measurements $\mathbf{D}$, and that the network parameters are constructed such that $\mathbf{D} = \mathbf{P} \cdot \mathbf{I}$. If the independent variables are driven with an arbitrary set of stimuli $\mathbf{S}$, the system obeys the equation

$$\begin{pmatrix} \mathbb{I} & 0 \\ -\mathbf{P} & \mathbb{I} \end{pmatrix} \cdot \begin{pmatrix} \mathbf{i} \\ \mathbf{d} \end{pmatrix} = \begin{pmatrix} \mathbf{s} \\ \mathbf{0} \end{pmatrix}$$

or

$$\begin{pmatrix} \mathbb{I} & 0 \\ \mathbf{P} & \mathbb{I} \end{pmatrix} \cdot \begin{pmatrix} \mathbf{s} \\ \mathbf{0} \end{pmatrix} = \begin{pmatrix} \mathbf{i} \\ \mathbf{d} \end{pmatrix} = \begin{pmatrix} \mathbf{s} \\ \mathbf{P} \cdot s \end{pmatrix}.$$

To make a conversion from an initial network parameter $\mathbf{P}_i$ to another network parameter $\mathbf{P}_f$, there are two sets of equations:

$$\begin{pmatrix} \mathbb{I} & 0 \\ \mathbf{P}_i & \mathbb{I} \end{pmatrix} \cdot \begin{pmatrix} \mathbf{s}_i \\ \mathbf{0} \end{pmatrix} = \begin{pmatrix} \mathbf{i}_i \\ \mathbf{d}_i \end{pmatrix} = \begin{pmatrix} \mathbf{s}_i \\ \mathbf{P}_i \cdot \mathbf{s}_i \end{pmatrix}$$

and

$$\begin{pmatrix} \mathbb{I} & 0 \\ \mathbf{P}_f & \mathbb{I} \end{pmatrix} \cdot \begin{pmatrix} \mathbf{s}_f \\ \mathbf{0} \end{pmatrix} = \begin{pmatrix} \mathbf{i}_f \\ \mathbf{d}_f \end{pmatrix} = \begin{pmatrix} \mathbf{s}_f \\ \mathbf{P}_f \cdot \mathbf{s}_f \end{pmatrix}.$$

Since the goal is that $\mathbf{P}_i$ is behaviorally equivalent to $\mathbf{P}_f$, there exists, for some arbitrary $\begin{pmatrix} \mathbf{i}_i \\ \mathbf{d}_i \end{pmatrix}$, a conversion matrix $\mathbf{C}$ that can convert this $\begin{pmatrix} \mathbf{i}_i \\ \mathbf{d}_i \end{pmatrix}$ into $\begin{pmatrix} \mathbf{i}_f \\ \mathbf{d}_f \end{pmatrix}$ according to

$$\mathbf{C} \cdot \begin{pmatrix} \mathbf{i}_i \\ \mathbf{d}_i \end{pmatrix} = \begin{pmatrix} \mathbf{i}_f \\ \mathbf{d}_f \end{pmatrix}.$$

The truth of this statement depends on $\mathbf{S}_i$ and $\mathbf{S}_f$, and this dependency is decoupled by stating that the job of $\mathbf{C}$ is to convert to the right types of parameters and the right ordering of parameters, and that $\mathbf{s}_i$ and $\mathbf{s}_f$ are left unconstrained right now – simply by realizing that there are such variables that can be chosen such that the equation holds.

Working with voltages and currents, $\mathbf{C}$ serves only to *permute* or reorder $\begin{pmatrix} \mathbf{i}_i \\ \mathbf{d}_i \end{pmatrix}$, and since it multiplies from the left it can only rearrange the rows of $\begin{pmatrix} \mathbf{i}_i \\ \mathbf{d}_i \end{pmatrix}$ and thus the order of the current and voltage variables. $\mathbf{C}$ is therefore calculated specifically to convert from one variable ordering to another; this is done by hand, examining the final variable ordering and choosing rows from the initial variable ordering to accomplish the objective.

After this $\mathbf{C}$ is calculated, it is partitioned into four quadrants:

$$\mathbf{C} = \begin{pmatrix} \mathbf{C}_{11} & \mathbf{C}_{12} \\ \mathbf{C}_{21} & \mathbf{C}_{22} \end{pmatrix}.$$

For $P$-port network parameters, $\mathbf{C}$ is a $2 \cdot P \times 2 \cdot P$ matrix, and the block matrices $\mathbf{C}_{11}$, $\mathbf{C}_{12}$, $\mathbf{C}_{21}$, and $\mathbf{C}_{22}$ are $P \times P$.

Applying this $\mathbf{C}$ leads to the following identity:

$$\begin{aligned} \mathbf{C} \cdot \begin{pmatrix} \mathbb{I} & 0 \\ \mathbf{P}_i & \mathbb{I} \end{pmatrix} \cdot \begin{pmatrix} \mathbf{s}_i \\ \mathbf{0} \end{pmatrix} &= \mathbf{C} \cdot \begin{pmatrix} \mathbf{i}_i \\ \mathbf{d}_i \end{pmatrix} = \mathbf{C} \cdot \begin{pmatrix} \mathbf{s}_i \\ \mathbf{P}_i \cdot \mathbf{s}_i \end{pmatrix} \\ &= \begin{pmatrix} \mathbb{I} & 0 \\ \mathbf{P}_f & \mathbb{I} \end{pmatrix} \cdot \begin{pmatrix} \mathbf{s}_f \\ \mathbf{0} \end{pmatrix} = \begin{pmatrix} \mathbf{i}_f \\ \mathbf{d}_f \end{pmatrix} = \begin{pmatrix} \mathbf{s}_f \\ \mathbf{P}_f \cdot \mathbf{s}_f \end{pmatrix}, \end{aligned}$$

which means that

$$\mathbf{C} \cdot \begin{pmatrix} \mathbf{s}_i \\ \mathbf{P}_i \cdot \mathbf{s}_i \end{pmatrix} = \begin{pmatrix} \mathbf{C}_{11} & \mathbf{C}_{12} \\ \mathbf{C}_{21} & \mathbf{C}_{22} \end{pmatrix} \cdot \begin{pmatrix} \mathbf{s}_i \\ \mathbf{P}_i \cdot \mathbf{s}_i \end{pmatrix} = \begin{pmatrix} \mathbf{s}_f \\ \mathbf{P}_f \cdot \mathbf{s}_f \end{pmatrix}$$

or

$$\begin{pmatrix} (\mathbf{C}_{11} + \mathbf{C}_{12} \cdot \mathbf{P}_i) \cdot \mathbf{s}_i \\ (\mathbf{C}_{21} + \mathbf{C}_{22} \cdot \mathbf{P}_i) \cdot \mathbf{s}_i \end{pmatrix} = \begin{pmatrix} \mathbf{s}_f \\ \mathbf{P}_f \cdot \mathbf{s}_f \end{pmatrix}$$

or

$$\begin{pmatrix} \mathbf{C}_{11} + \mathbf{C}_{12} \cdot \mathbf{P}_i \\ \mathbf{C}_{21} + \mathbf{C}_{22} \cdot \mathbf{P}_i \end{pmatrix} = \begin{pmatrix} \mathbf{s}_f \cdot \mathbf{s}_i^{-1} \\ \mathbf{P}_f \cdot \mathbf{s}_f \cdot \mathbf{s}_i^{-1} \end{pmatrix}.$$

Substituting the top into the bottom and solving for $\mathbf{P}_f$ yields

$$\mathbf{P}_f = (\mathbf{C}_{21} + \mathbf{C}_{22} \cdot \mathbf{P}_i) \cdot (\mathbf{C}_{11} + \mathbf{C}_{12} \cdot \mathbf{P}_i)^{-1}. \quad (1.7)$$

Note that in (1.7) the specific choice of $\mathbf{s}_i$ and $\mathbf{s}_f$ vanished from the solution.

#### *Steps for Converting Network Parameters*

1. Identify the independent and dependent port variables in the network parameter to be converted from $\mathbf{P}_i$ and stack these on top of each other, starting with the independent variable list followed by the dependent list.
2. Identify the independent and dependent port variables in the network parameter to be converted to $\mathbf{P}_f$ and stack these on top of each other, starting with the independent variable list followed by the dependent list.

3. Generate a permutation or conversion matrix $\mathbf{C}$ that converts the first variable list to the second variable list.

4. Partition $\mathbf{C}$ into four equal size block matrices such that $\mathbf{C} = \begin{pmatrix} \mathbf{C_{11}} & \mathbf{C_{12}} \\ \mathbf{C_{21}} & \mathbf{C_{22}} \end{pmatrix}$.

5. Generate the new network parameters as $\mathbf{P}_f = (\mathbf{C_{21}} + \mathbf{C_{22}} \cdot \mathbf{P}_i) \cdot (\mathbf{C_{11}} + \mathbf{C_{12}} \cdot \mathbf{P}_i)^{-1}$.

### 1.3.1 Conversion from Z-Parameters to Y-Parameters

Here is a simple example to convert Z-parameters to Y-parameters, even though the conversion is already known and obvious. For Z-parameters, the independent and dependent variables are $\left(\begin{smallmatrix} \mathbf{I} \\ \mathbf{V} \end{smallmatrix}\right)$; for Y-parameters, they are $\left(\begin{smallmatrix} \mathbf{V} \\ \mathbf{I} \end{smallmatrix}\right)$. To convert from Z- to Y-parameters, $\mathbf{C} = \left(\begin{smallmatrix} \mathbf{0} & \mathbb{I} \\ \mathbb{I} & \mathbf{0} \end{smallmatrix}\right)$, and therefore the conversion is

$$\mathbf{Y} = (\mathbf{C_{21}} + \mathbf{C_{22}} \cdot \mathbf{P}_i) \cdot (\mathbf{C_{11}} + \mathbf{C_{12}} \cdot \mathbf{P}_i)^{-1} = (\mathbb{I} + \mathbf{0} \cdot \mathbf{Z}) \cdot (\mathbf{0} + \mathbb{I} \cdot \mathbf{Z})^{-1} = \mathbf{Z}^{-1}.$$

This is as expected. The Y-parameters are the inverse of the Z-parameters, which also implies that the Z-parameters are the inverse of the Y-parameters.

### 1.3.2 Conversion from Z-Parameters to ABCD Parameters

For Z-parameters, the independent and dependent variables are $\left(\begin{smallmatrix} i_1 \\ i_2 \\ v_1 \\ v_2 \end{smallmatrix}\right)$; for ABCD parameters, using (1.2), they are $\left(\begin{smallmatrix} v_2 \\ i_2 \\ v_1 \\ -i_1 \end{smallmatrix}\right)$, and therefore $\mathbf{C}$ is constructed such that

$$\begin{pmatrix} 0 & 0 & 0 & 1 \\ 0 & 1 & 0 & 0 \\ 0 & 0 & 1 & 0 \\ -1 & 0 & 0 & 0 \end{pmatrix} \cdot \begin{pmatrix} i_1 \\ i_2 \\ v_1 \\ v_2 \end{pmatrix} = \begin{pmatrix} v_2 \\ i_2 \\ v_1 \\ -i_1 \end{pmatrix}.$$

Therefore, the conversion can be written as follows:

$$\begin{aligned} \mathbf{A} &= (\mathbf{C_{21}} + \mathbf{C_{22}} \cdot \mathbf{Z}) \cdot (\mathbf{C_{11}} + \mathbf{C_{12}} \cdot \mathbf{Z})^{-1} \\ &= \left[\left(\begin{smallmatrix} 0 & 0 \\ -1 & 0 \end{smallmatrix}\right) + \left(\begin{smallmatrix} 1 & 0 \\ 0 & 0 \end{smallmatrix}\right) \cdot \left(\begin{smallmatrix} Z_{11} & Z_{12} \\ Z_{21} & Z_{22} \end{smallmatrix}\right)\right] \cdot \left[\left(\begin{smallmatrix} 0 & 0 \\ 0 & 1 \end{smallmatrix}\right) + \left(\begin{smallmatrix} 0 & 1 \\ 0 & 0 \end{smallmatrix}\right) \cdot \left(\begin{smallmatrix} Z_{11} & Z_{12} \\ Z_{21} & Z_{22} \end{smallmatrix}\right)\right]^{-1} \\ &= \begin{pmatrix} Z_{11} & Z_{12} \\ -1 & 0 \end{pmatrix} \cdot \begin{pmatrix} Z_{21} & Z_{22} \\ 0 & 1 \end{pmatrix}^{-1} \\ &= \frac{1}{Z_{21}} \cdot \begin{pmatrix} Z_{11} & -|\mathbf{Z}| \\ -1 & Z_{22} \end{pmatrix}. \end{aligned} \tag{1.8}$$

### 1.3.3 Conversion from Y-Parameters to ABCD Parameters

For Y-parameters, the independent and dependent variables are $\begin{pmatrix} v_1 \\ v_2 \\ i_1 \\ i_2 \end{pmatrix}$; for ABCD parameters, using (1.2), they are $\begin{pmatrix} v_2 \\ i_2 \\ v_1 \\ -i_1 \end{pmatrix}$, and therefore $\mathbf{C}$ is constructed such that

$$\begin{pmatrix} 0 & 1 & 0 & 0 \\ 0 & 0 & 0 & 1 \\ 1 & 0 & 0 & 0 \\ 0 & 0 & -1 & 0 \end{pmatrix} \cdot \begin{pmatrix} v_1 \\ v_2 \\ i_1 \\ i_2 \end{pmatrix} = \begin{pmatrix} v_2 \\ i_2 \\ v_1 \\ -i_1 \end{pmatrix}.$$

Therefore, the conversion can be written as follows:

$$\begin{aligned} \mathbf{A} &= (\mathbf{C_{21}} + \mathbf{C_{22}} \cdot \mathbf{Y}) \cdot (\mathbf{C_{11}} + \mathbf{C_{12}} \cdot \mathbf{Y})^{-1} \\ &= \left[\left(\begin{smallmatrix} 1 & 0 \\ 0 & 0 \end{smallmatrix}\right) + \left(\begin{smallmatrix} 0 & 0 \\ -1 & 0 \end{smallmatrix}\right) \cdot \left(\begin{smallmatrix} Y_{11} & Y_{12} \\ Y_{21} & Y_{22} \end{smallmatrix}\right)\right] \cdot \left[\left(\begin{smallmatrix} 0 & 1 \\ 0 & 0 \end{smallmatrix}\right) + \left(\begin{smallmatrix} 0 & 0 \\ 0 & 1 \end{smallmatrix}\right) \cdot \left(\begin{smallmatrix} Y_{11} & Y_{12} \\ Y_{21} & Y_{22} \end{smallmatrix}\right)\right]^{-1} \\ &= \begin{pmatrix} 1 & 0 \\ -Y_{11} & -Y_{12} \end{pmatrix} \cdot \begin{pmatrix} 0 & 1 \\ Y_{21} & Y_{22} \end{pmatrix}^{-1} \\ &= \frac{1}{Y_{21}} \cdot \begin{pmatrix} -Y_{22} & 1 \\ |\mathbf{Y}| & -Y_{11} \end{pmatrix}. \end{aligned} \tag{1.9}$$

### 1.3.4 Conversion from ABCD Parameters to Z-Parameters

For ABCD parameters, using (1.2), the independent and dependent variables are $\begin{pmatrix} v_2 \\ i_2 \\ v_1 \\ -i_1 \end{pmatrix}$; for Z-parameters, they are $\begin{pmatrix} i_1 \\ i_2 \\ v_1 \\ v_2 \end{pmatrix}$, and therefore $\mathbf{C}$ is constructed such that

$$\begin{pmatrix} 0 & 0 & 0 & -1 \\ 0 & 1 & 0 & 0 \\ 0 & 0 & 1 & 0 \\ 1 & 0 & 0 & 0 \end{pmatrix} \cdot \begin{pmatrix} v_2 \\ i_2 \\ v_1 \\ -i_1 \end{pmatrix} = \begin{pmatrix} i_1 \\ i_2 \\ v_1 \\ v_2 \end{pmatrix}.$$

Therefore, the conversion can be written as follows:

$$\begin{aligned} \mathbf{Z} &= (\mathbf{C_{21}} + \mathbf{C_{22}} \cdot \mathbf{A}) (\mathbf{C_{11}} + \mathbf{C_{12}} \cdot \mathbf{A})^{-1} \\ &= \left[\left(\begin{smallmatrix} 0 & 0 \\ 1 & 0 \end{smallmatrix}\right) + \left(\begin{smallmatrix} 1 & 0 \\ 0 & 0 \end{smallmatrix}\right) \cdot \left(\begin{smallmatrix} A & B \\ C & D \end{smallmatrix}\right)\right] \cdot \left[\left(\begin{smallmatrix} 0 & 0 \\ 0 & 1 \end{smallmatrix}\right) + \left(\begin{smallmatrix} 0 & -1 \\ 0 & 0 \end{smallmatrix}\right) \cdot \left(\begin{smallmatrix} A & B \\ C & D \end{smallmatrix}\right)\right]^{-1} \\ &= \begin{pmatrix} A & B \\ 1 & 0 \end{pmatrix} \cdot \begin{pmatrix} -C & -D \\ 0 & 1 \end{pmatrix}^{-1} \\ &= -\frac{1}{C} \cdot \begin{pmatrix} A & A \cdot D - B \cdot C \\ 1 & D \end{pmatrix}. \end{aligned} \tag{1.10}$$

### 1.3.5 Conversion from ABCD Parameters to Y-Parameters

For ABCD parameters, using (1.2), the independent and dependent variables are $\begin{pmatrix} v_2 \\ i_2 \\ v_1 \\ -i_1 \end{pmatrix}$; for Y-parameters, they are $\begin{pmatrix} v_1 \\ v_2 \\ i_1 \\ i_2 \end{pmatrix}$, and therefore $\mathbf{C}$ is constructed such that

$$\begin{pmatrix} 0 & 0 & 1 & 0 \\ 1 & 0 & 0 & 0 \\ 0 & 0 & 0 & -1 \\ 0 & 1 & 0 & 0 \end{pmatrix} \cdot \begin{pmatrix} v_2 \\ i_2 \\ v_1 \\ -i_1 \end{pmatrix} = \begin{pmatrix} v_1 \\ v_2 \\ i_1 \\ i_2 \end{pmatrix}.$$

Therefore, the conversion can be written as follows:

$$\begin{aligned} \mathbf{Y} &= (\mathbf{C_{21}} + \mathbf{C_{22}} \cdot \mathbf{A}) \cdot (\mathbf{C_{11}} + \mathbf{C_{12}} \cdot \mathbf{A})^{-1} \\ &= \left[\left(\begin{smallmatrix} 0 & 0 \\ 0 & 1 \end{smallmatrix}\right) + \left(\begin{smallmatrix} 0 & -1 \\ 0 & 0 \end{smallmatrix}\right) \cdot \left(\begin{smallmatrix} A & B \\ C & D \end{smallmatrix}\right)\right] \cdot \left[\left(\begin{smallmatrix} 0 & 0 \\ 1 & 0 \end{smallmatrix}\right) + \left(\begin{smallmatrix} 1 & 0 \\ 0 & 0 \end{smallmatrix}\right) \cdot \left(\begin{smallmatrix} A & B \\ C & D \end{smallmatrix}\right)\right]^{-1} \\ &= \begin{pmatrix} -C & -D \\ 0 & 1 \end{pmatrix} \cdot \begin{pmatrix} A & B \\ 1 & 0 \end{pmatrix}^{-1} \\ &= -\frac{1}{B} \cdot \begin{pmatrix} D & C \cdot B - D \cdot A \\ -1 & A \end{pmatrix}. \end{aligned} \tag{1.11}$$

## 1.4 Network Simulation

Network simulation involves the solution of a network consisting of interconnected network devices and connected forcing functions or independent sources for the port voltages and currents. Given such a network, simulation involves the following steps:

1. Draw the network with labels for each device defining its name and the type of parameters used, numbering each port of the device and connecting ports of devices together at nodes. Drive some nodes with voltage and/or current sources.
2. Label the port voltages and port currents, remembering that generally port currents point into the device.[2]
3. Count the total number of ports in the network and multiply by two. This is the total number of independent equations required to solve the network.
4. Write an equation for each dependent variable at each port in terms of the independent variables and the network parameters for the device.
5. Write equations that set port voltages equal at all nodes (i.e. connected device ports) and equal to any driving voltage sources.
6. Write equations that set the sum of all currents into (or out of) all nodes to zero or equal to any driving current sources.

[2]Some network parameters specify the currents for one or more ports pointing out of the device. In these cases, the port current should be written in the direction specified by the network parameter definition.

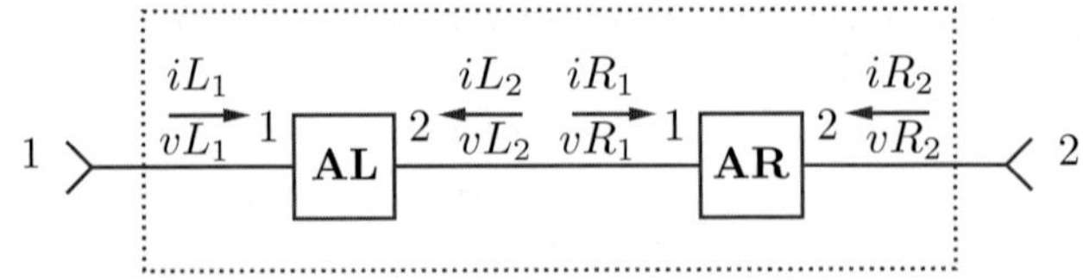


**Figure 1.12** Cascaded ABCD parameter networks

7. Check that the required number of equations have been obtained. Write all equations with linear combinations of the port voltages and currents on the left and all forcing functions on the right.
8. Determine the ordering of the node vector by listing all of the port voltages and currents in a vector.
9. Using the required equations, fill in a system characteristics matrix and a stimulus vector such that the system characteristics matrix multiplied from the left by the node vector equals the stimulus vector.
10. Solve the system for the node vector values by inverting the system characteristics matrix and multiplying by the stimulus vector.

Since network parameters are only described in order to get to the main topic, s-parameters, an example of this method is shown for s-parameters of a transmission line in §7.2; more general simulation solutions, again using s-parameters, are provided in Chapter 9.

## 1.5 Cascading Networks

Cascading is a method of attaching multiple elements together in a certain order. It is a special case of the more general problem of combining multiple network elements. Cascading is the only special case of network parameter combination that is discussed in this text, and it utilizes special network parameters designed for this purpose: ABCD parameters. The definition of ABCD parameters was given in (1.2), and is repeated here:

$$\begin{pmatrix} v_1 \\ -i_1 \end{pmatrix} = \begin{pmatrix} A & B \\ C & D \end{pmatrix} \cdot \begin{pmatrix} v_2 \\ i_2 \end{pmatrix}.$$

Consider now Figure 1.12, which contains two cascaded ABCD parameter two-port devices designated with ABCD parameters **AL** and **AR**. The equations for the two devices are:

$$\begin{pmatrix} vR_1 \\ -iR_1 \end{pmatrix} = \mathbf{AR} \cdot \begin{pmatrix} vR_2 \\ iR_2 \end{pmatrix} \tag{1.12}$$

and

$$\begin{pmatrix} vL_1 \\ -iL_1 \end{pmatrix} = \mathbf{AL} \cdot \begin{pmatrix} vL_2 \\ iL_2 \end{pmatrix}. \tag{1.13}$$

The voltages at port 2 of **AL** and port 1 of **AR** are equal, as is the current out of port 2 of **AL** and into port 1 of **AR**:

$$vL_2 = vR_1,$$
$$iL_2 = -iR_1.$$

Therefore,

$$\begin{pmatrix} vL_2 \\ iL_2 \end{pmatrix} = \begin{pmatrix} vR_1 \\ -iR_1 \end{pmatrix}. \tag{1.14}$$

Substituting (1.14) into (1.13), one obtains

$$\begin{pmatrix} vL_1 \\ -iL_1 \end{pmatrix} = \mathbf{AL} \cdot \begin{pmatrix} vR_1 \\ -iR_1 \end{pmatrix}, \tag{1.15}$$

and on substituting (1.12) into (1.15) one obtains

$$\begin{pmatrix} vL_1 \\ -iL_1 \end{pmatrix} = \mathbf{AL} \cdot \mathbf{AR} \cdot \begin{pmatrix} vR_2 \\ iR_2 \end{pmatrix}.$$

$\mathbf{AL} \cdot \mathbf{AR}$ simply describes the ABCD parameters of the cascaded network. Thus, it can be seen that ABCD parameters have a special property and are designed with cascading in mind: cascaded ABCD parameter network systems are solved by multiplying the ABCD parameters of the networks.

A reasonable strategy for cascading network parameters of other types is to convert the network parameters to ABCD parameters, multiply the ABCD parameter matrices together, and convert back to the network parameters of interest.

### 1.5.1 Cascaded Z-Parameters Using ABCD Parameters

Consider a network containing two cascaded two-port Z-parameter devices **ZL** and **ZR**. The ABCD parameters, using (1.8), are

$$\mathbf{AL} = \frac{1}{ZL_{21}} \cdot \begin{pmatrix} ZL_{11} & -|\mathbf{ZL}| \\ -1 & ZL_{22} \end{pmatrix},$$

$$\mathbf{AR} = \frac{1}{ZR_{21}} \cdot \begin{pmatrix} ZR_{11} & -|\mathbf{ZR}| \\ -1 & ZR_{22} \end{pmatrix}.$$

Multiplying these together results in

$$\mathbf{AL} \cdot \mathbf{AR} = \frac{1}{ZL_{21} \cdot ZR_{21}} \cdot \begin{pmatrix} ZL_{11} \cdot ZR_{11} + |\mathbf{ZL}| & -(ZL_{11} \cdot |\mathbf{ZR}| + |\mathbf{ZL}| \cdot ZR_{22}) \\ -(ZR_{11} + ZL_{22}) & ZL_{22} \cdot ZR_{22} + |\mathbf{ZR}| \end{pmatrix}.$$

Utilizing (1.10),

$$\begin{aligned} \mathbf{Z} &= -\frac{1}{C} \cdot \begin{pmatrix} A & A \cdot D - B \cdot C \\ 1 & D \end{pmatrix} \\ &= \frac{1}{ZR_{11} + ZL_{22}} \cdot \begin{pmatrix} ZL_{11} \cdot ZR_{11} + |\mathbf{ZL}| & ZL_{12} \cdot ZR_{12} \\ ZL_{21} \cdot ZR_{21} & ZL_{22} \cdot ZR_{22} + |\mathbf{ZR}| \end{pmatrix}. \end{aligned}$$

### 1.5.2 Cascaded Y-Parameters Using ABCD Parameters

Consider a network containing two cascaded two-port Y-parameter devices **YL** and **YR**. The ABCD parameters, using (1.9), are

$$\mathbf{AL} = \frac{1}{YL_{21}} \cdot \begin{pmatrix} -YL_{22} & 1 \\ |\mathbf{YL}| & -YL_{11} \end{pmatrix},$$

$$\mathbf{AR} = \frac{1}{YR_{21}} \cdot \begin{pmatrix} -YR_{22} & 1 \\ |\mathbf{YR}| & -YR_{11} \end{pmatrix}.$$

Multiplying these together results in

$$\mathbf{AL} \cdot \mathbf{AR} = \frac{1}{YL_{21} \cdot YR_{21}} \cdot \begin{pmatrix} YL_{22} \cdot YR_{22} + |\mathbf{YR}| & -(YR_{11} + YL_{22}) \\ -(YL_{11} \cdot |\mathbf{YR}| + |\mathbf{YL}| \cdot YR_{22}) & YL_{11} \cdot YR_{11} + |\mathbf{YL}| \end{pmatrix}.$$

Utilizing (1.11),

$$\begin{aligned} \mathbf{Y} &= -\frac{1}{B} \cdot \begin{pmatrix} D & C \cdot B - D \cdot A \\ -1 & A \end{pmatrix} \\ &= \frac{1}{YR_{11} + YL_{22}} \cdot \begin{pmatrix} YL_{11} \cdot YR_{11} + |\mathbf{YL}| & -YL_{12} \cdot YR_{12} \\ -YL_{21} \cdot YR_{21} & YL_{22} \cdot YR_{22} + |\mathbf{YR}| \end{pmatrix}. \end{aligned}$$

## 1.6 Network Parameter Summary

Network parameter devices define the relationship of independent port variables to other dependent port variables, where here these variables are voltages and currents. Network parameter models are behavioral as they hide the inner workings of an underlying circuit, as opposed to topological models which constrain a system to fit a particular circuit model with defined elements, interconnections, and values. Network parameter models are a matrix, where each matrix element contains an equation. Network parameters are traditionally supplied as a list of matrices, where each matrix corresponds to a given frequency. In other words, they are sampled network parameter models.

Network parameter models can be computed given a circuit, with the understanding that certain types of network parameters might not exist depending on the circuit topology and/or values. Once computed, they can be converted to other types of network parameters, both types representing the same circuit behavior. Different types of network parameters are useful in calculating interconnected networks, but these calculation methods are not used in this text; here, absolute network parameters are used, which are different from classical network parameters.

# 2

# Waves

CIRCUIT THEORY uses the concept of voltage and current, with the conventions for the sign of voltage drop and current flow as described in Chapter 1. Circuit analysis relies on Ohm's law and Kirchhoff's voltage and current laws.

Basic circuit analysis can be used for high-speed circuit analysis, but there are certain drawbacks of basic methods and improvements that can be made. One such drawback is that, at high frequencies, even relatively short wires cannot be considered as ideal. In reality a wire has some inductance and capacitance depending on its geometry, and the reactances associated with a wire are *distributed.* This distributed nature means that interconnects are actually *transmission lines*, as discussed extensively in Chapter 7. Transmission lines cannot be described perfectly by circuits with a finite number of lumped elements, and approximating them in this way tends to bog down simulators. A transmission line is best described with network parameters.

Furthermore, transmission lines are most easily described and understood by considering them as the means for the propagation of waves, as opposed to how they affect voltages and currents. Fortunately, there is a direct relationship between voltages, currents, and waves. This relationship is discussed in this chapter.

## 2.1 Wave Relationships to Voltage and Current

In 1965, Kurokawa [5] expanded on concepts that relate current and voltage to the concept of *power waves.* Power waves define a relationship between voltage and current across a load to incident and reflected power waves as (2.1) and (2.2):

$$a = \frac{1}{2\sqrt{|\mathrm{Re}\,(Z_l)|}} \cdot v + \frac{Z_l}{2\sqrt{|\mathrm{Re}\,(Z_l)|}} \cdot i; \tag{2.1}$$

$$b = \frac{1}{2\sqrt{|\mathrm{Re}\,(Z_l)|}} \cdot v - \frac{Z_l^*}{2\sqrt{|\mathrm{Re}\,(Z_l)|}} \cdot i. \tag{2.2}$$

In (2.1) and (2.2), $v$ is the voltage present at a port of the load, $i$ is the current entering a port of the load, and $Z_l$ is the impedance of the load. Note that $a$ refers to the power wave *incident* on the load and $b$ refers to the power wave *reflected* from the load. The units for power waves are $\mathrm{V}/\sqrt{\Omega}$, and when multiplied by $\sqrt{Z_l}$ are in V. These equations imply expressions for voltage and current as a function of power waves as follows:

$$v = \frac{\sqrt{|\mathrm{Re}(Z_l)|}}{\mathrm{Re}\,(Z_l)} \cdot \left(Z_l^* \cdot a + Z_l \cdot b\right), \tag{2.3}$$

$$i = \frac{\sqrt{|\mathrm{Re}(Z_l)|}}{\mathrm{Re}\,(Z_l)} \cdot (a - b)\,. \tag{2.4}$$

The concept here is that power waves are defined such that the power $p$ absorbed by the load is related to the incident and reflected power as, for $(Z_l \in \mathbb{R}) > 0$,

$$|a|^2 - |b|^2 = \mathrm{Re}\,(p)\,.$$

While the power wave concept is useful for determinations involving power, it presents certain problems in high-speed circuit analysis. A major problem arises from the usage of $Z_l^*$ in (2.3) for the description of voltage. This is discussed later.

An alternative wave description is that of traveling waves, as described in [6], which assumes that $Z0$, the reference impedance, is positive real (i.e. $(Z0 \in \mathbb{R}) > 0$):

$$a = \frac{1}{2 \cdot \sqrt{Z0}} \cdot (v + i \cdot Z0)\,; \tag{2.5}$$

$$b = \frac{1}{2 \cdot \sqrt{Z0}} \cdot (v - i \cdot Z0)\,. \tag{2.6}$$

Equations (2.5) and (2.6) imply the following relationships for voltage and current:

$$i = \frac{1}{\sqrt{Z0}} \cdot (a - b) = \frac{\sqrt{Z0}}{Z0} \cdot (a - b)\,,$$

$$v = \frac{Z0}{\sqrt{Z0}} \cdot (a + b) = \sqrt{Z0} \cdot (a + b)\,.$$

Finally, [7] proposes a variant on traveling waves that is suitable as an abstract concept to bridge the gap between wave theory and circuit theory called *pseudo-waves*. Pseudo-waves are particularly useful in the sense that they introduce a concept of arbitrary reference impedance $Z0$ that can be complex. Pseudo-waves are defined as follows:

$$a = \frac{|v_0|}{v_0} \cdot \frac{\sqrt{\mathrm{Re}\,(Z0)}}{2 \cdot |Z0|} \cdot (v + i \cdot Z0)\,,$$

$$b = \frac{|v_0|}{v_0} \cdot \frac{\sqrt{\mathrm{Re}\,(Z0)}}{2 \cdot |Z0|} \cdot (v - i \cdot Z0)\,.$$

These imply the following relationships for voltages and currents:

$$i = \frac{v_0}{|v_0|} \cdot \frac{|Z0|}{\sqrt{\mathrm{Re}\,(Z0)}} \cdot \frac{(a - b)}{Z0}\,,$$

$$v = \frac{v_0}{|v_0|} \cdot \frac{|Z0|}{\sqrt{\mathrm{Re}\,(Z0)}} \cdot (a + b)\,.$$

The nomenclature $\sqrt{Z0}$ is used to define the normalizing factor, which can be unrelated to $Z0$. In other words, $\sqrt{Z0}$ is a simple variable. This normalizing factor is constructed with the understanding that, when a source with a given source impedance is driving a given terminating impedance $Z0$, the constraints $v_0 = i_0 \cdot Z0$, and $p_0 = v_0 \cdot i_0^*$ can be used to rewrite the factor in the pseudo-wave equations as

$$\frac{1}{2 \cdot \sqrt{\mathcal{Z}0}} = \frac{|v_0|}{v_0} \cdot \frac{\sqrt{\mathrm{Re}\,(Z0)}}{2 \cdot |Z0|}.$$

This enables the definitions of the normalizing factor for pseudo-waves:

$$\sqrt{\mathcal{Z}0} = \frac{v_0}{|v_0|} \cdot \frac{|Z0|}{\sqrt{\mathrm{Re}\,(Z0)}} = \frac{v_0}{\sqrt{\mathrm{Re}\,(p_0)}} = \frac{v_0}{\sqrt{\mathrm{Re}\,(v_0 \cdot i_0^*)}} = \frac{v_0}{\sqrt{\mathrm{Re}\,(Z0 \cdot i_0 \cdot i_0^*)}} = \frac{v_0}{|i_0| \cdot \sqrt{\mathrm{Re}\,(Z0)}}.$$

Thus there are a number of ways in which to express the normalizing factor $\sqrt{\mathcal{Z}0}$, depending on metrology, which is part of the point of [7]. Note that when $Z0 \in \mathbb{R}$ and when the phase of $v_0$ is set to zero, then $\sqrt{\mathcal{Z}0} = \sqrt{Z0}$ and all of the wave types – power waves, traveling waves, and pseudo-waves – are all equivalent. Problems in using or assuming any of these wave types are encountered only when $Z0$ is complex, the normalizing factor $\sqrt{\mathcal{Z}0}$ is complex, and $\sqrt{\mathcal{Z}0} \neq \sqrt{Z0}$.

## 2.2 Wave Definition Requirements

Multiple definitions of waves were described in §2.1. In order to eliminate any confusion, this section considers some requirements on the definition of waves. These requirements are not physical, but are useful within the context of other things expressed in this book. For most purposes, there are only two requirements. These are:

1. Wave definitions follow the *reversal property.*
2. Wave definitions allow the determination of the meaning of scattering parameters.

### 2.2.1 The Reversal Property of the Wave Transformation

The reversal property is required for the interconnection of devices. It relates to the conditions at the boundary or interface between two connected devices. Stated simply:

*Incident waves on one device are the reflected waves from the other device on the other side of the boundary. Similarly, reflected waves from one device are the same as the incident waves from the other device on the other side of the boundary. The choice of which set of waves are incident and reflected waves at a boundary dictate only the reference direction for current flow.*

The reversal property means that, given a boundary between two devices, there are two sets of waves flowing: forward and reverse. The forward wave is incident on the rightmost device and considered reflected from the leftmost device. The reverse wave is incident on the leftmost device and considered reflected from the rightmost device. The forward wave flow convention is shown in Figure 2.1(a). Here, the forward wave value is chosen as $a$ and the reverse wave value as $b$ (see Figure 2.1(b)), as is typical. When these values are substituted into an equation for voltage, the voltage at the interface is obtained; if these

(a) Voltage/current (b) Waves

**Figure 2.1** Forward wave flow convention

(a) Voltage/current (b) Waves

**Figure 2.2** Reverse wave flow convention

values are substituted into an equation for current, the current *into the rightmost device* is obtained, as shown in Figure 2.1. A positive current indicates that current is indeed flowing into the rightmost device, and a negative sign indicates it is actually flowing out of it into the leftmost device. If the reversal property holds, this calculation can be repeated for a reverse wave convention, as shown in Figure 2.2, choosing the reverse wave value as $a$ and the forward wave value as $b$ (see Figure 2.2(b)). When these new values are substituted into an equation for voltage, the same value must be obtained. In other words, the voltage calculated must be independent of the choice of direction of $a$ or $b$. Furthermore, when these new values are substituted into the equation for current, in order to obtain the current into the leftmost device, it must be found to be the negative of the current flowing into the rightmost device calculated previously, as shown in Figure 2.2(a).

An arbitrary transform is defined that converts current and voltage to waves:

$$\begin{pmatrix} a \\ b \end{pmatrix} = \begin{pmatrix} x & y \\ w & z \end{pmatrix} \cdot \begin{pmatrix} v \\ i \end{pmatrix}.$$

The reversal property means that

$$\begin{pmatrix} b \\ a \end{pmatrix} = \begin{pmatrix} x & y \\ w & z \end{pmatrix} \cdot \begin{pmatrix} v \\ -i \end{pmatrix}.$$

This implies that

$$\begin{pmatrix} x & y \\ w & z \end{pmatrix} = \begin{pmatrix} 0 & 1 \\ 1 & 0 \end{pmatrix} \cdot \begin{pmatrix} x & y \\ w & z \end{pmatrix} \cdot \begin{pmatrix} 1 & 0 \\ 0 & -1 \end{pmatrix} = \begin{pmatrix} w & -z \\ x & -y \end{pmatrix} = \begin{pmatrix} x & y \\ x & -y \end{pmatrix}.$$

In other words, given a selection of $x$ and $y$, $w$ and $z$ can no longer be chosen arbitrarily: $w = x$ and $z = -y$.

### 2.2.2 Scattering Parameter Requirement on the Wave Transformation

Scattering parameters have not yet been mentioned, but these are network parameters that characterize devices using the ratio of reflected to incident waves (just as Z-parameters characterize devices using the ratio of voltage to current). For simplification, consider a single-port device consisting of an impedance to ground. The voltage at the input node is related to the current into the input node as $v = i \cdot Z$. In other words, given a known current $i$, the voltage is known. For s-parameters, there is a similar relationship. For the same one-port device, there is a relationship between the incident and reflected wave as $b = \Gamma \cdot a$. The requirement is that, given an s-parameter measurement $\Gamma$, the wave equations must be constructed in such a way that it is possible to obtain a unique value of the impedance $Z$. Using an arbitrary transform that converts current and voltage to waves,

$$\begin{pmatrix} a \\ b \end{pmatrix} = \begin{pmatrix} x & y \\ w & z \end{pmatrix} \cdot \begin{pmatrix} v \\ i \end{pmatrix}.$$

For this example,

$$\begin{pmatrix} a \\ \Gamma \cdot a \end{pmatrix} = \begin{pmatrix} x & y \\ w & z \end{pmatrix} \cdot \begin{pmatrix} i \cdot Z \\ i \end{pmatrix} = \begin{pmatrix} x \cdot i \cdot Z + y \cdot i \\ w \cdot i \cdot Z + z \cdot i \end{pmatrix}.$$

Solving for $\Gamma$,

$$\Gamma = \frac{w \cdot Z + z}{x \cdot Z + y} = \frac{w}{x} \cdot \frac{Z - z/w}{Z + y/x}.$$

This means that, in order to obtain a unique value for $\Gamma$ for a given value of $Z$, three ratios must be known: $w/x$, $z/w$, and $y/x$.

### 2.2.3 Wave Definition Requirement Combination

Given the reversal property and the scattering parameter requirement, the wave transformation is rewritten from the reversal property requirement:

$$\begin{pmatrix} a \\ b \end{pmatrix} = \begin{pmatrix} x & y \\ x & -y \end{pmatrix} \cdot \begin{pmatrix} v \\ i \end{pmatrix} = x \cdot \begin{pmatrix} 1 & y/x \\ 1 & -y/x \end{pmatrix} \cdot \begin{pmatrix} v \\ i \end{pmatrix}.$$

Then, from the scattering parameter requirement, only the ratio $y/x$ survives. This ratio is defined as the reference impedance $Z0 = y/x$, and the other remaining free variable is defined in terms of the normalizing factor as $x = 1/\left(2 \cdot \sqrt{Z0}\right)$.

So, the requirements specify that a transformation from voltage and current to waves is given by

$$\begin{pmatrix} a \\ b \end{pmatrix} = \frac{1}{2 \cdot \sqrt{\mathscr{Z}0}} \cdot \begin{pmatrix} 1 & Z0 \\ 1 & -Z0 \end{pmatrix} \cdot \begin{pmatrix} v \\ i \end{pmatrix}, \tag{2.7}$$

where the choice of $\sqrt{\mathscr{Z}0}$ and $Z0$ is arbitrary.

There are the following important implications of these requirements:

1. Power waves are not suitable as they cannot obey the reversal property because of the existence of $Z_l^*$, unless the reference impedance is real. As mentioned previously, all wave transformations are identical (with a possible difference in normalization factor) when the reference impedance is real.

2. Pseudo-waves and traveling waves are suitable because they obey the reversal property and differ only by the normalization factor.

3. Scattering parameters computed with wave transformations that meet the requirements can always be utilized because they can always be converted back to a set of network parameters that rely on the physical quantities of voltage and current, given only the knowledge of $Z0$ and $\sqrt{\mathscr{Z}0}$. In fact, for the single-port example used in §2.2.2, the value of $\sqrt{\mathscr{Z}0}$ is irrelevant. Although not shown here, it is also irrelevant if the same values of $Z0$ and $\sqrt{\mathscr{Z}0}$ are used for all ports in a multi-port situation.

4. Circuits and systems created with interconnected scattering parameter devices will all work properly providing the wave transformation used meets the requirements stated and all sources of waves in the system (current and voltage sources) utilize the same normalization factor. Devices (including current and voltage sources) with different reference impedances can be interconnected, given that the reference impedance by which the scattering parameters for the device were calculated is known and impedance transformers are utilized (see Chapter 5). The absolute quantities of the waves throughout the system will be dependent on the normalization factors employed and the known reference impedance at the various locations in the circuit.

## 2.3 Power and the Normalization Factor

In §2.2, and in the definition of pseudo-waves, one finds that there is no specific requirement on the specification of the reference impedance $Z0$ and the normalization factor $\sqrt{\mathscr{Z}0}$. Decoupling these two values probably just causes confusion, so there is a general preference to couple them as $\sqrt{\mathscr{Z}0} = \sqrt{Z0}$.

It was stated earlier that the power wave construct makes power calculations simple. To see this, using (2.3) and (2.4),

$$p = v \cdot i^* = \frac{\sqrt{|\mathrm{Re}(Z_l)|}}{\mathrm{Re}\,(Z_l)} \cdot (Z_l^* \cdot a + Z_l \cdot b) \cdot \left[ \frac{\sqrt{|\mathrm{Re}(Z_l)|}}{\mathrm{Re}\,(Z_l)} \cdot (a - b) \right]^* .$$

For $(Z_l \in \mathbb{R}) > 0$,

$$\begin{aligned} p &= \frac{1}{\mathrm{Re}\,(Z_l)} \cdot \left(Z_l^* \cdot a + Z_l \cdot b\right) \cdot (a-b)^* \\ &= \frac{1}{\mathrm{Re}\,(Z_l)} \cdot \left(Z_l^* \cdot a \cdot a^* - Z_l^* \cdot a \cdot b^* + Z_l \cdot b \cdot a^* - Z_l \cdot b \cdot b^*\right). \end{aligned}$$

This expands to

$$\begin{aligned} p = \frac{1}{\mathrm{Re}\,(Z_l)} \cdot \Big(&\mathrm{Re}\,(Z_l) \cdot |a|^2 - j \cdot \mathrm{Im}\,(Z_l) \cdot |a|^2 - \mathrm{Re}\,(Z_l) \cdot a \cdot b^* + j \cdot \mathrm{Im}\,(Z_l) \cdot a \cdot b^* \\ &+ \mathrm{Re}\,(Z_l) \cdot b \cdot a^* + j \cdot \mathrm{Im}\,(Z_l) \cdot b \cdot a^* - \mathrm{Re}\,(Z_l) \cdot |b|^2 - j \cdot \mathrm{Im}\,(Z_l) \cdot |b|^2\Big). \end{aligned}$$

Recognize that

$$\mathrm{Re}\,(Z_l) \cdot b \cdot a^* - \mathrm{Re}\,(Z_l) \cdot a \cdot b^* = j \cdot 2 \cdot \mathrm{Re}\,(Z_l) \cdot \mathrm{Im}\,(b \cdot a^*)$$

and

$$j \cdot \mathrm{Im}\,(Z_l) \cdot a \cdot b^* + j \cdot \mathrm{Im}\,(Z_l) \cdot b \cdot a^* = j \cdot 2 \cdot \mathrm{Im}\,(Z_l) \cdot \mathrm{Re}\,(a \cdot b^*),$$

which are both entirely imaginary quantities. The real and imaginary parts are separated as follows:

$$p = |a|^2 - |b|^2 + j \cdot \left[\frac{\mathrm{Im}\,(Zl)}{\mathrm{Re}\,(Zl)} \cdot \left(2 \cdot \mathrm{Re}\,(a \cdot b^*) - |a|^2 - |b|^2\right) + 2 \cdot \mathrm{Im}\,(b \cdot a^*)\right].$$

Finally, a nice relationship can be seen between the real power delivered and the power waves:[1]

$$\mathrm{Re}\,(p) = |a|^2 - |b|^2.$$

Despite this nice power relationship, it has already been established that power waves are unsuitable.

The power relationship is now calculated with the pseudo-wave definition (2.13):

$$p = v \cdot i^* = \frac{\sqrt{Z0} \cdot \sqrt{Z0}^*}{Z0^*} \cdot (a+b) \cdot (a-b)^* = \frac{\sqrt{Z0} \cdot \sqrt{Z0}^*}{Z0^*} \cdot \left(a \cdot a^* - a \cdot b^* + b \cdot a^* - b \cdot b^*\right).$$

Recognize that

$$b \cdot a^* - a \cdot b^* = j \cdot 2 \cdot \mathrm{Im}\,(b \cdot a^*),$$

which is entirely imaginary, and

$$a \cdot a^* - b \cdot b^* = |a|^2 - |b|^2,$$

[1]Since the units for waves are $\mathrm{V}/\sqrt{\Omega}$, the units for power are $\mathrm{V}^2/\Omega = \mathrm{W}$.

**Listing 2.1** Wave helper function for reference impedances and normalization factors

```
def Z0KHelper(Z0K,P):
    (Z0,K)=Z0K
    if Z0 is None:
        Z0=matrix(diag([50.0]*P))
    elif isinstance(Z0,list):
        Z0=matrix(diag([float(i.real)+float(i.imag)*1j for i in Z0]))
    elif isinstance(Z0.real,float) or isinstance(Z0.real,int):
        Z0=matrix(diag([float(Z0.real)+float(Z0.imag)*1j]*P))
    if K is None:
        K=sqrt(Z0)
    elif isinstance(K,list):
        K=matrix(diag([float(i.real)+float(i.imag)*1j for i in K]))
    elif isinstance(K.real,float) or isinstance(K.real,int):
        K=matrix(diag([float(K.real)+float(K.imag)*1j]*P))
    return (Z0,K)
```

**Listing 2.2** Power wave helper function for reference impedances

```
def Z0KHelperPW(Z0,P):
    if Z0 is None:
        Z0=matrix(diag([50.0]*P))
    elif isinstance(Z0,list):
        Z0=matrix(diag([float(i.real)+float(i.imag)*1j for i in Z0]))
    elif isinstance(Z0.real,float) or isinstance(Z0.real,int):
        Z0=matrix(diag([float(Z0.real)+float(Z0.imag)*1j]*P))
    K=sqrt(abs(Z0.real))
    return (Z0,K)
```

which is entirely real. Since $\frac{\sqrt{\mathcal{Z}0}\cdot\sqrt{\mathcal{Z}0}^*}{Z0^*}$ may be complex, separating real and imaginary parts yields

$$p = \mathrm{Re}\left(\frac{\sqrt{\mathcal{Z}0}\cdot\sqrt{\mathcal{Z}0}^*}{Z0^*}\right)\cdot\left(|a|^2-|b|^2\right) - \mathrm{Im}\left(\frac{\sqrt{\mathcal{Z}0}\cdot\sqrt{\mathcal{Z}0}^*}{Z0^*}\right)\cdot 2\cdot\mathrm{Im}\left(b\cdot a^*\right)$$
$$+\,j\cdot\left[\mathrm{Im}\left(\frac{\sqrt{\mathcal{Z}0}\cdot\sqrt{\mathcal{Z}0}^*}{Z0^*}\right)\cdot\left(|a|^2-|b|^2\right) + \mathrm{Re}\left(\frac{\sqrt{\mathcal{Z}0}\cdot\sqrt{\mathcal{Z}0}^*}{Z0^*}\right)\cdot 2\cdot\mathrm{Im}\left(b\cdot a^*\right)\right].$$

This means that, for the type of waves being specified, there is a complicated relationship between the power and the waves. One can see that if the reference impedance $Z0$ is real and if $\sqrt{\mathcal{Z}0}=\sqrt{Z0}$, then the power relationship holds. Thus, traveling waves are both suitable and maintain a simple power relationship when $Z0$ is real.

Despite the pseudo-wave definition, a commonly used convention is simply to define $\sqrt{\mathcal{Z}0}=\sqrt{Z0}$, whether $Z0$ is real or complex. This avoids confusion. That being said, the definitions and derivations in this text have been made carefully to allow for the usage of other conventions. Thus the reader will see the symbol $\sqrt{\mathcal{Z}0}$ referring to the normalization factor, written slightly differently from $Z0$ to imply that they need not be coupled. However, in the interest of not confusing things too much, the symbol is written such that one may simply consider $\sqrt{\mathcal{Z}0}=\sqrt{Z0}$ if preferred.

Listing 2.1 provides a helper function used in the accompanying software for resolving the reference impedance and normalization factor specification in a concentrated location.

**Table 2.1** General wave definitions

| | | |
|---|---|---|
| $v$ | $v = \sqrt{\mathcal{Z}0} \cdot (a + b)$ | (2.8) |
| $i$ | $i = \frac{\sqrt{\mathcal{Z}0}}{Z0} \cdot (a - b)$ | (2.9) |
| $\left(\begin{smallmatrix} v \\ i \end{smallmatrix}\right)$ | $\left(\begin{smallmatrix} v \\ i \end{smallmatrix}\right) = \sqrt{\mathcal{Z}0} \cdot \begin{pmatrix} 1 & 1 \\ \frac{1}{Z0} & -\frac{1}{Z0} \end{pmatrix} \cdot \left(\begin{smallmatrix} a \\ b \end{smallmatrix}\right)$ | (2.10) |
| $a$ | $a = \frac{1}{2\sqrt{\mathcal{Z}0}} \cdot (v + i \cdot Z0)$ | (2.11) |
| $b$ | $b = \frac{1}{2\sqrt{\mathcal{Z}0}} \cdot (v - i \cdot Z0)$ | (2.12) |
| $\left(\begin{smallmatrix} a \\ b \end{smallmatrix}\right)$ | $\left(\begin{smallmatrix} a \\ b \end{smallmatrix}\right) = \frac{1}{2\sqrt{\mathcal{Z}0}} \cdot \left(\begin{smallmatrix} 1 & Z0 \\ 1 & -Z0 \end{smallmatrix}\right) \cdot \left(\begin{smallmatrix} v \\ i \end{smallmatrix}\right)$ | (2.13) |

Listing 2.2 is also provided for power waves, but this is only used when converting between power wave definitions and pseudo-wave definitions. In Listing 2.1, if a reference impedance is not specified, it is 50 Ω for all ports, and, if the normalization factor is not provided, then $\sqrt{\mathcal{Z}0} = \sqrt{Z0}$ for all ports. But the capability exists to specify reference impedances other than 50 Ω (even complex), to specify different reference impedances on different ports, and to decouple the per-port normalization factor from the per-port reference impedance, while most easily handling the most common situations.

For power waves, $\sqrt{\mathcal{Z}0} = \sqrt{|\mathrm{Re}(Z0)|}$ is always used according to Kurokawa's definition.

## 2.4 Wave Equations

With the requirements for wave definitions provided in (2.7), all of the required equations involving waves are provided in Table 2.1 and Table 2.2. In these tables, single waves, voltages, and currents are shown in lower case, standard font: $a$ for the incident wave at a port (or defined forward propagating wave), $b$ as the reflected wave at a port (or defined reverse propagating wave), $i$ as the current into a port (or the defined forward flowing current), and $v$ as the voltage. Bold lower-case variables (e.g. $\mathbf{a}$, $\mathbf{b}$, $\mathbf{v}$, and $\mathbf{i}$) represent vector quantities, where each element is the value of the variable for a given port (i.e. $\mathbf{a}$ is a vector of incident waves such that, for a $P$-port device, $a_p$ is the incident wave on port $p$ for $p \in 1 \ldots P$). Bold upper-case variables (e.g. $\mathbf{A}$, $\mathbf{B}$, $\mathbf{V}$, and $\mathbf{I}$) represent matrix quantities, where each row represents a variable for a given port and each column represents a given measurement condition (i.e. $\mathbf{B}$ is a matrix of reflected waves such that, for a $P$-port device and $M$ measurements, $B_{pm}$ is the reflected wave from port $p$ in measurement condition $m$).

The variations in these equations depend on whether the normalization factor $\sqrt{\mathcal{Z}0}$ and/or the port reference impedance $Z0$ is constant or varies for each port. The most general case is that both $\sqrt{\mathcal{Z}0}$ and $Z0$ vary for each port. In this case, there are matrices represented as $\sqrt{\boldsymbol{\mathcal{Z}0}}$ and $\mathbf{Z0}$, where the port normalization factor and reference impedance, respectively, are listed for each port along the diagonal. Thus, for a $P$-port device, there is a normalization factor for the port at element $\sqrt{\mathcal{Z}0}_{pp}$ in $\sqrt{\boldsymbol{\mathcal{Z}0}}$ and a reference impedance at element $Z0_{pp}$ of $\mathbf{Z0}$, where $p \in 1 \ldots P$. The least general, but most common, situation

**Table 2.2** Multi-port wave definitions with variable or constant $Z0$, $\sqrt{\mathcal{Z}0}$

| | Variable per port | Constant value |
|---|---|---|
| $\mathbf{v}$ | $\mathbf{v} = \sqrt{\boldsymbol{\mathcal{Z}}\mathbf{0}} \cdot (\mathbf{a} + \mathbf{b})$ | $\mathbf{v} = \sqrt{\mathcal{Z}0} \cdot (\mathbf{a} + \mathbf{b})$ |
| $\mathbf{i}$ | $\mathbf{i} = \sqrt{\boldsymbol{\mathcal{Z}}\mathbf{0}} \cdot \mathbf{Z0}^{-1} \cdot (\mathbf{a} - \mathbf{b})$ | $\mathbf{i} = \frac{\sqrt{\mathcal{Z}0}}{Z0} \cdot (\mathbf{a} - \mathbf{b})$ |
| $\mathbf{a}$ | $\mathbf{a} = \frac{1}{2} \cdot \sqrt{\boldsymbol{\mathcal{Z}}\mathbf{0}}^{-1} \cdot (\mathbf{v} + \mathbf{Z0} \cdot \mathbf{i})$ | $\mathbf{a} = \frac{1}{2\sqrt{\mathcal{Z}0}} \cdot (\mathbf{v} + Z0 \cdot \mathbf{i})$ |
| $\mathbf{b}$ | $\mathbf{b} = \frac{1}{2} \cdot \sqrt{\boldsymbol{\mathcal{Z}}\mathbf{0}}^{-1} \cdot (\mathbf{v} - \mathbf{Z0} \cdot \mathbf{i})$ | $\mathbf{b} = \frac{1}{2\sqrt{\mathcal{Z}0}} \cdot (\mathbf{v} - Z0 \cdot \mathbf{i})$ |
| $\mathbf{V}$ | $\mathbf{V} = \sqrt{\boldsymbol{\mathcal{Z}}\mathbf{0}} \cdot (\mathbf{A} + \mathbf{B})$ | $\mathbf{V} = \sqrt{\mathcal{Z}0} \cdot (\mathbf{A} + \mathbf{B})$ |
| $\mathbf{I}$ | $\mathbf{I} = \sqrt{\boldsymbol{\mathcal{Z}}\mathbf{0}} \cdot \mathbf{Z0}^{-1} \cdot (\mathbf{A} - \mathbf{B})$ | $\mathbf{I} = \frac{\sqrt{\mathcal{Z}0}}{Z0} \cdot (\mathbf{A} - \mathbf{B})$ |
| $\mathbf{A}$ | $\mathbf{A} = \frac{1}{2} \cdot \sqrt{\boldsymbol{\mathcal{Z}}\mathbf{0}}^{-1} \cdot (\mathbf{V} + \mathbf{Z0} \cdot \mathbf{I}) \quad (2.14)$ | $\mathbf{A} = \frac{1}{2\sqrt{\mathcal{Z}0}} \cdot (\mathbf{V} + Z0 \cdot \mathbf{I})$ |
| $\mathbf{B}$ | $\mathbf{B} = \frac{1}{2} \cdot \sqrt{\boldsymbol{\mathcal{Z}}\mathbf{0}}^{-1} \cdot (\mathbf{V} - \mathbf{Z0} \cdot \mathbf{I}) \quad (2.15)$ | $\mathbf{B} = \frac{1}{2\sqrt{\mathcal{Z}0}} \cdot (\mathbf{V} - Z0 \cdot \mathbf{I})$ |
| $\binom{\mathbf{V}}{\mathbf{I}}$ | $\binom{\mathbf{V}}{\mathbf{I}} = \begin{pmatrix} \sqrt{\boldsymbol{\mathcal{Z}}\mathbf{0}} & \sqrt{\boldsymbol{\mathcal{Z}}\mathbf{0}} \\ \sqrt{\boldsymbol{\mathcal{Z}}\mathbf{0}} \cdot \mathbf{Z0}^{-1} & -\sqrt{\boldsymbol{\mathcal{Z}}\mathbf{0}} \cdot \mathbf{Z0}^{-1} \end{pmatrix} \cdot \binom{\mathbf{A}}{\mathbf{B}} \quad (2.16)$ | $\binom{\mathbf{V}}{\mathbf{I}} = \sqrt{\mathcal{Z}0} \cdot \begin{pmatrix} \mathbb{I} & \mathbb{I} \\ \frac{1}{Z0} \cdot \mathbb{I} & -\frac{1}{Z0} \cdot \mathbb{I} \end{pmatrix} \cdot \binom{\mathbf{A}}{\mathbf{B}}$ |
| $\binom{\mathbf{A}}{\mathbf{B}}$ | $\binom{\mathbf{A}}{\mathbf{B}} = \frac{1}{2} \cdot \begin{pmatrix} \sqrt{\boldsymbol{\mathcal{Z}}\mathbf{0}}^{-1} & \sqrt{\boldsymbol{\mathcal{Z}}\mathbf{0}}^{-1} \cdot \mathbf{Z0} \\ \sqrt{\boldsymbol{\mathcal{Z}}\mathbf{0}}^{-1} & -\sqrt{\boldsymbol{\mathcal{Z}}\mathbf{0}}^{-1} \cdot \mathbf{Z0} \end{pmatrix} \cdot \binom{\mathbf{V}}{\mathbf{I}} \quad (2.17)$ | $\binom{\mathbf{A}}{\mathbf{B}} = \frac{1}{2\sqrt{\mathcal{Z}0}} \cdot \begin{pmatrix} \mathbb{I} & Z0 \cdot \mathbb{I} \\ \mathbb{I} & -Z0 \cdot \mathbb{I} \end{pmatrix} \cdot \binom{\mathbf{V}}{\mathbf{I}}$ |

is that each port has the same normalization factor and reference impedance, with $\sqrt{\mathcal{Z}0} \cdot \mathbb{I}$ substituted for $\sqrt{\boldsymbol{\mathcal{Z}}\mathbf{0}}$ and $Z0 \cdot \mathbb{I}$ substituted for $\mathbf{Z0}$.

As a final note, the definition of a traveling wave has the normalization factor and reference impedance connected such that if a port has a reference impedance $Z0$, the normalization factor is $\sqrt{\mathcal{Z}0} = \sqrt{Z0}$. For devices with different port reference impedances, $\mathbf{Z0}$ represents the port reference impedances and $\sqrt{\boldsymbol{\mathcal{Z}}\mathbf{0}} = \sqrt{\mathbf{Z0}}$ represents the port normalization factors for the traveling wave case.

## 2.5 Power Wave Equations

Despite the fact that power waves do not meet the wave definition requirements because of their inability to conform to the reversal property, they are sometimes used as the basis for s-parameters, especially those calculated from field solvers. The power wave equations are tabulated in Table 2.3 and Table 2.4 and provide methods to convert back and forth between these types of waves in subsequent sections. These equations are the basis for the s-parameter conversions provided in §3.5.

**Table 2.3** General power wave definitions ($\sqrt{\mathcal{Z}0} = \sqrt{|\mathrm{Re}\,(Z0)|}$)

| | |
|---|---|
| $v$ | $v = \frac{\sqrt{\mathcal{Z}0}}{\mathrm{Re}(Z0)} \cdot (Z0^* \cdot a + Z0 \cdot b)$ |
| $i$ | $i = \frac{\sqrt{\mathcal{Z}0}}{\mathrm{Re}(Z0)} \cdot (a - b)$ |
| $\binom{v}{i}$ | $\binom{v}{i} = \frac{\sqrt{\mathcal{Z}0}}{\mathrm{Re}(Z0)} \cdot \begin{pmatrix} Z0^* & Z0 \\ 1 & -1 \end{pmatrix} \cdot \binom{a}{b}$ |
| $a$ | $a = \frac{1}{2 \cdot \sqrt{\mathcal{Z}0}} \cdot (v + Z0 \cdot i)$ |
| $b$ | $b = \frac{1}{2 \cdot \sqrt{\mathcal{Z}0}} \cdot (v - Z0^* \cdot i)$ |
| $\binom{a}{b}$ | $\binom{a}{b} = \frac{1}{2\sqrt{\mathcal{Z}0}} \cdot \begin{pmatrix} 1 & Z0 \\ 1 & -Z0^* \end{pmatrix} \cdot \binom{v}{i}$ |

**Table 2.4** Multi-port power wave definitions: variable port $Z0$ ($\sqrt{\mathcal{Z}0} = \sqrt{|\mathrm{Re}\,(Z0)|}$)

| | | |
|---|---|---|
| $\mathbf{v}$ | $\mathbf{v} = \sqrt{\boldsymbol{\mathcal{Z}0}} \cdot \mathrm{Re}\,(\mathbf{Z0})^{-1} \cdot \left(\mathbf{Z0}^{*'} \cdot \mathbf{a} + \mathbf{Z0} \cdot \mathbf{b}\right)$ | |
| $\mathbf{i}$ | $\mathbf{i} = \sqrt{\boldsymbol{\mathcal{Z}0}} \cdot \mathrm{Re}\,(\mathbf{Z0})^{-1} \cdot (\mathbf{a} - \mathbf{b})$ | |
| $\mathbf{a}$ | $\mathbf{a} = \frac{1}{2} \cdot \sqrt{\boldsymbol{\mathcal{Z}0}}^{-1} \cdot (\mathbf{v} + \mathbf{Z0} \cdot \mathbf{i})$ | |
| $\mathbf{b}$ | $\mathbf{b} = \frac{1}{2} \cdot \sqrt{\boldsymbol{\mathcal{Z}0}}^{-1} \cdot (\mathbf{v} - \mathbf{Z0}^* \cdot \mathbf{i})$ | |
| $\mathbf{V}$ | $\mathbf{V} = \sqrt{\boldsymbol{\mathcal{Z}0}} \cdot \mathrm{Re}\,(\mathbf{Z0})^{-1} \cdot \left(\mathbf{Z0}^{*'} \cdot \mathbf{A} + \mathbf{Z0} \cdot \mathbf{B}\right)$ | |
| $\mathbf{I}$ | $\mathbf{I} = \sqrt{\boldsymbol{\mathcal{Z}0}} \cdot \mathrm{Re}\,(\mathbf{Z0})^{-1} \cdot (\mathbf{A} - \mathbf{B})$ | |
| $\mathbf{A}$ | $\mathbf{A} = \frac{1}{2} \cdot \sqrt{\boldsymbol{\mathcal{Z}0}}^{-1} \cdot (\mathbf{V} + \mathbf{Z0} \cdot \mathbf{I})$ | |
| $\mathbf{B}$ | $\mathbf{B} = \frac{1}{2} \cdot \sqrt{\boldsymbol{\mathcal{Z}0}}^{-1} \cdot (\mathbf{V} - \mathbf{Z0}^* \cdot \mathbf{I})$ | |
| $\binom{\mathbf{V}}{\mathbf{I}}$ | $\binom{\mathbf{V}}{\mathbf{I}} = \begin{pmatrix} \sqrt{\boldsymbol{\mathcal{Z}0}} \cdot \mathrm{Re}(\mathbf{Z0})^{-1} & \mathbf{0} \\ \mathbf{0} & \sqrt{\boldsymbol{\mathcal{Z}0}} \cdot \mathrm{Re}(\mathbf{Z0})^{-1} \end{pmatrix} \cdot \begin{pmatrix} \mathbf{Z0}^* & \mathbf{Z0} \\ \mathbb{I} & -\mathbb{I} \end{pmatrix} \cdot \binom{\mathbf{A}}{\mathbf{B}}$ | (2.18) |
| $\binom{\mathbf{A}}{\mathbf{B}}$ | $\binom{\mathbf{A}}{\mathbf{B}} = \frac{1}{2} \cdot \begin{pmatrix} \sqrt{\boldsymbol{\mathcal{Z}0}}^{-1} & \mathbf{0} \\ \mathbf{0} & \sqrt{\boldsymbol{\mathcal{Z}0}}^{-1} \end{pmatrix} \cdot \begin{pmatrix} \mathbb{I} & \mathbf{Z0} \\ \mathbb{I} & -\mathbf{Z0}^* \end{pmatrix} \cdot \binom{\mathbf{V}}{\mathbf{I}}$ | (2.19) |

### 2.5.1 General Waves In Terms Of Power Waves

Substituting (2.18) into (2.17) provides an equation for general waves in terms of power waves:

$$\begin{aligned}
\begin{pmatrix}\mathbf{A}_w\\ \mathbf{B}_w\end{pmatrix} &= \frac{1}{2}\cdot\begin{pmatrix}\sqrt{\mathbf{Z0}}_w^{-1} & \mathbf{0}\\ \mathbf{0} & \sqrt{\mathbf{Z0}}_w^{-1}\end{pmatrix}\cdot\begin{pmatrix}\mathbb{I} & \mathbf{Z0}_w\\ \mathbb{I} & -\mathbf{Z0}_w\end{pmatrix}\\
&\cdot\begin{pmatrix}\sqrt{\mathbf{Z0}}_p\cdot\mathrm{Re}(\mathbf{Z0}_p)^{-1} & \mathbf{0}\\ \mathbf{0} & \sqrt{\mathbf{Z0}}_p\cdot\mathrm{Re}(\mathbf{Z0}_p)^{-1}\end{pmatrix}\cdot\begin{pmatrix}\mathbf{Z0}_p^* & \mathbf{Z0}_p\\ \mathbb{I} & -\mathbb{I}\end{pmatrix}\cdot\begin{pmatrix}\mathbf{A}_p\\ \mathbf{B}_p\end{pmatrix}\\
&= \frac{1}{2}\cdot\begin{pmatrix}\sqrt{\mathbf{Z0}}_w^{-1}\cdot\sqrt{\mathbf{Z0}}_p\cdot\mathrm{Re}(\mathbf{Z0}_p)^{-1} & \mathbf{0}\\ \mathbf{0} & \sqrt{\mathbf{Z0}}_w^{-1}\cdot\sqrt{\mathbf{Z0}}_p\cdot\mathrm{Re}(\mathbf{Z0}_p)^{-1}\end{pmatrix}\\
&\cdot\begin{pmatrix}\mathbb{I} & \mathbf{Z0}_w\\ \mathbb{I} & -\mathbf{Z0}_w\end{pmatrix}\cdot\begin{pmatrix}\mathbf{Z0}_p^* & \mathbf{Z0}_p\\ \mathbb{I} & -\mathbb{I}\end{pmatrix}\cdot\begin{pmatrix}\mathbf{A}_p\\ \mathbf{B}_p\end{pmatrix}\\
&= \frac{1}{2}\cdot\begin{pmatrix}\sqrt{\mathbf{Z0}}_w^{-1}\cdot\sqrt{\mathbf{Z0}}_p\cdot\mathrm{Re}(\mathbf{Z0}_p)^{-1} & \mathbf{0}\\ \mathbf{0} & \sqrt{\mathbf{Z0}}_w^{-1}\cdot\sqrt{\mathbf{Z0}}_p\cdot\mathrm{Re}(\mathbf{Z0}_p)^{-1}\end{pmatrix}\\
&\cdot\begin{pmatrix}\mathbf{Z0}_p^*+\mathbf{Z0}_w & \mathbf{Z0}_p-\mathbf{Z0}_w\\ \mathbf{Z0}_p^*-\mathbf{Z0}_w & \mathbf{Z0}_p+\mathbf{Z0}_w\end{pmatrix}\cdot\begin{pmatrix}\mathbf{A}_p\\ \mathbf{B}_p\end{pmatrix}.
\end{aligned} \tag{2.20}$$

One can verify that if $\mathbf{Z0} = \mathbf{Z0}_p = \mathbf{Z0}_w \in \mathbb{R}$, and $\sqrt{\mathbf{Z0}} = \sqrt{\mathbf{Z0}}_p = \sqrt{\mathbf{Z0}}_w$, then

$$\begin{pmatrix}\mathbf{A}_w\\ \mathbf{B}_w\end{pmatrix} = \frac{1}{2}\cdot\begin{pmatrix}\mathbf{Z0}^{-1} & \mathbf{0}\\ \mathbf{0} & \mathbf{Z0}^{-1}\end{pmatrix}\cdot\begin{pmatrix}2\cdot\mathbf{Z0} & \mathbf{0}\\ \mathbf{0} & 2\cdot\mathbf{Z0}\end{pmatrix}\cdot\begin{pmatrix}\mathbf{A}_p\\ \mathbf{B}_p\end{pmatrix} = \begin{pmatrix}\mathbf{A}_p\\ \mathbf{B}_p\end{pmatrix}.$$

### 2.5.2 Power Waves In Terms Of General Waves

Substituting (2.16) into (2.19) provides an equation for power waves in terms of general waves:

$$\begin{aligned}
\begin{pmatrix}\mathbf{A}_p\\ \mathbf{B}_p\end{pmatrix} &= \frac{1}{2}\cdot\begin{pmatrix}\sqrt{\mathbf{Z0}}_p^{-1} & \mathbf{0}\\ \mathbf{0} & \sqrt{\mathbf{Z0}}_p^{-1}\end{pmatrix}\cdot\begin{pmatrix}\mathbb{I} & \mathbf{Z0}_p\\ \mathbb{I} & -\mathbf{Z0}_p^*\end{pmatrix}\cdot\begin{pmatrix}\sqrt{\mathbf{Z0}}_w & \mathbf{0}\\ \mathbf{0} & \sqrt{\mathbf{Z0}}_w\end{pmatrix}\cdot\begin{pmatrix}\mathbb{I} & \mathbb{I}\\ \mathbf{Z0}_w^{-1} & -\mathbf{Z0}_w^{-1}\end{pmatrix}\cdot\begin{pmatrix}\mathbf{A}_w\\ \mathbf{B}_w\end{pmatrix}\\
&= \frac{1}{2}\cdot\begin{pmatrix}\sqrt{\mathbf{Z0}}_p^{-1}\cdot\sqrt{\mathbf{Z0}}_w\cdot\mathbf{Z0}_w^{-1} & \mathbf{0}\\ \mathbf{0} & \sqrt{\mathbf{Z0}}_p^{-1}\cdot\sqrt{\mathbf{Z0}}_w\cdot\mathbf{Z0}_w^{-1}\end{pmatrix}\cdot\begin{pmatrix}\mathbb{I} & \mathbf{Z0}_p\\ \mathbb{I} & -\mathbf{Z0}_p^*\end{pmatrix}\cdot\begin{pmatrix}\mathbf{Z0}_w & \mathbf{Z0}_w\\ \mathbb{I} & -\mathbb{I}\end{pmatrix}\cdot\begin{pmatrix}\mathbf{A}_w\\ \mathbf{B}_w\end{pmatrix}\\
&= \frac{1}{2}\cdot\begin{pmatrix}\sqrt{\mathbf{Z0}}_p^{-1}\cdot\sqrt{\mathbf{Z0}}_w\cdot\mathbf{Z0}_w^{-1} & \mathbf{0}\\ \mathbf{0} & \sqrt{\mathbf{Z0}}_p^{-1}\cdot\sqrt{\mathbf{Z0}}_w\cdot\mathbf{Z0}_w^{-1}\end{pmatrix}\cdot\begin{pmatrix}\mathbf{Z0}_w+\mathbf{Z0}_p & \mathbf{Z0}_w-\mathbf{Z0}_p\\ \mathbf{Z0}_w-\mathbf{Z0}_p^* & \mathbf{Z0}_w+\mathbf{Z0}_p^*\end{pmatrix}\cdot\begin{pmatrix}\mathbf{A}_w\\ \mathbf{B}_w\end{pmatrix}.
\end{aligned}$$

One can verify that if $\mathbf{Z0} = \mathbf{Z0}_p = \mathbf{Z0}_w \in \mathbb{R}$, and $\sqrt{\mathbf{Z0}} = \sqrt{\mathbf{Z0}}_p = \sqrt{\mathbf{Z0}}_w$, then

$$\begin{pmatrix}\mathbf{A}_p\\ \mathbf{B}_p\end{pmatrix} = \frac{1}{2}\cdot\begin{pmatrix}\mathbf{Z0}^{-1} & \mathbf{0}\\ \mathbf{0} & \mathbf{Z0}^{-1}\end{pmatrix}\cdot\begin{pmatrix}2\cdot\mathbf{Z0} & \mathbf{0}\\ \mathbf{0} & 2\cdot\mathbf{Z0}\end{pmatrix}\cdot\begin{pmatrix}\mathbf{A}_w\\ \mathbf{B}_w\end{pmatrix} = \begin{pmatrix}\mathbf{A}_w\\ \mathbf{B}_w\end{pmatrix}.$$

# 3

# Scattering Parameters

HAVING COVERED NETWORK PARAMETER MODELS (albeit network models for voltages and currents) and the concept of waves, the goal of this chapter is to combine the two concepts to describe the type of network parameter of most interest in this book – that being scattering parameters, or s-parameters.

S-parameters are network parameters that, instead of defining port voltage and current relationships, define the relationship of reflected waves to incident waves.

Methods are provided for determining s-parameters of simple networks, but the reader should recognize that this is covered comprehensively in the next chapter, Chapter 4.

Transmission parameters, or T-parameters, the cascadable version of s-parameters, are introduced, and some basic cascade and de-embedding methods are shown for two-port networks.

The chapter ends with a general form of T-parameters and a list of network parameters for various circuit elements that will be useful later on, and finally with a discussion of network parameter file formats, specifically the Touchstone format for s-parameters.

## 3.1 S-Parameter Definition

S-parameters define reflected waves in terms of incident waves. An s-parameter matrix **S** for a given frequency specifies the vector containing the port reflected waves **b** with respect to a vector containing the port incident waves **a** as follows:

$$\mathbf{b} = \mathbf{S} \cdot \mathbf{a}. \tag{3.1}$$

For example, for a two-port network, the s-parameters at a given frequency comprise the matrix shown in (3.2):

$$\begin{pmatrix} b_1 \\ b_2 \end{pmatrix} = \begin{pmatrix} S_{11} & S_{12} \\ S_{21} & S_{22} \end{pmatrix} \cdot \begin{pmatrix} a_1 \\ a_2 \end{pmatrix}. \tag{3.2}$$

Equation (3.2) implies that the reflected power waves in the two-port device are related by (3.3) and (3.4):

$$b_1 = S_{11} \cdot a_1 + S_{12} \cdot a_2, \tag{3.3}$$

$$b_2 = S_{21} \cdot a_1 + S_{22} \cdot a_2. \tag{3.4}$$

Equations (3.3) and (3.4) show that the definition of a given s-parameter $S_{xy}$ can be defined as the ratio of the reflected wave at port $x$ to the incident wave applied at port $y$

while the incident wave at all other ports is set to zero. This concept is not as intuitive as with circuits, because it might not be clear to the reader how this is accomplished: how one might drive a port with an incident wave at one port and not another.

## 3.2 Method of Determining S-Parameters of Circuits

Here, a method is put forth for determining the s-parameters of circuits. This method is the s-parameter version of what was outlined in §1.2:

1. The exposed terminals of the circuit model are labeled with port numbers, and the voltages at and currents into the terminals are labeled such that they are associated with the port number.
2. The circuit model or element is instrumented with voltage sources driving exposed terminals through a series impedance equal to the reference impedance for that port. Each source must be labeled with the port number for the network parameters to be obtained. Note that the reference impedance is arbitrary and need not be the same for each port, but one must keep track of the reference impedances as the interpretation of the resulting s-parameters depends on this knowledge. Similarly, note that the normalization factor is also arbitrary.
3. Each voltage source is turned on alternately (leaving all of the others set to zero) and all of the port voltages and currents are calculated for each circumstance. The measurements of the port currents and voltages are labeled as $i_{xy}$ and $v_{xy}$, where $x$ refers to the port number and $y$ is the measurement number (usually, which port is on for the circumstance).
4. All of the voltage measurements are arranged in a matrix $\mathbf{V}$, where each element $V_{xy} = v_{xy}$.
5. Similarly, all of the current measurements are arranged in a matrix $\mathbf{I}$, where each element $\mathbf{I}_{xy} = i_{xy}$.
6. Using (2.14) and (2.15), realize that the measurement of incident waves is given by $\mathbf{A} = \frac{1}{2} \cdot \sqrt{\boldsymbol{\mathcal{Z}0}}^{-1} \cdot (\mathbf{V} + \mathbf{Z0} \cdot \mathbf{I})$ and the measurement of the reflected waves is given by $\mathbf{B} = \frac{1}{2} \cdot \sqrt{\boldsymbol{\mathcal{Z}0}}^{-1} \cdot (\mathbf{V} - \mathbf{Z0} \cdot \mathbf{I})$ and that from (3.1) one can calculate $\mathbf{S} = \mathbf{B} \cdot \mathbf{A}^{-1}$. Therefore, the s-parameters are calculated as $\mathbf{S} = \sqrt{\boldsymbol{\mathcal{Z}0}}^{-1} \cdot (\mathbf{V} - \mathbf{Z0} \cdot \mathbf{I}) \cdot (\mathbf{V} + \mathbf{Z0} \cdot \mathbf{I})^{-1} \cdot \sqrt{\boldsymbol{\mathcal{Z}0}}$. Note that $\mathbf{Z0}$ is a matrix containing all of the port reference impedances on the diagonal. If, as is common, there is only one value of reference impedance $Z0$, then $\mathbf{Z0} = Z0 \cdot \mathbb{I}$. Furthermore, if there is only one normalization factor $\sqrt{\mathcal{Z}0}$, then $\sqrt{\boldsymbol{\mathcal{Z}0}} = \sqrt{\mathcal{Z}0} \cdot \mathbb{I}$.

It is useful to observe that if the Z-parameters of the circuit model are known, then $\mathbf{V} = \mathbf{Z} \cdot \mathbf{I}$. Substituting into the equation for $\mathbf{S}$ results in $\mathbf{S} = \sqrt{\boldsymbol{\mathcal{Z}0}}^{-1} \cdot (\mathbf{Z} - \mathbf{Z0}) \cdot (\mathbf{Z} + \mathbf{Z0})^{-1} \cdot \sqrt{\boldsymbol{\mathcal{Z}0}}$, and if the Y-parameters are known then $\mathbf{I} = \mathbf{Y} \cdot \mathbf{V}$. Substitution would result in $\mathbf{S} = \sqrt{\boldsymbol{\mathcal{Z}0}}^{-1} \cdot (\mathbb{I} - \mathbf{Z0} \cdot \mathbf{Y}) \cdot (\mathbb{I} + \mathbf{Z0} \cdot \mathbf{Y})^{-1} \cdot \sqrt{\boldsymbol{\mathcal{Z}0}}$. This compares with the results provided for conversions from Z- and Y-parameters to s-parameters provided in (3.17) and (3.19).

### 3.2.1 Alternative Form of S-Parameter Determination

Here, an alternative form of s-parameter determination based on matrix algebra methods is provided. The assumption is made that, when generating equations, those equations for the circuit or circuit elements are in terms of the port voltages and port currents only; the voltage sources are completely disregarded. This generates one equation for every port of the system. Then, a generic set of equations, which sets the port voltage as a function of the port current, the reference impedance, and the voltage source, is applied.

Since there is no voltage source involved in the first set of equations, it can be written in a form where there is a linear combination of port voltages and currents that add to zero:

$$\sum_{p=1}^{P} xi_{mp} \cdot i_p + \sum_{p=1}^{P} xv_{mp} \cdot v_p = 0,$$

where $p \in 1 \ldots P$ and $m \in 1 \ldots P$ is the equation number.

Then, the second set of equations has the form, for a given port $p \in 1 \ldots P$,

$$v_p - i_p \cdot Z0 = VS_p,$$

where $v_p$ is the voltage at port $p$, $i_p$ is the current into port $p$, and $VS_p$ is the voltage source driving port $p$. This system of equations can be written in the following form:

$$\begin{pmatrix} xi_{11} & xi_{12} & \cdots & xi_{1P} & xv_{11} & xv_{12} & \cdots & xv_{1P} \\ xi_{21} & xi_{22} & \cdots & xi_{2P} & xv_{21} & xv_{22} & \cdots & xv_{2P} \\ \vdots & \vdots & \ddots & \vdots & \vdots & \vdots & \ddots & \vdots \\ xi_{P1} & xi_{P2} & \cdots & xi_{PP} & xv_{P1} & xv_{P2} & \cdots & xv_{PP} \\ Z0 & 0 & \cdots & 0 & 1 & 0 & \cdots & 0 \\ 0 & Z0 & \cdots & 0 & 0 & 1 & \cdots & 0 \\ \vdots & \vdots & \ddots & \vdots & \vdots & \vdots & \ddots & \vdots \\ 0 & 0 & \cdots & Z0 & 0 & 0 & \cdots & 1 \end{pmatrix} \cdot \begin{pmatrix} i_1 \\ i_2 \\ \vdots \\ i_P \\ v_1 \\ v_2 \\ \vdots \\ v_P \end{pmatrix} = \begin{pmatrix} 0 \\ 0 \\ \vdots \\ 0 \\ VS_1 \\ VS_2 \\ \vdots \\ VS_P \end{pmatrix}$$

or in block form as

$$\begin{pmatrix} \mathbf{XI} & \mathbf{XV} \\ \mathbf{Z0} & \mathbb{I} \end{pmatrix} \cdot \begin{pmatrix} \mathbf{I} \\ \mathbf{V} \end{pmatrix} = \begin{pmatrix} \mathbf{0} \\ \mathbf{VS} \end{pmatrix}.$$

Here, all blocks are $P \times P$, even $\mathbf{I}$, $\mathbf{V}$, and $\mathbf{VS}$, where $I_{mp}$, $V_{mp}$, and $VS_{mp}$ contain the port current, port voltage, and value of the voltage source for driving condition $m$ at port $p$. It is advantageous to alternately set voltage source $m$ to $2 \cdot \sqrt{Z0}_m$ and all others to zero in driving condition $m$, which makes $\mathbf{VS} = 2 \cdot \sqrt{\mathcal{Z}\mathbf{0}}$. This is because, under this condition,

$$\begin{pmatrix} \mathbf{XI} & \mathbf{XV} \\ \mathbf{Z0} & \mathbb{I} \end{pmatrix} \cdot \begin{pmatrix} \mathbf{I} \\ \mathbf{V} \end{pmatrix} = \begin{pmatrix} \mathbf{0} \\ 2 \cdot \sqrt{\mathcal{Z}\mathbf{0}} \end{pmatrix}. \tag{3.5}$$

In block form, this represents two equations. One of them is

$$\mathbf{V} + \mathbf{Z0} \cdot \mathbf{I} = 2 \cdot \sqrt{\mathcal{Z}\mathbf{0}}. \tag{3.6}$$

Using (2.14),

$$\mathbf{A} = \frac{1}{2} \cdot \sqrt{\mathbf{Z0}}^{-1} \cdot (\mathbf{V} + \mathbf{Z0} \cdot \mathbf{I}) = \mathbb{I}.$$

This is helpful since the s-parameters are $\mathbf{S} = \mathbf{B} \cdot \mathbf{A}^{-1}$. Using $\mathbf{A} = \mathbb{I}$ avoids the determination of $\mathbf{A}$ and $\mathbf{A}^{-1}$, and $\mathbf{S} = \mathbf{B}$:

$$\mathbf{S} = \mathbf{B} = \frac{1}{2} \cdot \sqrt{\mathbf{Z0}}^{-1} \cdot (\mathbf{V} - \mathbf{Z0} \cdot \mathbf{I}). \tag{3.7}$$

This arrangement is equivalent to the situation where the incident waves from the source are unity and the system is driven through the reference impedance of the port (which is by design), which means that the s-parameters are read directly from the reflected waves only.

Using (3.6) enables the solution for the s-parameters in terms of either voltage or current only. Solving (3.6) for $\mathbf{V}$ and substituting for $\mathbf{V}$ in (3.7) produces

$$\mathbf{S} = \mathbf{B} = \frac{1}{2} \cdot \sqrt{\mathbf{Z0}}^{-1} \cdot \left(2 \cdot \sqrt{\mathbf{Z0}} - \mathbf{Z0} \cdot \mathbf{I} - \mathbf{Z0} \cdot \mathbf{I}\right) = \mathbb{I} - \sqrt{\mathbf{Z0}}^{-1} \cdot \mathbf{Z0} \cdot \mathbf{I}. \tag{3.8}$$

So, the strategy is to solve (3.5) for $\mathbf{I}$ and substitute this $\mathbf{I}$ into (3.8) to solve finally for the s-parameters.

The other equation in (3.5) is

$$\mathbf{XI} \cdot \mathbf{I} + \mathbf{XV} \cdot \mathbf{V} = \mathbf{0}.$$

This is just a matrix form of the $P$ equations for the port voltages in terms of the port currents mentioned earlier.

It is tempting to simply plug one of the two solutions implied by this equation into the other. For example,

$$\mathbf{I} = -\mathbf{XI}^{-1} \cdot \mathbf{XV} \cdot \mathbf{V}.$$

This is, it turns out, a poor starting point because the solution is predicated on the invertibility of $\mathbf{XI}$. In fact, careful examination reveals that the product $-\mathbf{XI}^{-1} \cdot \mathbf{XV}$ simply represents the Y-parameters of the circuit, and the Y-parameters may or may not exist. Similarly, one could try

$$\mathbf{V} = -\mathbf{XV}^{-1} \cdot \mathbf{XI} \cdot \mathbf{I}.$$

This has a similar problem in that the solution is predicated on the invertibility of $\mathbf{XV}$. The product $-\mathbf{XV}^{-1} \cdot \mathbf{XI}$ represents the Z-parameters of the circuit, and these also may or may not exist.

In problems like this, the order of solution will determine the outcome. These problems are best solved using traditional matrix algebra methods. To solve this system, one must make a wise choice of the first *pivot* element. This is the element that appears in the upper left corner of the matrix and is the element that will need to be inverted to solve the system (a $2 \times 2$ matrix will require two matrix inversions). In this case, one matrix inversion is eliminated and a solution is guaranteed by choosing $\mathbb{I}$ as the pivot element. This is done by first permuting the rows and columns of the system to put $\mathbb{I}$ in the first pivot position:

$$\begin{pmatrix}\mathbf{0} & \mathbb{I}\\ \mathbb{I} & \mathbf{0}\end{pmatrix}\cdot\begin{pmatrix}\mathbf{XI} & \mathbf{XV}\\ \mathbf{Z0} & \mathbb{I}\end{pmatrix}\cdot\begin{pmatrix}\mathbf{0} & \mathbb{I}\\ \mathbb{I} & \mathbf{0}\end{pmatrix}\cdot\begin{pmatrix}\mathbf{0} & \mathbb{I}\\ \mathbb{I} & \mathbf{0}\end{pmatrix}\cdot\begin{pmatrix}\mathbf{I}\\ \mathbf{V}\end{pmatrix}=\begin{pmatrix}\mathbf{0} & \mathbb{I}\\ \mathbb{I} & \mathbf{0}\end{pmatrix}\cdot\begin{pmatrix}\mathbf{0}\\ 2\cdot\sqrt{\boldsymbol{\mathcal{Z}}\mathbf{0}}\end{pmatrix}. \tag{3.9}$$

The operations in (3.9) do not affect the outcome or the validity of the equation because the matrix on the left is multiplied by both sides and the inner product $\begin{pmatrix}\mathbf{0} & \mathbb{I}\\ \mathbb{I} & \mathbf{0}\end{pmatrix}\cdot\begin{pmatrix}\mathbf{0} & \mathbb{I}\\ \mathbb{I} & \mathbf{0}\end{pmatrix}$ is the identity. Simplifying (3.9), one obtains

$$\begin{pmatrix}\mathbb{I} & \mathbf{Z0}\\ \mathbf{XV} & \mathbf{XI}\end{pmatrix}\cdot\begin{pmatrix}\mathbf{V}\\ \mathbf{I}\end{pmatrix}=\begin{pmatrix}2\cdot\sqrt{\boldsymbol{\mathcal{Z}}\mathbf{0}}\\ \mathbf{0}\end{pmatrix},$$

and now $\mathbb{I}$ appears in the upper left corner as desired.

The lower left element is set to zero by a matrix multiplication on both sides:

$$\begin{pmatrix}\mathbb{I} & 0\\ -\mathbf{XV} & \mathbb{I}\end{pmatrix}\cdot\begin{pmatrix}\mathbb{I} & \mathbf{Z0}\\ \mathbf{XV} & \mathbf{XI}\end{pmatrix}\cdot\begin{pmatrix}\mathbf{V}\\ \mathbf{I}\end{pmatrix}=\begin{pmatrix}\mathbb{I} & 0\\ -\mathbf{XV} & \mathbb{I}\end{pmatrix}\cdot\begin{pmatrix}2\cdot\sqrt{\boldsymbol{\mathcal{Z}}\mathbf{0}}\\ \mathbf{0}\end{pmatrix}$$

to obtain

$$\begin{pmatrix}\mathbb{I} & \mathbf{Z0}\\ \mathbf{0} & \mathbf{XI}-\mathbf{XV}\cdot\mathbf{Z0}\end{pmatrix}\cdot\begin{pmatrix}\mathbf{V}\\ \mathbf{I}\end{pmatrix}=\begin{pmatrix}2\cdot\sqrt{\boldsymbol{\mathcal{Z}}\mathbf{0}}\\ -2\cdot\mathbf{XV}\cdot\sqrt{\boldsymbol{\mathcal{Z}}\mathbf{0}}\end{pmatrix}.$$

Ordinarily, this step involves the inversion of the pivot element, but here the pivot element is $\mathbb{I}$.

The solution for $\mathbf{I}$ is therefore

$$\mathbf{I}=-2\cdot(\mathbf{XI}-\mathbf{XV}\cdot\mathbf{Z0})^{-1}\cdot\mathbf{XV}\cdot\sqrt{\boldsymbol{\mathcal{Z}}\mathbf{0}}.$$

Plugging this into (3.8), the s-parameters are obtained:

$$\mathbf{S}=\mathbb{I}-\sqrt{\boldsymbol{\mathcal{Z}}\mathbf{0}}^{-1}\cdot\mathbf{Z0}\cdot\mathbf{I}=\mathbb{I}+2\cdot\sqrt{\boldsymbol{\mathcal{Z}}\mathbf{0}}^{-1}\cdot\mathbf{Z0}\cdot(\mathbf{XI}-\mathbf{XV}\cdot\mathbf{Z0})^{-1}\cdot\mathbf{XV}\cdot\sqrt{\boldsymbol{\mathcal{Z}}\mathbf{0}}. \tag{3.10}$$

If the normalization factor is the same for all ports, then $\sqrt{\boldsymbol{\mathcal{Z}}\mathbf{0}}=\sqrt{\mathcal{Z}0}\cdot\mathbb{I}$ and the s-parameters are calculated as

$$\mathbf{S}=\mathbb{I}+2\cdot\mathbf{Z0}\cdot(\mathbf{XI}-\mathbf{XV}\cdot\mathbf{Z0})^{-1}\cdot\mathbf{XV}.$$

Thus, the normalization factor is irrelevant under the situation where a single normalization factor is utilized.

As a final note here, understand that the choice for driving voltages of $\mathbf{VS}=2\cdot\sqrt{\mathcal{Z}0}$ is used to place unity voltage at the port voltage $\mathbf{V}$, and, since this voltage appears driven through the reference impedance, $\mathbf{A}=\mathbb{I}$. Note that $\mathbf{VS}$ does not appear in the final result; this choice merely simplified the derivation, but does nothing for the actual calculation, which is now performed using (3.10). In subsequent examples, $\mathbf{VS}=\mathbb{I}$ is used, realizing that this does not invalidate the result in (3.10).

This is the preferred solution to the problem. Therefore, the new steps using this method are provided as follows:

**Table 3.1** Computation of s-parameters of a circuit

| Situation | Equation |
|---|---|
| Port normalization factor and/or reference impedances are different | $\mathbf{S} = \mathbb{I} + 2 \cdot \sqrt{\mathbf{\mathcal{Z}0}}^{-1} \cdot \mathbf{Z0} \cdot (\mathbf{XI} - \mathbf{XV} \cdot \mathbf{Z0})^{-1} \cdot \mathbf{XV} \cdot \sqrt{\mathbf{\mathcal{Z}0}}$ |
| Normalization factors and reference impedances are the same on all ports | $\mathbf{S} = \mathbb{I} + 2 \cdot Z0 \cdot (\mathbf{XI} - \mathbf{XV} \cdot Z0)^{-1} \cdot \mathbf{XV}$ |
| Traveling wave case (required only when port reference impedances differ) | $\mathbf{S} = \mathbb{I} + 2 \cdot \sqrt{\mathbf{Z0}} \cdot (\mathbf{XI} - \mathbf{XV} \cdot \mathbf{Z0})^{-1} \cdot \mathbf{XV} \cdot \sqrt{\mathbf{Z0}}$ |

1. Label the exposed terminals of the circuit model with port numbers and label the voltages at and currents into the terminals, associating them with port numbers.
2. Choose a reference impedance and normalization factor for each port and arrange them along the diagonal of $\mathbf{Z0}$ and $\sqrt{\mathbf{\mathcal{Z}0}}$, respectively.
3. Write equations for each port current and voltage in terms of each other. Arrange the equations such that they are a sum of linear combinations of the port voltages and currents and such that they sum to zero as follows:

$$\sum_{p=1}^{P} xi_{mp} \cdot i_p + \sum_{p=1}^{P} xv_{mp} \cdot v_p = 0,$$

   where, for $P$ ports, $p \in 1 \ldots P$ and $m \in 1 \ldots P$, $i_p$ represents the current into port $p$, $v_p$ represents the voltage at port $p$, and $m$ represents the equation number.
4. Arrange the equations in matrix form such that $\mathbf{XI} \cdot \mathbf{i} = \sum_{p=1}^{P} xi_{mp} \cdot i_p$ and $\mathbf{XV} \cdot \mathbf{v} = \sum_{p=1}^{P} xv_{mp} \cdot v_p$.
5. Calculate the s-parameters using the formulas in Table 3.1, depending on the situation.

## 3.3 Example S-Parameter Circuit Calculations

Two procedures are outlined in §3.2. The first is the most straightforward, but it can be cumbersome as it involves the inverse of a larger matrix. The second is the preferred matrix algebra style. Both methods are provided for the examples in this section; however, some teaching of the matrix algebra method will be included for the first (more straightforward) method. When demonstrating the first method, one:

1. Writes all equations in the form of port voltages and currents, and writes the equations that relate the port voltage to the port current, reference impedance, and voltage sources.

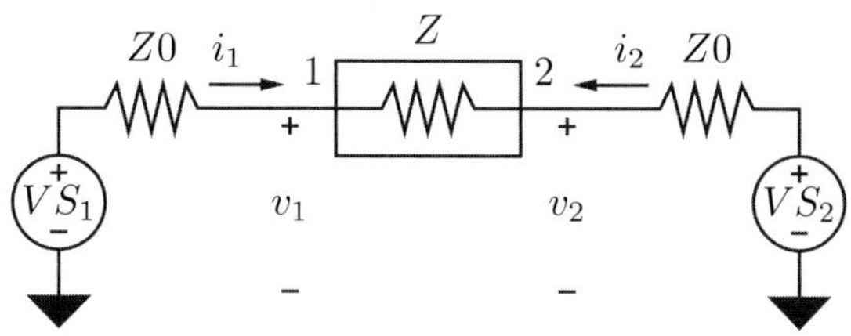


**Figure 3.1** A series impedance instrumented for s-parameter measurement

2. Writes the system of equations in matrix form (this is a good habit and helps in the identification of the form of the matrices when using the alternative method).
3. Chooses to drive the voltage sources alternately with unity voltage. This will result in $\mathbf{V} + \mathbf{Z0} \cdot \mathbf{I} = \mathbb{I}$.

The procedures outlined here are very simple, but there is a minor improvement that can often be employed to make things easier. The preference is to write all of the equations for the port voltages and currents and to solve the entire system in matrix form, as will be seen in the examples. When this is done, there is a minor issue in that the port voltage is not the same as the voltage source voltage because of the series reference impedance. So the preference is to use the port voltages in all equations (ignoring the voltage source) and then add the equations that relate the port voltage to the port voltage source through the port current and reference impedance. When this is carried out, a certain symmetry forms in the equations, but the reader is warned not to depend too much on this symmetry for the solution because the ability to utilize it is related to the existence of Z- or Y-parameters for the device being measured. As there are two equations in the preceding paragraph, one relating s-parameters to Z-parameters and the other relating them to Y-parameters, if Z-parameters do not exist, one obviously cannot use the Z-parameter equation (and the same goes for Y-parameters). There are some devices where neither Z- nor Y-parameters exist (a simple wire is one). If certain symmetry is depended on for a solution, this situation must be considered. This warning is demonstrated in the following example.

### 3.3.1 S-Parameter Model of a Series Impedance

A series impedance is shown instrumented for s-parameter measurement in Figure 3.1. The first step is to write all equations in terms of the port voltages and currents. In this case, two equations are generated. The first is for the single mesh through the resistor whose current is specified as $i_1$. Applying Kirchhoff's voltage law (KVL),[1]

$$-v_1 + i_1 \cdot Z + v_2 = 0.$$

Then, set $i_1$ equal to the negative of $i_2$:

$$i_1 + i_2 = 0.$$

[1]It is a common mistake to write the first equation as $-v_1 + (i_1 - i_2) \cdot Z + v_2 = 0$, mistaking the current through the impedance element as $i_1 - i_2$. It is not. It is $i_1$ or $-i_2$.

Now, all of the port voltages are specified in terms of the port voltage sources and port reference impedances:

$$v_1 + i_1 \cdot Z0_1 = VS_1,$$
$$v_2 + i_2 \cdot Z0_2 = VS_2.$$

This provides four equations with four unknowns, which can be written as

$$\begin{pmatrix} 1 & 1 & 0 & 0 \\ Z & 0 & -1 & 1 \\ Z0_1 & 0 & 1 & 0 \\ 0 & Z0_2 & 0 & 1 \end{pmatrix} \cdot \begin{pmatrix} i_1 \\ i_2 \\ v_1 \\ v_2 \end{pmatrix} = \begin{pmatrix} 0 \\ 0 \\ VS_1 \\ VS_2 \end{pmatrix}.$$

The two sets of voltage and current measurements can be performed by setting $VS_1$ and $VS_2$ alternately to 1:

$$\begin{pmatrix} \mathbf{I} \\ \mathbf{V} \end{pmatrix} = \begin{pmatrix} i_{11} & i_{12} \\ i_{21} & i_{22} \\ v_{11} & v_{12} \\ v_{21} & v_{22} \end{pmatrix} = \begin{pmatrix} 1 & 1 & 0 & 0 \\ Z & 0 & -1 & 1 \\ Z0_1 & 0 & 1 & 0 \\ 0 & Z0_2 & 0 & 1 \end{pmatrix}^{-1} \cdot \begin{pmatrix} 0 & 0 \\ 0 & 0 \\ 1 & 0 \\ 0 & 1 \end{pmatrix}$$
$$= \frac{1}{Z + Z0_1 + Z0_2} \cdot \begin{pmatrix} 1 & -1 \\ -1 & 1 \\ Z + Z0_2 & Z0_1 \\ Z0_2 & Z + Z0_1 \end{pmatrix}.$$

The s-parameters are solved as follows:

$$\mathbf{S}_{series\ Z} = \sqrt{\mathbf{Z0}}^{-1} \cdot (\mathbf{V} - \mathbf{Z0} \cdot \mathbf{I}) \cdot (\mathbf{V} + \mathbf{Z0} \cdot \mathbf{I})^{-1} \cdot \sqrt{\mathbf{Z0}}$$
$$= \begin{pmatrix} \frac{1}{\sqrt{Z0}_1} & 0 \\ 0 & \frac{1}{\sqrt{Z0}_2} \end{pmatrix} \cdot \left[ \left(\begin{smallmatrix} Z+Z0_2 & Z0_1 \\ Z0_2 & Z+Z0_1 \end{smallmatrix}\right) - \left(\begin{smallmatrix} Z0_1 & 0 \\ 0 & Z0_2 \end{smallmatrix}\right) \cdot \left(\begin{smallmatrix} 1 & -1 \\ -1 & 1 \end{smallmatrix}\right) \right]$$
$$\cdot \left[ \left(\begin{smallmatrix} Z+Z0_2 & Z0_1 \\ Z0_2 & Z+Z0_1 \end{smallmatrix}\right) + \left(\begin{smallmatrix} Z0_1 & 0 \\ 0 & Z0_2 \end{smallmatrix}\right) \cdot \left(\begin{smallmatrix} 1 & -1 \\ -1 & 1 \end{smallmatrix}\right) \right]^{-1} \cdot \begin{pmatrix} \sqrt{Z0}_1 & 0 \\ 0 & \sqrt{Z0}_2 \end{pmatrix}$$
$$= \frac{1}{Z + Z0_1 + Z0_2} \cdot \begin{pmatrix} Z + Z0_2 - Z0_1 & 2 \cdot \frac{\sqrt{Z0}_2}{\sqrt{Z0}_1} \cdot Z0_1 \\ 2 \cdot \frac{\sqrt{Z0}_1}{\sqrt{Z0}_2} \cdot Z0_2 & Z + Z0_1 - Z0_2 \end{pmatrix}.$$

Using the second method,

$$\mathbf{S}_{series\ Z} = \mathbb{I} + 2 \cdot \sqrt{\mathbf{Z0}}^{-1} \cdot \mathbf{Z0} \cdot (\mathbf{XI} - \mathbf{XV} \cdot \mathbf{Z0})^{-1} \cdot \mathbf{XV} \cdot \sqrt{\mathbf{Z0}}$$
$$= \left(\begin{smallmatrix} 1 & 0 \\ 0 & 1 \end{smallmatrix}\right) + 2 \cdot \begin{pmatrix} \frac{1}{\sqrt{Z0}_1} & 0 \\ 0 & \frac{1}{\sqrt{Z0}_2} \end{pmatrix} \cdot \left(\begin{smallmatrix} Z0_1 & 0 \\ 0 & Z0_2 \end{smallmatrix}\right)$$
$$\cdot \left[ \left(\begin{smallmatrix} 1 & 1 \\ Z & 0 \end{smallmatrix}\right) - \left(\begin{smallmatrix} 0 & 0 \\ -1 & 1 \end{smallmatrix}\right) \cdot \left(\begin{smallmatrix} Z0_1 & 0 \\ 0 & Z0_2 \end{smallmatrix}\right) \right]^{-1} \cdot \left(\begin{smallmatrix} 0 & 0 \\ -1 & 1 \end{smallmatrix}\right) \cdot \begin{pmatrix} \sqrt{Z0}_1 & 0 \\ 0 & \sqrt{Z0}_2 \end{pmatrix}$$
$$= \frac{1}{Z + Z0_1 + Z0_2} \cdot \begin{pmatrix} Z + Z0_2 - Z0_1 & 2 \cdot \frac{\sqrt{Z0}_2}{\sqrt{Z0}_1} \cdot Z0_1 \\ 2 \cdot \frac{\sqrt{Z0}_1}{\sqrt{Z0}_2} \cdot Z0_2 & Z + Z0_1 - Z0_2 \end{pmatrix}.$$

**Listing 3.1** S-parameters of series impedance

```
def SeriesZ(Z,Z0=50.):
    return [[Z/(Z+2.*Z0),2.*Z0/(Z+2.*Z0)],[2*Z0/(Z+2.*Z0),Z/(Z+2.*Z0)]]
```

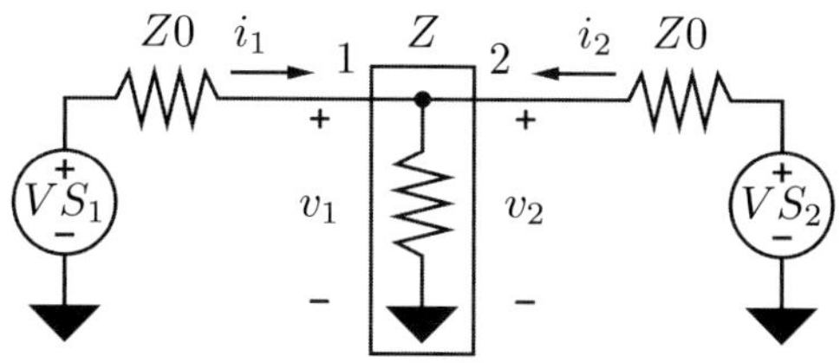


**Figure 3.2** A shunt impedance instrumented for s-parameter measurement

For a wire (i.e. $Z = 0$), this becomes

$$\mathbf{S}_{wire} = \frac{1}{Z0_1 + Z0_2} \cdot \begin{pmatrix} Z0_2 - Z0_1 & 2 \cdot \frac{\sqrt{Z0}_2}{\sqrt{Z0}_1} \cdot Z0_1 \\ 2 \cdot \frac{\sqrt{Z0}_1}{\sqrt{Z0}_2} \cdot Z0_2 & Z0_1 - Z0_2 \end{pmatrix}. \quad (3.11)$$

In both cases, if there is a single reference impedance $Z0$ such that $\mathbf{Z0} = Z0 \cdot \mathbb{I}$, and a single normalization factor $\sqrt{\mathbf{Z0}} = \sqrt{Z0} \cdot \mathbb{I}$,

$$\mathbf{S}_{series\,Z} = \frac{1}{Z + 2 \cdot Z0} \cdot \begin{pmatrix} Z & 2 \cdot Z0 \\ 2 \cdot Z0 & Z \end{pmatrix}.$$

The code for the series impedance is provided in Listing 3.1. Finally, for the case where the impedance is $Z = 0$,

$$\mathbf{S}_{wire} = \begin{pmatrix} 0 & 1 \\ 1 & 0 \end{pmatrix}. \quad (3.12)$$

Note that the s-parameters of a wire do not depend on the reference impedance, as long as the reference impedance at each port is the same. Further, they do not depend on the normalization factor as long as the normalization factor is the same at each port. The curious and very important equation for $\mathbf{S}_{wire}$ whereby the port impedances and normalization factors are different on each end is discussed further in Chapter 5.

### 3.3.2 S-Parameter Model of a Shunt Impedance

A shunt impedance is shown instrumented for s-parameter measurement in Figure 3.2. For a shunt impedance, the voltage at the center node is used to solve the system. All equations are written in terms of the port voltages and currents. In this case, there are two equations. The first is for the single node voltage specified as $v_1$:

$$v_1 = (i_1 + i_2) \cdot Z.$$

Then, two port voltages are set equal:

$$v_1 = v_2.$$

Now, all of the port voltages are specified in terms of the port voltage sources and port reference impedances:

$$v_1 + i_1 \cdot Z0_1 = VS_1,$$
$$v_2 + i_2 \cdot Z0_2 = VS_2.$$

This provides four equations with four unknowns, which can be written as follows:

$$\begin{pmatrix} -Z & -Z & 1 & 0 \\ 0 & 0 & 1 & -1 \\ Z0_1 & 0 & 1 & 0 \\ 0 & Z0_2 & 0 & 1 \end{pmatrix} \cdot \begin{pmatrix} i_1 \\ i_2 \\ v_1 \\ v_2 \end{pmatrix} = \begin{pmatrix} 0 \\ 0 \\ VS_1 \\ VS_2 \end{pmatrix}.$$

Setting $VS_1$ and $VS_2$ alternately to 1 performs the two sets of voltage and current measurements:

$$\begin{pmatrix} \mathbf{I} \\ \mathbf{V} \end{pmatrix} = \begin{pmatrix} i_{11} & i_{12} \\ i_{21} & i_{22} \\ v_{11} & v_{12} \\ v_{21} & v_{22} \end{pmatrix} = \begin{pmatrix} -Z & -Z & 1 & 0 \\ 0 & 0 & 1 & -1 \\ Z0_1 & 0 & 1 & 0 \\ 0 & Z0_2 & 0 & 1 \end{pmatrix}^{-1} \cdot \begin{pmatrix} 0 & 0 \\ 0 & 0 \\ 1 & 0 \\ 0 & 1 \end{pmatrix}$$
$$= \frac{1}{Z0_1 \cdot Z0_2 + Z \cdot (Z0_1 + Z0_2)} \cdot \begin{pmatrix} Z + Z0_2 & -Z \\ -Z & Z + Z0_1 \\ Z \cdot Z0_2 & Z \cdot Z0_1 \\ Z \cdot Z0_2 & Z \cdot Z0_1 \end{pmatrix}.$$

The s-parameters are solved as follows:

$$\mathbf{S}_{shunt\ Z} = \sqrt{\mathbf{Z0}}^{-1} \cdot (\mathbf{V} - \mathbf{Z0} \cdot \mathbf{I}) \cdot (\mathbf{V} + \mathbf{Z0} \cdot \mathbf{I})^{-1} \cdot \sqrt{\mathbf{Z0}}$$
$$= \begin{pmatrix} \frac{1}{\sqrt{Z0_1}} & 0 \\ 0 & \frac{1}{\sqrt{Z0_2}} \end{pmatrix} \cdot \left[ \begin{pmatrix} Z \cdot Z0_2 & Z \cdot Z0_1 \\ Z \cdot Z0_2 & Z \cdot Z0_1 \end{pmatrix} - \begin{pmatrix} Z0_1 & 0 \\ 0 & Z0_2 \end{pmatrix} \cdot \begin{pmatrix} Z+Z0_2 & -Z \\ -Z & Z+Z0_1 \end{pmatrix} \right]$$
$$\cdot \left[ \begin{pmatrix} Z \cdot Z0_2 & Z \cdot Z0_1 \\ Z \cdot Z0_2 & Z \cdot Z0_1 \end{pmatrix} + \begin{pmatrix} Z0_1 & 0 \\ 0 & Z0_2 \end{pmatrix} \cdot \begin{pmatrix} Z+Z0_2 & -Z \\ -Z & Z+Z0_1 \end{pmatrix} \right]^{-1} \cdot \begin{pmatrix} \sqrt{Z0_1} & 0 \\ 0 & \sqrt{Z0_2} \end{pmatrix}$$
$$= \frac{1}{Z0_1 \cdot Z0_2 + Z \cdot (Z0_1 + Z0_2)} \cdot \begin{pmatrix} Z \cdot (Z0_2 - Z0_1) - Z0_1 \cdot Z0_2 & 2 \cdot \frac{\sqrt{Z0_2}}{\sqrt{Z0_1}} \cdot Z0_1 \cdot Z \\ 2 \cdot \frac{\sqrt{Z0_1}}{\sqrt{Z0_2}} \cdot Z0_2 \cdot Z & Z \cdot (Z0_1 - Z0_2) - Z0_1 \cdot Z0_2 \end{pmatrix}.$$

The second method produces

$$\mathbf{S}_{shunt\ Z} = \mathbb{I} + 2 \cdot \sqrt{\mathbf{Z0}}^{-1} \cdot \mathbf{Z0} \cdot (\mathbf{XI} - \mathbf{XV} \cdot \mathbf{Z0})^{-1} \cdot \mathbf{XV} \cdot \sqrt{\mathbf{Z0}}$$
$$= \begin{pmatrix} 1 & 0 \\ 0 & 1 \end{pmatrix} + 2 \cdot \begin{pmatrix} \frac{1}{\sqrt{Z0_1}} & 0 \\ 0 & \frac{1}{\sqrt{Z0_2}} \end{pmatrix} \cdot \begin{pmatrix} Z0_1 & 0 \\ 0 & Z0_2 \end{pmatrix}$$
$$\cdot \left[ \begin{pmatrix} -Z & -Z \\ 0 & 0 \end{pmatrix} - \begin{pmatrix} 1 & 0 \\ 1 & -1 \end{pmatrix} \cdot \begin{pmatrix} Z0_1 & 0 \\ 0 & Z0_2 \end{pmatrix} \right]^{-1} \cdot \begin{pmatrix} 1 & 0 \\ 1 & -1 \end{pmatrix} \cdot \begin{pmatrix} \sqrt{Z0_1} & 0 \\ 0 & \sqrt{Z0_2} \end{pmatrix}$$
$$= \frac{1}{Z0_1 \cdot Z0_2 + Z \cdot (Z0_1 + Z0_2)} \cdot \begin{pmatrix} Z \cdot (Z0_2 - Z0_1) - Z0_1 \cdot Z0_2 & 2 \cdot \frac{\sqrt{Z0_2}}{\sqrt{Z0_1}} \cdot Z0_1 \cdot Z \\ 2 \cdot \frac{\sqrt{Z0_1}}{\sqrt{Z0_2}} \cdot Z0_2 \cdot Z & Z \cdot (Z0_1 - Z0_2) - Z0_1 \cdot Z0_2 \end{pmatrix}.$$

**Listing 3.2** S-parameters of shunt impedance

```
def ShuntZTwoPort(Z,Z0=50.):
    D=(Z0+2.*Z)
    return [[-Z0/D,2.*Z/D],
        [2.*Z/D,-Z0/D]]
```

For a short (i.e. $Z = 0$), one obtains

$$\mathbf{S}_{short} = \begin{pmatrix} -1 & 0 \\ 0 & -1 \end{pmatrix}.$$

The s-parameters of a short are independent of the reference impedance and the normalization factor.

In both cases, a single reference impedance $Z0$ such that $\mathbf{Z0} = Z0 \cdot \mathbb{I}$, and a single normalization factor $\sqrt{\boldsymbol{\mathcal{Z}}\mathbf{0}} = \sqrt{\mathcal{Z}0} \cdot \mathbb{I}$, results in

$$\mathbf{S}_{shunt\ Z} = \frac{1}{2 \cdot Z + Z0} \cdot \begin{pmatrix} -Z0 & 2 \cdot Z \\ 2 \cdot Z & -Z0 \end{pmatrix}.$$

The code for the shunt impedance device is shown in Listing 3.2. For $Z = \infty$, one obtains the s-parameters of a wire, as expected.

### 3.3.3 Mutual Inductance

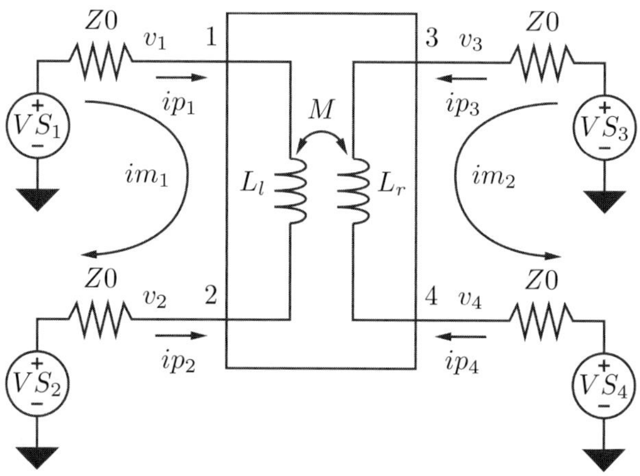


**Figure 3.3** Mutual inductance in test circuit

Consider the mutual inductance in Figure 3.3, which is shown in a test circuit for determining s-parameters. Since the concern here is the generation of network parameters,

**Listing 3.3** S-parameters of a mutual inductance

```
def Mutual(L1,Lr,M,f,Z0=None,K=None):
    s=1j*2.*math.pi*f
    D=s*s*(L1*Lr-M*M)+2*Z0*s*(L1+Lr)+4*Z0*Z0
    S11=(s*s*(L1*Lr-M*M)+2*s*L1*Z0)/D
    S12=(2*Z0*(s*Lr+2*Z0))/D
    S13=(2*s*M*Z0)/D
    S14=-S13
    S21=S12
    S22=S11
    S23=S14
    S24=S13
    S31=S13
    S32=S23
    S33=(s*s*(L1*Lr-M*M)+2*s*Lr*Z0)/D
    S34=2*Z0*(s*L1+2*Z0)/D
    S41=S14
    S42=S24
    S43=S34
    S44=S33
    return [[S11,S12,S13,S14],
            [S21,S22,S23,S24],
            [S31,S32,S33,S34],
            [S41,S42,S43,S44]]
```

the meaning of this device is not discussed. For a more complete understanding of this device, see [8].

A single reference impedance $Z0$ is used in this solution.

Four equations are needed to describe the operation of this transformer. The first two describe the behavior of the self and mutual inductance:

$$-v_1 + i_1 \cdot s \cdot L_l + i_3 \cdot s \cdot M + v_2 = 0,$$

$$-v_3 + i_3 \cdot s \cdot L_r + i_1 \cdot s \cdot M + v_4 = 0.$$

The second two equate the other two port currents:

$$i_1 + i_2 = 0,$$

$$i_3 + i_4 = 0.$$

This is written in matrix form as follows:

$$\begin{pmatrix} s \cdot L_l & 0 & s \cdot M & 0 & -1 & 1 & 0 & 0 \\ s \cdot M & 0 & s \cdot L_r & 0 & 0 & 0 & -1 & 1 \\ 1 & 1 & 0 & 0 & 0 & 0 & 0 & 0 \\ 0 & 0 & 1 & 1 & 0 & 0 & 0 & 0 \\ Z0 & 0 & 0 & 0 & 1 & 0 & 0 & 0 \\ 0 & Z0 & 0 & 0 & 0 & 1 & 0 & 0 \\ 0 & 0 & Z0 & 0 & 0 & 0 & 1 & 0 \\ 0 & 0 & 0 & Z0 & 0 & 0 & 0 & 1 \end{pmatrix} \cdot \begin{pmatrix} i_1 \\ i_2 \\ i_3 \\ i_4 \\ v_1 \\ v_2 \\ v_3 \\ v_4 \end{pmatrix} = \begin{pmatrix} 0 \\ 0 \\ 0 \\ 0 \\ VS_1 \\ VS_2 \\ VS_3 \\ VS_4 \end{pmatrix}.$$

Using the alternative form of solution outlined in §3.2.1 for expediency, one obtains

$$\begin{aligned}\mathbf{S}_{mutual+self} &= \mathbb{I} + 2\cdot\mathbf{Z0}\cdot\left(\mathbf{XI}-\mathbf{XV}\cdot\mathbf{Z0}\right)^{-1}\cdot\mathbf{XV}\\ &= \mathbb{I} + 2\cdot Z0\cdot\left[\begin{pmatrix} s\cdot L_l & 0 & s\cdot M & 0\\ s\cdot M & 0 & s\cdot L_r & 0\\ 1 & 1 & 0 & 0\\ 0 & 0 & 1 & 1\end{pmatrix} - \begin{pmatrix} -1 & 1 & 0 & 0\\ 0 & 0 & -1 & 1\\ 0 & 0 & 0 & 0\\ 0 & 0 & 0 & 0\end{pmatrix}\cdot Z0\right]^{-1}\cdot\begin{pmatrix} -1 & 1 & 0 & 0\\ 0 & 0 & -1 & 1\\ 0 & 0 & 0 & 0\\ 0 & 0 & 0 & 0\end{pmatrix}\\ &= \frac{1}{s^2\cdot(L_l\cdot L_r - M^2) + 2\cdot Z0\cdot s\cdot(L_l+L_r) + 4\cdot Z0^2}\\ &\quad\cdot\begin{pmatrix} \begin{matrix}s^2\cdot\left(L_l\cdot L_r - M^2\right)\\ +2\cdot s\cdot L_1 Z0\end{matrix} & 2\cdot Z0\cdot(s\cdot L_r+2\cdot Z0) & 2\cdot s\cdot M\cdot Z0 & -2\cdot s\cdot M\cdot Z0\\ 2\cdot Z0\cdot(s\cdot L_r+2\cdot Z0) & \begin{matrix}s^2\cdot\left(L_l\cdot L_r - M^2\right)\\ +2\cdot s\cdot L_l\cdot Z0\end{matrix} & -2\cdot s\cdot M\cdot Z0 & 2\cdot s\cdot M\cdot Z0\\ 2\cdot s\cdot M\cdot Z0 & -2\cdot s\cdot M\cdot Z0 & \begin{matrix}s^2\cdot\left(L_l\cdot L_r - M^2\right)\\ +2\cdot s\cdot L_r\cdot Z0\end{matrix} & 2\cdot Z0\cdot(s\cdot L_l+2\cdot Z0)\\ -2\cdot s\cdot M\cdot Z0 & 2\cdot s\cdot M\cdot Z0 & 2\cdot Z0\cdot(s\cdot L_l+2\cdot Z0) & \begin{matrix}s^2\cdot\left(L_l\cdot L_r - M^2\right)\\ +2\cdot s\cdot L_r\cdot Z0\end{matrix}\end{pmatrix}.\end{aligned}$$

The s-parameters of the mutual inductance device are provided in Listing 3.3. The Y-parameters are much simpler:

$$\mathbf{Y}_{mutual+self} = \frac{\begin{pmatrix} -L_r & L_r & M & -M\\ L_r & -L_r & -M & M\\ M & -M & -L_l & L_l\\ -M & M & L_l & -L_l\end{pmatrix}}{s\cdot\left(M^2 - L_l\cdot L_r\right)},$$

but note that they are undefined at zero frequency.

### 3.3.4 S-Parameter Model of a Circuit Node

One of the key concepts in analyzing s-parameter systems is the behavior at a circuit node. It turns out that a circuit node has s-parameters and it is improper to connect s-parameter networks to more than one other port of another device without considering the behavior at the node. This topic is covered completely in §4.1.4.

Consider $P$ wires joined at a node. In circuits, there is a dot at this point. To analyze the behavior of this junction using s-parameters, consider that there are $P$ equations at the node in addition to the three equations that determine the voltages at the port as a function of the port currents and the voltage sources. One of the equations is Kirchhoff's current law (KCL) – that the sum of the port currents into the node must sum to zero:

$$i_1 + i_2 + \ldots + i_P = 0.$$

There are $P - 1$ equations that set the port voltages equal at the node:

$$\begin{aligned} v_1 - v_2 &= 0,\\ &\vdots\\ v_1 - v_P &= 0.\end{aligned}$$

The equations are therefore written as

$$\begin{pmatrix} 1 & 1 & \cdots & 1 & 0 & 0 & \cdots & 0 \\ 0 & 0 & \cdots & 0 & 1 & -1 & \cdots & 0 \\ \vdots & \vdots & \ddots & \vdots & \vdots & \vdots & \ddots & 0 \\ 0 & 0 & \cdots & 0 & 1 & 0 & \cdots & -1 \\ Z0_1 & 0 & \cdots & 0 & 1 & 0 & \cdots & 0 \\ 0 & Z0_2 & \cdots & 0 & 0 & 1 & \cdots & 0 \\ \vdots & \vdots & \ddots & \vdots & \vdots & \vdots & \ddots & 0 \\ 0 & 0 & \cdots & Z0_P & 0 & 0 & \cdots & 1 \end{pmatrix} \cdot \begin{pmatrix} i_1 \\ i_2 \\ \vdots \\ i_P \\ v_1 \\ v_2 \\ \vdots \\ v_P \end{pmatrix} = \begin{pmatrix} 0 \\ 0 \\ \vdots \\ 0 \\ VS_1 \\ VS_2 \\ \vdots \\ VS_P \end{pmatrix}.$$

The solution is

$$\begin{aligned} \mathbf{S}_{P\text{-}port\,node} &= \mathbb{I} + 2 \cdot \mathbf{Z0} \cdot \left(\mathbf{XI} - \mathbf{XV} \cdot \mathbf{Z0}\right)^{-1} \cdot \mathbf{XV} \\ &= \mathbb{I} + 2 \cdot \begin{pmatrix} Z0_1 & 0 & \cdots & 0 \\ 0 & Z0_2 & \cdots & 0 \\ \vdots & \vdots & \ddots & \vdots \\ 0 & 0 & \cdots & Z0_P \end{pmatrix} \\ &\quad \cdot \left[ \begin{pmatrix} 1 & 1 & \cdots & 1 \\ 0 & 0 & \cdots & 0 \\ \vdots & \vdots & \ddots & \vdots \\ 0 & 0 & \cdots & 0 \end{pmatrix} - \begin{pmatrix} 0 & 0 & \cdots & 0 \\ 1 & -1 & \cdots & 0 \\ \vdots & \vdots & \ddots & 0 \\ 1 & 0 & \cdots & -1 \end{pmatrix} \cdot \begin{pmatrix} Z0_1 & 0 & \cdots & 0 \\ 0 & Z0_2 & \cdots & 0 \\ \vdots & \vdots & \ddots & \vdots \\ 0 & 0 & \cdots & Z0_P \end{pmatrix} \right]^{-1} \cdot \begin{pmatrix} 0 & 0 & \cdots & 0 \\ 1 & -1 & \cdots & 0 \\ \vdots & \vdots & \ddots & 0 \\ 1 & 0 & \cdots & -1 \end{pmatrix} \\ &= \frac{2}{\sum_p \frac{1}{Z0_p}} \cdot \begin{pmatrix} \frac{1}{Z0_1} & \frac{1}{Z0_2} & \cdots & \frac{1}{Z0_P} \\ \frac{1}{Z0_1} & \frac{1}{Z0_2} & \cdots & \frac{1}{Z0_P} \\ \vdots & \vdots & \ddots & \vdots \\ \frac{1}{Z0_1} & \frac{1}{Z0_2} & \cdots & \frac{1}{Z0_P} \end{pmatrix} - \mathbb{I}. \end{aligned} \tag{3.13}$$

Setting all of the reference impedances to the same value such that $Z0_p = Z0$ results in

$$\mathbf{S}_{P\text{-}port\,node} = \frac{1}{P} \cdot \begin{pmatrix} 2-P & 2 & \cdots & 2 \\ 2 & 2-P & \cdots & 2 \\ \vdots & \vdots & \ddots & \vdots \\ 2 & 2 & \cdots & 2-P \end{pmatrix}. \tag{3.14}$$

These results are valid for $P \geq 2$ ($P = 2$ produces the familiar s-parameters of a wire). This result is important because it means that if multiple s-parameter devices are connected together, and more than two devices connect to a node, then this device must be inserted into the network to facilitate the interconnection.

For the ideal tee, $P = 3$ and

$$\mathbf{S}_{ideal\,tee} = \begin{pmatrix} -1/3 & 2/3 & 2/3 \\ 2/3 & -1/3 & 2/3 \\ 2/3 & 2/3 & -1/3 \end{pmatrix}. \tag{3.15}$$

The code for a general $P$-port tee is provided in Listing 3.4.

**Listing 3.4** S-parameters of a tee

```
def Tee(P=None):
    if P is None: P=3
    return [[(2.-P)/P if r==c else 2./P for c in range(P)] for r in range(P)]
```

**Table 3.2** S-parameters in terms of Z-parameters

| Situation | Conversion equation |
|---|---|
| General conversion formula when port normalization factors and reference impedances are different | $\mathbf{S} = \sqrt{\mathbb{Z}\mathbb{0}}^{-1} \cdot (\mathbf{Z} - \mathbf{Z0})\,(\cdot\mathbf{Z} + \mathbf{Z0})^{-1} \cdot \sqrt{\mathbb{Z}\mathbb{0}}$ |
| Applicable conversion formula when normalization factors and reference impedances are the same on all ports | $\mathbf{S} = (\mathbf{Z} - \mathbb{I} \cdot Z0) \cdot (\mathbf{Z} + \mathbb{I} \cdot Z0)^{-1}$ (3.16) |
| Traveling wave case (required only when port reference impedances differ) | $\mathbf{S} = \sqrt{\mathbf{Z0}}^{-1} \cdot (\mathbf{Z} - \mathbf{Z0}) \cdot (\mathbf{Z} + \mathbf{Z0})^{-1} \cdot \sqrt{\mathbf{Z0}}$ |
| Single-port device | $\Gamma = \frac{Z-Z0}{Z+Z0}$ |

**Listing 3.5** General conversion from Z-parameters to s-parameters

```
def Z2S(Z,Z0=None,K=None):
    (Z0,K)=Z0KHelper((Z0,K),len(Z))
    Z=matrix(Z)
    return (K.getI()*(Z-Z0)*(Z+Z0).getI()*K).tolist()
```

## 3.4 S-Parameter Conversions

In this section, the network parameter conversion methods outlined in §1.3 are used to perform all of the necessary conversions between other network parameters and s-parameters.

### 3.4.1 S-Parameters In Terms Of Z-Parameters

To convert from Z-parameters to s-parameters, the voltage and current to wave conversion formula in (2.17) is used along with a permutation matrix and the method in §1.3:

$$\begin{pmatrix} \mathbf{A} \\ \mathbf{B} \end{pmatrix} = \frac{1}{2} \cdot \begin{pmatrix} \sqrt{\mathbb{Z}\mathbb{0}}^{-1} & \mathbf{0} \\ \mathbf{0} & \sqrt{\mathbb{Z}\mathbb{0}}^{-1} \end{pmatrix} \cdot \begin{pmatrix} \mathbb{I} & \mathbf{Z0} \\ \mathbb{I} & -\mathbf{Z0} \end{pmatrix} \cdot \begin{pmatrix} \mathbf{0} & \mathbb{I} \\ \mathbb{I} & \mathbf{0} \end{pmatrix} \cdot \begin{pmatrix} \mathbf{I} \\ \mathbf{V} \end{pmatrix},$$

and therefore

$$\mathbf{C} = \frac{1}{2} \cdot \begin{pmatrix} \sqrt{\mathbb{Z}\mathbb{0}}^{-1} & \mathbf{0} \\ \mathbf{0} & \sqrt{\mathbb{Z}\mathbb{0}}^{-1} \end{pmatrix} \cdot \begin{pmatrix} \mathbb{I} & \mathbf{Z0} \\ \mathbb{I} & -\mathbf{Z0} \end{pmatrix} \cdot \begin{pmatrix} \mathbf{0} & \mathbb{I} \\ \mathbb{I} & \mathbf{0} \end{pmatrix} = \begin{pmatrix} \frac{1}{2} \cdot \sqrt{\mathbb{Z}\mathbb{0}}^{-1} \cdot \mathbf{Z0} & \frac{1}{2} \cdot \sqrt{\mathbb{Z}\mathbb{0}}^{-1} \\ -\frac{1}{2} \cdot \sqrt{\mathbb{Z}\mathbb{0}}^{-1} \cdot \mathbf{Z0} & \frac{1}{2} \cdot \sqrt{\mathbb{Z}\mathbb{0}}^{-1} \end{pmatrix}.$$

**Table 3.3** Z-parameters in terms of s-parameters

| Situation | Conversion equation |
|---|---|
| General conversion formula when port normalization factors and reference impedances are different | $\mathbf{Z} = \sqrt{\mathcal{Z}\mathbf{0}} \cdot (\mathbb{I} + \mathbf{S}) \cdot (\mathbb{I} - \mathbf{S})^{-1} \cdot \mathbf{Z0} \cdot \sqrt{\mathcal{Z}\mathbf{0}}^{-1}$ |
| Applicable conversion formula when normalization factors and reference impedances are the same on all ports | $\mathbf{Z} = (\mathbb{I} + \mathbf{S}) \cdot (\mathbb{I} - \mathbf{S})^{-1} \cdot Z0$ |
| Traveling wave case (required only when port reference impedances differ) | $\mathbf{Z} = \sqrt{\mathbf{Z0}} \cdot (\mathbb{I} + \mathbf{S}) \cdot (\mathbb{I} - \mathbf{S})^{-1} \cdot \sqrt{\mathbf{Z0}}$ |
| Single-port device | $Z = Z0 \cdot \frac{1+\Gamma}{1-\Gamma}$ |

**Listing 3.6** General conversion from s-parameters to Z-parameters

```
def S2Z(S,Z0=None,K=None):
    (Z0,K)=Z0KHelper((Z0,K),len(S))
    I=matrix(identity(len(S)))
    S=matrix(S)
    return (K*(I+S)*(I-S).getI()*Z0*K.getI()).tolist()
```

Thus,

$$
\begin{aligned}
\mathbf{S} &= (\mathbf{C_{21}} + \mathbf{C_{22}} \cdot \mathbf{Z}) \cdot (\mathbf{C_{11}} + \mathbf{C_{12}} \cdot \mathbf{Z})^{-1} \\
&= \left(-\frac{1}{2} \cdot \sqrt{\mathcal{Z}\mathbf{0}}^{-1} \cdot \mathbf{Z0} + \frac{1}{2} \cdot \sqrt{\mathcal{Z}\mathbf{0}}^{-1} \cdot \mathbf{Z}\right) \cdot \left(\frac{1}{2} \cdot \sqrt{\mathcal{Z}\mathbf{0}}^{-1} \cdot \mathbf{Z0} + \frac{1}{2} \cdot \sqrt{\mathcal{Z}\mathbf{0}}^{-1} \cdot \mathbf{Z}\right)^{-1} \\
&= \sqrt{\mathcal{Z}\mathbf{0}}^{-1} \cdot (\mathbf{Z} - \mathbf{Z0}) \cdot \left[\sqrt{\mathcal{Z}\mathbf{0}}^{-1} \cdot (\mathbf{Z} + \mathbf{Z0})\right]^{-1} \\
&= \sqrt{\mathcal{Z}\mathbf{0}}^{-1} \cdot (\mathbf{Z} - \mathbf{Z0}) \cdot (\mathbf{Z} + \mathbf{Z0})^{-1} \cdot \sqrt{\mathcal{Z}\mathbf{0}}.
\end{aligned}
\tag{3.17}
$$

The conversion from Z-parameters to s-parameters under a variety of circumstances is shown in Table 3.2 and the code is provided in Listing 3.5.

### 3.4.2 Z-Parameters In Terms Of S-Parameters

To convert from s-parameters to Z-parameters, the wave to voltage and current conversion formula in (2.16) is used along with a permutation matrix and the method in §1.3:

$$
\begin{pmatrix} \mathbf{I} \\ \mathbf{V} \end{pmatrix} = \begin{pmatrix} \mathbf{0} & \mathbb{I} \\ \mathbb{I} & \mathbf{0} \end{pmatrix} \cdot \begin{pmatrix} \sqrt{\mathcal{Z}\mathbf{0}} & \mathbf{0} \\ \mathbf{0} & \sqrt{\mathcal{Z}\mathbf{0}} \end{pmatrix} \cdot \begin{pmatrix} \mathbb{I} & \mathbb{I} \\ \mathbf{Z0}^{-1} & -\mathbf{Z0}^{-1} \end{pmatrix} \cdot \begin{pmatrix} \mathbf{A} \\ \mathbf{B} \end{pmatrix},
$$

**Table 3.4** S-parameters in terms of Y-parameters

| **Situation** | **Conversion equation** |
|---|---|
| General conversion formula when port normalization factors and reference impedances are different | $\mathbf{S} = \sqrt{\mathcal{Z}\mathbf{0}}^{-1} \cdot (\mathbb{I} + \mathbf{Z0} \cdot \mathbf{Y})^{-1} \cdot (\mathbb{I} - \mathbf{Z0} \cdot \mathbf{Y}) \cdot \sqrt{\mathcal{Z}\mathbf{0}}$ |
| Applicable conversion formula when normalization factors and reference impedances are the same on all ports | $\mathbf{S} = (\mathbb{I} + Z0 \cdot \mathbf{Y})^{-1} \cdot (\mathbb{I} - Z0 \cdot \mathbf{Y}) \quad (3.18)$ |
| Traveling wave case (required only when port reference impedances differ) | $\mathbf{S} = \sqrt{\mathbf{Z0}}^{-1} \cdot (\mathbb{I} + \mathbf{Z0} \cdot \mathbf{Y})^{-1} \cdot (\mathbb{I} - \mathbf{Z0} \cdot \mathbf{Y}) \cdot \sqrt{\mathbf{Z0}}$ |
| Single-port device | $\Gamma = \frac{1 - Z0 \cdot Y}{1 + Z0 \cdot Y}$ |

or

$$\mathbf{C} = \begin{pmatrix} \mathbf{0} & \mathbb{I} \\ \mathbb{I} & \mathbf{0} \end{pmatrix} \cdot \begin{pmatrix} \sqrt{\mathcal{Z}\mathbf{0}} & \mathbf{0} \\ \mathbf{0} & \sqrt{\mathcal{Z}\mathbf{0}} \end{pmatrix} \cdot \begin{pmatrix} \mathbb{I} & \mathbb{I} \\ \mathbf{Z0}^{-1} & -\mathbf{Z0}^{-1} \end{pmatrix} = \begin{pmatrix} \sqrt{\mathcal{Z}\mathbf{0}} \cdot \mathbf{Z0}^{-1} & -\sqrt{\mathcal{Z}\mathbf{0}} \cdot \mathbf{Z0}^{-1} \\ \sqrt{\mathcal{Z}\mathbf{0}} & \sqrt{\mathcal{Z}\mathbf{0}} \end{pmatrix},$$

and therefore

$$\begin{aligned} \mathbf{Z} &= (\mathbf{C_{21}} + \mathbf{C_{22}} \cdot \mathbf{S}) \cdot (\mathbf{C_{11}} + \mathbf{C_{12}} \cdot \mathbf{S})^{-1} \\ &= \left(\sqrt{\mathcal{Z}\mathbf{0}} + \sqrt{\mathcal{Z}\mathbf{0}} \cdot \mathbf{S}\right) \cdot \left(\sqrt{\mathcal{Z}\mathbf{0}} \cdot \mathbf{Z0}^{-1} - \sqrt{\mathcal{Z}\mathbf{0}} \cdot \mathbf{Z0}^{-1} \cdot \mathbf{S}\right)^{-1} \\ &= \sqrt{\mathcal{Z}\mathbf{0}} \cdot (\mathbb{I} + \mathbf{S}) \cdot \left[\sqrt{\mathcal{Z}\mathbf{0}} \cdot \mathbf{Z0}^{-1} \cdot (\mathbb{I} - \mathbf{S})\right]^{-1} \\ &= \sqrt{\mathcal{Z}\mathbf{0}} \cdot (\mathbb{I} + \mathbf{S}) \cdot (\mathbb{I} - \mathbf{S})^{-1} \cdot \mathbf{Z0} \cdot \sqrt{\mathcal{Z}\mathbf{0}}^{-1}. \end{aligned}$$

Because of (C.13), one can also write

$$\mathbf{Z} = \sqrt{\mathcal{Z}\mathbf{0}} \cdot (\mathbb{I} - \mathbf{S})^{-1} \cdot (\mathbb{I} + \mathbf{S}) \cdot \mathbf{Z0} \cdot \sqrt{\mathcal{Z}\mathbf{0}}^{-1}.$$

The existence of Z-parameters therefore depends on the invertibility of $(\mathbb{I} - \mathbf{S})$. The conversion from s-parameters to Z-parameters under a variety of circumstances is shown in Table 3.3 and the code is provided in Listing 3.6.

### 3.4.3 S-Parameters In Terms Of Y-Parameters

To convert from Y-parameters to s-parameters, the voltage and current to wave conversion formula in (2.17) is used along with the method in §1.3:

$$\begin{pmatrix} \mathbf{A} \\ \mathbf{B} \end{pmatrix} = \frac{1}{2} \cdot \begin{pmatrix} \sqrt{\mathcal{Z}\mathbf{0}}^{-1} & \mathbf{0} \\ \mathbf{0} & \sqrt{\mathcal{Z}\mathbf{0}}^{-1} \end{pmatrix} \cdot \begin{pmatrix} \mathbb{I} & \mathbf{Z0} \\ \mathbb{I} & -\mathbf{Z0} \end{pmatrix} \cdot \begin{pmatrix} \mathbf{V} \\ \mathbf{I} \end{pmatrix},$$

**Listing 3.7** General conversion from Y-parameters to s-parameters

```
def Y2S(Y,Z0=None,K=None):
    (Z0,K)=Z0KHelper((Z0,K),len(Y))
    I=matrix(identity(len(Y)))
    Y=matrix(Y)
    return (K.getI()*(I+Z0*Y).getI()*(I-Z0*Y)*K).tolist()
```

**Listing 3.8** General conversion from s-parameters to Y-parameters

```
def S2Y(S,Z0=None,K=None):
    (Z0,K)=Z0KHelper((Z0,K),len(S))
    I=matrix(identity(len(S)))
    S=matrix(S)
    return (K*Z0.getI()*(I-S)*(I+S).getI()*K.getI()).tolist()
```

or

$$\mathbf{C} = \frac{1}{2}\cdot\begin{pmatrix} \sqrt{\mathcal{Z}\mathbf{0}}^{-1} & \mathbf{0} \\ \mathbf{0} & \sqrt{\mathcal{Z}\mathbf{0}}^{-1} \end{pmatrix}\cdot\begin{pmatrix} \mathbb{I} & \mathbf{Z0} \\ \mathbb{I} & -\mathbf{Z0} \end{pmatrix} = \begin{pmatrix} \frac{1}{2}\cdot\sqrt{\mathcal{Z}\mathbf{0}}^{-1} & \frac{1}{2}\cdot\sqrt{\mathcal{Z}\mathbf{0}}^{-1}\cdot\mathbf{Z0} \\ \frac{1}{2}\cdot\sqrt{\mathcal{Z}\mathbf{0}}^{-1} & -\frac{1}{2}\cdot\sqrt{\mathcal{Z}\mathbf{0}}^{-1}\cdot\mathbf{Z0} \end{pmatrix}.$$

Therefore,

$$\begin{aligned}\mathbf{S} &= (\mathbf{C_{21}} + \mathbf{C_{22}}\cdot\mathbf{Y})\cdot(\mathbf{C_{11}} + \mathbf{C_{12}}\cdot\mathbf{Y})^{-1} \\ &= \left(\frac{1}{2}\cdot\sqrt{\mathcal{Z}\mathbf{0}}^{-1} - \frac{1}{2}\cdot\sqrt{\mathcal{Z}\mathbf{0}}^{-1}\cdot\mathbf{Z0}\cdot\mathbf{Y}\right)\cdot\left(\frac{1}{2}\cdot\sqrt{\mathcal{Z}\mathbf{0}}^{-1} + \frac{1}{2}\cdot\sqrt{\mathcal{Z}\mathbf{0}}^{-1}\mathbf{Z0}\cdot\mathbf{Y}\right)^{-1} \\ &= \sqrt{\mathcal{Z}\mathbf{0}}^{-1}\cdot(\mathbb{I} - \mathbf{Z0}\cdot\mathbf{Y})\cdot\left[\sqrt{\mathcal{Z}\mathbf{0}}^{-1}\cdot(\mathbb{I} + \mathbf{Z0}\cdot\mathbf{Y})\right]^{-1} \\ &= \sqrt{\mathcal{Z}\mathbf{0}}^{-1}\cdot(\mathbb{I} - \mathbf{Z0}\cdot\mathbf{Y})\cdot(\mathbb{I} + \mathbf{Z0}\cdot\mathbf{Y})^{-1}\cdot\sqrt{\mathcal{Z}\mathbf{0}}. \end{aligned} \tag{3.19}$$

Because of (C.13), one can also write

$$\mathbf{S} = \sqrt{\mathcal{Z}\mathbf{0}}^{-1}\cdot(\mathbb{I} + \mathbf{Z0}\cdot\mathbf{Y})^{-1}\cdot(\mathbb{I} - \mathbf{Z0}\cdot\mathbf{Y})\cdot\sqrt{\mathcal{Z}\mathbf{0}}.$$

The conversion from Y-parameters to s-parameters under a variety of circumstances is shown in Table 3.4 and the code is provided in Listing 3.7.

### 3.4.4 Y-Parameters In Terms Of S-Parameters

To Convert from s-parameters to Y-parameters, the wave to voltage and current conversion formula in (2.16) is used along with a permutation matrix and the method in §1.3:

$$\begin{pmatrix} \mathbf{V} \\ \mathbf{I} \end{pmatrix} = \begin{pmatrix} \sqrt{\mathcal{Z}\mathbf{0}} & \mathbf{0} \\ \mathbf{0} & \sqrt{\mathcal{Z}\mathbf{0}} \end{pmatrix}\cdot\begin{pmatrix} \mathbb{I} & \mathbb{I} \\ \mathbf{Z0}^{-1} & -\mathbf{Z0}^{-1} \end{pmatrix}\cdot\begin{pmatrix} \mathbf{A} \\ \mathbf{B} \end{pmatrix},$$

or

$$\mathbf{C} = \begin{pmatrix} \sqrt{\mathcal{Z}\mathbf{0}} & \mathbf{0} \\ \mathbf{0} & \sqrt{\mathcal{Z}\mathbf{0}} \end{pmatrix}\cdot\begin{pmatrix} \mathbb{I} & \mathbb{I} \\ \mathbf{Z0}^{-1} & -\mathbf{Z0}^{-1} \end{pmatrix} = \begin{pmatrix} \sqrt{\mathcal{Z}\mathbf{0}} & \sqrt{\mathcal{Z}\mathbf{0}} \\ \sqrt{\mathcal{Z}\mathbf{0}}\cdot\mathbf{Z0}^{-1} & -\sqrt{\mathcal{Z}\mathbf{0}}\cdot\mathbf{Z0}^{-1} \end{pmatrix}.$$

**Table 3.5** Y-parameters in terms of s-parameters

| Situation | Conversion equation |
|---|---|
| General conversion formula when port normalization factors and reference impedances are different | $\mathbf{Y} = \sqrt{\mathcal{Z}\mathbf{0}} \cdot \mathbf{Z0}^{-1} \cdot (\mathbb{I} - \mathbf{S}) \cdot (\mathbb{I} + \mathbf{S})^{-1} \cdot \sqrt{\mathcal{Z}\mathbf{0}}^{-1}$ |
| Applicable conversion formula when normalization factors and reference impedances are the same on all ports | $\mathbf{Y} = \frac{1}{Z0} \cdot (\mathbb{I} - \mathbf{S}) \cdot (\mathbb{I} + \mathbf{S})^{-1}$ |
| Traveling wave case (required only when port reference impedances differ) | $\mathbf{Y} = \sqrt{\mathbf{Z0}} \cdot \mathbf{Z0}^{-1} \cdot (\mathbb{I} - \mathbf{S}) \cdot (\mathbb{I} + \mathbf{S})^{-1} \cdot \sqrt{\mathbf{Z0}}^{-1}$ |
| Single-port device | $Y = \frac{1}{Z0} \cdot \frac{1-\Gamma}{1+\Gamma}$ |

Therefore,

$$\begin{aligned}
\mathbf{Y} &= (\mathbf{C_{21}} + \mathbf{C_{22}} \cdot \mathbf{S}) \cdot (\mathbf{C_{11}} + \mathbf{C_{12}} \cdot \mathbf{S})^{-1} \\
&= \left(\sqrt{\mathcal{Z}\mathbf{0}} \cdot \mathbf{Z0}^{-1} - \sqrt{\mathcal{Z}\mathbf{0}} \cdot \mathbf{Z0}^{-1} \cdot \mathbf{S}\right) \cdot \left(\sqrt{\mathcal{Z}\mathbf{0}} + \sqrt{\mathcal{Z}\mathbf{0}} \cdot \mathbf{S}\right)^{-1} \\
&= \sqrt{\mathcal{Z}\mathbf{0}} \cdot \mathbf{Z0}^{-1} \cdot (\mathbb{I} - \mathbf{S}) \cdot \left[\sqrt{\mathcal{Z}\mathbf{0}} \cdot (\mathbb{I} + \mathbf{S})\right]^{-1} \\
&= \sqrt{\mathcal{Z}\mathbf{0}} \cdot \mathbf{Z0}^{-1} \cdot (\mathbb{I} - \mathbf{S}) \cdot (\mathbb{I} + \mathbf{S})^{-1} \cdot \sqrt{\mathcal{Z}\mathbf{0}}^{-1}.
\end{aligned}$$

Because of (C.13), one can also write

$$\mathbf{Y} = \sqrt{\mathcal{Z}\mathbf{0}} \cdot \mathbf{Z0}^{-1} \cdot (\mathbb{I} + \mathbf{S})^{-1} \cdot (\mathbb{I} - \mathbf{S}) \cdot \sqrt{\mathcal{Z}\mathbf{0}}^{-1}.$$

The existence of Y-parameters therefore depends on the invertibility of $(\mathbb{I} + \mathbf{S})$. The conversion from s-parameters to Y-parameters under a variety of circumstances is shown in Table 3.5 and the code is provided in Listing 3.8.

### 3.4.5 S-Parameters in Terms of ABCD Parameters

To convert from ABCD parameters to s-parameters, the voltage and current to wave conversion formulas in (2.11) and (2.12) are used along with the method in §1.3:

$$a_1 = \frac{1}{2 \cdot \sqrt{\mathcal{Z}0_1}} \cdot (v_1 + i_1 \cdot Z0_1), \qquad b_1 = \frac{1}{2 \cdot \sqrt{\mathcal{Z}0_1}} \cdot (v_1 - i_1 \cdot Z0_1),$$

$$a_2 = \frac{1}{2 \cdot \sqrt{\mathcal{Z}0_2}} \cdot (v_2 + i_2 \cdot Z0_2), \qquad b_2 = \frac{1}{2 \cdot \sqrt{\mathcal{Z}0_2}} \cdot (v_2 - i_2 \cdot Z0_2).$$

**Table 3.6** S-parameters in terms of ABCD parameters

| **Situation** | **Conversion equation** |
|---|---|
| General conversion formula when port normalization factors and reference impedances are different | $\mathbf{S} = \frac{1}{Z0_2\cdot A - B - Z0_1\cdot Z0_2\cdot C + Z0_1\cdot D} \cdot \begin{pmatrix} Z0_2\cdot A - B + Z0_1\cdot Z0_2 C - Z0_1\cdot D & 2\cdot Z0_1\cdot \frac{\sqrt{Z0_2}}{\sqrt{Z0_1}}\cdot (A\cdot D - C\cdot B) \\ 2\cdot Z0_2\cdot \frac{\sqrt{Z0_1}}{\sqrt{Z0_2}} & -Z0_2\cdot A - B + Z0_1\cdot Z0_2\cdot C + Z0_1\cdot D \end{pmatrix}$ |
| Applicable conversion formula when normalization factors and reference impedances are the same on all ports | $\mathbf{S} = \frac{1}{Z0\cdot A - B - Z0^2\cdot C + Z0\cdot D} \cdot \begin{pmatrix} Z0\cdot A - B + Z0^2\cdot C - Z0\cdot D & 2\cdot Z0\cdot (A\cdot D - C\cdot B) \\ 2\cdot Z0 & -Z0\cdot A - B + Z0^2\cdot C + Z0\cdot D \end{pmatrix} \quad (3.20)$ |
| Traveling wave case (required only when port reference impedances differ) | $\mathbf{S} = \frac{1}{Z0_2\cdot A - B - Z0_1\cdot Z0_2\cdot C + Z0_1\cdot D} \cdot \begin{pmatrix} Z0_2\cdot A - B + Z0_1\cdot Z0_2\cdot C - Z0_1\cdot D & 2\cdot \sqrt{Z0_1}\cdot \sqrt{Z0_2}\cdot (A\cdot D - C\cdot B) \\ 2\cdot \sqrt{Z0_1}\cdot \sqrt{Z0_2} & -Z0_2\cdot A - B + Z0_1\cdot Z0_2\cdot C + Z0_1 D \end{pmatrix}$ |

**Listing 3.9** General conversion from ABCD parameters to s-parameters

```
def ABCD2S(ABCD,Z0=None,K=None):
    (Z0,K)=Z0KHelper((Z0,K),len(ABCD))
    Z01=Z0.item(0,0)
    Z02=Z0.item(1,1)
    K1=K.item(0,0)
    K2=K.item(1,1)
    C11=matrix([[0,0],[1.0/(2.0*K2),Z02/(2.0*K2)]])
    C12=matrix([[1.0/(2.0*K1),-Z01/(2.0*K1)],[0,0]])
    C21=matrix([[0,0],[1.0/(2.0*K2),-Z02/(2.0*K2)]])
    C22=matrix([[1.0/(2.0*K1),Z01/(2.0*K1)],[0,0]])
    return array((C21+C22*ABCD)*((C11+C12*ABCD).getI())).tolist()
```

Ordering the waves and voltages in order of independent and dependent variables, based on the definition of s-parameters and ABCD parameters, produces

$$\begin{pmatrix} 0 & 0 & \frac{1}{2\cdot\sqrt{Z0_1}} & -\frac{1}{2\cdot\sqrt{Z0_1}}\cdot Z0_1 \\ \frac{1}{2\cdot\sqrt{Z0_2}} & \frac{1}{2\cdot\sqrt{Z0_2}}\cdot Z0_2 & 0 & 0 \\ 0 & 0 & \frac{1}{2\cdot\sqrt{Z0_1}} & \frac{1}{2\cdot\sqrt{Z0_1}}\cdot Z0_1 \\ \frac{1}{2\cdot\sqrt{Z0_2}} & -\frac{1}{2\cdot\sqrt{Z0_2}}\cdot Z0_2 & 0 & 0 \end{pmatrix} \cdot \begin{pmatrix} v_2 \\ i_2 \\ v_1 \\ -i_1 \end{pmatrix} = \begin{pmatrix} a_1 \\ a_2 \\ b_1 \\ b_2 \end{pmatrix},$$

and therefore

$$\mathbf{C_{11}} = \begin{pmatrix} 0 & 0 \\ \frac{1}{2\cdot\sqrt{Z0_2}} & \frac{1}{2\cdot\sqrt{Z0_2}}\cdot Z0_2 \end{pmatrix}, \qquad \mathbf{C_{12}} = \begin{pmatrix} \frac{1}{2\cdot\sqrt{Z0_1}} & -\frac{1}{2\cdot\sqrt{Z0_1}}\cdot Z0_1 \\ 0 & 0 \end{pmatrix},$$

$$\mathbf{C_{21}} = \begin{pmatrix} 0 & 0 \\ \frac{1}{2\cdot\sqrt{Z0_2}} & -\frac{1}{2\cdot\sqrt{Z0_2}}\cdot Z0_2 \end{pmatrix}, \qquad \mathbf{C_{22}} = \begin{pmatrix} \frac{1}{2\cdot\sqrt{Z0_1}} & \frac{1}{2\cdot\sqrt{Z0_1}}\cdot Z0_1 \\ 0 & 0 \end{pmatrix}.$$

**Table 3.7** ABCD parameters in terms of s-parameters

| Situation | Conversion equation |
|---|---|
| General conversion formula when port normalization factors and reference impedances are different | $\begin{pmatrix} A & B \\ C & D \end{pmatrix} = \frac{1}{2 \cdot S_{21}} \cdot \frac{\sqrt{\mathcal{Z}0_1}}{\sqrt{\mathcal{Z}0_2}} \cdot \begin{pmatrix} 1-S_{22}+S_{11}-\lvert\mathbf{S}\rvert & -Z0_2 \cdot (1+S_{22}+S_{11}+\lvert\mathbf{S}\rvert) \\ \frac{1}{Z0_1} \cdot (-1+S_{22}+S_{11}-\lvert\mathbf{S}\rvert) & \frac{Z0_2}{Z0_1} \cdot (1+S_{22}-S_{11}-\lvert\mathbf{S}\rvert) \end{pmatrix}$ |
| Applicable conversion formula when normalization factors and reference impedances are the same on all ports. | $\begin{pmatrix} A & B \\ C & D \end{pmatrix} = \frac{1}{2 \cdot S_{21}} \cdot \begin{pmatrix} 1-S_{22}+S_{11}-\lvert\mathbf{S}\rvert & -Z0 \cdot (1+S_{22}+S_{11}+\lvert\mathbf{S}\rvert) \\ \frac{1}{Z0} \cdot (-1+S_{22}+S_{11}-\lvert\mathbf{S}\rvert) & 1+S_{22}-S_{11}-\lvert\mathbf{S}\rvert \end{pmatrix}$ |
| Traveling wave case (required only when port reference impedances differ) | $\begin{pmatrix} A & B \\ C & D \end{pmatrix} = \frac{1}{2 \cdot S_{21}} \cdot \frac{\sqrt{Z0_1}}{\sqrt{Z0_2}} \cdot \begin{pmatrix} 1-S_{22}+S_{11}-\lvert\mathbf{S}\rvert & -Z0_2 \cdot (1+S_{22}+S_{11}+\lvert\mathbf{S}\rvert) \\ \frac{1}{Z0_1} \cdot (-1+S_{22}+S_{11}-\lvert\mathbf{S}\rvert) & \frac{Z0_2}{Z0_1} \cdot (1+S_{22}-S_{11}-\lvert\mathbf{S}\rvert) \end{pmatrix}$ |

Thus,

$$
\begin{aligned}
\mathbf{S} &= (\mathbf{C_{21}} + \mathbf{C_{22}} \cdot \mathbf{A}) \cdot (\mathbf{C_{11}} + \mathbf{C_{12}} \cdot \mathbf{A})^{-1} \\
&= \left[ \begin{pmatrix} 0 & 0 \\ \frac{1}{2 \cdot \sqrt{\mathcal{Z}0_2}} & -\frac{1}{2 \cdot \sqrt{\mathcal{Z}0_2}} \cdot Z0_2 \end{pmatrix} + \begin{pmatrix} \frac{1}{2 \cdot \sqrt{\mathcal{Z}0_1}} & \frac{1}{2 \cdot \sqrt{\mathcal{Z}0_1}} \cdot Z0_1 \\ 0 & 0 \end{pmatrix} \cdot \begin{pmatrix} A & B \\ C & D \end{pmatrix} \right] \\
&\quad \cdot \left[ \begin{pmatrix} 0 & 0 \\ \frac{1}{2 \cdot \sqrt{\mathcal{Z}0_2}} & \frac{1}{2 \cdot \sqrt{\mathcal{Z}0_2}} Z0_2 \end{pmatrix} + \begin{pmatrix} \frac{1}{2 \cdot \sqrt{\mathcal{Z}0_1}} & -\frac{1}{2 \cdot \sqrt{\mathcal{Z}0_1}} \cdot Z0_1 \\ 0 & 0 \end{pmatrix} \cdot \begin{pmatrix} A & B \\ C & D \end{pmatrix} \right]^{-1} \\
&= \frac{\begin{pmatrix} Z0_2 \cdot A - B + Z0_1 \cdot Z0_2 \cdot C - Z0_1 \cdot D & 2 \cdot Z0_1 \cdot \frac{\sqrt{\mathcal{Z}0_2}}{\sqrt{\mathcal{Z}0_1}} \cdot (A \cdot D - C \cdot B) \\ 2 \cdot Z0_2 \cdot \frac{\sqrt{\mathcal{Z}0_1}}{\sqrt{\mathcal{Z}0_2}} & -Z0_2 \cdot A - B + Z0_1 \cdot Z0_2 \cdot C + Z0_1 \cdot D \end{pmatrix}}{Z0_2 \cdot A - B - Z0_1 \cdot Z0_2 \cdot C + Z0_1 \cdot D} .
\end{aligned}
$$

The conversion from s-parameters to ABCD parameters under a variety of circumstances is shown in Table 3.6 and the code is provided in Listing 3.9.

### 3.4.6 ABCD Parameters In Terms Of S-Parameters

To convert from s-parameters to ABCD parameters, the wave to voltage and current conversion formulas in (2.8) and (2.9) are used along with the method in §1.3:

$$
v_1 = \sqrt{\mathcal{Z}0_1} \cdot (a_1 + b_1) , \qquad i_1 = \frac{\sqrt{\mathcal{Z}0_1}}{Z0_1} \cdot (a_1 - b_1) ,
$$

$$
v_2 = \sqrt{\mathcal{Z}0_2} \cdot (a_2 + b_2) , \qquad i_2 = \frac{\sqrt{\mathcal{Z}0_2}}{Z0_2} \cdot (a_2 - b_2) .
$$

**Listing 3.10** General conversion from s-parameters to ABCD parameters

```
def S2ABCD(S,Z0=None,K=None):
    (Z0,K)=Z0KHelper((Z0,K),len(S))
    Z01=Z0.item(0,0)
    Z02=Z0.item(1,1)
    K1=K.item(0,0)
    K2=K.item(1,1)
    C11=matrix([[0,K2],[0,K2/Z02]])
    C12=matrix([[0,K2],[0,-K2/Z02]])
    C21=matrix([[K1,0],[-K1/Z01,0]])
    C22=matrix([[K1,0],[K1/Z01,0]])
    return array((C21+C22*S)*((C11+C12*S).getI())).tolist()
```

Ordering the waves and voltages in order of independent and dependent variables, based on the definition of s-parameters and ABCD parameters, produces

$$\begin{pmatrix} 0 & \sqrt{Z0_2} & 0 & \sqrt{Z0_2} \\ 0 & \frac{\sqrt{Z0_2}}{Z0_2} & 0 & -\frac{\sqrt{Z0_2}}{Z0_2} \\ \sqrt{Z0_1} & 0 & \sqrt{Z0_1} & 0 \\ -\frac{\sqrt{Z0_1}}{Z0_1} & 0 & \frac{\sqrt{Z0_1}}{Z0_1} & 0 \end{pmatrix} \cdot \begin{pmatrix} a_1 \\ a_2 \\ b_1 \\ b_2 \end{pmatrix} = \begin{pmatrix} v_2 \\ i_2 \\ v_1 \\ -i_1 \end{pmatrix},$$

and therefore

$$\mathbf{C_{11}} = \begin{pmatrix} 0 & \sqrt{Z0_2} \\ 0 & \frac{\sqrt{Z0_2}}{Z0_2} \end{pmatrix}, \qquad \mathbf{C_{12}} = \begin{pmatrix} 0 & \sqrt{Z0_2} \\ 0 & -\frac{\sqrt{Z0_2}}{Z0_2} \end{pmatrix},$$

$$\mathbf{C_{21}} = \begin{pmatrix} \sqrt{Z0_1} & 0 \\ -\frac{\sqrt{Z0_1}}{Z0_1} & 0 \end{pmatrix}, \qquad \mathbf{C_{22}} = \begin{pmatrix} \sqrt{Z0_1} & 0 \\ \frac{\sqrt{Z0_1}}{Z0_1} & 0 \end{pmatrix}.$$

Thus,

$$\begin{aligned} \left(\begin{smallmatrix} A & B \\ C & D \end{smallmatrix}\right) &= (\mathbf{C_{21}} + \mathbf{C_{22}} \cdot \mathbf{S}) \cdot (\mathbf{C_{11}} + \mathbf{C_{12}} \cdot \mathbf{S})^{-1} \\ &= \left[ \begin{pmatrix} \sqrt{Z0_1} & 0 \\ -\frac{\sqrt{Z0_1}}{Z0_1} & 0 \end{pmatrix} + \begin{pmatrix} \sqrt{Z0_1} & 0 \\ \frac{\sqrt{Z0_1}}{Z0_1} & 0 \end{pmatrix} \cdot \left(\begin{smallmatrix} S_{11} & S_{12} \\ S_{21} & S_{22} \end{smallmatrix}\right) \right] \cdot \left[ \begin{pmatrix} 0 & \sqrt{Z0_2} \\ 0 & \frac{\sqrt{Z0_2}}{Z0_2} \end{pmatrix} + \begin{pmatrix} 0 & \sqrt{Z0_2} \\ 0 & -\frac{\sqrt{Z0_2}}{Z0_2} \end{pmatrix} \cdot \left(\begin{smallmatrix} S_{11} & S_{12} \\ S_{21} & S_{22} \end{smallmatrix}\right) \right]^{-1} \\ &= \frac{1}{2 \cdot S_{21}} \cdot \frac{\sqrt{Z0_1}}{\sqrt{Z0_2}} \cdot \begin{pmatrix} 1 - S_{22} + S_{11} - |\mathbf{S}| & -Z0_2 \cdot (1 + S_{22} + S_{11} + |\mathbf{S}|) \\ \frac{1}{Z0_1} \cdot (-1 + S_{22} + S_{11} - |\mathbf{S}|) & \frac{Z0_2}{Z0_1} \cdot (1 + S_{22} - S_{11} - |\mathbf{S}|) \end{pmatrix}. \end{aligned}$$

The conversion from ABCD parameters to s-parameters under a variety of circumstances is shown in Table 3.7 and the code is provided in Listing 3.10.

## 3.5 Power Wave Based S-Parameters

Different wave types were discussed in Chapter 2, resulting in wave definitions that met certain requirements, specifically the reversal property as outlined in §2.2.1. As such, the power wave concept was discarded because its definition violated this reversal property.

**Listing 3.11** General conversion from power wave based to wave based s-parameters

```
def Sp2Sw(Sp,Z0w=None,Z0p=None,Kw=None):
    if Z0p is None:
        Z0p=Z0w
    (Z0w,Kw)=Z0KHelper((Z0w,Kw),len(Sp))
    (Z0p,Kp)=Z0KHelperPW(Z0p,len(Sp))
    Sp=matrix(Sp)
    Sw=(Kw.getI()*Kp*Z0p.real.getI()*((Z0p.conjugate()-Z0w)+(Z0p+Z0w)*Sp)*
        ((Z0p.conjugate()+Z0w)+(Z0p-Z0w)*Sp).getI()*Kw*Kp.getI()*Z0p.real)
    return Sw
```

There are still cases, especially with regard to field solvers, that utilize power waves as the basis of the waves used to form s-parameters. This section provides conversions between network parameters using the power wave definition.

In the remainder of this book (except when specifically distinguishing between the various wave definitions in Chapter 2), the term *wave* refers to the wave definitions that are outlined in §2.4. In this section, waves are still defined in this manner as simply *waves* and power waves are referred to as *power waves*. As such, sets of forward and reverse propagating waves and wave defined s-parameters have $w$ subscripts and all power waves and power wave defined s-parameters have $p$ subscripts. In the remainder of this book, these subscripts are removed and all waves are defined as the general wave variety defined in §2.4.

## 3.5.1 Wave Based S-Parameters In Terms Of Power Wave Based S-Parameters

Referring to (2.20), waves defined in terms of power waves are given by

$$\begin{pmatrix}\mathbf{A}_w\\\mathbf{B}_w\end{pmatrix}=\frac{1}{2}\cdot\begin{pmatrix}\sqrt{\mathcal{Z}\mathbf{0}}_w^{-1}\cdot\sqrt{\mathcal{Z}\mathbf{0}}_p\cdot\mathrm{Re}(\mathbf{Z0}_p)^{-1} & \mathbf{0}\\ \mathbf{0} & \sqrt{\mathcal{Z}\mathbf{0}}_w^{-1}\cdot\sqrt{\mathcal{Z}\mathbf{0}}_p\cdot\mathrm{Re}(\mathbf{Z0}_p)^{-1}\end{pmatrix}\cdot\begin{pmatrix}\mathbf{Z0}_p^*+\mathbf{Z0}_w & \mathbf{Z0}_p-\mathbf{Z0}_w\\ \mathbf{Z0}_p^*-\mathbf{Z0}_w & \mathbf{Z0}_p+\mathbf{Z0}_w\end{pmatrix}\cdot\begin{pmatrix}\mathbf{A}_p\\\mathbf{B}_p\end{pmatrix}.$$

Using the method in §1.3 defines $\mathbf{C}$ as

$$\mathbf{C}=\frac{1}{2}\cdot\begin{pmatrix}\sqrt{\mathcal{Z}\mathbf{0}}_w^{-1}\cdot\sqrt{\mathcal{Z}\mathbf{0}}_p\cdot\mathrm{Re}(\mathbf{Z0}_p)^{-1}\cdot\left(\mathbf{Z0}_p^*+\mathbf{Z0}_w\right) & \sqrt{\mathcal{Z}\mathbf{0}}_w^{-1}\cdot\sqrt{\mathcal{Z}\mathbf{0}}_p\cdot\mathrm{Re}(\mathbf{Z0}_p)^{-1}\cdot(\mathbf{Z0}_p-\mathbf{Z0}_w)\\ \sqrt{\mathcal{Z}\mathbf{0}}_w^{-1}\cdot\sqrt{\mathcal{Z}\mathbf{0}}_p\cdot\mathrm{Re}(\mathbf{Z0}_p)^{-1}\cdot\left(\mathbf{Z0}_p^*-\mathbf{Z0}_w\right) & \sqrt{\mathcal{Z}\mathbf{0}}_w^{-1}\cdot\sqrt{\mathcal{Z}\mathbf{0}}_p\cdot\mathrm{Re}(\mathbf{Z0}_p)^{-1}\cdot(\mathbf{Z0}_p+\mathbf{Z0}_w)\end{pmatrix},$$

and therefore

$$\begin{aligned}\mathbf{S}_w&=(\mathbf{C_{21}}+\mathbf{C_{22}}\cdot\mathbf{S}_p)\cdot(\mathbf{C_{11}}+\mathbf{C_{12}}\cdot\mathbf{S}_p)^{-1}\\&=\left(\sqrt{\mathcal{Z}\mathbf{0}}_w^{-1}\cdot\sqrt{\mathcal{Z}\mathbf{0}}_p\cdot\mathrm{Re}\left(\mathbf{Z0}_p\right)^{-1}\cdot\left[\left(\mathbf{Z0}_p^*-\mathbf{Z0}_w\right)+\left(\mathbf{Z0}_p+\mathbf{Z0}_w\right)\cdot\mathbf{S}_p\right]\right)\\&\quad\cdot\left(\sqrt{\mathcal{Z}\mathbf{0}}_w^{-1}\cdot\sqrt{\mathcal{Z}\mathbf{0}}_p\cdot\mathrm{Re}\left(\mathbf{Z0}_p\right)^{-1}\cdot\left[\left(\mathbf{Z0}_p^*+\mathbf{Z0}_w\right)+\left(\mathbf{Z0}_p-\mathbf{Z0}_w\right)\cdot\mathbf{S}_p\right]\right)^{-1}\\&=\sqrt{\mathcal{Z}\mathbf{0}}_w^{-1}\cdot\sqrt{\mathcal{Z}\mathbf{0}}_p\cdot\mathrm{Re}\left(\mathbf{Z0}_p\right)^{-1}\cdot\left[\left(\mathbf{Z0}_p^*-\mathbf{Z0}_w\right)+\left(\mathbf{Z0}_p+\mathbf{Z0}_w\right)\cdot\mathbf{S}_p\right]\\&\quad\cdot\left[\left(\mathbf{Z0}_p^*+\mathbf{Z0}_w\right)+\left(\mathbf{Z0}_p-\mathbf{Z0}_w\right)\cdot\mathbf{S}_p\right]^{-1}\cdot\sqrt{\mathcal{Z}\mathbf{0}}_w\cdot\sqrt{\mathcal{Z}\mathbf{0}}_p^{-1}\cdot\mathrm{Re}\left(\mathbf{Z0}_p\right).\end{aligned}$$

**Listing 3.12** General conversion from wave based to power wave based s-parameters

```
def Sw2Sp(Sp,Z0w=None,Z0p=None,Kw=None):
    if Z0p is None:
        Z0p=Z0w
    (Z0w,Kw)=Z0KHelper((Z0w,Kw),len(Sp))
    (Z0p,Kp)=Z0KHelperPW(Z0p,len(Sp))
    Sp=matrix(Sp)
    Sw=(Kp.getI()*Kw*Z0w.getI()*((Z0w-Z0p.conjugate())+(Z0w+Z0p.conjugate())*Sp)*
        ((Z0w+Z0p)+(Z0w-Z0p)*Sp).getI()*Kp*Kw.getI()*Z0w)
    return Sw
```

A conversion from power waves to waves, *without simultaneously making an impedance transformation* (i.e. $\mathbf{Z0}_p = \mathbf{Z0}_w = \mathbf{Z0}$), results in

$$\begin{aligned}
\mathbf{S}_w &= \sqrt{\mathcal{Z}\mathbf{0}}_w^{-1} \cdot \sqrt{\mathcal{Z}\mathbf{0}}_p \cdot \mathrm{Re}\left(\mathbf{Z0}\right)^{-1} \cdot \left[\left(\mathbf{Z0}^* - \mathbf{Z0}\right) + \left(\mathbf{Z0} + \mathbf{Z0}\right) \cdot \mathbf{S}_p\right] \\
&\quad \cdot \left[\left(\mathbf{Z0}^* + \mathbf{Z0}\right) + \left(\mathbf{Z0} - \mathbf{Z0}\right) \cdot \mathbf{S}_p\right]^{-1} \cdot \sqrt{\mathcal{Z}\mathbf{0}}_w \cdot \sqrt{\mathcal{Z}\mathbf{0}}_p^{-1} \cdot \mathrm{Re}\left(\mathbf{Z0}\right) \\
&= \sqrt{\mathcal{Z}\mathbf{0}}_w^{-1} \cdot \sqrt{\mathcal{Z}\mathbf{0}}_p \cdot \mathrm{Re}\left(\mathbf{Z0}\right)^{-1} \cdot \left[2 \cdot \mathbf{Z0} \cdot \mathbf{S}_p - 2 \cdot j \cdot \mathrm{Im}\left(\mathbf{Z0}\right)\right] \\
&\quad \cdot \left[2 \cdot \mathrm{Re}\left(\mathbf{Z0}\right)\right]^{-1} \cdot \sqrt{\mathcal{Z}\mathbf{0}}_w \cdot \sqrt{\mathcal{Z}\mathbf{0}}_p^{-1} \cdot \mathrm{Re}\left(\mathbf{Z0}\right) \\
&= \sqrt{\mathcal{Z}\mathbf{0}}_w^{-1} \cdot \sqrt{\mathcal{Z}\mathbf{0}}_p \cdot \mathrm{Re}\left(\mathbf{Z0}\right)^{-1} \cdot \left[\mathbf{Z0} \cdot \mathbf{S}_p - j \cdot \mathrm{Im}\left(\mathbf{Z0}\right)\right] \cdot \sqrt{\mathcal{Z}\mathbf{0}}_w \cdot \sqrt{\mathcal{Z}\mathbf{0}}_p^{-1} \\
&= \sqrt{\mathcal{Z}\mathbf{0}}_w^{-1} \cdot \sqrt{\mathcal{Z}\mathbf{0}}_p \cdot \left[\mathbf{S}_p + j \cdot \mathrm{Im}\left(\mathbf{Z0}\right) \cdot \mathrm{Re}\left(\mathbf{Z0}\right)^{-1} \cdot \left(\mathbf{S}_p - \mathbb{I}\right)\right] \cdot \sqrt{\mathcal{Z}\mathbf{0}}_w \cdot \sqrt{\mathcal{Z}\mathbf{0}}_p^{-1}.
\end{aligned}$$

If the power wave normalization factor $\sqrt{\mathcal{Z}\mathbf{0}}_p = \sqrt{\left|\mathrm{Re}\left(\mathbf{Z0}\right)\right|}$ and the traveling wave normalization factor $\sqrt{\mathcal{Z}\mathbf{0}}_w = \sqrt{\mathbf{Z0}}$ are used, the result is

$$\mathbf{S}_w = \sqrt{\mathbf{Z0}}^{-1} \cdot \sqrt{\left|\mathrm{Re}\left(\mathbf{Z0}\right)\right|} \cdot \left[\mathbf{S}_p + j \cdot \mathrm{Im}\left(\mathbf{Z0}\right) \cdot \mathrm{Re}\left(\mathbf{Z0}\right)^{-1} \cdot \left(\mathbf{S}_p - \mathbb{I}\right)\right] \cdot \sqrt{\mathbf{Z0}} \cdot \sqrt{\left|\mathrm{Re}\left(\mathbf{Z0}\right)\right|}^{-1}.$$

As a check, if the wave based s-parameters use the traveling wave normalization factors and if the reference impedance is real (i.e. $\mathbf{Z0} \in \mathbb{R}$), then both types of s-parameters are identical. The code for converting power wave based to wave based s-parameters is provided in Listing 3.11.

### 3.5.2 Power Wave Based S-Parameters In Terms Of Wave Based S-Parameters

Referring to (2.20), waves defined in terms of power waves are given by

$$\begin{pmatrix} \mathbf{A}_p \\ \mathbf{B}_p \end{pmatrix} = \frac{1}{2} \cdot \begin{pmatrix} \sqrt{\mathcal{Z}\mathbf{0}}_p^{-1} \cdot \sqrt{\mathcal{Z}\mathbf{0}}_w \cdot \mathbf{Z0}_w^{-1} & \mathbf{0} \\ \mathbf{0} & \sqrt{\mathcal{Z}\mathbf{0}}_p^{-1} \cdot \sqrt{\mathcal{Z}\mathbf{0}}_w \cdot \mathbf{Z0}_w^{-1} \end{pmatrix} \cdot \begin{pmatrix} \mathbf{Z0}_w + \mathbf{Z0}_p & \mathbf{Z0}_w - \mathbf{Z0}_p \\ \mathbf{Z0}_w - \mathbf{Z0}_p^* & \mathbf{Z0}_w + \mathbf{Z0}_p^* \end{pmatrix} \cdot \begin{pmatrix} \mathbf{A}_w \\ \mathbf{B}_w \end{pmatrix}.$$

Using the method in §1.3 defines $\mathbf{C}$ as

$$\mathbf{C} = \frac{1}{2} \cdot \begin{pmatrix} \sqrt{\mathcal{Z}\mathbf{0}}_p^{-1} \cdot \sqrt{\mathcal{Z}\mathbf{0}}_w \cdot \mathbf{Z0}_w^{-1} \cdot \left(\mathbf{Z0}_w + \mathbf{Z0}_p\right) & \sqrt{\mathcal{Z}\mathbf{0}}_p^{-1} \cdot \sqrt{\mathcal{Z}\mathbf{0}}_w \cdot \mathbf{Z0}_w^{-1} \cdot \left(\mathbf{Z0}_w - \mathbf{Z0}_p\right) \\ \sqrt{\mathcal{Z}\mathbf{0}}_p^{-1} \cdot \sqrt{\mathcal{Z}\mathbf{0}}_w \cdot \mathbf{Z0}_w^{-1} \cdot \left(\mathbf{Z0}_w - \mathbf{Z0}_p^*\right) & \sqrt{\mathcal{Z}\mathbf{0}}_p^{-1} \cdot \sqrt{\mathcal{Z}\mathbf{0}}_w \cdot \mathbf{Z0}_w^{-1} \cdot \left(\mathbf{Z0}_w + \mathbf{Z0}_p^*\right) \end{pmatrix},$$

and therefore

$$
\begin{aligned}
\mathbf{S}_p &= (\mathbf{C_{21}} + \mathbf{C_{22}}{\cdot}\mathbf{S}_w) \cdot (\mathbf{C_{11}} + \mathbf{C_{12}} \cdot \mathbf{S}_w)^{-1} \\
&= \left( \sqrt{\mathcal{Z}\mathbf{0}}_p^{-1} \cdot \sqrt{\mathcal{Z}\mathbf{0}}_w \cdot \mathbf{Z0}_w^{-1} \cdot \left[ \left(\mathbf{Z0}_w - \mathbf{Z0}_p^*\right) + \left(\mathbf{Z0}_w + \mathbf{Z0}_p^*\right) \cdot \mathbf{S}_w \right] \right) \\
&\quad \cdot \left( \sqrt{\mathcal{Z}\mathbf{0}}_p^{-1} \cdot \sqrt{\mathcal{Z}\mathbf{0}}_w \cdot \mathbf{Z0}_w^{-1} \cdot \left[ (\mathbf{Z0}_w + \mathbf{Z0}_p) + (\mathbf{Z0}_w - \mathbf{Z0}_p) \cdot \mathbf{S}_w \right] \right)^{-1} \\
&= \sqrt{\mathcal{Z}\mathbf{0}}_p^{-1} \cdot \sqrt{\mathcal{Z}\mathbf{0}}_w \cdot \mathbf{Z0}_w^{-1} \cdot \left[ \left(\mathbf{Z0}_w - \mathbf{Z0}_p^*\right) + \left(\mathbf{Z0}_w + \mathbf{Z0}_p^*\right) \cdot \mathbf{S}_w \right] \\
&\quad \cdot \left[ (\mathbf{Z0}_w + \mathbf{Z0}_p) + (\mathbf{Z0}_w - \mathbf{Z0}_p) \cdot \mathbf{S}_w \right]^{-1} \cdot \sqrt{\mathcal{Z}\mathbf{0}}_p \cdot \sqrt{\mathcal{Z}\mathbf{0}}_w^{-1} \cdot \mathbf{Z0}_w .
\end{aligned}
$$

A conversion from power waves to waves, *without simultaneously making an impedance transformation* (i.e. $\mathbf{Z0}_p = \mathbf{Z0}_w = \mathbf{Z0}$), results in

$$
\begin{aligned}
\mathbf{S}_p &= \sqrt{\mathcal{Z}\mathbf{0}}_p^{-1} \cdot \sqrt{\mathcal{Z}\mathbf{0}}_w \cdot \mathbf{Z0}^{-1} \cdot \left[ (\mathbf{Z0} - \mathbf{Z0}^*) + (\mathbf{Z0} + \mathbf{Z0}^*) \cdot \mathbf{S}_w \right] \\
&\quad \cdot \left[ (\mathbf{Z0} + \mathbf{Z0}) + (\mathbf{Z0} - \mathbf{Z0}) \cdot \mathbf{S}_w \right]^{-1} \cdot \sqrt{\mathcal{Z}\mathbf{0}}_p \cdot \sqrt{\mathcal{Z}\mathbf{0}}_w^{-1} \cdot \mathbf{Z0} \\
&= \sqrt{\mathcal{Z}\mathbf{0}}_p^{-1} \cdot \sqrt{\mathcal{Z}\mathbf{0}}_w \cdot \mathbf{Z0}^{-1} \cdot \left[ j \cdot \mathrm{Im}\,(\mathbf{Z0}) + \mathrm{Re}\,(\mathbf{Z0}) \cdot \mathbf{S}_w \right] \cdot \sqrt{\mathcal{Z}\mathbf{0}}_p \cdot \sqrt{\mathcal{Z}\mathbf{0}}_w^{-1} .
\end{aligned}
$$

If the power wave normalization factor $\sqrt{\mathcal{Z}\mathbf{0}}_p = \sqrt{|\mathrm{Re}\,(\mathbf{Z0})|}$ and the traveling wave normalization factor $\sqrt{\mathcal{Z}\mathbf{0}}_w = \sqrt{\mathbf{Z0}}$ are used, the result is

$$
\mathbf{S}_p = \sqrt{|\mathrm{Re}\,(\mathbf{Z0})|}^{-1} \cdot \sqrt{\mathbf{Z0}} \cdot \mathbf{Z0}^{-1} \cdot \left[ j \cdot \mathrm{Im}\,(\mathbf{Z0}) + \mathrm{Re}\,(\mathbf{Z0}) \cdot \mathbf{S}_w \right] \cdot \sqrt{|\mathrm{Re}\,(\mathbf{Z0})|} \cdot \sqrt{\mathbf{Z0}}^{-1} .
$$

As a check, if the wave based s-parameters use the traveling wave normalization factors and the reference impedance is real (i.e. $\mathbf{Z0} \in \mathbb{R}$), then both types of s-parameters are identical.

The code for converting wave based to power wave based s-parameters is provided in Listing 3.12.

## 3.6 T-Parameters

In §1.5, ABCD parameters were used to cascade networks, and it was shown that these parameters are useful in determining cascaded Z- and Y-parameter sections. T-parameters are the equivalent cascading counterpart to s-parameters. T-parameters are defined, for a two-port network, as follows:

$$
\begin{pmatrix} b_1 \\ a_1 \end{pmatrix} = \begin{pmatrix} T_{11} & T_{12} \\ T_{21} & T_{22} \end{pmatrix} \cdot \begin{pmatrix} a_2 \\ b_2 \end{pmatrix} . \tag{3.21}
$$

Consider now a network containing two cascaded two-port devices designated with T-parameters $\mathbf{TL}$ and $\mathbf{TR}$. The equations for the two devices are

$$
\begin{pmatrix} b_{r1} \\ a_{r1} \end{pmatrix} = \mathbf{TR} \cdot \begin{pmatrix} a_{r2} \\ b_{r2} \end{pmatrix} , \tag{3.22}
$$

$$\begin{pmatrix} b_{l1} \\ a_{l1} \end{pmatrix} = \mathbf{TL} \cdot \begin{pmatrix} a_{l2} \\ b_{l2} \end{pmatrix}. \tag{3.23}$$

A goal is to equate the forward propagating wave at the right side of the first device to the backward propagating wave at the left side of the second device, and the backward propagating wave at the right side of the first device to the forward propagating wave at the left side of the second device:

$$b_{l2} = a_{r1}, \qquad a_{l2} = b_{r1}.$$

This assumption of equality of the reverse and forward propagating waves is only correct if the two sets of waves are in the same reference impedance. To understand this, the voltage at and current into port 1 of the right device are written using (2.10):

$$\begin{pmatrix} v \\ i \end{pmatrix} = \begin{pmatrix} 1 & 0 \\ 0 & -1 \end{pmatrix} \cdot \begin{pmatrix} v_{r1} \\ i_{r1} \end{pmatrix} = \begin{pmatrix} 1 & 0 \\ 0 & -1 \end{pmatrix} \cdot \sqrt{Z0}_{r1} \cdot \begin{pmatrix} 1 & 1 \\ \frac{1}{Z0_{r1}} & -\frac{1}{Z0_{r1}} \end{pmatrix} \cdot \begin{pmatrix} a_{r1} \\ b_{r1} \end{pmatrix}, \tag{3.24}$$

and the voltage at and current into port 2 of the left device are written as

$$\begin{pmatrix} v \\ i \end{pmatrix} = \begin{pmatrix} v_{l2} \\ i_{l2} \end{pmatrix} = \sqrt{Z0}_{l2} \cdot \begin{pmatrix} 1 & 1 \\ \frac{1}{Z0_{l2}} & -\frac{1}{Z0_{ls}} \end{pmatrix} \cdot \begin{pmatrix} 0 & 1 \\ 1 & 0 \end{pmatrix} \cdot \begin{pmatrix} b_{l2} \\ a_{l2} \end{pmatrix}. \tag{3.25}$$

Note that the current in (3.24) is negated to define a reference forward current $i$ and that the order of the waves in (3.25) is rearranged.

Equating the two produces

$$\begin{pmatrix} a_{l2} \\ b_{l2} \end{pmatrix} = \frac{1}{2 \cdot Z0_{r1}} \cdot \frac{\sqrt{Z0}_{r1}}{\sqrt{Z0}_{l2}} \cdot \begin{pmatrix} Z0_{r1} + Z0_{l2} & Z0_{r1} - Z0_{l2} \\ Z0_{r1} - Z0_{l2} & Z0_{r1} + Z0_{l2} \end{pmatrix} \cdot \begin{pmatrix} b_{r1} \\ a_{r1} \end{pmatrix}. \tag{3.26}$$

When one defines

$$\rho = \frac{Z0_{r1} - Z0_{l2}}{Z0_{r1} + Z0_{l2}}$$

and substitutes this into (3.26),

$$Z0_{r1} = Z0_{l2} \cdot \frac{1+\rho}{1-\rho},$$

then (3.26) can be written as

$$\begin{pmatrix} a_{l2} \\ b_{l2} \end{pmatrix} = \frac{\sqrt{Z0}_{r1}}{\sqrt{Z0}_{l2}} \cdot \frac{1}{\rho+1} \cdot \begin{pmatrix} 1 & \rho \\ \rho & 1 \end{pmatrix} \cdot \begin{pmatrix} b_{r1} \\ a_{r1} \end{pmatrix}. \tag{3.27}$$

Substituting (3.22) and (3.23) into (3.27) produces

$$\begin{pmatrix} b_{l1} \\ a_{l1} \end{pmatrix} = \mathbf{TL} \cdot \frac{\sqrt{Z0}_{r1}}{\sqrt{Z0}_{l2}} \cdot \frac{1}{\rho+1} \cdot \begin{pmatrix} 1 & \rho \\ \rho & 1 \end{pmatrix} \cdot \mathbf{TR} \cdot \begin{pmatrix} a_{r2} \\ b_{r2} \end{pmatrix}. \tag{3.28}$$

Examining (3.28), it is seen that if $\rho = 0$ (meaning that $Z0_{r1} = Z0_{l2}$) and if $\sqrt{Z0}_{r1} = \sqrt{Z0}_{l2}$, or, in other words, if the port reference impedance and normalization factor for port 1 of the device on the right and port 2 of the device on the left are equal, then (3.28) reduces to

$$\begin{pmatrix} b_{l1} \\ a_{l1} \end{pmatrix} = \mathbf{TL} \cdot \mathbf{TR} \cdot \begin{pmatrix} a_{r2} \\ b_{r2} \end{pmatrix}.$$

This is the main use of T-parameters – to cascade two-port devices by simply multiplying the T-parameter matrices in the order that they are connected in a network. Generally, in order to perform this cascade, the port impedances must be the same, otherwise the complication surfaces in (3.28). This is discussed further in Chapter 5. For the moment, observe that the complication in the middle of (3.28) can be understood by defining a two-port device with T-parameters given by

$$\mathbf{TM} = \frac{\sqrt{\mathcal{Z}0}_{r1}}{\sqrt{\mathcal{Z}0}_{l2}} \cdot \frac{1}{1+\rho} \cdot \begin{pmatrix} 1 & \rho \\ \rho & 1 \end{pmatrix}, \tag{3.29}$$

and therefore

$$\begin{pmatrix} b_{l1} \\ a_{l1} \end{pmatrix} = \mathbf{TL} \cdot \mathbf{TM} \cdot \mathbf{TR} \cdot \begin{pmatrix} a_{r2} \\ b_{r2} \end{pmatrix}. \tag{3.30}$$

Equation (3.30) simply demonstrates further the cascading capability of T-parameters and highlights what to do when port impedances differ. The device **TM** is simply a reference impedance transformer, and **TM** is simply a matrix that represents the T-parameters of a reference impedance transformer that resolves the differences between the two port impedances.

The port impedance problem only shows up at ports that are joined. The result $\mathbf{TL} \cdot \mathbf{TM} \cdot \mathbf{TR}$ represents the T-parameters of a device that has a reference impedance for port 1 as the reference impedance of port 1 of the left device and a reference impedance for port 2 as the reference impedance of port 2 of the right device. In other words, the inner, connected, ports are resolved with impedance transformers, and the outer, unconnected, ports stay the same.

### 3.6.1 T-Parameters In Terms Of S-Parameters

Here, s-parameters are converted to T-parameters. For s-parameters, there are independent and dependent variables $\begin{pmatrix} a_1 \\ a_2 \\ b_1 \\ b_2 \end{pmatrix}$ and for T-parameters, using (3.21), $\begin{pmatrix} a_2 \\ b_2 \\ b_1 \\ a_1 \end{pmatrix}$. Therefore **C** is constructed such that

$$\begin{pmatrix} 0 & 1 & 0 & 0 \\ 0 & 0 & 0 & 1 \\ 0 & 0 & 1 & 0 \\ 1 & 0 & 0 & 0 \end{pmatrix} \cdot \begin{pmatrix} a_1 \\ a_2 \\ b_1 \\ b_2 \end{pmatrix} = \begin{pmatrix} a_2 \\ b_2 \\ b_1 \\ a_1 \end{pmatrix}.$$

The conversion can be written as follows:

$$\begin{aligned}\mathbf{T} &= (\mathbf{C_{21}} + \mathbf{C_{22}} \cdot \mathbf{S}) \cdot (\mathbf{C_{11}} + \mathbf{C_{12}} \cdot \mathbf{S})^{-1} \\ &= \left[\begin{pmatrix} 0 & 0 \\ 1 & 0 \end{pmatrix} + \begin{pmatrix} 1 & 0 \\ 0 & 0 \end{pmatrix} \cdot \begin{pmatrix} S_{11} & S_{12} \\ S_{21} & S_{22} \end{pmatrix}\right] \cdot \left[\begin{pmatrix} 0 & 1 \\ 0 & 0 \end{pmatrix} + \begin{pmatrix} 0 & 0 \\ 0 & 1 \end{pmatrix} \cdot \begin{pmatrix} S_{11} & S_{12} \\ S_{21} & S_{22} \end{pmatrix}\right]^{-1} \\ &= \begin{pmatrix} S_{11} & S_{12} \\ 1 & 0 \end{pmatrix} \begin{pmatrix} 0 & 1 \\ S_{21} & S_{22} \end{pmatrix}^{-1} = \frac{1}{S_{21}} \cdot \begin{pmatrix} -|\mathbf{S}| & S_{11} \\ -S_{22} & 1 \end{pmatrix}. \end{aligned} \tag{3.31}$$

For a two-port device, T-parameters with respect to s-parameters are given by

$$\mathbf{T} = \frac{\begin{pmatrix} -|\mathbf{S}| & S_{11} \\ -S_{22} & 1 \end{pmatrix}}{S_{21}}. \tag{3.32}$$

### 3.6.2 S-Parameters In Terms Of T-Parameters

Here, T-parameters are converted to s-parameters. For T-parameters, using (3.21), there are independent and dependent variables $\begin{pmatrix} a_2 \\ b_2 \\ b_1 \\ a_1 \end{pmatrix}$ and for s-parameters, $\begin{pmatrix} a_1 \\ a_2 \\ b_1 \\ b_2 \end{pmatrix}$. Therefore **C** is constructed such that

$$\begin{pmatrix} 0 & 0 & 0 & 1 \\ 1 & 0 & 0 & 0 \\ 0 & 0 & 1 & 0 \\ 0 & 1 & 0 & 0 \end{pmatrix} \cdot \begin{pmatrix} a_2 \\ b_2 \\ b_1 \\ a_1 \end{pmatrix} = \begin{pmatrix} a_1 \\ a_2 \\ b_1 \\ b_2 \end{pmatrix}.$$

The conversion can be written as follows:

$$\begin{aligned}\mathbf{S} &= (\mathbf{C_{21}} + \mathbf{C_{22}} \cdot \mathbf{T}) \cdot (\mathbf{C_{11}} + \mathbf{C_{12}} \cdot \mathbf{T})^{-1} \\ &= \left[\begin{pmatrix} 0 & 0 \\ 0 & 1 \end{pmatrix} + \begin{pmatrix} 1 & 0 \\ 0 & 0 \end{pmatrix} \cdot \begin{pmatrix} T_{11} & T_{12} \\ T_{21} & T_{22} \end{pmatrix}\right] \cdot \left[\begin{pmatrix} 0 & 0 \\ 1 & 0 \end{pmatrix} + \begin{pmatrix} 0 & 1 \\ 0 & 0 \end{pmatrix} \cdot \begin{pmatrix} T_{11} & T_{12} \\ T_{21} & T_{22} \end{pmatrix}\right]^{-1} \\ &= \begin{pmatrix} T_{11} & T_{12} \\ 0 & 1 \end{pmatrix} \cdot \begin{pmatrix} T_{21} & T_{22} \\ 1 & 0 \end{pmatrix}^{-1} = \frac{1}{T_{22}} \cdot \begin{pmatrix} T_{12} & |\mathbf{T}| \\ 1 & -T_{21} \end{pmatrix}. \end{aligned} \tag{3.33}$$

For a two-port device, s-parameters with respect to T-parameters are given by

$$\mathbf{S} = \frac{\begin{pmatrix} T_{12} & |\mathbf{T}| \\ 1 & -T_{21} \end{pmatrix}}{T_{22}}. \tag{3.34}$$

## 3.7 Cascading

Analogous with the usage of ABCD parameters to find Z-parameters of cascaded Z-parameter devices and Y-parameters of cascaded Y-parameter devices, here s-parameters are found for cascaded s-parameter devices using T-parameters. A reasonable strategy for cascading network parameters of other types is to convert the network parameters to T-parameters, multiply the T-parameter matrices together, and convert back to the network parameters of interest.

### 3.7.1 S-Parameters of Two Cascaded Two-Port S-Parameter Devices Using T-Parameters

Consider a network containing two cascaded two-port T-parameter devices **SL** and **SR**. The T-parameters, using (3.31), are

$$\mathbf{TL} = \frac{1}{SL_{21}} \cdot \begin{pmatrix} -|\mathbf{SL}| & SL_{11} \\ -SL_{22} & 1 \end{pmatrix}, \qquad \mathbf{TR} = \frac{1}{SR_{21}} \cdot \begin{pmatrix} -|\mathbf{SR}| & SR_{11} \\ -SR_{22} & 1 \end{pmatrix}.$$

Multiplying these together results in

$$\mathbf{T} = \mathbf{TL} \cdot \mathbf{TR} = \frac{1}{SL_{21} \cdot SR_{21}} \cdot \begin{pmatrix} |\mathbf{SL}| \cdot |\mathbf{SR}| - SL_{11} \cdot SR_{22} & SL_{11} - |\mathbf{SL}| \cdot SR_{11} \\ |\mathbf{SR}| \cdot SL_{22} - SR_{22} & 1 - SL_{22} \cdot SR_{11} \end{pmatrix}.$$

Utilizing (3.33),

$$\begin{aligned} \mathbf{S} &= \frac{1}{T_{22}} \cdot \begin{pmatrix} T_{12} & |\mathbf{T}| \\ 1 & -T_{21} \end{pmatrix} \\ &= \frac{1}{1 - SL_{22} \cdot SR_{11}} \cdot \begin{pmatrix} SL_{11} - SR_{11} \cdot |\mathbf{SL}| & SL_{12} \cdot SR_{12} \\ SL_{21} \cdot SR_{21} & SR_{22} - SL_{22} \cdot |\mathbf{SR}| \end{pmatrix}. \end{aligned} \tag{3.35}$$

When applying (3.35), one should remember that the reference impedance at port 1 of the result is equal to the reference impedance at port 1 of **SL**, the reference impedance at port 2 of the result is equal to the reference impedance at port 2 of **SR**, and that the equation is only valid if the reference impedance at port 2 of **SL** is the same as the reference impedance at port 1 of **SR**.

## 3.8 Inverse and Identity Sections

Consider an identity system given by T-parameters:

$$\mathbf{T}_{identity} = \begin{pmatrix} 1 & 0 \\ 0 & 1 \end{pmatrix}. \tag{3.36}$$

Equation (3.36) is the identity matrix. Since cascaded sections given by T-parameters are formed by multiplying T-parameter matrices, it is clear that multiplying any set of T-parameters by (3.36) leaves the system unchanged.

Given a set of T-parameters, the inverse matrix, when multiplied, forms the identity matrix. In other words, the de-embedding problem with T-parameter matrices is solved simply as follows:

$$\mathbf{T}_R = \mathbf{T}_L^{-1} \cdot \mathbf{T}.$$

The s-parameters corresponding to the identity matrix, utilizing (3.34), are

$$\mathbf{S}_{identity} = \begin{pmatrix} 0 & 1 \\ 1 & 0 \end{pmatrix}. \tag{3.37}$$

Note that $S_{identity}$ is not the identity matrix and represents the s-parameters of an ideal wire, as provided in (3.12).

The s-parameters for the *inverse* of a given section are defined as the section that, when cascaded with the given section, forms (3.37). Note that for a given section $\mathbf{S}$ it is not $\mathbf{S}^{-1}$ as this only works for T-parameters. In order to calculate the inverse of a section, consider that a section $\mathbf{S}$ has T-parameters given by (3.34), whose inverse is

$$\mathbf{T}^{-1} = \frac{\begin{pmatrix} 1 & -S_{11} \\ S_{22} & -|S| \end{pmatrix}}{S_{12}}. \tag{3.38}$$

Using (3.34), (3.38) becomes

$$\mathbf{S}_{inv} = \frac{1}{|\mathbf{S}|} \cdot \begin{pmatrix} S_{11} & -S_{21} \\ -S_{12} & S_{22} \end{pmatrix}. \tag{3.39}$$

Thus (3.39) defines the inverse section $\mathbf{S}_{inv}$ such that $\mathbf{S}$ cascaded with $\mathbf{S}_{inv}$ equals the identity section given by (3.37).

## 3.9 De-embedding S-Parameters

De-embedding s-parameters involves the removal of sections from a system. Consider a two-port system with known s-parameters given as $\mathbf{S}$ with a known section $\mathbf{SL}$ and an unknown section $\mathbf{SR}$, whereby $\mathbf{SR}$ is the result of removing $\mathbf{SL}$ from $\mathbf{S}$. In other words, the goal is to find the section $\mathbf{SR}$ such that $\mathbf{SL}$ cascaded with $\mathbf{SR}$ equals $\mathbf{S}$. This is found by cascading the inverse of $\mathbf{SL}$ as provided in (3.39) with $\mathbf{S}$ using (3.35):

$$\mathbf{SR} = \frac{1}{SL_{22} \cdot S_{11} - |\mathbf{SL}|} \cdot \begin{pmatrix} S_{11} - SL_{11} & SL_{21} \cdot S_{12} \\ SL_{12} \cdot S_{21} & SL_{22} \cdot |\mathbf{S}| - S_{22} \cdot |\mathbf{SL}| \end{pmatrix}. \tag{3.40}$$

Thus, (3.40) defines the section to the right, $\mathbf{SR}$, formed by removing a section to the left, $\mathbf{SL}$, from $\mathbf{S}$.

Similarly, given a two-port system with known s-parameters $\mathbf{S}$ with a known section $\mathbf{SR}$ and an unknown section $\mathbf{SL}$, whereby $\mathbf{SL}$ is the result of removing $\mathbf{SR}$ from $\mathbf{S}$, the goal is to find the section $\mathbf{SL}$ such that $\mathbf{SL}$ cascaded with $\mathbf{SR}$ equals $\mathbf{S}$. This is found by cascading $\mathbf{S}$ with the inverse of $\mathbf{SR}$ as provided in (3.39) using (3.35):

$$\mathbf{SL} = \frac{1}{S_{22} \cdot SR_{11} - |\mathbf{SR}|} \cdot \begin{pmatrix} SR_{11} \cdot |\mathbf{S}| - S_{11} \cdot |\mathbf{SR}| & S_{12} \cdot SR_{21} \\ S_{21} \cdot SR_{12} & S_{22} - SR_{22} \end{pmatrix}.$$

While the basic concept of de-embedding is provided here, it is covered comprehensively in Chapter 10.

## 3.10 Network Parameters of Common Elements

This section summarizes the network parameters of the most common elements.

### 3.10.1 Single-Port Impedance to Ground

Consider a single-port impedance $Z$ to ground. To determine the Z-parameters, a current source with a value of 1 ampere is connected to the port and the voltage, equal to $Z$ volts, is measured. Therefore the Z-parameters are simply $Z$. To determine the Y-parameters, a voltage source with a value of 1 volt is connected to the port and the current, equal to $1/Z$ amperes, is measured. Therefore the Y-parameters are $1/Z$. The s-parameters can be calculated using (3.16):

Network parameters of a single-port impedance $Z$ to ground

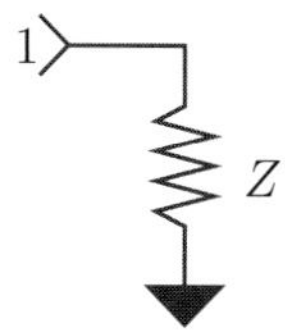


$$Z_{11} = Z,$$
$$Y_{11} = \frac{1}{Z},$$
$$\Gamma = \frac{Z - Z0}{Z + Z0}. \tag{3.41}$$

### 3.10.2 Two-Port Series Impedance

Consider a two-port impedance $Z$. The Z-parameters cannot be determined since a current source cannot be connected to one port with the other port open. To determine the Y-parameters, a voltage source with a value of 1 volt is connected to port 1, and port 2 is shorted to ground. The current into port 1 is measured as $1/Z$ amperes and the current into port 2 is $-1/Z$ amperes, i.e. $Y_{11} = 1/Z$ and $Y_{21} = -1/Z$. Next, the voltage source is connected to port 2 with port 1 shorted to ground. The current into port 2 is measured as $1/Z$ amperes and the current into port 1 is measured as $-1/Z$ amperes. Equation (3.18) is used to compute the s-parameters:

Network parameters of a two-port series impedance $Z$ (*Z-parameters do not exist)*

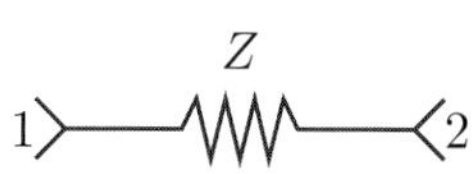


$$\mathbf{Y} = \frac{\begin{pmatrix} 1 & -1 \\ -1 & 1 \end{pmatrix}}{Z},$$
$$\mathbf{S} = \frac{\begin{pmatrix} Z & 2 \cdot Z0 \\ 2 \cdot Z0 & Z \end{pmatrix}}{Z + 2 \cdot Z0}. \tag{3.42}$$

### 3.10.3 Two-Port Shunt Impedance to Ground

Consider a two-port wire with an impedance $Z$ connected to ground and to the wire. To determine the Z-parameters, a current source with a value of 1 ampere is connected to port 1 and port 2 is left open. The voltage measured at both ports is $Z$ volts. The element is symmetric, so the Z-parameters are $Z$ in all rows and columns. The Y-parameters do not exist because the Z-parameter matrix is not invertible. This is because, in order to determine the Y-parameters, one must connect a voltage source to one port while the other is shorted to ground, which leads to infinite current.

Equation (3.16) is used to compute the s-parameters:

Network parameters of a two-port shunt impedance $Z$ (*Y-parameters do not exist*)

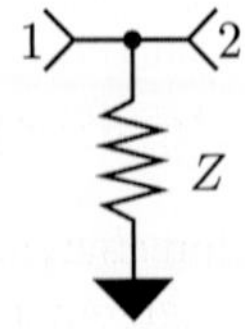


$$\mathbf{Z} = \begin{pmatrix} Z & Z \\ Z & Z \end{pmatrix},$$

$$\mathbf{S} = \frac{\begin{pmatrix} -Z0 & 2 \cdot Z \\ 2 \cdot Z & -Z0 \end{pmatrix}}{2 \cdot Z + Z0}. \tag{3.43}$$

### 3.10.4 Series/Shunt Impedance

To generate the common series/shunt impedance combination, consider a series impedance $Zs$ followed by a shunt impedance $Zh$. The s-parameters of the series impedance $SL$ are given by (3.42) as

$$SL = \begin{pmatrix} Zs & 2 \cdot Z0 \\ 2 \cdot Z0 & Zs \end{pmatrix} \cdot \frac{1}{Zs + 2 \cdot Z0}.$$

The s-parameters of the shunt impedance $SR$ are given by (3.43) as

$$SR = \begin{pmatrix} -Z0 & 2 \cdot Zh \\ 2 \cdot Zh & -Z0 \end{pmatrix} \cdot \frac{1}{2 \cdot Zh + Z0}.$$

Using the s-parameters of two cascaded sections provided by (3.35) results in

$$\mathbf{S} = \frac{\begin{pmatrix} SL_{11} - SR_{11} \cdot |\mathbf{SL}| & SL_{12} \cdot SR_{12} \\ SL_{21} \cdot SR_{21} & SR_{22} - SL_{22} \cdot |\mathbf{SR}| \end{pmatrix}}{1 - SL_{22} \cdot SR_{11}}.$$

This simplifies to (3.44).

S-parameters of series/shunt impedance to ground

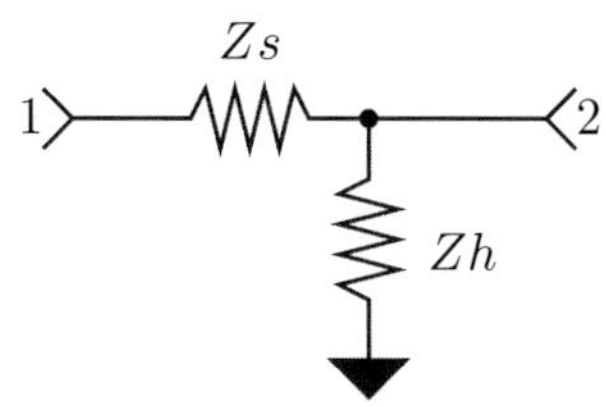


$$\mathbf{S} = \frac{\begin{pmatrix} Zs \cdot Zh + Zs \cdot Z0 - Z0^2 & 2 \cdot Zh \cdot Z0 \\ 2 \cdot Zh \cdot Z0 & Zs \cdot Zh - Zs \cdot Z0 - Z0^2 \end{pmatrix}}{Z0^2 + (2 \cdot Zh + Zs) \cdot Z0 + Zs \cdot Zh}. \tag{3.44}$$

## 3.11 Advanced Cascade Parameters – Multi-Port T-Parameters

In general, cascade parameters and hybrid parameters are ill defined for elements with more than two ports. That being said, there can be generally accepted agreement if one thinks in terms of left-hand and right-hand ports for the purpose of cascading. While much more advanced and reliable techniques for cascading are developed in Chapter 4, multi-port cascading is useful for relatively simple networks, meaning ones that are connected in a repetitive manner.

For the purposes of cascading, a generalized T-parameter matrix is defined in terms of left-hand and right-hand ports as

$$\mathbf{l}_{\mathbf{a}}^{\mathbf{b}} = \mathbf{T} \cdot \mathbf{r}_{\mathbf{b}}^{\mathbf{a}},$$

where $\mathbf{l}$ and $\mathbf{r}$ are lists of wave values for the left and right ports, respectively. The notation $\mathbf{r}_{\mathbf{b}}^{\mathbf{a}}$ means that the waves are ordered as incident followed by reflected for the right-hand ports, and $\mathbf{l}_{\mathbf{a}}^{\mathbf{b}}$ denotes the reverse ordering for the left-hand ports. For cascadability, the right-hand nodes must represent left-hand nodes of a downstream device, so the wave order must be reversed.

The T-parameters of the aggregate cascaded elements are calculated by multiplying the T-parameter matrices. All that is needed here is an ability to convert back and forth between s- and T-parameters. In other words, a reasonable strategy for cascading two devices whose s-parameters are known would be to convert both to T-parameters, multiply the two T-parameter matrices, and convert back to s-parameters.

A general T-parameter network device is shown in Figure 3.4. The device is shown with ports numbered according to numbers held in two vectors: $\mathbf{lp}$ and $\mathbf{rp}$. These vectors are presumed to contain a list of port numbers where, for a $P$-port device, $p \in 1 \ldots P$ is held uniquely in one of the locations of $\mathbf{lp}$ or $\mathbf{rp}$, meaning that every possible value of $p$ is in exactly one element of the two vectors. Thus, in Figure 3.4, the numbers $lp[1], lp[2], \ldots, lp[L]$ are the port numbers of the left-hand ports and the numbers $rp[1], rp[2], \ldots, rp[R]$ are the port numbers of the right-hand ports. Furthermore, if a $P$-port device with s-parameters designated as $\mathbf{S}$ has wave values held in a vector $\mathbf{a}$ incident on the device, and wave values

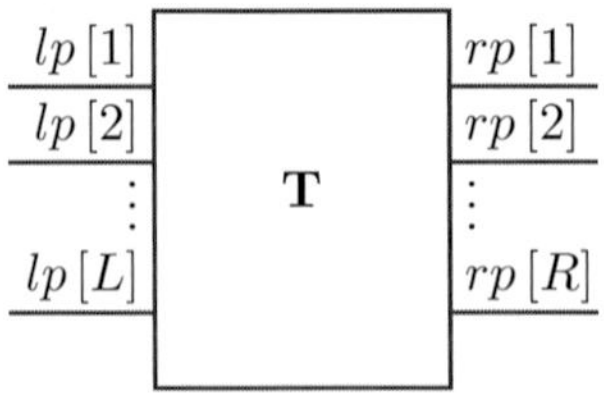


**Figure 3.4** Generalized T-parameters

held in a vector **b** reflected from the device, then the s-parameters define the relationship $\mathbf{b} = \mathbf{S} \cdot \mathbf{a}$, and the corresponding T-parameters for this device would obey the relationship

$$\begin{pmatrix} b\left[lp\left[1\right]\right] \\ a\left[lp\left[1\right]\right] \\ b\left[lp\left[2\right]\right] \\ a\left[lp\left[2\right]\right] \\ \vdots \\ a\left[lp\left[L\right]\right] \end{pmatrix} = \begin{pmatrix} T_{11} & T_{12} & T_{13} & T_{14} & \cdots & T_{1R} \\ T_{21} & T_{22} & T_{23} & T_{24} & \cdots & T_{2R} \\ T_{31} & T_{32} & T_{33} & T_{34} & \cdots & T_{3R} \\ T_{41} & T_{42} & T_{43} & T_{44} & \cdots & T_{4R} \\ \vdots & \vdots & \vdots & \vdots & \ddots & \vdots \\ T_{L1} & T_{L2} & T_{L3} & T_{L4} & \cdots & T_{LR} \end{pmatrix} \cdot \begin{pmatrix} a\left[rp\left[1\right]\right] \\ b\left[rp\left[1\right]\right] \\ a\left[rp\left[2\right]\right] \\ b\left[rp\left[2\right]\right] \\ \vdots \\ b\left[rp\left[R\right]\right] \end{pmatrix}.$$

To convert from s-parameters to T-parameters, the methods provided in §1.3 are used, stacking the input node vector (in this case the right-hand nodes) on top of the output node vector (in this case the left-hand nodes) and defining a permutation matrix that converts the stacked input node vector (the incident power waves) on top of the output node vector (the reflected power waves) into the first vector as

$$\mathbf{C} \cdot \begin{pmatrix} a_1 \\ a_2 \\ a_3 \\ a_4 \\ \vdots \\ a_P \\ b_1 \\ b_2 \\ b_3 \\ b_4 \\ \vdots \\ b_P \end{pmatrix} = \begin{pmatrix} a\left[rp\left[1\right]\right] \\ b\left[rp\left[1\right]\right] \\ a\left[rp\left[2\right]\right] \\ b\left[rp\left[2\right]\right] \\ \vdots \\ b\left[rp\left[R\right]\right] \\ b\left[lp\left[1\right]\right] \\ a\left[lp\left[1\right]\right] \\ b\left[lp\left[2\right]\right] \\ a\left[lp\left[2\right]\right] \\ \vdots \\ a\left[lp\left[L\right]\right] \end{pmatrix}. \tag{3.45}$$

In (3.45), $L + R = P$ is the total number of ports in the device; generally, $P$ is even and $L = R = P/2$.

In §1.3, after forming the permutation matrix, the matrix is partitioned and the conversion is calculated as

$$\mathbf{T} = \Big(\underset{2\cdot L\times P}{\mathbf{C_{21}}} + \underset{2\cdot L\times P}{\mathbf{C_{22}}} \cdot \underset{P\times P}{\mathbf{S}}\Big) \cdot \Big(\underset{2\cdot R\times P}{\mathbf{C_{11}}} + \underset{2\cdot R\times P}{\mathbf{C_{12}}} \cdot \underset{P\times P}{\mathbf{S}}\Big)^{-1}, \tag{3.46}$$

where each matrix is labeled with its dimensions to expose how to partition $\mathbf{C}$ to make this work out correctly.

While (3.46) is perfectly fine, there are some observations about this equation that can allow for economization. Because the incident and reflected waves for each port are grouped together, all of the odd rows of $\mathbf{C_{11}}$ contain a 1 in the column pertaining to a right-hand port number (i.e. the 1 is located to choose one of the incident waves). To be specific, for a given right-hand port $r \in 1 \ldots R$, $C_{11}[2 \cdot r - 1][rp[r]] = 1$. Because of the partitioning of $\mathbf{C}$, there are zeros in all of the even numbered rows of $\mathbf{C_{11}}$, as the even numbered rows of the first $R$ rows of $\mathbf{C}$ select reflected waves and these don't make it into $\mathbf{C_{11}}$. Note that $\mathbf{C_{12}}$ has the opposite situation in that the even rows contain a 1 in the column pertaining to a right-hand port number (i.e. the 1 is located to choose one of the reflected waves). To be specific, for a given right-hand port $r$, $C_{12}[2 \cdot r][rp[r]] = 1$. Because of the partitioning of $\mathbf{C}$, there are zeros in all of the odd numbered rows of $\mathbf{C_{12}}$.

$\mathbf{C_{12}}$ chooses a row of the s-parameters from $\mathbf{S}$ corresponding to the port number $rp[r]$, and $\mathbf{C_{11}}$ can be thought to choose a row of the similarly sized $P \times P$ identity matrix corresponding to the port number $rp[r]$. Since $\mathbf{C_{11}}$ and the product of $\mathbf{C_{12}}$ and $\mathbf{S}$ are added, the result is an interleaving of a row of the identity matrix and a row of the s-parameters.

A similar line of examination can be performed in relation to the matrices $\mathbf{C_{21}}$ and $\mathbf{C_{22}}$, and the T-parameters can be calculated as

$$\mathbf{T} = \begin{pmatrix} S[lp[1]][*] \\ \mathbb{I}[lp[1]][*] \\ S[lp[2]][*] \\ \mathbb{I}[lp[2]][*] \\ \vdots \\ S[lp[L]][*] \\ \mathbb{I}[lp[L]][*] \end{pmatrix} \cdot \begin{pmatrix} \mathbb{I}[rp[1]][*] \\ S[rp[1]][*] \\ \mathbb{I}[rp[2]][*] \\ S[rp[2]][*] \\ \vdots \\ \mathbb{I}[rp[R]][*] \\ S[rp[R]][*] \end{pmatrix}^{-1}. \tag{3.47}$$

Python code for performing this conversion is provided in Listing 3.13. The Python code has zero-based indexing and the algorithm described is one based. Also, this code has the option of not supplying any port ordering. In the case that port ordering is not supplied, it assumes that the first half of the ports are on the left and the second half are on the right.

Conversion from T-parameters back to s-parameters can be performed by solving (3.46) for $\mathbf{S}$:

$$\mathbf{S} = (\mathbf{C_{22}} - \mathbf{T} \cdot \mathbf{C_{12}})(\mathbf{T} \cdot \mathbf{C_{11}} - \mathbf{C_{21}})^{-1}.$$

Alternatively, (3.47) can be reversed. This is much more complicated because of the nature of $\mathbf{C}^T$. It is so complicated that it is best described programmatically rather than

**Listing 3.13** General conversion from s-parameters to T-parameters

```
def S2T(S,lp=None,rp=None):
    P=len(S)
    if not isinstance(lp,list):
        lp=range(1,P//2+1)
        rp=range(P//2+1,P+1)
    I=identity(P).tolist()
    TL=[]
    for r in range(len(lp)):
        TL.append(S[lp[r]-1])
        TL.append(I[lp[r]-1])
    TR=[]
    for r in range(len(rp)):
        TR.append(I[rp[r]-1])
        TR.append(S[rp[r]-1])
    return array(matrix(TL)*matrix(TR).getI()).tolist()
```

mathematically:

$$\mathbf{S} = \left(\begin{matrix} \begin{cases} \mathbb{I}\left[2\cdot\text{index}\left(1,\mathbf{rp}\right)\right][*] & \text{if } 1\in\mathbf{rp} \\ T\left[2\cdot\text{index}\left(1,\mathbf{lp}\right)-1\right][*] & \text{if } 1\in\mathbf{lp} \end{cases} \\ \begin{cases} \mathbb{I}\left[2\cdot\text{index}\left(2,\mathbf{rp}\right)\right][*] & \text{if } 2\in\mathbf{rp} \\ T\left[2\cdot\text{index}\left(2,\mathbf{lp}\right)-1\right][*] & \text{if } 2\in\mathbf{lp} \end{cases} \\ \vdots \\ \begin{cases} \mathbb{I}\left[2\cdot\text{index}\left(P,\mathbf{rp}\right)\right][*] & \text{if } P\in\mathbf{rp} \\ T\left[2\cdot\text{index}\left(P,\mathbf{lp}\right)-1\right][*] & \text{if } P\in\mathbf{lp} \end{cases} \end{matrix}\right)\cdot\left(\begin{matrix} \begin{cases} \mathbb{I}\left[2\cdot\text{index}\left(1,\mathbf{rp}\right)-1\right][*] & \text{if } 1\in\mathbf{rp} \\ T\left[2\cdot\text{index}\left(1,\mathbf{lp}\right)\right][*] & \text{if } 1\in\mathbf{lp} \end{cases} \\ \begin{cases} \mathbb{I}\left[2\cdot\text{index}\left(2,\mathbf{rp}\right)-1\right][*] & \text{if } 2\in\mathbf{rp} \\ T\left[2\cdot\text{index}\left(2,\mathbf{lp}\right)\right][*] & \text{if } 2\in\mathbf{lp} \end{cases} \\ \vdots \\ \begin{cases} \mathbb{I}\left[2\cdot\text{index}\left(P,\mathbf{rp}\right)-1\right][*] & \text{if } P\in\mathbf{rp} \\ T\left[2\cdot\text{index}\left(P,\mathbf{lp}\right)\right][*] & \text{if } P\in\mathbf{lp} \end{cases} \end{matrix}\right)^{-1}. \tag{3.48}$$

Previously it was mentioned that $L+R=P$ is the total number of ports in the network and that, generally, $P$ is even and $L=R=P/2$. Remember that $\mathbf{S}$ is a square $P\times P$ matrix and that $\mathbf{T}$ is a $2\cdot L\times 2\cdot R$ matrix. The right-hand side matrix in (3.47) is a $2\cdot R\times P$ matrix; therefore $R=P/2$ and the matrix is square. A square right-hand matrix in (3.47) is the normal situation and implies that half the ports are on the left and half are on the right.

Python code for performing this conversion is provided in Listing 3.14. Again, the code uses zero-based indexing.

As a final test, using a simple two-port device with port 1 on the left and port 2 on the right, the T-parameters are calculated as (3.49), which can be verified against (3.34):

$$\mathbf{T} = \begin{pmatrix} \mathbf{S}_{1*} \\ \mathbb{I}_{1*} \end{pmatrix}\cdot\begin{pmatrix} \mathbb{I}_{2*} \\ \mathbf{S}_{2*} \end{pmatrix}^{-1} = \begin{pmatrix} S_{11} & S_{12} \\ 1 & 0 \end{pmatrix}\cdot\begin{pmatrix} 0 & 1 \\ S_{21} & S_{22} \end{pmatrix}^{-1} = \frac{\begin{pmatrix} -|\mathbf{S}| & S_{11} \\ -S_{22} & 1 \end{pmatrix}}{S_{21}}. \tag{3.49}$$

**Listing 3.14** General conversion from T-parameters to s-parameters

```
def T2S(T,lp=None,rp=None):
    P=len(T)//2+len(T[0])//2
    if not isinstance(lp,list):
        lp=range(1,P//2+1)
        rp=range(P//2+1,P+1)
    I=identity(P).tolist()
    SL=[]
    SR=[]
    for p in range(P):
        if (p+1) in rp:
            SL.append(I[2*rp.index(p+1)+1])
            SR.append(I[2*rp.index(p+1)])
        else:
            SL.append(T[2*lp.index(p+1)])
            SR.append(T[2*lp.index(p+1)+1])
    return array(matrix(SL)*matrix(SR).getI()).tolist()
```

Similarly, the T-parameters are converted back to s-parameters as (3.50), which can be verified against (3.32):

$$\mathbf{S} = \begin{pmatrix} \mathbf{T}_{1*} \\ \mathbb{I}_{2*} \end{pmatrix} \cdot \begin{pmatrix} \mathbf{T}_{2*} \\ \mathbb{I}_{1*} \end{pmatrix}^{-1} = \begin{pmatrix} T_{11} & T_{12} \\ 0 & 1 \end{pmatrix} \cdot \begin{pmatrix} T_{21} & T_{22} \\ 1 & 0 \end{pmatrix}^{-1} = \frac{\begin{pmatrix} T_{12} & |\mathbf{T}| \\ 1 & -T_{21} \end{pmatrix}}{T_{22}}. \quad (3.50)$$

## 3.12 S-Parameter File Format

S-parameters are commonly exchanged in a file format that has become the standard for s-parameters and other network parameters called Touchstone.[2] Traditionally, the Touchstone format used has been version 1.1, which has been in use since 2002 [9]. Although a new version of the standard [10] has been created to improve various limitations of the version 1.1 standard, it is not widely used. The limitations of the version 1.1 standard are not huge and the version 2.0 standard is designed to be compatible with version 1.1 files (i.e. a version 2.0 parser must also support parsing of 1.1); currently, it is safest to just use version 1.1 to ensure compatibility. Also, the version 1.1 standard is very simple.

Touchstone 1.1 format is such that an s-parameter file is a list of numbers. Each frequency is listed along with the complex matrix representing the s-parameters at that frequency. Thus, for a device with $P$ ports and $N$ frequency points, there will be $(2 \cdot P^2 + 1) \cdot N$ numbers.

The stream of numbers is such that there are $N$ groups of $2 \cdot P^2 + 1$ numbers for each frequency. For each frequency, one number is for the frequency itself, and $2 \cdot P^2$ numbers are for the s-parameter matrix for that frequency. The s-parameter matrix is specified as two numbers per element, filling the row first (except for the two-port case, which is discussed later). The two numbers per element describe a complex number. If the two numbers are denoted $A$ and $B$, the complex number is expressed as

- real and imaginary (i.e. $A + j \cdot B$);

[2]Touchstone® is a trademark of Agilent Corporation.

- magnitude and angle (in degrees) (i.e. $A \cdot e^{j \cdot \frac{\pi}{180} \cdot B}$);
- decibels and angle (in degrees) (i.e. $10^{\frac{A}{20}} \cdot e^{j \cdot \frac{\pi}{180} \cdot B}$).

The frequencies are given in Hz, kHz, MHz, or GHz.

There are two special characters in the ASCII stream of numbers, the "#" and the "!":

- Whenever a "!" appears, everything beyond it on a line is ignored. These are used for comments.
- The "#" at the beginning of a line denotes that the line is an option line that can contain several keywords:
  - Hz, kHz, MHz, and GHz – determines the multiplier to apply to all frequencies;
  - RI, MA, dB – determines the format as explained above (angles are always in degrees);
  - S, Y, Z, G, or H – determines the type of network parameters;
  - R followed by a number, where the number is $Z0$, the real reference impedance for all ports.

All Touchstone 1.1 files have a file extension .sXp, where X is the number of ports in the device and the number of frequency points is not provided – the file must be read to the end to find out how many there are. Both of these items, along with the single real reference impedance for all ports, are some of the major drawbacks addressed in the new standard.

Listing 3.15 shows a piece of Python code that reads s-parameter files. It is mostly self-explanatory, but there are some salient points to note.

First, the overall strategy is to read each line and to separate the line into tokens delimited by exclamation points. The first token is chosen as the valid part of the line. This valid portion is again separated into tokens delimited by spaces. If the first token contains the pound sign, then it is the options line, and this line is parsed to determine whether it has any of the keywords above; these keywords are used to remember the frequency multiplier, the format of the complex numbers provided, the reference impedance, and the type of network parameter. Only s-parameters are of interest. Otherwise, the tokens are appended to a running list of tokens and each is assumed to represent a number.

At the end of this first pass, the number of tokens is utilized to determine the number of frequency points:

$$N = \frac{NumTokens}{1 + 2 \cdot P^2}.$$

Then, each of the two elements of the complex number are placed in their correct locations, for $n \in 0 \ldots N-1$, $r, c \in 0 \ldots P-1$:

$$\begin{aligned} f[n] &= TokenList\left[n \cdot \left(1 + 2 \cdot P^2\right)\right], \\ A[n][r][c] &= TokenList\left[\left(1 + 2 \cdot P^2\right) \cdot n + 1 + (r \cdot P + c) \cdot 2\right], \\ B[n][r][c] &= TokenList\left[\left(1 + 2 \cdot P^2\right) \cdot n + 1 + (r \cdot P + c) \cdot 2 + 1\right]. \end{aligned}$$

Each of the two numbers is converted to become complex according to the format read on the options line.

**Listing 3.15** S-parameters **SParameterFile** class

```
class SParameterFile(SParameters):
    def __init__(self,name,Z0=None):
        self.m_sToken='S'
        self.m_Z0=Z0
        self.m_P=int(str.lower(name).split('.')[-1].split('s')[1].split('p')[0])
        freqMul = 1e6
        complexType = 'MA'
        Z0=50.
        sp=True
        f=[]
        self.m_f=[]
        numbersList=[]
        spfile=open(name,'rU' if sys.version_info.major < 3 else 'r')
        for line in spfile:
            lineList = str.lower(line).split('!')[0].split()
            if len(lineList)>0:
                if lineList[0] == '#':
                    if 'hz' in lineList: freqMul = 1.0
                    if 'khz' in lineList: freqMul = 1e3
                    if 'mhz' in lineList: freqMul = 1e6
                    if 'ghz' in lineList: freqMul = 1e9
                    if 'ma' in lineList: complexType = 'MA'
                    if 'ri' in lineList: complexType = 'RI'
                    if 'db' in lineList: complexType = 'DB'
                    if 'r' in lineList:
                        Z0=float(lineList[lineList.index('r')+1])
                    if not self.m_sToken.lower() in lineList:
                        sp=False
                else: numbersList.extend(lineList)
        spfile.close()
        if not sp: return
        if self.m_Z0==None: self.m_Z0=Z0
        frequencies = len(numbersList)//(1+self.m_P*self.m_P*2)
        P=self.m_P
        self.m_d=[empty([P,P]).tolist() for fi in range(frequencies)]
        for fi in range(frequencies):
            f.append(float(numbersList[(1+P*P*2)*fi])*freqMul)
            for r in range(P):
                for c in range(P):
                    n1=float(numbersList[(1+P*P*2)*fi+1+(r*P+c)*2])
                    n2=float(numbersList[(1+P*P*2)*fi+1+(r*P+c)*2+1])
                    if complexType == 'RI':
                        self.m_d[fi][r][c]=n1+1j*n2
                    else:
                        expangle=cmath.exp(1j*math.pi/180.*n2)
                        if complexType == 'MA':
                            self.m_d[fi][r][c]=n1*expangle
                        elif complexType == 'DB':
                            self.m_d[fi][r][c]=math.pow(10.,n1/20)*expangle
            if P == 2:
                self.m_d[fi]=array(self.m_d[fi]).transpose().tolist()
            if Z0 != self.m_Z0:
                self.m_d[fi]=ReferenceImpedance(self.m_d[fi],self.m_Z0,Z0)
        self.m_f=GenericFrequencyList(f)
```

**Listing 3.16 SParameters class**

```
class SParameters(SParameterManipulation):
    def __init__(self,f,data,Z0=50.0):
        self.m_sToken='S'; self.m_d=data; self.m_Z0=Z0
        self.m_f=FrequencyList(f)
        if not data is None:
            if len(data)>0: self.m_P=len(data[0])
        else:
            mat=self[0]
            if not mat is None: self.m_P=len(mat[0])
    def __getitem__(self,item): return self.m_d[item]
    def __len__(self): return len(self.m_f)
    def f(self): return self.m_f
    def Response(self,ToP,FromP): return [mat[ToP-1][FromP-1] for mat in self]
    def FrequencyResponse(self,ToP,FromP):
        return FrequencyResponse(self.f(),self.Response(ToP,FromP))
    def WriteToFile(self,name,formatString=None):
        freqMul = 1e6; fToken = 'MHz'; cpxType = 'MA'; Z0 = 50.0
        if not formatString is None:
            lineList = str.lower(formatString).split('!')[0].split()
            if len(lineList)>0:
                if 'hz' in lineList: fToken = 'Hz'; freqMul = 1.0
                if 'khz' in lineList: fToken = 'KHz'; freqMul = 1e3
                if 'mhz' in lineList: fToken = 'MHz'; freqMul = 1e6
                if 'ghz' in lineList: fToken = 'GHz'; freqMul = 1e9
                if 'ma' in lineList: cpxType = 'MA'
                if 'ri' in lineList: cpxType = 'RI'
                if 'db' in lineList: cpxType = 'DB'
                if 'r' in lineList: Z0=float(lineList[lineList.index('r')+1])
        spfile=open(name,'w')
        for lin in self.header: spfile.write(('! '+lin if lin[0] != '!' else lin)+'\n')
        spfile.write('# '+fToken+' '+cpxType+' '+self.m_sToken+' R '+str(Z0)+'\n')
        for n in range(len(self.m_f)):
            line=[str(self.m_f[n]/freqMul)]
            mat=self[n]
            if Z0 != self.m_Z0: mat=ReferenceImpedance(mat,Z0,self.m_Z0)
            if self.m_P == 2: mat=array(mat).transpose().tolist()
            for r in range(self.m_P):
                for c in range(self.m_P):
                    val = mat[r][c]
                    if cpxType == 'MA':
                        line.append(str(round(abs(val),6)))
                        line.append(str(round(cmath.phase(val)*180./math.pi,6)))
                    elif cpxType == 'RI':
                        line.append(str(round(val.real,6)))
                        line.append(str(round(val.imag,6)))
                    elif cpxType == 'DB':
                        line.append(str(round(20*math.log10(abs(val)),6)))
                        line.append(str(round(cmath.phase(val)*180./math.pi,6)))
            pline = ' '.join(line)+'\n'
            spfile.write(pline)
        spfile.close()
        return self
...
```

There is one particular oddity of the Touchstone 1.1 format. The writers of the standard determined that two-port devices should be read in order as $S_{11}$, $S_{21}$, $S_{12}$, $S_{22}$. This means that, for a two-port matrix, the transpose is applied after following the rules stated for other devices.

Finally, the s-parameters should be converted from the reference impedance specified on the options line to whatever reference impedance is desired, usually $50\,\Omega$.

When writing the s-parameters, there really is no special need to write them in any of the particular formats, but an option is provided in Listing 3.16 to write them with whatever option line desired.

# 4

## S-Parameter System Models

THE PRECEDING CHAPTERS focused on the generation of network parameter models from circuits. This resulted from the application of KVL or KCL to a circuit instrumented with the proper sources and terminations depending on the desired type of network parameter involved. S-parameters, the main subject of this book, were obtained either by the determination of Y- or Z- network parameters and subsequent conversion to s-parameters, or directly through the use of a voltage source with the reference impedance $Z0$ as the source impedance and termination impedances. Sometimes relatively simple sets of s-parameters were combined through cascading using T-parameters. All of the methods used so far required analysis of circuit topology. In other words, a knowledge of circuit theory was relied upon to solve s-parameter model problems.

This chapter leaves circuit theory behind and turns its attention to working with s-parameter models in their rawest forms. An s-parameter system model is defined as a system containing interconnected s-parameter network models. The goal of this chapter is to provide ways of reducing a system to a single s-parameter model that describes the system.

This all starts with the introduction of the signal-flow diagram representation, which is the analog to a circuit where s-parameters are involved. The teaching builds on an important foundation that facilitates the solution of all s-parameter system problems, that being the duality between the signal-flow diagram and the matrix representation of the system and its equations. The *ideal tee* is introduced as an element critical to the employment of these methods. Armed with this understanding, one can actually solve all s-parameter model problems. The chapter builds upon this information to produce some refinements on s-parameter system models and produces some helpful equations for common situations that arise, similar to how T-parameters facilitate some simple cascading problems.

This material is very different from traditional methods taught in existing material. One widely taught method (Mason's non-touching loop rule) is excluded from this text because it is overly complicated and error prone and is easily avoided through maintenance of the signal-flow diagram and matrix representation duality.

The various node removal rules are provided, but should be used only for small systems.

Mason's non-touching loop rule was derived by Samuel J. Mason [11] and can be read about in various sources [6, 12, 13]. Node removal rules can be read about in [14]. Mason's rule is replaced here by matrix representations. Node reduction rules are supplanted by the graphical equation method and block matrix methods presented in this chapter.

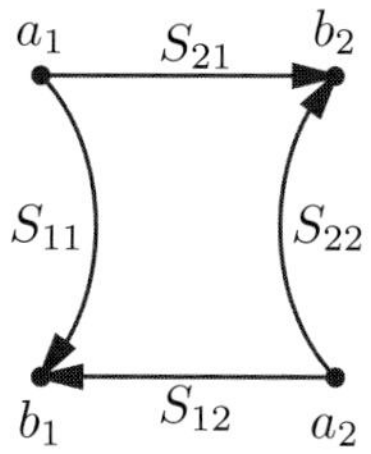


**Figure 4.1** Signal-flow diagram representation of a two-port s-parameter network device

## 4.1 Interconnection of S-Parameter Networks

S-parameter network elements are connected together in much the same way as a circuit or system of other kinds of network elements, but there are a few caveats that must be carefully considered and honored:

1. All device ports must be connected to exactly one other port of another device. If the desire is to have a device port open, then the device port must be connected to the port of an *ideal open* network device. Multiple device port connections must be handled with an *ideal tee* network device.
2. A connection between two device ports implies that the incident node of one port is connected to the reflected node of the other, and vice versa. Incident nodes are also called $a$ nodes and reflected nodes are called $b$ nodes. Internally, a device has exactly one $a$ and one $b$ node per port, and the s-parameters provide the equation for the $b$ nodes in terms of the $a$ nodes. Once two device ports are connected together, the type of node loses its meaning because each device port's $a$ node is connected directly to another device port's $b$ node, so there is no concept of a node being incident or reflected unless considered with respect to a port of a device.
3. Since s-parameters, unlike other types of network parameter, have a concept of a port reference impedance, only ports with the same reference impedance can be connected together. Otherwise, the network devices to be interconnected must either be put in the same reference impedance through a reference impedance transformation or the device ports can be connected through a reference impedance transformer; this is covered in Chapter 5.

### 4.1.1 Signal-Flow Representation of S-Parameter Network Devices

The definition of s-parameters is that they define the reflected waves in terms of the incident waves according to (3.2). Thus, a $P$-port s-parameter element has $2 \cdot P$ values of waves defined at the interface. Pairs of incident and reflected waves are grouped together, and this pair is called a *port*, just as a pair of voltages and currents represent the behavior at the device of other network parameter devices. The paired incident and reflected waves at a port represent a pair of values in a given row of vectors $\mathbf{a}$ and $\mathbf{b}$ in the definition $\mathbf{S} \cdot \mathbf{a} = \mathbf{b}$.

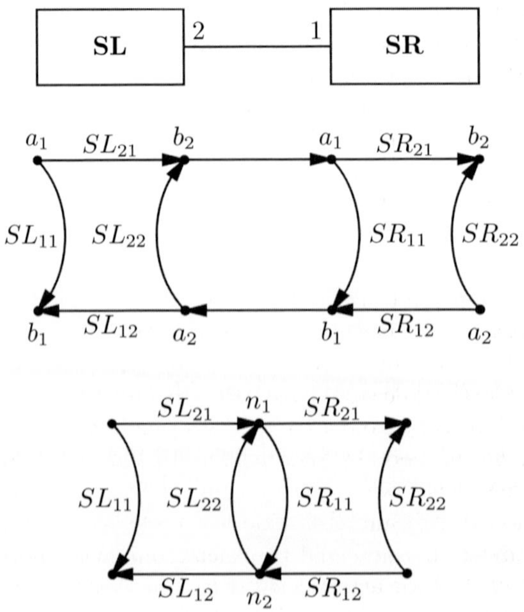


**Figure 4.2** Connection of two ports of two s-parameter devices

When s-parameter network devices are expressed as a signal-flow diagram, each value of incident and reflected wave is called a *node* and is represented by a dot in a diagram. Each of the s-parameter values therefore represents the weight of an arrow that can be drawn between the nodes to represent the nodal relationships. This is shown for a two-port device in Figure 4.1. On the left, nodes $a_1$ and $b_1$ represent the waves incident on and reflected from port 1 of the two-port device. Similarly, nodes $a_2$ and $b_2$ on the right represent the waves incident on and reflected from port 2 of the two-port device. It is customary to put the waves incident on a network on top of the reflected waves for a port when the port is on the left and to reverse the order when the port is on the right. This convention makes it easier to draw the signal-flow diagrams and indicates general forward wave flow along the top and general reverse wave flow along the bottom.

### 4.1.2 Port Connections

When generating an s-parameter system, all ports must connect to something. In other words, all ports of a network device must connect to exactly one port of another network device. When a pair of ports each from two network devices are connected, two node connections are made, as shown in Figure 4.2. The $a$ or incident wave on one port of one device also represents the $b$ or reflected waves from a port of the other device, and vice versa. This constraint is a function of the definition of current as a function of waves and is based on the reversal property explained in §2.2.1. With regard to s-parameter network element connection, it states that waves must be defined in a manner such that the current entering a device at a port is the same as the current leaving the other device at the connected port. When device ports are connected, the nodes are shared when drawing the signal-flow diagram. In other words, while technically two pairs of nodes are connected together for a total of four nodes, the connected nodes are merged, leaving two nodes per device port connection. This is shown in the bottom diagram of Figure 4.2. When this is done, there is no longer any concept of $a$ and $b$ nodes, so it is preferred to label these nodes generically as $n_1$, $n_2$, etc.

In circuit theory, a node in a circuit represents a location at which a certain voltage is specified and all wires connected to the dot representing a node acquire the same potential. In s-parameter networks, a node is a location with a certain wave value. In circuit theory, all wires leading away from a node are considered to carry current in a manner such that KCL is obeyed at the node (i.e. the sum of all currents into the node is zero). In s-parameter networks, all arrows leading away from a node have a weight that represents a multiplication of the wave at the originating node by that weight. In the signal-flow representation of an s-parameter device, arrows lead away from all $a$ nodes internal to the device and terminate on $b$ nodes. In a signal-flow diagram, the value of a node is defined as the weighted sum of the values at all other nodes. If a node does not depend on another node's value, there is no arrow. If it does, the weight of the arrow determines the weight in the sum. All s-parameter devices are such that waves enter a device through its $a$ nodes and exit through its $b$ nodes.

Unlike circuit theory, there is no inherent concept in a signal-flow representation of multiple wires joining at a single node (in the circuit theory sense), and this concept must be enforced through a special network element device when interconnecting multiple s-parameter device ports together. This is one of the most confusing aspects regarding s-parameters and is an absolutely key concept. It was seen in working with network parameters involving voltages and currents that certain network parameters were not defined for things as simple as a wire. A series resistance did not have Z-parameters defined, and has Y-parameters defined only for non-zero resistance according to (1.3). A shunt impedance to ground did not have Y-parameters, but had Z-parameters defined only for finite impedance according to (1.5). Thus, simple concepts like a wire and an open circuit cannot be expressed with either Y- or Z-parameters. It might therefore be surprising that s-parameters exist for wires (as provided in (3.11) and (3.12)) and open and short circuits (as provided in (3.41)). In summary, network parameters involving voltages and currents state that wires and open and short circuits do not exist as *network parameter elements*, but obviously they do exist and their existence is based on how one deals with the interconnection of networks. In other words, in voltage and current networks, wires don't exist as network elements, but are dealt with by how the voltages and currents are defined at the ports of interconnected devices. In s-parameters and signal-flow diagrams, this thinking must be altered. In s-parameters, these network elements do exist and the behavior of voltage nodes and wires is not handled automatically in the analysis.

### 4.1.3 Opens, Shorts, and Terminations of S-Parameter Device Ports

In circuits, it is easy to understand what it means to open, short, or terminate a device. With network parameters the rules are more complicated in that a Z-parameter device port

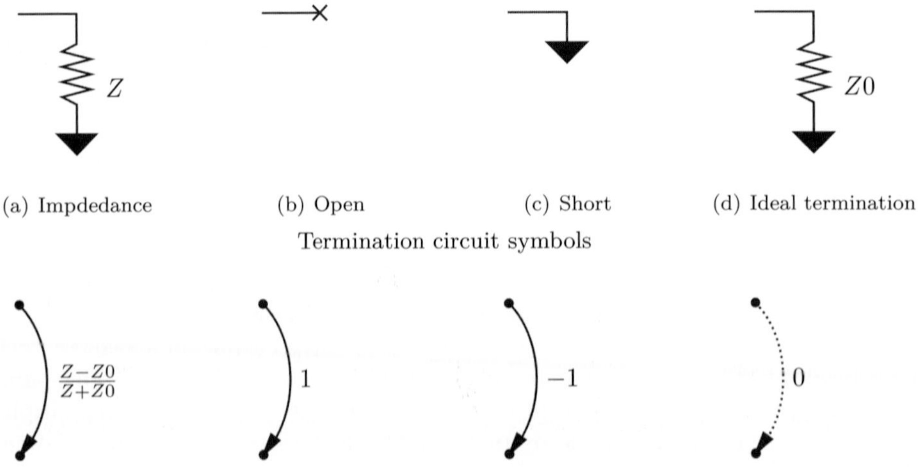


(a) Impdedance (b) Open (c) Short (d) Ideal termination

Termination circuit symbols

(e) Impedance (f) Open (g) Short (h) Ideal termination

Termination signal-flow diagrams

**Figure 4.3** Terminations

**Listing 4.1** S-parameters of a termination to ground

```
def TerminationZ(Z,Z0=None,K=None):
    (Z0,K)=Z0KHelper((Z0,K),1)
    Z0=Z0.item(0,0)
    Z=float(Z.real)+float(Z.imag)*1j
    return [[(Z-Z0)/(Z+Z0)]]
```

can be left open, implying that no current is entering the port, but might not be able to be shorted. Furthermore, a Y-parameter device can be shorted but might not be able to be left open.

Regarding s-parameters, things are further complicated regarding what "short" and "open" mean. A single-port impedance to ground was provided in (3.41). Noting that $S_{11}$, the s-parameters of a single-port device, are usually written as $\Gamma$, they approach 1 as the impedance $Z$ goes to infinity, $-1$ when $Z = 0$, and 0 when $Z = Z0$. These situations are shown in Figure 4.3, where the circuit element for a general impedance to ground is shown in Figure 4.3(a), with the corresponding signal-flow diagram representation given in Figure 4.3(e). The open and the short in Figure 4.3(b) and Figure 4.3(c) correspond to arrows with weights of 1 and $-1$ in Figure 4.3(f) and Figure 4.3(g), respectively.

Figure 4.3(d) represents an *ideal* termination. It is so-called because the signal-flow diagram in Figure 4.3(h) has no arrow. This is the analogous situation to the open and short circuit in circuit theory, because here any wave incident on the termination is not reflected back. In other words, waves reaching this termination exit the system and do not return.

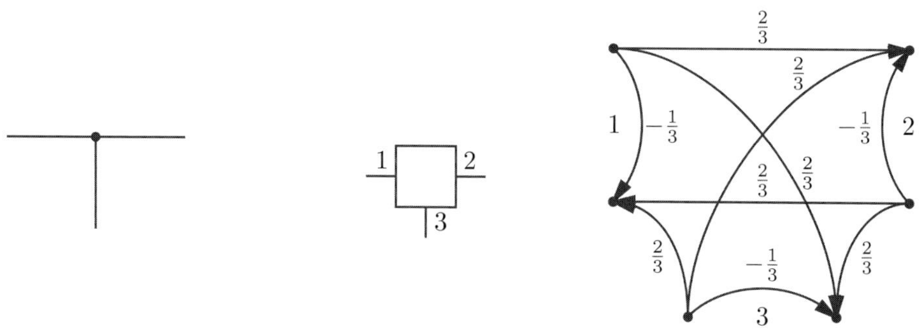


(a) Circuit (b) Network element (c) Signal-flow diagram

**Figure 4.4** Multiple device connections in circuit and s-parameter view

Sometimes, the ideal termination is referred to as matched, but, remember, it is matched to $Z0$, which is an arbitrary number, so the matched condition exists only for waves in that reference impedance.

The Python code for a termination is provided in Listing 4.1. The Python code for the open and the short are too trivial to show.

### 4.1.4 Multiple Device Port Connections

The rules are that every port of every s-parameter device is connected to exactly one other port of another s-parameter device, so one wonders how to handle multiple device port connections. Note that a node in a circuit sense is really a multi-port s-parameter device and that to realize multiple device port connections one must instantiate this multi-port s-parameter device and connect each of its ports to each of the interconnected device ports. The s-parameters for the ideal tee are provided in (3.15) and are provided generally for a $P$-port connection in (3.13) and (3.14).

The implications of multiple device connections are shown in Figure 4.4. Here, the benign connection of multiple devices at a circuit node in Figure 4.4(a) is shown to be represented by an actual s-parameter element in Figure 4.4(b). The internals of this network element are shown in Figure 4.4(c). If more than three devices are joined as shown, the element can be expanded to multiple ports using the equations referenced, or multiple tee devices can be joined together.

## 4.2 Signal-Flow Diagram Representation of Systems

While Figure 4.1 is a signal-flow diagram corresponding to an s-parameter model, Figure 4.5(b) shows a signal-flow diagram representation of the circuit or system in Figure 4.5(a). This system originated from a circuit schematic consisting of a voltage source with a given series impedance $Z_s$ driving a series impedance $Z_r$ which is terminated at a load $Z_l$. The

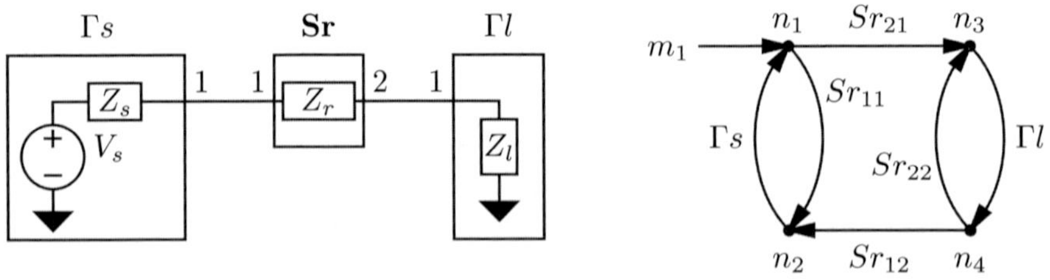


(a) Circuit (b) Signal-flow diagram

**Figure 4.5** Simple example circuit

original circuit is shown inside the s-parameter network devices, but this circuit is only a picture – the circuit meaning is entirely lost and the behavior of the system is dependent only on the reflection coefficient of the source $\Gamma s$, the stimuli emitted from the source $m_1$, the s-parameters of the series element **Sr**, and the reflection coefficient of the load $\Gamma l$. Later, one could substitute the stimulus and the s-parameters of the devices. To generate the signal-flow diagram, the previously outlined connection rules were followed. Note that all device ports are connected to exactly one other device port. There was no intent to connect multiple device ports, so the ideal tee concept was not required. There is an implicit assumption here that each connected device port is in the same reference impedance (it is most common that a single reference impedance is used for an entire system). In the diagram, the arrows and weights containing the s-parameters of the elements can be seen, where the weight of each arrow is given by an s-parameter value that is written with a subscript notation such as $S_{xy}$, where the $x$ refers to the reflected port and the $y$ corresponds to the incident port. All arrows terminate on a port which is a reflected port for a device, and all arrows originate from an incident port of a device according to the representation of each element as dictated by Figure 4.1. Here, the terminology used is that waves are called reflected whether they are truly reflected from the port on which the wave is incident or whether they are simply exiting the port. In Figure 4.5, the $a$ and $b$ naming conventions used when referring to individual devices as seen in Figure 4.1 have been dropped because they can no longer be referred to as incident or reflected nodes. For example, the node containing the incident wave on the termination is $n_3$, but this is the reflected wave from port 2 of the two-port device.

There is a duality between the signal-flow diagram and the system of equations describing the system. This is seen by writing the equations for each node in node order, where each node's equation is the weighted sum of the arrows from all other nodes with any stimulus applied directly to the node added:

$$n_1 = m_1 + \Gamma s \cdot n_2,$$
$$n_2 = Sr_{11} \cdot n_1 + Sr_{12} \cdot n_4,$$
$$n_3 = Sr_{21} \cdot n_1 + Sr_{22} \cdot n_4,$$
$$n_4 = \Gamma l \cdot n_3.$$

This is written in matrix form as

$$\begin{pmatrix} n_1 \\ n_2 \\ n_3 \\ n_4 \end{pmatrix} = \begin{pmatrix} 0 & \Gamma s & 0 & 0 \\ Sr_{11} & 0 & 0 & Sr_{12} \\ Sr_{21} & 0 & 0 & Sr_{22} \\ 0 & 0 & \Gamma l & 0 \end{pmatrix} \cdot \begin{pmatrix} n_1 \\ n_2 \\ n_3 \\ n_4 \end{pmatrix} + \begin{pmatrix} m_1 \\ 0 \\ 0 \\ 0 \end{pmatrix}.$$

This can be written in a preferred form as follows:

$$\left[ \mathbb{I} - \begin{pmatrix} 0 & \Gamma s & 0 & 0 \\ Sr_{11} & 0 & 0 & Sr_{12} \\ Sr_{21} & 0 & 0 & Sr_{22} \\ 0 & 0 & \Gamma l & 0 \end{pmatrix} \right] \cdot \begin{pmatrix} n_1 \\ n_2 \\ n_3 \\ n_4 \end{pmatrix} = \begin{pmatrix} m_1 \\ 0 \\ 0 \\ 0 \end{pmatrix}. \tag{4.1}$$

When written in this preferred form, this is called the *system equation*. The large matrix is called the *weights* matrix because it contains the weight of every arrow in the signal-flow diagram. The weights matrix is generally designated as **W**. The vector containing the nodes is called the *node vector* for obvious reasons and is generally designated as **n**. The vector on the right is called the *stimulus* vector because it contains all sources of stimuli to the system and is generally designated as **m**.

Therefore, the system equation is given by

$$(\mathbb{I} - \mathbf{W}) \cdot \mathbf{n} = \mathbf{m}.$$

Any matrix to the left of the node vector with the stimulus vector on the right is referred to as the *system characteristics* matrix and is generally designated as **SC**. Here, the system characteristics matrix is

$$\mathbf{SC} = \mathbb{I} - \mathbf{W}.$$

and therefore the system equation is also written as

$$\mathbf{SC} \cdot \mathbf{n} = \mathbf{m}. \tag{4.2}$$

This definition of the system characteristics matrix as a function of the weights matrix is not always correct, meaning that the implied value of the weights matrix as $\mathbf{W} = \mathbb{I} - \mathbf{SC}$ is not always meaningful. What is meant by this is that while a matrix **W** can be formed from a system characteristics matrix **SC** in this manner, the weights matrix might not be able to represent a correct signal-flow diagram of a system. This improper determination of **W** occurs whenever there are zeros on the diagonal of the system characteristics matrix. Therefore, if a system characteristics matrix is obtained by means other than subtracting a weights matrix from the identity matrix, whereby the weights matrix came from a signal-flow diagram, the matrix might require manipulation to get values onto its diagonal. This is performed by row and column permutations on the system equation. In the example where the system equation was generated directly from the signal-flow diagram, the following important steps were taken to formulate the proper weights matrix:

1. An equation was written for each node.
2. The equations for each node were listed in the same order as the nodes are ordered in the node vector.

The absolute node order in the node vector is not important (although some benefits to specific node ordering are seen later), the only requirements being that the equations were listed in the same order. When the equations are written in the proper order, it can be said that each row of the system equation is an equation for the node listed in each corresponding row of the node vector. In other words, the first row is the equation for node $n_1$, the second is an equation for node $n_2$, etc.

When the system characteristics matrix is constructed in this manner, it is said to be in *canonical* form. It is not the only canonical form; there are as many canonical forms as there are possibilities in which to order the node vector.

When the system characteristics matrix is in canonical form, it can be subtracted from the identity matrix and the weights of the arrows in a signal-flow diagram can be read directly. In this way, signal-flow diagrams are easily drawn that visualize a system equation. When a system equation is in canonical form, one never puts it in a non-canonical form. In other words, one never reorders the rows of the system equation without correspondingly reordering the columns.

Returning to the weights matrix, any given element at row $x$ and column $y$ of the weights matrix $W_{xy}$ is the weight of an arrow that originates from the node listed at row $y$ in the node vector and terminates on the node listed at row $x$ of the node vector. This is understood clearly by understanding that row $x$ of the equation corresponds to the equation for node $n_x$, and therefore all of the elements in row $x$ of $\mathbf{W}$ are the weights of the arrows that terminate at node $n_x$. Similarly, all elements in column $y$ of $\mathbf{W}$ multiply by the node listed at row $y$ of the node vector and are therefore the weights of the arrows leaving that node.

Once the system equation corresponding to the signal-flow diagram written in the form (4.2) is found, the goal is usually to solve the solution for the node values as a function of the stimulus as $\mathbf{n} = \mathbf{SC}^{-1} \cdot \mathbf{m}$.

Up to now, the system equation was formed by writing all of the nodal equations and putting these in matrix form. The ideal way that guarantees a canonical form is to take the following steps to write the system equation directly from the signal-flow diagram:

1. For an $N$-node system, label the nodes in the system from $1 \ldots N$.
2. List the node names that have been numbered (it helps if the nodes are named such as $n_1$, $n_2$, etc.) and list them in that order in the node vector.
3. Identify and label the sources of stimuli. These are the arrows that enter a node but do not come from another node (i.e. are set to a value external to the system).

4. Write an empty $N \times N$ weights matrix subtracted from an $N \times N$ identity matrix multiplied by the nodes vector (the identity matrix minus the weights matrix is the system characteristics matrix).
5. Set this equal to an empty $N$-element vector with an empty list of stimuli names.
6. Then, for every row in the node vector, identify the node at that row in the node vector. Find all the arrows that terminate on that node and enter the weight of the arrow in the corresponding row of the weights matrix in the column corresponding to the row in the node vector containing the name of the node from which each arrow originates. If the arrow does not originate from a node, then enter the weight in the stimulus vector at the corresponding row. If, when entering the arrow weights in the weights matrix or the stimulus vector, an element is occupied, add the weight to the weight already occupying the element.
7. Set all unused locations in the stimulus vector and weights matrix to zero.

For the example provided in Figure 4.5, the setup would start with

$$\left[\mathbb{I} - \begin{pmatrix} \square & \square & \square & \square \\ \square & \square & \square & \square \\ \square & \square & \square & \square \\ \square & \square & \square & \square \end{pmatrix}\right] \cdot \begin{pmatrix} n_1 \\ n_2 \\ n_3 \\ n_4 \end{pmatrix} = \begin{pmatrix} \square \\ \square \\ \square \\ \square \end{pmatrix}$$

and would end in the form[1]

$$\left[\mathbb{I} - \begin{pmatrix} W_{11} & W_{12} & W_{13} & W_{14} \\ W_{21} & W_{22} & W_{23} & W_{24} \\ W_{31} & W_{32} & W_{33} & W_{34} \\ W_{41} & W_{42} & W_{43} & W_{44} \end{pmatrix}\right] \cdot \begin{pmatrix} n_1 \\ n_2 \\ n_3 \\ n_4 \end{pmatrix} = \begin{pmatrix} m_1 \\ m_2 \\ m_3 \\ m_4 \end{pmatrix},$$

where for a system with $P$ total device ports, with $r, c \in 1 \dots P$ and $W_{rc}$ representing the sum of the weights of the arrows from $n_c$ to node $n_r$ and $m_r$ represents the sum of the stimuli terminating on $n_r$. These criteria are satisfied by

$$\mathbf{W} = \begin{pmatrix} 0 & \Gamma s & 0 & 0 \\ Sr_{11} & 0 & 0 & Sr_{12} \\ Sr_{21} & 0 & 0 & Sr_{22} \\ 0 & 0 & \Gamma l & 0 \end{pmatrix}, \qquad \mathbf{m} = \begin{pmatrix} m_1 \\ 0 \\ 0 \\ 0 \end{pmatrix}.$$

### 4.2.1 Generating a Signal-Flow Diagram from a System Equation

Generally, one is given a system equation in the form

$$\mathbf{SC} \cdot \mathbf{n} = \mathbf{m},$$

where $\mathbf{n}$ is a list of node variables for the system. Any system like this has a viable signal-flow diagram representation provided the matrix $\mathbf{SC}$ is invertible. In other words, it can be

[1] It is helpful to read $W_{xy}$ as the effect on node $n_x$ due to node $n_y$.

solved as

$$\mathbf{n} = \mathbf{SC}^{-1} \cdot \mathbf{m}.$$

In its solved form, all of the node values in $\mathbf{n}$ are expressed as a weighted sum of all of the values in $\mathbf{m}$, where the weights are held in $\mathbf{SC}^{-1}$. The solution to the system in (4.1) is

$$\begin{aligned}
\begin{pmatrix} n_1 \\ n_2 \\ n_3 \\ n_4 \end{pmatrix} &= \left[ \mathbb{I} - \begin{pmatrix} 0 & \Gamma s & 0 & 0 \\ Sr_{11} & 0 & 0 & Sr_{12} \\ Sr_{21} & 0 & 0 & Sr_{22} \\ 0 & 0 & \Gamma l & 0 \end{pmatrix} \right]^{-1} \begin{pmatrix} m_1 \\ 0 \\ 0 \\ 0 \end{pmatrix} \\
&= \begin{pmatrix} 1 & -\Gamma s & 0 & 0 \\ -Sr_{11} & 1 & 0 & -Sr_{12} \\ -Sr_{21} & 0 & 1 & -Sr_{22} \\ 0 & 0 & -\Gamma l & 1 \end{pmatrix}^{-1} \begin{pmatrix} m_1 \\ 0 \\ 0 \\ 0 \end{pmatrix} \\
&= \frac{1}{1 - Sr_{22} \cdot \Gamma l - Sr_{11} \cdot \Gamma s + \Gamma l \cdot \Gamma s \cdot |\mathbf{Sr}|} \cdot \begin{pmatrix} 1 - Sr_{22} \cdot \Gamma l \\ Sr_{11} - \Gamma l \cdot |\mathbf{Sr}| \\ Sr_{21} \\ Sr_{21} \cdot \Gamma l \end{pmatrix} \cdot m_1. \quad (4.3)
\end{aligned}$$

While this will not be drawn, it is seen that a signal-flow diagram representation of this equation is simply all of the nodes listed with stimuli pointing into each node. Often, this is the desired result, but this laborious equation shows the lack of insight obtained. The process of solving the equation is almost irreversible, so once the desired equation is solved the original diagram is lost. Also, the solution required a matrix inverse, or, technically speaking, a matrix-vector solution to the system. The point of signal-flow diagrams is to help solve the system without the math and to provide insight into how the system works.

Returning to (4.1) and keeping within the spirit of this text, one is usually provided with a system equation in the system characteristics matrix form. In the example, this would be as follows:

$$\begin{pmatrix} 1 & -\Gamma s & 0 & 0 \\ -Sr_{11} & 1 & 0 & -Sr_{12} \\ -Sr_{21} & 0 & 1 & -Sr_{22} \\ 0 & 0 & -\Gamma l & 1 \end{pmatrix} \cdot \begin{pmatrix} n_1 \\ n_2 \\ n_3 \\ n_4 \end{pmatrix} = \begin{pmatrix} m_1 \\ 0 \\ 0 \\ 0 \end{pmatrix}. \quad (4.4)$$

The system characteristics matrix is the most common form provided for a system of equations. All systems of equations have an equivalent weights matrix form. The weights matrix form is obtained by subtracting the system characteristics matrix from the identity matrix. Once in weights matrix form, the signal-flow diagram is easily drawn; the duality has been demonstrated between the signal-flow diagram and the system equation drawn in weights matrix form. That being said, it is important to pause and consider some points.

The canonical form of the system equation has been discussed. To reiterate, a canonical form occurs when the equation ordering and the node ordering are the same. An indication of this is all ones on the diagonal of the system characteristics matrix (or equivalently, all zeros on the diagonal of the weights matrix). The example equation (4.4) is in just that form, so it is safe to proceed immediately to the generation of a signal-flow diagram. But

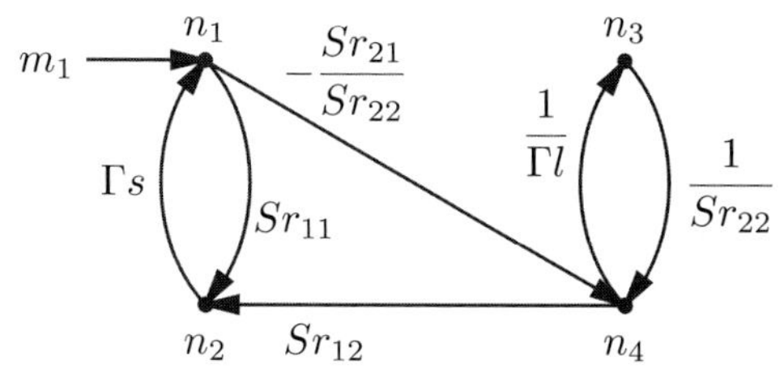


**Figure 4.6** A signal-flow diagram resulting from undesired canonical form of system characteristics matrix

this system equation could just as well have been provided with rows interchanged or entire rows scaled. Sometimes, this can still result in a canonical form. An example of another equivalent form of (4.4) is

$$\begin{pmatrix} 1 & -\Gamma s & 0 & 0 \\ -Sr_{11} & 1 & 0 & -Sr_{12} \\ -Sr_{21} & 0 & 1 & -\frac{1}{\Gamma l} \\ -\frac{Sr_{21}}{Sr_{22}} & 0 & -\frac{1}{Sr_{22}} & 1 \end{pmatrix} \cdot \begin{pmatrix} n_1 \\ n_2 \\ n_3 \\ n_4 \end{pmatrix} = \begin{pmatrix} m_1 \\ 0 \\ 0 \\ 0 \end{pmatrix}. \tag{4.5}$$

Equations (4.4) and (4.5) can be proven to be equivalent by solving them. Both have the solution in (4.3) and are equivalent representations of the same system. Notice that (4.5) has ones on the diagonal, so it is in a canonical form. This means that each row of (4.5) is an equation for the node at the same row in the node vector. The signal-flow diagram corresponding to this equation is shown in Figure 4.6. This is a very strange looking flow diagram and some observations are worth noting:

1. When a system is created from interconnected devices, it is expected that the weight of each non-zero arrow is an s-parameter of a device. In this case one finds combinations of s-parameters and reciprocals of s-parameters.
2. If an attempt is made to untangle this flow diagram to try to identify the interconnected devices, the interconnections would violate the interconnection rules of incident and reflected nodes of one port connected to the reflected and incident nodes of the other port, respectively. The devices themselves also would not have s-parameters because s-parameters are required to express reflected nodes of the device in terms of incident nodes.

It might be surprising, but this signal-flow diagram, with all of its problems, is completely valid; however, it has a nonsensical physical interpretation. While signal-flow diagrams are used to represent the flow of signals in a system and the cause-and-effect relationships, all valid signal-flow representations of a system equation do not have these characteristics. This is because signal-flow diagrams, at their essence, are really simultaneity diagrams or state diagrams. They express a mathematical relationship or constraints between all of the nodes and the stimuli that are met simultaneously. There is much to say about all of this that lead to philosophical considerations of the relationship between the signal-flow diagram and the system equation, but it suffices to recognize that the signal-flow diagram in Figure 4.6 is not very useful.

Furthermore, not much advice can be offered for getting a system equation into the right form; there are some valid operations or tools that can be employed, but mostly it involves trial and error. This topic is somewhat academic because the use of signal-flow diagrams presented in this text will always start with a system of interconnected s-parameter devices and, as long as they are interconnected properly, the correct signal-flow diagram is achieved. Furthermore, the ability to put diagrams back into the form of properly interconnected devices is retained, as long as the rows in the equivalent system equation are not reordered without reordering the columns equivalently.

### 4.2.2 System Equation Permutation and Scaling

For every initial vector $\mathbf{n_i}$ and every final vector $\mathbf{n_f}$, where $\mathbf{n_f}$ is a node vector with the nodes in $\mathbf{n_i}$ in a different order, there exists a *permutation* matrix that, when multiplied from the left of $\mathbf{n_i}$ creates $\mathbf{n_f}$. This matrix is designated as $\mathbf{P}$:

$$\mathbf{P} \cdot \mathbf{n_i} = \mathbf{n_f}. \tag{4.6}$$

$\mathbf{P}$ is constructed such that there is a single 1 in every row and every column. When multiplied from the left, for each row, the column containing the 1 determines the row chosen from the vector or matrix to the right for that row. In mathematics terms, if $P_{rc} = 1$, then it chooses row $c$ of the vector or matrix to the right of $\mathbf{P}$ and places it at row $r$ in the result of the matrix multiplication.

A permutation matrix as constructed produces a vector $\mathbf{n_f}$ that contains reordered values of $\mathbf{n_i}$. In other words, each element of $\mathbf{n_i}$ is placed somewhere at a different row in $\mathbf{n_f}$. Given an $R$ element reordering vector $\mathbf{v}$, where each element $v[r]$ for $r \in 1 \ldots R$ contains the row in $\mathbf{n_i}$ containing the element to place at row $r$ in $\mathbf{n_f}$ (i.e. $n_f[r] = n_i[v[r]]$), the row permutation matrix is an $R \times R$ element matrix $\mathbf{P}$ containing all zeros except that $P[r][v[r]] = 1$. Given the two lists of node names $\mathbf{n_i}$ and $\mathbf{n_f}$, the reordering vector $\mathbf{v}$ is formed such that, for each $r$, $v[r] = \text{index}(n_f[r], \mathbf{n_i})$, where the index function finds the row in $\mathbf{n_i}$ containing the element $n_f[r]$. Thus, the permutation matrix that reorders $\mathbf{n_i}$ as $\mathbf{n_f}$ is again an $R \times R$ element matrix $\mathbf{P}$ containing all zeros, except that for each $r \in 1 \ldots R$ there is a 1 at $P[r][\text{index}(n_f[r], \mathbf{n_i})]$. A matrix formed in this manner produces a matrix $\mathbf{P}$ that satisfies the equality in (4.6). Finally, if $\mathbf{n_i}$ contains a list of numbers $1 \ldots R$ in order, and $\mathbf{n_f}$ simply contains this list in a different order (i.e. contains the numbers $1 \ldots R$ in a different order), then the index function is not needed (because $\text{index}(n_f[r], \mathbf{n_i}) = n_f[r]$) and $P[r][n_f[r]] = 1$ is used. Said differently, in this situation, $\mathbf{v} = \mathbf{n_f}$.

For the purposes here, a row permutation matrix is designed to operate on a matrix to the right, where one thinks of the matrix on the right as containing rows that represent equations corresponding to the node at the same row in $\mathbf{n_i}$. If the row permutation matrix is generated to reorder nodes $\mathbf{n_i}$ to a different node order $\mathbf{n_f}$, then the row permutation matrix created would reorder the equations in a matrix to the right in a similar manner. As such, this matrix is capable of *equation reordering* (where its equation is synonymous with the row).

If a permutation matrix is multiplied from the right, then it operates somewhat opposite from a row permutation matrix. The matrix $\mathbf{P}$ is constructed such that there is a single 1 in every row and in every column. When multiplied from the right, for each column, the row containing the 1 determines the column chosen from the vector or matrix to the left for that column. In mathematics terms, if $P_{rc} = 1$, then it chooses column $r$ of the matrix to the left of $\mathbf{P}$ and places it at column $c$ in the result of the matrix multiplication. A permutation matrix designed to multiply from the right is called a *column permutation matrix.*

For the purposes here, a column permutation matrix is designed to operate on a matrix to the left, where one thinks of the matrix on the left as containing columns that represent nodes in a node vector. In other words, each column lists the weights applied to a given node at a column given by the row in the node vector. If a column permutation matrix is generated to reorder the *rows* of $\mathbf{n_i}$ to a different order in $\mathbf{n_f}$, then the column permutation matrix created would reorder the nodes in a matrix to the left in a similar manner. As such, this matrix is capable of *node reordering* (where its node is synonymous with the column).

### Row and Column Permutation Duality

The inverse of a permutation matrix is its transpose:

$$\mathbf{P}^T \cdot \mathbf{P} = \mathbf{P} \cdot \mathbf{P}^T = \mathbb{I}.$$

A row permutation matrix $\mathbf{P}$ created such that a row reordered matrix $\mathbf{A_r}$ is created by multiplying from the left of an initial matrix $\mathbf{A}$ (i.e. $\mathbf{A_r} = \mathbf{P} \cdot \mathbf{A}$) has the same column ordering effect when its transpose is multiplied from the right. In other words, given an initial node vector $\mathbf{n_i}$ and a final reordered node vector $\mathbf{n_f}$, and a row permutation matrix $\mathbf{P}$ such that $\mathbf{P} \cdot \mathbf{n_i} = \mathbf{n_f}$, then the column reordered matrix is $\mathbf{A_c} = \mathbf{A} \cdot \mathbf{P}^T$. And, if one wanted to reorder the rows (equations) and columns (nodes) of a matrix $\mathbf{A}$, one would perform this reordering by applying $\mathbf{P} \cdot \mathbf{A} \cdot \mathbf{P}^T$.

### Row and Column Permutation for Equation and Node Reordering

Regarding permutation, two rules in linear algebra are used:

1. It is allowed to multiply each side of an equation by the same matrix (with correct dimensions) without changing its validity.
2. It is allowed to insert the identity matrix (with correct dimensions) anywhere in an equation.

Using these rules, given a system equation $\mathbf{SC} \cdot \mathbf{n} = \mathbf{m}$, one can write

$$\begin{aligned}
\{\mathbf{P} \cdot \mathbf{SC}\} \cdot \mathbf{n} &= \mathbf{P} \cdot \mathbf{m}, \\
\left\{\mathbf{SC} \cdot \mathbf{P}^T\right\} \cdot \{\mathbf{P} \cdot \mathbf{n}\} &= \mathbf{m}, \\
\left\{\mathbf{P} \cdot \mathbf{SC} \cdot \mathbf{P}^T\right\} \cdot \{\mathbf{P} \cdot \mathbf{n}\} &= \mathbf{P} \cdot \mathbf{m}.
\end{aligned}$$

In these equations, braces are drawn around areas of the equation that are gathered together.

The first equation reorders the equations (i.e. the rows) in the system equation and does not touch the order of the node vector. It can be used to move equations around in a system equation to move the equation for a node into the correct position. It should not be used on a system equation in the correct canonical form.

The second equation is used less, but can reorder the node vector without touching the equation ordering (i.e. the row ordering). It can also be used to get the system equation in the proper form. It should not be used on a system equation in the correct canonical form.

The third equation is used very often as it reorders the equation ordering (the row ordering) while simultaneously reordering the nodes. This equation is used to reorder the nodes of the node vector, keeping the equation in the correct canonical form through commensurate reordering of the equations. Signal-flow diagrams drawn from this equation and the original will look identical. To solve problems of getting a system of equations in the correct canonical form, the first equation is used exclusively since each row is thought of as an equation; its use is crucial in lining up the equation and its associated node.

Aside from reordering equations, it is also helpful to scale the rows of the system. This is sometimes needed to put a 1 on the diagonal of the system characteristics matrix (which means a zero in the weights matrix, which means that no arrows loop back on a node). Loopbacks are annoying because they do not naturally occur through the interconnection of s-parameter network devices and they interfere with other things, as will be seen in later sections. Loopbacks are removed by dividing them out of a row, and rows are scaled by using a row scaling matrix $\mathbf{G}$. A row scaling matrix is a diagonal matrix containing the scaling factor for a given row in its row element on the diagonal. In other terms, a diagonal matrix $\mathbf{G}$ is constructed such that $G_{rr}$ contains the scale factor to apply to row $r$. It is fairly obvious that if a combination of matrices $\mathbf{P}$ and $\mathbf{G}$ are contrived to put the system matrix in a better form, they are both multiplied from the left of the system equation; if $\mathbf{G}$ is to the right of $\mathbf{P}$, the original row is scaled, and if it is to the left, the final row is scaled.

Note from (4.5) that if rows three and four are interchanged, having scaled the original rows three and four by $-\Gamma l$ and $-Sr_{22}$, respectively, one can get the system equation into its original form:

$$\mathbf{G} = \begin{pmatrix} 1 & 0 & 0 & 0 \\ 0 & 1 & 0 & 0 \\ 0 & 0 & -\Gamma l & 0 \\ 0 & 0 & 0 & -Sr_{22} \end{pmatrix}, \qquad \mathbf{P} = \begin{pmatrix} 1 & 0 & 0 & 0 \\ 0 & 1 & 0 & 0 \\ 0 & 0 & 0 & 1 \\ 0 & 0 & 1 & 0 \end{pmatrix},$$

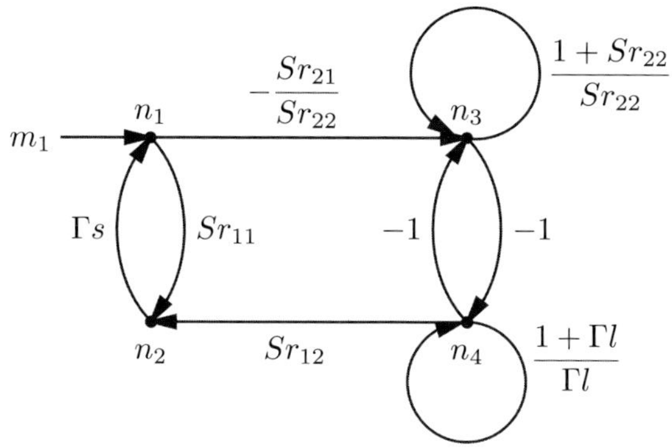


**Figure 4.7** A signal-flow diagram resulting from non-zero elements on the diagonal of the weights matrix

and

$$
\begin{aligned}
\mathbf{P}\cdot\mathbf{G}\cdot\mathbf{SC}\cdot\mathbf{n} &= \mathbf{P}\cdot\mathbf{G}\cdot\mathbf{m} \\
&= \begin{pmatrix} 1&0&0&0\\ 0&1&0&0\\ 0&0&0&1\\ 0&0&1&0 \end{pmatrix} \cdot \begin{pmatrix} 1&0&0&0\\ 0&1&0&0\\ 0&0&-\Gamma l&0\\ 0&0&0&-Sr_{22} \end{pmatrix} \cdot \begin{pmatrix} 1 & -\Gamma s & 0 & 0 \\ -Sr_{11} & 1 & 0 & -Sr_{12} \\ -Sr_{21} & 0 & 1 & -\frac{1}{\Gamma l} \\ -\frac{Sr_{21}}{Sr_{22}} & 0 & -\frac{1}{Sr_{22}} & 1 \end{pmatrix} \cdot \begin{pmatrix} n_1\\ n_2\\ n_3\\ n_4 \end{pmatrix} \\
&= \begin{pmatrix} 1&0&0&0\\ 0&1&0&0\\ 0&0&0&1\\ 0&0&1&0 \end{pmatrix} \cdot \begin{pmatrix} 1&0&0&0\\ 0&1&0&0\\ 0&0&-\Gamma l&0\\ 0&0&0&-Sr_{22} \end{pmatrix} \cdot \begin{pmatrix} m_1\\ 0\\ 0\\ 0 \end{pmatrix} \\
&= \begin{pmatrix} 1 & -\Gamma s & 0 & 0 \\ -Sr_{11} & 1 & 0 & -Sr_{12} \\ -Sr_{21} & 0 & 1 & -Sr_{22} \\ 0 & 0 & -\Gamma l & 1 \end{pmatrix} \cdot \begin{pmatrix} n_1\\ n_2\\ n_3\\ n_4 \end{pmatrix} = \begin{pmatrix} m_1\\ 0\\ 0\\ 0 \end{pmatrix}.
\end{aligned}
\tag{4.7}
$$

The result of (4.7) is that the equation is restored to the form in (4.4) and can be drawn as the desired signal-flow diagram shown in Figure 4.5.

In this example, where the form of the system equation was modified, the original equation provided is in a canonical form. There are some other variations that are worth looking at. One is the situation where the system equation has values on the diagonal, but the diagonal is not all ones. An example of this would be if the row interchange were performed, but not the row scaling. In this case, one obtains

$$
\begin{pmatrix} 1 & -\Gamma s & 0 & 0 \\ -Sr_{11} & 1 & 0 & -Sr_{12} \\ -\frac{Sr_{21}}{Sr_{22}} & 0 & -\frac{1}{Sr_{22}} & 1 \\ 0 & 0 & 1 & -\frac{1}{\Gamma l} \end{pmatrix} \cdot \begin{pmatrix} n_1\\ n_2\\ n_3\\ n_4 \end{pmatrix} = \begin{pmatrix} m_1\\ 0\\ 0\\ 0 \end{pmatrix}. \tag{4.8}
$$

The signal-flow diagram corresponding to (4.8) is shown in Figure 4.7, where there are loopbacks on nodes $n_3$ and $n_4$ due to the fact that the diagonal of the system characteristics

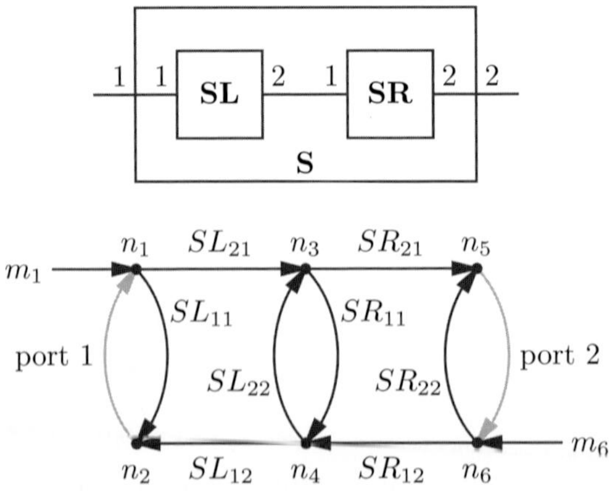


**Figure 4.8** Signal-flow diagram for cascaded two-port network example

matrix is not all ones. These loopbacks should always be removed. Finally, if the weight of a loopback element in a signal-flow diagram is unity, this indicates that there was a zero on the diagonal of the system characteristics matrix. This indicates that the row containing the equation for the node with the unity weight loopback did not contain an equation for that node at all. This is a very bad situation because, when this occurs, the loopback cannot be removed without removing other nodes from the network.

## 4.3 S-Parameters of Systems

The problem of determining s-parameters of a system starts with a signal-flow diagram representing the system with a plurality of interconnected nodes. The system exposes ports as pairs of nodes representing incident and reflected waves at that port. The nodes that are exposed represent *system ports*. The nodes that are not exposed are called *internal nodes*. The general nature of the problem is to restate the system with no internal nodes as a relationship between only incident and reflected waves at the system ports. For example, consider the problem of determining the s-parameters of a cascaded pair of two-port networks. In order to determine the s-parameters of this system, the interconnected system is drawn, with all devices labeled and all ports of the devices numbered, as the solution will be valid only for a system interconnected in the exact manner it is drawn. A box is placed around the entire network with the exposed ports identified and numbered. In drawing the signal-flow diagram corresponding to this system, the system ports are driven through the port reference impedance, which means there is no arrow connecting incident and reflected nodes at the system port. Furthermore, an arbitrary stimulus is driven into the incident node at the system ports. To illustrate the concept, the problem of determining the s-parameters of two two-port devices in cascade is shown in Figure 4.8.

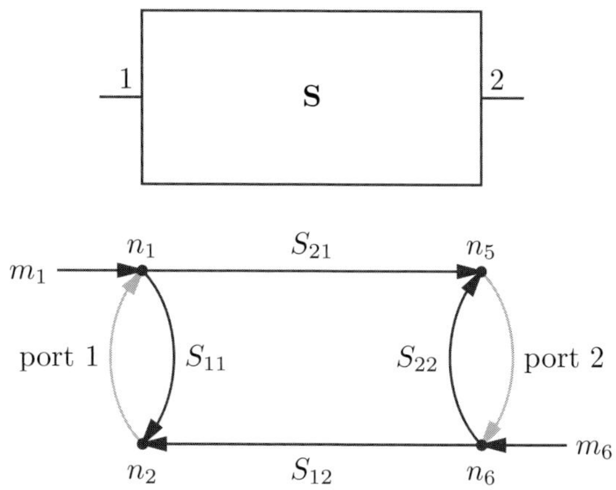


**Figure 4.9** Example resulting signal-flow diagram for s-parameters of cascaded network

In Figure 4.8, two cascaded two-port networks are shown with the accompanying signal-flow diagram drawn for determining the s-parameters of the aggregate system. Here, the connection between port 2 of the left device and port 1 of the right device form the internal nodes $n_3$ and $n_4$. Port 1 of the new device comprises the exposed nodes $n_1$ and $n_2$, which are the incident and reflected waves from port 1, respectively. Port 2 of the new device comprises the exposed nodes $n_5$ and $n_6$, which are the reflected and incident waves from port 2, respectively. In instrumenting the device combination for the aggregate s-parameter determination, two stimuli are connected: $m_1$ at node $n_1$, which is the incident wave on system port 1, and $m_2$ at node $n_6$, which is the incident wave on system port 2. The terminations for the exposed ports 1 and 2 are shown by the gray arrow in Figure 4.9, whose weight is actually zero. In other words, these arrows don't exist.

The determination of the s-parameters involves the determination of the ratio of the incident to the reflected waves at the system ports. Specifically:

- $S_{11}$ is found by determining the value of $n_2$ when $m_1$ is set to 1 and $m_2$ is set to 0.
- $S_{21}$ is found by determining the value of $n_5$ when $m_1$ is set to 1 and $m_2$ is set to 0.
- $S_{22}$ is found by determining the value of $n_5$ when $m_2$ is set to 1 and $m_1$ is set to 0.
- $S_{12}$ is found by determining the value of $n_2$ when $m_2$ is set to 1 and $m_1$ is set to 0.

The final solution looks like Figure 4.9.

The system equation corresponding to Figure 4.8 containing the weights matrix is provided in (4.9) and is shown in simplified form in (4.10). The solution for the node vector is provided in (4.11):

$$\left[ \mathbb{I} - \begin{pmatrix} 0 & 0 & 0 & 0 & 0 & 0 \\ SL_{11} & 0 & 0 & SL_{21} & 0 & 0 \\ SL_{21} & 0 & 0 & SL_{22} & 0 & 0 \\ 0 & 0 & SR_{11} & 0 & 0 & SR_{12} \\ 0 & 0 & SR_{21} & 0 & 0 & SR_{22} \\ 0 & 0 & 0 & 0 & 0 & 0 \end{pmatrix} \right] \cdot \begin{pmatrix} n_1 \\ n_2 \\ n_3 \\ n_4 \\ n_5 \\ n_6 \end{pmatrix} = \begin{pmatrix} m_1 \\ 0 \\ 0 \\ 0 \\ 0 \\ m_2 \end{pmatrix}, \tag{4.9}$$

$$\begin{pmatrix} 1 & 0 & 0 & 0 & 0 & 0 \\ -SL_{11} & 1 & 0 & -SL_{21} & 0 & 0 \\ -SL_{21} & 0 & 1 & -SL_{22} & 0 & 0 \\ 0 & 0 & -SR_{11} & 1 & 0 & -SR_{12} \\ 0 & 0 & -SR_{21} & 0 & 1 & -SR_{22} \\ 0 & 0 & 0 & 0 & 0 & 1 \end{pmatrix} \cdot \begin{pmatrix} n_1 \\ n_2 \\ n_3 \\ n_4 \\ n_5 \\ n_6 \end{pmatrix} = \begin{pmatrix} m_1 \\ 0 \\ 0 \\ 0 \\ 0 \\ m_2 \end{pmatrix}, \tag{4.10}$$

$$\begin{pmatrix} n_1 \\ n_2 \\ n_3 \\ n_4 \\ n_5 \\ n_6 \end{pmatrix} = \begin{pmatrix} 1 & 0 & 0 & 0 & 0 & 0 \\ -SL_{11} & 1 & 0 & -SL_{21} & 0 & 0 \\ -SL_{21} & 0 & 1 & -SL_{22} & 0 & 0 \\ 0 & 0 & -SR_{11} & 1 & 0 & -SR_{12} \\ 0 & 0 & -SR_{21} & 0 & 1 & -SR_{22} \\ 0 & 0 & 0 & 0 & 0 & 1 \end{pmatrix}^{-1} \cdot \begin{pmatrix} m_1 \\ 0 \\ 0 \\ 0 \\ 0 \\ m_2 \end{pmatrix}. \tag{4.11}$$

Only the values of nodes $n_2$ and $n_5$ are needed when the system is driven under two specific conditions: $m_1 = 1$ only and $m_2 = 1$ only. This is accomplished by writing a new equation for (4.11) as (4.12):

$$\begin{pmatrix} 1 & 0 \\ S_{11} & S_{12} \\ \cdots & \cdots \\ \cdots & \cdots \\ S_{21} & S_{22} \\ 0 & 1 \end{pmatrix} = \begin{pmatrix} n_{11} & n_{12} \\ n_{21} & n_{22} \\ n_{31} & n_{32} \\ n_{41} & n_{42} \\ n_{51} & n_{52} \\ n_{61} & n_{62} \end{pmatrix}$$
$$= \begin{pmatrix} 1 & 0 & 0 & 0 & 0 & 0 \\ -SL_{11} & 1 & 0 & -SL_{21} & 0 & 0 \\ -SL_{21} & 0 & 1 & -SL_{22} & 0 & 0 \\ 0 & 0 & -SR_{11} & 1 & 0 & -SR_{12} \\ 0 & 0 & -SR_{21} & 0 & 1 & -SR_{22} \\ 0 & 0 & 0 & 0 & 0 & 1 \end{pmatrix}^{-1} \cdot \begin{pmatrix} 1 & 0 \\ 0 & 0 \\ 0 & 0 \\ 0 & 0 \\ 0 & 0 \\ 0 & 1 \end{pmatrix}. \tag{4.12}$$

The explanation of (4.12) is as follows: The stimulus vector was first replaced by a matrix representing two driving conditions. The first driving condition, where only $m_1 = 1$, is represented by the first column of the matrix. The second column of the matrix represents the condition where only $m_2 = 1$. Thus the two driving conditions for the determination of the s-parameters have been satisfied. Because of these two driving conditions, the node vector expands to two columns. The first column represents the first driving condition and has its node names with a 1 added. The second column represents the second driving condition and has its node names with a 2 added. Thus, a node value $n_{xy}$ indicates the value of node $n_x$ under driving condition $y$. By driving the system this way, the s-parameters of the system are located in the rows corresponding to the port reflected wave nodes $n_2$ and $n_5$ and are shown in the matrix to the left of the new node matrix. The values corresponding to node values $n_{31}$, $n_{32}$, $n_{41}$, and $n_{42}$ have been left blank – these are "don't care" values as they are the values of waves at internal nodes. The node values $n_{11}$, $n_{12}$, $n_{61}$, and $n_{62}$ have been filled in because they are known from inspection. The values of these nodes are simply the values of the stimuli applied.

Therefore, the s-parameters of the system are calculated for the conditions in (4.12) as

$$\mathbf{S} = \begin{pmatrix} n_{21} & n_{22} \\ n_{51} & n_{52} \end{pmatrix}.$$

It is left to the reader to see that if, for some reason, it was not possible to drive the system under the two conditions where only $m_1 = 1$ and only $m_2 = 1$, but instead with unknown values $m_{11}$ and $m_{21}$ corresponding to driving condition 1 and $m_{12}$ and $m_{22}$ corresponding to driving condition 2, then the result would be

$$\begin{pmatrix} n_{11} & n_{12} \\ n_{21} & n_{22} \\ n_{31} & n_{32} \\ n_{41} & n_{42} \\ n_{51} & n_{52} \\ n_{61} & n_{62} \end{pmatrix} = \begin{pmatrix} 1 & 0 & 0 & 0 & 0 & 0 \\ -SL_{11} & 1 & 0 & -SL_{21} & 0 & 0 \\ -SL_{21} & 0 & 1 & -SL_{22} & 0 & 0 \\ 0 & 0 & -SR_{11} & 1 & 0 & -SR_{12} \\ 0 & 0 & -SR_{21} & 0 & 1 & -SR_{22} \\ 0 & 0 & 0 & 0 & 0 & 1 \end{pmatrix}^{-1} \cdot \begin{pmatrix} m_{11} & m_{12} \\ 0 & 0 \\ 0 & 0 \\ 0 & 0 \\ 0 & 0 \\ m_{21} & m_{22} \end{pmatrix}.$$

The s-parameters of the aggregate system are calculated as follows:

$$\mathbf{S} = \begin{pmatrix} n_{21} & n_{22} \\ n_{51} & n_{52} \end{pmatrix} \cdot \begin{pmatrix} n_{11} & n_{12} \\ n_{61} & n_{62} \end{pmatrix}^{-1},$$

where

$$\begin{pmatrix} n_{11} & n_{12} \\ n_{61} & n_{62} \end{pmatrix} = \begin{pmatrix} m_{11} & m_{12} \\ m_{21} & m_{22} \end{pmatrix}.$$

These patterns in the solution are interesting and are certainly the most basic way of solving for the s-parameters of systems of interconnected devices, but there is one final observation that makes the solution almost trivial. This observation begins with the fact that, given the system equation as in (4.10), one has the system characteristics matrix, node vector, and stimulus vector corresponding to the following equation:

$$\mathbf{SC} \cdot \mathbf{n} = \mathbf{m}.$$

Designating the matrix $\mathbf{Si} = \mathbf{SC}^{-1}$, then, for the example provided, the result is

$$\begin{pmatrix} Si_{11} & Si_{12} & Si_{13} & Si_{14} & Si_{15} & Si_{16} \\ Si_{21} & Si_{22} & Si_{23} & Si_{24} & Si_{25} & Si_{26} \\ Si_{31} & Si_{32} & Si_{33} & Si_{34} & Si_{35} & Si_{36} \\ Si_{41} & Si_{42} & Si_{43} & Si_{44} & Si_{45} & Si_{46} \\ Si_{51} & Si_{52} & Si_{53} & Si_{54} & Si_{55} & Si_{56} \\ Si_{61} & Si_{62} & Si_{63} & Si_{64} & Si_{65} & Si_{66} \end{pmatrix} \cdot \begin{pmatrix} 1 & 0 \\ 0 & 0 \\ 0 & 0 \\ 0 & 0 \\ 0 & 0 \\ 0 & 1 \end{pmatrix} = \begin{pmatrix} n_{11} & n_{12} \\ n_{21} & n_{22} \\ n_{31} & n_{32} \\ n_{41} & n_{42} \\ n_{51} & n_{52} \\ n_{61} & n_{62} \end{pmatrix}. \tag{4.13}$$

Examining (4.13) shows that the placement of the ones in the stimulus vector has the effect of choosing the first and sixth columns from **Si** such that

$$\begin{pmatrix} Si_{11} & Si_{16} \\ Si_{21} & Si_{26} \\ Si_{31} & Si_{36} \\ Si_{41} & Si_{46} \\ Si_{51} & Si_{56} \\ Si_{61} & Si_{66} \end{pmatrix} = \begin{pmatrix} n_{11} & n_{12} \\ n_{21} & n_{22} \\ n_{31} & n_{32} \\ n_{41} & n_{42} \\ n_{51} & n_{52} \\ n_{61} & n_{62} \end{pmatrix}, \tag{4.14}$$

and the placement of the nodes containing the s-parameters in the node vector has the effect of choosing the second and fifth columns from the result in (4.14) such that

$$\begin{pmatrix} Si_{21} & Si_{26} \\ Si_{51} & Si_{56} \end{pmatrix} = \begin{pmatrix} n_{21} & n_{22} \\ n_{51} & n_{52} \end{pmatrix}.$$

The columns and rows chosen from **Si** depend on the location of the nodes containing incident and reflected waves from the system ports in the node vector. While the choice of columns in **Si** depends on the locations of the ones in the stimulus vector, these locations are themselves a function of the node containing the incident waves on the system. Thus, the columns and rows chosen from **Si** are purely a function of the relationship between nodes containing the incident and reflected waves on the system and their locations in the node vector.

If one were to generate a list of two sets of nodes **a** and **b** corresponding to the nodes containing the waves incident on and reflected from the system ports *in port order*, then one could use these two lists in conjunction with the node vector and the inverse of the system characteristics matrix to pick out the s-parameters from **Si**. In the example provided, listing the incident and reflected waves from the system ports in port order provides $\mathbf{a} = \left(\begin{smallmatrix} n_1 \\ n_6 \end{smallmatrix}\right)$ and $\mathbf{b} = \left(\begin{smallmatrix} n_2 \\ n_5 \end{smallmatrix}\right)$. Defining the function that returns the one-based index of an element $x$ in a vector $\mathbf{y}$ as $\text{index}(x, \mathbf{y})$, then, for a system with $P$ ports, and for rows and columns of the system s-parameters $r, c \in 1 \ldots P$, the s-parameters of the system are given by

$$S[r][c] = Si[\text{index}(b[r], \mathbf{n})][\text{index}(a[c], \mathbf{n})].$$

This concludes the description of the most basic method for finding aggregate s-parameters of systems of interconnected devices. The remainder of the chapter will present some ways of doing this differently or methods that can simplify conceptually the inversion of the system characteristics matrix. But when all else fails, what was presented in this section is actually the most reliable method to use.

## 4.4 Block Matrix Solution of S-Parameter Systems

When solving a combination of networks for overall s-parameters, note that one starts with system equations of

$$\left(\mathbb{I} - \widehat{\mathbf{W}}\right) \cdot \widehat{\mathbf{n}} = \widehat{\mathbf{m}}. \tag{4.15}$$

The nodes listed in $\widehat{\mathbf{n}}$ can be classified into three broad categories:

1. Nodes that are waves incident on the entire system. These nodes are listed in order of system port number and are designated as **a**.
2. Nodes that are waves reflected from the entire system. These nodes are listed in order of system port number and are designated as **b**.
3. All other nodes are designated as **x**. These nodes are *internal* nodes.

Since all nodes must fall into one of the three categories, a new node vector can be created in block vector form as $\begin{pmatrix} \mathbf{a} \\ \mathbf{b} \\ \mathbf{x} \end{pmatrix}$.

This new node vector is simply a reordering of the original node vector **n**, and, since both node vectors are listings of nodes in each row, a row permutation matrix can be formed that is multiplied from the left such that

$$\mathbf{P} \cdot \widehat{\mathbf{n}} = \begin{pmatrix} \mathbf{a} \\ \mathbf{b} \\ \mathbf{x} \end{pmatrix}.$$

Recognizing that the transpose of a permutation matrix that simply reorders rows is the inverse of such a matrix, the original system equation can be rewritten as follows:

$$\mathbf{P} \cdot \left(\mathbb{I} - \widehat{\mathbf{W}}\right) \cdot \mathbf{P}^T \cdot \mathbf{P} \cdot \widehat{\mathbf{n}} = \mathbf{P} \cdot \widehat{\mathbf{m}}.$$

Note that multiplying **P** from the left of each side has no effect on the original equation and that multiplying $\mathbf{P}^T \cdot \mathbf{P} = \mathbb{I}$ inside also has no effect. Although these permutations have no effect on the original equation, they do reorder things if things are grouped a certain way. Using these permutations, the groupings are:

$$\mathbf{m} = \mathbf{P} \cdot \widehat{\mathbf{m}},$$

$$\mathbf{n} = \mathbf{P} \cdot \widehat{\mathbf{n}},$$

$$\mathbb{I} - \mathbf{W} = \mathbf{P} \cdot \left(\mathbb{I} - \widehat{\mathbf{W}}\right) \cdot \mathbf{P}^T = \mathbb{I} - \mathbf{P} \cdot \widehat{\mathbf{W}} \cdot \mathbf{P}^T.$$

Thus, the stimulus vector is reordered to reflect the new node ordering as well as the node vector. The weights matrix is both row and column reordered *in the same way* because the column permutation applied on the right is the same as the row permutation applied on the left. This type of reordering therefore has no effect on the identity matrix, and the permutations are applied directly to the weights matrix. The new system equation can therefore be written as

$$(\mathbb{I} - \mathbf{W}) \cdot \mathbf{n} = \mathbf{m}.$$

Again, this system equation is equivalent to that in (4.15), only the node and stimulus vector are in a different order (and the weights matrix is rearranged to reflect this new ordering). The system is still in a canonical form.

Because of this categorization and reordering of the nodes, this system can be partitioned as

$$\left[\mathbb{I}-\begin{pmatrix}\mathbf{W_{aa}} & \mathbf{W_{ab}} & \mathbf{W_{ax}}\\ \mathbf{W_{ba}} & \mathbf{W_{bb}} & \mathbf{W_{bx}}\\ \mathbf{W_{xa}} & \mathbf{W_{xb}} & \mathbf{W_{xx}}\end{pmatrix}\right]\cdot\begin{pmatrix}\mathbf{a}\\ \mathbf{b}\\ \mathbf{x}\end{pmatrix}=\begin{pmatrix}\mathbf{m_a}\\ \mathbf{m_b}\\ \mathbf{m_x}\end{pmatrix}. \tag{4.16}$$

This partitioning separates the weights matrix and stimulus vector along the same lines as the desired node ordering and grouping. The naming of the block weights matrices is such that subscripts are added, with the first letter representing sets of terminal nodes and the second letter representing originating nodes for the weights in the matrix. For example, $\mathbf{W_{ba}}$ contains all of the weights for the arrows in the signal-flow diagram originating from nodes designated **a** and terminating at nodes designated **b**. The stimulus block vectors are named such that the subscript represents the designation for the nodes that the group of stimuli points into. For example, $\mathbf{m_x}$ represents the group of stimuli that points into, or terminate on, the internal nodes.

The interconnection rules for devices were provided in §4.1.2 and it was explained that devices are interconnected by connecting incident nodes of a device to reflected nodes of another, and reflected nodes of a device to incident nodes of another at exactly one port. This means that for a system with nodes designated as **a** and **b** nodes these are exposed system ports that will be connected to sources to determine the system s-parameters and, when driven through the reference impedance, will have no interconnection possible between certain of these **a** and **b** nodes. For example, since waves enter the system at **a** nodes from external stimuli, there is no interconnection possible among the **a** nodes. Furthermore, since waves exit the system from **b** nodes, there is no interconnection possible among the **b** nodes. These constraints on interconnection possibilities cause whole blocks of the weights matrix to become zero and thus simplify the solution. Here is a list of restricted interconnections and their implications on the weights matrix and stimulus vector:

1. No arrows can terminate on an **a** node, except for stimuli applied to the system, therefore all block weights matrix named with a subscript that starts with **a** are zero. Therefore, $\mathbf{W_{aa}}=\mathbf{0}$, $\mathbf{W_{ab}}=\mathbf{0}$, and $\mathbf{W_{ax}}=\mathbf{0}$.
2. No arrows can originate from a **b** node since **b** nodes are waves leaving the system and the system is being driven through its reference impedance. Therefore, $\mathbf{W_{ab}}=\mathbf{0}$, $\mathbf{W_{bb}}=\mathbf{0}$, and $\mathbf{W_{xb}}=\mathbf{0}$.
3. Since external stimuli are applied to the system only at the **a** nodes, all other stimuli are equal to zero. Therefore, $\mathbf{m_b}=\mathbf{0}$ and $\mathbf{m_x}=\mathbf{0}$.
4. Since **a** and **b** are listed in order of system ports, and because there is a choice as to how to drive the system, $\mathbf{m_a}=\mathbb{I}$ is set arbitrarily.

Applying these observations, (4.16) is rewritten as

$$\left[\mathbb{I}-\begin{pmatrix}\mathbf{0} & \mathbf{0} & \mathbf{0}\\ \mathbf{W_{ba}} & \mathbf{0} & \mathbf{W_{bx}}\\ \mathbf{W_{xa}} & \mathbf{0} & \mathbf{W_{xx}}\end{pmatrix}\right]\cdot\begin{pmatrix}\mathbf{A}\\ \mathbf{B}\\ \mathbf{X}\end{pmatrix}=\begin{pmatrix}\mathbb{I}\\ \mathbf{0}\\ \mathbf{0}\end{pmatrix} \tag{4.17}$$

or

$$\begin{pmatrix} \mathbb{I} & \mathbf{0} & \mathbf{0} \\ -\mathbf{W_{ba}} & \mathbb{I} & -\mathbf{W_{bx}} \\ -\mathbf{W_{xa}} & \mathbf{0} & \mathbb{I}-\mathbf{W_{xx}} \end{pmatrix} \cdot \begin{pmatrix} \mathbf{A} \\ \mathbf{B} \\ \mathbf{X} \end{pmatrix} = \begin{pmatrix} \mathbb{I} \\ \mathbf{0} \\ \mathbf{0} \end{pmatrix}. \tag{4.18}$$

In order to solve this, the top two equations are written with $\mathbf{X}$ separated out:

$$\begin{pmatrix} \mathbb{I} & \mathbf{0} \\ -\mathbf{W_{ba}} & \mathbb{I} \end{pmatrix} \cdot \begin{pmatrix} \mathbf{A} \\ \mathbf{B} \end{pmatrix} + \begin{pmatrix} \mathbf{0} \\ -\mathbf{W_{bx}} \end{pmatrix} \cdot \mathbf{X} = \begin{pmatrix} \mathbb{I} \\ \mathbf{0} \end{pmatrix}. \tag{4.19}$$

Then, the bottom equation of (4.18) is written and solved for $\mathbf{X}$:

$$\begin{gathered} \begin{pmatrix} -\mathbf{W_{xa}} & \mathbf{0} \end{pmatrix} \cdot \begin{pmatrix} \mathbf{A} \\ \mathbf{B} \end{pmatrix} + (\mathbb{I}-\mathbf{W_{xx}}) \cdot \mathbf{X} = \mathbf{0}, \\ \mathbf{X} = (\mathbb{I}-\mathbf{W_{xx}})^{-1} \cdot \begin{pmatrix} \mathbf{W_{xa}} & \mathbf{0} \end{pmatrix} \cdot \begin{pmatrix} \mathbf{A} \\ \mathbf{B} \end{pmatrix}. \end{gathered} \tag{4.20}$$

Substituting (4.20) into (4.19) yields

$$\begin{pmatrix} \mathbb{I} & \mathbf{0} \\ -\mathbf{W_{ba}} & \mathbb{I} \end{pmatrix} \cdot \begin{pmatrix} \mathbf{A} \\ \mathbf{B} \end{pmatrix} + \begin{pmatrix} \mathbf{0} \\ -\mathbf{W_{bx}} \end{pmatrix} \cdot (\mathbb{I}-\mathbf{W_{xx}})^{-1} \cdot \begin{pmatrix} \mathbf{W_{xa}} & \mathbf{0} \end{pmatrix} \cdot \begin{pmatrix} \mathbf{A} \\ \mathbf{B} \end{pmatrix} = \begin{pmatrix} \mathbb{I} \\ \mathbf{0} \end{pmatrix}$$

or

$$\left[\begin{pmatrix} \mathbb{I} & \mathbf{0} \\ -\mathbf{W_{ba}} & \mathbb{I} \end{pmatrix} + \begin{pmatrix} \mathbf{0} \\ -\mathbf{W_{bx}} \end{pmatrix} \cdot (\mathbb{I}-\mathbf{W_{xx}})^{-1} \cdot \begin{pmatrix} \mathbf{W_{xa}} & \mathbf{0} \end{pmatrix}\right] \cdot \begin{pmatrix} \mathbf{A} \\ \mathbf{B} \end{pmatrix} = \begin{pmatrix} \mathbb{I} \\ \mathbf{0} \end{pmatrix}. \tag{4.21}$$

Equation (4.21) is simplified to

$$\begin{pmatrix} \mathbb{I} & \mathbf{0} \\ -\mathbf{W_{ba}} - \mathbf{W_{bx}} \cdot (\mathbb{I}-\mathbf{W_{xx}})^{-1} \cdot \mathbf{W_{xa}} & \mathbb{I} \end{pmatrix} \cdot \begin{pmatrix} \mathbf{A} \\ \mathbf{B} \end{pmatrix} = \begin{pmatrix} \mathbb{I} \\ \mathbf{0} \end{pmatrix}. \tag{4.22}$$

The s-parameters are calculated as $\mathbf{S} = \mathbf{B} \cdot \mathbf{A}^{-1}$, and one sees that $\mathbf{A} = \mathbb{I}$, so therefore, solving for $\mathbf{B}$,

$$\mathbf{S} = \mathbf{W_{ba}} + \mathbf{W_{bx}} \cdot (\mathbb{I}-\mathbf{W_{xx}})^{-1} \cdot \mathbf{W_{xa}}. \tag{4.23}$$

Observe that the negated s-parameters appear directly in the lower left quadrant of (4.22). This is because there are only two sets of nodes remaining – those incident on the system and those reflected from the system. Another way of seeing this is by expressing (4.22) as

$$\left[\mathbb{I} - \begin{pmatrix} \mathbf{0} & \mathbf{0} \\ \mathbf{W_{ba}} + \mathbf{W_{bx}} \cdot (\mathbb{I}-\mathbf{W_{xx}})^{-1} \cdot \mathbf{W_{xa}} & \mathbf{0} \end{pmatrix}\right] \cdot \begin{pmatrix} \mathbf{A} \\ \mathbf{B} \end{pmatrix} = \begin{pmatrix} \mathbb{I} \\ \mathbf{0} \end{pmatrix}.$$

### 4.4.1 Reduction of Systems With Stimuli

In somewhat rare instances, one is not looking for the s-parameters of a system, but rather the reduction of a system. The equation in (4.23) provides a reduction of a system, but this is only for a system with no stimuli – the only stimuli is supplied at the **a** nodes in the determination of the s-parameters. So, while a system with stimuli already embedded does not have s-parameters per se, it is often desirable to reduce the system such that it looks like a set of s-parameters, but with stimuli emanating from it. This is the case whenever systems containing internal sources are dealt with (see Chapter 6). Thus, the equation given in (4.16) reduces to

$$\left[\mathbb{I} - \begin{pmatrix} \mathbf{0} & \mathbf{0} & \mathbf{0} \\ \mathbf{W_{ba}} & \mathbf{0} & \mathbf{W_{bx}} \\ \mathbf{W_{xa}} & \mathbf{0} & \mathbf{W_{xx}} \end{pmatrix}\right] \cdot \begin{pmatrix} \mathbf{a} \\ \mathbf{b} \\ \mathbf{x} \end{pmatrix} = \begin{pmatrix} \mathbf{m_a} \\ \mathbf{m_b} \\ \mathbf{m_x} \end{pmatrix}, \tag{4.24}$$

where in (4.24), the simplifications shown in (4.17) have been retained, but the stimuli in the system are allowed to remain.

Using the algebra of the previous section, (4.24) can be rewritten as

$$\left[\mathbb{I} - \left[\begin{pmatrix} \mathbf{0} & \mathbf{0} \\ \mathbf{W_{ba}} & \mathbf{0} \end{pmatrix} + \begin{pmatrix} \mathbf{0} \\ \mathbf{W_{bx}} \end{pmatrix} \cdot (\mathbb{I} - \mathbf{W_{xx}})^{-1} \cdot \begin{pmatrix} \mathbf{W_{xa}} & \mathbf{0} \end{pmatrix}\right]\right] \cdot \begin{pmatrix} \mathbf{a} \\ \mathbf{b} \end{pmatrix}$$
$$= \begin{pmatrix} \mathbf{m_a} \\ \mathbf{m_b} \end{pmatrix} + \begin{pmatrix} \mathbf{0} \\ \mathbf{W_{bx}} \end{pmatrix} \cdot (\mathbb{I} - \mathbf{W_{xx}})^{-1} \cdot \mathbf{m_x}.$$

The s-parameters of this reduced system (again, the weights, connecting the **a** and **b** nodes) are therefore

$$\mathbf{S} = \mathbf{W_{ba}} + \mathbf{W_{bx}} \cdot (\mathbb{I} - \mathbf{W_{xx}})^{-1} \cdot \mathbf{W_{xa}}.$$

This is the same as (4.23). As for the stimuli, in principle, systems will not have stimuli pointing into **a** nodes, neither in the initial or final reduced system, so the stimuli emanating from the remaining **b** nodes $\mathbf{m_b'}$ are therefore calculated as

$$\mathbf{m_b'} = \mathbf{m_b} + \mathbf{W_{bx}} \cdot (\mathbb{I} - \mathbf{W_{xx}})^{-1} \cdot \mathbf{m_x}.$$

### 4.4.2 Solution to the Cascaded Two-Port Problem with the Block Solution

In the cascaded two-port example, there are the following groups of nodes:

$$\mathbf{a} = \begin{pmatrix} n_1 \\ n_6 \end{pmatrix}, \qquad \mathbf{b} = \begin{pmatrix} n_2 \\ n_5 \end{pmatrix}, \qquad \mathbf{x} = \begin{pmatrix} n_3 \\ n_4 \end{pmatrix}.$$

The nodes for **a** and **b** have been written in system port order. The order of the nodes in **x** is arbitrary.

Table 4.1 Block weights matrices for cascading example

| Matrix | Arrows originate from | Arrows terminate at | Form | Value |
|---|---|---|---|---|
| $\mathbf{W_{ba}}$ | $\mathbf{a} = \begin{pmatrix} n_1 \\ n_6 \end{pmatrix}$ | $\mathbf{b} = \begin{pmatrix} n_2 \\ n_5 \end{pmatrix}$ | $\begin{pmatrix} n_1 \to n_2 & n_6 \to n_2 \\ n_1 \to n_5 & n_6 \to n_5 \end{pmatrix}$ | $\begin{pmatrix} SL_{11} & 0 \\ 0 & SR_{22} \end{pmatrix}$ |
| $\mathbf{W_{bx}}$ | $\mathbf{x} = \begin{pmatrix} n_3 \\ n_4 \end{pmatrix}$ | $\mathbf{b} = \begin{pmatrix} n_2 \\ n_5 \end{pmatrix}$ | $\begin{pmatrix} n_3 \to n_2 & n_4 \to n_2 \\ n_3 \to n_5 & n_4 \to n_5 \end{pmatrix}$ | $\begin{pmatrix} 0 & SL_{12} \\ SR_{21} & 0 \end{pmatrix}$ |
| $\mathbf{W_{xx}}$ | $\mathbf{x} = \begin{pmatrix} n_3 \\ n_4 \end{pmatrix}$ | $\mathbf{x} = \begin{pmatrix} n_3 \\ n_4 \end{pmatrix}$ | $\begin{pmatrix} n_3 \to n_3 & n_4 \to n_3 \\ n_3 \to n_4 & n_4 \to n_4 \end{pmatrix}$ | $\begin{pmatrix} 0 & SL_{22} \\ SR_{11} & 0 \end{pmatrix}$ |
| $\mathbf{W_{xa}}$ | $\mathbf{a} = \begin{pmatrix} n_1 \\ n_6 \end{pmatrix}$ | $\mathbf{x} = \begin{pmatrix} n_3 \\ n_4 \end{pmatrix}$ | $\begin{pmatrix} n_1 \to n_3 & n_6 \to n_3 \\ n_1 \to n_4 & n_6 \to n_4 \end{pmatrix}$ | $\begin{pmatrix} SL_{21} & 0 \\ 0 & SR_{12} \end{pmatrix}$ |

There are four block matrices to be determined. They are shown in Table 4.1. Using the block weights matrices shown in Table 4.1, the solution is written as

$$\begin{aligned}\mathbf{S} &= \mathbf{W_{ba}} + \mathbf{W_{bx}} \cdot (\mathbb{I} - \mathbf{W_{xx}})^{-1} \cdot \mathbf{W_{xa}} \\ &= \begin{pmatrix} SL_{11} & 0 \\ 0 & SR_{22} \end{pmatrix} + \begin{pmatrix} 0 & SL_{12} \\ SR_{21} & 0 \end{pmatrix} \cdot \begin{pmatrix} 1 & -SL_{22} \\ -SR_{11} & 1 \end{pmatrix}^{-1} \cdot \begin{pmatrix} SL_{21} & 0 \\ 0 & SR_{12} \end{pmatrix} \\ &= \left(\begin{smallmatrix} SL_{11} & 0 \\ 0 & SR_{22} \end{smallmatrix}\right) + \frac{1}{1 - SL_{22} \cdot SR_{21}} \cdot \left(\begin{smallmatrix} 0 & SL_{12} \\ SR_{21} & 0 \end{smallmatrix}\right) \cdot \left(\begin{smallmatrix} 1 & SL_{22} \\ SR_{11} & 1 \end{smallmatrix}\right) \cdot \left(\begin{smallmatrix} SL_{21} & 0 \\ 0 & SR_{12} \end{smallmatrix}\right). \end{aligned} \tag{4.25}$$

The cascade equation is kept in the form of (4.25) because it explicitly states the solution in a simple matrix form (one weight per matrix element).

Equation (4.25) exposes how the solution is formed. One can see that the network starts with some connections interacting already at the system ports ($\mathbf{W_{ba}}$). To these arrows are added combinations of interactions between the system ports and the internal nodes ($\mathbf{W_{bx}}$ and $\mathbf{W_{xa}}$) and interactions within the internal nodes themselves ($\mathbf{W_{xx}}$).

### 4.4.3 Tip Embedding Using Block Weights

Consider, as shown in Figure 4.10, a $P$-port s-parameter network device where to each port $p$, a two-port network is connected. Each two-port network is defined as having s-parameters $\mathbf{ST_p}$, where $p$ is the port number to which it connects. As a convention, port 1 of the two-port device is connected to the port of the network device and port 2 is exposed as a system port. The $\mathbf{a}$ nodes are listed as nodes 1 through $P$ and the $\mathbf{b}$ nodes as nodes $P+1$ through $2 \cdot P$. Also, for simplicity, the $\mathbf{x}$ nodes are separated into two sets, with the incident nodes on the network device designated as nodes $2 \cdot P + 1$ through nodes $3 \cdot P$ and the reflected nodes of the network device designated as nodes $3 \cdot P + 1$ through $4 \cdot P$.

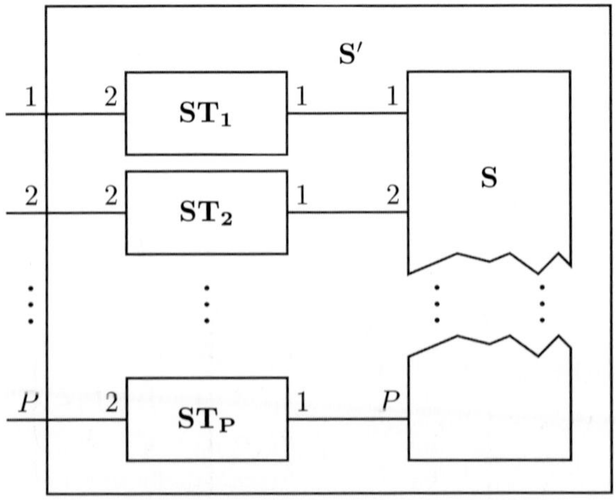


**Figure 4.10** Tip embedding

Therefore, the node vectors are defined as

$$\mathbf{a} = \begin{pmatrix} n_1 \\ n_2 \\ \vdots \\ n_P \end{pmatrix}, \qquad \mathbf{b} = \begin{pmatrix} n_{P+1} \\ n_{P+2} \\ \vdots \\ n_{2 \cdot P} \end{pmatrix}, \qquad \mathbf{x} = \begin{pmatrix} n_{2 \cdot P+1} \\ n_{2 \cdot P+2} \\ \vdots \\ n_{3 \cdot P} \\ n_{3 \cdot P+1} \\ n_{3 \cdot P+2} \\ \vdots \\ n_{4 \cdot P} \end{pmatrix}.$$

The weights matrices take the following form:

$$\mathbf{W_{ba}} = \begin{pmatrix} ST_{1_{22}} & 0 & \cdots & 0 \\ 0 & ST_{2_{22}} & \cdots & 0 \\ \vdots & \vdots & \ddots & \vdots \\ 0 & 0 & \vdots & ST_{P_{22}} \end{pmatrix} = \mathbf{ST_{22}},$$

$$\mathbf{W_{bx}} = \begin{pmatrix} 0 & 0 & \cdots & 0 & ST_{1_{21}} & 0 & \cdots & 0 \\ 0 & 0 & \cdots & 0 & 0 & ST_{2_{21}} & \cdots & 0 \\ \vdots & \vdots & \ddots & \vdots & \vdots & \vdots & \ddots & \vdots \\ 0 & 0 & \cdots & 0 & 0 & 0 & \vdots & ST_{P_{21}} \end{pmatrix} = \begin{pmatrix} \mathbf{0} & \mathbf{ST_{21}} \end{pmatrix},$$

$$\mathbf{W_{xx}} = \begin{pmatrix} 0 & 0 & \cdots & 0 & ST_{1_{11}} & 0 & \cdots & 0 \\ 0 & 0 & \cdots & 0 & 0 & ST_{2_{11}} & \cdots & 0 \\ \vdots & \vdots & \ddots & \vdots & \vdots & \vdots & \ddots & \vdots \\ 0 & 0 & \cdots & 0 & 0 & 0 & \vdots & ST_{P_{11}} \\ S_{11} & S_{12} & \cdots & S_{1P} & 0 & 0 & \cdots & 0 \\ S_{21} & S_{22} & \vdots & S_{2P} & 0 & 0 & \cdots & 0 \\ \vdots & \vdots & \ddots & \vdots & \vdots & \vdots & \ddots & \vdots \\ S_{P1} & S_{P2} & \cdots & S_{PP} & 0 & 0 & \cdots & 0 \end{pmatrix} = \begin{pmatrix} 0 & \mathbf{ST_{11}} \\ \mathbf{S} & \mathbf{0} \end{pmatrix},$$

$$\mathbf{W_{xa}} = \begin{pmatrix} ST_{1_{12}} & 0 & \cdots & 0 \\ 0 & ST_{2_{12}} & \cdots & 0 \\ \vdots & \vdots & \ddots & \vdots \\ 0 & 0 & \vdots & ST_{P_{12}} \\ 0 & 0 & \cdots & 0 \\ 0 & 0 & \cdots & 0 \\ \vdots & \vdots & \ddots & \vdots \\ 0 & 0 & \cdots & 0 \end{pmatrix} = \begin{pmatrix} \mathbf{ST_{12}} \\ \mathbf{0} \end{pmatrix}.$$

Partitioning the weights matrices produces the s-parameter solution as

$$\begin{aligned} \mathbf{S'} &= \mathbf{W_{ba}} + \mathbf{W_{bx}} \cdot (\mathbb{I} - \mathbf{W_{xx}})^{-1} \cdot \mathbf{W_{xa}} \\ &= \mathbf{ST_{22}} + \begin{pmatrix} \mathbf{0} & \mathbf{ST_{21}} \end{pmatrix} \cdot \begin{pmatrix} \mathbb{I} & -\mathbf{ST_{11}} \\ -\mathbf{S} & \mathbb{I} \end{pmatrix}^{-1} \cdot \begin{pmatrix} \mathbf{ST_{12}} \\ \mathbf{0} \end{pmatrix} \\ &= \mathbf{ST_{22}} + \mathbf{ST_{21}} \cdot (\mathbb{I} - \mathbf{S} \cdot \mathbf{ST_{11}})^{-1} \cdot \mathbf{S} \cdot \mathbf{ST_{12}}. \end{aligned} \tag{4.26}$$

Equation (4.26) can be used for a variety of purposes, including port reference impedance changes, as will be described in Chapter 5.

## 4.5 System Reduction Through Node Removal

A system equation can be written in the form

$$(\mathbb{I} - \mathbf{W}) \cdot \mathbf{n} = \mathbf{m}.$$

Regarding node removal, consider this system as consisting of a list of nodes $\mathbf{k}$ that it is desirable to keep and a list of nodes $\mathbf{x}$ that it is desirable to remove. Therefore, the vector of nodes can be partitioned into two block vectors:

$$\mathbf{n} = \begin{pmatrix} \mathbf{k} \\ \mathbf{x} \end{pmatrix}.$$

Partitioning the entire system to reflect this node partitioning results in

$$\left[\mathbb{I} - \begin{pmatrix} \mathbf{W_{kk}} & \mathbf{W_{kx}} \\ \mathbf{W_{xk}} & \mathbf{W_{xx}} \end{pmatrix}\right] \cdot \begin{pmatrix} \mathbf{k} \\ \mathbf{x} \end{pmatrix} = \begin{pmatrix} \mathbf{m_k} \\ \mathbf{m_x} \end{pmatrix}$$

or

$$\begin{pmatrix} \mathbb{I} - \mathbf{W_{kk}} & -\mathbf{W_{kx}} \\ -\mathbf{W_{xk}} & \mathbb{I} - \mathbf{W_{xx}} \end{pmatrix} \cdot \begin{pmatrix} \mathbf{k} \\ \mathbf{x} \end{pmatrix} = \begin{pmatrix} \mathbf{m_k} \\ \mathbf{m_x} \end{pmatrix}.$$

The two equations for this system are written as

$$(\mathbb{I} - \mathbf{W_{kk}}) \cdot \mathbf{k} - \mathbf{W_{kx}} \cdot \mathbf{x} = \mathbf{m_k},$$

$$-\mathbf{W_{xk}} \cdot \mathbf{k} + (\mathbb{I} - \mathbf{W_{xx}}) \cdot \mathbf{x} = \mathbf{m_x}.$$

Solving the second for $\mathbf{x}$:

$$\mathbf{x} = (\mathbb{I} - \mathbf{W_{xx}})^{-1} \cdot [\mathbf{m_x} + \mathbf{W_{xk}} \cdot \mathbf{k}].$$

Substituting this into the first,

$$(\mathbb{I} - \mathbf{W_{kk}}) \cdot \mathbf{k} - \mathbf{W_{kx}} \cdot (\mathbb{I} - \mathbf{W_{xx}})^{-1} \cdot [\mathbf{m_x} + \mathbf{W_{xk}} \cdot \mathbf{k}] = \mathbf{m_k}.$$

Collecting $\mathbf{k}$ yields

$$\left[(\mathbb{I} - \mathbf{W_{kk}}) - \mathbf{W_{kx}} \cdot (\mathbb{I} - \mathbf{W_{xx}})^{-1} \cdot \mathbf{W_{xk}}\right] \cdot \mathbf{k} = \mathbf{m_k} + \mathbf{W_{kx}} \cdot (\mathbb{I} - \mathbf{W_{xx}})^{-1} \cdot \mathbf{m_x}$$

or finally

$$\left[\mathbb{I} - \left(\mathbf{W_{kk}} + \mathbf{W_{kx}} \cdot (\mathbb{I} - \mathbf{W_{xx}})^{-1} \cdot \mathbf{W_{xk}}\right)\right] \cdot \mathbf{k} = \mathbf{m_k} + \mathbf{W_{kx}} \cdot (\mathbb{I} - \mathbf{W_{xx}})^{-1} \cdot \mathbf{m_x}. \quad (4.27)$$

Thus, the steps of node removal can be written as follows:

1. From a node list $\mathbf{n}$ identify the nodes to keep $\mathbf{k}$ and the nodes to be removed $\mathbf{x}$, and group this with the stimuli to keep $\mathbf{m_k}$ and the stimuli to be removed $\mathbf{m_x}$.
2. Generate four weights matrices either by partitioning the system weights matrix or separately such that:
   a) $\mathbf{W_{kk}}$ is a weights matrix containing the weights of arrows interacting with the nodes in $\mathbf{k}$ (i.e. the arrows go between these nodes in the list) such that a matrix element $\mathbf{W_{kk}}[r][c]$ contains the weight of the arrow originating at the node listed at $\mathbf{k}[c]$ and terminating at the node listed at $\mathbf{k}[r]$.
   b) $\mathbf{W_{kx}}$ is a weights matrix containing the weights of arrows going from the nodes to be removed listed in $\mathbf{x}$ to the nodes to keep listed in $\mathbf{k}$ such that a matrix element $\mathbf{W_{kx}}[r][c]$ contains the weight of the arrow originating at the node listed at $\mathbf{x}[c]$ and terminating at the node listed at $\mathbf{k}[r]$.
   c) $\mathbf{W_{xk}}$ is a weights matrix containing the weights of arrows going from the nodes to keep listed in $\mathbf{k}$ to the nodes to be removed listed in $\mathbf{x}$ such that a matrix element $\mathbf{W_{xk}}[r][c]$ contains the weight of the arrow originating at the node listed at $\mathbf{k}[c]$ and terminating at the node listed at $\mathbf{x}[r]$.
   d) $\mathbf{W_{xx}}$ is a weights matrix containing the weights of arrows interacting with the nodes in $\mathbf{x}$ (i.e. the arrows go between these nodes in the list) such that a matrix element $\mathbf{W_{xx}}[r][c]$ contains the weight of the arrow originating at the node listed at $\mathbf{x}[c]$ and terminating at the node listed at $\mathbf{x}[r]$.
3. Generate the weights matrix of the new system by adding to the matrix $\mathbf{W_{kk}}$ such that the new weights matrix is

$$\mathbf{W'} = \mathbf{W_{kk}} + \mathbf{W_{kx}} \cdot (\mathbb{I} - \mathbf{W_{xx}})^{-1} \cdot \mathbf{W_{xk}}.$$

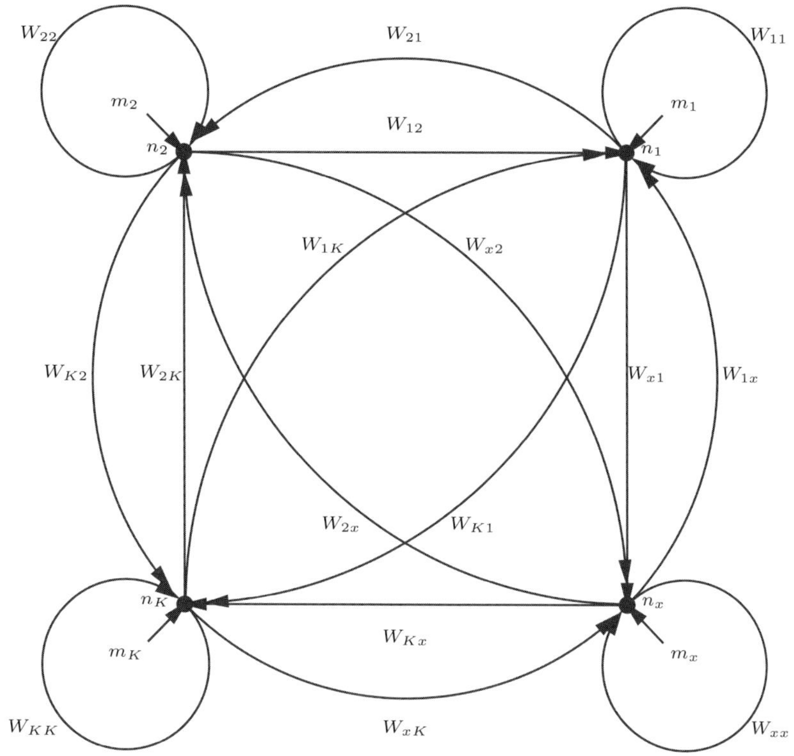


**Figure 4.11** Single node removal example

4. Generate the stimulus vector of the new system by adding to the vector $\mathbf{m_k}$ such that the new stimulus vector is

$$\mathbf{m}' = \mathbf{m_k} + \mathbf{W_{kx}} \cdot (\mathbb{I} - \mathbf{W_{xx}})^{-1} \cdot \mathbf{m_x}.$$

5. With the new list of nodes $\mathbf{n}' = \mathbf{k}$, write the new system equation as

$$(\mathbb{I} - \mathbf{W}') \cdot \mathbf{n}' = \mathbf{m}'.$$

6. After this step, the signal-flow diagram can be redrawn.

### 4.5.1 Removal of a Single Node in a Signal-Flow Diagram

Often, especially for small systems, one is tempted to reduce or solve the system without resorting to a lot of math. For very small systems, or when one or a few nodes are being removed, this can be a satisfactory method, but, from experience, node removal is an error-prone activity. That being said, this section will explain the steps of reducing a signal-flow diagram by hand by removing nodes. Generally, the nodes that are removed are the nodes not exposed to the system ports (i.e. the internal nodes). When internal nodes are removed properly, the result is a signal-flow diagram representative of the s-parameters of a system.

Node removal is performed by removing a single node at a time. In other words, if there are many nodes to be removed, one chooses a first node to remove, then follows the node removal steps and redraws the signal-flow diagram with this first node removed. The process continues for the second node, and so on. Therefore, the process of node removal is best understood by understanding what is happening in (4.27) when there is only one node removed.

One is given a system, containing $K$ nodes listed 1 through $K$, that comprises a list of nodes to keep and a single node $n_x$, which is a node to be removed. The system equation can be written as

$$(\mathbb{I} - \mathbf{W}) \cdot \mathbf{n} = \mathbf{m} = \left[ \mathbb{I} - \begin{pmatrix} W_{11} & W_{12} & \cdots & W_{1K} & W_{1x} \\ W_{21} & W_{22} & \cdots & W_{2K} & W_{2x} \\ \vdots & \vdots & \ddots & \vdots & \vdots \\ W_{K1} & W_{K2} & \cdots & W_{KK} & W_{Kx} \\ W_{x1} & W_{x2} & \cdots & W_{xK} & W_{xx} \end{pmatrix} \right] \cdot \begin{pmatrix} n_1 \\ n_2 \\ \vdots \\ n_K \\ n_x \end{pmatrix} = \begin{pmatrix} m_1 \\ m_2 \\ \vdots \\ m_K \\ m_x \end{pmatrix}.$$

The signal-flow diagram for this system with all possible arrows shown is provided in Figure 4.11. The goal here is to find the resulting equivalent signal-flow diagram with node $n_x$ removed. Following the steps just given, the list of nodes and stimuli to keep and remove is

$$\mathbf{k} = \begin{pmatrix} n_1 \\ n_2 \\ \vdots \\ n_K \end{pmatrix}, \qquad \mathbf{m_k} = \begin{pmatrix} m_1 \\ m_2 \\ \vdots \\ m_K \end{pmatrix}, \qquad \mathbf{x} = (n_x), \qquad \mathbf{m_x} = (m_x),$$

and the new partitioned weights matrices are

$$\mathbf{W_{kk}} = \begin{pmatrix} W_{11} & W_{12} & \cdots & W_{1K} \\ W_{21} & W_{22} & \cdots & W_{2K} \\ \vdots & \vdots & \ddots & \vdots \\ W_{K1} & W_{K2} & \cdots & W_{KK} \end{pmatrix}, \quad \mathbf{W_{kx}} = \begin{pmatrix} W_{1x} \\ W_{2x} \\ \vdots \\ W_{Kx} \end{pmatrix}, \quad \mathbf{W_{xk}} = \begin{pmatrix} W_{x1} \\ W_{x2} \\ \vdots \\ W_{xK} \end{pmatrix}^T,$$

$$\mathbf{W_{xx}} = (W_{xx}).$$

Following the steps, a new weights matrix $\mathbf{W}'$ is created by adding to the weights in $\mathbf{W_{kk}}$:

$$\begin{aligned}
\mathbf{W}' &= \mathbf{W_{kk}} + \mathbf{W_{kx}} \cdot (\mathbb{I} - \mathbf{W_{xx}})^{-1} \cdot \mathbf{W_{xk}} \\
&= \begin{pmatrix} W_{11} & W_{12} & \cdots & W_{1K} \\ W_{21} & W_{22} & \cdots & W_{2K} \\ \vdots & \vdots & \ddots & \vdots \\ W_{K1} & W_{K2} & \cdots & W_{KK} \end{pmatrix} + \begin{pmatrix} W_{1x} \\ W_{2x} \\ \vdots \\ W_{Kx} \end{pmatrix} \cdot \frac{1}{1 - W_{xx}} \cdot \begin{pmatrix} W_{x1} & W_{x2} & \cdots & W_{xK} \end{pmatrix} \\
&= \begin{pmatrix} W_{11} & W_{12} & \cdots & W_{1K} \\ W_{21} & W_{22} & \cdots & W_{2K} \\ \vdots & \vdots & \ddots & \vdots \\ W_{K1} & W_{K2} & \cdots & W_{KK} \end{pmatrix} \\
&\quad + \frac{1}{1 - W_{xx}} \cdot \begin{pmatrix} W_{1x} \cdot W_{x1} & W_{1x} \cdot W_{x2} & \cdots & W_{1x} \cdot W_{xK} \\ W_{2x} \cdot W_{x1} & W_{2x} \cdot W_{x2} & \cdots & W_{2x} \cdot W_{xK} \\ \vdots & \vdots & \ddots & \vdots \\ W_{Kx} \cdot W_{x1} & W_{Kx} \cdot W_{x2} & \cdots & W_{Kx} \cdot W_{xK} \end{pmatrix} \\
&= \begin{pmatrix} W_{11} + \frac{W_{1x} \cdot W_{x1}}{1 - W_{xx}} & W_{12} + \frac{W_{1x} \cdot W_{x2}}{1 - W_{xx}} & \cdots & W_{1K} + \frac{W_{1x} \cdot W_{xK}}{1 - W_{xx}} \\ W_{21} + \frac{W_{2x} \cdot W_{x1}}{1 - W_{xx}} & W_{22} + \frac{W_{2x} \cdot W_{x2}}{1 - W_{xx}} & \cdots & W_{2K} + \frac{W_{2x} \cdot W_{xK}}{1 - W_{xx}} \\ \vdots & \vdots & \ddots & \vdots \\ W_{K1} + \frac{W_{Kx} \cdot W_{x1}}{1 - W_{xx}} & W_{K2} + \frac{W_{Kx} \cdot W_{x2}}{1 - W_{xx}} & \cdots & W_{KK} + \frac{W_{Kx} \cdot W_{xK}}{1 - W_{xx}} \end{pmatrix}.
\end{aligned} \tag{4.28}$$

A new stimulus vector $\mathbf{m}'$ is created:

$$\begin{aligned}
\mathbf{m}' = \mathbf{m_k} + \mathbf{W_{kx}} \cdot (\mathbb{I} - \mathbf{W_{xx}})^{-1} \cdot \mathbf{m_x} &= \begin{pmatrix} m_1 \\ m_2 \\ \vdots \\ m_K \end{pmatrix} + \begin{pmatrix} W_{1x} \\ W_{2x} \\ \vdots \\ W_{Kx} \end{pmatrix} \cdot \left( \frac{1}{1 - W_{xx}} \right) \cdot (m_x) \\
&= \begin{pmatrix} m_1 \\ m_2 \\ \vdots \\ m_K \end{pmatrix} + \frac{1}{1 - W_{xx}} \cdot \begin{pmatrix} W_{1x} \cdot m_x \\ W_{2x} \cdot m_x \\ \vdots \\ W_{Kx} \cdot m_x \end{pmatrix} = \begin{pmatrix} m_1 + \frac{W_{1x} \cdot m_x}{1 - W_{xx}} \\ m_2 + \frac{W_{2x} \cdot m_x}{1 - W_{xx}} \\ \vdots \\ m_K + \frac{W_{Kx} \cdot m_x}{1 - W_{xx}} \end{pmatrix}.
\end{aligned} \tag{4.29}$$

The resulting changes to the weights and stimuli are shown graphically in the new signal-flow diagram shown in Figure 4.12. Here, the removed node, arrows, and stimulus are grayed out. The removed weights are shown, but grayed out, to aid in understanding the process. Note that every surviving arrow is affected by adding the product of three weights that were removed. For a surviving arrow going from an originating node number $o$ to a terminating node number $t$, one adds the product of:

1. $W_{tx}$ – the weight of a removed arrow that originated at node $n_x$ and terminated at node $n_t$.

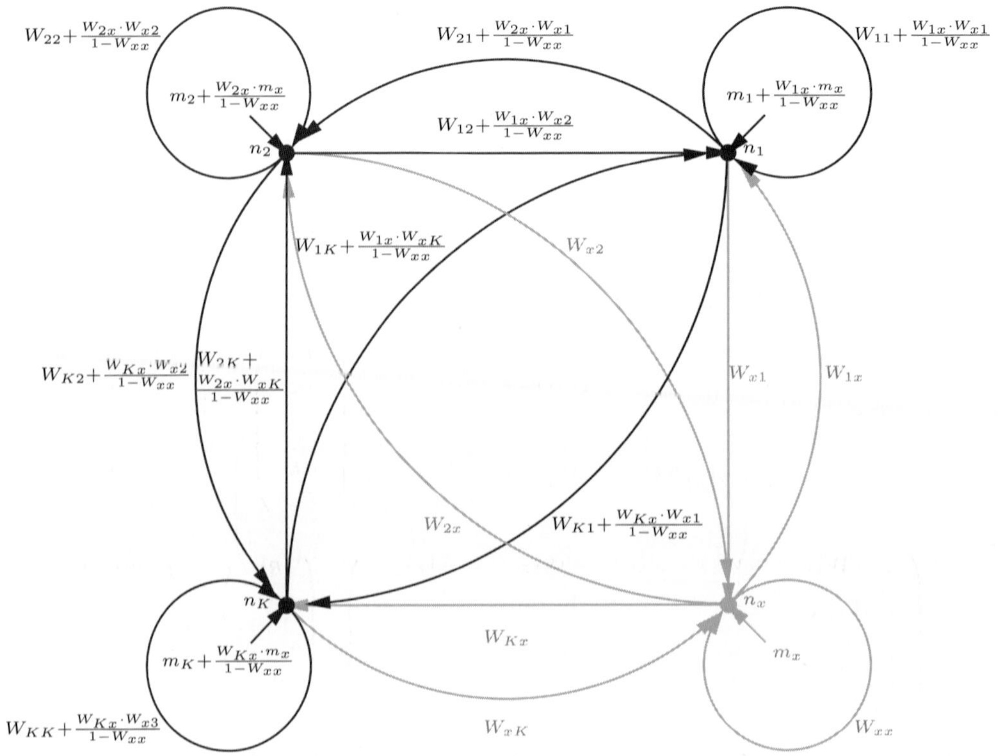


**Figure 4.12** Single node removal example with node $n_x$ removed

2. $W_{xo}$ – the weight of a removed arrow that originated at node $n_o$ and terminated at node $n_x$.

3. $\frac{1}{1-W_{xx}}$ – a factor that accounts for the removed arrow that looped back on node $n_x$.

In other words, disregarding the effect of the loopback on node $n_x$ for a moment, one adds the product of all path weights that originated at the same originating node and terminated at the same terminating node, which went through the removed node. Examining Figure 4.12 carefully should make this clear.

For every surviving stimulus arrow that terminates at a node number $t$, one adds the product of three weights:

1. $W_{tx}$ – the weight of a removed arrow that originated at node $n_x$ and terminated at node $n_t$.

2. $m_x$ – the removed stimulus that terminated at node $n_x$.

3. $\frac{1}{1-W_{xx}}$ – a factor that accounts for the removed arrow that looped back on node $n_x$.

In other words, again disregarding the effect of the loopback on node $n_x$ for a moment, for a given surviving stimulus, one adds the product of the removed stimulus and all path weights from the removed node that the removed stimulus terminated at and the node that the given surviving stimulus terminates at. This should be understood by carefully examining Figure 4.12.

Returning to the effect of the loopback arrow on $n_x$ (with a weight of $W_{xx}$): this loopback arrow causes some complications and disturbs the understanding of single node removal. It is desirable to have a recipe for node removal that does not involve this consideration. This is possible if loopbacks are removed from nodes to be removed, prior to removing the node. As mentioned previously, this loopback arrow is the result of a non-zero value on the diagonal of the weights matrix.

The original system equation is in the following form:

$$\left[\mathbb{I} - \begin{pmatrix} W_{11} & W_{12} & \cdots & W_{1K} & W_{1x} \\ W_{21} & W_{22} & \cdots & W_{2K} & W_{2x} \\ \vdots & \vdots & \ddots & \vdots & \vdots \\ W_{K1} & W_{K2} & \cdots & W_{KK} & W_{Kx} \\ W_{x1} & W_{x2} & \cdots & W_{xK} & W_{xx} \end{pmatrix}\right] \cdot \begin{pmatrix} n_1 \\ n_2 \\ \vdots \\ n_K \\ n_x \end{pmatrix} = \begin{pmatrix} m_1 \\ m_2 \\ \vdots \\ m_K \\ m_x \end{pmatrix}$$

or

$$\begin{pmatrix} 1-W_{11} & -W_{12} & \cdots & -W_{1K} & -W_{1x} \\ -W_{21} & 1-W_{22} & \cdots & -W_{2K} & -W_{2x} \\ \vdots & \vdots & \ddots & \vdots & \vdots \\ -W_{K1} & -W_{K2} & \cdots & 1-W_{KK} & -W_{Kx} \\ -W_{x1} & -W_{x2} & \cdots & -W_{xK} & 1-W_{xx} \end{pmatrix} \cdot \begin{pmatrix} n_1 \\ n_2 \\ \vdots \\ n_K \\ n_x \end{pmatrix} = \begin{pmatrix} m_1 \\ m_2 \\ \vdots \\ m_K \\ m_x \end{pmatrix}.$$

One can multiply each side of this equation to get rid of the loopback on node $n_x$:

$$\begin{pmatrix} 1 & 0 & 0 & \cdots & 0 \\ 0 & 1 & 0 & \cdots & 0 \\ \vdots & \vdots & \ddots & \vdots & \vdots \\ 0 & 0 & 0 & \cdots & 0 \\ 0 & 0 & 0 & \cdots & \frac{1}{1-W_{xx}} \end{pmatrix} \cdot \begin{pmatrix} 1-W_{11} & -W_{12} & \cdots & -W_{1K} & -W_{1x} \\ -W_{21} & 1-W_{22} & \cdots & -W_{2K} & -W_{2x} \\ \vdots & \vdots & \vdots & \vdots & \vdots \\ -W_{K1} & -W_{K2} & \cdots & 1-W_{KK} & -W_{Kx} \\ -W_{x1} & -W_{x2} & \cdots & -W_{xK} & 1-W_{xx} \end{pmatrix} \cdot \begin{pmatrix} n_1 \\ n_2 \\ \vdots \\ n_K \\ n_x \end{pmatrix}$$

$$= \begin{pmatrix} 1 & 0 & 0 & \cdots & 0 \\ 0 & 1 & 0 & \cdots & 0 \\ \vdots & \vdots & \ddots & \vdots & \vdots \\ 0 & 0 & 0 & \cdots & 0 \\ 0 & 0 & 0 & \cdots & \frac{1}{1-W_{xx}} \end{pmatrix} \cdot \begin{pmatrix} m_1 \\ m_2 \\ \vdots \\ m_K \\ m_x \end{pmatrix},$$

which results in

$$\begin{pmatrix} 1-W_{11} & -W_{12} & \cdots & -W_{1K} & -W_{1x} \\ -W_{21} & 1-W_{22} & \cdots & -W_{2K} & -W_{2x} \\ \vdots & \vdots & \ddots & \vdots & \vdots \\ -W_{K1} & -W_{K2} & \cdots & 1-W_{KK} & -W_{Kx} \\ -\frac{W_{x1}}{1-W_{xx}} & -\frac{W_{x2}}{1-W_{xx}} & \cdots & -\frac{W_{xK}}{1-W_{xx}} & 1 \end{pmatrix} \cdot \begin{pmatrix} n_1 \\ n_2 \\ \vdots \\ n_K \\ n_x \end{pmatrix} = \begin{pmatrix} m_1 \\ m_2 \\ \vdots \\ m_K \\ \frac{m_x}{1-W_{xx}} \end{pmatrix}$$

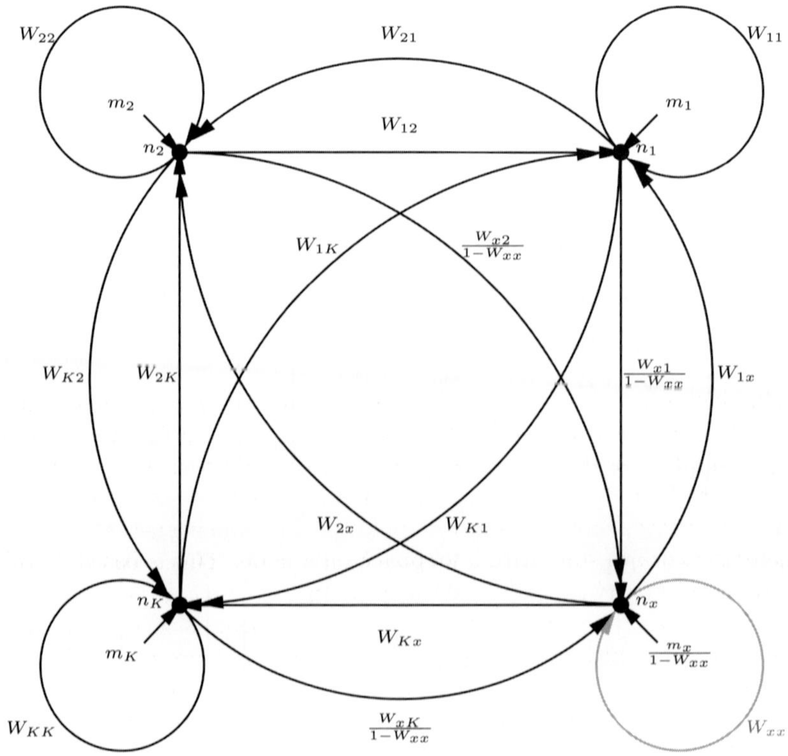


**Figure 4.13** Single node removal example with loopback on node $n_x$ removed

or

$$\left[ \mathbb{I} - \begin{pmatrix} W_{11} & W_{12} & \cdots & W_{1K} & W_{1x} \\ W_{21} & W_{22} & \cdots & W_{2K} & W_{2x} \\ \vdots & \vdots & \ddots & \vdots & \vdots \\ W_{K1} & W_{K2} & \cdots & W_{KK} & W_{Kx} \\ \frac{W_{x1}}{1-W_{xx}} & \frac{W_{x2}}{1-W_{xx}} & \cdots & \frac{W_{xK}}{1-W_{xx}} & 0 \end{pmatrix} \right] \cdot \begin{pmatrix} n_1 \\ n_2 \\ \vdots \\ n_K \\ n_x \end{pmatrix} = \begin{pmatrix} m_1 \\ m_2 \\ \vdots \\ m_K \\ \frac{m_x}{1-W_{xx}} \end{pmatrix}. \tag{4.30}$$

Examining (4.30) reveals that the loopback is removed and that every element in the row of the node containing the removed loopback is divided by $1 - W_{xx}$. This effect is shown graphically in Figure 4.13. Here one sees that the effect of scaling a row of the system equation is to scale every arrow (including the stimulus) pointing into the node defined by the row in the system equation. In other words, since a row in the system equation is an equation for a node, the effect of scaling the row is to scale arrows pointing at the node defined by that row.

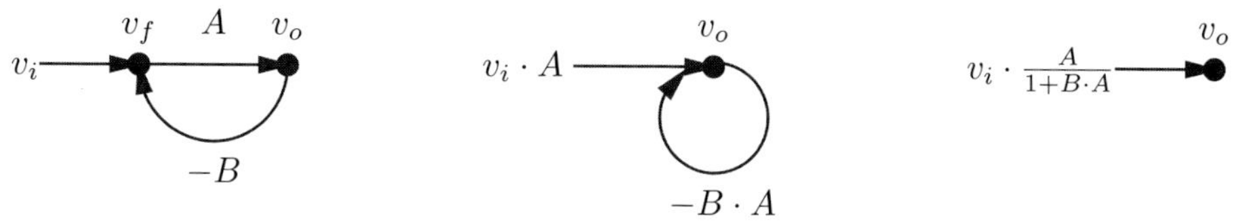


(a) Original network (b) Node $v_f$ removed (c) Loopback on $v_o$ removed

**Figure 4.14** Node removal in feedback analysis

The scaling factor is $1/(1 - W_{xx})$ and not $W_{xx}$. This is borne out by the math, but has been seen before in an area with which the reader might be familiar: feedback analysis. In Figure 4.14, a simple feedback example is shown with node removal performed. The step of removing the final loopback removes the feedback path.[2]

Figure 4.14 illustrates another important point: when reducing networks, it is unusual for an s-parameter network to start with a loopback on a node. This is based on the definition of s-parameters and on the rules for interconnecting s-parameter devices. However, it is not unusual for loopbacks to appear after a node is removed, as occurred in the removal of $v_f$ in Figure 4.14. Generally, these loopbacks, if they appear, are removed after each node removal step because generally it is desirable that each row of the system equation is an equation of a node in terms of all other nodes and stimuli. When a loopback exists, the equation for the node is also in terms of the node itself.

If the plan is that nodes are never removed until a loopback is removed first (where the rules for removing one are known), then the steps of node removal are simplified. To remove a node $n_x$:

1. If there is a loopback node at node $n_x$, perform step 4 to remove it and return to this step.
2. Draw a new diagram which is a copy of the original, but with $n_x$ and all the arrows that originate from it or terminate at it, including the stimulus, removed.
3. For each arrow to be removed that originates from the node to remove $n_x$ and terminates on a surviving node $n_t$:
    a) For each arrow to be removed that originates at a surviving node $n_o$ and terminates at the node to be removed $n_x$:
        i. Add the product of the weight of the arrow from $n_o$ to $n_x$ and the weight of the arrow from $n_x$ to $n_t$ to the weight of the arrow from $n_o$ to $n_t$ in the new diagram. If no such arrow exists, create it.
    b) Add the product of the stimulus applied to node $n_x$ and the weight of the arrow from $n_x$ to $n_t$ to the stimulus applied to node $n_t$ in the new diagram.
4. For each loopback created by step 3 at a surviving node $n_l$:

[2]The feedback effect is not removed as all of the networks are equivalent, only the ability to see it is.

a) Divide the weight of every arrow terminating on node $n_l$ (except for the loopback itself) by $1 - W_{ll}$, where $W_{ll}$ is the weight of the loopback arrow on node $n_l$. Be sure to include the stimulus that points at node $n_l$.
b) Remove the loopback arrow from node $n_l$.

Some notes about these steps:

- Step 1 is rarely needed when reducing signal-flow diagrams starting with interconnected s-parameter network devices because of the definition of s-parameters (reflected nodes in terms of incident nodes) and the interconnection rules (an incident node of one device port connects to one reflected node of another device). This is not to say loopbacks will not appear, only that they will not appear at the outset and step 4 removes them after each node is removed.
- Steps 3 and 3a comprise a looping process consisting of an outer loop looping over the $N_o$ arrows originating from the node to be removed and an inner loop looping over the $N_i$ arrows terminating on the node to be removed. Therefore, the total number of arrows (excluding stimuli) affected is known (i.e. added to) in the flow diagram as $N_o \cdot N_i$.
- Steps 3 and 3b comprise a loop over the $N_o$ arrows originating from the node to be removed, and therefore the total number of stimuli affected (i.e. added to) is known in the signal-flow diagram as $N_o$, presuming the node to be removed has a stimulus applied to it.
- Steps 3 and 3b are rarely needed when reducing signal-flow diagrams starting with interconnected network devices because of the usual goal for reduction. Usually, stimuli are only applied to system ports, and these ports contain nodes that are not usually removed from the signal-flow diagram.
- Steps 2 and 3 are performed in this order in practice since manual node removal typically involves redrawing the diagram as each step is performed and this is the safest and most foolproof way. That being said, one could just as well alter and add to the arrows in the original diagram in step 3 and then remove $n_x$ and all stimuli and arrows that terminate on it or originate from it.

These steps appear complicated (and they are), but after some practice they can often be performed fairly automatically. That being said, it is easy to make mistakes. Most of the mistakes involve not accounting for all of the paths or forgetting an arrow, as the process generally involves repeatedly redrawing the signal-flow diagram as each node is removed.

### 4.5.2 Signal-Flow Diagram Node Removal Example

A simple example that encompasses all of the rules specified for node removal is provided. While this example is relatively simple in the number of nodes, things still get very complicated. Most of the complication stems from the fact that, in order to make this an all encompassing example, the node to be removed has both a looped back arrow on it and stimulus applied to it.

The example starts in Figure 4.15, a three-node system with stimuli applied to all nodes, and all possible node-to-node connections and a looped back arrow on node 3. This example will follow the steps given in §4.5.1 to remove node 3.

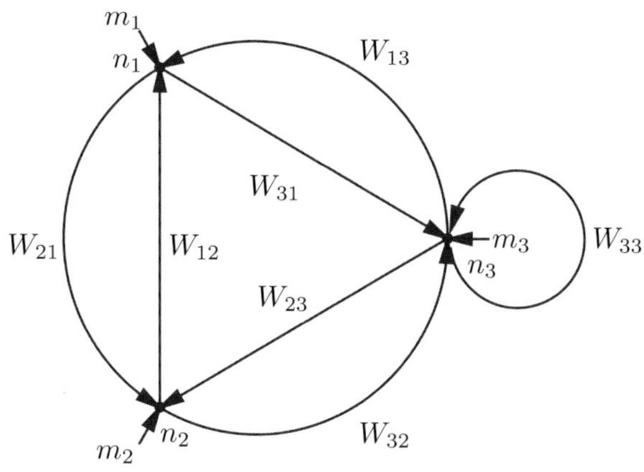


**Figure 4.15** Node removal example

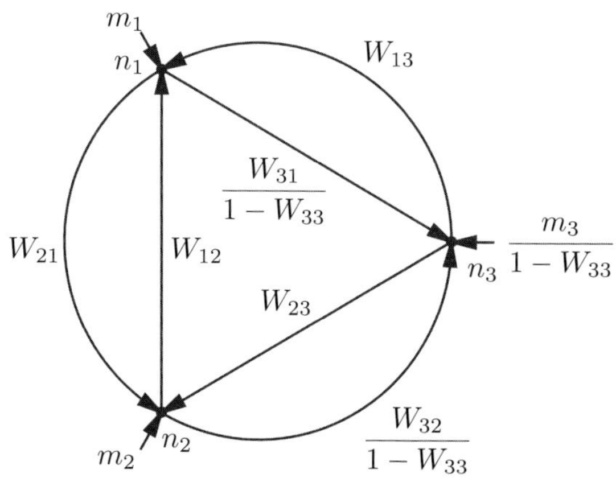


**Figure 4.16** Node removal example after step 1

The first step is to remove the loopback on node $n_3$ with weight $W_{33}$. This takes us immediately to step 4. According to step 4, one identifies all of the arrows including stimulus $m_3$ that terminate on $n_3$ and divides each of the weights of these arrows by $(1 - W_{33})$. This result can be seen in Figure 4.16, where this modification of the stimulus $M_3$ and the weights $W_{31}$ and $W_{32}$ has been made and the loopback on $n_3$ has been removed.

In step 2, the signal-flow diagram is redrawn with node $n_3$ removed and with all nodes, stimuli, and arrows that will survive already drawn in. This is shown in Figure 4.17, where it is seen that only the stimuli $m_1$ and $m_2$ along with the arrows $W_{21}$ and $W_{12}$ appear from the original diagram. According to step 1 in the instructions, all of the arrows and stimuli

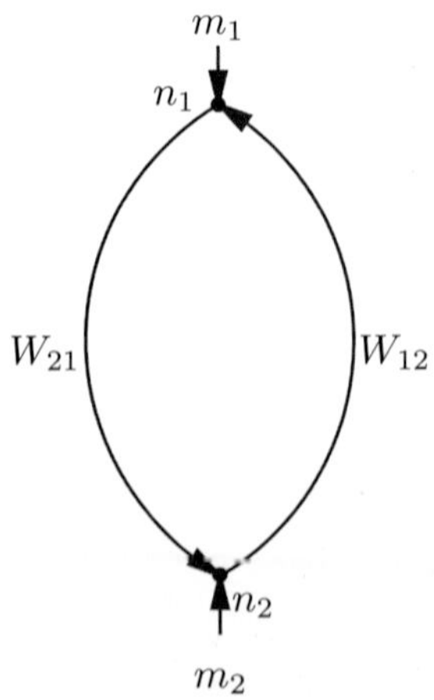


**Figure 4.17** Node removal example after step 2

involved with $n_3$ are identified. There are two arrows from the surviving nodes $n_1$ and $n_2$ that terminate on $n_3$ and two arrows that originate from $n_3$ and terminate on nodes $n_1$ and $n_2$.

In step 3, in an outer loop, one loops over the set of all arrows that originate from $n_3$ and terminate on surviving nodes. These are the two arrows $n_3 \rightarrow n_1$ and $n_3 \rightarrow n_2$. For the inner loop, one loops over the set of all arrows that originate from a surviving node and terminate on $n_3$. These are the two arrows $n_1 \rightarrow n_3$ and $n_2 \rightarrow n_3$. There will be six arrows (four arrows between nodes and two stimuli) either affected or added to in the new diagram. These steps are listed here as:

- For the first arrow that originates on $n_3$ and terminates on surviving node $n_1$:
    - For the first arrow that originates on $n_1$ and terminates on $n_3$, one adds the product of the weight of the arrow from $n_1$ to $n_3$ and the arrow from $n_3$ to $n_1$ to the arrow that goes from $n_1$ to $n_1$ in the new diagram. Since there is not such an arrow, a loopback arrow on $n_1$ with weight $W_{31} \cdot W_{13}/\left(1 - W_{33}\right)$ is created.
    - For the second arrow that originates on $n_2$ and terminates on $n_3$, one adds the product of the weight of the arrow from $n_2$ to $n_3$ and the arrow from $n_3$ to $n_1$ to the arrow that goes from $n_2$ to $n_1$. There is already an arrow with weight $W_{12}$ connecting these nodes, so $W_{32} \cdot W_{13}/\left(1 - W_{33}\right)$ is added to this weight.
    - Not forgetting the stimuli, one adds to the stimulus applied at node $n_1$ the stimulus applied at $n_3$ multiplied by the path weight from $n_3$ to $n_1$. Therefore, the stimulus at $n_1$ becomes $m_1 + W_{13} \cdot m_3/\left(1 - W_{33}\right)$.
- For the second arrow that originates on $n_3$ and terminates on surviving node $n_2$:
    - For the first arrow that originates on $n_1$ and terminates on $n_3$, one adds the product of the weight of the arrow from $n_1$ to $n_3$ and the arrow from $n_3$ to $n_2$ to

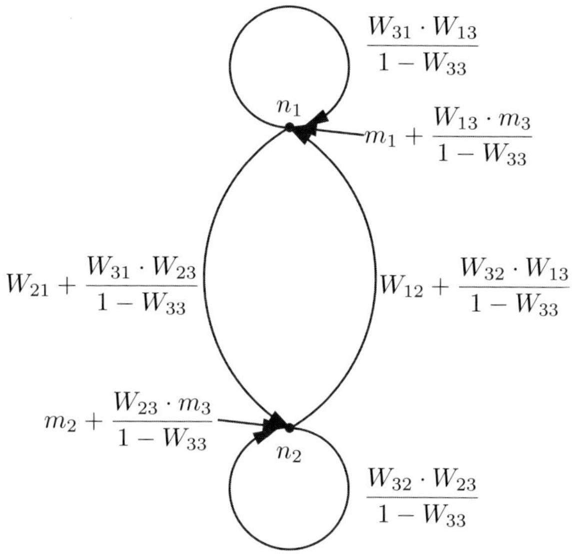


**Figure 4.18** Node removal example after step 3

the arrow that goes from $n_1$ to $n_2$. There is already an arrow with weight $W_{21}$ connecting these nodes, so $W_{31} \cdot W_{23}/(1 - W_{33})$ is added to this weight.

- For the second arrow that originates on $n_2$ and terminates on $n_3$, one adds the product of the weight of the arrow from $n_2$ to $n_3$ and the arrow from $n_3$ to $n_2$ to the arrow that goes from $n_2$ to $n_2$. Since there is not such an arrow, a loopback arrow on $n_2$ with weight $W_{32} \cdot W_{23}/(1 - W_{33})$ is created.
- For the stimulus applied at node $n_2$, one adds the stimulus applied at $n_3$ multiplied by the path weight from $n_3$ to $n_2$. Therefore, the stimulus at $n_2$ becomes $m_2 + W_{23} \cdot m_3/(1 - W_{33})$.

The result of this is shown in Figure 4.18. Check that node $n_x$ has been removed along with two arrows that terminate on the node from surviving nodes, two arrows that originated from $n_x$ and terminated on surviving nodes, and the stimuli applied to $n_x$. In the process, four arrows were affected that went between nodes and two stimuli that had arrows from $n_x$.

Finally, in step 4, the loopbacks that appeared on nodes $n_1$ and $n_2$ are removed. To remove the loopback on node $n_1$, one divides the stimulus applied to $n_1$ and the weight of all arrows that terminate on $n_1$ (the single arrow from $n_2$ to $n_1$) by $1 - \frac{W_{31} \cdot W_{13}}{1 - W_{33}}$. To remove the loopback on node $n_2$, one divides the stimulus applied to $n_2$ and the weight of all arrows that terminate on $n_2$ (the single arrow from node $n_1$ to $n_2$) by $1 - \frac{W_{32} \cdot W_{23}}{1 - W_{33}}$. The result is shown in Figure 4.19.

The new diagram with node $n_3$ removed shown in Figure 4.19 is behaviorally equivalent to Figure 4.15 in that the values at nodes 1 and 2 are identical.

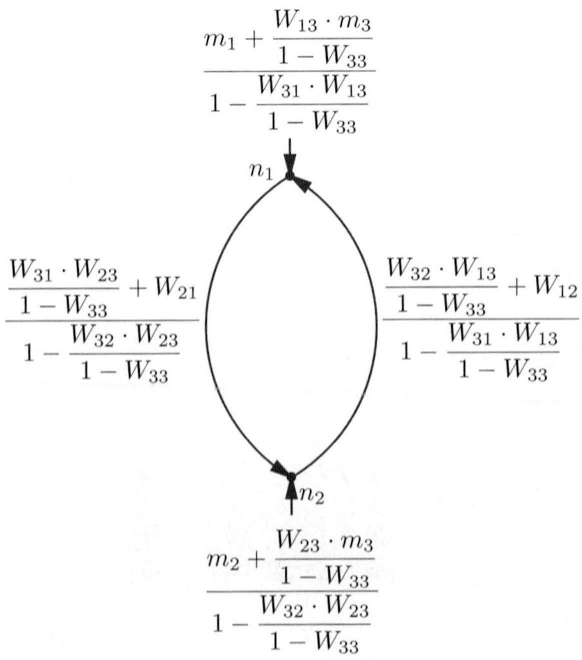


**Figure 4.19** Node removal example after step 4

## 4.6 Node Removal Using Graphical Equation Methods

In §4.5.1 the math was derived for removal of a single node in a signal-flow diagram and the results of the derivation were applied directly to the signal-flow diagram itself. As such, this method of node removal supplied an alternative to the matrix algebra method provided at the beginning of §4.5 and allowed for the direct manipulation of the signal-flow diagram. However, it required that only a single node is removed at a time. For simple systems, this is not a big drawback. The author's opinion of direct removal of a node in signal-flow diagrams is that it is a tedious and error-prone process. From experience, the wrong answer is often obtained and one ends up resorting to the matrix algebra methods to figure out what went wrong. In the end, one regrets that matrix algebra methods were not used in the first place. This being said, the matrix algebra methods are a bit heavy handed for simple problems and here is where a compromise is desired.

The compromise is to put the signal-flow diagram in matrix algebra form (or keep it there if equations were provided originally) and to remove nodes directly from the matrices and vectors in the equations by recognizing the patterns that have formed in the math derivations. Often, manipulation of the matrices and vectors is easier and less error prone than manipulation of the signal-flow diagram. In the end, it is easy to express the final

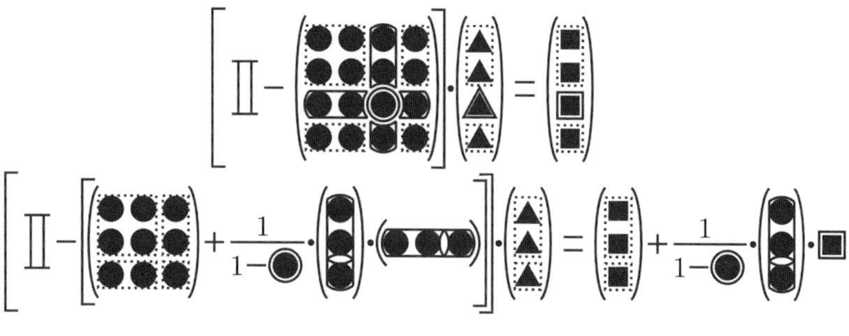

**Figure 4.20** Full node removal equation (graphical)

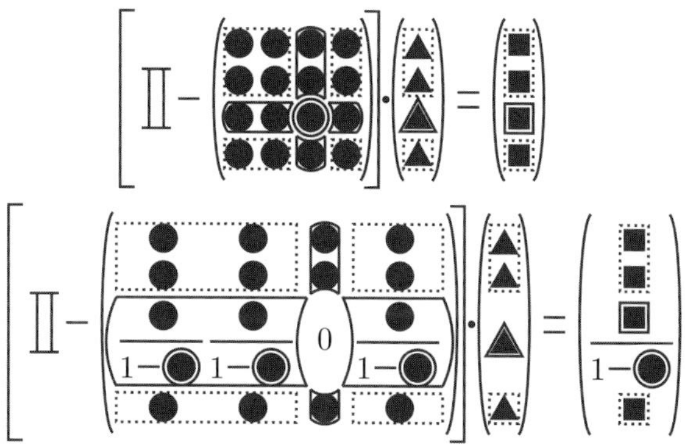

**Figure 4.21** Loopback removal equation (graphical)

result in signal-flow diagram form. The compromise method is called the *graphical equation* method because, to use it, one relies only on a graphical representation of the equations.

The equation for node removal can be shown graphically as a pattern to follow, as shown in Figure 4.20. Basically, the top equation shows the form of the system equation, with circles representing weights, triangles representing nodes, and squares representing stimuli. The matrix is shown partitioned for the removal of the node represented by the outlined triangle. Dotted lines are shown surrounding the surviving weights, nodes, and stimuli, and solid lines are shown surrounding the weights, nodes, and stimuli that will be removed from the system. Below this is the full solution for the removal of the node. Basically, it is a graphical representation of the resulting weights matrix derived in (4.28) and (4.29). This pattern can be followed to directly remove a single node from the original equation. The pattern for full node removal looks a bit complicated, but that is because it is still retaining the possibility for a loopback arrow on the node to be removed. It is easier if these loopbacks are removed first. The graphical equivalent of loopback removal is shown in Figure 4.21, where one sees that loopbacks are removed by dividing out the loopback element in the row

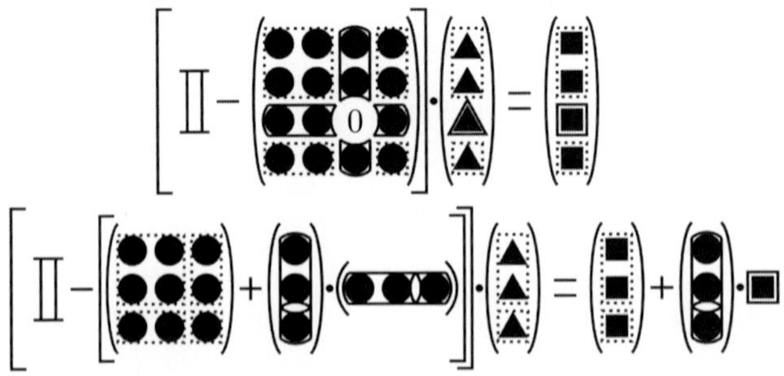

**Figure 4.22** Node removal equation with no loopback (graphical)

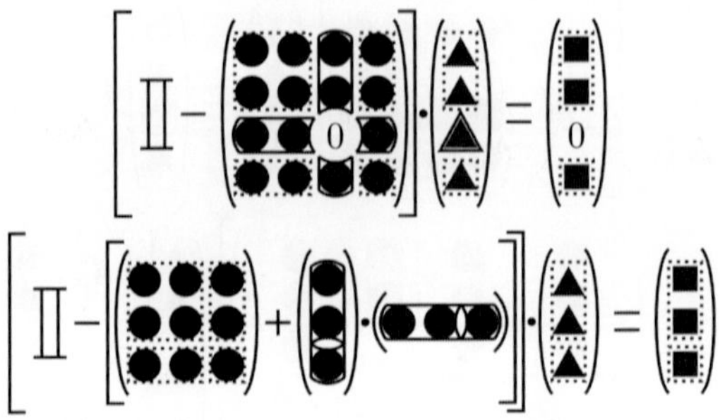

**Figure 4.23** Node removal equation with neither loopback nor stimuli (graphical)

containing the loopback. This has already been discussed mathematically. Here it is shown graphically.

Node removal without a loopback involved is shown graphically in Figure 4.22. Things now look much simpler. Finally, in the removal of nodes from a system, one does not generally remove nodes with stimuli applied, so an even simpler pattern is shown graphically in Figure 4.23.

In a somewhat more complicated fashion, this method can be used to remove entire sets of nodes at a time. Consider the fact that one can partition the matrices and vectors in the following form, where $R_1$, $R_2$, and $R_3$ and $C_1$, $C_2$, and $C_3$ are the rows and columns of the partitioned matrices with the understanding that $R_2 = C_2$ and $R_1+R_2+R_3 = C_1+C_2+C_3$:

$$\left[\underset{R_1+R_2+R_3}{\mathbb{I}} - \begin{pmatrix} \underset{R_1\times C_1}{\mathbf{A}_{11}} & \underset{R_1\times C_2}{\mathbf{A}_{12}} & \underset{R_1\times C_3}{\mathbf{A}_{13}} \\ \underset{R_2\times C_1}{\mathbf{A}_{21}} & \underset{R_2\times C_2}{\mathbf{A}_{22}} & \underset{R_2\times C_3}{\mathbf{A}_{23}} \\ \underset{R_3\times C_1}{\mathbf{A}_{31}} & \underset{R_3\times C_2}{\mathbf{A}_{32}} & \underset{R_3\times C_3}{\mathbf{A}_{33}} \end{pmatrix}\right] \cdot \begin{pmatrix} \underset{R_1}{\mathbf{n}_1} \\ \underset{R_2}{\mathbf{n}_2} \\ \underset{R_3}{\mathbf{n}_3} \end{pmatrix} = \begin{pmatrix} \underset{R_1}{\mathbf{m}_1} \\ \underset{R_2}{\mathbf{m}_2} \\ \underset{R_3}{\mathbf{m}_3} \end{pmatrix}.$$

If the desire is to remove all of the nodes in the vector $\mathbf{n}_2$, the first step is to zero out the $\mathbf{A}_{22}$ location. This is performed by making use of the equality following:

$$\left[\underset{R_1+R_2+R_3}{\mathbb{I}} - \left(\begin{matrix} \mathbf{A}_{11} & \mathbf{A}_{12} & \mathbf{A}_{13} \\ \left[\underset{R_2}{\mathbb{I}} - \mathbf{A}_{22}\right]^{-1} \cdot \mathbf{A}_{21} & \underset{R_2}{\mathbf{0}} & \left[\underset{R_2}{\mathbb{I}} - \mathbf{A}_{22}\right]^{-1} \cdot \mathbf{A}_{23} \\ \mathbf{A}_{31} & \mathbf{A}_{32} & \mathbf{A}_{33} \end{matrix}\right)\right] \cdot \begin{pmatrix} \mathbf{n}_1 \\ \mathbf{n}_2 \\ \mathbf{n}_3 \end{pmatrix}$$

$$= \left(\begin{matrix} \mathbf{m}_1 \\ \left[\underset{R_2}{\mathbb{I}} - \mathbf{A}_{22}\right]^{-1} \cdot \mathbf{m}_2 \\ \mathbf{m}_3 \end{matrix}\right).$$

Then, the nodes in the vector $\mathbf{n}_2$ are removed using

$$\left[\underset{R_1+R_3}{\mathbb{I}} - \left[\begin{pmatrix} \mathbf{A}_{11} & \mathbf{A}_{13} \\ \mathbf{A}_{31} & \mathbf{A}_{33} \end{pmatrix} + \begin{pmatrix} \mathbf{A}_{12} \\ \mathbf{A}_{32} \end{pmatrix} \cdot \left[\underset{R_2}{\mathbb{I}} - \mathbf{A}_{22}\right]^{-1} \cdot \begin{pmatrix} \mathbf{A}_{21} & \mathbf{A}_{23} \end{pmatrix}\right]\right] \cdot \begin{pmatrix} \mathbf{n}_1 \\ \mathbf{n}_3 \end{pmatrix}$$

$$= \begin{pmatrix} \mathbf{m}_1 \\ \mathbf{m}_3 \end{pmatrix} + \begin{pmatrix} \mathbf{A}_{12} \\ \mathbf{A}_{32} \end{pmatrix} \cdot \left[\underset{R_2}{\mathbb{I}} - \mathbf{A}_{22}\right]^{-1} \cdot \mathbf{m}_2.$$

All of this assumes, of course, that $\mathbb{I} - \mathbf{A}_{22}$ is invertible.

These graphical equation representations don't require much explanation, and are best demonstrated with examples. The examples used for single node removal are repeated using this graphical method.

### 4.6.1 Signal-Flow Diagram Node Removal Example

The removal of node $n_3$ from Figure 4.15 is repeated using the graphical equation method. The first step is to write the system equation for this diagram:

$$\left[\mathbb{I} - \begin{pmatrix} 0 & W_{12} & W_{13} \\ W_{21} & 0 & W_{23} \\ W_{31} & W_{32} & W_{33} \end{pmatrix}\right] \cdot \begin{pmatrix} n_1 \\ n_2 \\ n_3 \end{pmatrix} = \begin{pmatrix} m_1 \\ m_2 \\ m_3 \end{pmatrix}.$$

To remove $n_3$, one consults Figure 4.20 and finds that there is the inconvenient non-zero element $W_{33}$. One can apply the full formula in Figure 4.20, or the loopback can be removed first. It is removed first by consulting Figure 4.21, where it is apparent that one must divide the entirety of row 3 in the equation by $1/(1 - W_{33})$ and place a zero where $W_{33}$ used to be:

$$\left[\mathbb{I} - \begin{pmatrix} 0 & W_{12} & W_{13} \\ W_{21} & 0 & W_{23} \\ \frac{W_{31}}{1-W_{33}} & \frac{W_{32}}{1-W_{33}} & 0 \end{pmatrix}\right] \cdot \begin{pmatrix} n_1 \\ n_2 \\ n_3 \end{pmatrix} = \begin{pmatrix} m_1 \\ m_2 \\ \frac{m_3}{1-W_{33}} \end{pmatrix}.$$

There is a stimulus applied to node $n_3$, so Figure 4.22 is used to write

$$\left[\mathbb{I}-\left\{\begin{pmatrix} 0 & W_{12} \\ W_{21} & 0 \end{pmatrix}+\begin{pmatrix} W_{13} \\ W_{23} \end{pmatrix}\cdot\begin{pmatrix} \frac{W_{31}}{1-W_{33}} & \frac{W_{32}}{1-W_{33}} \end{pmatrix}\right\}\right]\cdot\begin{pmatrix} n_1 \\ n_2 \end{pmatrix}$$
$$=\begin{pmatrix} m_1 \\ m_2 \end{pmatrix}+\begin{pmatrix} W_{13} \\ W_{23} \end{pmatrix}\cdot\frac{m_3}{1-W_{33}}.$$

Simplifying this, one can now see the additions to the arrows and stimuli in the signal-flow diagram when this example was solved directly:

$$\left[\mathbb{I}-\left\{\begin{pmatrix} 0 & W_{12} \\ W_{21} & 0 \end{pmatrix}+\begin{pmatrix} \frac{W_{13}\cdot W_{31}}{1-W_{33}} & \frac{W_{13}\cdot W_{32}}{1-W_{33}} \\ \frac{W_{23}\cdot W_{31}}{1-W_{33}} & \frac{W_{23}\cdot W_{32}}{1-W_{33}} \end{pmatrix}\right\}\right]\cdot\begin{pmatrix} n_1 \\ n_2 \end{pmatrix}=\begin{pmatrix} m_1 \\ m_2 \end{pmatrix}+\begin{pmatrix} \frac{W_{13}\cdot m_3}{1-W_{33}} \\ \frac{W_{23}\cdot m_3}{1-W_{33}} \end{pmatrix}.$$

Simplifying further results in

$$\left[\mathbb{I}-\begin{pmatrix} \frac{W_{13}\cdot W_{31}}{1-W_{33}} & W_{12}+\frac{W_{13}\cdot W_{32}}{1-W_{33}} \\ W_{21}+\frac{W_{23}\cdot W_{31}}{1-W_{33}} & \frac{W_{23}\cdot W_{32}}{1-W_{33}} \end{pmatrix}\right]\cdot\begin{pmatrix} n_1 \\ n_2 \end{pmatrix}=\begin{pmatrix} m_1+\frac{W_{13}\cdot m_3}{1-W_{33}} \\ m_2+\frac{W_{23}\cdot m_3}{1-W_{33}} \end{pmatrix}.$$

Finally, Figure 4.21 is used to remove both loopbacks (non-zero elements on the diagonal of the weights matrix) in one shot:

$$\left[\mathbb{I}-\begin{pmatrix} 0 & \frac{W_{12}+\frac{W_{13}\cdot W_{32}}{1-W_{33}}}{1-\frac{W_{13}\cdot W_{31}}{1-W_{33}}} \\ \frac{W_{21}+\frac{W_{23}\cdot W_{31}}{1-W_{33}}}{1-\frac{W_{23}\cdot W_{32}}{1-W_{33}}} & 0 \end{pmatrix}\right]\cdot\begin{pmatrix} n_1 \\ n_2 \end{pmatrix}=\begin{pmatrix} \frac{m_1+\frac{W_{13}\cdot m_3}{1-W_{33}}}{1-\frac{W_{13}\cdot W_{31}}{1-W_{33}}} \\ \frac{m_2+\frac{W_{23}\cdot m_3}{1-W_{33}}}{1-\frac{W_{23}\cdot W_{32}}{1-W_{33}}} \end{pmatrix}.$$

The result can be placed back into signal-flow diagram form, if desired, as shown in Figure 4.19.

## 4.7 Examples

To conclude this chapter, two simple solution examples are provided:

- a terminated two-port device;
- a cascade of two two-port devices.

These solutions are provided using four solution methods:

1. signal-flow diagram node removal, as proposed in §4.5;
2. direct solution, as proposed in §4.3;
3. node removal with the graphical equation method block form, as proposed in §4.6;
4. node removal with the graphical equation method one node at a time, as proposed in §4.6.

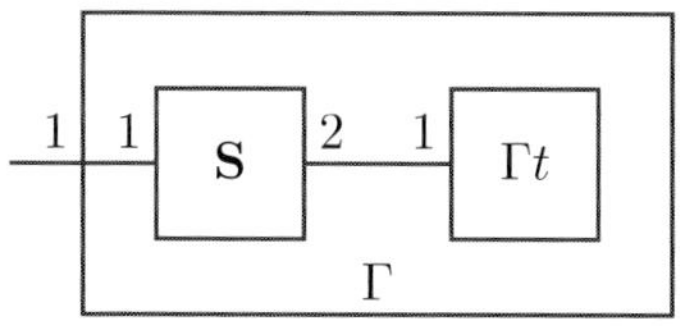


(a) Network block diagram

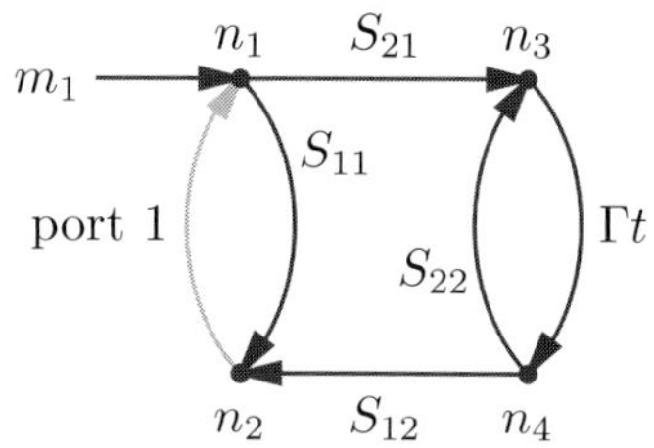


(b) Signal-flow diagram

**Figure 4.24** Terminated two-port example

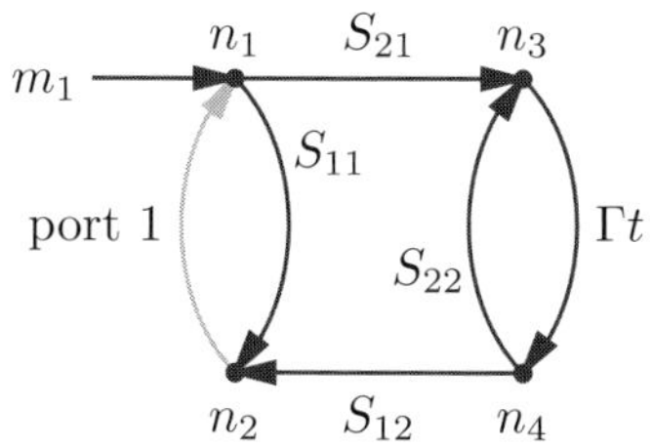


(a) Starting point

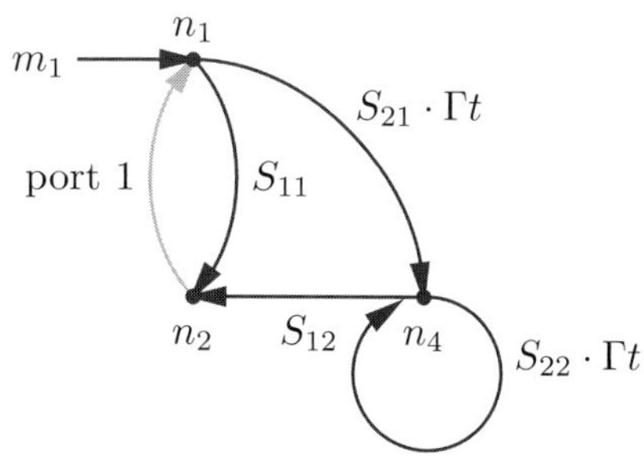


(b) Node 3 removal

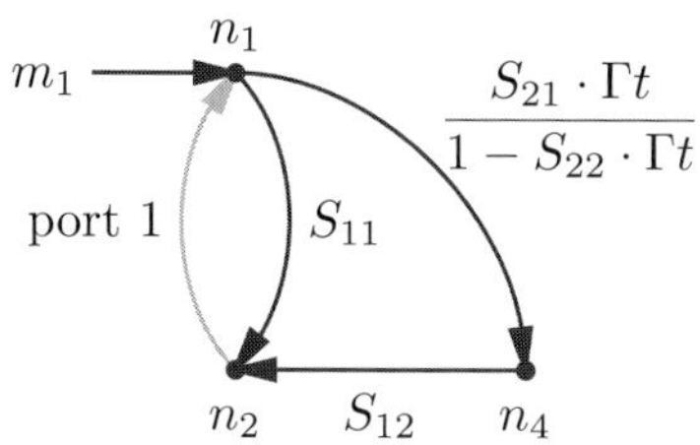


(c) Node 4 loopback removal

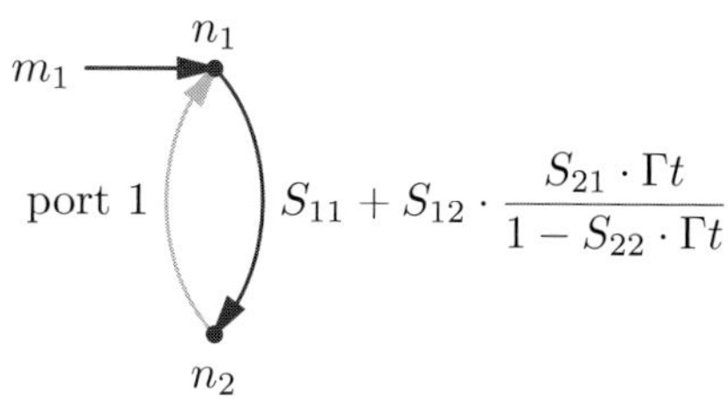


(d) Node 4 removal

**Figure 4.25** Node removal to solve terminated two-port example

### 4.7.1 Terminated Two-Port Device Example

Consider the terminated two-port device example in Figure 4.24, where the network block diagram is shown in Figure 4.24(a) and its corresponding signal-flow diagram is provided in Figure 4.24(b).

Some of the methods require the system equation, which is

$$\left[\mathbb{I}-\begin{pmatrix}0&0&0&0\\S_{11}&0&0&S_{12}\\S_{21}&0&0&S_{22}\\0&0&\Gamma t&0\end{pmatrix}\right]\cdot\begin{pmatrix}n_1\\n_2\\n_3\\n_4\end{pmatrix}=\begin{pmatrix}m_1\\0\\0\\0\end{pmatrix}. \tag{4.31}$$

#### *Node Removal*

The node removal method is shown in Figure 4.25. The starting point in Figure 4.25(a) is just a copy of Figure 4.24(b). Node $n_3$ is removed first in Figure 4.25(b) and the resulting loopback is removed in Figure 4.25(c). Finally, node $n_4$ is removed in Figure 4.25(d), which provides the result.

#### *Direct Solution*

$$\Gamma=\left(\left[\mathbb{I}-\begin{pmatrix}0&0&0&0\\S_{11}&0&0&S_{12}\\S_{21}&0&0&S_{22}\\0&0&\Gamma t&0\end{pmatrix}\right]^{-1}\cdot\begin{pmatrix}1\\0\\0\\0\end{pmatrix}\right)_2=\frac{(S_{11}\cdot S_{22}-S_{21}\cdot S_{12})\cdot\Gamma t-S_{11}}{S_{22}\cdot\Gamma t-1}=\frac{|\mathbf{S}|\cdot\Gamma t-S_{11}}{S_{22}\cdot\Gamma t-1}.$$

#### *Block Matrix Solution*

$$\begin{aligned}\mathbf{S}&=\mathbf{W_{ba}}+\mathbf{W_{bx}}\cdot(\mathbb{I}-\mathbf{W_{xx}})^{-1}\cdot\mathbf{W_{xa}}\\&=S_{11}+\begin{pmatrix}0&S_{12}\end{pmatrix}\cdot\left[\mathbb{I}-\begin{pmatrix}0&S_{22}\\\Gamma t&0\end{pmatrix}\right]^{-1}\cdot\begin{pmatrix}S_{21}\\0\end{pmatrix}=\frac{|\mathbf{S}|\cdot\Gamma t-S_{11}}{S_{22}\cdot\Gamma t-1}.\end{aligned} \tag{4.32}$$

#### *Node Removal Solution – Block Form*

$$\begin{aligned}\mathbf{W}'&=\mathbf{W_{kk}}+\mathbf{W_{kx}}\cdot(\mathbb{I}-\mathbf{W_{xx}})^{-1}\cdot\mathbf{W_{xk}}\\&=\begin{pmatrix}0&0\\S_{11}&0\end{pmatrix}+\begin{pmatrix}0&0\\0&S_{12}\end{pmatrix}\cdot\left[\mathbb{I}-\begin{pmatrix}0&S_{22}\\\Gamma t&0\end{pmatrix}\right]^{-1}\cdot\begin{pmatrix}S_{21}&0\\0&0\end{pmatrix}=\begin{pmatrix}0&0\\\frac{|\mathbf{S}|\cdot\Gamma t-S_{11}}{S_{22}\cdot\Gamma t-1}&0\end{pmatrix},\end{aligned}$$

$$\left[\mathbb{I}-\begin{pmatrix}0&0\\\frac{|\mathbf{S}|\cdot\Gamma t-S_{11}}{S_{22}\cdot\Gamma t-1}&0\end{pmatrix}\right]\cdot\begin{pmatrix}n_1\\n_2\end{pmatrix}=\begin{pmatrix}m_1\\0\end{pmatrix}.$$

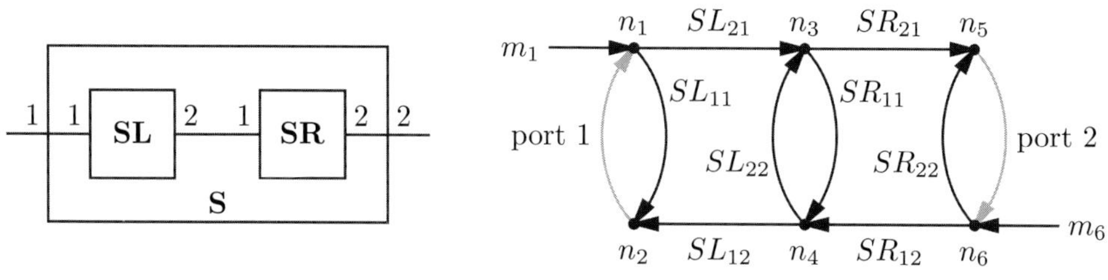


(a) Network block diagram (b) Signal-flow diagram

**Figure 4.26** Two cascaded two-port devices example

***Node Removal – Graphical Equation Method***

Node 3 removal, starting with (4.31):

$$\left[\mathbb{I} - \left\{\begin{pmatrix} 0 & 0 & 0 \\ S_{11} & 0 & S_{12} \\ 0 & 0 & 0 \end{pmatrix} + \begin{pmatrix} 0 \\ 0 \\ \Gamma t \end{pmatrix} \cdot \begin{pmatrix} S_{21} & 0 & S_{22} \end{pmatrix}\right\}\right] \cdot \begin{pmatrix} n_1 \\ n_2 \\ n_4 \end{pmatrix} = \begin{pmatrix} m_1 \\ 0 \\ 0 \end{pmatrix},$$

$$\left[\mathbb{I} - \begin{pmatrix} 0 & 0 & 0 \\ S_{11} & 0 & S_{12} \\ \Gamma t \cdot S_{21} & 0 & \Gamma t \cdot S_{22} \end{pmatrix}\right] \cdot \begin{pmatrix} n_1 \\ n_2 \\ n_4 \end{pmatrix} = \begin{pmatrix} m_1 \\ 0 \\ 0 \end{pmatrix},$$

$$\left[\mathbb{I} - \begin{pmatrix} 0 & 0 & 0 \\ S_{11} & 0 & S_{12} \\ \frac{\Gamma t \cdot S_{21}}{1 - \Gamma t \cdot S_{22}} & 0 & 0 \end{pmatrix}\right] \cdot \begin{pmatrix} n_1 \\ n_2 \\ n_4 \end{pmatrix} = \begin{pmatrix} m_1 \\ 0 \\ 0 \end{pmatrix}.$$

Node 4 removal:

$$\left[\mathbb{I} - \begin{pmatrix} 0 & 0 & 0 \\ S_{11} & 0 & S_{12} \\ \frac{\Gamma t \cdot S_{21}}{1 - \Gamma t \cdot S_{22}} & 0 & 0 \end{pmatrix}\right] \cdot \begin{pmatrix} n_1 \\ n_2 \\ n_4 \end{pmatrix} = \begin{pmatrix} m_1 \\ 0 \\ 0 \end{pmatrix},$$

$$\left[\mathbb{I} - \left\{\begin{pmatrix} 0 & 0 \\ S_{11} & 0 \end{pmatrix} + \begin{pmatrix} 0 \\ S_{12} \end{pmatrix} \cdot \begin{pmatrix} \frac{\Gamma t \cdot S_{21}}{1 - \Gamma t \cdot S_{22}} & 0 \end{pmatrix}\right\}\right] \cdot \begin{pmatrix} n_1 \\ n_2 \end{pmatrix} = \begin{pmatrix} m_1 \\ 0 \end{pmatrix},$$

$$\left[\mathbb{I} - \begin{pmatrix} 0 & 0 \\ S_{11} + S_{12} \cdot \frac{\Gamma t \cdot S_{21}}{1 - \Gamma t \cdot S_{22}} & 0 \end{pmatrix}\right] \cdot \begin{pmatrix} n_1 \\ n_2 \end{pmatrix} = \begin{pmatrix} m_1 \\ 0 \end{pmatrix}.$$

### 4.7.2 Two Cascaded Two-Port Devices Example

A two cascaded two-port devices example is provided in Figure 4.26, where the network block diagram is shown in Figure 4.26(a) and its corresponding signal-flow diagram is provided in Figure 4.26(b).

Some of the methods require the system equation, which is

$$\left[\mathbb{I} - \begin{pmatrix} 0 & 0 & 0 & 0 & 0 & 0 \\ SL_{11} & 0 & 0 & SL_{12} & 0 & 0 \\ SL_{21} & 0 & 0 & SL_{22} & 0 & 0 \\ 0 & 0 & SR_{11} & 0 & 0 & SR_{12} \\ 0 & 0 & SR_{21} & 0 & 0 & SR_{22} \\ 0 & 0 & 0 & 0 & 0 & 0 \end{pmatrix}\right] \cdot \begin{pmatrix} n_1 \\ n_2 \\ n_3 \\ n_4 \\ n_5 \\ n_6 \end{pmatrix} = \begin{pmatrix} m_1 \\ 0 \\ 0 \\ 0 \\ 0 \\ m_6 \end{pmatrix}. \tag{4.33}$$

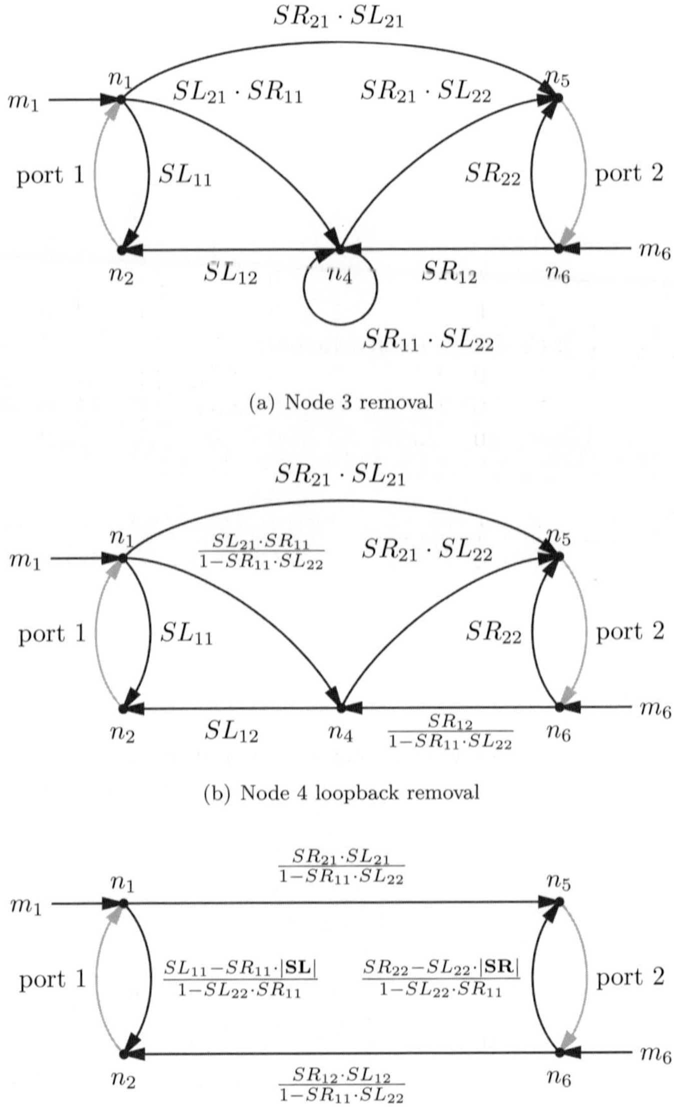


(a) Node 3 removal



(b) Node 4 loopback removal



(c) Node 4 removal

**Figure 4.27** Node removal to solve cascaded two-port example

### *Node Removal*

The node removal method is provided in Figure 4.27. Node $n_3$ is shown removed in Figure 4.27(a) and the resulting loopback at node $n_4$ is removed in Figure 4.27(b). Node $n_4$ is removed in Figure 4.27(c), which shows the final result.

### *Direct Solution*

Starting with (4.33),

$$\mathbf{N} = \left[ \mathbb{I} - \begin{pmatrix} 0 & 0 & 0 & 0 & 0 & 0 \\ SL_{11} & 0 & 0 & SL_{12} & 0 & 0 \\ SL_{21} & 0 & 0 & SL_{22} & 0 & 0 \\ 0 & 0 & SR_{11} & 0 & 0 & SR_{12} \\ 0 & 0 & SR_{21} & 0 & 0 & SR_{22} \\ 0 & 0 & 0 & 0 & 0 & 0 \end{pmatrix} \right]^{-1} \cdot \begin{pmatrix} 1 & 0 \\ 0 & 0 \\ 0 & 0 \\ 0 & 0 \\ 0 & 0 \\ 0 & 1 \end{pmatrix},$$

$$\mathbf{S} = \begin{pmatrix} N_{21} & N_{22} \\ N_{51} & N_{52} \end{pmatrix} = \frac{1}{1 - SL_{22} \cdot SR_{11}} \cdot \begin{pmatrix} SL_{11} - SR_{11} \cdot |\mathbf{SL}| & SL_{12} \cdot SR_{12} \\ SL_{21} \cdot SR_{21} & SR_{22} - SL_{22} \cdot |\mathbf{SR}| \end{pmatrix}.$$

### *Block Matrix Solution*

The remaining methods require row and column permutations to regroup the nodes in the node vector so that the matrix partitioning can be read directly from the weights matrix:

$$\mathbf{P} = \begin{pmatrix} 1&0&0&0&0&0 \\ 0&0&0&0&0&1 \\ 0&1&0&0&0&0 \\ 0&0&0&0&1&0 \\ 0&0&1&0&0&0 \\ 0&0&0&1&0&0 \end{pmatrix},$$

$$\mathbf{P} \cdot [\mathbb{I} - \mathbf{W}] \cdot \mathbf{P}^T \cdot \mathbf{P} \cdot \mathbf{n} = \mathbf{P} \cdot \mathbf{m} = \left[ \mathbb{I} - \begin{pmatrix} 0 & 0 & 0 & 0 & 0 & 0 \\ 0 & 0 & 0 & 0 & 0 & 0 \\ SL_{11} & 0 & 0 & 0 & 0 & SL_{12} \\ 0 & SR_{22} & 0 & 0 & SR_{21} & 0 \\ SL_{21} & 0 & 0 & 0 & 0 & SL_{22} \\ 0 & SR_{12} & 0 & 0 & SR_{11} & 0 \end{pmatrix} \right] \cdot \begin{pmatrix} n_1 \\ n_6 \\ n_2 \\ n_5 \\ n_3 \\ n_4 \end{pmatrix} = \begin{pmatrix} m_1 \\ m_6 \\ 0 \\ 0 \\ 0 \\ 0 \end{pmatrix}.$$

$$\begin{aligned} \mathbf{S} &= \mathbf{W_{ba}} + \mathbf{W_{bx}} \cdot (\mathbb{I} - \mathbf{W_{xx}})^{-1} \cdot \mathbf{W_{xa}} \\ &= \begin{pmatrix} SL_{11} & 0 \\ 0 & SR_{22} \end{pmatrix} + \begin{pmatrix} 0 & SL_{12} \\ SR_{21} & 0 \end{pmatrix} \cdot \left[ \mathbb{I} - \begin{pmatrix} 0 & SL_{22} \\ SR_{11} & 0 \end{pmatrix} \right]^{-1} \cdot \begin{pmatrix} SL_{21} & 0 \\ 0 & SR_{12} \end{pmatrix} \\ &= \begin{pmatrix} SL_{11} & 0 \\ 0 & SR_{22} \end{pmatrix} + \frac{1}{1 - SL_{22} \cdot SR_{21}} \cdot \begin{pmatrix} 0 & SL_{12} \\ SR_{21} & 0 \end{pmatrix} \cdot \begin{pmatrix} 1 & SL_{22} \\ SR_{11} & 1 \end{pmatrix} \cdot \begin{pmatrix} SL_{21} & 0 \\ 0 & SR_{12} \end{pmatrix}. \end{aligned} \quad (4.34)$$

### Node Removal Solution – Block Form

$$\begin{aligned}
\mathbf{W}' &= \mathbf{W_{kk}} + \mathbf{W_{kx}} \cdot (\mathbb{I} - \mathbf{W_{xx}})^{-1} \cdot \mathbf{W_{xk}} \\
&= \begin{pmatrix} 0 & 0 & 0 & 0 \\ 0 & 0 & 0 & 0 \\ SL_{11} & 0 & 0 & 0 \\ 0 & SR_{22} & 0 & 0 \end{pmatrix} + \begin{pmatrix} 0 & 0 \\ 0 & 0 \\ 0 & SL_{12} \\ SR_{21} & 0 \end{pmatrix} \cdot \left[\mathbb{I} - \begin{pmatrix} 0 & SL_{22} \\ SR_{11} & 0 \end{pmatrix}\right]^{-1} \cdot \begin{pmatrix} SL_{21} & 0 & 0 & 0 \\ 0 & SR_{12} & 0 & 0 \end{pmatrix} \\
&= \begin{pmatrix} 0 & 0 & 0 & 0 \\ 0 & 0 & 0 & 0 \\ SL_{11} & 0 & 0 & 0 \\ 0 & SR_{22} & 0 & 0 \end{pmatrix} + \frac{1}{1 - SL_{22} \cdot SR_{21}} \cdot \begin{pmatrix} 0 & 0 \\ 0 & 0 \\ 0 & SL_{12} \\ SR_{21} & 0 \end{pmatrix} \cdot \begin{pmatrix} 1 & SL_{22} \\ SR_{11} & 1 \end{pmatrix} \cdot \begin{pmatrix} SL_{21} & 0 & 0 & 0 \\ 0 & SR_{12} & 0 & 0 \end{pmatrix} \\
&= \frac{1}{1 - SL_{22} \cdot SR_{11}} \cdot \begin{pmatrix} 0 & 0 & 0 & 0 \\ 0 & 0 & 0 & 0 \\ SL_{11} - SR_{11} \cdot |\mathbf{SL}| & SL_{12} \cdot SR_{12} & 0 & 0 \\ SL_{21} \cdot SR_{21} & SR_{22} - SL_{22} \cdot |\mathbf{SR}| & 0 & 0 \end{pmatrix} .
\end{aligned}$$

$$\left[\mathbb{I} - \frac{1}{1 - SL_{22} \cdot SR_{11}} \cdot \begin{pmatrix} 0 & 0 & 0 & 0 \\ 0 & 0 & 0 & 0 \\ SL_{11} - SR_{11} \cdot |\mathbf{SL}| & SL_{12} \cdot SR_{12} & 0 & 0 \\ SL_{21} \cdot SR_{21} & SR_{22} - SL_{22} \cdot |\mathbf{SR}| & 0 & 0 \end{pmatrix}\right] \cdot \begin{pmatrix} n_1 \\ n_6 \\ n_2 \\ n_5 \end{pmatrix} = \begin{pmatrix} m_1 \\ m_6 \\ 0 \\ 0 \end{pmatrix} .$$

### Node Removal – Graphical Equation Method

Node 3 removal, starting with (4.33):

$$\left[\mathbb{I} - \left\{\begin{pmatrix} 0 & 0 & 0 & 0 & 0 \\ SL_{11} & 0 & SL_{12} & 0 & 0 \\ 0 & 0 & 0 & 0 & SR_{12} \\ 0 & 0 & 0 & 0 & SR_{22} \\ 0 & 0 & 0 & 0 & 0 \end{pmatrix} + \begin{pmatrix} 0 \\ 0 \\ SR_{11} \\ SR_{21} \\ 0 \end{pmatrix} \cdot \begin{pmatrix} SL_{21} & 0 & SL_{22} & 0 & 0 \end{pmatrix}\right\}\right] \cdot \begin{pmatrix} n_1 \\ n_2 \\ n_4 \\ n_5 \\ n_6 \end{pmatrix} = \begin{pmatrix} m_1 \\ 0 \\ 0 \\ 0 \\ m_6 \end{pmatrix} ,$$

$$\left[\mathbb{I} - \begin{pmatrix} 0 & 0 & 0 & 0 & 0 \\ SL_{11} & 0 & SL_{12} & 0 & 0 \\ SR_{11} \cdot SL_{21} & 0 & SR_{11} \cdot SL_{22} & 0 & SR_{12} \\ SR_{21} \cdot SL_{21} & 0 & SR_{21} \cdot SL_{22} & 0 & SR_{22} \\ 0 & 0 & 0 & 0 & 0 \end{pmatrix}\right] \cdot \begin{pmatrix} n_1 \\ n_2 \\ n_4 \\ n_5 \\ n_6 \end{pmatrix} = \begin{pmatrix} m_1 \\ 0 \\ 0 \\ 0 \\ m_6 \end{pmatrix} ,$$

$$\left[\mathbb{I} - \begin{pmatrix} 0 & 0 & 0 & 0 & 0 \\ SL_{11} & 0 & SL_{12} & 0 & 0 \\ \frac{SR_{11} \cdot SL_{21}}{1 - SR_{11} \cdot SL_{22}} & 0 & 0 & 0 & \frac{SR_{12}}{1 - SR_{11} \cdot SL_{22}} \\ SR_{21} SL_{21} & 0 & SR_{21} SL_{22} & 0 & SR_{22} \\ 0 & 0 & 0 & 0 & 0 \end{pmatrix}\right] \cdot \begin{pmatrix} n_1 \\ n_2 \\ n_4 \\ n_5 \\ n_6 \end{pmatrix} = \begin{pmatrix} m_1 \\ 0 \\ 0 \\ 0 \\ m_6 \end{pmatrix} .$$

Node 4 removal:

$$\left[\mathbb{I} - \begin{pmatrix} 0 & 0 & 0 & 0 & 0 \\ SL_{11} & 0 & SL_{12} & 0 & 0 \\ \frac{SR_{11} \cdot SL_{21}}{1 - SR_{11} \cdot SL_{22}} & 0 & 0 & 0 & \frac{SR_{12}}{1 - SR_{11} \cdot SL_{22}} \\ SR_{21} \cdot SL_{21} & 0 & SR_{21} SL_{22} & 0 & SR_{22} \\ 0 & 0 & 0 & 0 & 0 \end{pmatrix}\right] \cdot \begin{pmatrix} n_1 \\ n_2 \\ n_4 \\ n_5 \\ n_6 \end{pmatrix} = \begin{pmatrix} m_1 \\ 0 \\ 0 \\ 0 \\ m_6 \end{pmatrix} ,$$

$$\left[\mathbb{I} - \left\{\begin{pmatrix} 0 & 0 & 0 & 0 \\ SL_{11} & 0 & 0 & 0 \\ SR_{21} \cdot SL_{21} & 0 & 0 & SR_{22} \\ 0 & 0 & 0 & 0 \end{pmatrix} + \begin{pmatrix} \frac{SR_{11} \cdot SL_{21}}{1 - SR_{11} \cdot SL_{22}} & 0 & 0 & \frac{SR_{12}}{1 - SR_{11} \cdot SL_{22}} \end{pmatrix} \cdot \begin{pmatrix} 0 \\ SL_{12} \\ SR_{21} \cdot SL_{22} \\ 0 \end{pmatrix}\right\}\right] \cdot \begin{pmatrix} n_1 \\ n_2 \\ n_5 \\ n_6 \end{pmatrix} = \begin{pmatrix} m_1 \\ 0 \\ 0 \\ m_6 \end{pmatrix} ,$$

$$\left[\mathbb{I} - \frac{1}{1 - SL_{22} \cdot SR_{11}} \cdot \begin{pmatrix} 0 & 0 & 0 & 0 \\ SL_{11} - SR_{11} \cdot |\mathbf{SL}| & 0 & 0 & SL_{12} SR_{12} \\ SL_{21} \cdot SR_{21} & 0 & 0 & SR_{22} - SL_{22} \cdot |\mathbf{SR}| \\ 0 & 0 & 0 & 0 \end{pmatrix}\right] \cdot \begin{pmatrix} n_1 \\ n_2 \\ n_5 \\ n_6 \end{pmatrix} = \begin{pmatrix} m_1 \\ 0 \\ 0 \\ m_6 \end{pmatrix} .$$

## 4.8 Summary

In this chapter, many methods have been explored for working with interconnected s-parameter networks, mostly for the purpose of determining the overall s-parameters. While these methods are systematic, one sees from the two simple examples, that things become complicated very quickly. This might lead the reader to believe that these methods are not actually useful, that the complexity might deter their usage. Chapter 8 solves this complexity by applying programmatic methods for assembling the system equation in order to perform the matrix partitioning, thus allowing for relatively simple program based solutions while still practicing the underlying methods presented in this chapter.

# 5

# Reference Impedance

THE USE OF REFERENCE IMPEDANCE with waves causes confusion regarding the interpretation of wave values. This chapter introduces the concept of reference impedance domains in signal-flow diagrams by employing an abstract element called the *impedance transformer*. The impedance transformer is used to conceptualize a wave and is used to convert waves from one impedance domain to another. Finally, the conversion of s-parameters from one reference impedance to another is shown.

## 5.1 Basic Reference Impedance Transformation

It is apparent from the equations that relate waves and voltages and currents in Table 2.1 that wave values depend on the choice of normalization factor, $\sqrt{\mathcal{Z}0}$, and reference impedance, $Z0$. Since s-parameters are ratios of incident and reflected waves, the s-parameter values themselves also depend on $\sqrt{\mathcal{Z}0}$ and $Z0$. This is entirely evident from the equations that relate absolute network parameters like Z- and Y-parameters to s-parameters, as provided in Table 3.2 and Table 3.4. When one is provided a set of s-parameters with the chosen reference impedance or normalization factor, it is sometimes necessary to convert these to another set with a new normalization factor and reference impedance. This is especially necessary when interconnecting s-parameter devices, as explained in §4.1.

A simple way to perform this transformation is first to convert the initial s-parameters $\mathbf{S}$, with normalization factor $\sqrt{\mathcal{Z}0}$ and reference impedance $Z0$, to Y- or Z-parameters using the equations provided in Table 3.3 and Table 3.5:

$$\mathbf{Z} = \sqrt{\boldsymbol{\mathcal{Z}0}} \cdot (\mathbb{I} + \mathbf{S}) \cdot (\mathbb{I} - \mathbf{S})^{-1} \cdot \mathbf{Z0} \cdot \sqrt{\boldsymbol{\mathcal{Z}0}}^{-1}$$

or

$$\mathbf{Y} = \sqrt{\boldsymbol{\mathcal{Z}0}} \cdot \mathbf{Z0}^{-1} \cdot (\mathbb{I} - \mathbf{S}) \cdot (\mathbb{I} + \mathbf{S})^{-1} \cdot \sqrt{\boldsymbol{\mathcal{Z}0}}^{-1}.$$

Then, convert these Y- or Z-parameters back to s-parameters $\mathbf{S}'$ with normalization factor $\sqrt{\mathcal{Z}0}'$ and reference impedance $Z0'$ using the equations provided in Table 3.2 and Table 3.4:

$$\mathbf{S}' = \sqrt{\boldsymbol{\mathcal{Z}0}}'^{-1} \cdot (\mathbf{Z} - \mathbf{Z0}') \cdot (\mathbf{Z} + \mathbf{Z0}')^{-1} \cdot \sqrt{\boldsymbol{\mathcal{Z}0}}'$$

or

$$\mathbf{S}' = \sqrt{\boldsymbol{\mathcal{Z}0}}'^{-1} \cdot (\mathbb{I} + \mathbf{Z0}' \cdot \mathbf{Y})^{-1} \cdot (\mathbb{I} - \mathbf{Z0}' \cdot \mathbf{Y}) \cdot \sqrt{\boldsymbol{\mathcal{Z}0}}'.$$

This works, provided the Y- or Z-parameters exist. Better formulas are considered later in this chapter.

There are some things to note about this process, the most important being that the values of the s-parameters depend on the normalization factor and their port reference impedances. But the normalization factor has an odd side-effect. If the normalization factor is the same on all ports of the device in both the initial and final s-parameters (i.e. $\sqrt{\boldsymbol{Z0}} = \sqrt{Z0} \cdot \mathbb{I}$ and $\sqrt{\boldsymbol{Z0}}' = \sqrt{Z0}' \cdot \mathbb{I}$), then the normalization factors drop out in both the conversion to and from Y- and Z-parameters. This can be seen from the tables provided in §3.4. In other words, while the wave values will depend on the normalization factor, normally the ratios of the waves (the s-parameters) do not.

Despite the fact that, under all conditions, the s-parameters are affected by the reference impedance, one should not think that the characteristics of the device described by the s-parameters are affected. In fact, the interpretation of s-parameters requires us to know the reference impedance. One set of s-parameters with a given reference impedance is no different from another set of s-parameters for the same device in another reference impedance, as long as they are related properly through a reference impedance transformation. This proper relationship is tested by verifying that the Z- or Y- parameter conversion produces the same Z- or Y-parameters. Devices with the same Z- or Y-parameters behave in the same way in a circuit or system. S-parameters with different s-parameter values might behave in the same way in a circuit or system depending on the reference impedance.

One question that may arise is: Why have different reference impedances? After all, Z-, Y-, and ABCD parameters don't have a reference impedance. Why would anyone want to have the problem of keeping the s-parameters and reference impedances aligned?

The reasons are as follows:

1. In circuits and systems, in combining devices with different s-parameters, or in simulation etc., there is no real reason to have different reference impedances. The behavior in the circuit is (or ought to be) the same, regardless of the reference impedance. But...
2. In interpreting, plotting, analyzing, and validating the performance of devices, the reference impedance is very important.

In other words, the reference impedance is important when making judgments about s-parameter values, but is not important regarding simulations and actual performance. That being said:

1. Some circuit simulators accidentally handle reference impedances improperly and assume that the reference impedance is real and is $Z0 = 50\,\Omega$ on every port of the device. The handling of this complexity is not so easy.
2. Some engineers always interpret s-parameters in the common, real $Z0 = 50\,\Omega$, which is often a mistake.

Here is some advice:

1. When using a simulator, or using s-parameters with software tools, or when giving them to your colleagues, always provide s-parameters that have a reference impedance $Z0 = 50\,\Omega$. Do this to avoid mistakes and to avoid the need for reference impedance transformations or transformers in the system.
2. When analyzing s-parameters, always consider the reference impedance relative to the source and termination impedances in the system.

Item 1 is really straightforward. Since the choice of reference impedance is arbitrary and does not affect the behavior of the device, it is easiest to stay on the safe side and use what everyone else is using. Most measurement instruments that measure s-parameters provide $50\,\Omega$ s-parameters anyway.

Item 2 needs some explaining. When plotting and analyzing s-parameters, one should keep in mind what the definition of s-parameters implies. For a $P$-port device with s-parameters $\mathbf{S}$ and a port reference impedance for $p \in 1 \ldots P$ of $Z0_p$, a given s-parameter element $S_{xy}$ is the ratio of the reflected wave from port $x$ to the incident wave at port $y$ when port $y$ is driven with all ports terminated in their reference impedance. In the case of the driven port, it is driven through its reference impedance. The simplest example is a one-port device with s-parameters given by $\Gamma$. The value of $\Gamma$ is the ratio of reflected waves to incident waves when driven through the reference impedance. The s-parameters of this one-port device are described by $\Gamma = (Z - Z0) / (Z + Z0)$, which is therefore zero when $Z = Z0$. Usually, this is the goal. When $Z = Z0$ and $\Gamma = 0$ the termination is said to be a perfect match. What one might omit or forget is that it is a perfect match to the reference impedance. If one has a $45\,\Omega$ driver and a perfect match is desired, then the perfect match is a $45\,\Omega$ termination. So, for this situation, one would want to know the s-parameters in a $45\,\Omega$ reference impedance and the desire would be for $\Gamma = 0$ under this circumstance. If the s-parameters of this termination are supplied to a simulator in $50\,\Omega$ reference impedance and simulated, one would indeed find that if it is a $45\,\Omega$ termination driven by a $45\,\Omega$ driver, it is a good match, but the plots of the s-parameters would show that it is not a good match because they would show the match to $50\,\Omega$, not $45\,\Omega$.

A simple example of this effect is supplied in Figure 5.1. Figure 5.1(a) shows an 8 GHz Chebyshev filter. The plots on the left of Figure 5.1 show plots of $S_{11}$ magnitude in Figure 5.1(b), $S_{21}$ magnitude in Figure 5.1(d), and the impedance profile in Figure 5.1(f) all plotted in a $50\,\Omega$ reference impedance. In Figure 5.1(b), the match (to $50\,\Omega$) is poor and in Figure 5.1(d) there is a lot of ripple in the response. The impedance profile in Figure 5.1(f) plots the impedance over time of this circuit (see Chapter 14 for more information about this) and shows that the impedance is more like $30\,\Omega$, which is what it was designed for. It is a circuit meant to be driven by a $30\,\Omega$ source and terminated in a $30\,\Omega$ load. So it is not really useful to examine the s-parameter plots in a $50\,\Omega$ reference impedance.

The plots on the right of Figure 5.1 show the same plots of $S_{11}$ magnitude in Figure 5.1(c), $S_{21}$ magnitude in Figure 5.1(e), and the impedance profile in Figure 5.1(g) all plotted in a $30\,\Omega$ reference impedance. All of these plots show a good match to $30\,\Omega$ and good performance. The impedance profile in Figure 5.1(g) shows that the impedance is more or less $30\,\Omega$, which again is what is expected.

In some sense, one can consider the plots of s-parameters to be the results of a simulation according to the s-parameter definition. One wants these plots to reflect the actual usage of the device in a system.

## 5.2 The Reference Impedance Transformer

The s-parameters of a wire were derived in §3.3.1; during the derivation, the normalization factor and reference impedance at each port of the wire were considered. The s-parameters

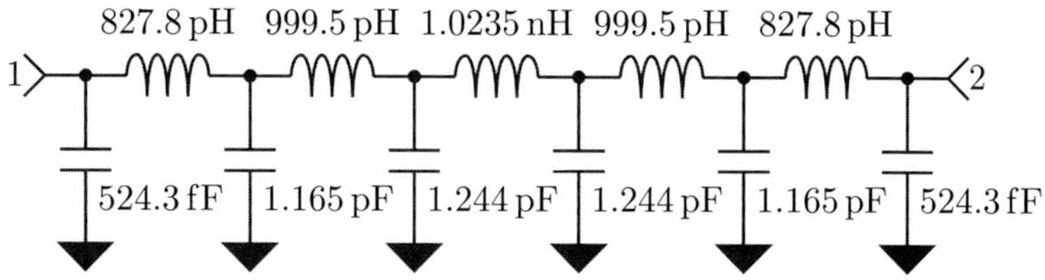


(a) 8 GHz Chebychev low-pass filter

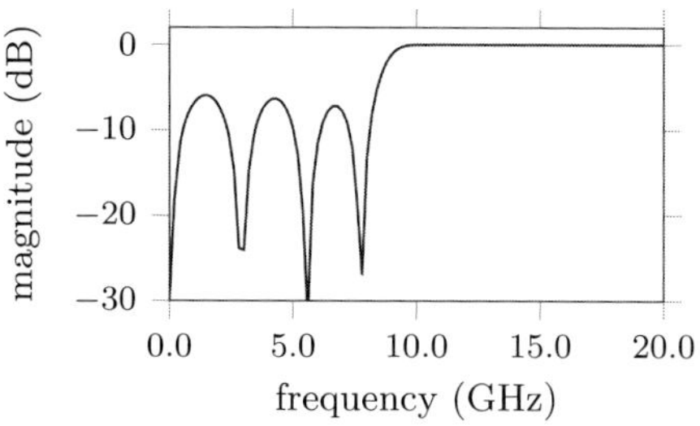


(b) $S_{11}$ magnitude in 50 Ω reference impedance

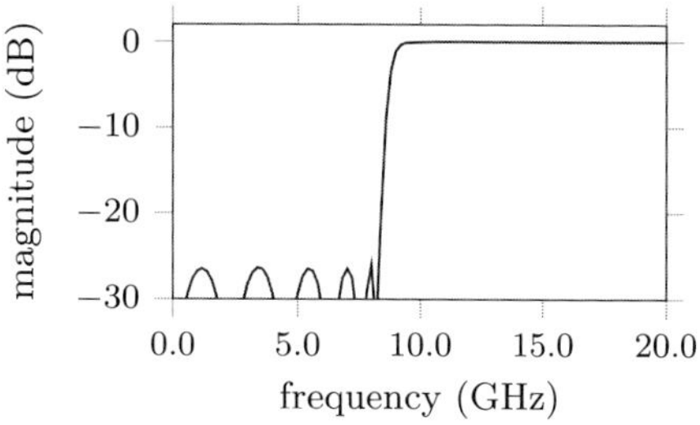


(c) $S_{11}$ magnitude in 30 Ω reference impedance

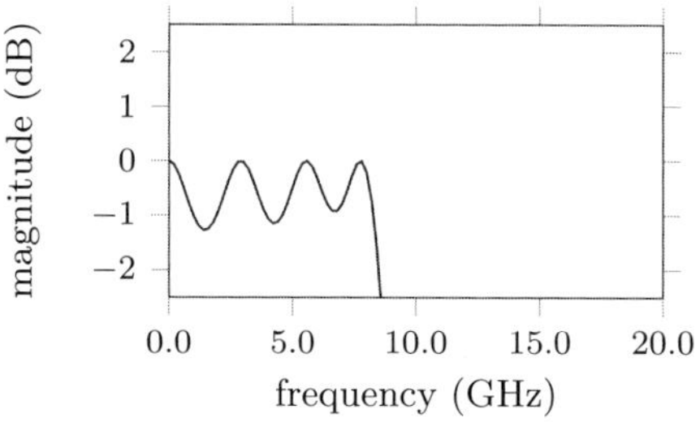


(d) $S_{21}$ magnitude in 50 Ω reference impedance

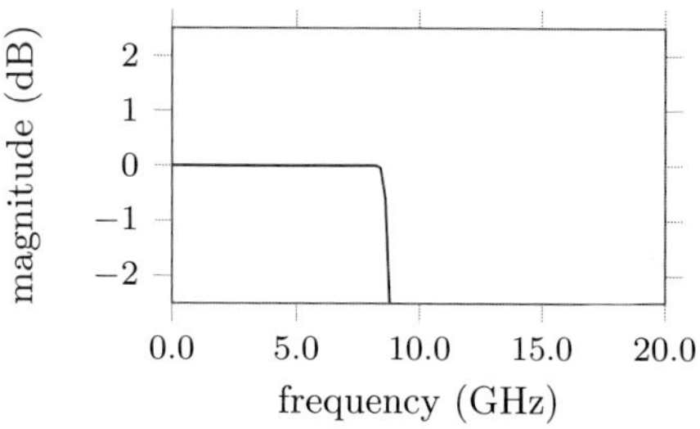


(e) $S_{21}$ magnitude in 30 Ω reference impedance

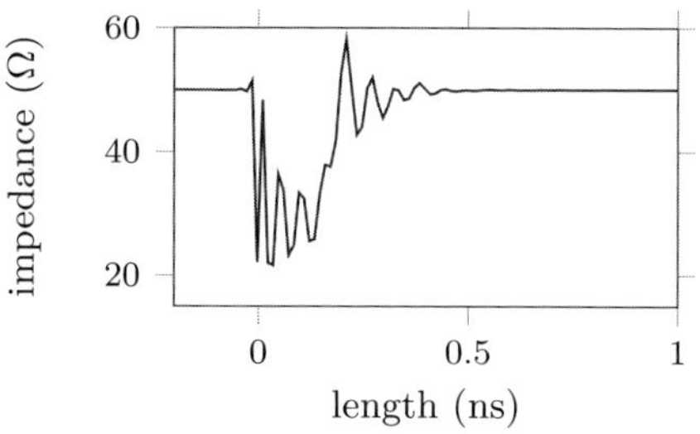


(f) Impedance profile in 50 Ω reference impedance

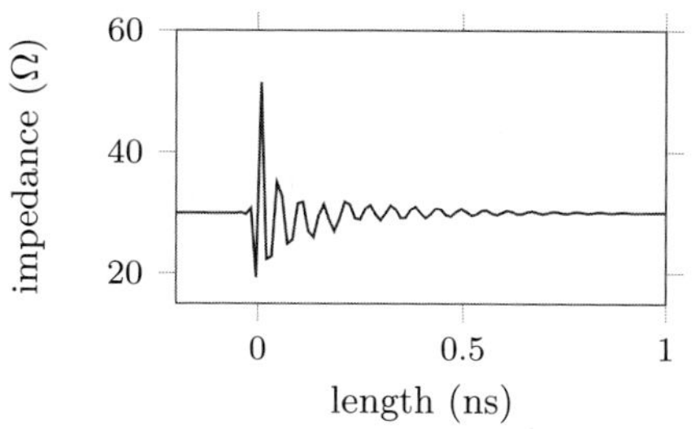


(g) Impedance profile in 30 Ω reference impedance

**Figure 5.1** Chebyshev filter viewed in different reference impedances

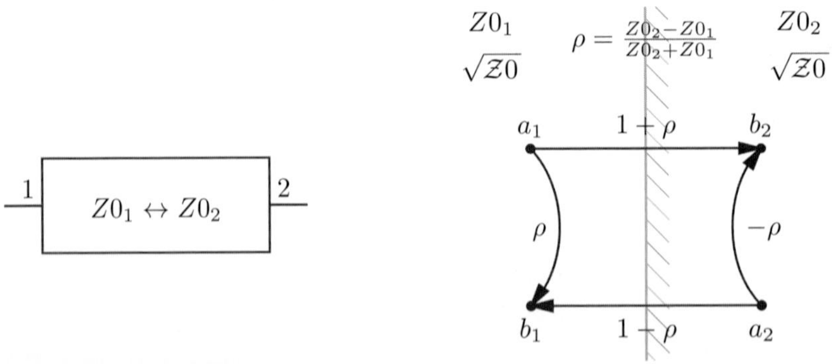


(a) Circuit representation (b) Signal-flow diagram representation

**Figure 5.2** Reference impedance transformer

were presented in (3.11), which is repeated below:

$$\mathbf{S}_{wire} = \frac{1}{Z0_1 + Z0_2} \cdot \begin{pmatrix} Z0_2 - Z0_1 & 2 \cdot \frac{\sqrt{Z0}_2}{\sqrt{Z0}_1} \cdot Z0_1 \\ 2 \cdot \frac{\sqrt{Z0}_1}{\sqrt{Z0}_2} \cdot Z0_2 & Z0_1 - Z0_2 \end{pmatrix}, \tag{5.1}$$

where $Z0_1$ is the reference impedance at port 1 and $Z0_2$ is the reference impedance at port 2. Similarly, $\sqrt{Z0}_1$ is the normalization factor applied at port 1 and $\sqrt{Z0}_2$ is the normalization factor applied at port 2. To alter this definition, one defines $\rho$ as an impedance change *going from port 1 to 2*:

$$\rho = \frac{Z0_2 - Z0_1}{Z0_2 + Z0_1}.$$

This allows (5.1) to be rewritten as

$$\mathbf{S}_{wire} = \mathbf{S}_{\substack{Z0_1 \leftrightarrow Z0_2 \\ \sqrt{Z0}_1 \leftrightarrow \sqrt{Z0}_2}} = \begin{pmatrix} \rho & (1-\rho) \cdot \frac{\sqrt{Z0}_2}{\sqrt{Z0}_1} \\ (1+\rho) \cdot \frac{\sqrt{Z0}_1}{\sqrt{Z0}_2} & -\rho \end{pmatrix}. \tag{5.2}$$

The device defined by (5.2) is actually two devices combined. If one sets $\sqrt{Z0}_1 = \sqrt{Z0}_2$ then one has purely a reference impedance transformer, as in Figure 5.2. The reference impedance transformer is a two-port device in Figure 5.2(a), with its corresponding signal-flow diagram shown in Figure 5.2(b). The Python code is shown in Listing 5.1. In Figure 5.2, the two reference impedance regions are displayed separated by an interface. On each side of the interface, the reference impedance is listed (as $Z0_1$ and $Z0_2$) along with a constant normalization factor $\sqrt{Z0} = \sqrt{Z0}_1 = \sqrt{Z0}_2$. For this reference impedance transformer, the assumption is that $\sqrt{Z0}$ is the same on each side of the boundary. Thus, the definition of a reference impedance transformer is

$$\mathbf{S}_{Z0_1 \leftrightarrow Z0_2} = \begin{pmatrix} \rho & 1-\rho \\ 1+\rho & -\rho \end{pmatrix}. \tag{5.3}$$

**Listing 5.1** Reference impedance transformer

```
def ReferenceImpedanceTransformer(Z0f,Z0i=None,Kf=None,Ki=None):
    Z0f=float(Z0f.real)+float(Z0f.imag)*1j
    if Z0i is None:
        Z0i=50.0
    Z0i=float(Z0i.real)+float(Z0i.imag)*1j
    if Kf is None:
        Kf=cmath.sqrt(Z0f)
    if Ki is None:
        Ki=cmath.sqrt(Z0i)
    Kf=float(Kf.real)+float(Kf.imag)*1j
    Ki=float(Ki.real)+float(Ki.imag)*1j
    p=(Z0f-Z0i)/(Z0f+Z0i)
    return [[p,(1.0-p)*Kf/Ki],[(1.0+p)*Ki/Kf,-p]]
```

If $\rho$ is defined as an impedance change going from port 2 to port 1, this is the same as replacing $\rho$ with $-\rho$, and the result is as if the two ports of the device were reordered in (5.2). In other words, it would be the same as turning the device around:

$$\begin{aligned}\mathbf{S}_{Z0_2 \leftrightarrow Z0_1} &= \left.\begin{pmatrix} \rho & 1-\rho \\ 1+\rho & -\rho \end{pmatrix}\right|_{\rho=-\rho} = \begin{pmatrix} 0 & 1 \\ 1 & 0 \end{pmatrix} \cdot \begin{pmatrix} \rho & 1-\rho \\ 1+\rho & -\rho \end{pmatrix} \cdot \begin{pmatrix} 0 & 1 \\ 1 & 0 \end{pmatrix} \\ &= \begin{pmatrix} -\rho & 1+\rho \\ 1-\rho & \rho \end{pmatrix}.\end{aligned}$$

A wire must have the same voltage and current on each side no matter the value of the port impedance. If the voltage and current at port 1 are defined as $v$ and $i$, according to (2.13), the definition of the waves at port 1 is given by

$$\begin{pmatrix} b_1 \\ a_1 \end{pmatrix} = \frac{1}{2 \cdot \sqrt{\mathcal{Z}0}_1} \cdot \begin{pmatrix} 1 & -Z0_1 \\ 1 & Z0_1 \end{pmatrix} \cdot \begin{pmatrix} v \\ i \end{pmatrix}, \tag{5.4}$$

where the row ordering has been swapped to put $b_1$ on top. Also, there is not only a different reference impedance at port 1 of $Z0_1$, but also an opportunity for a different normalization factor $\sqrt{\mathcal{Z}0}_1$.

At port 2, the same voltage and current are present, but the incident and reflected waves are reversed:

$$\begin{pmatrix} a_2 \\ b_2 \end{pmatrix} = \frac{1}{2 \cdot \sqrt{\mathcal{Z}0}_2} \cdot \begin{pmatrix} 1 & -Z0_2 \\ 1 & Z0_2 \end{pmatrix} \cdot \begin{pmatrix} v \\ i \end{pmatrix}. \tag{5.5}$$

Substituting the solution to (5.5) for $v$ and $i$ into (5.4) results in

$$\begin{aligned}\begin{pmatrix} b_1 \\ a_1 \end{pmatrix} &= \frac{1}{2 \cdot \sqrt{\mathcal{Z}0}_1} \cdot \begin{pmatrix} 1 & -Z0_1 \\ 1 & Z0_1 \end{pmatrix} \cdot \left[\frac{1}{2 \cdot \sqrt{\mathcal{Z}0}_2} \cdot \begin{pmatrix} 1 & -Z0_2 \\ 1 & Z0_2 \end{pmatrix}\right]^{-1} \cdot \begin{pmatrix} a_2 \\ b_2 \end{pmatrix} \\ &= \begin{pmatrix} \frac{1}{2} \cdot \frac{\sqrt{\mathcal{Z}0}_2}{\sqrt{\mathcal{Z}0}_1} \cdot \frac{Z0_2+Z0_1}{Z0_2} & \frac{1}{2} \cdot \frac{\sqrt{\mathcal{Z}0}_2}{\sqrt{\mathcal{Z}0}_1} \cdot \frac{Z0_2-Z0_1}{Z0_2} \\ \frac{1}{2} \cdot \frac{\sqrt{\mathcal{Z}0}_2}{\sqrt{\mathcal{Z}0}_1} \cdot \frac{Z0_2-Z0_1}{Z0_2} & \frac{1}{2} \cdot \frac{\sqrt{\mathcal{Z}0}_2}{\sqrt{\mathcal{Z}0}_1} \cdot \frac{Z0_2+Z0_1}{Z0_2} \end{pmatrix} \cdot \begin{pmatrix} a_2 \\ b_2 \end{pmatrix} \\ &= \begin{pmatrix} \frac{\sqrt{\mathcal{Z}0}_2}{\sqrt{\mathcal{Z}0}_1} \cdot \frac{1}{1+\rho} & \frac{\sqrt{\mathcal{Z}0}_2}{\sqrt{\mathcal{Z}0}_1} \cdot \frac{\rho}{1+\rho} \\ \frac{\sqrt{\mathcal{Z}0}_2}{\sqrt{\mathcal{Z}0}_1} \cdot \frac{\rho}{1+\rho} & \frac{\sqrt{\mathcal{Z}0}_2}{\sqrt{\mathcal{Z}0}_1} \cdot \frac{1}{1+\rho} \end{pmatrix} \cdot \begin{pmatrix} a_2 \\ b_2 \end{pmatrix}.\end{aligned} \tag{5.6}$$

Not only is (5.6) an equation that provides the relationships between waves on each side of a reference impedance interface, it also describes the T-parameters of a reference impedance transformer, and it can be checked that, if $\sqrt{Z0}_1 = \sqrt{Z0}_2$ and $Z0_1 = Z0_2$, the T-parameters are the identity matrix as one would expect. By the way, something similar was seen previously in (3.29).

Using (3.34) to convert (5.6) to s-parameters,

$$\begin{aligned}
\mathbf{S}_{\substack{Z0_1 \leftrightarrow Z0_2 \\ \sqrt{Z0}_1 \leftrightarrow \sqrt{Z0}_2}} &= \frac{\begin{pmatrix} T_{12} & |\mathbf{T}| \\ 1 & -T_{21} \end{pmatrix}}{T_{22}} \\
&= \begin{pmatrix} \frac{Z0_2 - Z0_1}{Z0_2 + Z0_1} & \frac{\sqrt{Z0}_2}{\sqrt{Z0}_1} \cdot \frac{2 \cdot Z0_1}{Z0_2 + Z0_1} \\ \frac{\sqrt{Z0}_1}{\sqrt{Z0}_2} \cdot \frac{2 \cdot Z0_2}{Z0_2 + Z0_1} & \frac{Z0_2 - Z0_1}{Z0_2 + Z0_1} \end{pmatrix} \\
&= \begin{pmatrix} \rho & (1-\rho) \cdot \frac{\sqrt{Z0}_2}{\sqrt{Z0}_1} \\ (1+\rho) \cdot \frac{\sqrt{Z0}_1}{\sqrt{Z0}_2} & -\rho \end{pmatrix}.
\end{aligned} \tag{5.7}$$

The definition in (5.7) is of an s-parameter device that performs a reference impedance and normalization factor transformation, and it is a direct result of equating voltages and currents on each side of the interface.

The reference impedance transformation can be differentiated from the normalization factor transformation, so the device defined in (5.7) is actually a cascade of two devices: the reference impedance transformer defined in (5.3) and another device called a normalization factor transformer defined by setting $\rho = 0$ in (5.2) and (5.7):

$$\mathbf{S}_{\sqrt{Z0}_1 \leftrightarrow \sqrt{Z0}_2} = \begin{pmatrix} 0 & \frac{\sqrt{Z0}_2}{\sqrt{Z0}_1} \\ \frac{\sqrt{Z0}_1}{\sqrt{Z0}_2} & 0 \end{pmatrix}.$$

The normalization factor transformer is shown schematically in Figure 5.3. Again, it is a two-port device, as shown in Figure 5.3(a), with a corresponding signal-flow diagram shown in Figure 5.3(b). To see the results of cascading a reference impedance transformer with a normalization factor transformer, consider Figure 5.4, where Figure 5.4(a) shows the two in cascade. Here there are three separate regions. Two of the three regions have the same normalization factor and another two regions have the same reference impedance. The two regions at the edges each have a distinct reference impedance $Z0_1$ or $Z0_2$ and an accompanying normalization factor $\sqrt{Z0}_1$ and $\sqrt{Z0}_2$. In Figure 5.4(b) and Figure 5.4(c) the node removal steps are shown to obtain the signal-flow diagram equivalent to the two in cascade. The signal-flow diagram in Figure 5.4(c) is equivalent to a device with s-parameters described by (5.7). Thus, the reference impedance and normalization factor transformer has been created, which is provided in Figure 5.5. The two-port device in Figure 5.5(a) has the signal-flow diagram shown in Figure 5.5(b).

As stated in Chapter 2, the commonly used traveling wave definition requires both a reference impedance and a normalization factor change, and this device accommodates that.

These transformation devices are utilized for several purposes, which are explored subsequently:

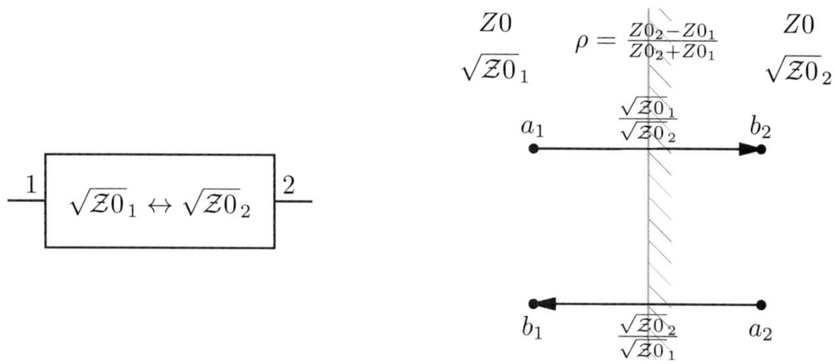


(a) Circuit representation (b) Signal-flow diagram representation

**Figure 5.3** Normalization factor transformer

1. They can be cascaded with the tips or ports of a device in the calculation of a reference impedance transformation.
2. They can be interconnected with elements in a circuit or system to join together devices with different reference impedances.
3. They are part of the internals of a transmission line model.
4. They can be positioned in a mirrored arrangement in a circuit or system to make measurements of what waves would be in a different reference impedance.

Items 1 and 4 will be covered in this chapter. Item 3 will be covered in Chapter 7, Transmission Lines.

## 5.3 Reference Impedance Transformers in Wave Measurements

Reference impedance transformers can be placed back-to-back in an arrangement for measuring waves in alternative reference impedance environments. This arrangement is shown in Figure 5.6. The idea here is that an s-parameter network is broken and these two back-to-back reference impedance transformers are inserted into the network, reconnecting the network. In this arrangement, there are waves at nodes $n_1$ and $n_2$ and nodes $n_5$ and $n_6$, all in the reference impedance $Z0_1$ and utilizing the normalization factor $\sqrt{Z0}_1$. In between the two networks at nodes $n_3$ and $n_4$ are nodes in reference impedance $Z0_2$ and normalization factor $\sqrt{Z0}_2$. In this manner, the waves at nodes $n_3$ and $n_4$ are transformed waves from the other nodes and provide for wave values in the new reference impedance.

The understanding of how this works begins with examining the network parameters of the entire arrangement, considering nodes $n_1$ and $n_2$ as the port 1 incident and reflected waves and nodes $n_6$ and $n_5$ as the port 2 incident and reflected waves, respectively, and nodes $n_3$ and $n_4$ as internal nodes. There is an easy way to calculate this by returning to the T-parameters provided in (5.6). These are the T-parameters for a reference impedance

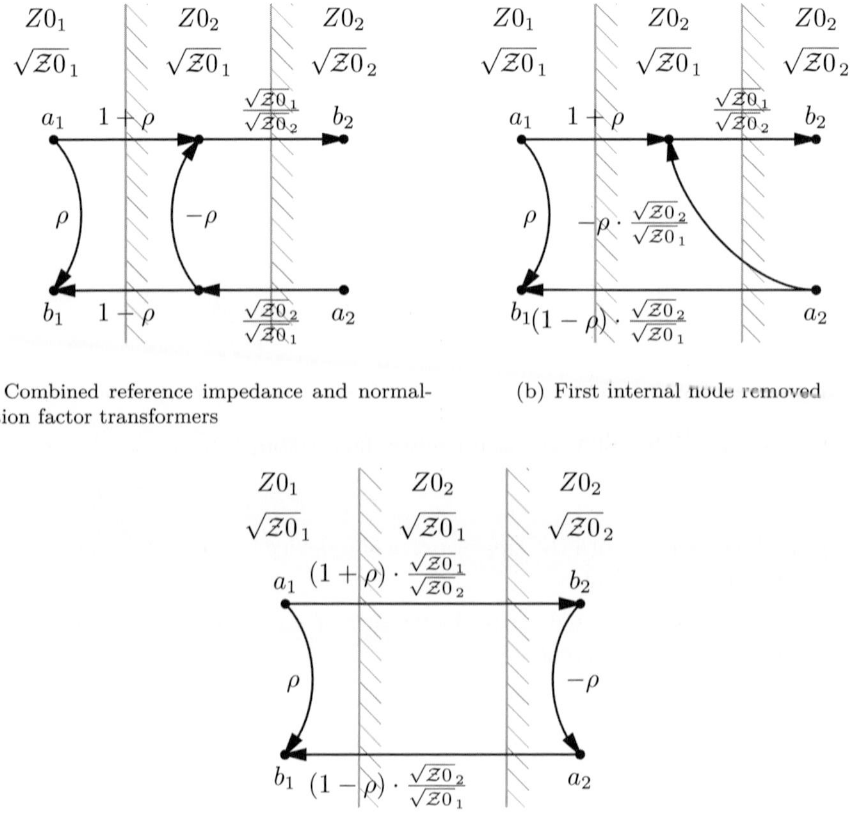


(a) Combined reference impedance and normalization factor transformers

(b) First internal node removed

(c) Last internal node removed

**Figure 5.4** Reference impedance transformer cascaded with normalization factor transformer

transformer from $Z0_1$ to $Z0_2$. The second reference impedance transformer is from $Z0_2$ to $Z0_1$, so the T-parameters for the second device are found by swapping $\sqrt{Z0}_1$ and $\sqrt{Z0}_2$ and replacing $\rho$ with $-\rho$. Therefore, the cascade devices have T-parameters as the product of the two T-parameter matrices:

$$\mathbf{T}_{Z0_1\leftrightarrow Z0_2\leftrightarrow Z0_1} = \begin{pmatrix} \frac{\sqrt{Z0}_2}{\sqrt{Z0}_1}\cdot\frac{1}{1+\rho} & \frac{\sqrt{Z0}_2}{\sqrt{Z0}_1}\cdot\frac{\rho}{1+\rho} \\ \frac{\sqrt{Z0}_2}{\sqrt{Z0}_1}\cdot\frac{\rho}{1+\rho} & \frac{\sqrt{Z0}_2}{\sqrt{Z0}_1}\cdot\frac{1}{1+\rho} \end{pmatrix} \cdot \begin{pmatrix} \frac{\sqrt{Z0}_1}{\sqrt{Z0}_2}\cdot\frac{1}{1-\rho} & \frac{\sqrt{Z0}_1}{\sqrt{Z0}_2}\cdot\frac{-\rho}{1-\rho} \\ \frac{\sqrt{Z0}_1}{\sqrt{Z0}_2}\cdot\frac{-\rho}{1-\rho} & \frac{\sqrt{Z0}_1}{\sqrt{Z0}_2}\cdot\frac{1}{1-\rho} \end{pmatrix} = \mathbb{I}. \tag{5.8}$$

The T-parameters in (5.8) indicate that the insertion of the two back-to-back impedance transformers have no effect on the system and that the node values $n_1 = n_5$ and $n_6 = n_2$. This is as expected. Because of this, one can redraw the equivalent diagram representing the

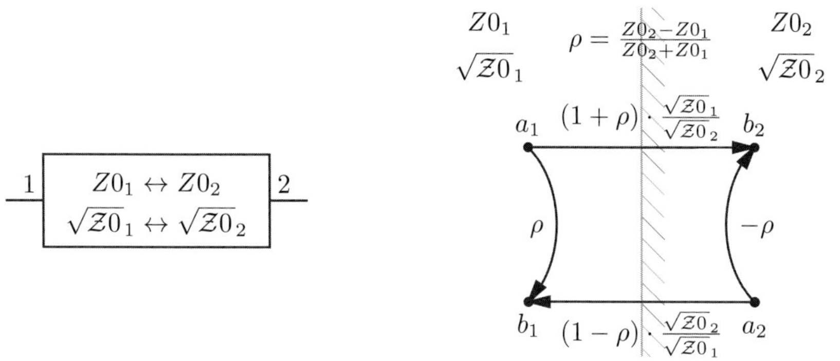


(a) Circuit representation (b) Signal-flow diagram representation

**Figure 5.5** Normalization factor and reference impedance transformer

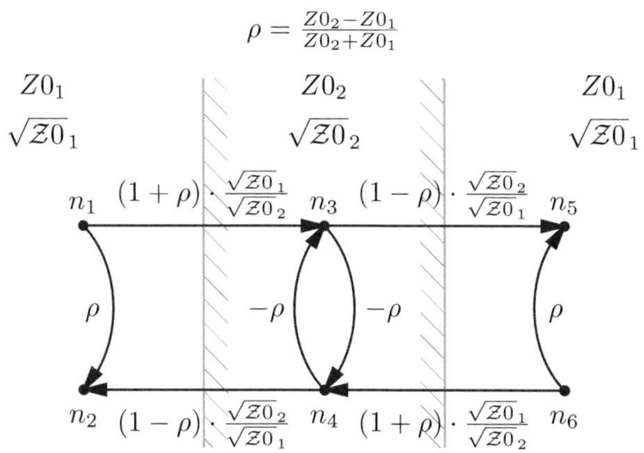


**Figure 5.6** Impedance transformers in a wave measuring setup

back-to-back reference impedance transformers, as shown in Figure 5.7. In this alternative arrangement, the equivalence of the node values at the perimeter is enforced, and nodes $n_3$ and $n_4$ are shown isolated internally.

Removing node $n_4$, the relationship between $n_3$ and nodes $n_1$ and $n_8$ can be generated. This process is shown in Figure 5.8, with node $n_4$ removal shown in Figure 5.8(a) and the final result shown in Figure 5.8(b):

$$n_3 = \frac{1}{1-\rho} \cdot \frac{\sqrt{Z0}_1}{\sqrt{Z0}_2} \cdot n_1 + \frac{-\rho}{1-\rho} \cdot \frac{\sqrt{Z0}_1}{\sqrt{Z0}_2} \cdot n_6.$$

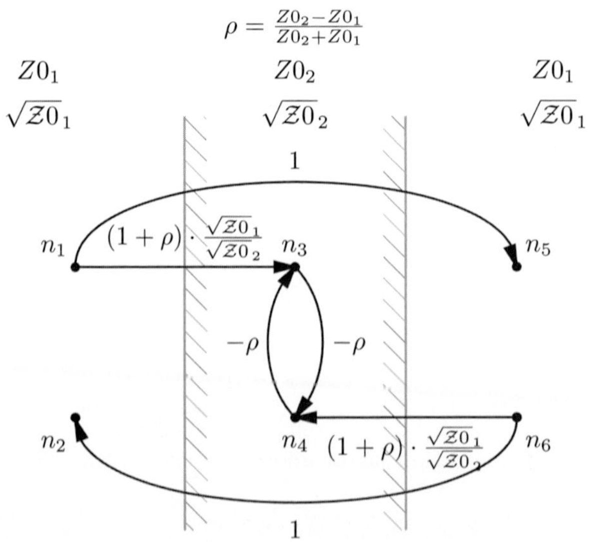


**Figure 5.7** Alternative system of impedance transformers in a wave measuring setup

Examining Figure 5.8, the result of removing node $n_3$ is seen by symmetry:

$$n_4 = -\frac{\rho}{1-\rho} \cdot \frac{\sqrt{Z0_1}}{\sqrt{Z0_2}} \cdot n_1 + \frac{1}{1-\rho} \cdot \frac{\sqrt{Z0_1}}{\sqrt{Z0_2}} \cdot n_6.$$

It should be understood that node $n_1$ represents a forward propagating wave and that $n_6$ is equal to $n_2$, which is the reverse propagating wave in the reference impedance $Z0_1$. Nodes $n_3$ and $n_4$ are the corresponding forward and reverse propagating waves in reference impedance $Z0_2$. Thus, if the forward propagating wave is designated as $a_1$ and the reverse propagating wave is designated as $b_1$ in reference impedance $Z0_1$, and $a_2$ and $b_2$ are designated as the corresponding waves in reference impedance $Z0_2$, then (5.9) can be written as follows:

$$\begin{pmatrix} a_2 \\ b_2 \end{pmatrix} = \begin{pmatrix} -\frac{\rho}{1-\rho} \cdot \frac{\sqrt{Z0_1}}{\sqrt{Z0_2}} & \frac{1}{1-\rho} \cdot \frac{\sqrt{Z0_1}}{\sqrt{Z0_2}} \\ \frac{1}{1-\rho} \cdot \frac{\sqrt{Z0_1}}{\sqrt{Z0_2}} & -\frac{\rho}{1-\rho} \cdot \frac{\sqrt{Z0_1}}{\sqrt{Z0_2}} \end{pmatrix} \cdot \begin{pmatrix} b_1 \\ a_1 \end{pmatrix}. \tag{5.9}$$

Equation (5.9) solved for $\binom{b_1}{a_1}$ and compared to (5.6), confirms that the back-to-back reference impedance transformers provide proper measurements of the waves in the system in reference impedance $Z0_2$ with no adverse effects on the rest of the system. In effect, this configuration, as shown in Figure 5.6, provides a peek into the system of the wave values in this different reference impedance. In another sense, these back-to-back devices can be viewed as performing a reference impedance transformation on each of the two ports of the connected devices. As stated previously, device ports can be interconnected as long as they have the same normalization factor and reference impedance. It does not matter what they

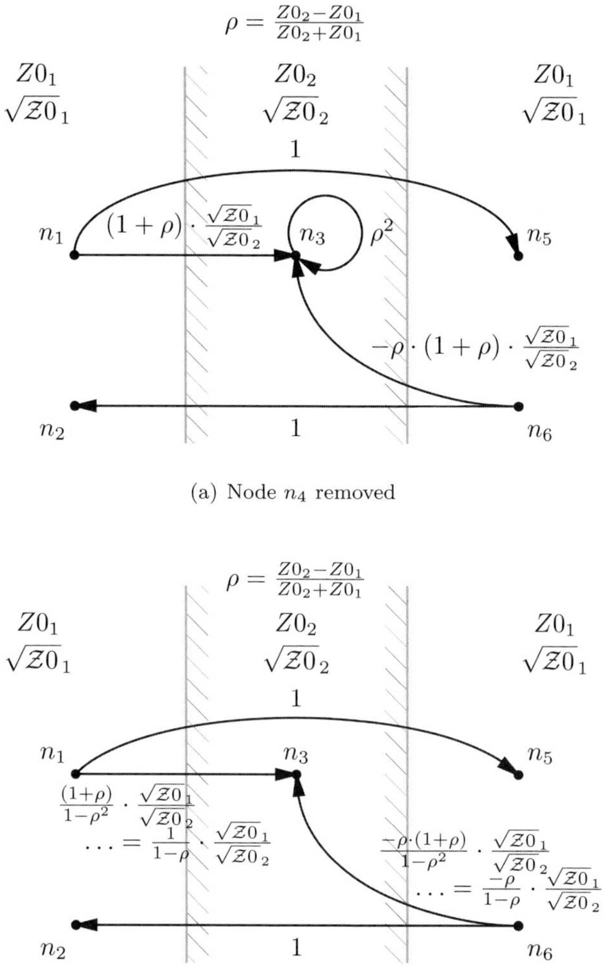


(a) Node $n_4$ removed



(b) Loopback on node $n_3$ removed

**Figure 5.8** Wave measurement calculation through node removal

are, they just have to be the same. So here, the back-to-back arrangement is like taking two interconnected device ports with the same reference impedance and normalization factor and transforming both of their reference impedances and normalization factors and reconnecting them. The waves now computed at the interface are in the new reference impedance.

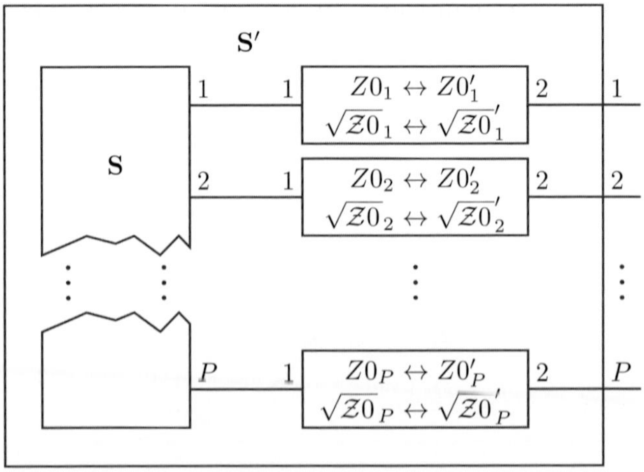


**Figure 5.9** Reference impedance transformation using tip embedding

## 5.4 Reference Impedance Transformation Using Transformers

In this section, general formulas are derived for reference impedance transformation using two methods:

1. Tip embedding: creating an s-parameter system by connecting an impedance transformer to each port of a device and solving for the aggregate system.

2. Tip de-embedding: solving the de-embedding problem by de-embedding a reference impedance transformer from each port of a device.

Both of these methods produce similar results, but the math in the tip de-embedding solution allows for the simplest final result, which is tabulated in Table 5.1.

### 5.4.1 Reference Impedance Transformation Through Tip Embedding

In §4.4.3, (4.26) was provided for embedding two-port devices at the tips or ports of a device. This will be used here to effect a reference impedance transformation by embedding reference impedance transformers. Consider a $P$-port device with an original reference impedance for each port contained on the diagonal of a matrix $\mathbf{Z0}$. In other words, for $p \in 1 \ldots P$, $Z0_{pp}$ is the original port reference impedance for port $p$. A matrix $\mathbf{Z0}'$, in a form similar to $\mathbf{Z0}$, contains a new reference impedance for each device port. For now, assume that normalization factors are similarly defined for each port as $\sqrt{\mathbf{Z0}}'$ and $\sqrt{\mathbf{Z0}}$. As mentioned in §4.4.3, the two-port devices that are embedded have port 1 connected to the device and expose port 2 as a new system port. This is shown schematically in Figure 5.9. Using the definition of impedance transformers, port 1 of device $p$ has a reference impedance

$Z0_p$ and port 2 has a reference impedance $Z0'_p$ and each device has s-parameters given by

$$\mathbf{S}_{\substack{Z0_p \leftrightarrow Z0'_p \\ \sqrt{Z0}_p \leftrightarrow \sqrt{Z0}'_p}} = \mathbf{ST}_p = \begin{pmatrix} \rho_{pp} & (1-\rho_{pp}) \cdot \frac{1}{\sqrt{Z0} f_{pp}} \\ (1+\rho_{pp}) \cdot \sqrt{Z0} f_{pp} & -\rho_{pp} \end{pmatrix},$$

where

$$\boldsymbol{\rho} = (\mathbf{Z0}' - \mathbf{Z0}) \cdot (\mathbf{Z0}' + \mathbf{Z0})^{-1}$$

and

$$\boldsymbol{\sqrt{Z0} f} = \boldsymbol{\sqrt{Z0}} \cdot \boldsymbol{\sqrt{Z0}}'^{-1}.$$

By the definition of (4.26), the following blocks are defined:

$$\mathbf{ST_{11}} = \boldsymbol{\rho}, \qquad \mathbf{ST_{12}} = (\mathbb{I} - \boldsymbol{\rho}) \cdot \boldsymbol{\sqrt{Z0} f}^{-1},$$
$$\mathbf{ST_{21}} = (\mathbb{I} + \boldsymbol{\rho}) \cdot \boldsymbol{\sqrt{Z0} f}, \qquad \mathbf{ST_{22}} = -\boldsymbol{\rho}.$$

The new s-parameters in the new reference impedance are defined as

$$\begin{aligned} \mathbf{S}' &= \mathbf{ST_{22}} + \mathbf{ST_{21}} \cdot (\mathbb{I} - \mathbf{S} \cdot \mathbf{ST_{11}}) \cdot \mathbf{S} \cdot \mathbf{ST_{12}} \\ &= -\boldsymbol{\rho} + (\mathbb{I} + \boldsymbol{\rho}) \cdot \boldsymbol{\sqrt{Z0} f} \cdot (\mathbb{I} - \mathbf{S} \cdot \boldsymbol{\rho})^{-1} \cdot \mathbf{S} \cdot (\mathbb{I} - \boldsymbol{\rho}) \cdot \boldsymbol{\sqrt{Z0} f}^{-1}. \end{aligned} \tag{5.10}$$

It is useful to examine this situation for a one-port device with s-parameters of $\Gamma$ in a reference impedance $Z0$ with a normalization factor of $\sqrt{Z0}$. The goal is to see the result in converting this to a new reference impedance $Z0'$ with a new normalization factor $\sqrt{Z0}'$. Using (5.10),

$$\begin{aligned} \Gamma' &= -\rho + (1+\rho) \cdot \sqrt{Z0} f \cdot (1 - \Gamma\rho)^{-1} \cdot \Gamma \cdot (1-\rho) \cdot \sqrt{Z0} f^{-1} \\ &= -\rho + \frac{(1+\rho) \cdot \Gamma \cdot (1-\rho)}{1 - \Gamma \cdot \rho} = \frac{\rho - \Gamma}{\Gamma \cdot \rho - 1}. \end{aligned} \tag{5.11}$$

Substituting $\rho = (Z0' - Z0) \,/\, (Z0' + Z0)$,

$$\Gamma' = \frac{\frac{Z0'-Z0}{Z0'+Z0} - \Gamma}{\Gamma \cdot \frac{Z0'-Z0}{Z0'+Z0} - 1} = \frac{Z0 \cdot \frac{1+\Gamma}{1-\Gamma} - Z0'}{Z0 \cdot \frac{1+\Gamma}{1-\Gamma} + Z0'}. \tag{5.12}$$

Both (5.11) and (5.12) are interesting in the sense that they do not depend on the normalization factors employed. Furthermore, the final form of (5.12) was constructed by merging the definition of $Z$ as a function of $\Gamma$ and $Z0$ and the definition of $\Gamma'$ as a function of $Z$ and $Z0'$.

The result of the one-port s-parameter reference impedance change might lead one to the wrong conclusion that the normalization factor is irrelevant in reference impedance transformations. For a two-port device,

$$\mathbf{S}' = -\begin{pmatrix} \rho_1 & 0 \\ 0 & \rho_2 \end{pmatrix} + \left[\mathbb{I} + \begin{pmatrix} \rho_1 & 0 \\ 0 & \rho_2 \end{pmatrix}\right] \cdot \begin{pmatrix} \sqrt{Z0}\,f_1 & 0 \\ 0 & \sqrt{Z0}\,f_2 \end{pmatrix} \cdot \left[\mathbb{I} - \begin{pmatrix} S_{11} & S_{12} \\ S_{21} & S_{22} \end{pmatrix} \cdot \begin{pmatrix} \rho_1 & 0 \\ 0 & \rho_2 \end{pmatrix}\right]^{-1}$$
$$\cdot \begin{pmatrix} S_{11} & S_{12} \\ S_{21} & S_{22} \end{pmatrix} \cdot \left[\mathbb{I} - \begin{pmatrix} \rho_1 & 0 \\ 0 & \rho_2 \end{pmatrix}\right] \cdot \begin{pmatrix} \sqrt{Z0}\,f_1 & 0 \\ 0 & \sqrt{Z0}\,f_2 \end{pmatrix}^{-1}$$
$$= \frac{1}{(S_{11} - |\mathbf{S}| \cdot \rho_2) \cdot \rho_1 + S_{22} \cdot \rho_2 - 1} \cdot \begin{pmatrix} (1-S_{22}\cdot\rho_2)\cdot\rho_1+|\mathbf{S}|\cdot\rho_2-S_{11} & \frac{\sqrt{Z0}\,f_1}{\sqrt{Z0}\,f_2}\cdot S_{12}\cdot(\rho_1+1)\cdot(\rho_2-1) \\ \frac{\sqrt{Z0}\,f_2}{\sqrt{Z0}\,f_1}\cdot S_{21}\cdot(\rho_2+1)\cdot(\rho_1-1) & (1-S_{11}\cdot\rho_1)\cdot\rho_2+|\mathbf{S}|\cdot\rho_1-S_{22} \end{pmatrix}.$$

Here it is seen that the s-parameters are, in fact, dependent on the normalization factor employed. Note that only the off-diagonal terms are dependent, which makes sense. The diagonal terms have incident waves entering the system scaled exactly opposite to the reflected waves leaving the system. So, all wave paths involving these normalization factors have canceling effects – even at the other port. However, when one port is driven and the other port is measured, the path from the incident wave at the driven port and the reflected wave at the other port must involve these different scaling effects. Note, however, that if the normalization factor is different in each of the two reference impedances, but the same at each port, there is no effect. Fortunately, this is the most common situation as most s-parameters are presented with all ports in a single reference impedance, and the most common operation involving reference impedances is to convert all ports in this one reference impedance to another reference impedance. In other words, usually the original and final reference impedances can be expressed as $\mathbf{Z0} = \mathbb{I} \cdot Z0$ and $\mathbf{Z0}' = \mathbb{I} \cdot Z0'$ (which leads as a side benefit to $\boldsymbol{\rho} = \mathbb{I} \cdot \rho$). This means hopefully that, if different normalization factors are employed, $\sqrt{\mathbf{Z0}} = \mathbb{I} \cdot \sqrt{Z0}$ and $\sqrt{\mathbf{Z0}}' = \mathbb{I} \cdot \sqrt{Z0}'$. Equation (5.10) can be solved generally for this situation:

$$\begin{aligned}
\mathbf{S}' &= -\mathbb{I} \cdot \rho + (\mathbb{I} + \mathbb{I} \cdot \rho) \cdot \sqrt{Z0}\,f \cdot (\mathbb{I} - \mathbf{S} \cdot \rho)^{-1} \cdot \mathbf{S} \cdot (\mathbb{I} - \mathbb{I} \cdot \rho) \cdot \sqrt{Z0}\,f^{-1} \\
&= -\mathbb{I} \cdot \rho + (\mathbb{I} - \mathbf{S} \cdot \rho)^{-1} \cdot \mathbf{S} \cdot (1 + \rho) \cdot (1 - \rho) \\
&= -(\mathbb{I} - \mathbf{S} \cdot \rho)^{-1} \cdot (\mathbb{I} - \mathbf{S} \cdot \rho) \cdot \rho + (\mathbb{I} - \mathbf{S} \cdot \rho)^{-1} \cdot \mathbf{S} \cdot (1 + \rho) \cdot (1 - \rho) \\
&= (\mathbb{I} - \mathbf{S} \cdot \rho)^{-1} \cdot \left[\mathbf{S} \cdot \rho^2 - \mathbb{I} \cdot \rho + \mathbf{S} - \mathbf{S} \cdot \rho^2\right] \\
&= (\mathbb{I} - \mathbf{S} \cdot \rho)^{-1} \cdot (\mathbf{S} - \mathbb{I} \cdot \rho).
\end{aligned}$$

With the normalization factors expressed in this manner, it is found that there is no effect and they can be ignored in reference impedance changes. Of course, the reason that the topic of normalization factors is brought up when talking about reference impedance transformations is because of the traveling wave case. In this case, one would have, for an impedance transformation from $Z0$ to $Z0'$, a corresponding normalization factor change from $\sqrt{Z0} = \sqrt{Z0}$ to $\sqrt{Z0}' = \sqrt{Z0'}$. Again, if all ports are in a given reference impedance and are to be transformed to another, one does not have to worry about normalization factors at all, even in the traveling wave case. If one is mixing port reference impedances, then more careful methods have to be utilized.

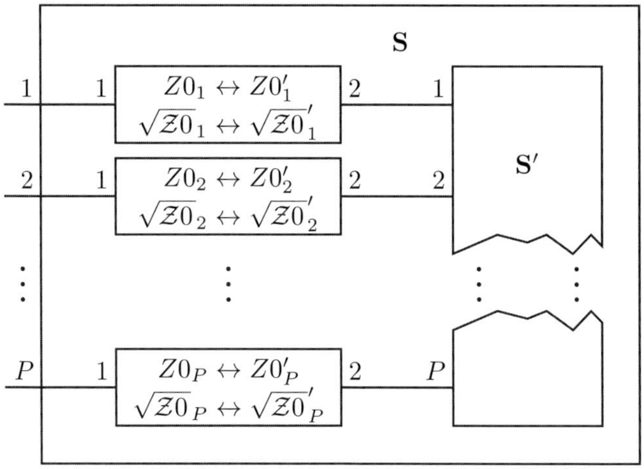


**Figure 5.10** Reference impedance transformation using tip de-embedding

**Listing 5.2** General reference impedance transformation

```
def ReferenceImpedance(S,Z0f,Z0i=None,Kf=None,Ki=None):
    (Z0f,Kf)=Z0KHelper((Z0f,Kf),len(S))
    (Z0i,Ki)=Z0KHelper((Z0i,Ki),len(S))
    I=matrix(identity(len(S)))
    p=(matrix(Z0f)-matrix(Z0i))*(matrix(Z0f)+matrix(Z0i)).getI()
    Kf=matrix(Ki)*matrix(Kf).getI()
    S=matrix(S)
    return (Kf*(I-p).getI()*(S-p)*(I-p*S).getI()*(I-p)*Kf.getI()).tolist()
```

### 5.4.2 Reference Impedance Transformation Through Tip De-embedding

In §10.4 a method is provided for de-embedding two-port devices at the tips or ports of a device. Here, this method is used for reference impedance transformation by de-embedding reference impedance transformers presumed to be at the tips of a device that are transforming the device from desired reference impedances into the given device with the given port reference impedances.

Consider diagonal matrices of reference impedances $\mathbf{Z0}$ and $\mathbf{Z0}'$ such that, for a $P$-port device, for $p \in 1 \ldots P$, $Z0'_{pp}$ is the new port reference impedance for port $p$ and $Z0_{pp}$ is the original port reference impedance for port $p$. It is assumed that there are normalization factors similarly defined for each port as $\sqrt{\mathbf{Z0}}'$ and $\sqrt{\mathbf{Z0}}$. As mentioned in §4.4.3, the two-port devices that are embedded have port 2 connected to the unknown device and expose port 1 as a system port on the given device. Using the definition of impedance transformers, port 1 of device $p$ has a reference impedance $Z0_p$ and port 2 has a reference

**Table 5.1** Reference impedance transformation

| **Situation** | **Conversion equation**[a] |
|---|---|
| General conversion formula when port normalization factors and reference impedances are different | $\mathbf{S}' = \sqrt{\mathcal{Z}\mathbf{0}}\boldsymbol{f} \cdot (\mathbb{I} - \boldsymbol{\rho})^{-1} \cdot (\mathbf{S} - \boldsymbol{\rho}) \cdot (\mathbb{I} - \boldsymbol{\rho} \cdot \mathbf{S})^{-1} \cdot (\mathbb{I} - \boldsymbol{\rho}) \cdot \sqrt{\mathcal{Z}\mathbf{0}}\boldsymbol{f}^{-1}$ |
| Applicable conversion formula when normalization factors and reference impedances are the same on all ports | $\mathbf{S}' = (\mathbf{S} - \mathbb{I} \cdot \rho) \cdot (\mathbb{I} - \rho \cdot \mathbf{S})^{-1}$ (5.13) |
| Traveling wave case (required only when port reference impedances differ) | $\mathbf{S}' = \sqrt{\mathbf{Z0}} \cdot \sqrt{\mathbf{Z0}'}^{-1} \cdot (\mathbb{I} - \boldsymbol{\rho})^{-1} \cdot (\mathbf{S} - \boldsymbol{\rho}) \cdot (\mathbb{I} - \boldsymbol{\rho} \cdot \mathbf{S})^{-1} \cdot (\mathbb{I} - \boldsymbol{\rho}) \cdot \sqrt{\mathbf{Z0}'} \cdot \sqrt{\mathbf{Z0}}^{-1}$ |

[a] The original reference impedance is $\mathbf{Z0}$ and the new, desired reference impedance is $\mathbf{Z0}'$, with $\boldsymbol{\rho} = (\mathbf{Z0}' - \mathbf{Z0}) \cdot (\mathbf{Z0}' + \mathbf{Z0})^{-1}$, $\sqrt{\mathcal{Z}\mathbf{0}}\boldsymbol{f} = \sqrt{\mathcal{Z}\mathbf{0}} \cdot \sqrt{\mathcal{Z}\mathbf{0}}'^{-1}$.

impedance $Z0'_p$ and each device has s-parameters given by

$$\mathbf{S}\,_{\substack{Z0_p \leftrightarrow Z0'_p \\ \sqrt{\mathcal{Z}0}_p \leftrightarrow \sqrt{\mathcal{Z}0}'_p}} = \mathbf{ST}_p = \begin{pmatrix} \rho_{pp} & (1 - \rho_{pp}) \cdot \frac{1}{\sqrt{\mathcal{Z}0}f_{pp}} \\ (1 + \rho_{pp}) \cdot \sqrt{\mathcal{Z}0}f_{pp} & -\rho_{pp} \end{pmatrix},$$

where

$$\boldsymbol{\rho} = (\mathbf{Z0}' - \mathbf{Z0}) \cdot (\mathbf{Z0}' + \mathbf{Z0})^{-1},$$

and

$$\sqrt{\mathcal{Z}\mathbf{0}}\boldsymbol{f} = \sqrt{\mathcal{Z}\mathbf{0}} \cdot \sqrt{\mathcal{Z}\mathbf{0}}'^{-1}.$$

A diagram depicting the arrangement of devices and reference impedance transformers is shown in Figure 5.10. Note that the orientation and definition of the reference impedance transformers are the same as in §5.4.1, only the location of $\mathbf{S}$ and $\mathbf{S}'$ are different.

By the methods provided in §10.4, and by comparing the de-embedding problem shown in Figure 5.10 to the solution provided in Figure 10.5, the following blocks are defined:

$$\mathbf{F_{11}} = \boldsymbol{\rho}, \qquad \mathbf{F_{12}} = (\mathbb{I} - \boldsymbol{\rho}) \cdot \sqrt{\mathcal{Z}\mathbf{0}}\boldsymbol{f}^{-1},$$
$$\mathbf{F_{21}} = (\mathbb{I} + \boldsymbol{\rho}) \cdot \sqrt{\mathcal{Z}\mathbf{0}}\boldsymbol{f}, \qquad \mathbf{F_{22}} = -\boldsymbol{\rho}.$$

Using (10.17), (10.18), and (10.19) leads to

$$\begin{aligned}\mathbf{S}' &= \mathbf{F_{12}}^{-1} \cdot (\mathbf{S_i} - \mathbf{F_{11}}) \cdot \left[\mathbf{F_{21}} + \mathbf{F_{22}} \cdot \mathbf{F_{12}}^{-1} \cdot (\mathbf{S_i} - \mathbf{F_{11}})\right]^{-1} \\ &= \sqrt{\mathcal{Z}\mathbf{0}}\boldsymbol{f} \cdot (\mathbb{I} - \boldsymbol{\rho})^{-1} \cdot (\mathbf{S} - \boldsymbol{\rho}) \cdot \left[(\mathbb{I} + \boldsymbol{\rho}) - \boldsymbol{\rho} \cdot (\mathbb{I} - \boldsymbol{\rho})^{-1} \cdot (\mathbf{S} - \boldsymbol{\rho})\right]^{-1} \cdot \sqrt{\mathcal{Z}\mathbf{0}}\boldsymbol{f}^{-1}.\end{aligned} \tag{5.14}$$

By recognizing that

$$(\mathbb{I} + \boldsymbol{\rho}) - \boldsymbol{\rho} \cdot (\mathbb{I} - \boldsymbol{\rho})^{-1} \cdot (\mathbf{S} - \boldsymbol{\rho}) = (\mathbb{I} - \boldsymbol{\rho})^{-1} \cdot (\mathbb{I} - \boldsymbol{\rho} \cdot \mathbf{S}),$$

(5.14) can be rewritten in its best and most general final form:

$$\mathbf{S}' = \sqrt{\boldsymbol{\mathcal{Z}}\mathbf{0}}\boldsymbol{f} \cdot (\mathbb{I} - \boldsymbol{\rho})^{-1} \cdot (\mathbf{S} - \boldsymbol{\rho}) \cdot (\mathbb{I} - \boldsymbol{\rho} \cdot \mathbf{S})^{-1} \cdot (\mathbb{I} - \boldsymbol{\rho}) \cdot \sqrt{\boldsymbol{\mathcal{Z}}\mathbf{0}}\boldsymbol{f}^{-1}.$$

The code that handles this general form is provided in Listing 5.2.

Given normalization factors that are the same for every port (i.e. $\sqrt{\boldsymbol{\mathcal{Z}}\mathbf{0}} = \mathbb{I} \cdot \sqrt{\mathcal{Z}0}$ and $\sqrt{\boldsymbol{\mathcal{Z}}\mathbf{0}}' = \mathbb{I} \cdot \sqrt{\mathcal{Z}0}'$), then these terms drop out, to yield

$$\mathbf{S}' = (\mathbb{I} - \boldsymbol{\rho})^{-1} \cdot (\mathbf{S} - \boldsymbol{\rho}) \cdot (\mathbb{I} - \boldsymbol{\rho} \cdot \mathbf{S})^{-1} \cdot (\mathbb{I} - \boldsymbol{\rho}).$$

Furthermore, for the common situation in which port impedances are the same for every port (i.e. $\mathbf{Z0} = \mathbb{I} \cdot Z0$ and $\mathbf{Z0}' = \mathbb{I} \cdot Z0'$, which leads as a side benefit to $\boldsymbol{\rho} = \mathbb{I} \cdot \rho$), the result is the following simple solution:

$$\mathbf{S}' = (\mathbf{S} - \boldsymbol{\rho}) \cdot (\mathbb{I} - \boldsymbol{\rho} \cdot \mathbf{S})^{-1}.$$

All of the reference impedance transformation possibilities are tabulated in Table 5.1.

# 6

# Sources

THIS CHAPTER CONCERNS ITSELF with the construction of the various building blocks necessary for simulation with s-parameters. Here, voltage and current sources are constructed along with all of the dependent source capabilities. Since dependent sources depend on voltage or current, devices are constructed to contain either a dependent voltage or current source along with a current or voltage sense element. By constructing devices in this way, s-parameters can be directly extracted for dependent source devices that are easily placed in circuits. The reason that a chapter is devoted to this topic is that sources and sense elements do not inherently have s-parameters (sources because they are also a source of stimuli, and sense elements because they relate waves to currents and voltages). Dependent sources have these elements inside, so these devices do not lend themselves well to the direct methods for s-parameter determination that were presented in Chapter 3 and Chapter 4. These solutions provide for a complete repertoire of elements.

## 6.1 Source Elements

### 6.1.1 The Current Source

The circuit symbol for the ideal current source is shown in Figure 6.1(a). The current source is a device that can be defined as an equality of the current entering the device from each port. Using the definition of current in (2.9),

$$\frac{\sqrt{Z0}}{Z0}\cdot(a_1-b_1)=i=-\frac{\sqrt{Z0}}{Z0}\cdot(a_2-b_2)\,.$$

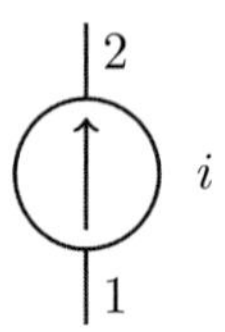


(a) Circuit representation

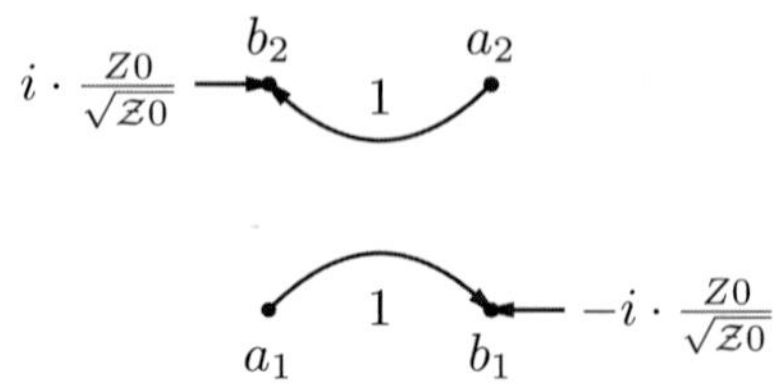


(b) Signal-flow diagram

**Figure 6.1** Ideal current source

(a) Circuit representation (b) Signal-flow diagram

**Figure 6.2** Ideal single-port current source

Solving for the $b$ nodes results in:

$$\begin{pmatrix} 1 & 0 \\ 0 & 1 \end{pmatrix} \cdot \begin{pmatrix} a_1 \\ a_2 \end{pmatrix} + \begin{pmatrix} -\frac{1}{\sqrt{Z0}} \cdot i \cdot Z0 \\ \frac{1}{\sqrt{Z0}} \cdot i \cdot Z0 \end{pmatrix} = \begin{pmatrix} b_1 \\ b_2 \end{pmatrix}. \tag{6.1}$$

Equation (6.1) clearly expresses the s-parameters of a source whose terminals look like a perfect open (i.e. a device with infinite output impedance) and with waves emanating from the $b$ nodes, as shown in Figure 6.1(b). Note that waves emanate from both ports.

Very often, one side of the current source is connected to ground, as shown in Figure 6.2(a). This new one-port current source is generated by connecting an arrow with weight $-1$ between nodes $a_1$ and $b_1$, thus terminating port 1 of the two-port source to ground. Port 2 of the former device is now port 1 of the new single-port device. Removing the old nodes $a_1$ and $b_1$, and renaming nodes $a_2$ and $b_2$ as the new $a_1$ and $b_1$ because they are the new port 1, results in the one-port current source diagram shown in Figure 6.2(b).

Another common configuration of the current source is with some output impedance. The ideal current source has infinite output impedance. To provide an output impedance, the ideal current source is cascaded with a shunt impedance, as shown in Figure 6.3(a). To calculate this, it is easiest to consider a generic one-port source cascaded with a generic two-port device. This allows this solution to be reused in §6.1.2. Node removal techniques are used because the stimulus entering one of the nodes to be removed precludes the use of the general cascading methods. The node removal steps are shown in Figure 6.4. The starting diagram is provided in Figure 6.4(a) as a device with reflection coefficient $\Gamma$ with an incident wave $m$ on a generic two-port device. The nodes are removed in Figure 6.4(b) and Figure 6.4(c), arriving at the general solution in Figure 6.4(d). Since it is a general solution, it will be reused for the voltage source calculations.

Equation (3.43) provides the s-parameters of a shunt impedance. The following substitutions are applied to Figure 6.4(d):

$$\Gamma = 1, \qquad \mathbf{S} = \frac{1}{2 \cdot Z + Z0} \cdot \begin{pmatrix} -Z0 & 2 \cdot Z \\ 2 \cdot Z & -Z0 \end{pmatrix}, \qquad m = \frac{1}{\sqrt{Z0}} \cdot i \cdot Z0.$$

This leads to a source reflection coefficient of

$$\Gamma_{cs} = \frac{Z - Z0}{Z + Z0}. \tag{6.2}$$

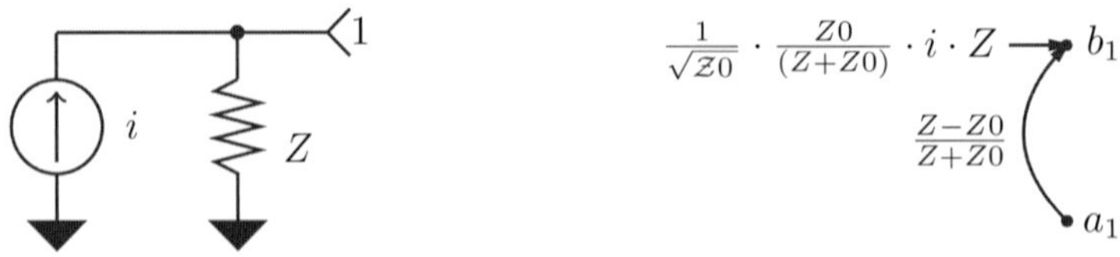


(a) Circuit representation (b) Signal-flow diagram

**Figure 6.3** Single-port current source with shunt impedance (according to (6.2) and (6.3))

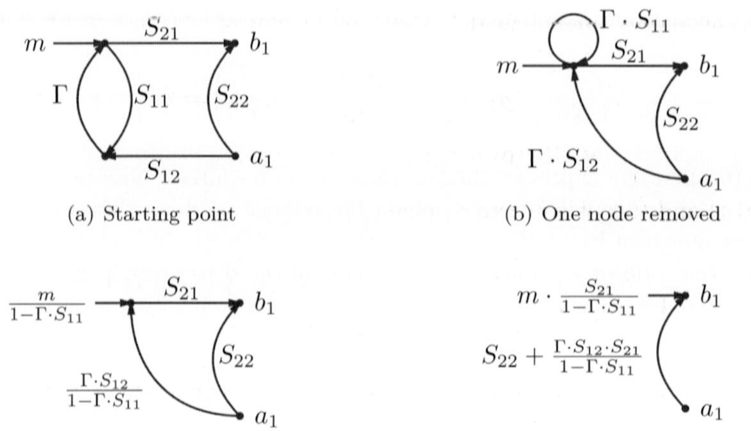


(a) Starting point (b) One node removed

(c) Loopback removed (d) Final signal-flow diagram

**Figure 6.4** One-port source cascaded with two-port impedance network

The driving stimulus becomes

$$m_{cs} = \frac{1}{\sqrt{Z0}} \cdot \frac{Z0}{Z + Z0} \cdot i \cdot Z. \tag{6.3}$$

The result is shown in Figure 6.3, with the signal-flow diagram solution in Figure 6.3(b) that is equivalent to the single-port current source with output impedance shown in Figure 6.3(a).

### 6.1.2 The Voltage Source

The circuit symbol for the ideal voltage source is shown in Figure 6.5(a). The voltage source is a device that can be defined as an equality of the current entering the device from each port, along with the fact that the voltage difference at the two ports must be the voltage

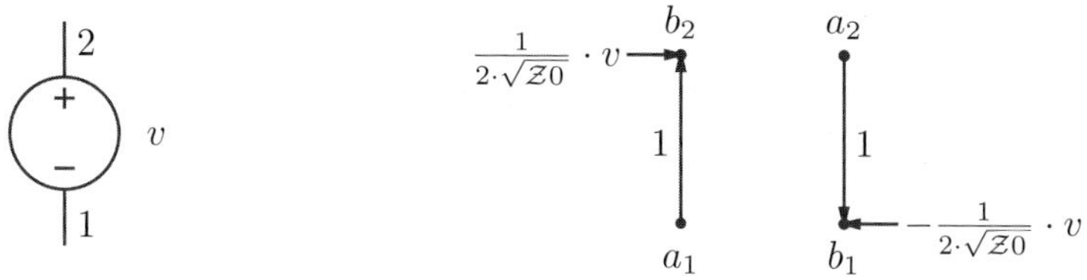


(a) Circuit representation (b) Signal-flow diagram

**Figure 6.5** Ideal voltage source

developed by the source. Using the definitions of current and voltage from (2.9) and (2.8),

$$\frac{\sqrt{Z0}}{Z0}\cdot(a_1-b_1)+\frac{\sqrt{Z0}}{Z0}\cdot(a_2-b_2)=0, \tag{6.4}$$

$$\sqrt{Z0}\cdot(a_1+b_1)+v=\sqrt{Z0}\cdot(a_2+b_2)\,. \tag{6.5}$$

Solving (6.4) and (6.5) for $b_1$ and $b_2$ results in

$$\begin{pmatrix}0 & 1\\ 1 & 0\end{pmatrix}\cdot\begin{pmatrix}a_1\\ a_2\end{pmatrix}+\begin{pmatrix}-\frac{1}{2\cdot\sqrt{Z0}}\cdot v\\ \frac{1}{2\cdot\sqrt{Z0}}\cdot v\end{pmatrix}=\begin{pmatrix}b_1\\ b_2\end{pmatrix}. \tag{6.6}$$

Equation (6.6) clearly expresses the s-parameters of a source whose terminals look like a perfectly matched thru device with waves emanating from the $b$ nodes, as shown in Figure 6.5(b). It might be surprising to see a voltage source looking like a perfect match to the reference impedance. Contrasting this with the ideal current source in §6.1.1, one might expect the output to look like a perfect short. But a voltage source has to have its port 1 connected to something, and what Figure 6.5(b) is really demonstrating is that the impedance seen at port 2 will be the impedance of the termination of port 1. Another way to look at comparing the current source to the voltage source is that setting the current to zero in the current source causes the circuit element to appear as open at each port. Similarly, setting the voltage to zero in the voltage source causes the circuit element to be a wire that shorts each port together.

As with the current source, voltage sources are popularly used with one side connected to ground, exposing only one port, as shown in Figure 6.6(a). The signal-flow diagram for this one-port voltage source is generated by connecting an arrow with weight $-1$ between nodes $a_1$ and $b_1$, thus terminating port 1 of the two-port source to ground. Port 2 of the former device becomes port 1 of the new single-port device. Removing nodes $a_1$ and $b_1$ using node removal techniques, and renaming nodes $a_2$ and $b_2$ as the new $a_1$ and $b_1$ because they are the new port 1, results in the one-port voltage source diagram shown in Figure 6.6(b).

Another common configuration of the voltage source is with some output impedance. The ideal one-port voltage source has zero impedance. To provide an output impedance, the ideal voltage source is cascaded with a series impedance, as shown in Figure 6.7(a).

(a) Circuit representation (b) Signal-flow diagram

**Figure 6.6** Ideal single-port voltage source

The equations in the generic one-port source and the two-port device cascade arrangement shown in Figure 6.4 are used to calculate this, along with the equation for a series impedance provided by (3.42). The following substitutions are made in Figure 6.4(d):

$$\Gamma = -1, \qquad \mathbf{S} = \frac{\begin{pmatrix} Z & 2 \cdot Z0 \\ 2 \cdot Z0 & Z \end{pmatrix}}{Z + 2 \cdot Z0}, \qquad m = \frac{1}{\sqrt{Z0}} \cdot v.$$

This leads to a source reflection coefficient of

$$\Gamma_{vs} = \frac{Z - Z0}{Z + Z0}. \tag{6.7}$$

The driving stimulus becomes

$$m_{vs} = \frac{1}{\sqrt{Z0}} \cdot \frac{Z0}{Z + Z0} \cdot v. \tag{6.8}$$

This is shown in Figure 6.7(b). It is interesting that in the world of waves and s-parameters the single-port current source and voltage source are identical if $v = i \cdot Z$; this is analogous to the fact that, for every Thévenin equivalent source, there is a Norton equivalent source. Here, the Thévenin and Norton equivalent is formed by interchanging $v$ and $i \cdot Z$.

## 6.2 Sense Elements

The use of s-parameters for circuit analysis and simulation is useful because circuits are reduced to signal-flow diagrams. Thus, meshes and nodes in circuits are converted to nodes only in signal-flow diagrams. This capability of s-parameter utilization would come with a limitation if it were not possible to sense and incorporate voltages and currents into controlled sources. This section deals with the methods for sensing voltages and currents that will be encapsulated into controlled sources. In a sense, hybrid network parameters are developed consisting of waves, voltages, and currents that, when combined with voltage and current sources, enable the creation of dependent sources for which s-parameters can exist. This enables dependent sources to be devices with s-parameters, which allows these devices

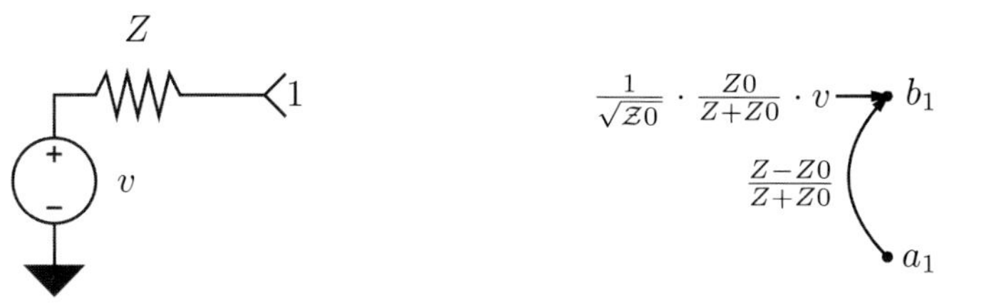


(a) Circuit representation (b) Signal-flow diagram

**Figure 6.7** Single-port voltage source with series impedance (according to (6.7) and (6.8))

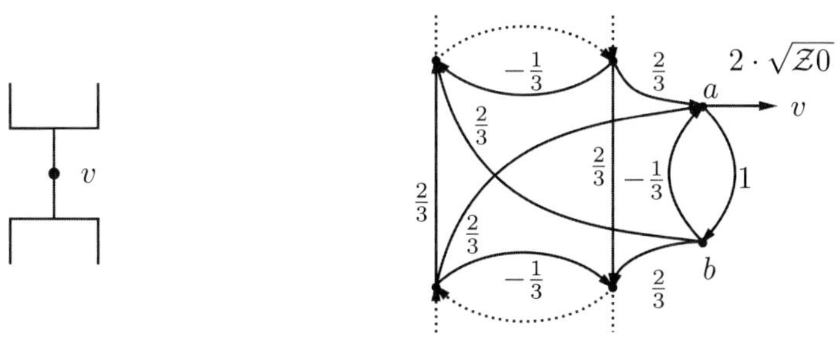


(a) Circuit representation (b) Signal-flow diagram with ideal tee

**Figure 6.8** Absolute voltage sense element

to be hooked into circuits in a normal manner and removes the need for any other use of voltage and current in a circuit (other than perhaps the final determination of a simulation result).

### 6.2.1 The Voltage Sensor

The voltage sensor is an element that senses either an absolute voltage, as shown in Figure 6.8(a), or preferably the voltage across an element. In order to connect the absolute voltage sense element into a circuit, the two terminals must appear as a wire (i.e. a short circuit) across the terminals, but must also appear as an open circuit (i.e. have infinite input impedance) at the sensor output. The voltage at a *voltage node* in a circuit is defined by the forward and reverse propagating waves making up the node and has no direction, as

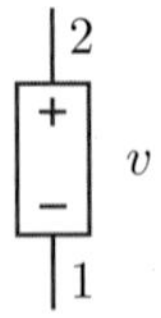


(a) Circuit representation

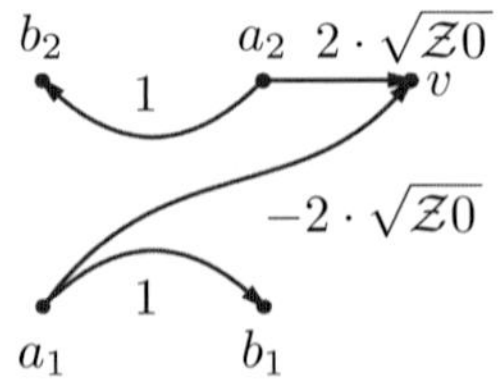


(b) Signal-flow diagram

**Figure 6.9** Differential voltage sense element

provided in (2.8):

$$v = \sqrt{Z0} \cdot (a + b) .$$

In order to detect this voltage, an open-circuit element is created connected to an ideal tee, as shown in Figure 6.8(b). Since an open-circuit element has a reflection coefficient $\Gamma = 1$, the $b$ node of the sensor will have the same value as the $a$ node, provided no waves are launched into the $b$ node, which is enforced. With this enforcement, the voltage is defined as

$$v = 2 \cdot \sqrt{Z0} \cdot a.$$

Thus, the voltage sense element shown in Figure 6.8 shows how voltage sensing is accomplished.

To sense the voltage *across* an element, or the voltage difference between two nodes, two of these sense elements are placed as in Figure 6.9, and the voltage sensed by the element labeled as the negative voltage is subtracted from the voltage sensed by the element labeled as the positive voltage. This is the same effect as connecting two terminals of an ideal voltmeter across a circuit. The element shown in Figure 6.9(a) measures the voltage across two nodes of a circuit by connecting port 2 as the positive terminal and port 1 as the negative terminal. Technically the voltage cannot be exported and must be utilized within the final device housing the sense element. The connection of this element has no effect on the circuit it connects to, and, because its terminals appear as open, as shown in Figure 6.9(b), it can be connected in the usual manner for any circuit connection.

### 6.2.2 The Current Sensor

The goal of a current sensor, as shown in Figure 6.10, is to detect the current through a device. In order to sense current, a device has to be physically inserted into the circuit. This device must have the notion of direction of current flow, and, since it must be inserted into a circuit, it should ideally have no effect on the circuit. It requires two ports, as shown in Figure 6.10(a). Using (2.9), the current flowing into port 1 is described as

$$i = \frac{\sqrt{Z0}}{Z0} \cdot (a_1 - b_1) .$$

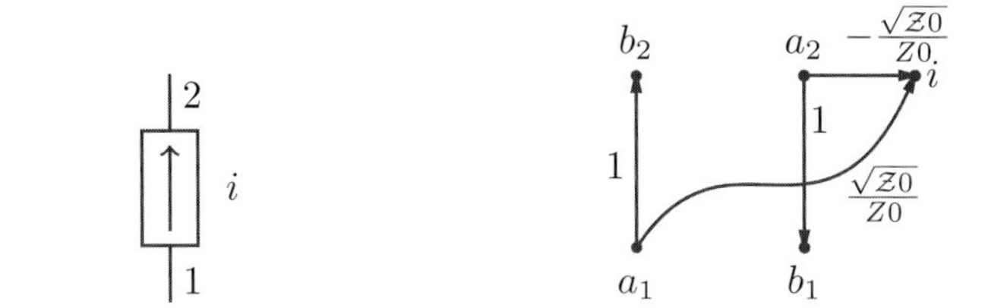


(a) Circuit representation (b) Signal-flow diagram

**Figure 6.10** Current sense element

The current flowing out of port 2 is described as

$$i = -\frac{\sqrt{Z0}}{Z0} \cdot (a_2 - b_2).$$

It is preferable not to tap off of the $b$ nodes as it complicates the determination of the s-parameters. This is not a problem because if the current sensor is not to disturb the circuit in which it is inserted, an ideal thru connection is placed between the ports. This means that $b_2 = a_1$ and $b_1 = a_2$, which means that the current sensed can be detected as the current entering port 1 as

$$i = \frac{\sqrt{Z0}}{Z0} \cdot (a_1 - a_2).$$

The element shown in Figure 6.10(a) measures current through the device flowing from port 1 to port 2. To utilize this element in a circuit, the circuit is broken and the device is inserted in the reference direction for current flow (i.e. current flowing from port 1 to port 2 will be measured as positive current). Like the voltage sense element, the current measured must be utilized with a device housing the sense element. Note that the insertion of this element into a circuit has no effect on the circuit, as can be seen in Figure 6.10(b).

## 6.3 Dependent Sources

To form dependent sources, the sense elements described in §6.2.1 and §6.2.2 are applied to the source elements described in §6.1.1 and §6.1.2 to form dependent sources. All four possible combinations are provided here. Since in dependent sources the sense elements are self-contained with the source element, the devices provided here have s-parameters and can be inserted into circuits with no special considerations.

### 6.3.1 Voltage Controlled Voltage Source

The voltage controlled voltage source is created by combining the voltage sense element, as shown in Figure 6.9, with the voltage source, as shown in Figure 6.5, resulting in the circuit and signal-flow representations shown in Figure 6.11(a) and Figure 6.11(b). Because the scaling factor of $2 \cdot \sqrt{Z0}$, applied to the arrows from the $a$ node of the sense element,

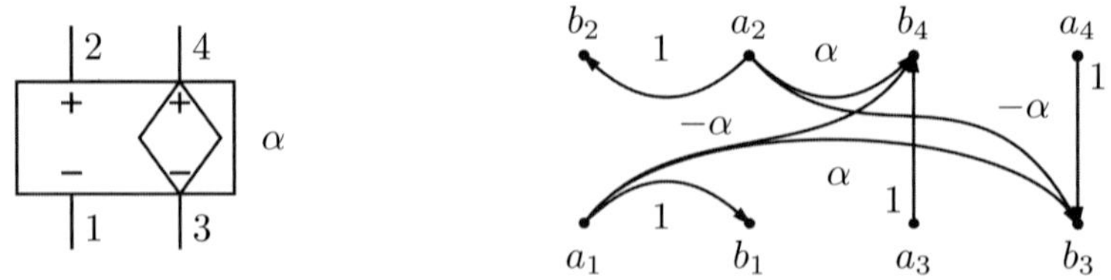


(a) Circuit representation (b) Signal-flow diagram

$$\mathbf{S} = \begin{pmatrix} 1 & 0 & 0 & 0 \\ 0 & 1 & 0 & 0 \\ \alpha & -\alpha & 0 & 1 \\ -\alpha & \alpha & 1 & 0 \end{pmatrix}$$

(c) S-parameters

**Figure 6.11** Voltage controlled voltage source

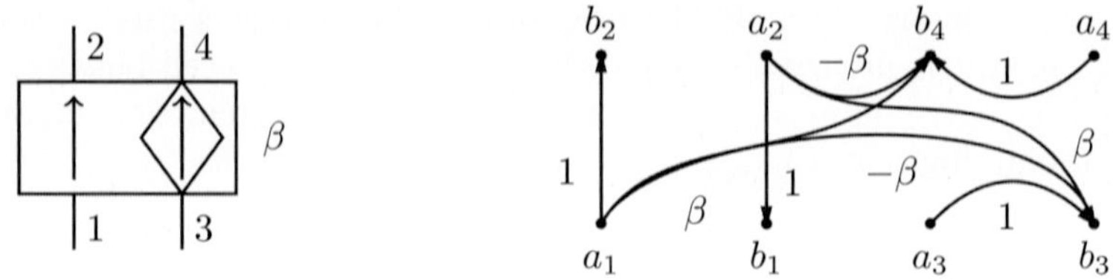


(a) Circuit representation (b) Signal-flow diagram

$$\mathbf{S} = \begin{pmatrix} 0 & 1 & 0 & 0 \\ 1 & 0 & 0 & 0 \\ -\beta & \beta & 1 & 0 \\ \beta & -\beta & 0 & 1 \end{pmatrix}$$

(c) S-parameters

**Figure 6.12** Current controlled current source

appears in the denominator of the stimulus in the voltage source, it allows the connection of the arrows with only the voltage gain $\alpha$ applied.

The s-parameters of the voltage controlled voltage source are easily determined by reading the weights of the arrows in Figure 6.11(b), shown in Figure 6.11(c).

### 6.3.2 Current Controlled Current Source

The current controlled current source is created by combining the current sense element, as shown in Figure 6.10, with the current source, as shown in Figure 6.1, resulting in the circuit and signal-flow representations shown in Figure 6.12(a) and Figure 6.12(b). Because the scaling factor of $\sqrt{\mathcal{Z}0}/Z0$, applied to the arrows from the $a$ node of the sense element,

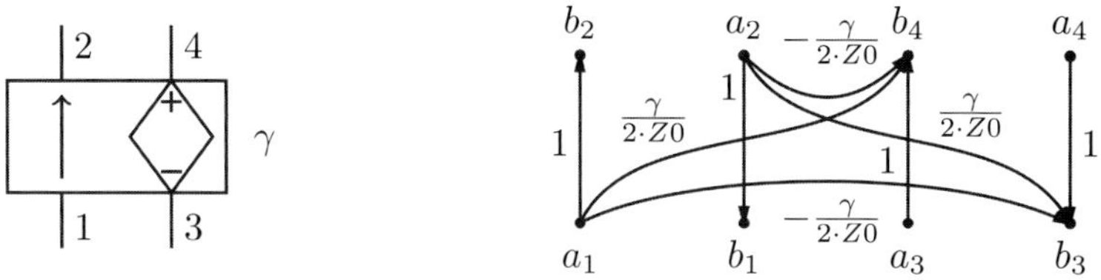


(a) Circuit representation (b) Signal-flow diagram

$$\mathbf{S} = \begin{pmatrix} 0 & 1 & 0 & 0 \\ 1 & 0 & 0 & 0 \\ -\frac{\gamma}{2\cdot Z0} & \frac{\gamma}{2\cdot Z0} & 0 & 1 \\ \frac{\gamma}{2\cdot Z0} & -\frac{\gamma}{2\cdot Z0} & 1 & 0 \end{pmatrix}$$

(c) S-parameters

**Figure 6.13** Current controlled voltage source

appears inverted in the stimulus of the current source, it allows the connection of the arrows with only the current gain $\beta$ applied.

The s-parameters of the current controlled current source are easily determined by reading the weights of the arrows connecting $a$ nodes to $b$ nodes in Figure 6.12(b) and are shown in Figure 6.12(c).

### 6.3.3 Current Controlled Voltage Source

The current controlled voltage source is created by combining the current sense element, as shown in Figure 6.10, with the voltage source, as shown in Figure 6.5, resulting in the circuit and signal-flow representations shown in Figure 6.13(a) and Figure 6.13(b). The scaling factor of $\sqrt{Z0}/Z0$ in the sense element combines with the factor of $1/\left(2\cdot\sqrt{Z0}\right)$ in the voltage source to form a scaling factor of $1/\left(2\cdot Z0\right)$; this is applied to the gain $\gamma$, which is in units of V/A, or ohms ($\Omega$).

The s-parameters of the current controlled voltage source are easily determined by reading the weights of the arrows in Figure 6.13(b) and are shown in Figure 6.13(c).

### 6.3.4 Voltage Controlled Current Source

The voltage controlled current source is created by combining the voltage sense element, as shown in Figure 6.9, with the current source, as shown in Figure 6.1, resulting in circuit and signal-flow representations shown in Figure 6.14(a) and Figure 6.14(b). The scaling factor of $2\cdot\sqrt{Z0}$ in the sense element combines with the factor of $Z0/\sqrt{Z0}$ in the source to form a scaling factor $2\cdot Z0$; this is applied to the gain $\delta$, which is in units of A/V, or siemens (S).

The s-parameters of the voltage controlled current source are easily determined by reading the weights of the arrows in Figure 6.14(b) and are shown in Figure 6.14(c).

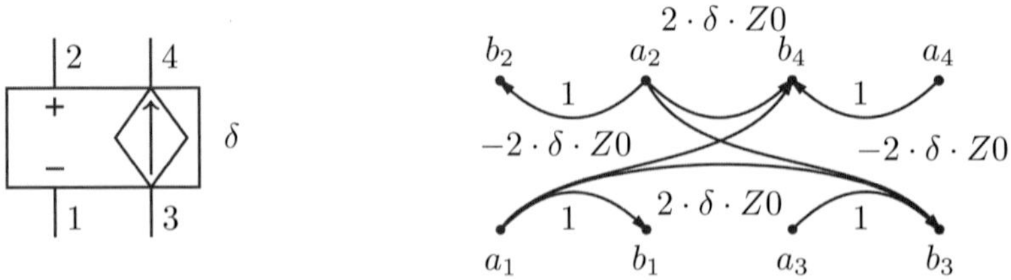


(a) Circuit representation (b) Signal-flow diagram

$$\mathbf{S} = \begin{pmatrix} 1 & 0 & 0 & 0 \\ 0 & 1 & 0 & 0 \\ 2 \cdot \delta \cdot Z0 & -2 \cdot \delta \cdot Z0 & 1 & 0 \\ -2 \cdot \delta \cdot Z0 & 2 \cdot \delta \cdot Z0 & 0 & 1 \end{pmatrix}$$

(c) S-parameters

**Figure 6.14** Voltage controlled current source

## 6.4 Amplifiers

Depending on their type, amplifiers are dependent sources with impedances either in series with or shunting the inputs or outputs. To simplify their usage, basic four-port amplifiers are developed and used to develop three- and two-port versions.

### 6.4.1 Voltage Amplifiers

Voltage amplifiers are formed by shunting the voltage sense inputs of a voltage controlled voltage source with a shunt input impedance $Z_i$ and adding an output impedance in series with the voltage output. In this section, the four-port voltage amplifier is derived first and the remaining amplifiers are derived by shorting ports together or to ground.

#### *Four-Port Voltage Amplifier*

The four-port voltage amplifier is shown in Figure 6.15. The circuit symbol is shown in Figure 6.15(a), with the internal circuit shown schematically in Figure 6.15(b), where there is a voltage controlled voltage source with a shunt input impedance $Z_i$ and a series output impedance $Z_o$. The addition of these elements leads to a complicated solution that is provided in Figure D.1 in Appendix D, with the results of this solution provided in Figure 6.15(c). Note the slightly different numbering on the ports from that in the voltage controlled voltage source, as seen in Figure 6.15(b).

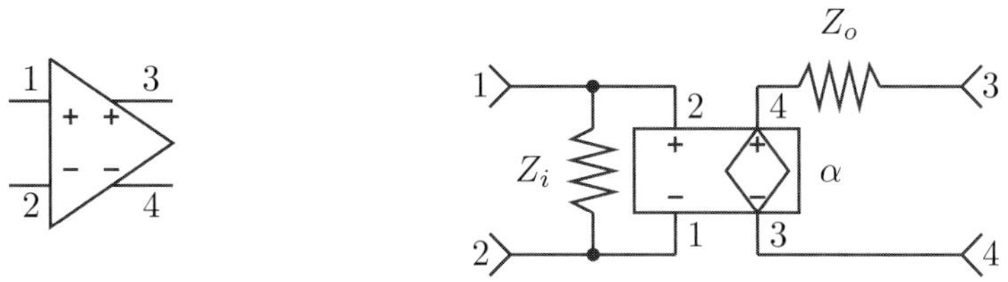


(a) Circuit symbol

(b) Circuit representation

$$\mathbf{S} = \begin{pmatrix} \frac{Z_i}{Z_i+2\cdot Z0} & \frac{2\cdot Z0}{Z_i+2\cdot Z0} & 0 & 0 \\ \frac{2\cdot Z0}{Z_i+2\cdot Z0} & \frac{Z_i}{Z_i+2\cdot Z0} & 0 & 0 \\ \frac{2\cdot Z_i\cdot Z0\cdot\alpha}{(Z_i+2\cdot Z0)\cdot(Z_o+2\cdot Z0)} & -\frac{2\cdot Z_i\cdot Z0\cdot\alpha}{(Z_i+2\cdot Z0)\cdot(Z_o+2\cdot Z0)} & \frac{Z_o}{Z_o+2\cdot Z0} & \frac{2\cdot Z0}{Z_o+2\cdot Z0} \\ -\frac{2\cdot Z_i\cdot Z0\cdot\alpha}{(Z_i+2\cdot Z0)\cdot(Z_o+2\cdot Z0)} & \frac{2\cdot Z_i\cdot Z0\cdot\alpha}{(Z_i+2\cdot Z0)\cdot(Z_o+2\cdot Z0)} & \frac{2\cdot Z0}{Z_o+2\cdot Z0} & \frac{Z_o}{Z_o+2\cdot Z0} \end{pmatrix}$$

(c) S-parameters

**Figure 6.15** Four-port voltage amplifier

As a check, opening the input impedance and shorting across the output impedance results in the same s-parameters as the voltage controlled voltage source in Figure 6.11(c):

$$\lim_{\substack{Z_i\to\infty \\ Z_o\to 0}} \mathbf{S} = \begin{pmatrix} 1 & 0 & 0 & 0 \\ 0 & 1 & 0 & 0 \\ \alpha & -\alpha & 0 & 1 \\ -\alpha & \alpha & 1 & 0 \end{pmatrix}.$$

The port ordering on the voltage controlled voltage source was changed from $\left(\begin{smallmatrix}1 & 2 & 3 & 4\end{smallmatrix}\right)^T$ to $\left(\begin{smallmatrix}2 & 1 & 4 & 3\end{smallmatrix}\right)^T$, but this is essentially the same as flipping the voltage controlled voltage source vertically and has no effect on its performance. In other words, swapping the input ports changes the sign of the input, but swapping the output ports reverses this sign change.

To see this, a permutation matrix is applied to the voltage controlled voltage source with this new pin numbering:

$$\begin{pmatrix} 0 & 1 & 0 & 0 \\ 1 & 0 & 0 & 0 \\ 0 & 0 & 0 & 1 \\ 0 & 0 & 1 & 0 \end{pmatrix} \cdot \begin{pmatrix} 1 & 0 & 0 & 0 \\ 0 & 1 & 0 & 0 \\ \alpha & -\alpha & 0 & 1 \\ -\alpha & \alpha & 1 & 0 \end{pmatrix} \cdot \begin{pmatrix} 0 & 1 & 0 & 0 \\ 1 & 0 & 0 & 0 \\ 0 & 0 & 0 & 1 \\ 0 & 0 & 1 & 0 \end{pmatrix}^T = \begin{pmatrix} 1 & 0 & 0 & 0 \\ 0 & 1 & 0 & 0 \\ \alpha & -\alpha & 0 & 1 \\ -\alpha & \alpha & 1 & 0 \end{pmatrix}.$$

This compares favorably with Figure 6.11(c).

### *Three-Port Voltage Amplifier*

A three-port voltage amplifier is shown schematically in Figure 6.16(a). It is formed by shorting ports 2 and 4 of the four-port voltage amplifier together, exposing this as port 3,

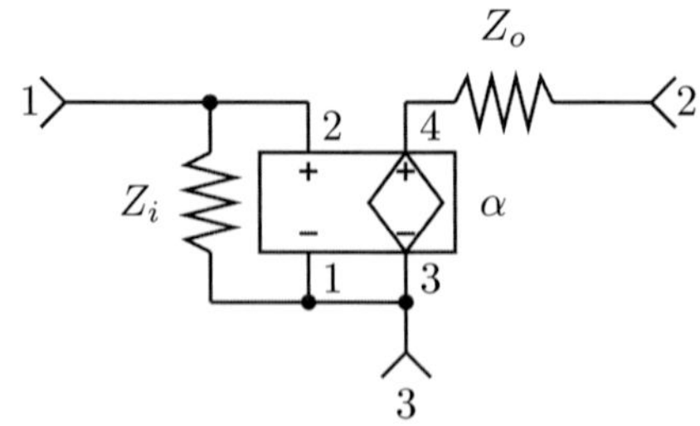


(a) Circuit representation

$$\mathbf{S} = \frac{1}{-Z_o \cdot Z_i - 2 \cdot Z_o \cdot Z0 - 2 \cdot Z_i \cdot Z0 - 3 \cdot Z0^2 + \alpha \cdot Z_i \cdot Z0} \cdot \begin{pmatrix} -Z_o \cdot Z_i - 2 \cdot Z_i \cdot Z0 + Z0^2 + \alpha \cdot Z_i \cdot Z0 & -2 \cdot Z0^2 & -2 \cdot Z0 \cdot (Z_o + Z0) \\ -2 \cdot Z0 \cdot (\alpha \cdot Z_i + Z0) & Z0^2 - 2 \cdot Z_o \cdot Z0 + \alpha \cdot Z_i \cdot Z0 - Z_o \cdot Z_i & 2 \cdot Z0 \cdot (\alpha \cdot Z_i - Z_i - Z0) \\ 2 \cdot Z0 \cdot (-Z0 + \alpha \cdot Z_i - Z_o) & -2 \cdot Z0 \cdot (Z_i + Z0) & -Z_o \cdot Z_i + Z0^2 - \alpha \cdot Z_i \cdot Z0 \end{pmatrix}$$

(b) S-parameters

**Figure 6.16** Three-port voltage amplifier

and renumbering the output port as port 2. This is a common configuration for three-port elements such as transistors.

The detailed solution to this problem is provided in Figure D.2 in Appendix D. The resulting s-parameters from this calculation are shown in Figure 6.16(b).

As a check, opening the input impedance and shorting across the output impedance results in the s-parameters of the voltage controlled voltage source with ports 1 and 3 shorted together and exposed:

$$\lim_{\substack{Z_i \to \infty \\ Z_o \to 0}} \mathbf{S} = \begin{pmatrix} 1 & 0 & 0 \\ \frac{2 \cdot \alpha}{2 - \alpha} & -\frac{\alpha}{2 - \alpha} & 2 \cdot \frac{1 - \alpha}{2 - \alpha} \\ -\frac{2 \cdot \alpha}{2 - \alpha} & \frac{2}{2 - \alpha} & \frac{\alpha}{2 - \alpha} \end{pmatrix}.$$

### *Two-Port Voltage Amplifier*

The two-port amplifier is provided in Figure 6.17, with the circuit symbol shown in Figure 6.17(a). This device is formed directly from the four-port voltage amplifier by shorting port 2 at the input and port 4 at the output to ground, as shown in Figure 6.17(b).

The detailed solution to this problem is provided in Figure D.3 in Appendix D. The resulting s-parameters from this calculation are shown in Figure 6.17(c).

Opening the input impedance and shorting across the output impedance results in the following s-parameters:

$$\lim_{\substack{Z_i \to \infty \\ Z_o \to 0}} = \begin{pmatrix} 1 & 0 \\ 2 \cdot \alpha & -1 \end{pmatrix}.$$

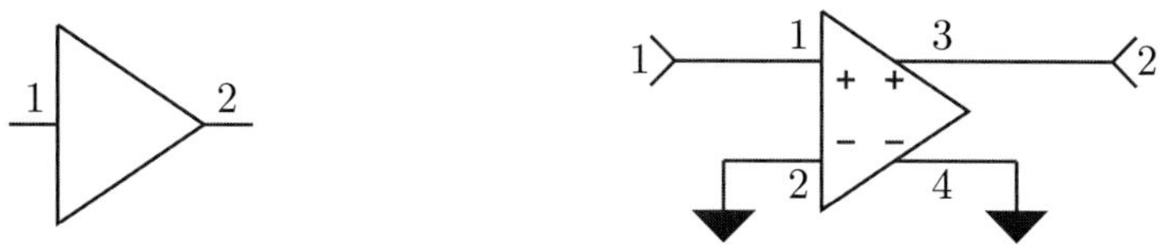


(a) Circuit symbol (b) Circuit representation

$$\mathbf{S} = \begin{pmatrix} \frac{Z_i - Z0}{Z_i + Z0} & 0 \\ \frac{2 \cdot \alpha \cdot Z_i \cdot Z0}{(Z_i + Z0) \cdot (Z_o + Z0)} & \frac{Z_o - Z0}{Z_o + Z0} \end{pmatrix}$$

(c) S-parameters

**Figure 6.17** Two-port voltage amplifier

The input is an open ($S_{11} = 1$), the output is a short ($S_{22} = -1$), there is perfect reverse isolation ($S_{12} = 0$), and the voltage gain is $\alpha$ ($s_{21} = 2 \cdot \alpha$). If it is unclear why $S_{21}$ has this factor of two, consider an ideal voltage source driving this amplifier, with the amplifier output driving a termination in the reference impedance. In this case, there is an arrow pointing into the system of $1/\left(2 \cdot \sqrt{Z0}\right)$, a loop of $-1$ formed by the combination of the output impedance of the ideal source and the input impedance of the amplifier, and an arrow of weight $2 \cdot \alpha$ driving the load. Since it is terminated in $Z0$, there is no wave incident on the amplifier output. Removing the loopback results in

$$b = \frac{1}{\sqrt{Z0}} \cdot v \cdot \frac{1}{1 - -1} \cdot 2 \cdot \alpha = \frac{\alpha}{\sqrt{Z0}} \cdot v,$$

and, since $a = 0$, the voltage at the output is computed as

$$\sqrt{Z0} \cdot (a + b) = \sqrt{Z0} \cdot \left(0 + \frac{\alpha}{\sqrt{Z0}} \cdot v\right) = \alpha \cdot v.$$

Another way to intuit this is to realize that the voltage developed at the input when the amplifier presents a high impedance is actually proportional to twice the wave incident on it (i.e. the incident wave is proportional to half the voltage), but s-parameters are the ratio of reflected to incident waves, not voltages.

### 6.4.2 Current Amplifiers

Current amplifiers are analogous to voltage amplifiers. They are formed by providing an input impedance in series with the current sense path from port 1 to port 2, and by shunting the output between ports 4 and ports 3 of a current controlled current source.

Exactly as in the voltage amplifier, simplified versions are provided by shorting various ports together and/or to ground. Limiting cases are also calculated here, as these limiting cases are popular and allow better sense to be made of the results.

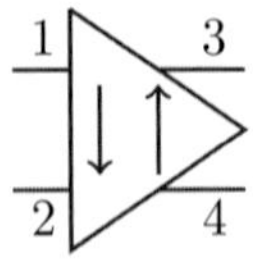


(a) Circuit symbol

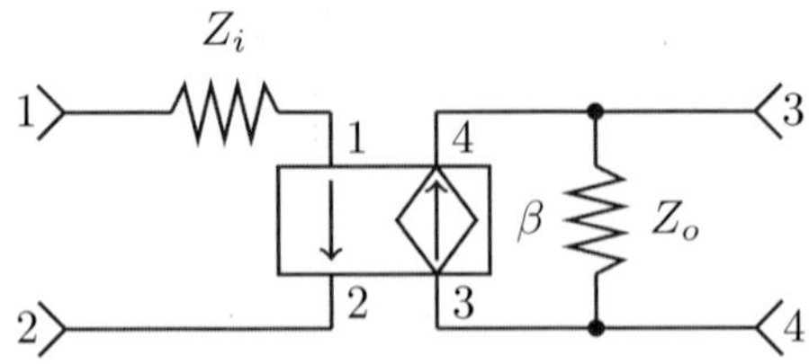


(b) Circuit representation

$$\mathbf{S} = \begin{pmatrix} \frac{Z_i}{Z_i+2\cdot Z0} & \frac{2\cdot Z0}{Z_i+2\cdot Z0} & 0 & 0 \\ \frac{2\cdot Z0}{Z_i+2\cdot Z0} & \frac{Z_i}{Z_i+2\cdot Z0} & 0 & 0 \\ \frac{2\cdot Z_o\cdot Z0\cdot\beta}{(Z_i+2\cdot Z0)\cdot(Z_o+2\cdot Z0)} & -\frac{2\cdot Z_o\cdot Z0\cdot\beta}{(Z_i+2\cdot Z0)\cdot(Z_o+2\cdot Z0)} & \frac{Z_o}{Z_o+2\cdot Z0} & \frac{2\cdot Z0}{Z_o+2\cdot Z0} \\ -\frac{2\cdot Z_o\cdot Z0\cdot\beta}{(Z_i+2\cdot Z0)\cdot(Z_o+2\cdot Z0)} & \frac{2\cdot Z_o\cdot Z0\cdot\beta}{(Z_i+2\cdot Z0)\cdot(Z_o+2\cdot Z0)} & \frac{2\cdot Z0}{Z_o+2\cdot Z0} & \frac{Z_o}{Z_o+2\cdot Z0} \end{pmatrix}$$

(c) S-parameters

**Figure 6.18** Four-port current amplifier

### *Four-Port Current Amplifier*

The four-port current amplifier is provided in Figure 6.18. Its schematic symbol is shown as a current controlled current source, but having input impedance $Z_i$ and output impedance $Z_o$ specified along with the current gain.

The circuit symbol is shown in Figure 6.18(a) and the circuit used to calculate this device is shown in Figure 6.18(b), where careful accounting of the port numbering is made. Since the preference is for the current amplifier to have current sensed as the current flowing into port 1 and exiting port 2, and the current output flowing into port 4 and out of port 3, some of the ports on the current controlled current source require rearrangement.

The details of the solution are shown in Figure D.4 in Appendix D, with the final, simplified s-parameters shown in Figure 6.18(c).

As a check, shorting across the input impedance and opening the output impedance results in the s-parameters of the current controlled current source, but with the pins re-ordered:

$$\lim_{\substack{Z_i\to 0\\ Z_o\to\infty}} \mathbf{S} = \begin{pmatrix} 0 & 1 & 0 & 0 \\ 1 & 0 & 0 & 0 \\ \beta & -\beta & 1 & 0 \\ -\beta & \beta & 0 & 1 \end{pmatrix}.$$

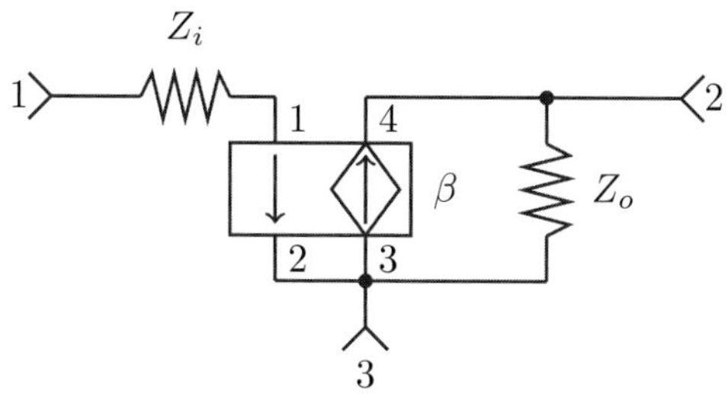


(a) Circuit representation

$$\mathbf{S} = \frac{1}{3 \cdot Z0^2 + (2 \cdot Z_o + 2 \cdot Z_i - \beta \cdot Z_o) \cdot Z0 + Z_o \cdot Z_i} \cdot \begin{pmatrix} Z_o \cdot Z_i + Z0 \cdot (2 \cdot Z_i - \beta \cdot Z_o) - Z0^2 & 2 \cdot Z0^2 & 2 \cdot Z0^2 + 2 \cdot Z_o \cdot Z0 \\ 2 \cdot Z0^2 + 2 \cdot \beta \cdot Z_o \cdot Z0 & Z_o \cdot Z_i + Z0 \cdot (2 \cdot Z_o - \beta \cdot Z_o) - Z0^2 & 2 \cdot Z0^2 + Z0 \cdot (2 \cdot Z_i - 2 \cdot \beta \cdot Z_o) \\ 2 \cdot Z0^2 + Z0 \cdot (2 \cdot Z_o - 2 \cdot \beta \cdot Z_o) & 2 \cdot Z0^2 + 2 \cdot Z_i \cdot Z0 & Z_o \cdot Z_i - Z0^2 + \beta \cdot Z_o \cdot Z0 \end{pmatrix}$$

(b) S-parameters

**Figure 6.19** Three-port current amplifier

Because the ports have been reordered from $\begin{pmatrix} 1 & 2 & 3 & 4 \end{pmatrix}^T$ to $\begin{pmatrix} 1 & 2 & 4 & 3 \end{pmatrix}^T$, there is a permutation matrix that performs this reordering multiplied as follows:

$$\begin{pmatrix} 1 & 0 & 0 & 0 \\ 0 & 1 & 0 & 0 \\ 0 & 0 & 0 & 1 \\ 0 & 0 & 1 & 0 \end{pmatrix} \cdot \begin{pmatrix} 0 & 1 & 0 & 0 \\ 1 & 0 & 0 & 0 \\ \beta & -\beta & 1 & 0 \\ -\beta & \beta & 0 & 1 \end{pmatrix} \cdot \begin{pmatrix} 1 & 0 & 0 & 0 \\ 0 & 1 & 0 & 0 \\ 0 & 0 & 0 & 1 \\ 0 & 0 & 1 & 0 \end{pmatrix}^T = \begin{pmatrix} 0 & 1 & 0 & 0 \\ 1 & 0 & 0 & 0 \\ -\beta & \beta & 1 & 0 \\ \beta & -\beta & 0 & 1 \end{pmatrix}.$$

This compares favorably with Figure 6.12(c).

### *Three-Port Current Amplifier*

The three-port current amplifier is provided schematically in Figure 6.19(a). It is formed by shorting together ports 2 and 4 of the four-port current amplifier and exposing this port as port 3. The output port becomes port 2. This is a common configuration for three-port elements such as transistors.

The detailed solution to this problem is provided in Figure D.5 in Appendix D. The resulting s-parameters from this calculation are shown in Figure 6.19(b).

As a check, shorting across the input impedance and opening the output impedance results in the s-parameters of the current controlled current source with ports 2 and 3 shorted together and exposed, and the output flipped:

$$\lim_{\substack{Z_i \to 0 \\ Z_o \to \infty}} \mathbf{S} = \frac{1}{2 - \beta} \cdot \begin{pmatrix} -\beta & 0 & 2 \\ 2 \cdot \beta & 2 - \beta & -2 \cdot \beta \\ 2 \cdot (1 - \beta) & 0 & \beta \end{pmatrix}.$$

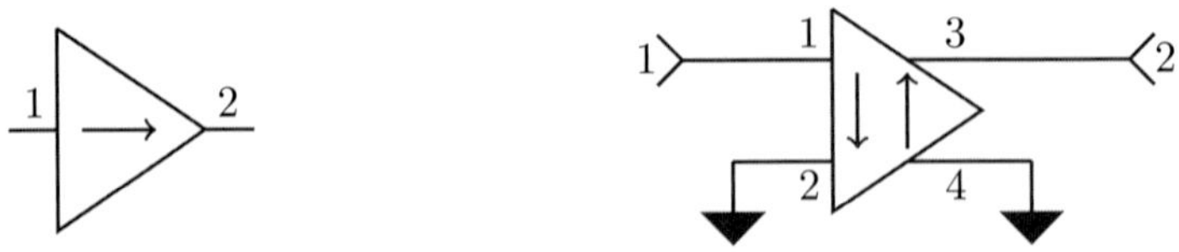


(a) Circuit symbol (b) Circuit representation

$$\mathbf{S} = \begin{pmatrix} \frac{Z_i - Z0}{Z_i + Z0} & 0 \\ \frac{2\cdot\beta\cdot Z_o\cdot Z0}{(Z_i+Z0)\cdot(Z_o+Z0)} & \frac{Z_o - Z0}{Z_o + Z0} \end{pmatrix}$$

(c) S-parameters

**Figure 6.20** Two-port current amplifier

### *Two-Port Current Amplifier*

The two-port current amplifier is provided in Figure 6.20, with its symbol given in Figure 6.20(a). Schematically, the two-port current amplifier is found by grounding ports 2 and 4 of the four-port current amplifier (and exposing port 3 as port 2), as seen in Figure 6.20(b), or alternatively by grounding port 3 of the three-port current amplifier.

The detailed solution to this system is provided in Figure D.6 in Appendix D, and the simplified s-parameters resulting from this symbolic solution are provided in Figure 6.20(c).

As a check, shorting the across the input impedance and opening the output impedance results in the s-parameters of the ideal two-port current amplifier:

$$\lim_{\substack{Z_i \to 0 \\ Z_o \to \infty}} \mathbf{S} = \begin{pmatrix} -1 & 0 \\ 2\cdot\beta & 1 \end{pmatrix}.$$

As another check, the amplifier is driven with an ideal current source, and the output current into a termination of $Z0$ is measured. An arrow with weight $\left(1/\sqrt{Z0}\right)\cdot i\cdot Z0$ drives a loopback formed by the interaction of the unity reflection coefficient of the current source and the shorted input impedance of the amplifier, with another arrow with weight $s_{21} = 2\cdot\beta$. Because of the ideal termination, the wave incident on the output is $a = 0$ and the wave emanating from the amplifier is

$$b = \frac{1}{\sqrt{Z0}}\cdot i\cdot Z0\cdot\frac{1}{1 - -1}\cdot 2\cdot\beta = \frac{1}{\sqrt{Z0}}\cdot i\cdot Z0\cdot\beta.$$

The current *exiting* the amplifier is therefore given by

$$-\frac{\sqrt{Z0}}{Z0}\cdot(a-b) = -\frac{\sqrt{Z0}}{Z0}\cdot\left(0 - \frac{1}{\sqrt{Z0}}\cdot i\cdot Z0\cdot\beta\right) = i\cdot\beta.$$

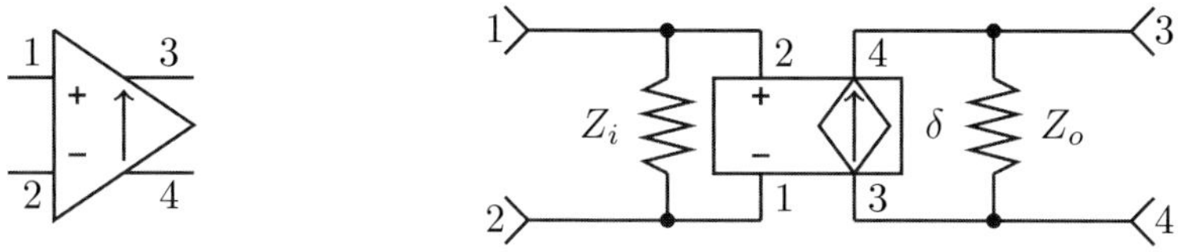


(a) Circuit symbol

(b) Circuit representation

$$\mathbf{S} = \begin{pmatrix} \frac{Z_i}{Z_i+2\cdot Z0} & \frac{2\cdot Z0}{Z_i+2\cdot Z0} & 0 & 0 \\ \frac{2\cdot Z0}{Z_i+2\cdot Z0} & \frac{Z_i}{Z_i+2\cdot Z0} & 0 & 0 \\ \frac{2\cdot Z_i\cdot Z_o\cdot Z0\cdot\delta}{(Z_i+2\cdot Z0)\cdot(Z_o+2\cdot Z0)} & -\frac{2\cdot Z_i\cdot Z_o\cdot Z0\cdot\delta}{(Z_i+2\cdot Z0)\cdot(Z_o+2\cdot Z0)} & \frac{Z_o}{Z_o+2\cdot Z0} & \frac{2\cdot Z0}{Z_o+2\cdot Z0} \\ -\frac{2\cdot Z_i\cdot Z_o\cdot Z0\cdot\delta}{(Z_i+2\cdot Z0)\cdot(Z_o+2\cdot Z0)} & \frac{2\cdot Z_i\cdot Z_o\cdot Z0\cdot\delta}{(Z_i+2\cdot Z0)\cdot(Z_o+2\cdot Z0)} & \frac{2\cdot Z0}{Z_o+2\cdot Z0} & \frac{Z_o}{Z_o+2\cdot Z0} \end{pmatrix}$$

(c) S-parameters

**Figure 6.21** Four-port transconductance amplifier

### 6.4.3 Transconductance Amplifiers

Transconductance amplifiers sense voltage and output current. They are the same as the voltage controlled current source except they have an input impedance $Z_i$ shunting the input and an output impedance $Z_o$ shunting the output. The gain is expressed as a conductance in A/V, or siemens (S).

#### *Four-Port Transconductance Amplifier*

The four-port transconductance amplifier is provided in Figure 6.21, with the circuit symbol for this amplifier shown in Figure 6.21(a). The circuit is shown in Figure 6.21(b), where there is a voltage controlled current source with input ports shunted by the input resistance $Z_i$ and output ports shunted by an output impedance $Z_o$. Normally, these impedances would be high. Note the port reordering of the voltage controlled current source, which has its input ports swapped.

The symbolic circuit solution is provided in Figure D.7 in Appendix D, and the simplified result is provided as the s-parameters for this device in Figure 6.21(c).

Opening the input and output impedance reduces the s-parameters in Figure 6.21(c) to those of the ideal transconductance amplifier, resulting in the same s-parameters of the voltage controlled current source:

$$\lim_{\substack{Z_i\to\infty \\ Z_o\to\infty}} \mathbf{S} = \begin{pmatrix} 1 & 0 & 0 & 0 \\ 0 & 1 & 0 & 0 \\ 2\cdot\delta\cdot Z0 & -2\cdot\delta\cdot Z0 & 1 & 0 \\ -2\cdot\delta\cdot Z0 & 2\cdot\delta\cdot Z0 & 0 & 1 \end{pmatrix}.$$

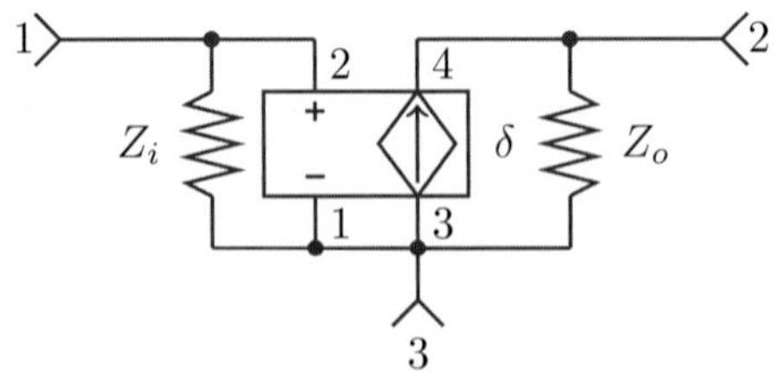


(a) Circuit representation

$$\mathbf{S} = \frac{1}{3 \cdot Z0^2 + (2 \cdot Z_o + 2 \cdot Z_i - \delta \cdot Z_i \cdot Z_o) \cdot Z0 + Z_o \cdot Z_i}$$

$$\cdot \begin{pmatrix} Z_o{\cdot}Z_i{+}Z0{\cdot}(2{\cdot}Z_i{-}\delta{\cdot}Z_i{\cdot}Z_o){-}Z0^2 & 2{\cdot}Z0^2 & 2{\cdot}Z0^2{+}2{\cdot}Z_o{\cdot}Z0 \\ 2{\cdot}Z0^2{+}2{\cdot}\delta{\cdot}Z_i{\cdot}Z_o{\cdot}Z0 & Z_o{\cdot}Z_i{+}Z0{\cdot}(2{\cdot}Z_o{-}\delta{\cdot}Z_i{\cdot}Z_o){-}Z0^2 & 2{\cdot}Z0^2{+}Z0{\cdot}(2{\cdot}Z_i{-}2{\cdot}\delta{\cdot}Z_i{\cdot}Z_o) \\ 2{\cdot}Z0^2{+}Z0{\cdot}(2{\cdot}Z_o{-}2{\cdot}\delta{\cdot}Z_i{\cdot}Z_o) & 2{\cdot}Z0^2{+}2{\cdot}Z_i{\cdot}Z0 & Z_o{\cdot}Z_i{-}Z0^2{+}\delta{\cdot}Z_i{\cdot}Z_o{\cdot}Z0 \end{pmatrix}$$

(b) S-parameters

**Figure 6.22** Three-port transconductance amplifier

The port reordering served to flip the input and output ports of this device, and therefore made the ideal device identical to the voltage controlled current source. This is seen by recognizing that the port ordering went from $\begin{pmatrix} 1 & 2 & 3 & 4 \end{pmatrix}^T$ to $\begin{pmatrix} 2 & 1 & 4 & 3 \end{pmatrix}^T$, and therefore

$$\begin{pmatrix} 0 & 1 & 0 & 0 \\ 1 & 0 & 0 & 0 \\ 0 & 0 & 0 & 1 \\ 0 & 0 & 1 & 0 \end{pmatrix} \cdot \begin{pmatrix} 1 & 0 & 0 & 0 \\ 0 & 1 & 0 & 0 \\ 2 \cdot \delta \cdot Z0 & -2 \cdot \delta \cdot Z0 & 1 & 0 \\ -2 \cdot \delta \cdot Z0 & 2 \cdot \delta \cdot Z0 & 0 & 1 \end{pmatrix} \cdot \begin{pmatrix} 0 & 1 & 0 & 0 \\ 1 & 0 & 0 & 0 \\ 0 & 0 & 0 & 1 \\ 0 & 0 & 1 & 0 \end{pmatrix}^T$$
$$= \begin{pmatrix} 1 & 0 & 0 & 0 \\ 0 & 1 & 0 & 0 \\ 2 \cdot \delta \cdot Z0 & -2 \cdot \delta \cdot Z0 & 1 & 0 \\ -2 \cdot \delta \cdot Z0 & 2 \cdot \delta \cdot Z0 & 0 & 1 \end{pmatrix}.$$

This compares favorably with Figure 6.14(c).

### *Three-Port Transconductance Amplifier*

The three-port transconductance amplifier is shown schematically in Figure 6.22(a), where it is formed by shorting together ports 2 and 4 of the four-port transconductance amplifier and exposing it as port 3, with the output port changed to port 2.

The circuit is solved symbolically in Figure D.8 in Appendix D, and the s-parameters resulting from simplifying these results are shown in Figure 6.22(b).

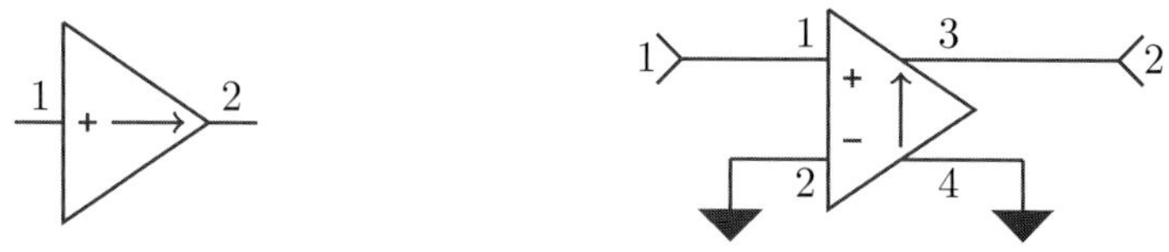


(a) Circuit symbol (b) Circuit representation

$$\mathbf{S} = \begin{pmatrix} \frac{Z_i - Z0}{Z_i + Z0} & 0 \\ \frac{2\cdot\delta\cdot Z_i\cdot Z_o\cdot Z0}{(Z_i+Z0)\cdot(Z_o+Z0)} & \frac{Z_o - Z0}{Z_o + Z0} \end{pmatrix}$$

(c) S-parameters

**Figure 6.23** Two-port transconductance amplifier

Opening the input and output impedance results in the s-parameters of the ideal three-port transconductance amplifier:

$$\lim_{\substack{Z_i\to\infty \\ Z_o\to\infty}} \mathbf{S} = \frac{1}{\delta\cdot Z0 - 1}\cdot\begin{pmatrix} \delta\cdot Z0 - 1 & 0 & 0 \\ -2\cdot\delta\cdot Z0 & \delta\cdot Z0 - 1 & 2\cdot\delta\cdot Z0 \\ 2\cdot\delta\cdot Z0 & 0 & -(\delta\cdot Z0 + 1) \end{pmatrix}.$$

### *Two-Port Transconductance Amplifier*

The two-port transconductance amplifier is provided in Figure 6.23, with the circuit symbol shown in Figure 6.23(a). Figure 6.23(b) provides a schematic of this device formed by shorting ports 2 and 4 of the four-port transconductance amplifier to ground and exposing port 3 as output port 2.

The symbolic solution is provided in Figure D.9 in Appendix D, and the simplified s-parameters are shown in Figure 6.23(c).

Opening the input and output impedance results in the s-parameters of the ideal two-port transconductance amplifier:

$$\lim_{\substack{Z_i\to\infty \\ Z_o\to\infty}} \mathbf{S} = \begin{pmatrix} 1 & 0 \\ 2\cdot\delta\cdot Z0 & 1 \end{pmatrix}.$$

## 6.4.4 Transresistance Amplifiers

The transresistance amplifier senses current, and outputs voltage. It is essentially a current controlled voltage source with series input and output impedance. The gain of the amplifier is expressed as a resistance in V/A, or ohms (Ω).

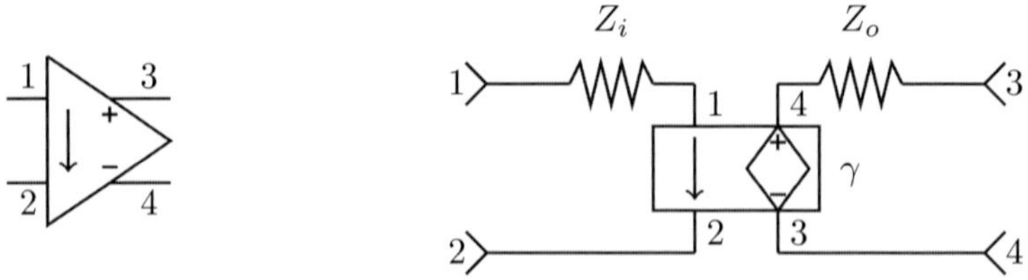


(a) Circuit symbol

(b) Circuit representation

$$\mathbf{S} = \begin{pmatrix} \frac{Z_i}{Z_i+2\cdot Z0} & \frac{2\cdot Z0}{Z_i+2\cdot Z0} & 0 & 0 \\ \frac{2\cdot Z0}{Z_i+2\cdot Z0} & \frac{Z_i}{Z_i+2\cdot Z0} & 0 & 0 \\ \frac{2\cdot\gamma\cdot Z0}{(Z_o+2\cdot Z0)\cdot(Z_i+2\cdot Z0)} & -\frac{2\cdot\gamma\cdot Z0}{(Z_o+2\cdot Z0)\cdot(Z_i+2\cdot Z0)} & \frac{Z_o}{Z_o+2\cdot Z0} & \frac{2\cdot Z0}{Z_o+2\cdot Z0} \\ -\frac{2\cdot\gamma\cdot Z0}{(Z_o+2\cdot Z0)\cdot(Z_i+2\cdot Z0)} & \frac{2\cdot\gamma\cdot Z0}{(Z_o+2\cdot Z0)\cdot(Z_i+2\cdot Z0)} & \frac{2\cdot Z0}{Z_o+2\cdot Z0} & \frac{Z_o}{Z_o+2\cdot Z0} \end{pmatrix}$$

(c) S-parameters

**Figure 6.24** Four-port transresistance amplifier

### *Four-Port Transresistance Amplifier*

The four-port transresistance amplifier is provided in Figure 6.24, with the circuit symbol shown in Figure 6.24(a).

The amplifier is shown schematically in Figure 6.24(b) as a current controlled voltage source with series impedance $Z_i$ and $Z_o$ at the input and output, respectively. Note that the output ports of the current controlled voltage source are swapped.

The symbolic solution is provided in Figure D.10 in Appendix D, and the simplified, resulting s-parameters are shown in Figure 6.24(c).

The ideal four-port transresistance amplifier has zero input impedance $Z_i$ and zero input impedance $Z_o$:

$$\lim_{\substack{Z_i \to 0 \\ Z_o \to 0}} \mathbf{S} = \begin{pmatrix} 0 & 1 & 0 & 0 \\ 1 & 0 & 0 & 0 \\ \frac{\gamma}{2\cdot Z0} & -\frac{\gamma}{2\cdot Z0} & 0 & 1 \\ -\frac{\gamma}{2\cdot Z0} & \frac{\gamma}{2\cdot Z0} & 1 & 0 \end{pmatrix}.$$

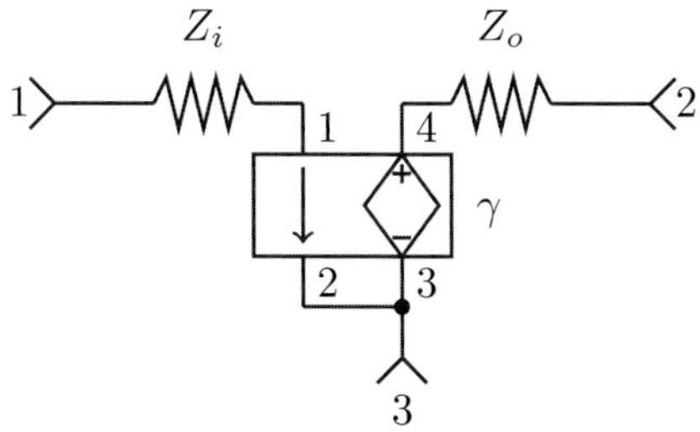


(a) Circuit representation

$$\mathbf{S} = \begin{pmatrix} \frac{Z_o \cdot Z_i + Z0 \cdot (2 \cdot Z_i - \gamma) - Z0^2}{3 \cdot Z0^2 + (2 \cdot Z_o + 2 \cdot Z_i - \gamma) \cdot Z0 + Z_o \cdot Z_i} & \frac{2 \cdot Z0^2}{3 \cdot Z0^2 + (2 \cdot Z_o + 2 \cdot Z_i - \gamma) \cdot Z0 + Z_o \cdot Z_i} & \frac{2 \cdot Z0^2 + 2 \cdot Z_o \cdot Z0}{3 \cdot Z0^2 + (2 \cdot Z_o + 2 \cdot Z_i - \gamma) \cdot Z0 + Z_o \cdot Z_i} \\ \frac{2 \cdot Z0^2 + 2 \cdot \gamma \cdot Z0}{3 \cdot Z0^2 + (2 \cdot Z_o + 2 \cdot Z_i - \gamma) \cdot Z0 + Z_o \cdot Z_i} & \frac{Z_o \cdot Z_i + Z0 \cdot (2 \cdot Z_o - \gamma) - Z0^2}{3 \cdot Z0^2 + (2 \cdot Z_o + 2 \cdot Z_i - \gamma) \cdot Z0 + Z_o \cdot Z_i} & \frac{2 \cdot Z0^2 + Z0 \cdot (2 \cdot Z_i - 2 \cdot \gamma)}{3 \cdot Z0^2 + (2 \cdot Z_o + 2 \cdot Z_i - \gamma) \cdot Z0 + Z_o \cdot Z_i} \\ \frac{2 \cdot Z0^2 + Z0 \cdot (2 \cdot Z_o - 2 \cdot \gamma)}{3 \cdot Z0^2 + (2 \cdot Z_o + 2 \cdot Z_i - \gamma) \cdot Z0 + Z_o \cdot Z_i} & \frac{2 \cdot Z0^2 + 2 \cdot Z_i \cdot Z0}{3 \cdot Z0^2 + (2 \cdot Z_o + 2 \cdot Z_i - \gamma) \cdot Z0 + Z_o \cdot Z_i} & \frac{Z_o \cdot Z_i - Z0^2 + \gamma \cdot Z0}{3 \cdot Z0^2 + (2 \cdot Z_o + 2 \cdot Z_i - \gamma) \cdot Z0 + Z_o \cdot Z_i} \end{pmatrix}$$

(b) S-parameters

**Figure 6.25** Three-port transresistance amplifier

This is, of course, the current controlled voltage source with its ports reordered. Since the ordering is from $(\,1\ 2\ 3\ 4\,)^T$ to $(\,1\ 2\ 4\ 3\,)^T$,

$$\begin{pmatrix} 1 & 0 & 0 & 0 \\ 0 & 1 & 0 & 0 \\ 0 & 0 & 0 & 1 \\ 0 & 0 & 1 & 0 \end{pmatrix} \cdot \begin{pmatrix} 0 & 1 & 0 & 0 \\ 1 & 0 & 0 & 0 \\ \frac{\gamma}{2 \cdot Z0} & -\frac{\gamma}{2 \cdot Z0} & 0 & 1 \\ -\frac{\gamma}{2 \cdot Z0} & \frac{\gamma}{2 \cdot Z0} & 1 & 0 \end{pmatrix} \cdot \begin{pmatrix} 1 & 0 & 0 & 0 \\ 0 & 1 & 0 & 0 \\ 0 & 0 & 0 & 1 \\ 0 & 0 & 1 & 0 \end{pmatrix}^T = \begin{pmatrix} 0 & 1 & 0 & 0 \\ 1 & 0 & 0 & 0 \\ -\frac{\gamma}{2 \cdot Z0} & \frac{\gamma}{2 \cdot Z0} & 0 & 1 \\ \frac{\gamma}{2 \cdot Z0} & -\frac{\gamma}{2 \cdot Z0} & 1 & 0 \end{pmatrix}.$$

This compares favorably with Figure 6.13(c).

### *Three-Port Transresistance Amplifier*

The three-port transresistance amplifier is provided in Figure 6.25 and is shown schematically in Figure 6.25(a). It is formed by shorting ports 2 and 4 of the four-port transresistance amplifier and exposing it as port 3, with the output port being numbered as port 2.

The symbolic solution is provided in Figure D.11 in Appendix D, where the simplified s-parameters are shown in Figure 6.25(b).

Shorting across the input and output impedance results in the ideal three-port transresistance amplifier:

$$\lim_{\substack{Z_i \to 0 \\ Z_o \to 0}} \mathbf{S} = \frac{1}{\gamma - 3 \cdot Z0} \cdot \begin{pmatrix} Z0 + \gamma & -2 \cdot Z0 & -2 \cdot Z0 \\ -2 \cdot Z0 - 2 \cdot \gamma & Z0 + \gamma & 2 \cdot \gamma - 2 \cdot Z0 \\ 2 \cdot \gamma - 2 \cdot Z0 & -2 \cdot Z0 & Z0 - \gamma \end{pmatrix}.$$

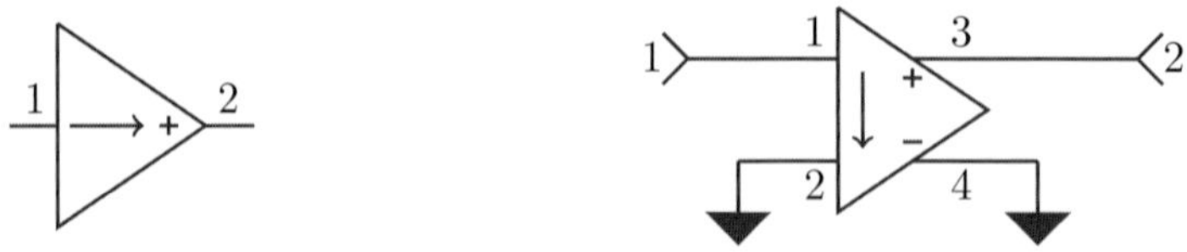


(a) Circuit symbol (b) Circuit representation

$$\mathbf{S} = \begin{pmatrix} \frac{Z_i - Z0}{Z_i + Z0} & 0 \\ \frac{2 \cdot \gamma \cdot Z0}{(Z_i + Z0) \cdot (Z_o + Z0)} & \frac{Z_o - Z0}{Z_o + Z0} \end{pmatrix}$$

(c) S-parameters

**Figure 6.26** Two-port transresistance amplifier

***Two-Port Transresistance Amplifier***

The two-port transresistance amplifier is provided in Figure 6.26, with its circuit symbol shown in Figure 6.26(a). The schematic representation in Figure 6.26(b) shows that it is created by shorting pins 2 and 4 of the of the four-port transresistance amplifier to ground and exposing port 3 as output port 2.

The symbolic solution is provided in Figure D.12 in Appendix D, and the simplified version of this solution is provided as the s-parameters in Figure 6.26(c).

Shorting across the input and output impedance produces the s-parameters of the ideal two-port transresistance amplifier:

$$\lim_{\substack{Z_i \to 0 \\ Z_o \to 0}} \mathbf{S} = \begin{pmatrix} -1 & 0 \\ \frac{2 \cdot \gamma}{Z0} & -1 \end{pmatrix}.$$

### 6.4.5 Operational Amplifier

The operational amplifier, whose symbol is shown in Figure 6.27(a), is formed from the four-port voltage amplifier in §6.4.1 by grounding the minus output, making it single ended and by connecting two input impedances to ground on the inputs (the voltage amplifier has an internal impedance across its inputs). The circuit for this is shown in Figure 6.27(b).

The details of the solution are provided in §D.15 in Appendix D, and the simplified solution is shown in Figure 6.27(c).

When looking at operational amplifiers in the literature, it is unclear how the input impedance is defined. There is an impedance $Z_i$ to ground on each input pin and an impedance $Z_d$ across the two inputs. For operational amplifiers that are specified with an

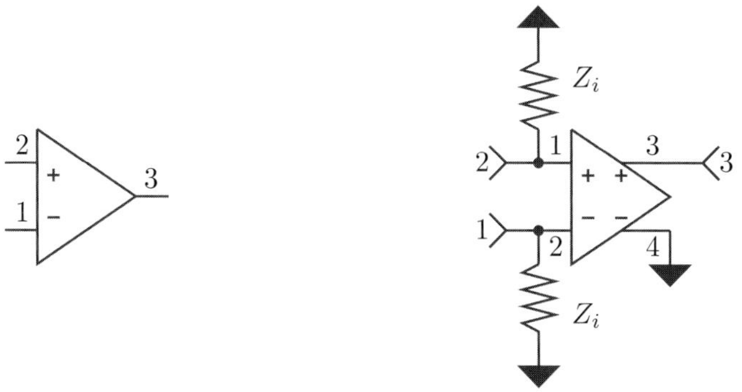


(a) Circuit symbol (b) Circuit representation

$$\mathbf{S} = \begin{pmatrix} \frac{(-2\cdot Z_i - Z_d)\cdot Z0^2 + Z_i^2\cdot Z_d}{(2\cdot Z_i+Z_d)\cdot Z0^2+(2\cdot Z_i\cdot Z_d+2\cdot Z_i^2)\cdot Z0+Z_i^2\cdot Z_d} & \frac{2\cdot Z_i^2\cdot Z0}{(2\cdot Z_i+Z_d)\cdot Z0^2+(2\cdot Z_i\cdot Z_d+2\cdot Z_i^2)\cdot Z0+Z_i^2\cdot Z_d} & 0 \\ \frac{2\cdot Z_i^2\cdot Z0}{(2\cdot Z_i+Z_d)\cdot Z0^2+(2\cdot Z_i\cdot Z_d+2\cdot Z_i^2)\cdot Z0+Z_i^2\cdot Z_d} & \frac{(-2\cdot Z_i - Z_d)\cdot Z0^2 + Z_i^2\cdot Z_d}{(2\cdot Z_i+Z_d)\cdot Z0^2+(2\cdot Z_i\cdot Z_d+2\cdot Z_i^2)\cdot Z0+Z_i^2\cdot Z_d} & 0 \\ -\frac{2\cdot G\cdot Z_i\cdot Z_d\cdot Z0}{(Z_o+Z0)\cdot((2\cdot Z_i+Z_d)\cdot Z0+Z_i\cdot Z_d)} & \frac{2\cdot G\cdot Z_i\cdot Z_d\cdot Z0}{(Z_o+Z0)\cdot((2\cdot Z_i+Z_d)\cdot Z0+Z_i\cdot Z_d)} & \frac{Z_o-Z0}{Z_o+Z0} \end{pmatrix}$$

(c) S-parameters

**Figure 6.27** Operational amplifier

impedance across the two inputs only, the following limit is taken:

$$\lim_{Z_i\to\infty} \mathbf{S} = \begin{pmatrix} \frac{Z_d}{Z_d+2\cdot Z0} & \frac{2\cdot Z0}{Z_d+2\cdot Z0} & 0 \\ \frac{2\cdot Z0}{Z_d+2\cdot Z0} & \frac{Z_d}{Z_d+2\cdot Z0} & 0 \\ -\frac{2\cdot G\cdot Z_d\cdot Z0}{(Z_d+2\cdot Z0)\cdot(Z_o+Z0)} & \frac{2\cdot G\cdot Z_d\cdot Z0}{(Z_d+2\cdot Z0)\cdot(Z_o+Z0)} & \frac{Z_o-Z0}{Z_o+Z0} \end{pmatrix}.$$

For operational amplifiers that are specified with input impedance to ground and no impedance across the input, the following limit is taken:

$$\lim_{Z_d\to\infty} \mathbf{S} = \begin{pmatrix} \frac{Z_i-Z0}{Z_i+Z0} & 0 & 0 \\ 0 & \frac{Z_i-Z0}{Z_i+Z0} & 0 \\ -\frac{2\cdot G\cdot Z_i\cdot Z0}{(Z_i+Z0)\cdot(Z_o+Z0)} & \frac{2\cdot G\cdot Z_i\cdot Z0}{(Z_i+Z0)\cdot(Z_o+Z0)} & \frac{Z_o-Z0}{Z_o+Z0} \end{pmatrix}.$$

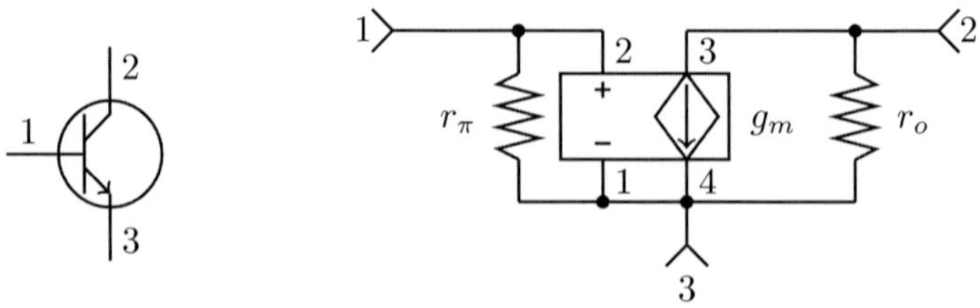


(a) Circuit symbol (b) Circuit representation

$$\mathbf{S} = \frac{1}{3 \cdot Z0^2 + (2 \cdot r_o + 2 \cdot r_\pi + g_m \cdot r_\pi \cdot r_o) \cdot Z0 + r_o \cdot r_\pi} \cdot \begin{pmatrix} r_o{\cdot}r_\pi{+}Z0{\cdot}(2{\cdot}r_\pi{+}g_m{\cdot}r_\pi{\cdot}r_o){-}Z0^2 & 2{\cdot}Z0^2 & 2{\cdot}Z0^2{+}2{\cdot}r_o{\cdot}Z0 \\ 2{\cdot}Z0^2{-}2{\cdot}g_m{\cdot}r_\pi{\cdot}r_o{\cdot}Z0 & r_o{\cdot}r_\pi{+}Z0{\cdot}(2{\cdot}r_o{+}g_m{\cdot}r_\pi{\cdot}r_o){-}Z0^2 & 2{\cdot}Z0^2{+}Z0{\cdot}(2{\cdot}r_\pi{+}2{\cdot}g_m{\cdot}r_\pi{\cdot}r_o) \\ 2{\cdot}Z0^2{+}Z0{\cdot}(2{\cdot}r_o{+}2{\cdot}g_m{\cdot}r_\pi{\cdot}r_o) & 2{\cdot}Z0^2{+}2{\cdot}r_\pi{\cdot}Z0 & r_o{\cdot}r_\pi{-}Z0^2{-}g_m{\cdot}r_\pi{\cdot}r_o{\cdot}Z0 \end{pmatrix}$$

(c) S-parameters

**Figure 6.28** Simple transistor

## 6.5 Transistors

Figure 6.28(a) shows an NPN transistor. The s-parameters of a simplified NPN transistor are calculated by reusing the three-port transconductance amplifier in §6.4.4, as shown in Figure 6.28(b), then substituting $r_\pi$ for $Z_i$ and $r_o$ for $Z_o$. Note that the pins on the transconductance amplifier have been swapped to change the reference direction of the collector current.

The result in Figure 6.28(c) is in closed form, but represents a very simple transistor model. A more complex model is provided in §D.17 in Appendix D.

## 6.6 Ideal Transformer

The ideal transformer shown in Figure 6.29(a) is modeled using dependent sources, specifically current controlled current sources and voltage controlled voltage sources, as shown in Figure 6.29(b).

The only variable for the ideal transformer is the turns ratio $a = N2/N1$, the ratio of the secondary windings $N2$ to the primary windings $N1$.

According to Figure 6.29(a), the primary comprises ports 1 and 2 and the secondary comprises ports 3 and 4. The dot convention has the dots placed on ports 1 and 3.

The ideal transformer has the effect of transforming an impedance at the secondary so it appears different to that at the primary:

$$Z_p = \frac{1}{a^2} \cdot Z_s,$$

where $Z_s$ is an impedance across the secondary and $Z_p$ is the effective impedance seen across the primary.

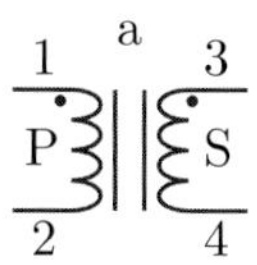


(a) Circuit symbol

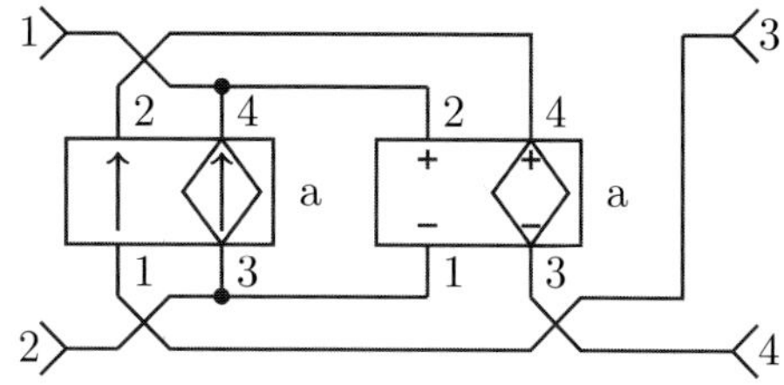


(b) Circuit representation

$$\boldsymbol{S} = \tfrac{1}{a^2+1} \cdot \begin{pmatrix} 1 & a^2 & a & -a \\ a^2 & 1 & -a & a \\ a & -a & a^2 & 1 \\ -a & a & 1 & a^2 \end{pmatrix}$$

(c) S-parameters

**Figure 6.29** Ideal transformer

The detailed calculation of the ideal transformer is provided in Figure D.16 of Appendix D, and the resulting s-parameters are shown in Figure 6.29(c).

An ideal transformer with unity turns ratio is

$$\lim_{a\to 1} \mathbf{S} = \frac{1}{2} \cdot \begin{pmatrix} 1 & 1 & 1 & -1 \\ 1 & 1 & -1 & 1 \\ 1 & -1 & 1 & 1 \\ -1 & 1 & 1 & 1 \end{pmatrix}.$$

# 7

# Transmission Lines

TRANSMISSION LINES FORM THE BASIS FOR SIGNAL INTEGRITY. After all, the job of signal integrity is to move signals from one place to another, and the channel through which the signals flow are transmission lines of some sort. To understand them fully requires a deep knowledge and understanding of electromagnetics. Most electrical engineers abstract the concepts of electromagnetics through the use of circuit elements. The transmission line defies circuit interpretation because the electromagnetic characteristics are distributed. This distributed nature leads naturally to network parameters in lieu of circuits. Finally, because transmission lines are best viewed as propagators of waves, as opposed to circuit elements with characteristics based on voltage and current, they lend themselves naturally to s-parameters. One can even say that the transmission line is the main s-parameter element, much like impedance is the main circuit element.

## 7.1 The Transmission Line Model

In circuit analysis, the basic circuit elements are the resistor, the inductor, and the capacitor. These elements are called *lumped elements* because they are lumps, or point elements, having no length or time associated with them. They obey Kirchhoff's laws such that the current into one port equals the current out at all times, and the equations that govern their voltages and currents are instantaneous at each port of the device.

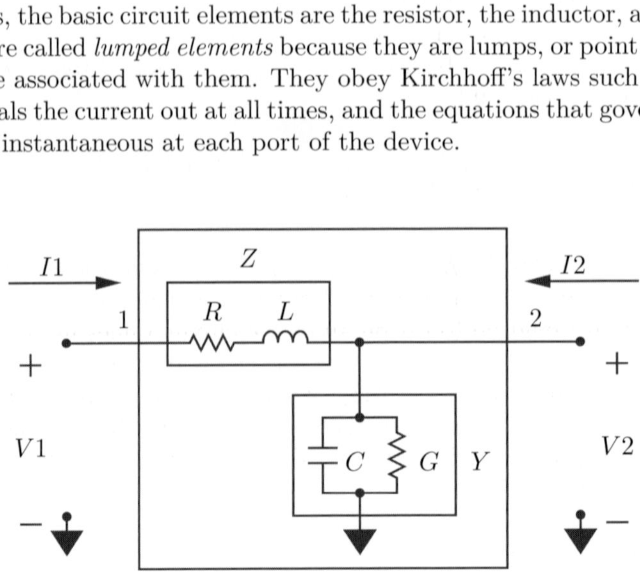


**Figure 7.1** Circuit representing telegrapher's equations

In signal integrity, the most basic element is the transmission line. A transmission line has *electrical length*,[1] which, despite the wording, implies that it has time associated with it. Because of the length associated with a transmission line, the current going into a port does not necessarily equal the current coming out. It will be seen that the transmission line tends to have internal ground connections; this might lead one to conclude that this causes the imbalances in the current, but it is, in fact, the distributed nature of the transmission line that causes this effect.

A single-ended two-port transmission line can be described and analyzed using the transmission line model used to develop the telegrapher's equations. This model is shown in Figure 7.1. A derivation of the telegrapher's equations is provided in Appendix B, §B.1.

The transmission line model consists of a series impedance $Z$, internally consisting of a series resistance $R$ and inductance $L$, followed by a shunt admittance $Y$, internally consisting of a conductance $G$ and a capacitance $C$. There are many physical reasons for this circuit representation that are not discussed here.

While the transmission line model shows lumped elements, it is intended to represent *distributed* values of inductance, resistance, conductance, and capacitance, as occur in a transmission line. As such, the circuit shown in Figure 7.1 can be thought of as an approximation of a small section of line. The first task here will be to convert this approximation using lumped elements into a network model of a transmission line with distributed elements.

The work begins with the network parameters of the section shown. Here, ABCD parameters are used as presented in §1.1.3. According to §1.5, ABCD parameters are cascadable, so this feature is used to form the ABCD parameters as a cascade of two devices: a series impedance and a shunt admittance.

The series impedance is obtained from (1.4), the shunt admittance is obtained from (1.6) (making the appropriate substitution for admittance), and the ABCD parameters of the series/shunt combination are

$$\mathbf{A}_{zy} = \begin{pmatrix} 1 & -Z \\ 0 & 1 \end{pmatrix} \cdot \begin{pmatrix} 1 & 0 \\ -Y & 1 \end{pmatrix} = \begin{pmatrix} 1+Z\cdot Y & -Z \\ -Y & 1 \end{pmatrix}. \tag{7.1}$$

Therefore, (7.1) provides the ABCD network parameters for the circuit shown in Figure 7.1. As mentioned, Figure 7.1 and now (7.1) are approximations of a transmission line. A better approximation would be to divide all of the circuit values in half and cascade two sections. An even better approximation would be to divide the circuit values by 1000 and cascade 1000 such sections. The exact representation of the transmission line is found by cascading an infinite number of infinitesimally small sections. If the values $R$, $L$, $G$, and $C$ represent the total resistance, inductance, conductance, and capacitance of a transmission line, respectively, and a value $K$ represents the number of sections that are desired (i.e. cascaded), then $Z = \Delta Z \cdot K$ and $Y = \Delta Y \cdot K$, where $\Delta Z = Z/K$ and $\Delta Y/K$ now represent the impedance and admittance in a small section. The ABCD parameters of this small

[1]In microwaves, electrical length refers to phase at a particular frequency, but in signal integrity, one prefers to think of electrical length as the propagation time.

section are

$$\begin{pmatrix} \Delta A & \Delta B \\ \Delta C & \Delta D \end{pmatrix} = \begin{pmatrix} 1 + \Delta Z \cdot \Delta Y & -\Delta Z \\ -\Delta Y & 1 \end{pmatrix} = \begin{pmatrix} 1 + \frac{Z \cdot Y}{K^2} & -\frac{Z}{K} \\ -\frac{Y}{K} & 1 \end{pmatrix}. \tag{7.2}$$

The ABCD parameters of the line can be represented by distributed impedance and admittance as the limit of cascading $K$ sections of line each having $1/K$ times the impedance and admittance in the limit as $K \to \infty$:

$$\begin{pmatrix} A & B \\ C & D \end{pmatrix}_{distributed} = \lim_{K \to \infty} \begin{pmatrix} \Delta A & \Delta B \\ \Delta C & \Delta D \end{pmatrix}^K = \lim_{K \to \infty} \begin{pmatrix} 1 + \frac{Z \cdot Y}{K^2} & -\frac{Z}{K} \\ -\frac{Y}{K} & 1 \end{pmatrix}^K. \tag{7.3}$$

Equation (7.3) is the differential element implied by the telegrapher's equations.[2] In (7.3) the cascading property of ABCD parameters is what enables the description of $K$ cascaded sections as the ABCD parameters of one section raised to the power $K$. Fortunately, one does not have to actually cascade an infinite number of sections, but instead can make use of a matrix property for computing the power of a matrix. This property utilizes eigenvalues and eigenvectors. The definition of an eigenvalue $\lambda$ and a corresponding eigenvector $\mathbf{v}$ of a matrix $\mathbf{A}$ is that

$$\mathbf{A} \cdot \mathbf{v} = \lambda \cdot \mathbf{v}. \tag{7.4}$$

This implies that, if all of the eigenvalues were placed along the diagonal of a matrix $\mathbf{\Lambda}$ and all of the eigenvectors were placed in a matrix $\mathbf{V}$ (i.e. each column of $\mathbf{V}$ is an eigenvector corresponding to the eigenvalue in that same row and column of $\mathbf{\Lambda}$), then

$$\mathbf{A} \cdot \mathbf{V} = \mathbf{V} \cdot \mathbf{\Lambda}$$

or, equivalently,

$$\mathbf{A} = \mathbf{V} \cdot \mathbf{\Lambda} \cdot \mathbf{V}^{-1}.$$

Therefore,

$$\mathbf{A}^2 = \mathbf{A} \cdot \mathbf{A} = \mathbf{V} \cdot \mathbf{\Lambda} \cdot \mathbf{V}^{-1} \cdot \mathbf{V} \cdot \mathbf{\Lambda} \cdot \mathbf{V}^{-1} = \mathbf{V} \cdot \mathbf{\Lambda}^2 \cdot \mathbf{V}^{-1}.$$

The power of a matrix can be calculated as

$$\mathbf{A}^P = \mathbf{V} \cdot \mathbf{\Lambda}^P \cdot \mathbf{V}^{-1}.$$

The plan, therefore, is to compute the two eigenvalues $\lambda_1$ and $\lambda_2$ and eigenvectors $\mathbf{v_1} = \left(\begin{smallmatrix} v_{11} \\ v_{21} \end{smallmatrix}\right)$ and $\mathbf{v_2} = \left(\begin{smallmatrix} v_{12} \\ v_{22} \end{smallmatrix}\right)$ of $\left(\begin{smallmatrix} \Delta A & \Delta B \\ \Delta C & \Delta D \end{smallmatrix}\right)$ and then compute the ABCD parameters of the distributed transmission line as

$$\begin{aligned} \begin{pmatrix} A & B \\ C & D \end{pmatrix}_{distributed} &= \lim_{K \to \infty} \begin{pmatrix} \Delta A & \Delta B \\ \Delta C & \Delta D \end{pmatrix}^K \\ &= \lim_{K \to \infty} \left[ \begin{pmatrix} v_{11} & v_{12} \\ v_{21} & v_{22} \end{pmatrix} \cdot \begin{pmatrix} \lambda_1^K & 0 \\ 0 & \lambda_2^K \end{pmatrix} \cdot \begin{pmatrix} v_{11} & v_{12} \\ v_{21} & v_{22} \end{pmatrix}^{-1} \right]. \end{aligned} \tag{7.5}$$

[2] This is shown in detail in Appendix B, §B.2.

The first step is to find the eigenvalues. Equation (7.4) implies that $(\mathbf{A} - \mathbb{I} \cdot \lambda) \cdot \mathbf{v} = 0$. Thus, for non-zero $\mathbf{v}$ there is the requirement

$$|\mathbf{A} - \mathbb{I} \cdot \lambda| = 0$$

or

$$\left| \begin{pmatrix} \Delta A - \lambda & \Delta B \\ \Delta C & \Delta D - \lambda \end{pmatrix} \right| = 0.$$

Therefore, each eigenvalue is the solution to

$$\lambda^2 - (\Delta A + \Delta D) \cdot \lambda + \Delta A \cdot \Delta D - \Delta B \cdot \Delta C = 0,$$

which can be rewritten, recognizing that the trace of the matrix (the sum of the diagonal elements) is given by

$$t = \Delta A + \Delta D \tag{7.6}$$

and the determinant of the matrix is

$$\delta = \Delta A \cdot \Delta D - \Delta B \cdot \Delta C, \tag{7.7}$$

and therefore,

$$\lambda^2 - t \cdot \lambda + \delta = 0. \tag{7.8}$$

The eigenvalues are therefore expressed as the two solutions to (7.8):

$$\lambda_1, \lambda_2 = \frac{t}{2} \pm \frac{1}{2}\sqrt{t^2 - 4 \cdot \delta}. \tag{7.9}$$

Regarding the eigenvectors, for each eigenvalue $\lambda$ and eigenvector $\mathbf{v}$,

$$\begin{pmatrix} \Delta A - \lambda & \Delta B \\ \Delta C & \Delta D - \lambda \end{pmatrix} \cdot \begin{pmatrix} v_1 \\ v_2 \end{pmatrix} = \begin{pmatrix} 0 \\ 0 \end{pmatrix}.$$

Using $A - \lambda$ as the pivot element:[3]

$$\begin{pmatrix} 1 & 0 \\ -(\Delta A - \lambda)^{-1} \cdot \Delta C & 1 \end{pmatrix} \cdot \begin{pmatrix} \Delta A - \lambda & \Delta B \\ \Delta C & \Delta D - \lambda \end{pmatrix} \cdot \begin{pmatrix} v_1 \\ v_2 \end{pmatrix} = \begin{pmatrix} 1 & 0 \\ -(\Delta A - \lambda)^{-1} \cdot \Delta C & 1 \end{pmatrix} \cdot \begin{pmatrix} 0 \\ 0 \end{pmatrix},$$

the following is obtained:

$$\begin{pmatrix} \Delta A - \lambda & \Delta B \\ 0 & \frac{\lambda^2 - t \cdot \lambda + \delta}{\Delta A - \lambda} \end{pmatrix} \cdot \begin{pmatrix} v_1 \\ v_2 \end{pmatrix} = \begin{pmatrix} 0 \\ 0 \end{pmatrix}.$$

The lower right element of the matrix is zero when an eigenvalue is substituted for $\lambda$, so

$$\begin{pmatrix} \Delta A - \lambda & \Delta B \\ 0 & 0 \end{pmatrix} \cdot \begin{pmatrix} v_1 \\ v_2 \end{pmatrix} = \begin{pmatrix} 0 \\ 0 \end{pmatrix}.$$

[3] There is no good justification for using $A - \lambda$ as the pivot element because it is not known whether it might or might not be non-zero in this symbolic calculation, but things do turn out okay in the end. Things might not have and another pivot might have been necessary.

Therefore, the eigenvector for a given eigenvalue $\begin{pmatrix} v_1 \\ \frac{\lambda-\Delta A}{\Delta B}\cdot v_1 \end{pmatrix}$, allowing $v_1$ to be arbitrarily chosen as 1, is

$$\mathbf{v} = \begin{pmatrix} 1 \\ \frac{\lambda-\Delta A}{\Delta B} \end{pmatrix},$$

and (7.5) can now be written as

$$\begin{aligned}\begin{pmatrix} A & B \\ C & D \end{pmatrix}_{distributed} &= \lim_{K\to\infty} \begin{pmatrix} \Delta A & \Delta B \\ \Delta C & \Delta D \end{pmatrix}^K \\ &= \lim_{K\to\infty} \left[ \begin{pmatrix} 1 & 1 \\ \frac{\lambda_1-\Delta A}{\Delta B} & \frac{\lambda_2-\Delta A}{\Delta B} \end{pmatrix} \cdot \begin{pmatrix} \lambda_1^K & 0 \\ 0 & \lambda_2^K \end{pmatrix} \cdot \begin{pmatrix} 1 & 1 \\ \frac{\lambda_1-\Delta A}{\Delta B} & \frac{\lambda_2-\Delta A}{\Delta B} \end{pmatrix}^{-1} \right].\end{aligned}$$

The limit is computed by first substituting (7.2) into (7.6) and (7.7) and the results into (7.9):

$$\lambda_1, \lambda_2 = \frac{t}{2} \pm \frac{1}{2}\sqrt{t^2 - 4\cdot\delta} = \frac{1}{2}\left(2 + \frac{Z\cdot Y}{K^2}\right) \pm \frac{1}{2}\cdot\sqrt{\left(2+\frac{Z\cdot Y}{K^2}\right)^2 - 4}.$$

Computing the limit of each eigenvalue raised to the power $K$ yields

$$\lim_{K\to\infty} \lambda_1^K, \lambda_2^K = \lim_{K\to\infty} \left[ \frac{1}{2}\cdot\left(2+\frac{Z\cdot Y}{K^2}\right) \pm \frac{1}{2}\cdot\sqrt{\left(2+\frac{Z\cdot Y}{K^2}\right)^2 - 4} \right]^K = e^{\pm\sqrt{Z\cdot Y}}.$$

Computing the limit of the eigenvector elements yields

$$\lim_{K\to\infty} \frac{\lambda_1, \lambda_2 - \Delta A}{\Delta B} = \lim_{K\to\infty} \left[ \frac{\frac{1}{2}\cdot(2+\frac{Z\cdot Y}{K^2}) \pm \frac{1}{2}\sqrt{(2+\frac{Z\cdot Y}{K^2})^2 - 4} - (1+\frac{Z\cdot Y}{K^2})}{\frac{Z}{K}} \right]^K = \mp\frac{1}{\sqrt{\frac{Z}{Y}}}.$$

Thus, the ABCD parameters of the distributed transmission line are

$$\begin{pmatrix} A & B \\ C & D \end{pmatrix}_{distributed} = \begin{pmatrix} 1 & 1 \\ -\frac{1}{\sqrt{\frac{Z}{Y}}} & \frac{1}{\sqrt{\frac{Z}{Y}}} \end{pmatrix} \cdot \begin{pmatrix} e^{\sqrt{Z\cdot Y}} & 0 \\ 0 & e^{-\sqrt{Z\cdot Y}} \end{pmatrix} \cdot \begin{pmatrix} 1 & 1 \\ -\frac{1}{\sqrt{\frac{Z}{Y}}} & \frac{1}{\sqrt{\frac{Z}{Y}}} \end{pmatrix}^{-1}.$$

To simplify things (and for further insight), the characteristics of a transmission line are defined as

$$\gamma = \sqrt{Z\cdot Y} \tag{7.10}$$

and

$$Z_c = \sqrt{Z/Y}, \tag{7.11}$$

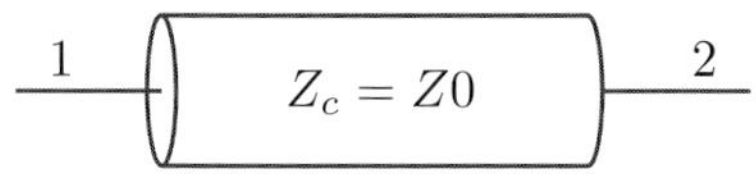


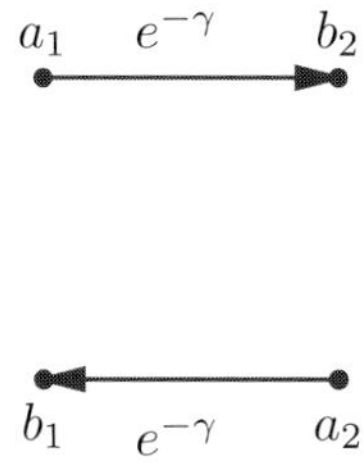


(a) Circuit representation (b) Signal-flow diagram representation

**Figure 7.2** Transmission line in reference impedance of $Z_c$

where $\gamma$ is the *propagation constant* and $Z_c$ is the *characteristic impedance* of the line:

$$\begin{pmatrix} A & B \\ C & D \end{pmatrix}_{distributed} = \begin{pmatrix} 1 & 1 \\ -\frac{1}{Z_c} & \frac{1}{Z_c} \end{pmatrix} \cdot \begin{pmatrix} e^{\gamma} & 0 \\ 0 & e^{-\gamma} \end{pmatrix} \cdot \begin{pmatrix} 1 & 1 \\ -\frac{1}{Z_c} & \frac{1}{Z_c} \end{pmatrix}^{-1}$$
$$= \begin{pmatrix} \frac{1}{2} \cdot (e^{\gamma} + e^{-\gamma}) & \frac{1}{2} \cdot Z_c \cdot (e^{-\gamma} - e^{\gamma}) \\ \frac{1}{2 \cdot Z_c} \cdot (e^{-\gamma} - e^{\gamma}) & \frac{1}{2} \cdot (e^{\gamma} + e^{-\gamma}) \end{pmatrix}. \tag{7.12}$$

The reflection coefficient $\rho$ is defined according to

$$\rho = \frac{Z_c - Z0}{Z_c + Z0}, \tag{7.13}$$

and therefore, solving for $Z_c$:

$$Z_c = -Z0 \cdot \frac{\rho + 1}{\rho - 1}.$$

Substituting for $Z_c$ in (7.12) and applying the conversion from ABCD parameters to s-parameters in (3.20) results in

$$\mathbf{S}_{distributed} = \begin{pmatrix} \frac{\rho \cdot \left(1 - e^{-2 \cdot \gamma}\right)}{1 - \rho^2 \cdot e^{-2 \cdot \gamma}} & \frac{\left(1 - \rho^2\right) \cdot e^{-\gamma}}{1 - \rho^2 \cdot e^{-2 \cdot \gamma}} \\ \frac{\left(1 - \rho^2\right) \cdot e^{-\gamma}}{1 - \rho^2 \cdot e^{-2 \cdot \gamma}} & \frac{\rho \cdot \left(1 - e^{-2 \cdot \gamma}\right)}{1 - \rho^2 \cdot e^{-2 \cdot \gamma}} \end{pmatrix}. \tag{7.14}$$

To understand (7.14), an important observation is made that is at the heart of the concept of s-parameters and reference impedance: that when $\rho = 0$, meaning that, from (7.13), $Z_c = Z0$, the transmission line s-parameters become

$$\mathbf{S}_{tline}\big|_{Z0 = Z_c} = \begin{pmatrix} 0 & e^{-\gamma} \\ e^{-\gamma} & 0 \end{pmatrix}. \tag{7.15}$$

Therefore, (7.15) defines the s-parameters of a transmission line when the reference impedance $Z0$ is the characteristic impedance of the transmission line $Z_c$, as shown in Figure 7.2. The transmission line symbol in Figure 7.2(a) shows that $Z_c = Z0$. The signal-flow representation in Figure 7.2(b) shows that, for the reference impedance chosen equal to the characteristic impedance of a transmission line, there is no wave reflected from the ports when driven in the reference impedance, and the transmission line serves to carry the incident wave at a port to the other port with weight $e^{-\gamma}$.

Using (5.13), an impedance transformation from $Z_c$ back to the reference impedance $Z0$ is made:[4]

$$\begin{aligned}\mathbf{S}_{tline} &= \left(\mathbf{S}_{tline}|_{Z0=Z_c} - \mathbb{I}\cdot\rho\right)\cdot\left(\mathbb{I} - \rho\cdot\mathbf{S}_{tline}|_{Z0=Z_c}\right)^{-1} \\ &= \left[\begin{pmatrix} 0 & e^{-\gamma} \\ e^{-\gamma} & 0 \end{pmatrix} + \begin{pmatrix} 1 & 0 \\ 0 & 1 \end{pmatrix}\cdot\rho\right]\cdot\left[\begin{pmatrix} 1 & 0 \\ 0 & 1 \end{pmatrix} + \begin{pmatrix} 0 & e^{-\gamma} \\ e^{-\gamma} & 0 \end{pmatrix}\cdot\rho\right]^{-1} \\ &= \begin{pmatrix} \frac{\rho\cdot\left(1-e^{-2\cdot\gamma}\right)}{1-\rho^2\cdot e^{-2\cdot\gamma}} & \frac{\left(1-\rho^2\right)\cdot e^{-\gamma}}{1-\rho^2\cdot e^{-2\cdot\gamma}} \\ \frac{\left(1-\rho^2\right)\cdot e^{-\gamma}}{1-\rho^2\cdot e^{-2\cdot\gamma}} & \frac{\rho\cdot\left(1-e^{-2\cdot\gamma}\right)}{1-\rho^2\cdot e^{-2\cdot\gamma}} \end{pmatrix}. \end{aligned} \tag{7.16}$$

Thus, (7.16) produces the same result as (7.14), and now it is understood that the complicated looking equation for the transmission line is due to the change in reference impedance.

Revisiting (7.10), for the resistance, inductance, conductance, and capacitance of the line:

$$\gamma = \sqrt{Z\cdot Y} = \sqrt{(R + j\cdot\omega\cdot L)\cdot(G + j\cdot\omega\cdot C)}.$$

When the line consists of only resistance and conductance, then $\gamma = \sqrt{R\cdot G}$ and is entirely real valued. Under this circumstance, the line has loss, but, most interestingly, has no phase shift or time delay: the wave incident on one port arrives immediately at the other port, albeit attenuated. A more important situation is when the line consists of only inductance and capacitance; then $\gamma = j\cdot\omega\cdot\sqrt{L\cdot C}$. Under this circumstance, the line has no loss and has constant group delay: the wave incident on one port arrives at the other port after the time $T_d = \sqrt{L\cdot C}$. The distributed inductance and capacitance result in a time delay only. Furthermore, for lines consisting of only inductance and capacitance, it is seen from (7.11) that the characteristic impedance of the line is the completely real-valued $Z_c = \sqrt{L/C}$. Actual transmission lines contain a mixture of real $\gamma$, producing loss due to the resistance of the conductor and dielectric losses, and imaginary $\gamma$, producing delay.

Returning to (7.15), if a reference impedance transformation from $Z_c$ to some arbitrary reference impedance $Z0$ were performed, then using the definition of $\rho$ in (7.13) and employing (5.13) results in the s-parameters in (7.14). A more interesting way to perform the

[4] In (5.13), the formula for reference impedance transformation provides $\mathbf{S}' = (\mathbf{S} - \mathbb{I}\cdot\rho)\cdot(\mathbb{I} - \rho\cdot\mathbf{S})^{-1}$, but in (7.16) there are plus signs. This is because the definition of $\rho$ used for the impedance transformation is $\boldsymbol{\rho} = (\mathbf{Z0}' - \mathbf{Z0})\cdot(\mathbf{Z0}' + \mathbf{Z0})^{-1}$, with $\mathbf{Z0}'$ being the new reference impedance ($Z0$ in this case) and $\mathbf{Z0}$ being the original reference impedance ($Z_c$ in this case). Applying this definition for the transformation results in a value of $\rho$ which is the negative of the $\rho$ defined in (7.13).

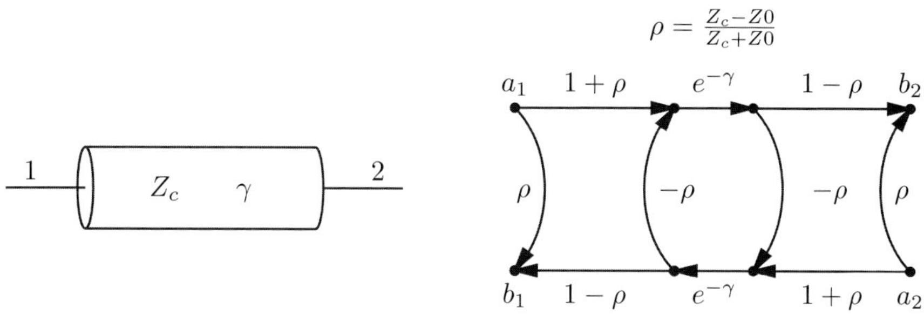


(a) Circuit representation

(b) Signal-flow diagram representation

**Figure 7.3** Transmission line

impedance transformation would be to cascade the network in (7.15) with a device that will transform the reference impedance. The s-parameters of a wire with arbitrary port impedances were found in (3.11), which is repeated below:

$$\mathbf{S}_{wire} = \frac{1}{Z0_1 + Z0_2} \cdot \begin{pmatrix} Z0_2 - Z0_1 & 2 \cdot Z0_1 \\ 2 \cdot Z0_2 & Z0_1 - Z0_2 \end{pmatrix}.$$

Let $\rho = (Z0_2 - Z0_1) / (Z0_2 + Z0_1)$ and realize that one can now write the s-parameters of a wire with a port reference impedance of $Z0_2$ on one port and a reference impedance of $Z0_1$ on the other port as

$$\mathbf{S}_{Z0_1 \rightarrow Z0_2} = \begin{pmatrix} \rho & 1-\rho \\ 1+\rho & -\rho \end{pmatrix}.$$

This is the reference impedance transformer that was discussed in Chapter 5 and shown in Figure 5.2. A transmission line is essentially a set of bidirectional arrows that transmit waves from one port to another while in the characteristic impedance of the line, with reference impedance transformers on the ends to adapt to the reference impedance of the system.

The circuit symbol for a transmission line is shown in Figure 7.3(a) and the best looking signal-flow diagram representation is given in Figure 7.3(b), exposing an inner portion that transmits waves with no interaction with impedance transformers attached, accounting for any mismatch between the characteristic impedance and the reference impedance.

### 7.1.1 Realistic Transmission Lines

While the resistance, inductance, conductance, capacitance (RLGC) transmission line model was used to produce equations for the distributed transmission line, there are two additional physical effects (circuit elements) that can be added to complete the model. Although this book will not go into detail on the physical nature of these additions, they will be used in the models.

**Listing 7.1** Two-port transmission line

```
def TLineTwoPort(Zc,gamma,Z0=50.):
    p=(Zc-Z0)/(Zc+Z0)
    L=cmath.exp(-gamma)
    S1=(p*(1.-L*L))/(1.-p*p*L*L)
    S2=((1.-p*p)*L)/(1.-p*p*L*L)
    return [[S1,S2],[S2,S1]]
```

**Listing 7.2** Lossless two-port transmission line

```
def TLineTwoPortLossless(Zc,Td,f,Z0=50.):
    return TLineTwoPort(Zc,1j*2.*math.pi*f*Td,Z0)
```

These additions account for:

- skin effect – the frequency dependent change in resistance due to current crowding on the edge of conductors;
- dissipation factor – also called loss tangent – an effect related to capacitors that causes them to dissipate real power.

Skin effect is modeled as a resistance defined in units of $\Omega/\sqrt{\text{Hz}}$ – a resistance that is proportional to the square root of the frequency. This resistance is put in series with the direct current (DC) resistance in the RLGC model.

The dissipation factor relates to the imaginary component of the capacitance. An ideal capacitor has frequency dependent susceptance $B = 2\pi \cdot f \cdot C$, which leads to a purely imaginary admittance. The dissipation factor $\tan\delta$ adds a frequency dependent conductance of $G = 2\pi \cdot f \cdot C \cdot \tan\delta$ such that the admittance of a capacitor is expressed as

$$Y = G + j \cdot B = 2\pi \cdot f \cdot C \cdot (j + \tan\delta) .$$

The dissipation factor is the same as the loss tangent,[5] which is written as $\tan\delta$ because there is an angle $\delta$ from the purely imaginary admittance. A $\delta = 0$ results in $\tan\delta = 0$, resulting in no real power dissipation in a capacitor. In the Python code, the effective series resistance (ESR) of a capacitor is also added, which is a constant resistance (i.e. it is not frequency dependent).

### 7.1.2 Software Transmission Line Models

Several models of the single-ended transmission line have been derived in this chapter and are available. These models can be separated into two code classes. The first class includes devices that provide a single s-parameter matrix for a given frequency. The second class provides a complete set of s-parameters over all desired frequencies. In Listing 7.1 and Listing 7.2 two devices are provided. Listing 7.1 provides the s-parameters at a single

[5] While these are the same effects, the dissipation factor usually refers to capacitors and applies to lower frequencies, whereas the loss tangent usually refers to dielectric properties in substrates and applies to higher frequencies.

**Listing 7.3** Lossless transmission line

```
class TLineLossless(SParameters):
    def __init__(self,f,P,Zc,Td,Z0=50.):
        self.m_Zc=Zc;    self.m_Td=Td;    self.m_P=P
        SParameters.__init__(self,f,None,Z0)
    def __getitem__(self,n):
        if self.m_P==2: return dev.TLineTwoPortLossless(
                self.m_Zc,self.m_Td,self.m_f[n],self.m_Z0)
        elif self.m_P==4: return dev.TLineFourPortLossless(
                self.m_Zc,self.m_Td,self.m_f[n],self.m_Z0)
```

**Listing 7.4** Two-port transmission line model

```
class TLineTwoPortRLGC(SParameters):
    def __init__(self,f,R,Rse,L,G,C,df,Z0=50.,K=0):
        if K==0: self.sp=TLineTwoPortRLGCAnalytic(f,R,Rse,L,G,C,df,Z0)
        else: self.sp=TLineTwoPortRLGCApproximate(f,R,Rse,L,G,C,df,Z0,K)
        SParameters.__init__(self,f,None,Z0)
    def __getitem__(self,n):
        return self.sp[n]
```

**Listing 7.5** Two-port transmission line approximation

```
class TLineTwoPortRLGCApproximate(SParameters):
    def __init__(self,f, R, Rse, L, G, C, df, Z0=50., K=0):
        if K==0:
            Td=math.sqrt(L*C)
            Rt=0.45/f[-1] # fastest risetime
            K=int(math.ceil(Td*2/(Rt*self.rtFraction)))

        self.m_K=K
        sdp=SystemDescriptionParser().AddLines(['device R 2','device Rse 2',
        'device L 2','device C 1','device G 1','connect R 2 Rse 1',
        'connect Rse 2 L 1','connect L 2 G 1 C 1','port 1 R 1 2 G 1'])
        self.m_sspn=SystemSParametersNumeric(sdp.SystemDescription())
        self.m_sspn.AssignSParameters('R',SeriesZ(R/K,Z0))
        self.m_sspn.AssignSParameters('G',TerminationG(G/K,Z0))
        self.m_spdl=[('Rse',SeriesRse(f,Rse/K,Z0)),
                     ('L',SeriesL(f,L/K,Z0)),
                     ('C',TerminationC(f,C/K,Z0,df))]
        SParameters.__init__(self,f,None,Z0)
    def __getitem__(self,n):
        for ds in self.m_spdl: self.m_sspn.AssignSParameters(ds[0],ds[1][n])
        sp=self.m_sspn.SParameters()
        return T2S(linalg.matrix_power(S2T(sp),self.m_K))
```

frequency with respect to a characteristic impedance $Z_c$ and $\gamma$, where $\rho$ is defined according to (7.13) and the remaining transmission line is defined according to (7.14). Given an impedance $Z$ and an admittance $Y$ at a given frequency, one can use (7.10) and (7.11) along with (7.14) to provide the s-parameter matrix.

Listing 7.2 provides the ideal lossless model in which an electrical length (time delay) $T_d$ along with the characteristic impedance $Z_c$ and frequency $f$ are specified.

Usually, the complete s-parameters of the transmission line are desired. Four models are provided for this purpose in Listing 7.3, Listing 7.4, Listing 7.5, and Listing 7.6. All of

**Listing 7.6** Two-port analytic transmission line model

```
class TLineTwoPortRLGCAnalytic(SParameters):
    def __init__(self,f, R, Rse, L, G, C, df, Z0=50.):
        self.R=R;    self.Rse=Rse; self.L=L
        self.G=G;    self.C=C;     self.df=df
        SParameters.__init__(self,f,None,Z0)
    def __getitem__(self,n):
        f=self.m_f[n]
        Z=self.R+self.Rse*math.sqrt(f)+1j*2*math.pi*f*self.L
        Y=self.G+2.*math.pi*f*self.C*(1j+self.df)
        try: Zc=cmath.sqrt(Z/Y)
        except: Zc=self.m_Z0
        gamma=cmath.sqrt(Z*Y)
        return TLineTwoPort(Zc,gamma,self.m_Z0)
```

these derive from the **SParameters** class in Listing 3.16 and utilize operator overloading of the `__getitem__()` function to allow for the mimicking of a complete stored s-parameter set, whereas in fact the s-parameters are calculated when accessed.

Listing 7.3 provides the lossless transmission line that employs the lossless device previously described in Listing 7.2. There is a two-port version and a four-port version. The four-port version is described in §7.1.3.

Listing 7.4 provides the two-port transmission line model that is specified by total constant resistance $R$, skin-effect resistance $R_{se}$, inductance $L$, conductance $G$, capacitance $C$, and dissipation factor (or loss tangent) $df$. The number of sections, denoted $K$, is specified for any approximation and defaults to zero. If the number of sections is specified as zero, Listing 7.6 is used, which utilizes the device previously described in Listing 7.1 and a completely analytic model is used. Otherwise, the approximation provided in Listing 7.5 is used. The approximation is not really needed, but is utilized for testing and educational purposes; it provides a good example of how to form complicated s-parameters using the techniques and classes described in Chapter 8. A complete netlist is assembled in Listing 7.5 for the transmission line built up from basic elements, and this circuit is solved on each s-parameter element access. Although it is not necessary to use this approximate model, it is analogous to the differential transmission line model provided in Figure 7.9(b), which is required under certain circumstances.

In Listing 7.5, the frequency independent variables are assigned during the construction of the circuit and a list of the frequency dependent s-parameters is assembled. The frequency dependent s-parameters are assigned prior to circuit solution on each element access.

### 7.1.3 The Four-Port Transmission Line Model

One oddity found in some circuit simulators is the four-port transmission line, whose symbol is shown in Figure 7.4(a). This element looks like a transmission line whose ground connections are exposed. In fact, if ports 3 and 4 in Figure 7.4(a) are grounded, this device functions exactly like a two-port transmission line. Unfortunately, the symbol itself is very misleading and the actual underlying model is shown in Figure 7.4(b), where two ideal transformers are shown on each side of the circuit with a single-ended transmission line in between.

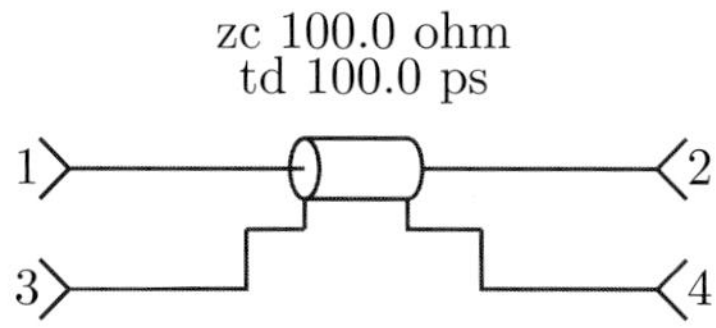


(a) Circuit symbol

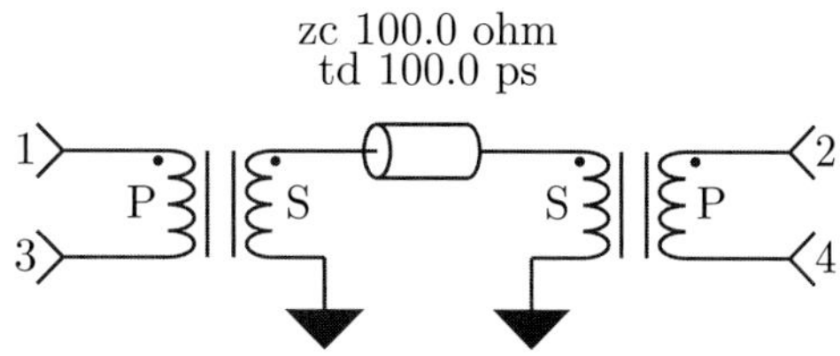


(b) Ideal transformer representation

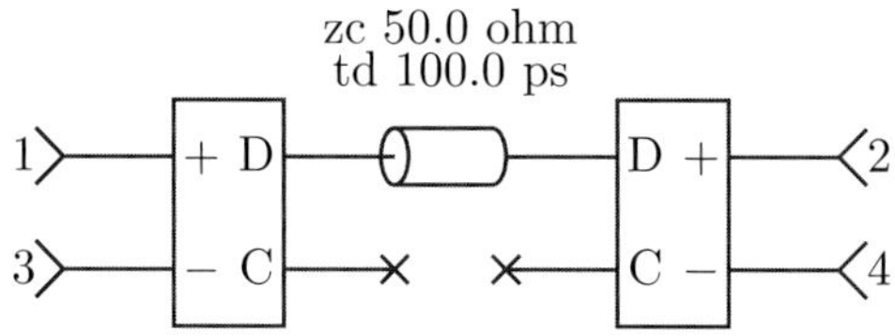


(c) Mixed-mode converter representation

```python
def TLineFourPort(Zc,gamma,Z0=50.):
    p=(Zc-Z0)/(Zc+Z0)
    a=(1.-3.*p)/(p-3.)
    Y=cmath.exp(-gamma)
    D=2.*(1-Y*Y*a*a)
    S1=(1.-Y*Y*a*a+a*(1.-Y*Y))/D
    S2=(1.-a*a)*Y/D
    S3=((1.-Y*Y*a*a)-a*(1.-Y*Y))/D
    return [[S1,S2,S3,-S2],
            [S2,S1,-S2,S3],
            [S3,-S2,S1,S2],
            [-S2,S3,S2,S1]]
```

(d) Device Python code

```python
def TLineFourPortLossless(Zc,Td,f,Z0=50.):
    return TLineFourPort(Zc,1j*2.*math.pi*f*Td,Z0)
```

(e) Lossless device Python code

**Figure 7.4** Four-port transmission line

Although the mixed-mode converter element has not yet been described (see §7.3.1), the behavior of the model in Figure 7.4(b) is best understood by the alternative model in Figure 7.4(c). This model shows that the differential mode (i.e. the differences between the voltages at ports 1 and 3 or 2 and 4) is transmitted back and forth through this model, but the common mode (the average value of the signals at port 1 and 3 or 2 and 4) is not; in fact there is an open circuit presented to the common mode. This is understood better in the context of the remainder of this chapter.

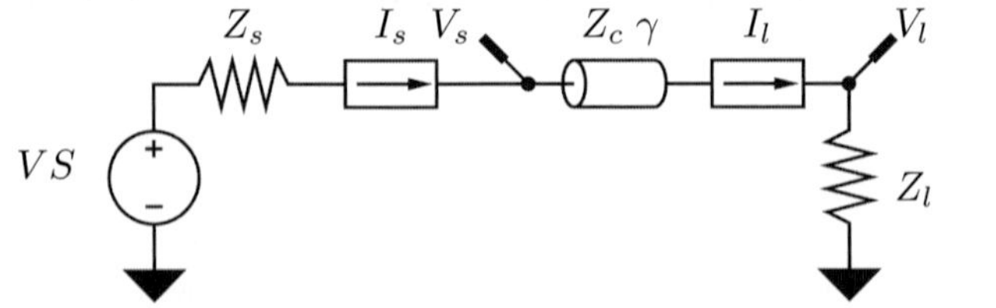


(a) Circuit representation

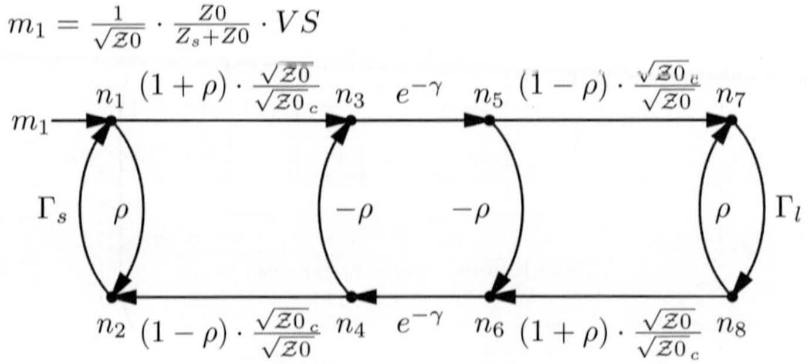


(b) Signal-flow diagram representation

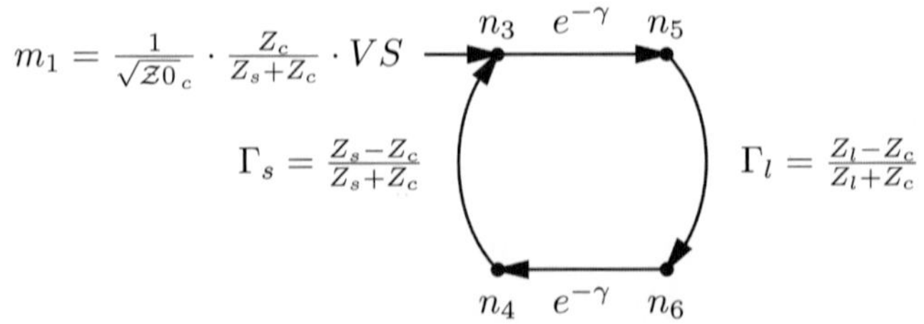


(c) Simplified signal-flow diagram representation in $Z0 = Z_c$

**Figure 7.5** Transmission line simulation

The four-port transmission line is utilized to build rather abstract representations of differential transmission lines, for example in [15], but the preferred method is to use mixed-mode converters, as put forth originally in [16]. The code for s-parameters of the four-port transmission line is provided in Figure 7.4(e), which uses the four-port transmission line device provided in Figure 7.4(d).

## 7.2 Simulation Example of Single-Ended Transmission Line

Figure 7.5 shows a simulation of a circuit (Figure 7.5(a)) containing a transmission line driven by a voltage source through a series impedance and terminated in a load impedance. The signal-flow representation of this circuit is shown in Figure 7.5(b), where there is a voltage source supplying a stimulus according to the calculations provided in §6.1.2. The solution begins with the system equation corresponding to Figure 7.5(b):

$$\left[\mathbb{I}-\begin{pmatrix} 0 & \Gamma_s & 0 & 0 & 0 & 0 & 0 & 0 \\ \rho & 0 & 0 & (1-\rho)\cdot\frac{\sqrt{Z0_c}}{\sqrt{Z0}} & 0 & 0 & 0 & 0 \\ (1+\rho)\cdot\frac{\sqrt{Z0}}{\sqrt{Z0_c}} & 0 & 0 & -\rho & 0 & 0 & 0 & 0 \\ 0 & 0 & 0 & 0 & 0 & e^{-\gamma} & 0 & 0 \\ 0 & 0 & e^{-\gamma} & 0 & 0 & 0 & 0 & 0 \\ 0 & 0 & 0 & 0 & -\rho & 0 & 0 & (1+\rho)\cdot\frac{\sqrt{Z0}}{\sqrt{Z0_c}} \\ 0 & 0 & 0 & 0 & (1-\rho)\cdot\frac{\sqrt{Z0_c}}{\sqrt{Z0}} & 0 & 0 & \rho \\ 0 & 0 & 0 & 0 & 0 & 0 & \Gamma_l & 0 \end{pmatrix}\right]\cdot\begin{pmatrix} n_1 \\ n_2 \\ n_3 \\ n_4 \\ n_5 \\ n_6 \\ n_7 \\ n_8 \end{pmatrix}=\begin{pmatrix} m_1 \\ 0 \\ 0 \\ 0 \\ 0 \\ 0 \\ 0 \\ 0 \end{pmatrix}. \tag{7.17}$$

The solution to (7.17) is provided in (7.18) in the form of waves:

$$\begin{pmatrix} n_1 \\ n_2 \\ n_3 \\ n_4 \\ n_5 \\ n_6 \\ n_7 \\ n_8 \end{pmatrix}=\frac{m_1}{\begin{bmatrix}(\rho-\Gamma_l)\cdot(\rho-\Gamma_s)\cdot e^{-2\cdot\gamma} \\ -(\Gamma_l\cdot\rho-1)\cdot(\rho\cdot\Gamma_s-1)\end{bmatrix}}\cdot\begin{pmatrix} \rho\cdot(\rho-\Gamma_l)\cdot e^{-2\cdot\gamma}+\Gamma_l\cdot\rho-1 \\ (\rho-\Gamma_l)\cdot e^{-2\cdot\gamma}+\rho\cdot(\Gamma_l\cdot\rho-1) \\ \frac{\sqrt{Z0}}{\sqrt{Z0_c}}\cdot(\rho+1)\cdot(\Gamma_l\cdot\rho-1) \\ \frac{\sqrt{Z0}}{\sqrt{Z0_c}}\cdot(\rho+1)\cdot(\rho-\Gamma_l)\cdot e^{-2\cdot\gamma} \\ \frac{\sqrt{Z0}}{\sqrt{Z0_c}}\cdot(\rho+1)\cdot(\Gamma_l\cdot\rho-1)\cdot e^{-\gamma} \\ \frac{\sqrt{Z0}}{\sqrt{Z0_c}}\cdot(\rho+1)\cdot(\rho-\Gamma_l)\cdot e^{-\gamma} \\ (\rho^2-1)\cdot e^{-\gamma} \\ (\rho^2-1)\cdot\Gamma_l\cdot e^{-\gamma} \end{pmatrix}. \tag{7.18}$$

The voltages in this system are computed using (2.8):

$$\begin{pmatrix} v_1 \\ v_1' \\ v_2' \\ v_2 \end{pmatrix} = \begin{pmatrix} \sqrt{Z0} & \sqrt{Z0} & 0 & 0 & 0 & 0 & 0 & 0 \\ 0 & 0 & \sqrt{Z0}_c & \sqrt{Z0}_c & 0 & 0 & 0 & 0 \\ 0 & 0 & 0 & 0 & \sqrt{Z0}_c & \sqrt{Z0}_c & 0 & 0 \\ 0 & 0 & 0 & 0 & 0 & 0 & \sqrt{Z0} & \sqrt{Z0} \end{pmatrix} \cdot \begin{pmatrix} n_1 \\ n_2 \\ n_3 \\ n_4 \\ n_5 \\ n_6 \\ n_7 \\ n_8 \end{pmatrix}$$

$$= \frac{m_1 \cdot \sqrt{Z0}}{\left[\begin{matrix} (\rho - \Gamma_l) \cdot (\rho - \Gamma_s) \cdot e^{-2 \cdot \gamma} \\ - (\Gamma_l \cdot \rho - 1) \cdot (\rho \cdot \Gamma_s - 1) \end{matrix}\right]} \cdot \begin{pmatrix} (\rho + 1) \cdot (\rho - \Gamma_l) \cdot e^{-2 \cdot \gamma} + (\rho + 1) \cdot (\Gamma_l \cdot \rho - 1) \\ (\rho + 1) \cdot (\rho - \Gamma_l) \cdot e^{-2 \cdot \gamma} + (\rho + 1) \cdot (\Gamma_l \cdot \rho - 1) \\ (\rho - 1) \cdot (\rho + 1) \cdot (1 + \Gamma_l) \cdot e^{-\gamma} \\ (\rho - 1) \cdot (\rho + 1) \cdot (1 + \Gamma_l) \cdot e^{-\gamma} \end{pmatrix}. \tag{7.19}$$

In (7.19), the voltages $v_1$ and $v_2$ labeled with the primes represent the voltages calculated inside the transmission line, and those without a prime represent the voltages calculated at the ports of the transmission line. As expected, the voltages at the ports and inside the transmission line match. The value of $\sqrt{Z0}$ cancels with the value used to define $m_1$. The usage of a different value of $\sqrt{Z0}$ in the interior of the transmission line is discussed at the end of this section.

The currents are calculated according to (2.9):

$$\begin{pmatrix} i_1 \\ i_1' \\ i_2' \\ i_2 \end{pmatrix} = \begin{pmatrix} \frac{\sqrt{Z0}}{Z0} & -\frac{\sqrt{Z0}}{Z0} & 0 & 0 & 0 & 0 & 0 & 0 \\ 0 & 0 & \frac{\sqrt{Z0}_c}{Z_c} & -\frac{\sqrt{Z0}_c}{Z_c} & 0 & 0 & 0 & 0 \\ 0 & 0 & 0 & 0 & \frac{\sqrt{Z0}_c}{Z_c} & -\frac{\sqrt{Z0}_c}{Z_c} & 0 & 0 \\ 0 & 0 & 0 & 0 & 0 & 0 & \frac{\sqrt{Z0}}{Z0} & -\frac{\sqrt{Z0}}{Z0} \end{pmatrix} \cdot \begin{pmatrix} n_1 \\ n_2 \\ n_3 \\ n_4 \\ n_5 \\ n_6 \\ n_7 \\ n_8 \end{pmatrix}.$$

For the port 1 current, two values are obtained:

$$i_1 = \frac{(\rho - 1) \cdot (\rho - \Gamma_l) \cdot e^{-2 \cdot \gamma} + (\rho - 1) \cdot (1 - \Gamma_l \cdot \rho)}{(\rho - \Gamma_l) \cdot (\rho - \Gamma_s) \cdot e^{-2 \cdot \gamma} - (\Gamma_l \cdot \rho - 1) \cdot (\rho \cdot \Gamma_s - 1)} \cdot \frac{\sqrt{Z0} \cdot m_1}{Z0},$$

$$i_1' = \frac{-(\rho + 1) \cdot (\rho - \Gamma_l) \cdot e^{-2 \cdot \gamma} + (\rho + 1) \cdot (\Gamma_l \cdot \rho - 1)}{(\rho - \Gamma_l) \cdot (\rho - \Gamma_s) \cdot e^{-2 \cdot \gamma} - (\Gamma_l \cdot \rho - 1) \cdot (\rho \cdot \Gamma_s - 1)} \cdot \frac{\sqrt{Z0} \cdot m_1}{Z_c};$$

in order to have $i_1 = i_1'$, $Z_c = Z0 \cdot (1 + \rho) / (1 - \rho)$ is substituted into the equation for $i_1'$.

This analysis looks very cumbersome. Things can be simplified greatly by choosing $Z0 = Z_c$ and performing the entire analysis with the reference impedance equal to the characteristic impedance of the transmission line; this is what the reference impedance is for in the first place. Then, the voltages and currents reduce to much simpler equations:

$$V_s = m_1 \cdot \sqrt{Z0} \cdot \frac{1 + \Gamma_l \cdot e^{-2\cdot\gamma}}{1 - \Gamma_l \cdot \Gamma_s \cdot e^{-2\cdot\gamma}}, \qquad V_l = m_1 \cdot \sqrt{Z0} \cdot \frac{(1 + \Gamma_l) \cdot e^{-\gamma}}{1 - \Gamma_l \cdot \Gamma_s \cdot e^{-2\cdot\gamma}},$$

$$I_s = m_1 \cdot \frac{\sqrt{Z0}}{Z_c} \cdot \frac{1 - \Gamma_l \cdot e^{-2\cdot\gamma}}{1 - \Gamma_l \cdot \Gamma_s \cdot e^{-2\cdot\gamma}}, \qquad I_l = m_1 \cdot \frac{\sqrt{Z0}}{Z_c} \cdot \frac{(1 - \Gamma_l) \cdot e^{-\gamma}}{1 - \Gamma_l \cdot \Gamma_s \cdot e^{-2\cdot\gamma}}.$$

With $Z0 = Z_c$, and therefore $\rho = 0$, the very simple diagram shown in Figure 7.5(c) results. This simplification removes the impedance transformer at the interface. There are new definitions of the stimulus, and source and load terminations due to the redefinition of $Z0$.

Now it is seen that the voltage source injects a forward propagating wave into the circuit:

$$m_1 = \frac{1}{\sqrt{Z0}_c} \cdot \frac{Z_c}{Z_s + Z_c} \cdot V,$$

and this wave races around the circle. Thus, the waves at the nodes can be defined as:

$$n_3 = m_1 \cdot \left[1 + \sum_{m=1}^{\infty} \left(\Gamma_l \cdot \Gamma_s \cdot e^{-2\cdot\gamma}\right)^m\right],$$

$$n_5 = n_3 \cdot e^{-\gamma} = m_1 \cdot e^{-\gamma} \cdot \left[1 + \sum_{m=1}^{\infty} \left(\Gamma_l \cdot \Gamma_s \cdot e^{-2\cdot\gamma}\right)^m\right],$$

$$n_6 = n_5 \cdot \Gamma_l = m_1 \cdot e^{-\gamma} \cdot \Gamma_l \cdot \left[1 + \sum_{m=1}^{\infty} \left(\Gamma_l \cdot \Gamma_s \cdot e^{-2\cdot\gamma}\right)^m\right],$$

$$n_4 = n_6 \cdot e^{-\gamma} = m_1 \cdot e^{-2\cdot\gamma} \cdot \Gamma_l \cdot \left[1 + \sum_{m=1}^{\infty} \left(\Gamma_l \cdot \Gamma_s \cdot e^{-2\cdot\gamma}\right)^m\right].$$

If $\Gamma_s = 0$, $n_3 = m_1$, and the waves reflected from the load end at $n_4$ on their return trip. Furthermore, if $\Gamma_l = 0$, $n_3 = m_1$, and $n_5 = m_1 \cdot e^{-\gamma}$, the wave ends at $n_5$. Ideally, $\Gamma_s = \Gamma_l = 0$, and $\gamma$ represent a lossless (purely imaginary) time delay (with propagation time constant over frequency), and waves would enter at $n_3$ and arrive at $n_5$ with no further travel.

### 7.2.1 Energy Storage

If a step waveform is applied in the circuit in Figure 7.5(a) and the line is lossless, consisting of only inductance and capacitance, then, at time zero, the source resistance $Z_s$ and the characteristic impedance $Z_c$ form a voltage divider. Then, a voltage appears at the interface of the transmission line,

$$V_s = \frac{Z_c}{Z_s + Z_c} \cdot VS,$$

and a current $I_s$ flows into the transmission line,

$$I_s = \frac{V_s}{Z_c} = \frac{VS}{Z_s + Z_c}.$$

Initially, no voltage appears at the load, nor does any current flow into it. Finally, some time later, a voltage $V_l$ appears at the load,

$$V_l = 2 \cdot \frac{Z_l}{Z_l + Z_c} \cdot V_s,$$

and a current $I_l$ flows in the load,

$$I_l = V_l / Z_l.$$

The electrical length of the line is the time delay. During the period of time between the application of the step at the source and its arrival at the load, an amount of energy was supplied to the line equal to

$$E = P \cdot T_d = \frac{V_s^2}{Z_c} \cdot T_d.$$

This energy was not lost, but is stored in the line. Consider that, in the lossless line, $T_d = \sqrt{L \cdot C}$ and $Z_c = \sqrt{L/C}$, where $L$ and $C$ are the total inductance and capacitance of the line; therefore,

$$\begin{aligned} E &= \frac{V_s^2}{Z_c} \cdot \sqrt{L \cdot C} \\ &= \frac{V_s^2}{Z_c} \cdot \left( \frac{1}{2} \cdot \sqrt{L \cdot C} + \frac{1}{2} \cdot \sqrt{L \cdot C} \right) = \frac{V_s^2}{Z_c} \cdot \left( \frac{1}{2} \cdot \frac{L}{\sqrt{\frac{L}{C}}} + \frac{1}{2} \cdot C \cdot \sqrt{\frac{L}{C}} \right) \\ &= \frac{V_s^2}{Z_c} \cdot \left( \frac{1}{2} \cdot \frac{L}{Z_c} + \frac{1}{2} \cdot C \cdot Z_c \right), \end{aligned}$$

which simplifies to

$$E = \frac{1}{2} \cdot L \cdot I_s^2 + \frac{1}{2} \cdot C \cdot V_s^2.$$

This is the familiar equation for energy stored in an inductor on the left and a capacitor on the right. Thus, a transmission line stores energy both in a magnetic field due to current flow and inductance and in an electric field due to charge storage and capacitance.

## 7.3 Differential Signaling

In signal integrity, signals are most often transmitted differentially. In other words, the signal content is contained in the difference between two voltages. There are two main reasons for doing this:

- Differential signals have higher immunity to external interference. This is because external interference tends to raise the voltage of two adjacent lines similarly and therefore the difference between the voltages on these adjacent lines remains mostly unaffected.
- Differential signals can have less radiation that interferes with other surrounding signals. This is because two adjacent lines driven differentially have a virtual ground between them where the electric fields terminate. Similarly, the current flowing on one line is the opposite of the current flowing on the adjacent line, causing the magnetic fields to cancel.

Also:

- Differential signals do not require a precise ground or absolute voltage reference. Since the signal is the difference between the voltages on the lines, the differential signal, for the most part, does not care what the absolute voltages are with respect to ground.

Given designated positive and negative voltages, $V_p$ and $V_n$, in a differential system, the differential-mode signal is defined electrically as

$$V_D = V_p - V_n. \tag{7.20}$$

Another less interesting but important signal is the common-mode signal. It is defined as the average of the voltage on the positive and negative terminals at each end of a transmission line:

$$V_C = \frac{V_p + V_n}{2}. \tag{7.21}$$

While the differential-mode signal carries the information, the common-mode signal must still be managed. This is to keep the absolute voltages within the ranges of the devices transmitting or receiving them, and to prevent radiated emissions.

The equations in (7.20) and (7.21) lead to the definition of the designated positive and negative voltages as

$$V_p = V_C + \frac{1}{2} \cdot V_D, \tag{7.22}$$

$$V_n = V_C - \frac{1}{2} \cdot V_D. \tag{7.23}$$

Some confusion immediately occurs when dealing with differential voltages. One source of confusion is in the description of the voltage levels. For example, a differential signal might be generated from two single-ended outputs of a transmitter, where each single-ended voltage swings from zero to one volt. One might say that the amplitude of each single-ended signal is therefore one volt. However, when the designated positive signal is one volt, the designated negative signal will ideally be zero, in this case leading to a differential output level of one volt. And when the designated positive signal is zero, the designated negative signal should swing to one volt, leading to a differential output level of minus one volt. One might therefore describe the output signal as plus or minus one volt differential amplitude, or as two volts differential peak–peak. Because these numbers are different by a factor of two, it is best always to include the plus/minus prefix, or the peak–peak suffix exclusive of each other. To avoid confusion, one should never say one volt or two volt differential amplitude alone, or "plus/minus one volt peak–peak."

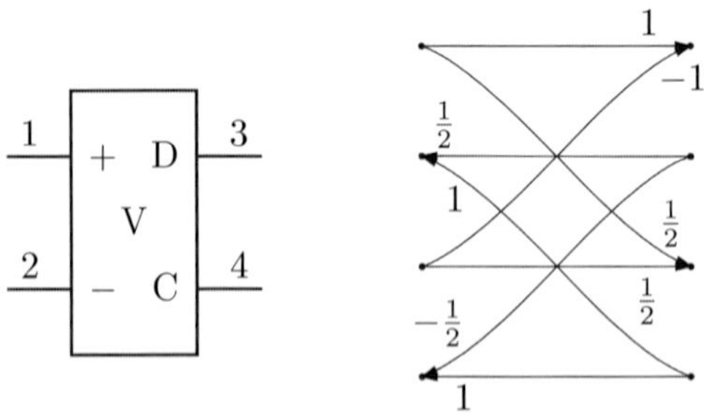


(a) Circuit symbol (b) Signal-flow diagram

$$\mathbf{S} = \begin{pmatrix} 0 & 0 & \frac{D}{2} & \frac{C}{2} \\ 0 & 0 & -\frac{D}{2} & \frac{C}{2} \\ \frac{1}{D} & -\frac{1}{D} & 0 & 0 \\ \frac{1}{C} & \frac{1}{C} & 0 & 0 \end{pmatrix}$$

(c) S-parameters; $D = 1$, $C = 2$

```
def MixedModeConverterVoltage():
    DF=1.; CF=2.
    return [[0.,0.,DF/2.,CF/2.],
            [0.,0.,-DF/2.,CF/2.],
            [1./DF,-1./DF,0.,0.],
            [1./CF,1./CF,0.,0.]]
```

(d) Python code

**Figure 7.6** Voltage mixed-mode converter

### 7.3.1 Differential and Common Modes and Mode Converters

A network device that performs the desired decomposition and superposition between single-ended voltages, designated as positive and negative, and the differential- and common-mode voltages is the voltage mixed-mode converter, as shown in Figure 7.6. The circuit symbol is shown in Figure 7.6(a), exposing the positive and negative waves at pins 1 and 2 and the differential- and common-mode waves at pins 3 and 4. The signal-flow diagram in Figure 7.6(b) shows the application of the four equations (7.20), (7.21), (7.22), and (7.23) to the terminals. There is no interaction between ports 1 and 2, nor between ports 3 and 4. The s-parameters of this device are provided in Figure 7.6(c) along with the Python code in Figure 7.6(d). The division factors are provided for conversion from single-ended to the differential- and common-mode waves; the common-mode wave is divided by 1 and the differential-mode wave is divided by 2.

If two of these mixed-mode converters are placed into a standard network parameter circuit in a mirrored, cascaded arrangement (i.e. the two differential-mode and common-mode ports are connected together) the arrangement has no effect on the circuit except to pass the signals between the two exposed positive terminals and the two exposed negative terminals with no interaction between them. A voltage probe between two such connected devices would pick off the differential- and common-mode voltages.

Unfortunately, the creators of mixed-mode s-parameters [17] did not like the difference in scaling factors for the differential and common modes, as found in what is termed a *voltage* mixed-mode converter, and preferred an equal scaling. This leads to definitions

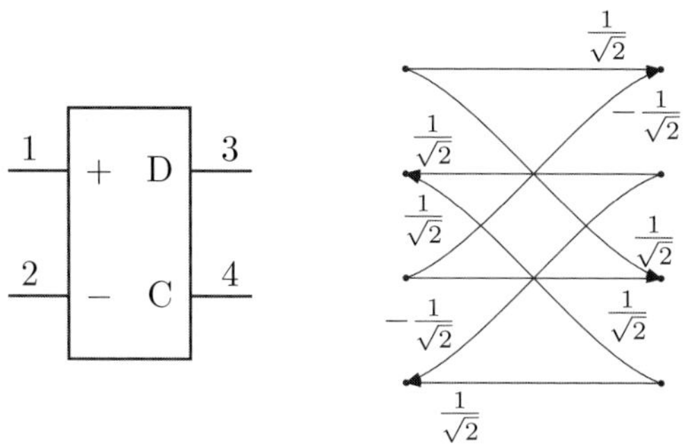


(a) Circuit symbol (b) Signal-flow diagram

$$\mathbf{S} = \begin{pmatrix} 0 & 0 & \frac{D}{2} & \frac{C}{2} \\ 0 & 0 & -\frac{D}{2} & \frac{C}{2} \\ \frac{1}{D} & -\frac{1}{D} & 0 & 0 \\ \frac{1}{C} & \frac{1}{C} & 0 & 0 \end{pmatrix}$$

(c) S-parameters; $D = \sqrt{2}$, $C = \sqrt{2}$

```
def MixedModeConverter():
    DF=math.sqrt(2.0); CF=math.sqrt(2.0)
    return [[0.,0.,DF/2.,CF/2.],
            [0.,0.,-DF/2.,CF/2.],
            [1./DF,-1./DF,0.,0.],
            [1./CF,1./CF,0.,0.]]
```

(d) Python code

**Figure 7.7** Standard mixed-mode converter

relating differential- and mixed-mode voltages whereby the differential mode is the difference between the positive and negative terminals divided by $\sqrt{2}$:

$$V_D = \frac{V_p - V_n}{\sqrt{2}}. \tag{7.24}$$

The common mode is the sum of the voltage on the positive and negative terminals divided by $\sqrt{2}$:

$$V_C = \frac{V_p + V_n}{\sqrt{2}}. \tag{7.25}$$

As such, what one normally thinks of as the differential-mode voltage is divided by $\sqrt{2}$ and what one normally thinks of as the common-mode voltage is multiplied by $\sqrt{2}$.

This leads to the definition of the designated positive and negative voltages as follows:

$$V_p = \frac{V_C + V_D}{\sqrt{2}}, \tag{7.26}$$

$$V_n = \frac{V_C - V_D}{\sqrt{2}}. \tag{7.27}$$

Even though there is symmetry in the relationships, these definitions can cause some confusion when interpreting results using mixed-mode s-parameters.

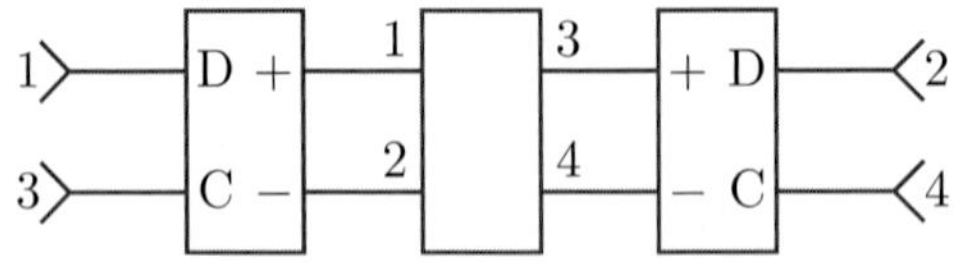


(a) Single-ended to mixed-mode conversion

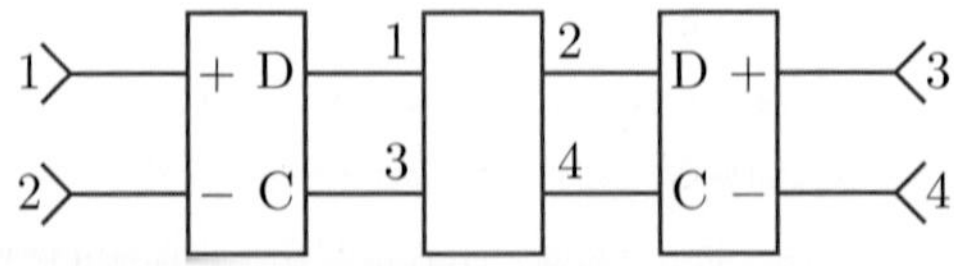


(b) Mixed-mode to single-ended conversion

$$\begin{pmatrix} s_{11} & s_{12} & s_{13} & s_{14} \\ s_{21} & s_{22} & s_{23} & s_{24} \\ s_{31} & s_{32} & s_{33} & s_{34} \\ s_{41} & s_{42} & s_{43} & s_{44} \end{pmatrix} \leftrightarrow \begin{pmatrix} s_{d1d1} & s_{d1d2} & s_{d1c1} & s_{d1c2} \\ s_{d2d1} & s_{d2d2} & s_{d2c1} & s_{d2c2} \\ s_{c1d1} & s_{c1d2} & s_{c1c1} & s_{c1c2} \\ s_{c2d1} & s_{c2d2} & s_{c2c1} & s_{c2c2} \end{pmatrix}$$

(c) Assumed s-parameter conversion

**Figure 7.8** Conversion between mixed-mode and single-ended s-parameters

The standard mixed-mode converter is shown in Figure 7.7 (for a generalized form, see [18]). The circuit symbol is shown in Figure 7.7(a), exposing the positive and negative waves at pins 1 and 2 and the differential- and common-mode waves at pins 3 and 4. The signal-flow diagram in Figure 7.7(b) shows the application of the four equations (7.24), (7.25), (7.26), and (7.27) to the terminals. There is no interaction between ports 1 and 2 nor between ports 3 and 4. The s-parameters of this device are provided in Figure 7.7(c) along with Python code in Figure 7.7(d). These are the same s-parameters as as in Figure 7.6(c), except here there is a common division factor for conversion from single-ended to the differential- and common-mode waves of $\sqrt{2}$.

Like the voltage mixed-mode converter, two of these mixed-mode converters placed into a standard network parameter circuit in a mirrored, cascaded arrangement have no effect on the circuit except to pass the signals between the two exposed positive terminals and the two exposed negative terminals.

The standard mixed-mode converter, henceforth called a mixed-mode converter (retaining the name voltage mixed-mode converter for the other type), is used to convert the s-parameters of systems between their single-ended and mixed-mode counterparts.

### 7.3.2 Mixed-Mode S-Parameters

Mixed-mode s-parameters [17] are used primarily to analyze and understand the performance of differential transmission lines in systems meant to transmit signals differentially. Based on the previous discussion, this means that the primary motive is to transmit information as the potential difference between two terminals, specifically as the voltage $V_D$. In other words, the primary goal is to transmit a differential signal $V_D$ from a transmitter to a receiver. In differential systems, the common-mode signal $V_C$ is not the signal of interest, although it is still important. This is because:

- External interference primarily induces changes to $V_C$ as opposed to $V_D$.
- Radiated emissions primarily come from $V_C$ as opposed to $V_D$.
- The common-mode voltage $V_C$ must stay within the operating range of all the devices. For example, if a receiver can tolerate voltages of zero to five volts, then, if the differential signal is two volts peak–peak, the common-mode signal should stay between one and four volt absolute levels for this circuit to operate.
- Imbalance in the source, line, and termination can lead to *mode conversion*, and hence produce the common mode (and the common mode can be converted to the differential mode).

Although many equations exist that generate mixed-mode s-parameters, the easiest way is to connect a single-ended circuit to mixed-mode converters, as shown in Figure 7.8. Four-port single-ended s-parameters are converted to mixed-mode s-parameters in Figure 7.8(a). There are many sources of confusion when performing this transformation, but the confusion is at least reduced by considering the graphical representation as opposed to using equations. The plus and minus terminals identify ports 1 and 2 of the single-ended device as one mixed-mode port, and ports 3 and 4 as another. It is customary to have both ports 1 and 2 of the mixed-mode result as the differential mode and to have 3 and 4 as the corresponding common mode. They are corresponding because port 1 is the differential mode for mixed-mode port 1 and 3 is the corresponding common mode for mixed-mode port 1.

The port ordering on the conversion from mixed mode back to single ended in Figure 7.8(b) is required in order to restore the proper port ordering.

The s-parameter matrices assumed for this conversion are shown in Figure 7.8(c). The mixed-mode s-parameters are labeled differently from their row and column ordering as traditionally used in the single-ended s-parameters.[6] Because of the convention of listing the differential ports first, the matrix could be partitioned as follows:

$$\begin{pmatrix} S_{d1d1} & S_{d1d2} & S_{d1c1} & S_{d1c2} \\ S_{d2d1} & S_{d2d2} & S_{d2c1} & S_{d2c2} \\ S_{c1d1} & S_{c1d2} & S_{c1c1} & S_{c1c2} \\ S_{c2d1} & S_{c2d2} & S_{c2c1} & S_{c2c2} \end{pmatrix} = \begin{pmatrix} \mathbf{S_{dd}} & \mathbf{S_{dc}} \\ \mathbf{S_{cd}} & \mathbf{S_{cc}} \end{pmatrix},$$

where each block represents a $2 \times 2$ matrix. In this manner, each block corresponds to a phenomenon of interest:

[6] Using $S_{d1c2}$ as opposed to the commonly used $S_{dc12}$ is preferred to avoid the ambiguity that occurs when devices with more than nine ports are analyzed.

- $\mathbf{S_{dd}}$ describes the propagation and characteristic impedance of a fictitious transmission line carrying the differential mode.
- $\mathbf{S_{cc}}$ describes the propagation and characteristic impedance of a fictitious transmission line carrying the common mode.
- $\mathbf{S_{cd}}$ describes the mode conversion occurring within the device that converts differential mode into common mode.
- $\mathbf{S_{dc}}$ describes the mode conversion occurring within the device that converts common mode into differential mode.

Each of these blocks describes interesting information in their own right. The $\mathbf{S_{dd}}$ block could be considered as the most important because it describes the propagation of the mode of interest within the device, but the $\mathbf{S_{cc}}$ block describing the propagation of the common mode is also important because the common mode will need to be terminated properly to avoid resonances and other bad behavior that could affect the ability of the line to carry the differential mode. The $\mathbf{S_{cd}}$ block is important because it describes how the line might radiate and interfere with other devices, and the $\mathbf{S_{dc}}$ block describes the susceptibility of the device to radiated interference from other devices and its sensitivity to imbalance in the transmitter or receiver termination.

Mixed-mode s-parameters have many pros and cons and are very confusing from the standpoint of port connections, but they are absolutely necessary for understanding the behavior of a differential transmission line. Here is some advice with regard to mixed-mode and single-ended s-parameters:

1. Don't exchange mixed-mode s-parameter files with others, especially for use in simulation, unless the only goal is to plot the s-parameters. Instead, exchange single-ended s-parameter files with the port numbering specified and deal with any mixed-mode conversions when they are plotted.
2. Don't plot single-ended s-parameter files and don't test against single-ended s-parameter masks for differential systems, especially if the lines are coupled. Single-ended s-parameters will look very confusing (perhaps broken) in the coupled line situation.
3. When examining mixed-mode s-parameters, as with any s-parameters, ensure that a reference impedance is used that is appropriate to the expected source and termination impedance, otherwise any interpretation of the results will be misleading.

## 7.4 Differential Transmission Lines

Differential signals are transmitted over a differential transmission line. The line is called differential even though it also transmits a common-mode signal; the intended signal is differential.

Figure 7.9(a) shows an extension of Figure 7.1 to the differential case. Here, there is a series resistance and inductance and a shunt capacitance and conductance, where, for each side of the line, the values are subscripted with a $p$ for the positive and an $n$ for the negative side of each of the single-ended legs making up the transmission line. In addition,

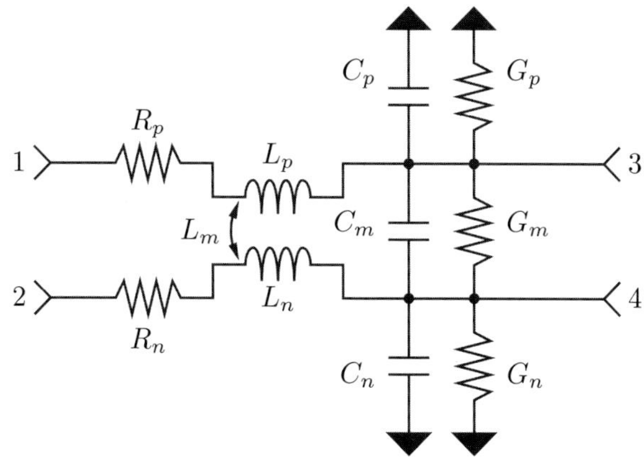


(a) Circuit model

```
class TLineDifferentialRLGCApproximate(SParameters):
    def __init__(self,f, Rp, Rsep, Lp, Gp, Cp, dfp, Rn, Rsen, Ln, Gn, Cn, dfn,
                      Cm, dfm, Gm, Lm, Z0=50., K=0):
        if K==0:
            Td=math.sqrt((max(Lp,Ln)+Lm)*(max(Cp,Cn)+2*Cm)); Rt=0.45/f[-1]
            K=int(math.ceil(Td*2/(Rt*self.rtFraction)))
        self.m_K=K
        sdp=SystemDescriptionParser().AddLines([
        'device rsep 2','device rp 2','device lp 2','device gp 1','device cp 1',
        'device rsen 2','device rn 2','device ln 2','device gn 1','device cn 1',
        'device lm 4','device gm 2','device cm 2','connect rp 2 rsep 1',
        'connect rsep 2 lp 1','connect rn 2 rsen 1','connect rsen 2 ln 1',
        'connect lp 2 lm 1','connect ln 2 lm 3','connect lm 2 gp 1 cp 1 gm 1 cm 1',
        'connect lm 4 gn 1 cn 1 gm 2 cm 2','port 1 rp 1 2 rn 1 3 lm 2 4 lm 4'])
        self.m_sspn=SystemSParametersNumeric(sdp.SystemDescription())
        self.m_sspn.AssignSParameters('rp',SeriesZ(Rp/K,Z0))
        self.m_sspn.AssignSParameters('gp',TerminationG(Gp/K,Z0))
        self.m_sspn.AssignSParameters('rn',SeriesZ(Rn/K,Z0))
        self.m_sspn.AssignSParameters('gn',TerminationG(Gn/K,Z0))
        self.m_sspn.AssignSParameters('gm',SeriesG(Gm/K,Z0))
        self.m_spdl=[('rsep',dev.SeriesRse(f,Rsep/K,Z0)),
            ('lp',dev.SeriesL(f,Lp/K,Z0)),('cp',dev.TerminationC(f,Cp/K,Z0,dfp)),
            ('rsen',dev.SeriesRse(f,Rsen/K,Z0)),('ln',dev.SeriesL(f,Ln/K,Z0)),
            ('cn',dev.TerminationC(f,Cn/K,Z0,dfn)),('lm',dev.Mutual(f,Lm/K,Z0)),
            ('cm',dev.SeriesC(f,Cm/K,Z0,dfm))]
        SParameters.__init__(self,f,None,Z0)
    def __getitem__(self,n):
        for ds in self.m_spdl: self.m_sspn.AssignSParameters(ds[0],ds[1][n])
        sp=self.m_sspn.SParameters()
        if sp == 1: return sp
        lp=[1,2]; rp=[3,4]
        return T2S(linalg.matrix_power(S2T(sp,lp,rp),self.m_K),lp,rp)
```

(b) S-parameters Python code

**Figure 7.9** Differential transmission line model

there are values subscripted with $m$ for mutual inductance, mutual capacitance, and mutual conductance, representing the coupling between the two lines.

In the analysis of differential transmission lines, there are several possibilities that may make the analysis method more or less difficult. The two main considerations are whether the line is balanced and whether the line is coupled.

For the circuit in Figure 7.9(a), coupling implies that there exists at least one of mutual inductance, mutual capacitance, or mutual conductance. In other words, at least one of the values $L_m$, $C_m$, or $G_m$ are non-zero. When a differential transmission line contains two single-ended sides that are uncoupled, they can be analyzed as two separate single-ended transmission lines.

For the circuit in Figure 7.9(a), balance implies that each of the circuit values subscripted with $p$ and with $n$ are equal. When a line is balanced, there are no interactions between the modes; the common mode sent by the transmitter arrives at the receiver with no differential-mode component, and similarly the differential mode sent by the transmitter arrives at the receiver with no common-mode component. When a transmission line can generate components of one mode from another, it is referred to as *mode conversion*; the only thing that can create mode conversion in this model is imbalance.

Balance and coupling are entirely separate concepts. An imbalanced, uncoupled line will have mode conversion. A good example of this is a transmission line consisting of two uncoupled lines of different lengths. Similarly, a balanced, coupled line cannot be analyzed as two separate single-ended lines as the two single-ended signals interact. All of these possibilities will be dealt with.

Generally speaking, the solution to a distributed transmission line described by Figure 7.9(a) is very complicated and is solved numerically. This solution is sort of a last resort and is used when no other method is possible. It must be used if the transmission line is unbalanced *and* coupled. If the line is either balanced *or* uncoupled, other means exist, even though this numerical approximation can still supply an accurate model and can be analyzed for educational purposes. In other words, although involving numerical approximation, the solution works in all possible cases.

There are at least two hindrances to solving the differential, distributed RLGC model analytically:

- There is no real definition of ABCD parameters for a four-port device. This can be overcome by utilizing T-parameters, which also don't have a strict definition, although a useful definition of multi-port T-parameters is supplied in §3.11.
- While the eigenvalues and eigenvectors of a $2 \times 2$ matrix can be described analytically, there are no generally usable formulas for the eigenvalues and eigenvectors of a $4 \times 4$ matrix. These equations do exist, but they are huge and not really usable.

The solution is created by piecing together a circuit consisting of the elements in Figure 7.9, and utilizing methods that will be discussed in §8.4. As before, all of the values are divided by the number of sections used for the approximation. The s-parameters of the system are converted to T-parameters using the equations in (3.47) and the algorithm in Listing 3.13. After raising the T-parameters to a power equal to the number of sections in the approximation, they are converted back to s-parameters using the equation in (3.48) and the algorithm in Listing 3.14. This is all shown in Figure 7.9(b), which is a surprisingly compact piece of code despite the complexity of the solution. The model includes skin-effect resistances for each side of the line and dissipation factors for all capacitors in the circuit.

### 7.4.1 Uncoupled Differential Transmission Lines

An uncoupled differential transmission line is shown in Figure 7.10. A differential transmission line is uncoupled when all of the mutual terms are zero, namely $L_m = C_m = G_m = 0$, as shown in Figure 7.10(a). When the lines are uncoupled, the analysis is greatly simplified in that the equivalent circuit model comprises simply two independent transmission lines, as shown in Figure 7.10(b). Not much more needs to be said about the analysis of this model.

### 7.4.2 Balanced Differential Transmission Lines

A balanced differential transmission line is one in which all of the values for the positive half of the line are equal to the values for the negative half of the line. Namely:

$$R_p = R_n = R, \qquad L_p = L_n = L, \qquad G_p = G_n = G, \qquad C_p = C_n = C.$$

The analysis of a balanced differential transmission line can be simplified by transforming the problem into how the differential and common modes are transmitted. Since the line is balanced, there is no mode conversion and the transfer of the differential and common modes is completely independent. There can still be mode conversion due to imbalance in the transmitter or the receiver, but, insofar as the transmission line is concerned, the two modes propagate independently once in the line.

To examine the mode propagation, two modes related to the differential and common mode are introduced. These are the even and odd modes. The even and odd modes are not actual modes that propagate, but are methods of analyzing port behavior under certain driving conditions. These driving conditions are shown in Figure 7.11, in which there are two voltage generators, each connected to ground at the transmitter driving the line with an impedance equal to the reference impedance $Z0$. The line is terminated with the reference impedance center tapped to ground. Thus, there are absolute voltages (referenced to ground) on each of the positive and negative sides of the system at both the receiver and the transmitter.

Even-mode analysis is related to the common mode; it is defined as analyzing the behavior on one side of the line when both sides are driven equivalently. Usually, one defines the even mode with a positive voltage on the side being analyzed. With the even mode, this is clear by simply driving both lines with a positive voltage.

Odd-mode analysis is related to the differential mode; it is defined as analyzing the behavior on one side of the line when each side is driven with an opposite voltage. Usually, one defines the odd mode by looking at the positive side, while the positive side is driven with a positive voltage and the negative side is driven with an equal, but opposite polarity, voltage.

As mentioned previously, the transmission line in Figure 7.9 is *balanced* if each side of the four-port transmission line is the same. This means that the analysis of this circuit can

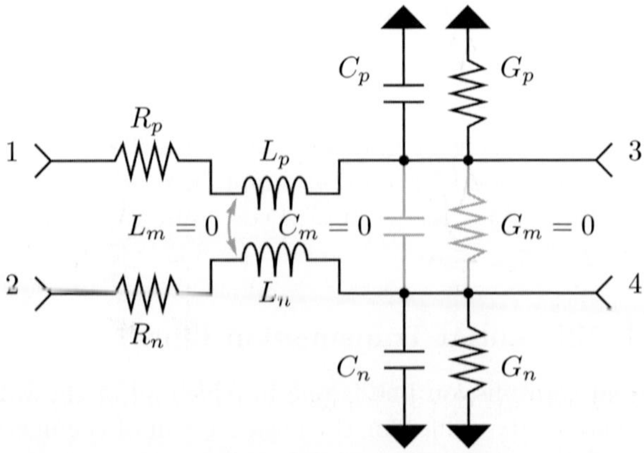


(a) Circuit model

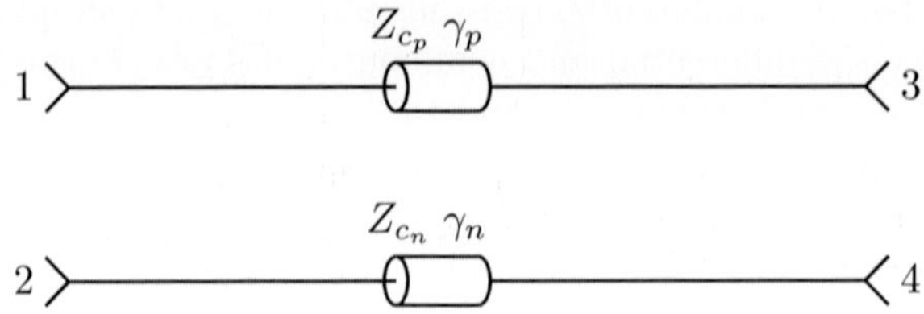


(b) Equivalent circuit

```
class TLineDifferentialRLGCUncoupled(SParameters):
    def __init__(self,f,Rp,Rsep,Lp,Gp,Cp,dfp,Rn,Rsen,Ln,Gn,Cn,dfn,Z0=50.,K=0):
        sdp=SystemDescriptionParser()
        sdp.AddLines(['device TP 2','device TN 2',
                      'port 1 TP 1 2 TN 1 3 TP 2 4 TN 2'])
        self.m_sspn=SystemSParametersNumeric(sdp.SystemDescription())
        self.m_spdl=[('TP',TLineTwoPortRLGC(f,Rp,Rsep,Lp,Gp,Cp,dfp,Z0,K)),
                     ('TN',TLineTwoPortRLGC(f,Rn,Rsen,Ln,Gn,Cn,dfn,Z0,K))]
        SParameters.__init__(self,f,None,Z0)
    def __getitem__(self,n):
        for ds in self.m_spdl:
            self.m_sspn.AssignSParameters(ds[0],ds[1][n])
        return self.m_sspn.SParameters()
```

(c) S-parameters Python code

**Figure 7.10** Circuit representing telegrapher's equations for uncoupled differential transmission line

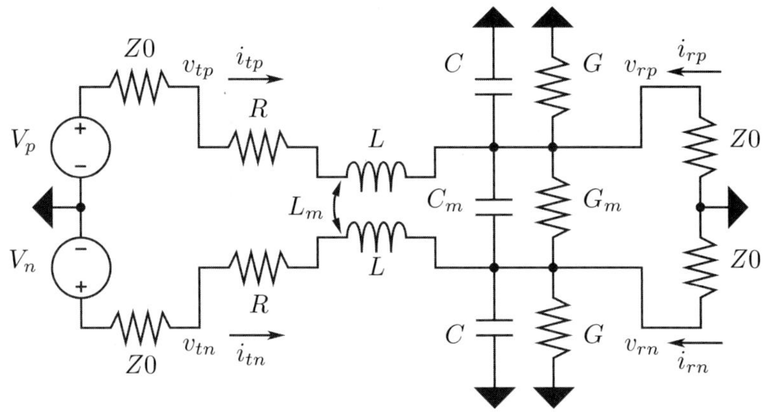


**Figure 7.11** Driving conditions for balanced transmission line circuit

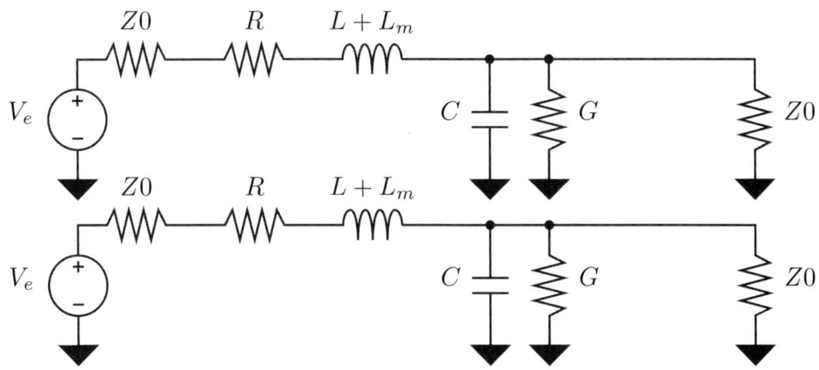


**Figure 7.12** Even-mode analysis of a balanced transmission line circuit

be simplified by decomposition and superposition. To perform this analysis, the circuit is driven and analyzed in each of the even and odd modes.

Even-mode analysis is performed by driving ports 1 and 2 with the same signal. This means that $V_p = V_n$. Because of the assumed balance of the line, this makes the voltages and currents on each side of the line identical. In other words,

$$v_{tn} = v_{tp}, \qquad v_{rn} = v_{rp}, \qquad i_{tn} = i_{tp}, \qquad i_{rn} = i_{rp}.$$

Under these conditions, there is no voltage across the mutual conductance and mutual capacitance and therefore no current flows through these elements; they can therefore be removed from the circuit. Similarly, because the current through each inductor is the same, the mutual inductance can be seen as adding to the self inductance of each of the positive and negative legs. This means that the circuit can be separated into two identical pieces under

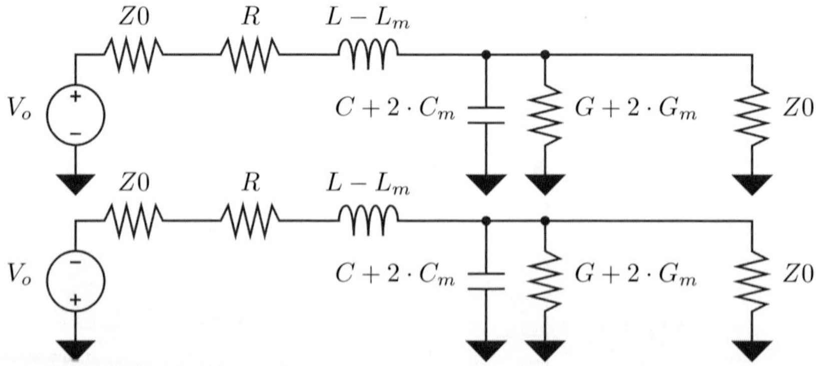


**Figure 7.13** Odd-mode analysis of a balanced transmission line circuit

this driving condition, as shown in Figure 7.12. There are now two equal and independent circuits that can utilize the two-port transmission line analysis previously derived. The series impedance and shunt admittance are

$$Z_{see} = R + j \cdot 2\pi \cdot f \cdot (L + L_m), \qquad Y_{she} = G + j \cdot 2\pi \cdot f \cdot C.$$

The characteristics of an even-mode transmission line are

$$\gamma_e = \sqrt{Z_{see} \cdot Y_{she}}, \qquad Z_e = \sqrt{Z_{see}/Y_{she}}.$$

Returning to Figure 7.11, odd-mode analysis is performed by driving ports 1 and 2 with an equal but opposite signal. This means that $V_n = -V_p$. Because of the assumed balance of the line, this makes the voltages and currents on each side of the line have an equal but opposite relationship. In other words,

$$v_{tn} = -v_{tp}, \qquad v_{rn} = -v_{rp}, \qquad i_{tn} = -i_{tp}, \qquad i_{rn} = -i_{rp}.$$

Under these conditions, there is twice the voltage across the mutual conductance and mutual capacitance and therefore these elements can be replaced with twice their values to ground. Similarly, because the current through each inductor has the opposite direction, the mutual inductance can be seen as subtracting from the self inductance of each of the positive and negative legs. This means that the circuit can be separated into two identical pieces under this driving condition, as shown in Figure 7.13. Again, there are now two independent circuits that can utilize the two-port transmission line analysis previously derived. The series impedance and shunt admittance are

$$Z_{seo} = R + j \cdot 2\pi \cdot f \cdot (L - L_m), \qquad Y_{sho} = G + 2 \cdot G_m + j \cdot 2\pi \cdot f \cdot (C + 2 \cdot C_m).$$

The characteristics of an odd-mode transmission line are

$$\gamma_o = \sqrt{Z_{seo} \cdot Y_{sho}}, \qquad Z_o = \sqrt{Z_{seo}/Y_{sho}}.$$

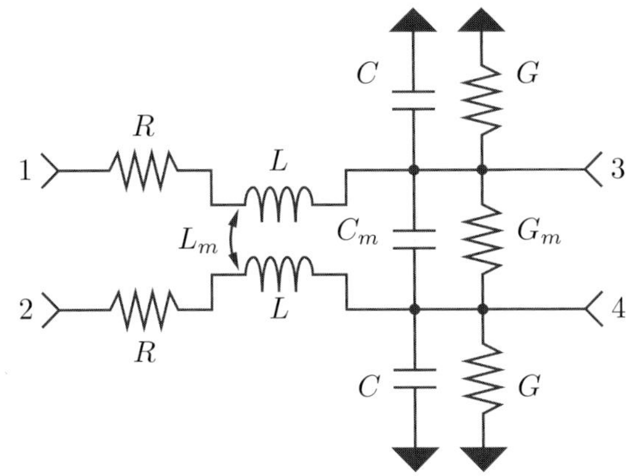


(a) RLGC circuit

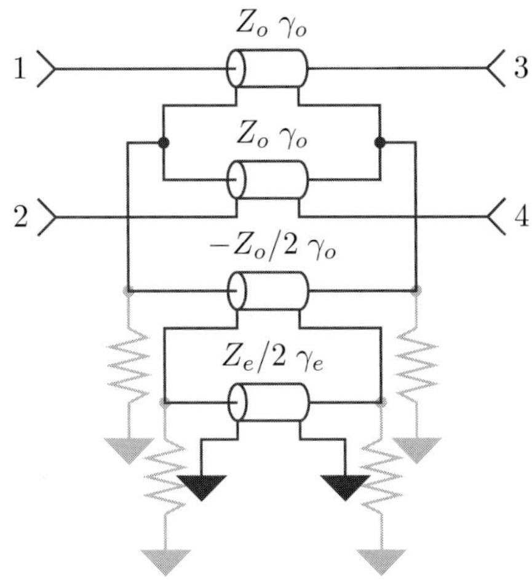


(b) Alternative model

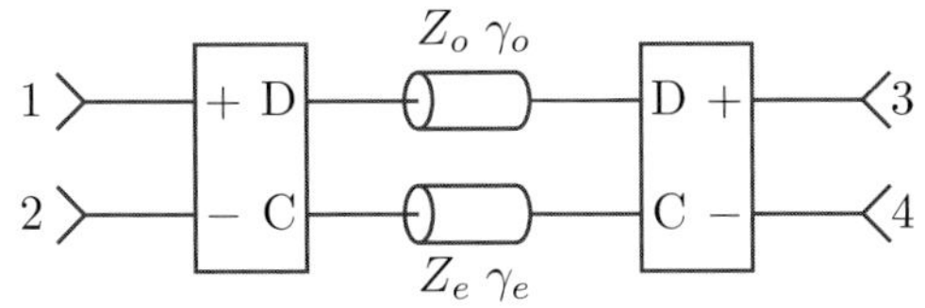


(c) Preferred model

```
class TLineDifferentialRLGCBalanced(SParameters):
    def __init__(self,f,R,Rse,L,G,C,df,Cm,dfm,Gm,Lm,Z0=50.,K=0):
        sdp=SystemDescriptionParser()
        sdp.AddLines(['device L 4 mixedmode','device R 4 mixedmode','device TE 2',
                      'device TO 2','port 1 L 1 2 L 2 3 R 1 4 R 2','connect L 3 TO 1',
                      'connect R 3 TO 2','connect L 4 TE 1','connect R 4 TE 2'])
        self.m_sspn=SystemSParametersNumeric(sdp.SystemDescription())
        self.m_spdl=[('TE',TLineTwoPortRLGC(f,R,Rse,L+Lm,G,C,df,Z0,K)),
                    ('TO',TLineTwoPortRLGC(f,R,Rse,L-Lm,G+2*Gm,C+2*Cm,
                                    (C*df+2*Cm*dfm)/(C+2*Cm),Z0,K))]
        SParameters.__init__(self,f,None,Z0)
    def __getitem__(self,n):
        for ds in self.m_spdl: self.m_sspn.AssignSParameters(ds[0],ds[1][n])
        return self.m_sspn.SParameters()
```

(d) Python code

**Figure 7.14** Balanced differential transmission line

Therefore, there is now an odd-mode transmission line and an even-mode transmission line, each transmitting the differential and common modes, respectively.

These transmission line definitions are used to provide a model of a balanced differential transmission line, as shown in Figure 7.14. Given the circuit model in Figure 7.14(a), with common parameter values in each leg of the transmission line, the traditional model using the even- and odd-mode definitions is provided in Figure 7.14(b) [15]. This model utilizes the four-port transmission line model provided in Figure 7.4. This model, although traditional, is labeled as an alternative because there is a better model that provides better insight. In any case, if one wants to use the model in Figure 7.4, it is helpful to place large resistors in the circuit (shown in gray) in order to help convergence.

A model that provides better insight is shown in Figure 7.14(c), where now the model is created using mixed-mode converters, as provided in Figure 7.7. The mixed-mode converters convert between the single-ended ports at the periphery of the model and the differential- and common-mode ports in the interior. The differential mode passes through a two-port odd-mode transmission line and the common mode passes through a two-port even-mode transmission line. The code for assembling this model is provided in Figure 7.14(d). As this model uses two-port models, there is some complexity in dealing with the dissipation factor for the odd mode. The dissipation factor $\tan\delta$ passed to the model relates to the capacitance and dissipation factor of each leg $C_p$ and $\tan\delta_p$, and the mutual capacitance and its dissipation factor $C_m$ and $\tan\delta_m$ as follows:

$$C_p \cdot (j + \tan\delta_p) + 2 \cdot C_m \cdot (j + \tan\delta_m) = (C_p + 2 \cdot C_m) \cdot (j + \tan\delta)\,,$$

and therefore

$$\tan\delta = \frac{C_p \cdot \tan\delta_p + 2 \cdot C_m \cdot \tan\delta_m}{C_p + 2 \cdot C_m}.$$

### 7.4.3 Summary of Differential Transmission Line Models

Listing 7.7 provides the final model of the s-parameters for the transmission line model. This model determines whether the line is balanced or uncoupled by examining the parameters sent to it and chooses the appropriate mode. The numerical approximation in Figure 7.9(b) is used if the number of sections is specified as non-zero. If the line is uncoupled, the model in Figure 7.10(c) is used. If the line is balanced, then the model in Figure 7.14(d) is utilized. If the line is coupled and unbalanced, there is no choice but to use the numerical approximation in Figure 7.9(b), although there has been some success in modeling imbalanced, coupled lines [19].

When zero sections are specified ($K = 0$), but the numerical approximation must be used, an appropriate value of the number of sections $K$ is calculated. The rule of thumb is that each cascaded section should be such that the round trip delay through the element is less than some fraction of the risetime of the signal. The approximate worst case delay through the section is calculated as

$$T_d = \sqrt{(\max(L_p, L_n) + L_m) \cdot (\max(C_p, C_n) + 2 \cdot C_m)}.$$

**Listing 7.7** Differential transmission line model

```
class TLineDifferentialRLGC(SParameters):
    def __init__(self,f, Rp, Rsep, Lp, Gp, Cp, dfp,
                        Rn, Rsen, Ln, Gn, Cn, dfn,
                        Cm, dfm, Gm, Lm, Z0=50., K=0):
        balanced = Rp==Rn and Rsep==Rsen and Lp==Ln and Gp==Gn and Cp==Cn
        uncoupled = Cm==0 and Gm==0 and Lm==0
        if K != 0 or (not balanced and not uncoupled):
            self.sp=TLineDifferentialRLGCApproximate(f,
                        Rp, Rsep, Lp, Gp, Cp, dfp,
                        Rn, Rsen, Ln, Gn, Cn, dfn,
                        Cm, dfm, Gm, Lm, Z0, K)
        elif uncoupled:
            self.sp=TLineDifferentialRLGCUncoupled(f,
                        Rp, Rsep, Lp, Gp, Cp, dfp,
                        Rn, Rsen, Ln, Gn, Cn, dfn,
                        Z0, K)
        elif balanced:
            self.sp=TLineDifferentialRLGCBalanced(f,
                        Rp, Rsep, Lp, Gp, Cp, dfp,
                        Cm, dfm, Gm, Lm, Z0, K)
        SParameters.__init__(self,f,None,Z0)
    def __getitem__(self,n):
        return self.sp[n]
```

This is really a worst case. Although the risetime is unknown, the bandwidth is hard limited to the last frequency point $f_{bw}$, meaning that the fastest risetime is approximately[7]

$$rt = 0.45/f_{bw}.$$

A fractional amount $frac$ of the risetime used by default in the class is 0.01, or 1 %,[8] which leads to

$$rt \cdot frac \leq 2 \cdot T_d/K,$$

and therefore

$$K = \lceil (2 \cdot T_d) \, / \, (rt \cdot frac) \rceil \, .$$

The value of $K$ does not need to be precisely chosen, and although the approximation is for $K$ sections in cascade, this does not lead to more or less work with higher or lower values of $K$; $K$ only needs to be large enough to be a good approximation, but not so large as to cause numerical problems.

## 7.5 Mixed-Mode Terminations

This chapter on transmission lines concludes with a brief discussion of mixed-mode terminations. Here, some useful models and s-parameter derivations are provided for terminating the differential transmission line.

[7] The risetime of a single-pole filter is such that the bandwidth $f_{bw}$ and risetime $rt$ are related as $f_{bw} \cdot rt = 0.35$. Here 0.45 is used since the system is brick-wall limited.

[8] Johnson and Graham [20] state that 33 % can be used to consider the transmission line as lumped (i.e. $K = 1$), but 10 % is commonly used; 1 % is used here for more accuracy in simulation.

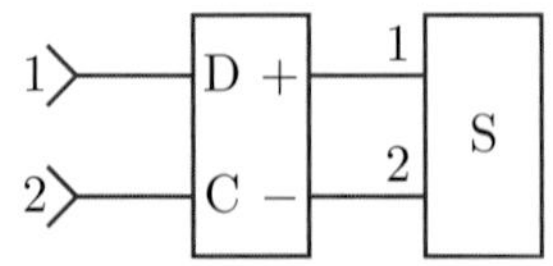


(a) Block diagram

$$\begin{pmatrix} S_{dd} & S_{dc} \\ S_{cd} & S_{cc} \end{pmatrix} = \frac{1}{2} \cdot \begin{pmatrix} S_{11} - S_{21} - S_{12} + S_{22} & S_{11} - S_{21} + S_{12} - S_{22} \\ S_{11} + S_{21} - S_{12} - S_{22} & S_{11} + S_{21} + S_{12} + S_{22} \end{pmatrix}$$

(b) S-parameters

**Figure 7.15** Mixed-mode termination

As mentioned previously, the goal of the differential transmission line is to transmit the differential mode in a distortionless manner through the line. Despite the importance of the differential mode, the common mode should not be ignored either, and proper circuit design has the source and load terminations designed such that they terminate both adequately.

Figure 7.15 provides a mixed-mode two-port termination. A schematic is shown in Figure 7.15(a), whereby a single-ended two-port device is connected to a mixed-mode converter with system port 1 as the differential-mode port and port 2 as the common-mode port. The mixed-mode s-parameters of this system are provided in Figure 7.15(b), which supplies the popular equations used for calculating mixed-mode terminations.[9]

While Figure 7.15 provides the most general case, there are two common types of terminations employed, and it is helpful to have the equations for these on hand. These are:

1. the tee termination, as shown in Figure 7.16;
2. the pi termination, as shown in Figure 7.17.

In the tee termination structure shown in Figure 7.16(a), there is a tee network with two of the legs tied to the plus and minus single-ended ports and one leg tied to ground. The s-parameters of this network are shown most generally in Figure 7.16(b), where any values of $Z_1$, $Z_2$, and $Z_3$ are allowed. Usually, it is desirable to terminate both the differential and common modes in a balanced fashion with no cross terms (i.e. no mode conversion at the termination). For the tee network, this means setting $Z_1 = Z_2$ and designating this as $Z$. Thus in Figure 7.16(c) the s-parameters of a balanced tee termination structure are provided.

These equations can be used in reverse to find the proper values for a desired termination. For a differential-mode termination of $Z_D$ and a common-mode termination of $Z_C$, remember that the mixed-mode s-parameters are for $Z_o = Z_D/2$ and for $Z_e = Z_C \cdot 2$. The desired values are

$$Z = Z_o = \frac{Z_D}{2},$$

[9]Although these equations are popular, it is much safer to convert single-ended s-parameters through the explicit connection of mixed-mode converters as shown. This cuts down on errors made in the equations with regard to port numbering.

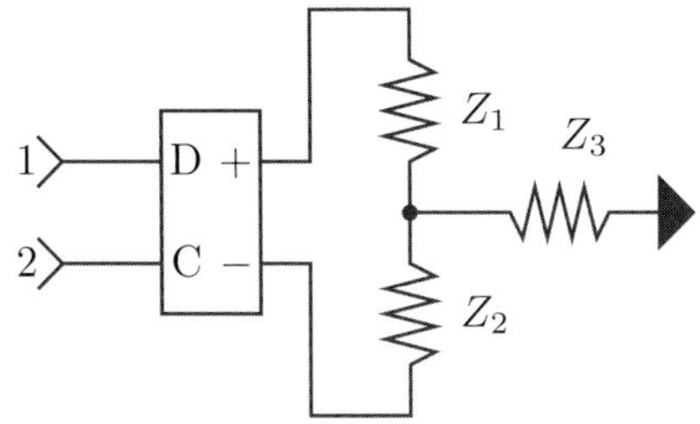


(a) Block diagram

$$\frac{\begin{pmatrix} Z_2 \cdot Z_1 + Z_3 \cdot Z_1 + Z_2 \cdot Z_3 - Z0^2 - 2 \cdot Z0 \cdot Z_3 & -Z0 \cdot (Z_2 - Z_1) \\ -Z0 \cdot (Z_2 - Z_1) & Z_2 \cdot Z_1 + Z_3 \cdot Z_1 + Z_2 \cdot Z_3 - Z0^2 + 2 \cdot Z0 \cdot Z_3 \end{pmatrix}}{Z_2 \cdot Z_3 + Z_2 \cdot Z_1 + Z_2 \cdot Z0 + Z_3 \cdot Z_1 + 2 \cdot Z0 \cdot Z_3 + Z0 \cdot Z_1 + Z0^2}$$

(b) S-parameters

$$\begin{pmatrix} \frac{Z-Z0}{Z+Z0} & 0 \\ 0 & \frac{2 \cdot Z_3 + Z - Z0}{2 \cdot Z_3 + Z + Z0} \end{pmatrix}$$

(c) S-parameters of balanced configuration; $Z_2 = Z_1 = Z$

**Figure 7.16** Mixed-mode tee termination

$$2 \cdot Z_3 + Z = Z_e = Z_C \cdot 2,$$

or

$$Z_3 = Z_C - \frac{Z_D}{4}.$$

As a check, if there is an uncoupled, balanced transmission line with each of the single-ended lines having a characteristic impedance of $50\,\Omega$, then $Z_D = 100\,\Omega$ and $Z_C = 25\,\Omega$, and therefore $Z = 50\,\Omega$ and $Z_3 = 0\,\Omega$. This is a common situation and is a grounded, center-tapped termination with each leg being $50\,\Omega$. In this situation it is sometimes common to terminate the DC component of the common mode in an open and the alternating current (AC) component of the common mode in $25\,\Omega$ through the use of a capacitor for $Z_3$.

In the pi termination structure, shown in Figure 7.17(a), there is a pi network, with the center of the pi connected across the plus and minus terminals and each leg of the pi tied to ground through an impedance. The s-parameters of this network are shown generally in Figure 7.17(b). As in the tee termination case, a balanced termination is usually desired with no mode conversion terms. Balance occurs only when $Z_1 = Z_2$, designated as $Z$. The s-parameters of the balanced pi termination are given in Figure 7.17(c). The even-mode termination is set by $Z$ and the odd-mode termination is determined by half of $Z_3$ in parallel with $Z$.

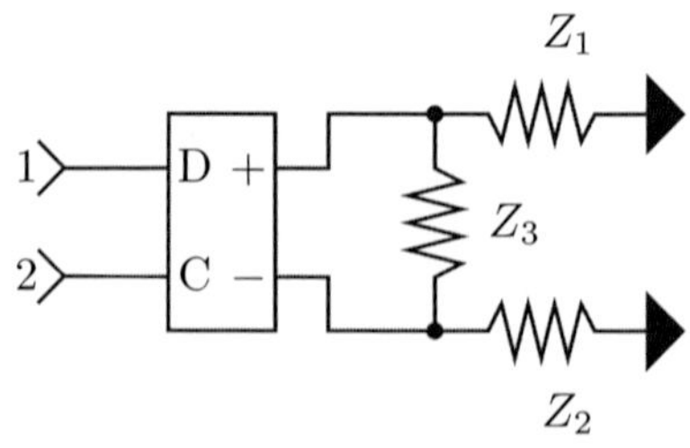


(a) Block diagram

$$\frac{\begin{pmatrix} \frac{Z_1\cdot Z_2\cdot Z_3-2\cdot Z0\cdot Z_1\cdot Z_2-Z0^2\cdot(Z_1+Z_2+Z_3)}{Z_3\cdot Z0\cdot(Z_1-Z_2)} & \frac{Z_3\cdot Z0\cdot(Z_1-Z_2)}{Z_1\cdot Z_2\cdot Z_3+2\cdot Z0\cdot Z_1\cdot Z_2-Z0^2\cdot(Z_1+Z_2+Z_3)} \end{pmatrix}}{Z_1\cdot Z_2\cdot Z_3+Z0\cdot(Z_1\cdot Z_3+2\cdot Z_1\cdot Z_2+Z_3\cdot Z_2)+Z0^2\cdot(Z_1+Z_2+Z_3)}$$

(b) S-parameters

$$\begin{pmatrix} \frac{\frac{Z_3}{2}\|Z-Z0}{\frac{Z_3}{2}\|Z+Z0} & 0 \\ 0 & \frac{Z-Z0}{Z+Z0} \end{pmatrix}$$

(c) S-parameters of balanced configuration; $Z_2 = Z_1 = Z$

**Figure 7.17** Mixed-mode pi termination

For a desired $Z_D$ and $Z_C$, one can calculate the impedances for the pi termination as

$$Z = Z_e = Z_C \cdot 2$$

and solving:

$$\frac{\frac{Z_3}{2}\cdot Z}{\frac{Z_3}{2}+Z} = Z_o = \frac{Z_D}{2}$$

or

$$Z_3 = \frac{Z_C \cdot Z_D}{Z_C - \frac{Z_D}{4}}.$$

Again, if there is an uncoupled, balanced transmission line as before, with both uncoupled lines having a characteristic impedance of $50\,\Omega$, and remembering that the common-mode impedance is half the single-ended impedance, then $Z = Z_C \cdot 2 = 50\,\Omega$ and $Z_3$ is infinity, or open, and the same structure results as for the tee structure solution.

Pi and tee terminations are provided in Table 7.1 for the most commonly used balanced configurations, where $Z = Z_1 = Z_2$. These terminations will appear matched at the D port of a mixed-mode converter in a reference impedance equal to the odd-mode impedance, and will appear matched at the C port in a reference impedance equal to the even-mode impedance.

**Table 7.1** Pi and tee terminations

| **Impedance** | **Tee** | **Pi** |
|---|---|---|
| $Z_o = Z_D/2$ | $Z$ | $(Z_3/2) \parallel Z$ |
| $Z_e = Z_C \cdot 2$ | $2 \cdot Z_3 + Z$ | $Z$ |
| $Z_D = Z_o \cdot 2$ | $2 \cdot Z$ | $Z_3 \parallel (2 \cdot Z)$ |
| $Z_C = Z_e/2$ | $Z_3 + Z/2$ | $Z/2$ |
| $Z$ | $Z_D/2 = Z_o$ | $2 \cdot Z_C = Z_e$ |
| $Z_3$ | $Z_C - Z_D/4 = (Z_e - Z_o)/2$ | $Z_C \cdot Z_D/(Z_C - Z_D/4) =$ $2 \cdot Z_e \cdot Z_o/(Z_e - Z_o)$ |

The complete details of the s-parameter calculations in Figures 7.15(b), 7.16(b), and 7.17(b) are provided in Figures D.20, D.18, and D.19 in Appendix D.

# Part II

## Applications

### Introduction

In the first part of this book, systematic methods were provided for computing s-parameters of circuits and systems containing multiple, interconnected devices. Despite the fact that the problems solved so far have been purposely small, they are still too complicated, and one expects the problems to grow to unwieldy proportions as more devices are added to a system. To solve real problems, systematic methods are needed. Fortunately, as will be seen, the system solutions taught earlier lend themselves well to very simple, programmatic solutions.

In this part, exactly four specific applications of s-parameters are covered to solve specific problems. The first problem was covered theoretically in Part I, where the construction of interconnected systems with s-parameter based elements to solve for the s-parameters of a system was provided. In Chapter 8, a base of software to achieve this is constructed and the fundamental concept of the system description is developed. Here, system descriptions are scripted and converted programmatically, using relatively small pieces of software, to provide system s-parameter solutions. Symbolic methods are shown, along with numeric methods, and finally parser methods that allow construction of a system through a simple text file describing the system.

The concept of a system description is used to solve linear simulation problems in Chapter 9. When simulation problems are solved, it is by generating transfer matrices that can be used to produce filters for processing waveforms, as described in Part III of the book, with the understanding that the main problem is the generation of these transfer matrices.

In Chapter 10, the system description concept is extended to solve de-embedding problems. Here, the classic de-embedding problem is described as solving for unknown elements in a system containing interconnected known and unknown elements, where the s-parameters of the entire system are known at the periphery.

Finally, in Chapter 11, the linear simulation techniques provided in Chapter 9 are expanded to a technique called *virtual probing*. This is a powerful tool that allows measurement and output probes to be placed in a schematic, and, with the identification of sources of waves entering the system, enables the conversion of measured waveforms to output waveforms in the system. This enables time-domain de-embedding and embedding of elements, removal of probing effects, and the ability to probe at inaccessible points in a system.

# 8

# System Descriptions

It was seen in Chapter 4 that the main problem concerning the solution of systems of interconnected s-parameter devices is the setup of the equations. After that, the math is quite simple. This chapter deals with this topic by providing a certain level of abstraction inherent in working with systems of s-parameters: that of converting circuit drawings and descriptions into signal-flow diagrams and more importantly into the systems of equations corresponding to the circuit.

Generally one works with graphical pictures of circuits. When dealing with systems of s-parameters, these pictures are a schematic that shows devices as blocks with numbered ports. Sometimes, such as in the case of inductors, capacitors, and other such simple elements found in electrical engineering, these blocks are shown as special symbols, but they always contain one or more ports. In the general case of a device in s-parameter systems, what is required are the s-parameters of the device, which are often specified as the name of a file containing them. Otherwise, if the device is of a special type, the schematic shows instead the devices along with various values that dictate the s-parameters. For example, a resistor is shown with its resistance value which determines its s-parameters. The schematic shows these devices with lines connecting the device ports in the system. These lines determine the interconnectedness of the devices. Finally, sometimes the end of an unconnected line is designated in a special way; for example as a numbered or named system port. This is shown, for example, in the signal-flow diagram for a cascaded two-port network of Figure 4.8.

Graphical circuit representations help humans work with circuits, but they are generally a representation of a more computer readable form called a *netlist*. A netlist is a computer readable, usually text based form of the description shown in graphical form. Since it is a somewhat trivial matter to convert a graphical description of a system to a netlist and vice versa, the netlist is the starting point for the description of solutions.

Based on the needs of a graphical schematic, a netlist requires a small amount of specialized information. This set of information is called a *system description*.

Figure 8.1 is a universal modeling language (UML) diagram that shows the internal structure of a system description. UML diagrams are used in object-oriented programming and are discussed in §17.2.

With a basic understanding of UML, in simple language, Figure 8.1 states that a system description is a list of devices and that a device has a name, a matrix (a list of lists of complex numbers representing the s-parameters of the device at a single frequency), and a type. Furthermore, a device is a list of ports. A port has an incident node name (`A`), a reflected node name (`B`), and a stimulus name (`M`).

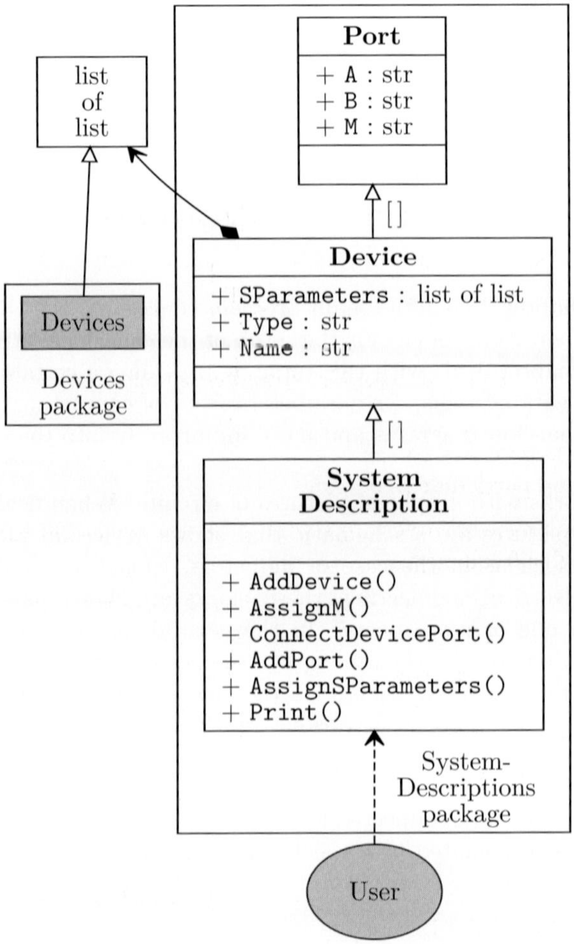


**Figure 8.1** System description UML diagram

## 8.1 System Descriptions

The Python class definitions for a **Port**, **Device**, and **SystemDescription** are provided in Listings 8.1, 8.2, 8.3, and 8.4. These are fewer than two pages of program listings that are capable of assembling a system description. This code will now be described in varying detail depending on the complexity of the implementation after examining the overall structure of these classes using a Python and C++ language style object-oriented nomenclature that defines the **SystemDescription** class structure.

Consider a system description declared as `sd = SystemDescription()`; the access to data in the system description is worth clarifying as follows:

- `sd` is the system description class structure or class instance.

- `len(sd)` is the number of devices.
- `sd[d]` refers to device $d$ (i.e. an instance of class type **Device**), which is also a list of type **Port**.
- `sd[d].Name` refers to the name of device $d$.
- `len(sd[d])` refers to the number of ports in device $d$.
- `sd[d][p]` refers to port $p$ of device $d$ (i.e. an instance of class type **Port**).
- `sd[d][p].A` refers to the name of the node of the wave incident on port $p$ of device $d$.
- `sd[d][p].B` refers to the name of the node of the wave reflected from port $p$ of device $d$.
- `sd[d][p].M` refers to the name of the stimulus emanating from the `B` node of port $p$ of device $d$.
- `sd[d].SParameters` refers to the complex matrix which represents the s-parameters of device $d$ at one particular frequency.
- `len(sd[d].SParameters)` refers to the number of rows in the s-parameters of device $d$ at one particular frequency.
- `sd[d].SParameters[o][i]` refers to the reflected wave from port $o$ due to an incident wave on port $i$ of device $d$ at at one particular frequency.

S-parameters are usually referred to using an indexing notation that begins with "1". For example, $S_{21}$ refers to the reflected wave from port 2 due to an incident wave on port 1. This document refers to s-parameters in this manner, but the data structures utilized are indexed using zero as the starting value. In other words, in the previous example, if the s-parameters at a given frequency are contained in a variable "`s`" then $s_{21}$ would be accessed as $s[1][0]$.

The class definition for a **Port** is shown in Listing 8.1, which is not worth much discussion as it is simply the holder of the three strings `A`, `B`, and `M` which can be read from and written to. A **Port** instance is initialized as empty strings.

The class definition for a **Device** is shown in Listing 8.2. Again, there is not much to note except that it is holder of a name variable called `Name`, an s-parameter matrix called `SParameters`, and a type called `Type`. Because of the class derivation, a **Device** is a list of **Port** instances. It contains a static member function called `SymbolicMatrix()` on line 10, which generates the s-parameter as a matrix formed from the name. Until s-parameters are provided to a device through an assignment of the s-parameter property, these s-parameters are initialized to a symbolic matrix containing a string representation of the device's s-parameters. For example, a two-port device named `'Name'` would have default s-parameters provided as a symbolic matrix `[['Name_11', 'Name_12'], ['Name21', 'Name_22']]`. This is so that the code provided here can be used to provide algebraic, symbolic results.

The main business is done in the **SystemDescription** class provided in Listing 8.3. A **SystemDescription** has three main functions to be accessed, from the perspective of a user, in the construction of a system description. These are the functions `AddDevice()` on line 3, `AddPort()` on line 26, and `ConnectDevicePort()` on line 8. The function

### Listing 8.1 **Port** class

```
class Port(object):
    def __init__(self):
        self.A=''
        self.B=''
        self.M=''
    def IsConnected(self):
        return self.A != ''
    def Print(self,level=0):
        if level==0:
            print('\n','Node','Name')
        for t in range(3):
            if not t==0:
                if level >= 2:
                    print(repr('').strip('\'').rjust(6), end=' ')
                if level >= 1:
                    print(repr('').strip('\'').rjust(4), end=' ')
                    print(repr('').strip('\'').rjust(4), end=' ')
            if t==0:
                print(repr('A').rjust(4),repr(self.A).rjust(4))
            elif t==1:
                print(repr('B').rjust(4),repr(self.B).rjust(4))
            else:
                print(repr('M').rjust(4),repr(self.M).rjust(4))
```

### Listing 8.2 **Device** class

```
class Device(list):
    def __init__(self,Name,Ports,Type='device'):
        self.Name = Name
        list.__init__(self,[Port() for _ in range(Ports)])
        self.SParameters = self.SymbolicMatrix(Name,Ports)
        self.Type=Type
    def AssignSParameters(self,SParameters):
        self.SParameters=SParameters
    @staticmethod
    def SymbolicMatrix(Name,Rows,Columns=-1):
        if Columns == -1:
            Columns = Rows
        if Rows == 1 and Columns == 1:
            return [[Name]]
        return [[Name+'_'+(str(r+1)+str(c+1) if r<9 and c<9 else str(r+1)+','+str(c+1))
                for c in range(Columns)] for r in range(Rows)]
    def Print(self,level=0):
        if level==0:
            print('\n','Name','Port','Node','Name')
        for p in range(len(self)):
            if p==0:
                print(repr(self.Name).rjust(4), end=' ')
            else:
                if level==1:
                    print(repr('').strip('\'').rjust(6), end=' ')
                print(repr('').strip('\'').rjust(4), end=' ')
            print(repr(p+1).rjust(4), end=' ')
            self[p].Print(level+1)
```

**Listing 8.3 SystemDescription** class (important functions)

```
class SystemDescription(list):
...
    def AddDevice(self,Name,Ports,SParams=None,Type='device'):
        self.append(Device(Name,Ports,Type))
        if isinstance(SParams,list):
            self.AssignSParameters(Name,SParams)
...
    def ConnectDevicePort(self,FromN,FromP,ToN,ToP):
        dfi = self.IndexOfDevice(FromN)
        dti = self.IndexOfDevice(ToN)
        if not self[dfi][FromP-1].IsConnected():
            if not self[dti][ToP-1].IsConnected():
                uN1,uN2 = (self.m_UniqueNode.Name(),self.m_UniqueNode.Name())
                self._InsertNodeName(FromN,FromP,uN2,uN1)
                self._InsertNodeName(ToN,ToP,uN1,uN2)
            else:
                self.ConnectDevicePort(ToN,ToP,FromN,FromP)
        else:
            TeeN = self.m_UniqueDevice.Name()
            self.AddDevice(TeeN,3)
            self.AssignSParameters(TeeN,Tee())
            self._InsertNodeName(TeeN,1,self[dfi][FromP-1].A,self[dfi][FromP-1].B)
            self._InsertNodeName(FromN,FromP,'','')
            self.ConnectDevicePort(FromN,FromP,TeeN,2)
            self.ConnectDevicePort(TeeN,3,ToN,ToP)
    def AddPort(self,DeviceName,DevicePort,SystemPort,AddThru=False):
        PortName = 'P'+str(SystemPort)
        self.AddDevice(PortName,1,[[0.0]])
        self.AssignM(PortName,1,'m'+str(SystemPort))
        if not AddThru:
            AddThru = self[self.IndexOfDevice(DeviceName)].Type == 'unknown'
        if AddThru:
            thruName=self.m_UniqueDevice.Name()
            self.AddDevice(thruName,2,Thru())
            self.ConnectDevicePort(PortName,1,thruName,1)
            self.ConnectDevicePort(thruName,2,DeviceName,DevicePort)
        else:
            self.ConnectDevicePort(PortName,1,DeviceName,DevicePort)
...
```

`AddDevice()` takes the device `Name`, the number of `Ports` and optional `SParams`. Since a **SystemDescription** is an array of devices, `AddDevice()` appends a new instance of a **Device** to the array. This triggers the instantiation of a **Device** with the given name which in turn instantiates an array of **Port**, with incident, reflected, and stimulus wave names defaulted to empty strings. If no s-parameters are provided, then the **Device** instance will be left initialized as the symbolic matrix defined by the `Name` and with numbers of rows and columns defined by the `Ports`. The function `AddPort()` takes the `DeviceName` and `DevicePort` to which the new system port is connected along with the number of the `SystemPort` being added. Ignoring the optional `AddThru` argument for a moment, this function adds a single-port device with a name formed by concatenating `'P'` with the `SystemPort` number and with s-parameters of 0, then connects this device to the `DeviceName` and `DevicePort` number specified through the `ConnectDevicePort()` function. It also assigns the stimulus on the system port device with a name associated with the system port number. The system port device is therefore perfectly matched to the reference

**Listing 8.4 SystemDescription** class (basic functions)

```
class SystemDescription(list):
    def __init__(self,sd=None):
        if not sd is None:
            list.__init__(self,list(sd))
            self.m_UniqueDevice=sd.m_UniqueDevice
            self.m_UniqueNode=sd.m_UniqueNode
        else:
            list.__init__(self,[])
            self.m_UniqueDevice=UniqueNameFactory('#')
            self.m_UniqueNode=UniqueNameFactory('n')
...
    def AssignM(self,DeviceN,DeviceP,MName):
        di = self.IndexOfDevice(DeviceN)
        self[di][DeviceP-1].M = MName
    def DeviceNames(self):
        return [self[d].Name for d in range(len(self))]
    def IndexOfDevice(self,DeviceName):
        return self.DeviceNames().index(DeviceName)
    def _InsertNodeName(self,DeviceName,Port,AName,BName):
        di = self.IndexOfDevice(DeviceName)
        self[di][Port-1].A = AName
        self[di][Port-1].B = BName
    def CheckConnections(self):
        if len(self)==0:
            raise SignalIntegrityExceptionSystemDescription('no devices')
        if not all([self[d][p].IsConnected()
            for d in range(len(self)) for p in range(len(self[d]))]):
            raise SignalIntegrityExceptionSystemDescription('unconnected device ports')
...
    def AssignSParameters(self,DeviceName,SParameters):
        self[self.IndexOfDevice(DeviceName)].AssignSParameters(SParameters)
    def Print(self):
        print('\n','Device','Name','Port','Node','Name')
        for d in range(len(self)):
            print(repr(d+1).rjust(6), end='  ')
            self[d].Print(1)
```

impedance of the system with a stimulus emanating from it incident on the `DeviceName` and `DevicePort` specified.

The `ConnectDevicePort()` function is the trickiest as it must handle four possibilities for device port connection and utilizes recursion. It takes as arguments `FromN` and `FromP`, which are the name and port number of the device from which the connection is made (i.e. the *from* device/port), and `ToN` and `ToP`, which are the name and port number to which the connection is made (i.e. the *to* device/port). In other words, the function connects *from* one device/port *to* another device/port. The four possibilities are: neither device/port has been previously connected; the *from* device/port has been connected but the *to* device/port has not; the *to* device/port has been connected but the *from* device/port has not; both device/ports have already been connected. If they are connected, the presumption is that they have already been connected to some other device/port and the desire is to connect them to another device/port as well, as outlined in §4.1.4.

The simplest case is addressed first: that of no previous connection on either device/port. In this case, two unique node names are first created, `uN1` and `uN2`, and, following the device connection rules, the *from* device/port's `A` and `B` nodes are assigned to `uN2` and `uN1`, while the *to* device/port's `A` and `B` nodes are assigned oppositely to `uN1` and `uN2`. This assigns the wave reflected from one device/port as the wave incident on the other device/port and vice versa, as outlined in §4.1.2. It turns out that the actual node names are unimportant, so they are generated automatically.

The next simplest case is that there is no previous connection on the *from* device/port but there is a previous connection on the *to* device/port. Since the assignment of one device/port as *from* and another as *to* is arbitrary, one can simply switch which device/port is *from* and which is *to* and call `ConnectDevicePort()` recursively. Since it has already been verified that the *from* device (soon to be the *to* device) is not connected and that the *to* device (soon to be the *from* device) is connected, on reentering the `ConnectDevicePort()` function, one finds that the *from* device is connected.

The remaining cases can now be handled in the context of the *from* device/port already being connected, with judgment suspended on whether the *to* device/port is connected or not.

If the *from* device was previously connected, one sees from the outer else clause of the `ConnectDevicePort()` function in Listing 8.3 that the action is to create a three-port tee device and add it to the system description. The node names of the reflected and incident waves on the *from* device/port are copied to port 1 of this new tee device, and the *from* device/port is disconnected by assigning the reflected and incident waves to empty nodes. This has the effect of connecting the device/port that the *from* device/port was connected to to the new tee device and disconnecting the *from* device/port from anything. Then, the *from* device/port is connected to port 2 of the tee through the `ConnectDevicePort()` function, called recursively. This will follow the first case of no device/ports connected because the *from* device/port was disconnected and port 2 of the newly created tee device is also not connected. Finally, the *to* device/port is connected to port 3 of the tee device, again through the `ConnectDevicePort()` recursive call. This can go two ways – either the *to* port is or is not already connected. If it is not already connected, it will follow again the case of two not previously connected ports, otherwise another tee device will be created and connected to the *to* port in the manner already described.

The result of the use of these three functions on a **SystemDescription** instance is to create an array of devices, each device being an array of ports, each port containing the names `A`, `B`, and `M`, with all of these names filled in based on the interconnectedness of the system. In fact, checking whether the system is connected is as simple as checking that all of the names contain strings, as seen in the `CheckConnections()` function on line 23 in Listing 8.4.

It is important to understand how the s-parameters of a system are calculated based on such a construction. Consider the class **SystemSParameters** provided in Listing 8.5. This class is derived from the **SystemDescription** class. This class derivation is provided to separate the function of computing s-parameters from the function of assembling the system description. The idea is that one constructs a system description from an instance of **SystemSParameters**, which inherits all of the functionality of the **SystemDescription** class or starts with an instance of **SystemDescription** and then "up-casts" to **SystemSParameters** by instantiating one with the **SystemDescription** class instance provided as an argument.

**Listing 8.5 SystemSParameters** class

```
1 class SystemSParameters(SystemDescription):
2     def __init__(self,sd=None):
3         SystemDescription.__init__(self,sd)
4     def PortANames(self):
5         return [x[1] for x in sorted
6                 ([(int(self[d].Name.strip('P')),self[d][0].A)
7                   for d in range(len(self)) if self[d].Name[0]=='P'])]
8     def PortBNames(self):
9         return [x[1] for x in sorted
10                ([(int(self[d].Name.strip('P')),self[d][0].B)
11                  for d in range(len(self)) if self[d].Name[0]=='P'])]
12    def OtherNames(self,K):
13        other=[]
14        for item in self.NodeVector():
15            if not item in K: other.append(item)
16        return other
17    def NodeVector(self):
18        return [self[d][p].B for d in range(len(self)) for p in range(len(self[d]))]
19    def StimulusVector(self):
20        return [self[d][p].M for d in range(len(self)) for p in range(len(self[d]))]
21    def WeightsMatrix(self,ToN=None,FromN=None):
22        if not isinstance(ToN,list):
23            nv = self.NodeVector()
24            ToN = nv
25        if not isinstance(FromN,list):
26            FromN=ToN
27        PWM = [[0]*len(FromN) for r in range(len(ToN))]
28        for d in range(len(self)):
29            for p in range(len(self[d])):
30                if self[d][p].B in ToN:
31                    r=ToN.index(self[d][p].B)
32                    for c in range(len(self[d])):
33                        if self[d][c].A in FromN:
34                            ci=FromN.index(self[d][c].A)
35                            PWM[r][ci]=self[d].SParameters[p][c]
36        return PWM
```

The **SystemSParameters** class contains the basic methods utilized by all other derived classes that make use of system descriptions; all of the other classes that make use of system descriptions derive from the **SystemSParameters** class. Three main methods are provided for obtaining the necessary information for a solution. These are, in Listing 8.5, `NodeVector()` on line 17, `StimulusVector()` on line 19, and `WeightsMatrix()` on line 21; these allow the generation of the complete system equation in either numeric or symbolic form. For the block matrix solutions, which require the categorization of the nodes and the generation of block matrices, the functions `PortANames()`, `PortBNames()`, and `OtherNames()` on lines 4, 8, and 12 are provided. Support for block matrices is built into the `WeightsMatrix()` function on line 21, which has been generalized to deal with all of the block weights matrix cases.

Understanding the operation of the **SystemSParameters** class begins with an understanding of the node vector generated by the `NodeVector()` function. It is known that there is nothing special about the ordering of this vector. Since it is arbitrary, this function simply loops over all of the devices and ports, extracts the `B` node name, and puts these names in a list. This is the choice made for the node vector, and it is very simple.

The equation ordering in the system equation is not arbitrary once a node vector has been determined; each row of the stimulus vector and weights matrix must correspond to the equation for the given node at that row of the node vector. In other words, the ordering of the node vector determines everything else about the system equation. Therefore, the stimulus vector is generated by the `StimulusVector()` function in the exact same way as the node vector; each device and port is looped over and the stimulus name is extracted and put in the stimulus vector list. This ensures that it is in an order consistent with the node vector.

The weights matrix generation by the `WeightsMatrix()` function is simple to describe, but difficult to understand. For every device and port in the system description, the `B` node name is obtained. This is the name of the node from which arrows in the signal-flow diagram emanate. For a given device and port containing the `B` node name, the source of all arrows terminating on the node can only be on the periphery of the device, and the only possibilities are the incident waves on the device (i.e. from nodes named by the `A` node names on every port of the device). Furthermore, the weights of the arrows in the signal-flow diagram can only be the s-parameters of the device as the arrows originate at the device incident nodes and terminate on the device reflected nodes. Therefore, the algorithm is fairly simple. Each device `d` and each port `p` of that device is looped over. For each `d` and `p`, the `B` node name is obtained and the row `r` found in the weights matrix from the corresponding row of the `B` node name in the node vector. Then each of the $P$ ports in device `d` are looped over, defining each port as `c`, the column in the s-parameter matrix for the weight. For each port `c`, the `A` node name and its index in the node vector is obtained as `ci`. This `ci` is the column in which the weight will be placed in the weights matrix. The value of the weight to place there is that of the device's s-parameter matrix at row `p` and column `c`.

The `WeightsMatrix()` function can be called with optional arguments defining `ToN` and `FromN`, which are the lists containing nodes that arrows terminate on and originate from, respectively. If `ToN` and `FromN` are not provided, then the node vector is used for `ToN` and `FromN`, and the weights matrix provided is the weights matrix for the entire system. Otherwise, `WeightsMatrix()` can be called with various lists corresponding to the nodes containing waves incident on the system, reflected from the system, or neither incident nor reflected from the system, as outlined in §4.4. Note that the waves incident on the system are found by calling `PortBNames()` as these are the waves emanating from the single-port, perfectly terminated, port device added through an `AddPort()` call.

## 8.2 System Description Example

The purpose of the simple example given here is to show how the **SystemDescription** and **SystemSParameters** classes automate the setup of the equations required for system solution. The capability for solving for the system s-parameters using this small amount of software will be demonstrated in §8.3.

Consider the Python code in Figure 8.2(a). This is a Python program duplicating the setup of the example provided in §4.7.2. On line 2 is the importation of the *SignalIntegrity* Python library, and on line 3 the instantiation of a **SystemDescription** class called `sd`.

```
from __future__ import print_function
import SignalIntegrity.Lib as si
sd = si.sd.SystemDescription()
sd.AddDevice('L', 2)  # add two-port left device
sd.AddDevice('R', 2)  # add two-port right device
sd.AddPort('L', 1, 1)  # add a port at port 1 of left device
sd.AddPort('R', 2, 2)  # add a port at port 2 of right device
sd.ConnectDevicePort('L', 2, 'R', 1)  # connect the other ports
sd.Print()  # print the system description
spc = si.sd.SystemSParameters(sd)
n = spc.NodeVector() # get the node vector
m = spc.StimulusVector()  # get the stimulus vector
W = spc.WeightsMatrix()  # get the weights matrix
# print out the vectors and matrices
print(('{0:' + str(5 * len(n)) + '}').format('Weights Matrix'), end=' ')
print('| {0:4}'.format('n'), end=' ')
print('| {0:4} |'.format('m'))
print('-------------------------------------------')
for r in range(len(W)):
    for c in range(len(W[r])):
        print('{0:4}'.format(str(W[r][c])), end=' ')
    print(' | {0:4}'.format(n[r]), end=' ')
    print('| {0:4} |'.format(m[r]))
print('-------------------------------------------')
```

(a) Python code

```
  Device Name Port Node Name
       1  'L'    1  'A'  'n1'
                    'B'  'n2'
                    'M'  ''
                 2  'A'  'n6'
                    'B'  'n5'
                    'M'  ''
       2  'R'    1  'A'  'n5'
                    'B'  'n6'
                    'M'  ''
                 2  'A'  'n3'
                    'B'  'n4'
                    'M'  ''
       3 'P1'    1  'A'  'n2'
                    'B'  'n1'
                    'M'  'm1'
       4 'P2'    1  'A'  'n4'
                    'B'  'n3'
                    'M'  'm2'
Weights Matrix                | n    | m    |
-------------------------------------------
0    0    L_12 0    L_11 0    | n2   |      |
0    0    L_22 0    L_21 0    | n5   |      |
0    R_11 0    0    0    R_12 | n6   |      |
0    R_21 0    0    0    R_22 | n4   |      |
0.0  0    0    0    0    0    | n1   | m1   |
0    0    0    0.0  0    0    | n3   | m2   |
-------------------------------------------
```

(b) Text output

**Figure 8.2** System description example

Lines 4 through 8 of Figure 8.2(a) contain calls on `sd` to describe the interconnected network. Lines 4 and 5 add two-port devices called `'L'` and `'R'`, just as in Figure 4.26; lines 6 and 7 add system ports to port 1 of `'L'` and to port 2 of `'R'`; and line 8 connects port 2 of `'L'` to port 1 of `'R'`. The description of the network is complete.

The system description is printed on line 9, the output of which is shown in the first part of Figure 8.2(b) and will be discussed subsequently. Line 10 converts the completed system description into an instance of the **SystemSParameters** class. For this example, it will be shown only that the tedious activity of setting up the equations has been taken care of. On lines 11 and 12, the node vector and the stimulus vector are extracted from the system description, and the weights matrix is obtained on line 13.

Because each device added is initialized symbolically, the weights matrix provided will remain symbolic (i.e. as text). If desired, one could have provided s-parameters for the `'L'` and `'R'` devices through a call to `AssignSParameters()` shown on line 30 of Listing 8.4, which allows the provision of the `DeviceName` and a complex matrix of `SParameters`, in which case, the weights matrix returned would be entirely numerical.

Finally, lines 14–24 show some tedium of printing out a formatted view of the system equation. Its output is shown as the latter part of Figure 8.2(b).

One of the outputs of Figure 8.2(a) was a printout of the system description. This was generated through a call on `Print()` on the **SystemDescription** class shown on line 32 of Listing 8.4, which can be seen to loop over `Print()` calls on the **Device** class shown on line 17 of Listing 8.2 that in turn loops over `Print()` calls on the **Port** class shown on line 8 of Listing 8.1. In other words, the act of printing system descriptions involves mostly printing each device in the system description, which involves printing the device number and name along with each of the ports in the device.

So, here the four devices are listed, two of which were added by the user called `'L'` and `'R'`, along with two which were added as a result of adding the two system ports called `'P1'` and `'P2'`. These port devices are a deviation from anything discussed previously so they deserve some explanation.

In Chapter 4, to solve this problem, one applied stimuli to the $a$ nodes corresponding to port 1 of device **L** and port 2 of device **R**; these $a$ nodes would be the source of incident waves on the system, and the $b$ nodes would be the source of reflected waves from the system, thus enabling the determination of the system s-parameters. Here, to solve the bookkeeping in an automated fashion, the convention is that the stimulus (`M`) for a given port emanates from the `B` node for that port. Therefore, in order to provide these stimuli on the `A` nodes, system port devices `'P1'` and `'P2'`, whose stimuli are emanating from their `B` nodes, must be generated and connected to the device ports in question. Thus the `B` nodes containing the stimuli in `'P1'` and `'P2'` are now incident on the desired device ports. Furthermore, there is a location in the system containing only the list of stimuli and system incident and reflected waves: all of the system ports. Of course, these system ports were initialized with s-parameters of zero so that they do not have any effect on the system (i.e. no arrow with any weight in the system). Unlike other devices added, there is no attempt to initialize the s-parameters of the port symbolically. These devices are set directly with single-port s-parameters of zero. These two devices with no effect added to the system do not change anything, as these two zero-weight arrows are implicit in the description of the problem, as shown by the faint gray zero-weight arrows in Figure 4.26. The use of these two system port devices enforces the notion that in order to solve the system, everything must be connected to something. In other words, each port must contain both an `A` and a `B` node connection and each of these `A`/`B` pairs must be connected to a reversed `B`/`A` pair on another device port. Thus, a check for a fully connected system is performed by calling `CheckConnections()` at line 23 of Listing 8.4, which loops over each device port calling `IsConnected()` on the **Port** class at line 6 of Listing 8.1, which in turn just checks its `A` node (because connecting any device ports through legal calls of `ConnectDevicePort()` or `AddPort()` on the **SystemDescription** class must always result in the assignment of `A` and `B` nodes simultaneously.

Finally, note that the devices appear in the order in which they were added to the system. The Python code began with two calls to `AddDevice()`, which added the two devices `'L'` and `'R'`, and then two calls to `AddPort()`, which added the two system ports and therefore determined the order of the devices in the system description as shown in Figure 8.2(b). Also note that, because there were no attempts to connect multiple device ports to any device port, there were no tee devices created automatically. If they were added, like the system ports, these tee devices would have numerical s-parameters added automatically according to (3.14).

Now an examination is made of the ports of the devices. For devices 1 and 2, these are two-port devices as specified in the `AddDevice()` calls, and for devices 3 and 4, these are one-port devices as explained for the operation of the `AddPort()` calls.

After the `AddDevice()` calls, the first system port added is port 1, defined as port 1 of device `'L'`. Therefore, nodes `'n1'` and `'n2'` are the `B` and `A` nodes on `'P1'` and are the `A` and `B` nodes on port 1 of device `'L'`. The next system port added is port 2, as defined as port 2 of device `'R'`. Therefore nodes `'n3'` and `'n4'` are the `B` and `A` nodes on `'P2'` and are the `A` and `B` nodes on port 2 of device `'R'`. Note that the two `AddPort()` calls also defined the stimulus on `'P1'` and `'P2'` as `'m1'` and `'m2'`.

Finally, the `ConnectDevicePort()` call adds nodes `'n5'` and `'n6'` as the `B` and `A` nodes of the *from* port 2 of device `'L'` and as the `A` and `B` nodes of the *to* port 1 of device `'R'`.

The result is six nodes for six total device ports in the system, each listed exactly twice: once as an `A` node of a device port and once as a `B` node of another device port. The listed system description in the first part of Figure 8.2(b) entirely defines the system and, as will be seen, the system equation.

The node vector, stimulus vector, and weights matrix were obtained in lines 11–13 of Figure 8.2(a), and were printed in lines 14–24 in a formatted manner. Having the system description means that it is a trivial matter (for a computer) to generate the required node and stimulus vector, and weights matrix that provide for a complete description of the system and the system equation.

In subsequent sections, the Python **SystemDescription** and **SystemSParameters** classes will be used to solve the system fully.

## 8.3 Symbolics

In §8.2, a system description was generated using the **SystemDescription** class and various information that described the system was output. In subsequent sections, similar

**Listing 8.6 Symbolic** class

```
class Symbolic():
    def __init__(self,**args):
        self.m_lines=[]
        size = args['size'] if 'size' in args else 'normal'
        self.m_docStart = args['docstart'] if 'docstart' in args else\
        '\\documentclass[10pt]{article}\n'+'\\usepackage{amsmath}\n'+\
        '\\usepackage{bbold}\n'+'\\begin{document}'
        self.m_docEnd = args['docend'] if 'docend' in args else '\\end{document}'
        self.m_eqPrefix = args['eqprefix'] if 'eqprefix' in args else '\\[ '
        self.m_eqSuffix = args['eqsuffix'] if 'eqsuffix' in args else ' \\]'
        self.m_identity = '\\boldsymbol{\\mathbbm{I}} '
        self.m_eqEnvironment = args['eqenv'] if 'eqenv' in args else True
        self.m_small= (size == 'small')
    def Clear(self):
        self.m_lines=[]
        return self
    def Emit(self):
        for line in self.m_lines: print(line)
        return self
    def DocStart(self):
        self._AddLine(self.m_docStart)
        return self
    def DocEnd(self):
        self._AddLine(self.m_docEnd)
        return self
    def _BeginEq(self):
        if self.m_eqEnvironment:
            return self.m_eqPrefix
        else: return ''
    def _EndEq(self):
        if self.m_eqEnvironment:
            return self.m_eqSuffix
        else: return ''
    def _AddLine(self,line):
        if len(line) == 0: return
        wlinelist=wrap(line)
        for wline in wlinelist: self.m_lines.append(wline)
        return self
    def _AddLines(self,lines):
        for line in lines: self._AddLine(line)
        return self
    def WriteToFile(self,name):
        with open(name,'w') as equationFile:
            for line in self.m_lines: equationFile.write(line+'\n')
    def _SmallMatrix(self):
        return self.m_small
    def _Identity(self):
        return self.m_identity
    def Get(self):
        lineBuffer=''
        for line in self.m_lines:
            lineBuffer=lineBuffer+line+'\n'
        return lineBuffer
...
    def _AddEq(self,text):
        self._AddLine(self._BeginEq() + text + self._EndEq())
        return self
    def _LaTeXMatrix(self,matrix):
        return Matrix2LaTeX(matrix,self._SmallMatrix())
```

methods are employed to output *symbolic* results. Symbolic means algebraic, mathematically rendered solutions that are readable as equations. All classes that produce symbolic results are derived from the **Symbolic** class.

The **Symbolic** class is shown in Listing 8.6 and contains odds and ends used to deal with symbolic information. On examining Listing 8.6, one finds that it is basically a collection of *lines* of text and, although the class does not contain any capability for determining what these lines of text ought to be, it supplies the base methods for dealing with these lines.

Symbolic information involves LaTeX code, and all of the classes that derive from **Symbolic** fill in LaTeX code in these lines. Although the entire details of LaTeX are probably unfamiliar to the reader and are really beyond the scope of this text, here are some important points:

- LaTeX is a powerful language for typesetting, especially of math oriented books; it was originated by the famous Stanford professor, Donald Knuth, himself the author of many seminal books on mathematics and computer programming. It should not be surprising that the current book was typeset in LaTeX and all of the symbolic classes in the *SignalIntegrity* package were utilized to test the math provided in this book and to provide some of the results.
- LaTeX provides arguably the only standard and somewhat official way of entering equations using ASCII characters. Perhaps the newer MathML standard, which has the specific goal of representing math between machines, will eventually take the place of LaTeX. LaTeX code is in a form that can be directly typeset by a LaTeX typesetting program or directly pasted into other math tools. For example, the program MathType[1] allows direct entry in LaTeX, and Maple[2] provides for direct equation entry in LaTeX.

Here are some basics of LaTeX code:

- A LaTeX document begins with code such as

  ```
  \documentclass[10pt]{article}
  \usepackage{amsmath}
  \begin{document}
  ```

  This tells a LaTeX typesetting program to begin a document based on the document class **article** to use a 10 point font, to use the American Mathematical Society (AMS) math package, and to start the document.
- A LaTeX document ends with `'\end{document}'`.
- A LaTeX equation can begin and end in a variety of ways. The simplest is with the `'\[ '` and `' \]'` strings. Equations delimited in this manner can be directly *included* into a LaTeX document and will be typeset as equations. Other forms are bracketing with '$' or `'\begin{equation}'` and `'\end{equation}'`. Primarily, the `'\[ '` and `' \]'` bracketing strings are used because they allow multiple equations to be brought into LaTeX.

[1]MathType® is a trademark of Design Science, Inc.

[2]Maple® is a trademarks of Waterloo Maplesoft, Inc.

- A LaTeX matrix appears as
  `'\[ \left(\begin{array}{cc}1&2\\3&4\end{array}\right)\]'`
  A line like this with a complete equation would typeset as $\left(\begin{array}{cc}1&2\\3&4\end{array}\right)$.
- LaTeX matrices can be made smaller using a small matrix environment. For example,
  `'\[\left(\begin{smallmatrix}1&2\\3&4\end{smallmatrix}\right)\]'`
  would typeset as $\left(\begin{smallmatrix}1&2\\3&4\end{smallmatrix}\right)$.
- A bold font variable name (as used for matrices) can be specified. For example, '`\mathbf{H}`' would typeset as $\mathbf{H}$.
- A multiplication symbol (a dot) uses the macro `'\cdot'`. For example, `'a\cdot b'` typesets as $a \cdot b$.
- Fractions use the macro `'\frac{}{}'`. For example, `'\frac{a}{b}'` typesets as $\frac{a}{b}$.
- Subscripts are achieved using an underscore. For example, `'S_{12}'` typesets as $S_{12}$.
- Superscripts are achieved using a caret. For example, `'\mathbf{A}^{-1}'` typesets as $\mathbf{A}^{-1}$.
- Parentheses, brackets, and braces are opened and closed using macros such as `'\left('`, `'\left['`, and `'\left{'` for opening and macros such as `'\right)'`, `'\right]'`, and `'\right}'` for closing. For example, `'a\cdot\left(b+c\right)'` typesets as $a \cdot (b + c)$.

These basics of LaTeX are thought to contain all of the information needed to understand the LaTeX output from any of the **Symbolic** derived classes. After all, one never needs to read the LaTeX code, just typeset it.

The **Symbolic** class contains many private and public methods. The private methods are intended for use internally or by the other derived classes (in C++ this would be called *protected*) and have their method names preceded by an underscore. The public methods are worth describing as they are intended for public use. These are:

- `__init__()` This function is private but is the constructor of the **Symbolic** class, and this function will be called by the constructors of all derived classes. This function takes a *dictionary* of arguments. A Python dictionary is a list of keys and values, which are assigned in the argument list of the function and specify various options regarding how symbolic results will be created. The possible keys and values and their explanations are shown in Table 8.1.
- `Clear()` clears the lines.
- `Emit()` prints the lines.
- `WriteToFile()` outputs the lines to a file.
- `DocStart()` adds the document start to the lines. This should be called first prior to any other function call that adds lines to ensure that this is on the first line. It adds the string in `m_docStart` to the line list which is set through the `docstart` key value in the constructor arguments.

**Table 8.1** Arguments for **Symbolic** class

| Key | Value | Description |
|---|---|---|
| eqenv | True (default) | Bracket all LaTeX equations in the equation environment (i.e. before each equation insert the member string m_eqPrefix and after each equation insert the member string m_eqSuffix). |
| | False | Do not bracket equations with any extra strings. |
| size | 'normal' (default) | Display LaTeX matrices using normal sizes – in LaTeX this means beginning matrices with the macro '\begin{array}...'. |
| | 'small' | Display LaTeX matrices using small sizes – in LaTeX this means beginning matrices with the macro '\begin{smallmatrix}...'. |
| eqprefix | | Defines the string to be used as the LaTeX equation prefix (i.e. sets the member string m_eqPrefix). The default value is '\[ '. |
| eqsuffix | | Defines the string to be used as the LaTeX equation suffix (i.e. sets the member string m_eqSuffix). The default value is ' \]'. |
| docstart | | Defines the string to be used as the LaTeX document start, if desired (i.e. set the member string m_docStart). The default value is '\documentclass[10pt]{article} \usepackage{amsmath} \usepackage{bbold} \begin{document}'. |
| docend | | Defines the string to be used as the LaTeX document end, if desired (i.e. set the member string m_docEnd). The default value is '\end{document}'. |

- `DocEnd()` adds the document end to the lines. This should be called at the very end of any functions that add to the lines prior to any call to `Emit()`, `WriteToFile()`, or `Get()`. It adds the string in the `m_docEnd` member variable, which is set through the `docend` key value in the constructor arguments.
- `Get()` gets the list of lines and returns them as one long string delimited by the newline `'\n'` character.

The **Symbolic** class provides methods that allow handling of the basic housekeeping operations of maintaining and outputting symbolic information in LaTeX format. The derived classes fill in this information.

A very simple example of the use of the class is shown in Figure 8.3, which shows a simple script in Figure 8.3(a), the output of the script in Figure 8.3(b), and the result of typesetting this output in Figure 8.3(c).

```
import SignalIntegrity.Lib as si
symbolic = si.sd.Symbolic()
symbolic.DocStart()
symbolic._AddEq('\\mathbf{S}='+symbolic._LaTeXMatrix(si.sy.SeriesZ('Z')))
symbolic.DocEnd()
symbolic.WriteToFile('Symbolic.tex')
```

(a) Python code

```
\documentclass[10pt]{article} \usepackage{amsmath} \usepackage{bbold}
\begin{document}
\[ \mathbf{S}=\left(\begin{array}{cc} \frac{Z}{Z+2\cdot Z0} &
\frac{2\cdot Z0}{Z+2\cdot Z0} \\ \frac{2\cdot Z0}{Z+2\cdot Z0} &
\frac{Z}{Z+2\cdot Z0} \end{array}\right) \]
\end{document}
```

(b) LaTeX output

$$\mathbf{S}=\left(\begin{array}{cc} \frac{Z}{Z+2\cdot Z0} & \frac{2\cdot Z0}{Z+2\cdot Z0} \\ \frac{2\cdot Z0}{Z+2\cdot Z0} & \frac{Z}{Z+2\cdot Z0} \end{array}\right)$$

(c) LaTeX processed equations

**Figure 8.3** Symbolic example

**Listing 8.7 SystemDescriptionSymbolic** class

```
class SystemDescriptionSymbolic(SystemSParameters,Symbolic):
    def __init__(self,sd=None,**args):
        SystemSParameters.__init__(self,sd)
        Symbolic.__init__(self,**args)
    def LaTeXSystemEquation(self):
        sW=self._LaTeXMatrix(self.WeightsMatrix())
        sn=self._LaTeXMatrix(SubscriptedVector(self.NodeVector()))
        sm=self._LaTeXMatrix(SubscriptedVector(self.StimulusVector()))
        self._AddEq('\\left[ '+self._Identity()+' - '+sW+'\\right] \\cdot '+\
            sn+' = '+sm)
        return self
```

### 8.3.1 Symbolic System Descriptions

While the **SystemDescription** class is capable of describing the interconnectedness of the system, derived classes can be created that make use of this information to produce algebraic results. The simplest such derived class is the **SystemDescriptionSymbolic** class, whose Python code is shown in Listing 8.7.

The **SystemDescriptionSymbolic** class derives from two classes, **SystemDescription** and **Symbolic**. This is called multiple inheritance, and the new **SystemDescription-Symbolic** class takes on the attributes of each of these two base classes. The **System-Description** class has been discussed previously in Listing 8.4 and Listing 8.3 and the **Symbolic** class was explained at the start of §8.3.

```
1 import SignalIntegrity.Lib as si
2 ssps = si.sd.SystemDescriptionSymbolic()
3 ssps.AddDevice('L', 2)  # add two-port left device
4 ssps.AddDevice('R', 2)  # add two-port right device
5 ssps.AddPort('L', 1, 1)  # add a port at port 1 of left device
6 ssps.AddPort('R', 2, 2)  # add a port at port 2 of right device
7 ssps.ConnectDevicePort('L', 2, 'R', 1)  # connect the other ports
8 ssps.LaTeXSystemEquation().Emit()
```

(a) Python code

```
\[ \left[ \boldsymbol{\mathbbm{I}}  - \left(\begin{array}{cccccc} 0 &
0 & L_{12} & 0 & L_{11} & 0 \\ 0 & 0 & L_{22} & 0 & L_{21} & 0 \\ 0 &
R_{11} & 0 & 0 & 0 & R_{12} \\ 0 & R_{21} & 0 & 0 & 0 & R_{22} \\ 0 &
0 & 0 & 0 & 0 & 0 \\ 0 & 0 & 0 & 0 & 0 & 0 \end{array}\right)\right]
\cdot \left(\begin{array}{c} n_{2} \\ n_{5} \\ n_{6} \\ n_{4} \\ n_{1}
\\ n_{3} \end{array}\right) = \left(\begin{array}{c} 0 \\ 0 \\ 0 \\ 0
\\ m_{1} \\ m_{2} \end{array}\right) \]
```

(b) LaTeX output

$$\left[ \mathbb{I} - \left(\begin{array}{cccccc} 0 & 0 & L_{12} & 0 & L_{11} & 0 \\ 0 & 0 & L_{22} & 0 & L_{21} & 0 \\ 0 & R_{11} & 0 & 0 & 0 & R_{12} \\ 0 & R_{21} & 0 & 0 & 0 & R_{22} \\ 0 & 0 & 0 & 0 & 0 & 0 \\ 0 & 0 & 0 & 0 & 0 & 0 \end{array}\right)\right] \cdot \left(\begin{array}{c} n_{2} \\ n_{5} \\ n_{6} \\ n_{4} \\ n_{1} \\ n_{3} \end{array}\right) = \left(\begin{array}{c} 0 \\ 0 \\ 0 \\ 0 \\ m_{1} \\ m_{2} \end{array}\right)$$

(c) LaTeX processed equations

**Figure 8.4** Symbolic system description example

The **SystemDescriptionSymbolic** class contains only two methods. The first is the `__init__()` function on line 2, which is a Python constructor. The constructor takes an instance of the **SystemDescription** class along with a *dictionary* of arguments. A Python dictionary is a list of keys and values. These keys and values are assigned in the argument list of the function. This dictionary is passed to the constructor of the **Symbolic** class from which all of the symbolic functions are derived from. The possible arguments are shown in Table 8.1.

All of the symbolic classes have the **Symbolic** class as their base either by directly deriving from this class or deriving from other classes that derive from it. The first step of the `__init__()` function is to initialize the base **SystemDescription** class with the system description optionally provided. It then initializes the **Symbolic** class with the argument list.

The second method is the `LaTeXSystemEquation()` method on line 5 of Listing 8.7. The job of this method is to assemble the system description into a LaTeX rendition of the system equation defining the system. It assembles this by making calls on the `WeightsMatrix()`, `NodeVector()`, and `StimulusVector()` functions in the base **SystemDescription** class and utilizing this information along with helper functions within the **Symbolic** class to turn the information into LaTeX equations.

An example is shown in Figure 8.4 for a system containing two cascaded two-port devices. In fact, Figure 8.4(a) is the same as Figure 8.2(a) with respect to the construction of the system description. Instead of printing out text information, however, on line 2 the **SystemDescriptionSymbolic** class is instantiated. Two calls are made on line 8, one to `LaTeXSystemEquation()` to produce the system equation and another to `Emit()` to print out the result.

The printed result is shown in Figure 8.4(b), which represents raw LaTeX code representing the system equation. One can examine the LaTeX code generated in Figure 8.4(b) and identify the elements of the LaTeX language as described in §8.3. Typesetting this equation with a LaTeX compiler produces the fully rendered equation in Figure 8.4(c). This equation is not only a more pleasing rendition of the equation shown textually in Figure 8.2(b), but also it can be pasted directly into many mathematics software packages as equation entry.

### 8.3.2 Symbolic Solutions

Up to this point, classes like the **SystemDescription** class have been found to simplify the conversion of netlists to information that describes the interconnectedness of systems and, in the case of the **SystemDescriptionSymbolic** class, provides symbolic algebraic descriptions of systems as described in Chapter 4. In this section, symbolic *solutions* will be produced. Specifically, the symbolic equivalent of the system description example in §8.2 will be produced and the two examples provided in §4.7.1 and §4.7.2 are repeated to solve the system completely.

These problems are repeated using the **SystemDescription** and **SystemSParameters** Python classes, using another derived class called **SystemSParametersSymbolic**, the listing of which is shown in Listing 8.8.

The **SystemSParametersSymbolic** class derives from the **SystemDescription-Symbolic** classes already discussed. The **SystemDescriptionSymbolic** class consists only of a constructor (the `__init__()` function on line 2) and a member function called `LaTeXSolution()` on line 9 that constructs LaTeX code for symbolic solutions for the s-parameters of a system, given the system description during construct time. The function `LaTeXSolution()` produces the main solution in a variety of ways depending on the input arguments and will be discussed after the first example in §8.3.3.

The constructor of the class **SystemSParametersSymbolic** takes a variable list of arguments, in addition to the instance of the **SystemDescription** class. This variable list contains the list of keys and values passed to the **Symbolic** class, as shown in Table 8.1.

#### *Symbolic Solution Example*

A repeat of the example in §4.7.1 is provided in Figure 8.5. On line 1 of Figure 8.5(a), the *SignalIntegrity* library is imported, and on line 2, an instance of the **SystemSParameters-Symbolic** class is created as `ssps`, on which all subsequent method calls are made. Lines 3–6 construct the system shown in Figure 4.24. Lines 3 and 4 call `AddDevice()` to add a two-port device called `'S'` and a one-port device called `'\Gamma t'`. Line 5 adds a system

### Listing 8.8 SystemSParametersSymbolic class

```
class SystemSParametersSymbolic(SystemDescriptionSymbolic):
    def __init__(self,sd=None,**args):
        SystemDescriptionSymbolic.__init__(self,sd,**args)
    def _LaTeXSi(self):
        sW=self._LaTeXMatrix(self.WeightsMatrix())
        self._AddEq('\mathbf{Si} = \\left[ '+self._Identity()+\
            ' - '+sW+' \\right]^{-1}')
        return self
    def LaTeXSolution(self,**args):
        solvetype = args['solvetype'] if 'solvetype' in args else 'block'
        size = args['size'] if 'size' in args else 'normal'
        AN=self.PortBNames(); BN=self.PortANames()
        if solvetype=='direct':
            self._LaTeXSi()
            BN=self.PortANames(); AN=self.PortBNames()
            n=self.NodeVector()
            SCI=Device.SymbolicMatrix('Si',len(n))
            B=[[0]*len(BN) for p in range(len(BN))]
            for r in range(len(BN)):
                for c in range(len(BN)):
                    B[r][c]=SCI[n.index(BN[r])][n.index(AN[c])]
            self._AddEq('\\mathbf{S} = '+self._LaTeXMatrix(B))
            return self
        XN=self.OtherNames(AN+BN)
        Wba=self.WeightsMatrix(BN,AN)
        sWba=self._LaTeXMatrix(Wba)
        Wxx=self.WeightsMatrix(XN,XN)
        if len(Wxx)==0:
            self._AddEq('\\mathbf{S} = '+sWba)
            return self
        Wbx=self.WeightsMatrix(BN,XN)
        Wxa=self.WeightsMatrix(XN,AN)
        if AllZeroMatrix(Wbx) or AllZeroMatrix(Wxa):
            self._AddEq('\\mathbf{S} = '+sWba)
            return self
        I=self._Identity()
...
        sWbx=self._LaTeXMatrix(Wbx)
        sWxa=self._LaTeXMatrix(Wxa)
        sWxx=self._LaTeXMatrix(Wxx)
        if size=='biggest':
            if len(Wba) != 0: self._AddEq('\\mathbf{W_{ba}} = '+sWba)
            if len(Wbx) != 0: self._AddEq('\\mathbf{W_{bx}} = '+sWbx)
            if len(Wxa) != 0: self._AddEq('\\mathbf{W_{xa}} = '+sWxa)
            if len(Wxx) != 0: self._AddEq('\\mathbf{W_{xx}} = '+sWxx)
            self._AddEq('\\mathbf{S}=\\mathbf{W_{ba}}+\\mathbf{W_{bx}}\\cdot'+\
                '\\left[ '+I+\
                ' -\\mathbf{W_{xx}}\\right]^{-1}\\cdot\\mathbf{W_{xa}}')
        elif size=='big':
            self._AddEq('\\mathbf{Wi} = '+' \\left[ '+I+' - '+sWxx+' \\right]^{-1} ')
            self._AddEq('\\mathbf{S} = '+sWba+' + '+sWbx+\
                ' \\cdot \\mathbf{Wi} \\cdot' +sWxa)
        else:
            self._AddEq('\\mathbf{S} = '+sWba+' + '+sWbx+' \\cdot \\left[ '+\
            I+' - '+sWxx+' \\right]^{-1} \\cdot'+sWxa)
        return self
```

```
import SignalIntegrity.Lib as si
ssps = si.sd.SystemSParametersSymbolic()
ssps.AddDevice('S', 2)  # add two-port left device
ssps.AddDevice('\\Gamma t', 1)  # add a termination
ssps.AddPort('S', 1, 1)  # add a port at port 1 of left device
ssps.ConnectDevicePort('S', 2, '\\Gamma t', 1)  # connect the other ports
ssps.LaTeXSystemEquation()
ssps.LaTeXSolution(solvetype='direct')
ssps.LaTeXSolution(solvetype='block').Emit()
```

(a) Python code

```
\[ \left[ \boldsymbol{\mathbbm{I}}  - \left(\begin{array}{cccc} 0 & 0
& S_{12} & S_{11} \\ 0 & 0 & S_{22} & S_{21} \\ 0 & \Gamma t & 0 & 0
\\ 0 & 0 & 0 & 0 \end{array}\right)\right] \cdot
\left(\begin{array}{c} n_{2} \\ n_{3} \\ n_{4} \\ n_{1}
\end{array}\right) = \left(\begin{array}{c} 0 \\ 0 \\ 0 \\ m_{1}
\end{array}\right) \]
\[ \mathbf{Si} = \left[ \boldsymbol{\mathbbm{I}}  -
\left(\begin{array}{cccc} 0 & 0 & S_{12} & S_{11} \\ 0 & 0 & S_{22} &
S_{21} \\ 0 & \Gamma t & 0 & 0 \\ 0 & 0 & 0 & 0 \end{array}\right)
\right]^{-1} \]
\[ \mathbf{S} = Si_{14} \]
\[ \mathbf{S} = S_{11} + \left(\begin{array}{cc} 0 & S_{12}
\end{array}\right) \cdot \left[ \boldsymbol{\mathbbm{I}}  -
\left(\begin{array}{cc} 0 & S_{22} \\ \Gamma t & 0 \end{array}\right)
\right]^{-1} \cdot\left(\begin{array}{c} S_{21} \\ 0
\end{array}\right) \]
```

(b) LaTeX output

$$\left[ \mathbb{I} - \left(\begin{array}{cccc} 0 & 0 & S_{12} & S_{11} \\ 0 & 0 & S_{22} & S_{21} \\ 0 & \Gamma t & 0 & 0 \\ 0 & 0 & 0 & 0 \end{array}\right)\right] \cdot \left(\begin{array}{c} n_{2} \\ n_{3} \\ n_{4} \\ n_{1} \end{array}\right) = \left(\begin{array}{c} 0 \\ 0 \\ 0 \\ m_{1} \end{array}\right)$$

$$\mathbf{Si} = \left[ \mathbb{I} - \left(\begin{array}{cccc} 0 & 0 & S_{12} & S_{11} \\ 0 & 0 & S_{22} & S_{21} \\ 0 & \Gamma t & 0 & 0 \\ 0 & 0 & 0 & 0 \end{array}\right) \right]^{-1}$$

$$\mathbf{S} = Si_{14}$$

$$\mathbf{S} = S_{11} + \left(\begin{array}{cc} 0 & S_{12} \end{array}\right) \cdot \left[ \mathbb{I} - \left(\begin{array}{cc} 0 & S_{22} \\ \Gamma t & 0 \end{array}\right) \right]^{-1} \cdot \left(\begin{array}{c} S_{21} \\ 0 \end{array}\right)$$

(c) LaTeX processed equations

**Figure 8.5** Symbolic solution example

port numbered port 1 defined as port 1 of device `'S'` in an `AddPort()` call. Finally, line 6, through a `ConnectDevicePort()` call, connects port 2 of `'S'` to port 1 of the termination.

Lines 7–9 emit LaTeX equations. A call to `LaTeXSystemEquation()` causes the LaTeX equations to be generated, and a subsequent call to `Emit()` causes these equations to be printed to the console.

The output is shown in Figure 8.5(b). Without knowing LaTeX, this looks like a barely comprehensible pile of gibberish, although some of the details can be picked out following the brief LaTeX information provided at the beginning of §8.3. The result of typesetting this LaTeX is shown in Figure 8.5(c), in which the equations take on their decoded, or typeset, form.

There are four equations in Figure 8.5(c). The first is the familiar system equation which, based on the system description, describes the relationship between the node vector, the weights matrix, and the stimulus vector. It was created by the call to `LaTeXSystemEquation()`. The second is the inverse of the system characteristics matrix assigned to **Si**. The third equation is the solution of the system s-parameters assigned to **S** in terms of the elements of **Si**. Thus, the second equation coupled with the third equation provides the solution to the system s-parameters in symbolic form according to the method provided in §4.3 and was created by a call to `LaTeXSolution()` with the argument `solvetype='direct'`. The fourth and final equation is the definition of the system s-parameters assigned to **S** using the block matrix method put forth in §4.4 and was created by the call to `LaTeXSolution()` with the argument `solvetype='block'`. Thus the fourth equation is a stand-alone equation that defines the s-parameters of the system.

These resulting equations can be entered, in LaTeX form, into a symbolic math solver such as Maple and simplified, where the simplification is basically the symbolic execution of the matrix inverse and the simplification of the resulting math. In numerical software providing a matrix inverse, the equations provided by **SystemSParametersSymbolic** are as good as any.

Comparing the equations in Figure 8.5(c) to those in §4.7.1 reveals some differences due to the arbitrary ordering of the node vector. Remember, there is no correct ordering of this vector, and solutions to these problems only insist that the equations are in a canonical form, as outlined in §4.2. The system equation, along with the system solution provided in the first three equations in Figure 8.5(c), is an alternative canonical form and solution. All such final solutions are the same numerically. These equations can be manipulated into other equivalent canonical forms by utilizing permutations provided in §4.2.2. The last equation in Figure 8.5(c), which is a direct solution using the block matrix method, matches that found in §4.7.1 because it is a single stand-alone equation and therefore must match.

### 8.3.3 System S-Parameter Solution Output Methods

Forms of the symbolic solution were output specifying direct and block form in the example in §8.3.2. There are a few other variations that can be employed depending on whether one wants the matrices all together in a single equation or for the equation to be split up into individual matrices. The single equation works well for normal or small size solutions, and the separate matrices work well for bigger solutions. As a reminder, the size of the individual matrices can be controlled using the size argument to symbolic, as shown in

$$\mathbf{Si} = \left[ \mathbb{I} - \begin{pmatrix} 0 & 0 & L_{12} & 0 & L_{11} & 0 \\ 0 & 0 & L_{22} & 0 & L_{21} & 0 \\ 0 & R_{11} & 0 & 0 & 0 & R_{12} \\ 0 & R_{21} & 0 & 0 & 0 & R_{22} \\ 0 & 0 & 0 & 0 & 0 & 0 \\ 0 & 0 & 0 & 0 & 0 & 0 \end{pmatrix} \right]^{-1}$$

$$\mathbf{S} = \begin{pmatrix} Si_{15} & Si_{16} \\ Si_{45} & Si_{46} \end{pmatrix}$$

(a) `solvetype='direct',size='normal'`

$$\mathbf{S} = \begin{pmatrix} L_{11} & 0 \\ 0 & R_{22} \end{pmatrix} + \begin{pmatrix} 0 & L_{12} \\ R_{21} & 0 \end{pmatrix} \cdot \left[ \mathbb{I} - \begin{pmatrix} 0 & L_{22} \\ R_{11} & 0 \end{pmatrix} \right]^{-1} \cdot \begin{pmatrix} L_{21} & 0 \\ 0 & R_{12} \end{pmatrix}$$

(b) `solvetype='block',size='normal'`

$$\mathbf{Wi} = \left[ \mathbb{I} - \begin{pmatrix} 0 & L_{22} \\ R_{11} & 0 \end{pmatrix} \right]^{-1}$$

$$\mathbf{S} = \begin{pmatrix} L_{11} & 0 \\ 0 & R_{22} \end{pmatrix} + \begin{pmatrix} 0 & L_{12} \\ R_{21} & 0 \end{pmatrix} \cdot \mathbf{Wi} \cdot \begin{pmatrix} L_{21} & 0 \\ 0 & R_{12} \end{pmatrix}$$

(c) `solvetype='block',size='big'`

$$\mathbf{W_{ba}} = \begin{pmatrix} L_{11} & 0 \\ 0 & R_{22} \end{pmatrix}$$

$$\mathbf{W_{bx}} = \begin{pmatrix} 0 & L_{12} \\ R_{21} & 0 \end{pmatrix}$$

$$\mathbf{W_{xa}} = \begin{pmatrix} L_{21} & 0 \\ 0 & R_{12} \end{pmatrix}$$

$$\mathbf{W_{xx}} = \begin{pmatrix} 0 & L_{22} \\ R_{11} & 0 \end{pmatrix}$$

$$\mathbf{S} = \mathbf{W_{ba}} + \mathbf{W_{bx}} \cdot [\mathbb{I} - \mathbf{W_{xx}}]^{-1} \cdot \mathbf{W_{xa}}$$

(d) `solvetype='block',size='biggest'`

**Figure 8.6** Result of function `LaTeXSolution()` in Figure 8.7

Table 8.1. Figure 8.6 provides examples corresponding to the example provided in Figure 8.7 (see §8.4) of the output of the various symbolic system s-parameter solutions:

- Figure 8.6(a) provides the direct solution. This is always provided as two equations. One equation provides a definition of a matrix **Si**, which is the inverse of the system characteristics matrix (or the inverse of the weights matrix subtracted from the identity matrix). The second equation provides the solution **S** shown as the elements chosen from **Si**, as put forth in §4.3.
- Figure 8.6(b) provides the normal sized block solution as provided by (4.23) in §4.4. Essentially, the solution provided is (4.23), but with the block matrices filled in based on the system description.
- Figure 8.6(c) provides the big sized block solution. This solution is the same as the normal sized solution, except the portion of the solution $(\mathbb{I} - \mathbf{W}_{\mathbf{xx}})^{-1}$ is split out. This is because this matrix becomes big as the system becomes big.
- Figure 8.6(d) provides the biggest sized block solution. In this solution, all of the matrices that make up (4.23) are assigned separately and (4.23) is simply displayed at the end. This allows for very large systems to be displayed.

## 8.4 The System Description Parser

Classes were provided to create the system description in §8.1. The **SystemDescription** class contains four main functions to describe the system. Three involve the interconnectedness of the system: the `AddDevice()`, `AddPort()`, and `ConnectDevicePort()` member functions. The fourth main function (which has not been used yet) is the `AssignSParameters()` function, which assigns s-parameters to a device. The system description example of §8.2 used these functions to assemble the system. In this section, a class is provided to assist further in making the assembly of a system a bit more terse and more easily readable in the **SystemDescriptionParser** class shown in Listing 8.9. This class derives from two classes (not shown) called **ParserArgs** and **ParserFile**. **ParserArgs** is a class that allows handling of variables so that netlists can be created with arguments. In this way, netlists can be generalized with certain arguments provided to change the circuit behavior. This feature is described in §8.6. **ParserFile** is a class that has two methods, `File()` and `WriteToFile()`, that allow reading and writing of netlist lines from and to a file. The **SystemDescriptionParser** class has only three main public functions. The first two functions, `AddLine()` and `AddLines()`, are used to add a line of text and a list of lines of text to the text version of a system description. The third function is `SystemDescription()`, which initiates the processing of the added lines and returns an instance of the **SystemDescription** class.

Looking at the internal function `_ProcessLines()`, one sees that it splits a line of text into space separated tokens and handles token lists whose first token is one of three keywords: `'device'`, `'port'`, and `'connect'`. These work as follows:

- `'device arg1 arg2 arg3 arg4 ...'` adds a device named `arg1` with `arg2` ports through a call that looks like `AddDevice(arg1,arg2)` on the **SystemDescription** class in Listing 8.3. The remaining tokens are passed to the **DeviceParser** class with `arg2` as the first argument reflecting the number of ports in the device. For

**Listing 8.9 SystemDescriptionParser class**

```
class SystemDescriptionParser(ParserFile,ParserArgs):
    def __init__(self,f=None,args=None):
        self.m_sd = None
        self.m_f=f
        self.m_lines=[]
        self.m_addThru = False
        self.AssignArguments(args)
        self.known=None
    def SystemDescription(self):
        if self.m_sd is None: self._ProcessLines()
        return self.m_sd
    def AddKnownDevices(self,known):
        self.known=known
        return self
    def AddLine(self,line):
        self.m_sd = None
        if len(line) == 0: return
        self.m_lines.append(line)
        return self
    def AddLines(self,lines):
        self.m_sd = None
        for line in lines:
            self.AddLine(line)
        return self
    def _ProcessLine(self,line,exclusionList):
        lineList=self.ReplaceArgs(LineSplitter(line))
        if len(lineList) == 0: return
        if self.ProcessVariables(lineList): return
        elif lineList[0] in exclusionList: self.m_ul.append(line)
        elif lineList[0] == 'device':
            argList = lineList[3:]
            if [lineList[2]]+argList in self.m_spcl:
                dev = DeviceParser(self.m_f,int(lineList[2]),None)
                dev.m_spf = self.m_spc[self.m_spcl.index([lineList[2]]+argList)][1]
            else: dev=DeviceParser(self.m_f,int(lineList[2]),argList)
            self.m_sd.AddDevice(lineList[1],int(lineList[2]),dev.m_sp)
            if not dev.m_spf is None:
                self.m_spc.append((lineList[1],dev.m_spf))
                self.m_spcl.append([lineList[2]]+argList)
        elif lineList[0] == 'connect':
            for i in range(3,len(lineList),2):
                self.m_sd.ConnectDevicePort(lineList[1],int(lineList[2]),
                    lineList[i],int(lineList[i+1]))
        elif lineList[0] == 'port':
            for i in range((len(lineList)-1)//3):
                self.m_sd.AddPort(lineList[i*3+2],int(lineList[i*3+3]),
                    int(lineList[i*3+1]),self.m_addThru)
        else: self.m_ul.append(line)
    def _ProcessLines(self,exclusionList=[]):
        self.m_sd=SystemDescription()
        self.m_spc=[]; self.m_spcl=[]; self.m_ul=[]
        if not self.known is None:
            for key in self.known.keys():
                self.m_spcl.append(LineSplitter(key))
                self.m_spc.append((None,self.known[key].Resample(self.m_f)))
        for line in self.m_lines: self._ProcessLine(line,exclusionList)
        return self
```

now, the examples that will be provided include only the first tokens, so discussion of the **DeviceParser** class is postponed until §8.5 with the organization of the software described in §17.3.5.

- `'connect arg1 arg2 arg3 arg4 arg5 arg6 ...'` makes device port connections through multiple calls on `ConnectDevicePort()` on the **SystemDescription** class in Listing 8.3. As described in §8.1, the `ConnectDevicePort()` method takes four arguments reflecting the *from* device/port and the *to* device/port, therefore, for each pair of tokens following the first two, calls are made with the pair of tokens as the *to* device/port and the first two tokens are always the *from* device/port. It follows a pattern like `ConnectDevicePort(arg1,arg2,arg3,arg4)`, `ConnectDevicePort(arg1,arg2,arg5,arg6)`, etc. Thus multiple device/port connections can be made in a single text line whereby all of the device/ports listed are all connected together.
- `'port arg1 arg2 arg3 arg4 arg5 arg6 ...'` adds system ports through multiple calls on `AddPort()` on the **SystemDescription** class in Listing 8.3. As described in §8.1, the `AddPort()` method takes three arguments: the device name, the device port, and the system port number in that order. The syntax for the text netlist is in a different order. For each triplet of tokens, calls are made using the first token as the system port number and the following two as the name of the device and the device port number. It follows a pattern like `AddPort(arg2,arg3,arg1)`, `AddPort(arg5,arg6,arg4)`, etc. Thus multiple system ports can be added in a single text line.

If the line to be processed is unrecognizable as not containing any of the keywords mentioned, then the unprocessed lines are appended to a member variable `m_ul`. This allows the derived parser classes to have the base **SystemDescriptionParser** class process the lines it recognizes and allows the derived class to process any lines that are left over.

### 8.4.1 Symbolic Solution Example Using the SystemDescriptionParser Class

The example in §4.7.2 that is shown graphically in Figure 4.26 is repeated in Figure 8.7, using the **SystemDescriptionParser** class.

The Python code for this example is shown in Figure 8.7(a). On line 2, the first task is to instantiate a **SystemDescriptionParser** class as `sdp`. Then, on line 3, a list of lines is added to the netlist. Each line is described as follows:

1. `'device L 2'` is the same as a call to `AddDevice('L',2)`, which adds a two-port device named `'L'` to the internally held system description.
2. `'device R 2'` is the same as a call to `AddDevice('R',2)`, which adds a two-port device named `'R'` to the internally held system description.
3. `'port 1 L 1 2 R 2'` is the same as two calls, one to `AddPort('L',1,1)` and the other to `AddPort('R',2,2)`. These add a system port numbered 1 to port 1 of the device named `'L'` and a system port numbered 2 to port 2 of the device named `'R'`. Note that each triplet of information has the system port number first.

```
1 import SignalIntegrity.Lib as si
2 sdp = si.p.SystemDescriptionParser()
3 sdp.AddLines(['device L 2','device R 2','port 1 L 1 2 R 2','connect L 2 R 1'])
4 sdp.WriteToFile('SymbolicSolution2.txt',False)
5 ssps = si.sd.SystemSParametersSymbolic(sdp.SystemDescription())
6 ssps.LaTeXSystemEquation()
7 ssps.LaTeXSolution(solvetype='direct')
8 ssps.LaTeXSolution(solvetype='block').Emit()
```

(a) Python code

$$\left[\mathbb{I}-\begin{pmatrix} 0 & 0 & L_{12} & 0 & L_{11} & 0\\ 0 & 0 & L_{22} & 0 & L_{21} & 0\\ 0 & R_{11} & 0 & 0 & 0 & R_{12}\\ 0 & R_{21} & 0 & 0 & 0 & R_{22}\\ 0 & 0 & 0 & 0 & 0 & 0\\ 0 & 0 & 0 & 0 & 0 & 0 \end{pmatrix}\right]\cdot\begin{pmatrix} n_2\\ n_5\\ n_6\\ n_4\\ n_1\\ n_3 \end{pmatrix}=\begin{pmatrix} 0\\ 0\\ 0\\ 0\\ m_1\\ m_2 \end{pmatrix}$$

$$\mathbf{Si}=\left[\mathbb{I}-\begin{pmatrix} 0 & 0 & L_{12} & 0 & L_{11} & 0\\ 0 & 0 & L_{22} & 0 & L_{21} & 0\\ 0 & R_{11} & 0 & 0 & 0 & R_{12}\\ 0 & R_{21} & 0 & 0 & 0 & R_{22}\\ 0 & 0 & 0 & 0 & 0 & 0\\ 0 & 0 & 0 & 0 & 0 & 0 \end{pmatrix}\right]^{-1}$$

$$\mathbf{S}=\begin{pmatrix} Si_{15} & Si_{16}\\ Si_{45} & Si_{46} \end{pmatrix}$$

$$\mathbf{S}=\begin{pmatrix} L_{11} & 0\\ 0 & R_{22} \end{pmatrix}+\begin{pmatrix} 0 & L_{12}\\ R_{21} & 0 \end{pmatrix}\cdot\left[\mathbb{I}-\begin{pmatrix} 0 & L_{22}\\ R_{11} & 0 \end{pmatrix}\right]^{-1}\cdot\begin{pmatrix} L_{21} & 0\\ 0 & R_{12} \end{pmatrix}$$

(b) LaTeX processed equations

**Figure 8.7** Symbolic solution example using the **SystemDescriptionParser** class

4. `'connect L 2 R 1'` is the same as a call to `ConnectDevicePort('L',2,'R',1)`, which connects port 2 of `'L'` to port 1 of `'R'`. If there were more device/ports listed, each call would be a call with `'L'` and 2 as the first two arguments.

This netlist is written to a file using a call to `WriteToFile()` on line 4.

On line 5, an instance of the class **SystemSParametersSymbolic** is instantiated with the **SystemDescription** instance produced from a call to `SystemDescription()` on `sdp`. This call causes the lines added to be processed and the system description returned. The remainder of the example proceeds like the example in §8.3.2, producing the raw LaTeX code with the typeset output provided in Figure 8.7(b).

The last equation in Figure 8.7(b) is exactly equivalent to the solution provided in (4.34), demonstrating that the **SystemDescriptionParser** class allows the description of

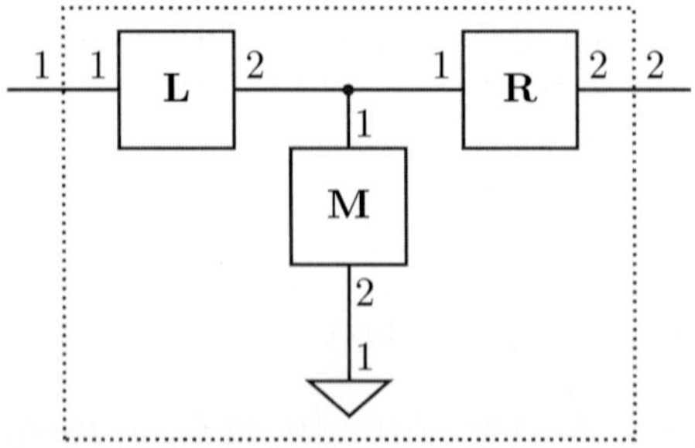


(a) Block diagram

```
device L 2
device R 2
device M 2
device G 1 ground
port 1 L 1 2 R 2
connect L 2 R 1 M 1
connect G 1 M 2
```

(b) Text netlist file

```
import SignalIntegrity.Lib as si
sdp = si.p.SystemDescriptionParser().File('SymbolicSolution3.txt')
ssps=si.sd.SystemSParametersSymbolic(sdp.SystemDescription(),size='small')
ssps.LaTeXSolution().Emit()
```

(c) Python code

$$\mathbf{S} = \begin{pmatrix} L_{11} & 0 \\ 0 & R_{22} \end{pmatrix} + \begin{pmatrix} 0&0&0&0&0&0&L_{12}&0 \\ 0&0&0&0&0&R_{21}&0&0 \end{pmatrix} \cdot \left[ \mathbb{I} - \begin{pmatrix} 0&0&0&0&0&0&L_{22}&0 \\ 0&0&0&0&0&R_{11}&0&0 \\ 0&0&0&0&M_{12}&0&0&M_{11} \\ 0&0&0&0&M_{22}&0&0&M_{21} \\ 0&0&0&-1&0&0&0&0 \\ \frac{2}{3}&-\frac{1}{3}&\frac{2}{3}&0&0&0&0&0 \\ -\frac{1}{3}&\frac{2}{3}&\frac{2}{3}&0&0&0&0&0 \\ \frac{2}{3}&\frac{2}{3}&-\frac{1}{3}&0&0&0&0&0 \end{pmatrix} \right]^{-1} \cdot \begin{pmatrix} L_{21}&0 \\ 0&R_{12} \\ 0&0 \\ 0&0 \\ 0&0 \\ 0&0 \\ 0&0 \\ 0&0 \end{pmatrix}$$

(d) LaTeX processed equations

**Figure 8.8** Symbolic solution example of a larger network

the network in simple terms for the **SystemSParametersSymbolic** class to solve the system symbolically.

### 8.4.2 Symbolic Solution Example of a Larger Network

Figure 8.8 shows a final symbolic example of the **SystemDescriptionParser** class utilized with a text file netlist and also with a larger network to highlight how it easily solves systems in a programmatic manner.

A block diagram showing a left device, a right device, and a middle device shunted to ground between the left and right devices is shown in Figure 8.8(a). In this system, there are three device ports connected together, so the system description is expected to become larger than initially expected due to the addition of a tee device.

The netlist in Figure 8.8(b) should be verified to describe the block diagram in Figure 8.8(a). This netlist is stored on the disk as `'SymbolicSolution3.txt'`.

The Python code designed to work with the file in Figure 8.8(b) is shown in Figure 8.8(c). On line 2, a **SystemDescriptionParser** class is instantiated as `sdp` with an immediate call to `File()` with `'SymbolicSolution3.txt'` specified as the file name. This has the effect of reading all of the lines in Figure 8.8(a) into the class instance. The call to `SystemDescription()` on `sdp` on line 3 has the effect of processing this netlist.

The raw LaTeX output of Figure 8.8(c) is shown typeset in Figure 8.8(d).

In Figure 8.8(c) the call to `LaTeXSolution()` on `ssps` was made with the specification of small matrices, which causes only the block matrix solution to be produced with small matrix format. This was to limit the shear size of all of the equations. The block matrix solution provides the solution to the s-parameters of the overall system.

The use of a tee device can be found inside the equation shown in Figure 8.8(d). This is exhibited by the $-1/3$ and $2/3$ s-parameter values found, and is expected. It is seen to be confined to the large matrix because all of the nodes connected to the tee are internal (i.e. they are not connected to system ports).

## 8.5 Numeric Solutions

The **SystemDescription** class has been shown to be useful for describing and connecting systems, and the **SystemSParameters** class has been shown to be capable of extracting the necessary information for the generation of the weights matrix, node vector, and stimulus vector. The **SystemSParameters** base class has been utilized to actually solve these systems (and other systems) in various ways. The **SystemSParametersSymbolic** class provided symbolic solutions.

The **SystemSParametersNumeric** class is shown in Listing 8.10 and is another derived class capable of numerical solutions of systems. Aside from the constructor, the only member function `SParameters()` provides the numeric solution.

The function `SParameters()` provides for a solution using the block matrix method given in §4.4 or the direct method from §4.3. The former method is preferred as it involves inverting smaller matrices and can handle some pathological cases, but both methods are provided for teaching purposes and both produce the same answer, as long as the matrices are invertible.

All solutions are at a single frequency and are calculated based on the s-parameters assigned to the devices. Thus, a method used to solve for a complete set of s-parameters of a system would be, by looping over all frequencies and at each frequency, supplying any frequency dependent s-parameters using calls to `AssignSParameters()`. Four examples of this are shown in Figure 8.9, which comprises a set of numeric solutions that duplicate the symbolic solution provided in Figure 8.7, albeit with two specific devices. Figures 8.9(a) and 8.9(b) both create an instance of the **SystemSParametersNumeric** class, and the system

```
import SignalIntegrity.Lib as si
sspn = si.sd.SystemSParametersNumeric()
sspn.AddDevice('L', 2)  # add two-port left device
sspn.AddDevice('R', 2)  # add two-port right device
sspn.AddPort('L', 1, 1)  # add a port at port 1 of left device
sspn.AddPort('R', 2, 2)  # add a port at port 2 of right device
sspn.ConnectDevicePort('L', 2, 'R', 1)  # connect the other ports
fl=[i*100.*1e6 for i in range(100+1)]; sp=[]
spl=si.sp.SParameterFile('cable.s2p',50.).Resample(fl)
spr=si.sp.SParameterFile('filter.s2p',50.).Resample(fl)
for n in range(len(fl)):
    sspn.AssignSParameters('L',spl[n])
    sspn.AssignSParameters('R',spr[n])
    sp.append(sspn.SParameters())
spres=si.sp.SParameters(fl,sp).WriteToFile('result.s2p')
```

(a) Using **SystemSParametersNumeric** class

```
import SignalIntegrity.Lib as si
sspn = si.sd.SystemSParametersNumeric()
sspn.AddDevice('L', 2)  # add two-port left device
sspn.AddDevice('R', 2)  # add two-port right device
sspn.AddPort('L', 1, 1)  # add a port at port 1 of left device
sspn.AddPort('R', 2, 2)  # add a port at port 2 of right device
sspn.ConnectDevicePort('L', 2, 'R', 1)  # connect the other ports
fl=[i*100.*1e6 for i in range(100+1)]; spdl=[]; sp=[]
spdl.append(('L',si.sp.SParameterFile('cable.s2p',50.).Resample(fl)))
spdl.append(('R',si.sp.SParameterFile('filter.s2p',50.).Resample(fl)))
for n in range(len(fl)):
    for ds in spdl: sspn.AssignSParameters(ds[0],ds[1][n])
    sp.append(sspn.SParameters())
spres=si.sp.SParameters(fl,sp).WriteToFile('result.s2p')
```

(b) Alternative using **SystemSParametersNumeric** class

```
import SignalIntegrity.Lib as si
sspnp = si.p.SystemSParametersNumericParser([i*100.*1e6 for i in range(100+1)])
sspnp.AddLines(['device L 2 file cable.s2p','device R 2 file filter.s2p',
    'port 1 L 1 2 R 2','connect L 2 R 1']).WriteToFile('example2.txt')
spres=sspnp.SParameters().WriteToFile('result.s2p')
```

(c) Using **SystemSParametersNumericParser** class

```
device L 2 file cable.s2p
device R 2 file filter.s2p
port 1 L 1 2 R 2
connect L 2 R 1
```

(d) Text netlist file

```
import SignalIntegrity.Lib as si
fl=[i*100.*1e6 for i in range(100+1)]
sspnp = si.p.SystemSParametersNumericParser(fl).File('example2.txt')
spres=sspnp.SParameters().WriteToFile('result.s2p')
```

(e) Using **SystemSParametersNumericParser** class with file

**Figure 8.9** Numeric solution example

**Listing 8.10 SystemSParametersNumeric** class

```
class SystemSParametersNumeric(SystemSParameters,Numeric):
    def __init__(self,sd=None):
        SystemSParameters.__init__(self,sd)
    def SParameters(self,**args):
        from numpy.linalg.linalg import LinAlgError
        solvetype = args['solvetype'] if 'solvetype' in args else 'block'
        AN=self.PortBNames()
        BN=self.PortANames()
        if solvetype == 'direct':
            n=self.NodeVector()
            PL=self.PermutationMatrix([n.index(BN[r])
                for r in range(len(BN))], len(n))
            PR=matrix(self.PermutationMatrix([n.index(AN[r])
                for r in range(len(AN))], len(n))).transpose()
            SCI=self.Dagger(matrix(identity(len(n)))-matrix(self.WeightsMatrix()),
                Left=PL,Right=PR).tolist()
            B=[[0]*len(BN) for p in range(len(BN))]
            for r in range(len(BN)):
                for c in range(len(BN)):
                    B[r][c]=SCI[n.index(BN[r])][n.index(AN[c])]
            return B
        # else solvetype assumed to be 'block'
        XN=self.OtherNames(AN+BN)
        Wba=self.WeightsMatrix(BN,AN)
        Wxx=self.WeightsMatrix(XN,XN)
        if len(Wxx)==0:
            return matrix(Wba).tolist()
        Wbx=self.WeightsMatrix(BN,XN)
        Wxa=self.WeightsMatrix(XN,AN)
        if AllZeroMatrix(Wbx) or AllZeroMatrix(Wxa):
            return matrix(Wba).tolist()
        I=matrix(identity(len(Wxx)))
        # Wba+Wbx*[(I-Wxx)^-1]*Wxa
        result = matrix(Wba)+self.Dagger(I-matrix(Wxx),Left=Wbx,Right=Wxa,Mul=True)
        return result.tolist()
```

is connected as before. Then a list of frequencies is created and any s-parameter files are read and resampled to this frequency list. For each frequency point, the s-parameters are assigned to the devices and the system is solved numerically with a call to `SParameters()`. Each of these results is aggregated with the frequency list into an instance of **SParameters** and written out to a file.

Figures 8.9(c) and 8.9(e) use the **SystemSParametersNumericParser** class, which will be described next. In Figure 8.9(c), the lines are added inside the program and written to a file. The netlist is shown in Figure 8.9(d) and read in Figure 8.9(e). The devices are specified as s-parameter files that are automatically read in and resampled. S-parameters are most often and most easily calculated using text file netlist files used with the **SystemSParametersNumericParser** class.

The **SystemSParametersNumericParser** class is shown in Listing 8.11. This class makes use of the functionality inherited from **SystemDescriptionParser** to parse the netlist. When the `'device'` keyword is encountered, the **DeviceParser** class is utilized to determine the type and arguments for the device specified. All device specifications are on a line that begins with `'device [name] [ports]'`, where `name` is the name given to the

**Listing 8.11 SystemSParametersNumericParser** class

```
class SystemSParametersNumericParser(SystemDescriptionParser,CallBacker,LinesCache):
    def __init__(self,f=None,args=None,callback=None,cacheFileName=None):
        SystemDescriptionParser.__init__(self,f,args)
        self.sf = None
    def SParameters(self,solvetype='block'):
        self.SystemDescription()
        self.m_sd.CheckConnections()
        spc=self.m_spc
        result = []
        for n in range(len(self.m_f)):
            for d in range(len(spc)):
                if not spc[d][0] is None:
                    self.m_sd.AssignSParameters(spc[d][0],spc[d][1][n])
            result.append(SystemSParametersNumeric(self.m_sd).SParameters(
                solvetype=solvetype))
        self.sf = SParameters(self.m_f, result)
        return self.sf
```

**Listing 8.12 DeviceParser** class

```
class DeviceParser():
    def __init__(self,f,ports,argsList):
        self.m_f=f
        self.m_sp=None
        self.m_spf=None
        if argsList is None:
            return
        if len(argsList) == 0:
            return
        if argsList[0] == 'subcircuit':
            self.m_spf=SubCircuit(self.m_f,argsList[1],
            ' '.join([x if len(x.split())==1 else "\'"+x+"\'" for x in argsList[2:]]))
            return
        if self.deviceFactory.MakeDevice(ports, argsList, f):
            if self.deviceFactory.frequencyDependent:
                self.m_spf=self.deviceFactory.dev
            else:
                self.m_sp=self.deviceFactory.dev
        else:
            #print 'device not found: '+' '.join(argsList)
            raise SignalIntegrityExceptionDeviceParser(
                'device not found: '+' '.join(argsList))
        return
```

device and `ports` denotes the number of ports in the device followed by other arguments that specify the device.

The **DeviceParser** is shown in Listing 8.12, which takes as arguments the number of ports, the device name, and an argument list. The **DeviceParser** class owns a static member of the class **DeviceFactory** and uses the `MakeDevice()` method shown in Listing 8.13 to make the device specified. Devices are made from a list of instances of the **ParserDevice** class shown in Listing 8.14. All of the valid instances of this class are instantiated at run-time and are shown in Listing 8.15. The **ParserDevice** class in Listing 8.15 and Listing 8.16 defines devices in terms of a name and number of ports (or range of ports, where appropriate) along with a list of arguments and defaults for the arguments. In some

**Listing 8.13 DeviceFactory** class `MakeDevice()` method

```
class DeviceFactory(list):
...
    def MakeDevice(self,ports,argsList,f):
        self.dev=None
        if len(argsList) == 0:
            return False
        name=argsList[0].lower()
        argsList=argsList[1:]
        for device in self:
            if device.ports is not None:
                if isinstance(device.ports,int):
                    if device.ports != ports:
                        continue
                elif isinstance(device.ports,str):
                    if '-' in device.ports:
                        (minPort,maxPort) = device.ports.split('-')
                        if ports < int(minPort):
                            continue
                        if ports > int(maxPort):
                            continue
                    else:
                        acceptablePorts = device.ports.split(',')
                        if not any(ports == int(acceptablePort)
                                   for acceptablePort in acceptablePorts):
                            continue
            if device.devicename != name: continue
            # this is the device, try to make it
            if device.arginname:
                if len(argsList) > 0:
                    argsList=['']+argsList
            argsProvidedDict = {argsList[i].lower():argsList[i+1]
                                for i in range(0,len(argsList),2)}
            arg=copy.copy(device.defaults)
            arg.update(argsProvidedDict)
            self.dev=eval(device.func)
            self.frequencyDependent=device.frequencyDependent
            return True
        return False
```

**Listing 8.14 ParserDevice** class

```
class ParserDevice(object):
    def __init__(self,devicename,ports,arginname,defaults,frequencyDependent,func):
        self.devicename=devicename
        self.ports=ports
        self.arginname=arginname
        self.defaults=defaults
        self.frequencyDependent=frequencyDependent
        self.func=func
```

**Listing 8.15 DeviceFactory** class initialization

```
class DeviceFactory(list):
    def __init__(self):
        list.__init__(self,[
        ParserDevice('file',None,True,{'':None},True,
            "SParameterFile(arg[''],50.).Resample(f)"),
        ParserDevice('c',1,True,{'':None,'df':0.,'esr':0.,'z0':50.},True,
            "TerminationC(f,float(arg['']),float(arg['z0']),\
            float(arg['df']),float(arg['esr']))"),
        ParserDevice('c',2,True,{'':None,'df':0.,'esr':0.,'z0':50.},True,
            "SeriesC(f,float(arg['']),float(arg['z0']),float(arg['df']),\
            float(arg['esr']))"),
        ParserDevice('l',1,True,{'':None},True,"TerminationL(f,float(arg['']))"),
        ParserDevice('l',2,True,{'':None},True,"SeriesL(f,float(arg['']))"),
        ParserDevice('r',1,True,{'':None},False,"TerminationZ(float(arg['']))"),
        ParserDevice('r',2,True,{'':None},False,"SeriesZ(float(arg['']))"),
        ParserDevice('rse',2,True,{'':None},False,"SeriesRse(float(arg['']))"),
        ParserDevice('shunt','2-4',True,{'':None},False,
            "ShuntZ(ports,float(arg['']))"),
        ParserDevice('m',4,True,{'':None},True,"Mutual(f,float(arg['']))"),
        ParserDevice('ground',1,False,{},False,"Ground()"),
        ParserDevice('open',1,False,{},False,"Open()"),
        ParserDevice('thru',2,False,{},False,"Thru()"),
        ParserDevice('directionalcoupler','3-4',False,{},False,
            "DirectionalCoupler(ports)"),
        ParserDevice('termination',None,False,{},False,
            "zeros(shape=(ports,ports)).tolist()"),
        ParserDevice('tee',None,False,{},False,"Tee(ports)"),
        ParserDevice('mixedmode',4,True,{'':'power'},False,
            "(MixedModeConverterVoltage() if arg[''] == 'voltage'\
            else MixedModeConverter())"),
        ParserDevice('idealtransformer',4,True,{'':1.},False,
            "IdealTransformer(float(arg['']))"),
        ParserDevice('voltagecontrolledvoltagesource',4,True,{'':None},False,
            "VoltageControlledVoltageSource(float(arg['']))"),
        ParserDevice('currentcontrolledcurrentsource',4,True,{'':None},False,
            "CurrentControlledCurrentSource(float(arg['']))"),
        ParserDevice('currentcontrolledvoltagesource',4,True,{'':None},False,
            "CurrentControlledVoltageSource(float(arg['']))"),
        ParserDevice('voltagecontrolledcurrentsource',4,True,{'':None},False,
            "VoltageControlledCurrentSource(float(arg['']))"),
        ParserDevice('voltageamplifier','2-4',False,{'gain':None,'zo':0,'zi':1e8,
            'z0':50.},False,"VoltageAmplifier(ports,float(arg['gain']),\
            float(arg['zi']),float(arg['zo']))"),
        ParserDevice('currentamplifier','2-4',False,{'gain':None,'zo':1e8,'zi':0,
            'z0':50.},False,"CurrentAmplifier(ports,float(arg['gain']),\
            float(arg['zi']),float(arg['zo']))"),
        ParserDevice('transresistanceamplifier','2-4',False,{'gain':None,'zo':0.,
            'zi':0.,'z0':50.},False,"TransresistanceAmplifier(ports,\
            float(arg['gain']),float(arg['zi']),float(arg['zo']))"),
        ParserDevice('transconductanceamplifier','2-4',False,{'gain':None,'zo':1e8,
            'zi':1e8,'z0':50.},False,"TransconductanceAmplifier(ports,\
            float(arg['gain']),float(arg['zi']),float(arg['zo']))"),
        ParserDevice('opamp',3,False,{'zi':1e8,'zd':1e8,'zo':0.,'gain':1e8,'z0':50.},
            False,"OperationalAmplifier(float(arg['zi']),float(arg['zd']),\
            float(arg['zo']),float(arg['gain']),float(arg['z0']))")])
...
```

**Listing 8.16 DeviceFactory** class initialization (continued)

```
class DeviceFactory(list):
...
    def __init__Contd(self):
        list.__init__(self,list(self+[
        ParserDevice('tline','2,4',False,{'zc':50.,'td':0.},True,
            "TLineLossless(f,ports,float(arg['zc']),float(arg['td']))"),
        ParserDevice('telegrapher',2,False,{'r':0.,'rse':0.,'l':0.,'c':0.,'df':0.,
            'g':0.,'z0':50.,'sect':0},True,"TLineTwoPortRLGC(f,\
            float(arg['r']),float(arg['rse']),float(arg['l']),float(arg['g']),\
            float(arg['c']),float(arg['df']),float(arg['z0']),int(arg['sect']))"),
        ParserDevice('telegrapher',4,False,{'rp':0.,'rsep':0.,'lp':0.,'cp':0.,'dfp':0.,
            'gp':0.,'rn':0.,'rsen':0.,'ln':0.,'cn':0.,'dfn':0.,'gn':0.,'lm':0.,
            'cm':0.,'dfm':0.,'gm':0.,'z0':50.,'sect':0},
            True,"TLineDifferentialRLGC(f, float(arg['rp']),float(arg['rsep']),\
            float(arg['lp']),float(arg['gp']),float(arg['cp']),float(arg['dfp']),\
            float(arg['rn']),float(arg['rsen']),float(arg['ln']),float(arg['gn']),\
            float(arg['cn']),float(arg['dfn']),float(arg['cm']),float(arg['dfm']),\
            float(arg['gm']),float(arg['lm']),float(arg['z0']),int(arg['sect']))"),
        ParserDevice('shortstd',1,False,{'od':0.,'oz0':50.,'ol':0.0,
            'l0':0.0,'l1':0.0,'l2':0.0,'l3':0.0},True,
            "ShortStandard(f,float(arg['od']),float(arg['oz0']),float(arg['ol']),\
            float(arg['l0']),float(arg['l1']),float(arg['l2']),float(arg['l3']))"),
        ParserDevice('openstd',1,False,{'od':0.,'oz0':50.,'ol':0.0,
            'c0':0.0,'c1':0.0,'c2':0.0,'c3':0.0},True,
            "OpenStandard(f,float(arg['od']),float(arg['oz0']),float(arg['ol']),\
            float(arg['c0']),float(arg['c1']),float(arg['c2']),float(arg['c3']))"),
        ParserDevice('loadstd',1,False,{'od':0.,'oz0':50.,'ol':0.0,'tz0':50.0},
            True,"LoadStandard(f,float(arg['od']),float(arg['oz0']),float(arg['ol']),\
            float(arg['tz0']))"),
        ParserDevice('thrustd',2,False,{'od':0.,'oz0':50.,'ol':0.0},
            True,"ThruStandard(f,float(arg['od']),float(arg['oz0']),float(arg['ol']))")
        ]))
...
```

cases, one of the arguments is specified without a keyword. Usually this is when there is only one argument, as in the resistance of a resistor. Otherwise, arguments come in keyword/value pairs, which can be seen in Listing 8.15. When a keyword/value pair does not have a default value (specified as `None` being the default value), then the value must be supplied when instantiating the device.

When the **SystemDescriptionParser** was used for parsing netlists used for symbolic solutions, the types of devices were restricted. Most devices were specified without any mention of their s-parameters, and the only ones that did have known s-parameters were ideal devices, like the tee, ground, open, etc., that have constant s-parameters independent of frequency. There are ways to use symbolic s-parameters for devices, but the s-parameters must be assigned after the netlist is read in or established through `AddLine()` calls and cannot be done through the netlist. An example of the assignment of symbolic s-parameters is shown in many of the examples in Appendix D, e.g. Figure D.1(a). When the **SystemDescriptionParser** is used for parsing netlists used in numeric solutions, the possible devices are provided in Tables 8.2, 8.3, 8.4, 8.5, and 8.6.

One aspect of the **SystemDescriptionParser** class shown in Listing 8.9 is that it maintains two lists:

**Table 8.2** Devices recognized by the **DeviceParser** class

| Device | Declared as 'device [name] [ports]' followed by ... |
|---|---|
| S-parameter file | 'file [filename]' – a device whose s-parameters are specified by an s-parameter file named filename. The file is read using Listing 3.15. |
| Capacitor | 'C [capacitance]' – a capacitor with given capacitance. The s-parameters are given by (3.42), where $Z = 1/(j \cdot 2\pi \cdot f \cdot C)$. |
| Inductor | 'L [inductance]' – an inductor with given inductance. The s-parameters are given by (3.42), where $Z = j \cdot 2\pi \cdot f \cdot L$. |
| Resistor | 'R [resistance]' – a resistor with given resistance. If the number of ports specified is one, then it is a resistor tied to ground with s-parameters given by (3.41). Otherwise it is a two-port series resistor with s-parameters given by (3.42). |
| Mutual inductor | 'M [inductance]' – a mutual inductance with given mutual inductance. The s-parameters are given by §3.3.3 with $L_l = L_r = 0$. The primary refers to ports 1 and 2 and the secondary refers to ports 3 and 4. The dot is on ports 1 and 3. |
| Ground | 'ground': s-parameters are −1. |
| Open | 'open': s-parameters are 1. |
| Ideal thru | 'thru' – an ideal thru with s-parameters given by $\left(\begin{smallmatrix}0 & 1\\ 1 & 0\end{smallmatrix}\right)$. |
| Ideal tee | 'tee [ports]' – an ideal tee with given number of ports. The s-parameters are as given by (3.14). |
| Directional coupler | 'directionalcoupler' – a directional coupler that directly connects ports 1 and 2. If three ports are specified, the wave incident on port 1 is sent to port 3. If four ports are specified, the wave incident on port 2 is additionally sent to port 4. This device is not passive. |
| Subcircuit | 'subcircuit [filename] <args>' – a subcircuit with a file containing a netlist called filename with given args to be provided to the netlist. See §8.6. |
| Mixed-mode converter | 'mixedmode <voltage>' – a mixed-mode converter with ports 1, 2, 3, and 4 being the +, −, D, and C ports. If the optional voltage keyword is specified, the differential mode is the + minus the − and the common mode is the average. Otherwise, it is the standard definition for s-parameters. |
| Ideal transformer | 'idealtransformer [ratio]' – an ideal transformer with given turns ratio, as given by Figure 6.29(c). The primary refers to ports 1 and 2 and the secondary refers to ports 3 and 4. The dot is on ports 1 and 3. The ratio is the secondary with respect to the primary. |

**Table 8.3** Dependent source devices recognized by the **DeviceParser** class

| **Device** | **Declared as `'device [name] [ports]'` followed by...** |
|---|---|
| Voltage controlled voltage source | `'voltagecontrolledvoltagesource [gain]'` – a voltage controlled voltage source with given voltage `gain` as shown in Figure 6.11. The measured voltage is across ports 1 and 2 (port 1 is positive). The source is between ports 3 and 4 (port 3 is positive). |
| Current controlled current source | `'currentcontrolledcurrentsource [gain]'` – a current controlled current source with given current `gain` as shown in Figure 6.12. The measured current is into and out of ports 1 and 2. The current is sunk and sourced at ports 3 and 4. |
| Voltage controlled current source | `'voltagecontrolledcurrentsource [transconductance]'` – a voltage controlled current source with given `transconductance` as shown in Figure 6.14. The measured voltage is across ports 1 and 2 (port 1 is positive). The current is sunk and sourced at ports 3 and 4. |
| Current controlled voltage source | `'currentcontrolledvoltagesource [transresistance]'` – a current controlled voltage source with given `transresistance` as shown in Figure 6.13. The measured current is into and out of ports 1 and 2. The source is between ports 3 and 4 (port 3 is positive). |

- `m_spcl`: a list of arguments specifying a device with frequency dependent s-parameters.
- `m_spc`: a dictionary of frequency dependent s-parameters according to the device name.

The elements in `m_spc` and `m_spcl` go together. For a given element $i$, `m_spc[i][0]` is the name of a device with frequency dependent s-parameters, `m_spc[i][1]` are the s-parameters of the device, and `m_spcl[i]` contains the arguments used to create those s-parameters.

In examining the `_ProcessLine()` function in the **SystemDescriptionParser** class shown in Listing 8.9, specifically the case where the first token on the line is the keyword `'device'`, one expects the name of the device to be in token 1, the number of ports to be in token 2, and the rest of the arguments to be from token 3 onward. These remaining arguments are loaded into `argList` to be passed to the **DeviceParser** class along with the number of ports. A check is made beforehand to see if the arguments are found in `m_spcl`. If they were not, a device is created by instantiating a **DeviceParser** class with the given arguments. The **DeviceParser** class produces frequency dependent s-parameters in a member `m_spf`. If the member is `None`, then there aren't any and the s-parameters are assumed frequency independent and are loaded permanently during the call to `AddDevice()`. Note that during the symbolic system description creation, this means that text s-parameters are loaded; otherwise, permanent numeric s-parameters are loaded. For example, the ground device produces s-parameters of $-1$. This value is loaded into the device in the `AddDevice()`

**Table 8.4** Amplifier devices recognized by the **DeviceParser** class

| Device | Declared as 'device [name] [ports] <parameters>'[a] |
|---|---|
| Voltage amplifier | 'voltageamplifier <gain> <zo> <zi>' – a voltage amplifier, like the voltage controlled voltage source with given input impedance zi and given output impedance zo. If not provided, zi defaults to $10^8$ and zo defaults to zero. It supports 4, 3, and 2 ports, as provided in §6.4.1. The gain is the voltage gain. |
| Current amplifier | 'currentamplifier <gain> <zo> <zi>' – a current amplifier, like the current controlled current source with given input impedance zi and given output impedance zo. If not provided, zi defaults to zero and zo defaults to $10^8$. It supports 4, 3, and 2 ports, as provided in §6.4.2. The gain is the current gain. |
| Transresistance amplifier | 'transresistanceamplifier <gain> <zo> <zi>' – a transresistance amplifier, like the current controlled voltage source with given input impedance zi and given output impedance zo. If not provided, zi and zo default to zero. It supports 4, 3, and 2 ports, as provided in §6.4.4. The gain is the transresistance. |
| Trans-conductance amplifier | 'transconductanceamplifier <gain> <zo> <zi>' – a transconductance amplifier, like the voltage controlled current source with given input impedance zi and given output impedance zo. If not provided, zi and zo default to $10^8$. It supports 4, 3, and 2 ports, as provided in §6.4.3. The gain is the transconductance. |
| Operational amplifier | 'opamp <zi> <zd> <zo> <gain>' – an operational amplifier. The minus and plus inputs are ports 1 and 2, respectively, and the output port is port 3. The impedance zd is the impedance across the input pins and defaults to $10^8$. The impedance zi is the impedance of each input pin to ground and defaults to $10^8$. The impedance zo is the output impedance and defaults to zero. The gain is the voltage gain of the device and defaults to unity. |

[a] The first parameter is the device name. The remaining parameters are specified as a keyword followed by value (e.g. <zo> is specified like 'zo 0'). The order of the keyword/value pairs is unimportant.

call and never needs to change again. Otherwise, for something like an s-parameter file, the frequency dependent s-parameters are contained in the m_spf member. If frequency dependent s-parameters are found, they are appended to the list m_spc along with the device name, and the arguments used for their creation are appended to m_spcl.

On subsequent calls to _ProcessLine(), the argument list for a device might be found in m_spcl. This happens when devices with the exact same characteristics are found in a schematic. In this case, a device is still created, but by instantiating a **DeviceParser** class without any arguments and adding through an AddDevice() call to get it into the system

**Table 8.5** Transmission line devices recognized by the **DeviceParser** class

| **Device** | **Declared as 'device [name] [ports] <parameters>'**[a] |
|---|---|
| Transmission line | 'tline <zc> <td>' – an ideal transmission line with characteristic impedance zc and propagation constant of td. The inner conductor goes from ports 1 to 2. If ports is 2, then the outer conductor is grounded. Otherwise, if ports is 4, then the outer conductor is exposed as ports 3 and 4. The two-port s-parameters are provided in (7.16). |
| Telegrapher (two-port) | 'telegrapher <r> <rse> <l> <g> <c> <df> <zo> <sect>' – a two-port transmission line defined by total DC resistance r in $\Omega$, skin-effect resistance rse in $\Omega/\sqrt{\text{Hz}}$, inductance l in H, DC conductance g in S, capacitance c in F, dissipation factor (or loss tangent) df, reference impedance zo in $\Omega$, and the number of sections sect. All values default to zero except for the reference impedance, which defaults to $50\,\Omega$. This device uses the model in Listing 7.4. |
| Telegrapher (four-port) | 'telegrapher <rp> <rsep> <lp> <gp> <cp> <dfp> <rn> <rsen> <ln> <gn> <cn> <dfn> <lm> <gm> <cm> <dfm> <zo> <sect>' – a four-port transmission line defined by parameters for assumed positive side (suffix 'p'), negative side (suffix 'n'), and mutual (suffix 'm') by total DC resistance rp and rn in $\Omega$, skin-effect resistance rsep and rsen in $\Omega/\sqrt{\text{Hz}}$, inductance lp, ln, and lm in H, DC conductance gp, gn, and gm in S, capacitance cp, cn, and cm in F, dissipation factor (or loss tangent) dfp, dfn, and dfm, reference impedance zo in $\Omega$, and the number of sections sect. All values default to zero except for the reference impedance, which defaults to $50\,\Omega$. The device uses the model in Listing 7.7. The ports are numbered 1, 2, 3, and 4 as left positive, right positive, left negative, and right negative, respectively. |

[a] The first parameter is the device name. The remaining parameters are specified as a keyword followed by value (e.g. <zc> is specified like 'zc 50'). The order of the keyword/value pairs is unimportant.

description. However, the frequency dependent s-parameters are loaded from m_spc and assigned to the device. In this way devices do not need to be duplicated.

At the end of processing all of the lines, m_spc contains a dictionary of all of the devices with frequency dependent s-parameters. Again, all of the devices without frequency independent s-parameters already have their s-parameters loaded into the device in the system description.

The **SystemSParametersNumericParser** automates the process of looping over the frequencies and assigns the s-parameters at each frequency step prior to solving the system. In Listing 8.11, the only member function is SParameters(). Its first job is to cause the system description to be generated by processing the lines and checking the connections for validity. Then, for each frequency point, it loads all of the frequency dependent s-

**Table 8.6** Calibration standard devices recognized by the **DeviceParser** class

| Device | Declared as 'device [name] [ports] <parameters>'[a] |
|---|---|
| Short standard | 'shortstd <od> <oz0> <ol> <l0> <l1> <l2> <l3>' – a short standard defined by an offset delay od in s, an offset loss ol in GΩ/s at 1 GHz, a real characteristic impedance of the offset oz0 in Ω, and a termination inductance given by a polynomial with coefficients l0, l1, l2, and l3, specified as provided in §15.5.3 and Listing 15.14. |
| Open standard | 'openstd <od> <oz0> <ol> <c0> <c1> <c2> <c3>' – an open standard defined by an offset delay od in s, an offset loss ol in GΩ/s at 1 GHz, a real characteristic impedance of the offset oz0 in Ω, and a termination capacitance given by a polynomial with coefficients c0, c1, c2, and c3, specified as provided in §15.5.4 and Listing 15.15. |
| Load standard | 'loadstd <od> <oz0> <ol> <tz0' – a load standard defined by an offset delay od in s, an offset loss ol in GΩ/s at 1 GHz, a real characteristic impedance of the offset oz0 in Ω, and a real termination impedance tz0 in Ω, as provided in §15.5.2 and Listing 15.13. |
| Thru standard | 'thrustd <od> <oz0> <ol>' – a thru standard defined by an offset delay od in s, an offset loss ol in GΩ/s at 1 GHz, and a real characteristic impedance of the offset oz0 in Ω, as provided in §15.5.1 and Listing 15.12. |

[a] The first parameter is the device name. The remaining parameters are specified as a keyword followed by value (e.g. <od> is specified like 'od 53e-12'). The order of the keyword/value pairs is unimportant.

**Listing 8.17 SubCircuit** class

```
class SubCircuit(SParameters):
    def __init__(self,f,fileName,args):
        sspnp=SystemSParametersNumericParser(f,args).File(fileName)
        SParameters.__init__(self,f,sspnp.SParameters())
```

parameters found in m_spc and loads them into the specified devices. After doing this, it instantiates a **SystemSParametersNumeric** class with the system description in its current frequency state, extracts the s-parameters, and appends them to a list. Finally, it returns the calculated s-parameters.

## 8.6 Subcircuits

A capability allowing for hierarchical design of problems is built into the **SystemDescriptionParser** class; this involves the **SubCircuit** class. The **SubCircuit** class in Listing 8.17 derives from **SParameters** and is essentially a way to read in s-parameters

**Listing 8.18 ParserArgs** class

```
class ParserArgs():
    def AssignArguments(self,args):
        self.m_vars=dict()
        if args is None:
            self.m_args=[]
        else:
            args=LineSplitter(args)
            self.m_args = dict([('$'+args[i]+'$',args[i+1])
                for i in range(0,len(args),2)])
    def ReplaceArgs(self,lineList):
        replacedOne=False
        for i in range(len(lineList)):
            if lineList[i] in self.m_vars:
                replacedOne=True
                lineList[i] = self.m_vars[lineList[i]]
        if replacedOne:
            lineList=' '.join(lineList).split()
        return lineList
    def ProcessVariables(self,lineList):
        if lineList[0] == 'var':
            variables=dict([(lineList[i*2+1],lineList[i*2+2])
                for i in range((len(lineList)-1)//2)])
            for key in self.m_args:
                if key in variables:
                    variables[key]=self.m_args[key]
            self.m_vars=dict(list(self.m_vars.items())+list(variables.items()))
            return True
        else:
            return False
```

not through a file, but rather by referencing a netlist file, for which the s-parameters are solved.

In Listing 8.17, the **SubCircuit** class is initialized with a frequency list, a file name, and some arguments. This causes a **SystemSParametersNumericParser** class to be instantiated with the frequency list and arguments, and for the file name to be read in as the netlist, just as in the previous examples. Then, the s-parameters are solved for in the netlist and the base **SParameters** class is initialized with these calculated s-parameters. At the end of instantiating a **SubCircuit** class, one is holding **SParameters** just as if an s-parameter file had been read in.

The key capability here is that of the arguments. These are handled by the **ParserArgs** class, as shown in Listing 8.18. The three member functions `AssignArguments()`, `ReplaceArgs()`, and `ProcessVariables()` are utilized at different stages of handling a subcircuit (although they can be utilized without the **SubCircuit** class). The `AssignArguments()` function is called during the construction of the **SystemDescriptionParser** class with any arguments specified. The arguments come in a space delimited string containing variable name/value pairs. These arguments are stored in the `m_args` member variable. The variable names supplied have `'$'` added to the front and back of the name as they are added to the dictionary. These are the arguments for the presumed subcircuit.

Each subcircuit text file ought to contain a special line starting with the keyword `'var'`. This keyword should be on the very first line of a subcircuit text file and declares the variables to be used in the subcircuit. Each of the remaining arguments comprise a variable

```
var $DL$ thru $DR$ thru
device L 2 $DL$
device R 2 $DR$
port 1 L 1 2 R 2
connect L 2 R 1
```

(a) Subcircuit netlist file 'cascade.sub'

```
device D 2 subcircuit cascade.sub DL 'file cable.s2p' DR 'file filter.s2p'
port 1 D 1 2 D 2
```

(b) Example netlist file 'SubCircuitExample2.txt'

```
import SignalIntegrity.Lib as si
fl=[i*100.*1e6 for i in range(100+1)]
sspnp = si.p.SystemSParametersNumericParser(fl).File('SubCircuitExample2.txt')
spres=sspnp.SParameters().WriteToFile('result.s2p')
```

(c) Example Python code using **SystemSParametersNumericParser** class with file

**Figure 8.10** Subcircuit example

name/value pair, where the name defines the variable name and the value defines the default value if an argument of that name is not specified. The variable name is always expected to be bracketed by '$'. During execution of the `_ProcessLine()` function in the **SystemDescriptionParser** class shown in Listing 8.9, a call is made to `ProcessVariables()` to handle such a line. When a variable declaration line is encountered, the names are first stored in a dictionary with the default values specified. Then the dictionary is compared with `m_args` and, for any argument specified during construction, the default value is replaced with the arguments provided and the dictionary of variables is placed in the member variable `m_vars`.

Also during processing of lines in the `_ProcessLine()` function, a call is made to `ReplaceArgs()` with each and every line processed. Here, any token on a line to be processed is searched for a variable found in `m_vars`, replacing any variable found with the value associated with the variable. In this manner, subcircuits can be created with variables provided to suit its usage.

Consider the cascaded two-port example in Figure 8.9. Suppose one wanted to cascade many pairs of two-port devices or had a frequently used system configuration and did not want to build a new netlist each time. An example of how this is achieved is provided in Figure 8.10.

There is a subcircuit netlist in Figure 8.10(a) that looks very much like the netlist provided in Figure 8.9(d), except that the two devices specified have been replaced by variables and these variables appear on the first line with the 'var' keyword. Each of these device types is specified to default to an ideal thru.

A netlist is shown in Figure 8.10(b) that makes use of this subcircuit. The subcircuit is specified as a two-port device named 'D' with the 'cascade.sub' file name and the arguments DL and DR specified. Arguments with spaces in them must be enclosed in single quotes. Also, the arguments supplied do not have the '$' added to them.

This netlist is read in and processed in Figure 8.10(c). Although this is a simple example, it contains all the necessary elements required to understand how to use subcircuits.

## 8.7 Summary of Python Code Arrangement

Numerous classes have been demonstrated in this chapter that provide for solutions to systems of interconnected s-parameter devices.

A similar structure is utilized in subsequent chapters, which cover de-embedding, simulation, and virtual probing, whereby a parser class, a symbolic class, and a numeric class provide for similar types of solutions. The complete organization of the system that derives from the **SystemDescription** class is provided in Chapter 17.

# 9

# Simulation

SIMULATION IN SIGNAL INTEGRITY is defined as the production of output waveforms in a system containing a circuit model and sources. Both the source and output waveforms represent currents or voltages with respect to time. Voltage waveforms are either the absolute potential of a node with respect to ground or are differences between nodes (i.e. differential measurements).

This chapter addresses a form of simulation that is purely linear and convolutional in nature. This means that the waveforms are processed with filters, and in this book, without loss of generality, these filters are finite impulse response (FIR) filters. Simulation is broken down into two separate steps with regard to the computations involved.

1. The first step is to compute the relationship between the specified sources and output waveforms in the system. This is found to involve calculations similar to those in Chapter 4, where nodes are solved with respect to stimuli. Here, the nodes are waves that are converted to voltages and currents, using methods to be shown, and the stimuli are waves provided by sources. The calculated relationship will be independent of the sources and will be valid for any source waveform provided.
2. The second step is to use the relationship calculated in the first step, along with the specification of source waveforms, to provide the output waveforms. Although an example will be shown at the end of this section that involves such waveform processing, the description of how waveforms are processed is not described until Chapter 13 – Waveforms and Filters.

In summary, this chapter will focus on the calculation of the relationships between source waveforms and output waveforms.

## 9.1 Simulation Solutions

Using the **SystemDescription** class developed in Chapter 8, a symbolic solution for a simulation problem is shown in Figure 9.1. The problem is shown schematically in Figure 9.1(a). According to the equivalent network diagram form in Figure 9.1(b), Figure 9.1(c) shows Python code that creates three devices: a one-port device named $\Gamma_s$ representative of a voltage source, a two-port device named **S** representative of any generic two-port device, such as a transmission line or filter, and a one-port device named $\Gamma_l$ representative of a termination. These devices are connected together such that port 1 of the voltage source is connected to port 1 of the generic device and port 2 of the generic device is connected to port 1 of the termination. The one-port voltage source is assumed to be a voltage source

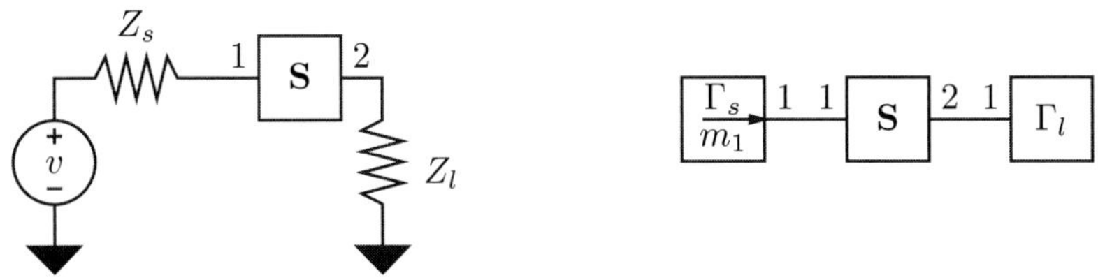


(a) Circuit diagram (b) Block diagram

```python
import SignalIntegrity.Lib as si
ssps=si.sd.SystemSParametersSymbolic()
ssps.AddDevice('S',2)
ssps.AddDevice('\\Gamma_l',1)
ssps.AddDevice('\\Gamma_s',1)
ssps.ConnectDevicePort('\\Gamma_s',1,'S',1)
ssps.ConnectDevicePort('S',2,'\\Gamma_l',1)
ssps.AssignM('\\Gamma_s',1,'m1')
ssps.LaTeXSystemEquation().Emit()
ssps.AssignSParameters('\\Gamma_s',si.sy.ShuntZ(1,'Z_s'))
ssps.AssignSParameters('\\Gamma_l',si.sy.ShuntZ(1,'Z_l'))
ssps.Clear().LaTeXSystemEquation().Emit()
```

(c) Python code

$$\left[\mathbb{I}-\left(\begin{array}{cccc} 0 & 0 & S_{12} & S_{11} \\ 0 & 0 & S_{22} & S_{21} \\ 0 & \Gamma_l & 0 & 0 \\ \Gamma_s & 0 & 0 & 0 \end{array}\right)\right]\cdot\left(\begin{array}{c} n_2 \\ n_3 \\ n_4 \\ n_1 \end{array}\right)=\left(\begin{array}{c} 0 \\ 0 \\ 0 \\ m_1 \end{array}\right)$$

$$\left[\mathbb{I}-\left(\begin{array}{cccc} 0 & 0 & S_{12} & S_{11} \\ 0 & 0 & S_{22} & S_{21} \\ 0 & \frac{Z_l-Z0}{Z_l+Z0} & 0 & 0 \\ \frac{Z_s-Z0}{Z_s+Z0} & 0 & 0 & 0 \end{array}\right)\right]\cdot\left(\begin{array}{c} n_2 \\ n_3 \\ n_4 \\ n_1 \end{array}\right)=\left(\begin{array}{c} 0 \\ 0 \\ 0 \\ m_1 \end{array}\right)$$

(d) LaTeX processed equations

**Figure 9.1** Symbolic simulation example

as provided in Figure 6.6, and the one-port termination is assumed to be a termination to ground as provided in (3.41). A stimulus is set as $m_1$ emanating from port 1 of $\Gamma_s$.

The instance of the **SystemDescription** class is converted to an instance of **SystemSParametersSymbolic** and the LaTeX system equation is emitted, which is shown in Figure 9.1(d).

The solution to the node values is expressed as:

$$\mathbf{Si} = \begin{pmatrix} Si_{11} & Si_{12} & Si_{13} & Si_{14} \\ Si_{21} & Si_{22} & Si_{23} & Si_{24} \\ Si_{31} & Si_{32} & Si_{33} & Si_{34} \\ Si_{41} & Si_{42} & Si_{43} & Si_{44} \end{pmatrix} = \left[ \mathbb{I} - \begin{pmatrix} 0 & 0 & S_{12} & S_{11} \\ 0 & 0 & S_{22} & S_{21} \\ 0 & \Gamma_l & 0 & 0 \\ \Gamma_s & 0 & 0 & 0 \end{pmatrix} \right]^{-1},$$

$$\begin{pmatrix} n_2 \\ n_3 \\ n_4 \\ n_1 \end{pmatrix} = \begin{pmatrix} Si_{11} & Si_{12} & Si_{13} & Si_{14} \\ Si_{21} & Si_{22} & Si_{23} & Si_{24} \\ Si_{31} & Si_{32} & Si_{33} & Si_{34} \\ Si_{41} & Si_{42} & Si_{43} & Si_{44} \end{pmatrix} \cdot \begin{pmatrix} 0 \\ 0 \\ 0 \\ m_1 \end{pmatrix}.$$

This expression can be improved by creating a matrix $\mathbf{Si}'$ and a corresponding $\mathbf{m}'$. The vector $\mathbf{m}'$ is formed by removing all of the zero elements of $\mathbf{m}$ and $\mathbf{Si}'$ is formed by removing all of the columns of $\mathbf{Si}$ corresponding to the zero rows of $\mathbf{m}$. Therefore,

$$\mathbf{m}' = m_1, \qquad \mathbf{Si}' = \begin{pmatrix} Si_{14} \\ Si_{24} \\ Si_{34} \\ Si_{44} \end{pmatrix},$$

and

$$\begin{pmatrix} n_2 \\ n_3 \\ n_4 \\ n_1 \end{pmatrix} = \begin{pmatrix} Si_{14} \\ Si_{24} \\ Si_{34} \\ Si_{44} \end{pmatrix} \cdot m_1.$$

It requires some inspection to see that nodes 1–4 represent the following:

- $n_1$ is the wave incident on port 1 of the generic two-port device;
- $n_2$ is the wave reflected from port 1 of the generic two-port device;
- $n_3$ is the wave reflected from port 2 of the generic two-port device;
- $n_4$ is the wave incident on port 2 of the generic two-port device.

Using the wave equation for voltage in (2.8), this means that the voltage at port 1 of the generic two-port device is proportional to the sum of $n_1$ and $n_2$, and that the voltage at port 2 of the generic two-port device is proportional to the sum of $n_3$ and $n_4$.

Thus, a voltage extraction matrix can be created that converts these waves to voltages as follows:

$$\begin{pmatrix} v_s \\ v_l \end{pmatrix} = \sqrt{Z0} \cdot \begin{pmatrix} 1 & 1 & 0 & 0 \\ 0 & 0 & 1 & 1 \end{pmatrix} \cdot \begin{pmatrix} n_1 \\ n_2 \\ n_3 \\ n_4 \end{pmatrix} = \sqrt{Z0} \cdot \begin{pmatrix} 1 & 0 & 0 & 1 \\ 0 & 1 & 1 & 0 \end{pmatrix} \cdot \begin{pmatrix} n_2 \\ n_3 \\ n_4 \\ n_1 \end{pmatrix}.$$

Therefore, the voltages can be solved as

$$\begin{pmatrix} v_s \\ v_l \end{pmatrix} = \sqrt{Z0} \cdot \begin{pmatrix} 1 & 0 & 0 & 1 \\ 0 & 1 & 1 & 0 \end{pmatrix} \cdot \begin{pmatrix} n_2 \\ n_3 \\ n_4 \\ n_1 \end{pmatrix} = \sqrt{Z0} \cdot \begin{pmatrix} 1 & 0 & 0 & 1 \\ 0 & 1 & 1 & 0 \end{pmatrix} \cdot \begin{pmatrix} Si_{14} \\ Si_{24} \\ Si_{34} \\ Si_{44} \end{pmatrix} \cdot m_1$$

$$= \sqrt{Z0} \cdot \begin{pmatrix} Si_{14} + Si_{44} \\ Si_{24} + Si_{34} \end{pmatrix} \cdot m_1.$$

Finally, from Figure 6.7,

$$m_1 = \frac{1}{\sqrt{Z0}} \cdot \frac{Z0}{Z_s + Z0} \cdot v,$$

resulting in

$$\begin{pmatrix} v_s \\ v_l \end{pmatrix} = \begin{pmatrix} Si_{14} + Si_{44} \\ Si_{24} + Si_{34} \end{pmatrix} \cdot \frac{Z0}{Z_s + Z0} \cdot v.$$

This is a *simulation* solution, so called because there is a system driven by sources (a single source in this case), outputs (the voltages at the source and load), and a system of equations that relate the output voltages to the source voltage. This means that frequency responses for filters can be generated that can be applied to a waveform supplied for the voltage source.

Note here that the normalization factor $\sqrt{Z0}$ canceled when combined with the source, and that if the output impedance $Z_s$ is zero, the reference impedance $Z0$ cancels as well. Here, all voltage sources will have zero output impedance, and any source resistance will be taken care of by an extra element in the circuit, allowing the removal of the reference impedance and the normalization factor from all equations.

There are several extensions here to the system description concept put forth in Chapter 8. These are:

1. The concept of the matrix $\mathbf{m}'$ and $\mathbf{Si}'$, where $\mathbf{m}'$ is formed by removing all of the zero elements of $\mathbf{m}$ and $\mathbf{Si}'$ is formed by removing all of the columns of $\mathbf{Si}$ corresponding to the zero rows of $\mathbf{m}$.
2. The concept of the voltage extraction matrix, which converts node wave values to absolute port voltages.
3. The addition of source devices formed by inserting appropriate s-parameters representative of the device and inserting stimuli emanating from the source device ports.
4. The concept of defining stimuli in terms of a source (voltage).

Thus, given a system equation

$$[\mathbb{I} - \mathbf{W}] \cdot \mathbf{n} = \mathbf{m}$$

and a list of output voltages associated with the nodes, such that there is a voltage extraction matrix $\mathbf{VE}$ such that

$$\mathbf{VE} \cdot \mathbf{n} = \mathbf{vo},$$

**Listing 9.1 Simulator** class

```
class Simulator(SystemSParameters,object):
    def __init__(self,sd=None):
        SystemSParameters.__init__(self,sd)
        self.m_ol = sd.m_ol if hasattr(sd, 'm_ol') else None
...
    def pOutputList(self):
        return self.m_ol
    @pOutputList.setter
    def pOutputList(self,ol=None):
        if not ol is None: self.m_ol = ol
        return self
    def AddVoltageSource(self,Name,Ports):
        if Ports == 1: self.AddDevice(Name,Ports,Ground(),'voltage source')
        elif Ports == 2: self.AddDevice(Name,Ports,Thru(),'voltage source')
        stimNumber=len(self.StimsPrime())+1
        for p in range(Ports): self.AssignM(Name,p+1,'m'+str(stimNumber+p))
    def AddCurrentSource(self,Name,Ports):
        if Ports == 1: self.AddDevice(Name,Ports,Open(),'current source')
        elif Ports == 2: self.AddDevice(Name,Ports,[[1.,0.],[0.,1.]],'current source')
        stimNumber=len(self.StimsPrime())+1
        for p in range(Ports): self.AssignM(Name,p+1,'m'+str(stimNumber+p))
...
```

the inverse of the system characteristics matrix is calculated as

$$\mathbf{Si} = [\mathbb{I} - \mathbf{W}]^{-1}.$$

The zero rows of $\mathbf{m}$ and the corresponding columns of $\mathbf{Si}$ are removed, forming $\mathbf{m}'$ and $\mathbf{Si}'$ such that

$$\mathbf{vo} = \mathbf{VE} \cdot \mathbf{n} = \mathbf{VE} \cdot \mathbf{Si}' \cdot \mathbf{m}'.$$

Finally, a vector of source voltages is associated with the stimuli such that

$$\mathbf{D} \cdot \mathbf{vs} = \mathbf{m}',$$

and the simulation solution is written as

$$\mathbf{vo} = \mathbf{VE} \cdot \mathbf{Si}' \cdot \mathbf{D} \cdot \mathbf{vs}.$$

The matrix that is multiplied by the source voltages to form the output voltages is called the *transfer matrix*

$$\mathbf{H} = \mathbf{VE} \cdot \mathbf{Si}' \cdot \mathbf{D}, \tag{9.1}$$

and is used in a simulation with input waveforms $\mathbf{vs}$ to form output waveforms $\mathbf{vo}$:

$$\mathbf{vo} = \mathbf{H} \cdot \mathbf{vs}.$$

## 9.2 The Simulator Class

The simulation solution can be accomplished programmatically with some extensions to the **SystemDescription** and **SystemSParameters** classes put forth in Listings 8.3, 8.4, and

**Listing 9.2 Simulator class (continued)**

```
class Simulator(SystemSParameters,object):
...
    def SourceVector(self):
        sv=[]
        for d in self:
            if d.Type == 'current source' or d.Type == 'voltage source':
                sv.append(d.Name)
        return sv
    def SourceToStimsPrimeMatrix(self,symbolic=False):
        sv=self.SourceVector()
        sp=self.StimsPrime()
        Z0='Z0' if symbolic else 50.
        sm = [[0]*len(sv) for r in range(len(sp))]
        for s in sv:
            d=self[self.IndexOfDevice(s)]
            if d.Type == 'current source':
                if len(d) == 1:
                    sm[sp.index(d[0].M)][sv.index(s)] = Z0
                elif len(d) == 2:
                    sm[sp.index(d[0].M)][sv.index(s)] = Z0
                    sm[sp.index(d[1].M)][sv.index(s)] = Z0
            elif d.Type == 'voltage source':
                if len(d) == 1:
                    sm[sp.index(d[0].M)][sv.index(s)] = 1.
                elif len(d) == 2:
                    sm[sp.index(d[0].M)][sv.index(s)] = -1./2.
                    sm[sp.index(d[1].M)][sv.index(s)] = 1./2.
        return sm
    def StimsPrime(self):
        sv=self.StimulusVector()
        sp=[]
        for s in range(len(sv)):
            sn='m'+str(s+1)
            if sn in sv: sp.append(sn)
            else: return sp
        return sp
    def SIPrime(self,symbolic=False,Left=None,Right=None):
        from numpy.linalg.linalg import LinAlgError
        n=self.NodeVector()
        m=self.StimulusVector()
        mprime=self.StimsPrime()
        if symbolic: SI=Device.SymbolicMatrix('Si',len(n))
        else:
            PR=matrix(self.PermutationMatrix([m.index('m'+str(c+1))
                        for c in range(len(mprime))], len(n))).transpose()
            if Right is None: Right = PR
            else: Right = matrix(PR)*matrix(Right)
            SI=self.Dagger(
                matrix(identity(len(n)))-matrix(self.WeightsMatrix()),
                Left=Left,Right=Right).tolist()
              SiPrime2=(SI*PR).tolist()
        SiPrime=[[0]*len(mprime) for r in range(len(n))]
        for c in range(len(mprime)):
            for r in range(len(n)):
                SiPrime[r][c]=SI[r][m.index('m'+str(c+1))]
        return SiPrime
    def VoltageExtractionMatrix(self,nl):
        n=self.NodeVector()
        result=[[0]*len(n) for r in range(len(nl))]
        for r in range(len(nl)):
            dp=self[self.DeviceNames().index(nl[r][0])][nl[r][1]-1]
            result[r][n.index(dp.A)]=1
            result[r][n.index(dp.B)]=1
        return result
```

8.5 by creating a derived **Simulator** class whose extensions to the **SystemDescription** are shown in Listing 9.1.

There are two classes of extensions. One is the ability to specify the member property `pOutputList`. This is a list of device name and port number tuples which define the desired voltages in the simulation. The second class of extensions involves the ability to add voltage and current sources using the member functions `AddVoltageSource()` and `AddCurrentSource()`. These functions add a device to the system description, where the name of the device represents the variable for the voltage or current in the source. One- and two-port devices are supported, where these devices correspond to the ideal current and voltage sources shown in Figures 6.1, 6.2, 6.5, and 6.6. Stimuli are assigned to emanate from each port of each source. As such, these extensions serve only to add functions which allow additional modification to the system description, or supply additional information that is needed for simulation only.

The remainder of the **Simulator** class that assists in the solution is shown in Listing 9.2. Corresponding to the discussion so far, the member functions are:

- `SourceVector()` gathers the list of source voltage or current names from the devices whose type is a current or voltage source.
- `SourceToStimsPrimeMatrix()` generates the matrix $\mathbf{D}$ (see §9.1) that converts the source voltages $\mathbf{vs}$ to the abbreviated stimulus vector $\mathbf{m}'$.
- `StimsPrime()` generates the abbreviated stimulus vector $\mathbf{m}'$.
- `SIPrime()` generates the matrix $\mathbf{Si}'$.
- `VoltageExtractionMatrix()` generates a voltage extraction matrix for translating the node vector $\mathbf{n}$ into the list of output voltages supplied.

## 9.3 Symbolic Simulation Solutions

As in §8.3.2, where symbolic s-parameter solutions were formed, one forms symbolic simulator solutions using the **SimulatorSymbolic** class, shown in Listing 9.3, which derives from the **Simulator** and **SystemSParametersSymbolic** classes.

A symbolic solution is provided in Figure 9.2. The solution corresponds to the situation in Figure 9.1, but here, instead of just generating the system equations, the simulation is performed. The code provided in Figure 9.1(c) is rewritten to provide the code in Figure 9.2(a), which now makes use of the **Simulator** and **SimulatorSymbolic** classes. The processed LaTeX output is shown in Figure 9.2(b).

## 9.4 The SimulatorParser Class

In this section, the capability generated in Chapter 8 is extended to the simulation concept. Thus, the capability of the **SystemDescriptionParser** class shown in Listing 8.9 is extended to provide the derived **SimulatorParser** class shown in Listing 9.4.

The arguments supplied to the **SimulatorParser** are used to initialize the base **SystemDescriptionParser** class. It is capable of processing lines, and in fact it overrides the **SystemDescriptionParser** function `_ProcessLines()`. In §8.4, it was seen

```
import SignalIntegrity.Lib as si
ss=si.sd.SimulatorSymbolic()
ss.AddDevice('S',2)
ss.AddDevice('\\Gamma_l',1)
ss.ConnectDevicePort('S',2,'\\Gamma_l',1)
ss.AddVoltageSource('V',1)
ss.AddDevice('Z_s',2)
ss.ConnectDevicePort('V',1,'Z_s',1)
ss.ConnectDevicePort('Z_s',2,'S',1)
ss.pOutputList = [('S',1),('S',2)]
ss.AssignSParameters('Z_s',si.sy.SeriesZ('Z_s'))
ss.AssignSParameters('\\Gamma_l',si.sy.ShuntZ(1,'Z_l'))
ss.LaTeXEquations().Emit()
```

(a) Python code

$$\left[\mathbb{I}-\left(\begin{array}{cccccc} 0 & 0 & S_{12} & 0 & 0 & S_{11} \\ 0 & 0 & S_{22} & 0 & 0 & S_{21} \\ 0 & \frac{Z_l-Z0}{Z_l+Z0} & 0 & 0 & 0 & 0 \\ 0 & 0 & 0 & 0 & -1 & 0 \\ \frac{2\cdot Z0}{Z_s+2\cdot Z0} & 0 & 0 & \frac{Z_s}{Z_s+2\cdot Z0} & 0 & 0 \\ \frac{Z_s}{Z_s+2\cdot Z0} & 0 & 0 & \frac{2\cdot Z0}{Z_s+2\cdot Z0} & 0 & 0 \end{array}\right)\right]\cdot\left(\begin{array}{c} n_6 \\ n_1 \\ n_2 \\ n_3 \\ n_4 \\ n_5 \end{array}\right)=\left(\begin{array}{c} 0 \\ 0 \\ 0 \\ m_1 \\ 0 \\ 0 \end{array}\right)$$

$$\mathbf{Si}=\left[\mathbb{I}-\left(\begin{array}{cccccc} 0 & 0 & S_{12} & 0 & 0 & S_{11} \\ 0 & 0 & S_{22} & 0 & 0 & S_{21} \\ 0 & \frac{Z_l-Z0}{Z_l+Z0} & 0 & 0 & 0 & 0 \\ 0 & 0 & 0 & 0 & -1 & 0 \\ \frac{2\cdot Z0}{Z_s+2\cdot Z0} & 0 & 0 & \frac{Z_s}{Z_s+2\cdot Z0} & 0 & 0 \\ \frac{Z_s}{Z_s+2\cdot Z0} & 0 & 0 & \frac{2\cdot Z0}{Z_s+2\cdot Z0} & 0 & 0 \end{array}\right)\right]^{-1}$$

$$\left(\begin{array}{c} S_1 \\ S_2 \end{array}\right)=\left(\begin{array}{c} Si_{14}+Si_{64} \\ Si_{24}+Si_{34} \end{array}\right)\cdot V$$

(b) LaTeX processed equations

**Figure 9.2** Symbolic simulation example (continued from Figure 9.1)

**Listing 9.3 SimulatorSymbolic** class

```
class SimulatorSymbolic(SystemSParametersSymbolic, Simulator):
    def __init__(self,sd=None,**args):
        SystemSParametersSymbolic.__init__(self,sd,**args)
        Simulator.__init__(self, sd)
    def LaTeXTransferMatrix(self):
        self.Check()
        self._LaTeXSi()
        veosi = MatrixMultiply(
            self.VoltageExtractionMatrix(self.pOutputList), self.SIPrime(True))
        veosil = Matrix2LaTeX(veosi, self._SmallMatrix())
        if len(veosi)==1:
            if len(veosi[0])==1:
                veosil='\\left('+veosil+'\\right)'
        on=Matrix2LaTeX([[D+'_{'+str(P)+'}'] for (D,P) in self.pOutputList])
        sv=Matrix2LaTeX(SubscriptedVector(self.SourceVector()))
        ssm=self.SourceToStimsPrimeMatrix(False)
        if len(ssm) == len(ssm[0]): # matrix is square
            isidentity=True
            for r in range(len(ssm)):
                for c in range(len(ssm[0])):
                    if r==c:
                        if ssm[r][c]!=1.:
                            isidentity=False
                            break
                    else:
                        if ssm[r][c]!=0.:
                            isidentity=False
                            break
        else: isidentity=False
        if isidentity:
            sm = ''
        else:
            sm=' \\cdot '+Matrix2LaTeX(self.SourceToStimsPrimeMatrix(True))
        line = on + '=' + veosil+sm+'\\cdot '+sv
        self._AddEq(line)
        return self
    def LaTeXEquations(self):
        self.LaTeXSystemEquation()
        self.LaTeXTransferMatrix()
        return self
```

that lines of a system description netlist were processed using commands to add devices and ports and to connect ports of devices together to form the system description. Here, the `_ProcessLines()` function first calls the base class `_ProcessLines()` function, with the commands `'connect'` and `'port'` excluded from the capability. Adding ports does not make sense in simulation as they are only for s-parameter determination. The `'connect'` commands are excluded because after processing the lines according to the **SystemDescriptionParser** base class, more devices are added in the form of potential sources. After processing these lines by making calls on `_ProcessSimulatorLine()` on any commands not recognized by the **SystemDescriptionParser** base class, it reprocesses the lines not recognized by the **SimulatorParser** class allowing the connections to be made.

The `_ProcessSimulatorLine()` function extends the capability of the parser to handle three new commands.

**Listing 9.4 SimulatorParser** class

```
class SimulatorParser(SystemDescriptionParser):
    def __init__(self, f=None, args=None):
        SystemDescriptionParser.__init__(self, f, args)
    def _ProcessSimulatorLine(self,line):
        lineList=self.ReplaceArgs(line.split())
        if len(lineList) == 0: return
        elif lineList[0] == 'output':
            if self.m_sd.pOutputList is None: self.m_sd.pOutputList = []
            for i in range(1,len(lineList),2):
                self.m_sd.pOutputList.append((lineList[i],int(lineList[i+1])))
        elif lineList[0] == 'voltagesource':
            self.m_sd.AddVoltageSource(lineList[1],int(lineList[2]))
        elif lineList[0] == 'currentsource':
            self.m_sd.AddCurrentSource(lineList[1],int(lineList[2]))
        else: self.m_ul.append(line)
    def _ProcessLines(self):
        SystemDescriptionParser._ProcessLines(self,['connect','port'])
        self.m_sd = Simulator(self.m_sd)
        lines=copy.deepcopy(self.m_ul); self.m_ul=[]
        for line in lines: self._ProcessSimulatorLine(line)
        lines=copy.deepcopy(self.m_ul); self.m_ul=[]
        for line in lines: SystemDescriptionParser._ProcessLine(self,line,['port'])
        return self
```

The internal function `_ProcessSimulatorLine()` splits a line of text into space separated tokens and handles token lists whose first token is one of three keywords: `'output'`, `'voltagesource'`, or `'currentsource'`. These tokens operate as follows:

- `'output arg1 arg2 arg3 arg4 ...'` adds a device named `arg1` and a port of that device `arg2` to a list of outputs. If more than one output is provided, it adds them in groups of two. The output list formed in this way looks just like that specified in Figure 9.2(a). Thus, a command such as `'output S 1 S 2'` to the **SimulatorParser** class looks just like a call to `pOutput=[('S',1),('S',2)]` on the **Simulator** class.
- `'voltagesource arg1 arg2'` adds a voltage source named `arg1` with a number of ports equal to `arg2`. It does this by making the call `AddVoltageSource()` on the **Simulator** class. Thus, the command `'voltagesource V 1'` processed by the **SimulatorParser** class is just like a call to `AddVoltageSource('V',1)` on the **Simulator** class.
- `'currentsource arg1 arg2'` adds a current source named `arg1` with a number of ports equal to `arg2`. It does this by making the call `AddCurrentSource()` on the **Simulator** class. Thus, the command `'currentsource I 2'` processed by the **SimulatorParser** class is just like a call to `AddCurrentSource('I',2)` on the **Simulator** class.

The **SimulatorParser** only automates the creation of a Simulator class through a netlist provided line-by-line, a set of lines, or a text file. It can therefore be used as a front-end processor for symbolic or numeric solutions. An example of this is shown in Figure 9.3. The Python code in Figure 9.3(b) connects a four-port device named `'X'` to two voltage sources connected through a series impedance and terminates the device in two impedances to ground, as shown in Figure 9.3(a). It creates the simulation solu-

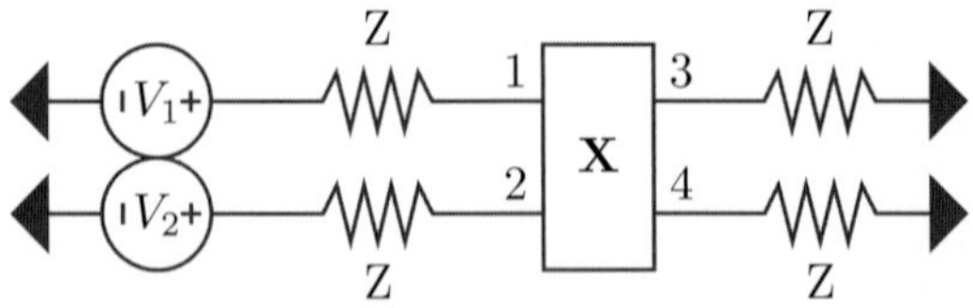


(a) Circuit diagram

```
import SignalIntegrity.Lib as si
sp = si.p.SimulatorParser()
sp.AddLine('device X 4')
sp.AddLine('device P 2')
sp.AddLine('device M 2')
sp.AddLine('device \\Gamma_P 1')
sp.AddLine('device \\Gamma_M 1')
sp.AddLine('voltagesource V1 1')
sp.AddLine('voltagesource V2 1')
sp.AddLine('connect V1 1 P 1')
sp.AddLine('connect V2 1 M 1')
sp.AddLine('connect P 2 X 1')
sp.AddLine('connect M 2 X 2')
sp.AddLine('connect X 3 \\Gamma_P 1')
sp.AddLine('connect X 4 \\Gamma_M 1')
sp.AddLine('output X 3 X 4 X 1 X 2')
ss=si.sd.SimulatorSymbolic(sp.SystemDescription(),size='small')
ss.AssignSParameters('P',si.sy.SeriesZ('Z'))
ss.AssignSParameters('M',si.sy.SeriesZ('Z'))
ss.AssignSParameters('\\Gamma_P',si.sy.ShuntZ(1,'Z'))
ss.AssignSParameters('\\Gamma_M',si.sy.ShuntZ(1,'Z'))
ss.LaTeXTransferMatrix().Emit()
```

(b) Python code

$$\mathbf{Si} = \left[ \mathbb{I} - \left( \begin{array}{cccccccccccc} 0 & 0 & 0 & 0 & 0 & X_{11} & 0 & X_{12} & X_{13} & X_{14} & 0 & 0 \\ 0 & 0 & 0 & 0 & 0 & X_{21} & 0 & X_{22} & X_{23} & X_{24} & 0 & 0 \\ 0 & 0 & 0 & 0 & 0 & X_{31} & 0 & X_{32} & X_{33} & X_{34} & 0 & 0 \\ 0 & 0 & 0 & 0 & 0 & X_{41} & 0 & X_{42} & X_{43} & X_{44} & 0 & 0 \\ \frac{2\cdot Z0}{Z+2\cdot Z0} & 0 & 0 & 0 & 0 & 0 & 0 & 0 & 0 & 0 & \frac{Z}{Z+2\cdot Z0} & 0 \\ \frac{Z}{Z+2\cdot Z0} & 0 & 0 & 0 & 0 & 0 & 0 & 0 & 0 & 0 & \frac{2\cdot Z0}{Z+2\cdot Z0} & 0 \\ 0 & \frac{2\cdot Z0}{Z+2\cdot Z0} & 0 & 0 & 0 & 0 & 0 & 0 & 0 & 0 & 0 & \frac{Z}{Z+2\cdot Z0} \\ 0 & \frac{Z}{Z+2\cdot Z0} & 0 & 0 & 0 & 0 & 0 & 0 & 0 & 0 & 0 & \frac{2\cdot Z0}{Z+2\cdot Z0} \\ 0 & 0 & \frac{Z-Z0}{Z+Z0} & 0 & 0 & 0 & 0 & 0 & 0 & 0 & 0 & 0 \\ 0 & 0 & 0 & \frac{Z-Z0}{Z+Z0} & 0 & 0 & 0 & 0 & 0 & 0 & 0 & 0 \\ 0 & 0 & 0 & 0 & -1 & 0 & 0 & 0 & 0 & 0 & 0 & 0 \\ 0 & 0 & 0 & 0 & 0 & 0 & -1 & 0 & 0 & 0 & 0 & 0 \end{array} \right) \right]^{-1}$$

$$\left( \begin{array}{c} X_3 \\ X_4 \\ X_1 \\ X_2 \end{array} \right) = \left( \begin{array}{cc} Si_{3,11}+Si_{9,11} & Si_{3,12}+Si_{9,12} \\ Si_{4,11}+Si_{10,11} & Si_{4,12}+Si_{10,12} \\ Si_{1,11}+Si_{6,11} & Si_{1,12}+Si_{6,12} \\ Si_{2,11}+Si_{8,11} & Si_{2,12}+Si_{8,12} \end{array} \right) \cdot \left( \begin{array}{c} V_1 \\ V_2 \end{array} \right)$$

(c) LaTeX processed equations

**Figure 9.3** Symbolic simulator parser example

**Listing 9.5 SimulatorNumeric** class

```
class SimulatorNumeric(Simulator,Numeric):
    def __init__(self,sd=None):
        Simulator.__init__(self,sd)
    def TransferMatrix(self):
        self.Check()
        VE_o=matrix(self.VoltageExtractionMatrix(self.m_ol))
        SIPrime=matrix(self.SIPrime(Left=VE_o))
        sm=matrix(self.SourceToStimsPrimeMatrix())
        Result=(VE_o*SIPrime*sm).tolist()
        return Result
```

**Listing 9.6 SimulatorNumericParser** class

```
class SimulatorNumericParser(SimulatorParser,CallBacker,LinesCache):
    def __init__(self, f=None, args=None,  callback=None, cacheFileName=None):
        SimulatorParser.__init__(self, f, args)
        self.transferMatrices = None
    def TransferMatrices(self):
        self.SystemDescription()
        self.m_sd.CheckConnections()
        spc=self.m_spc
        result=[]
        for n in range(len(self.m_f)):
            for d in range(len(self.m_spc)):
                self.m_sd.AssignSParameters(spc[d][0],spc[d][1][n])
            tm=SimulatorNumeric(self.m_sd).TransferMatrix()
            result.append(tm)
        self.transferMatrices=TransferMatrices(self.m_f,result)
        return self.transferMatrices
```

tion by putting the device together with a **SimulatorParser** and then converting this to a **SimulatorSymbolic**. Then, the symbolic s-parameters are assigned and **SimulatorSymbolic** produces the symbolic result shown in Figure 9.3(c).

## 9.5 Numeric Solutions

In the previous sections, the **Simulator** class was utilized to generate a simulation system and form a solution. The **SimulatorSymbolic** class was used to generate a symbolic solution. In Listing 9.5 the **SimulatorNumeric** class is shown as a derived class capable of numeric simulation solutions insofar as the transfer matrices are concerned.

It is initialized with an instance of a **Simulator** class and, once initialized, generates the numeric solution through a single member function `TransferMatrix()`.

The function `TransferMatrix()` is straightforward given the previous discussion. It forms the matrix necessary according to (9.1).

This solution is at a single frequency and is calculated based on the s-parameters assigned to the devices. A method for solving for a complete set of transfer matrices of a system would loop over all frequencies, and at each frequency it would supply any frequency dependent s-parameters using calls on `AssignSParameters()`. Even better is to utilize the **SimulatorNumericParser** class shown in Listing 9.6. As this class inherits

**Listing 9.7 TransferMatrices** class

```
class TransferMatrices(list):
    def __init__(self,f,d):
        self.f=FrequencyList(f)
        list.__init__(self,d)
        self.Inputs=len(d[0][0])
        self.Outputs=len(d[0])
    def SParameters(self):
        if self.Inputs == self.Outputs:
            return SParameters(self.f,self)
        else:
            squareMatrices=[]
            P=max(self.Inputs,self.Outputs)
            for transferMatrix in self:
                squareMatrix=zeros((P,P),complex).tolist()
                for r in range(len(transferMatrix)):
                    for c in range(len(transferMatrix[0])):
                        squareMatrix[r][c]=transferMatrix[r][c]
                squareMatrices.append(squareMatrix)
            return SParameters(self.f,squareMatrices)
...
```

from **SimulatorParser**, which inherits from **SystemDescriptionParser**, all of the capability utilized system description generation is applicable here. This means that netlists containing devices specified through s-parameter files or models can be solved just as for s-parameter solutions.

The solution result is an instance of the **TransferMatrices** class, as shown in Listing 9.7. This is a container for transfer functions used mostly for time-domain processing, which will be discussed in §13.5, but has an `SParameters()` function for saving transfer functions for use with s-parameter viewing tools.

### 9.5.1 Simulation Example

Up to this point, simulation solutions were shown as solving for a transfer matrix (or transfer matrices at various frequencies). Essentially, transfer matrices provide the frequency responses of filters used to convert input waveforms into output waveforms. Waveforms and filters are discussed in Chapter 13, but they are not conceptually difficult to imagine. The details of working with them are quite complicated, however. Despite the fact that these details have not yet been discussed, it would be wrong to end a chapter on simulation without showing any actual waveform processing results. If necessary, one might want to gloss over this section until Chapter 13 has been read. Furthermore, one might refer to Chapter 17 to understand all of the class relationships and package locations for the classes used in this example. This example makes much more sense in the context of those chapters.

A *parser* simulation example using most of the things already discussed is shown in Figure 9.4. The circuit in Figure 9.4(a) was drawn with the *SignalIntegrityApp* that is discussed in Chapter 18, and is shown here to visualize the netlist being assembled as shown in Figure 9.4(b). The Python code for manually assembling this netlist and solving for the transfer matrices is provided in Figure 9.4(c).

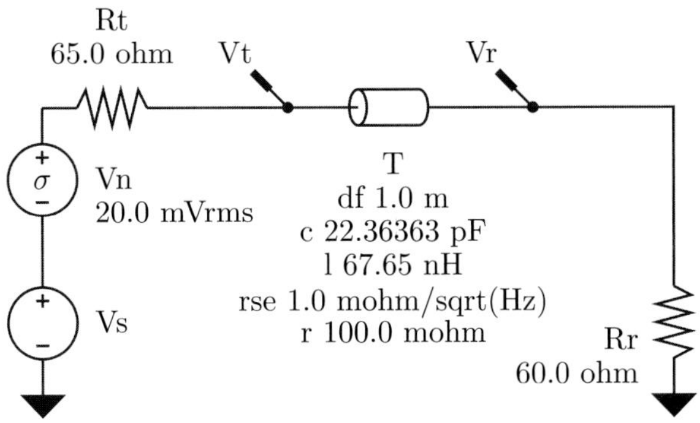


(a) Circuit

```
device T 2 telegrapher r 0.1 rse 0.001 l 6.765e-08 c 2.23636363636e-11 df 0.001
voltagesource Vs 1
voltagesource Vn 2
device Rt 2 R 65
device Rr 1 R 60
connect Vs 1 Vn 1
connect Vn 2 Rt 1
connect Rt 2 T 1
connect T 2 Rr 1
output T 1
output T 2
```

(b) Netlist

```
import SignalIntegrity.Lib as si

Td=1.23e-9; Zc=55; C=Td/Zc; L=Td*Zc
Rse=0.001; df=.001; R=.1;

netlist=['device T 2 telegrapher'+' r '+str(R)+' rse '+str(Rse)+' l '+str(L)+
         ' c '+str(C)+' df '+str(df),
         'voltagesource Vs 1','voltagesource Vn 2','device Rt 2 R 65',
         'device Rr 1 R 60','connect Vs 1 Vn 1','connect Vn 2 Rt 1',
         'connect Rt 2 T 1','connect T 2 Rr 1','output T 1','output T 2']
for line in netlist: print(line)

Fs=40e9; Ts=1./Fs; irlength=20e-9
fdtm=si.td.wf.TimeDescriptor(0,irlength/Ts,Fs).FrequencyList()
smp=si.p.SimulatorNumericParser(fdtm).AddLines(netlist)
tm=smp.TransferMatrices()
```

(c) Simulation solution

**Figure 9.4** Simulation example

Referring to Figure 9.4(c), on line 1 the *SignalIntegrity* toolbox is imported as `si`. This means that `si` is how one refers to the *SignalIntegrity* package; all of the sub-packages will be accessed with dots. These packages are provided in §17.1.

Here, the desire is to simulate serial data transmission over a single-ended transmission line. The desired electrical length is defined as 1.23 ns, and the desired characteristic impedance is 55 Ω. Using this, the capacitance and inductance of the transmission line are calculated. The series resistance is specified as 100 mΩ, the skin-effect resistance as $1\,\text{m}\Omega/\sqrt{\text{Hz}}$, and the loss tangent as 0.001. These values are specified without regard to realism and are used only to construct the example.

The netlist is assembled in lines 6 through 10, and printed on line 11. The printout is shown in Figure 9.4(b).

Waveform processing and filters have not yet been discussed, but, on line 13, the sample rate of the waveforms is specified as 40 GS/s, which would make the sample period of the waveforms 25 ps. It is assumed that, for this simulation, a reasonable impulse response length would be 20 ns. This is reasonable based on the short length and reasonably good match of the transmission line to the source impedance of 65 Ω and the termination impedance of 60 Ω.

The time descriptor for the impulse response waveform is specified as starting at time zero, having a number of samples equal to the impulse response length divided by the sample period, and the sample rate specified. From this, the frequencies required are extracted. See §12.4.2 for more on this subject.

On line 15, the **SimulatorNumericParser** class is instantiated and the netlist lines added to it.

On line 16, the transfer matrices solution is extracted. Most of this process has been described in this chapter.

The transfer matrices should be examined, so in Figure 9.5 the frequency responses of the transfer matrices are plotted using the code provided in Figure 9.5(c). On line 1, the output waveform names are listed, and on line 2 the input waveform names are listed. The input waveforms are the names of the sources in the system in the order listed in Figure 9.4(b). The output waveforms are the names of the outputs, also in the order the outputs are listed in Figure 9.4(b).

The frequency responses are extracted on line 3. If there are $O$ output waveforms and $I$ input waveforms, this is an $O \times I$ matrix of frequency responses.

These are plotted using matplotlib, which is the Python equivalent of MATLAB's plotting capability. To plot a matrix of frequency responses, matplotlib offers the concept of subplots. A subplot is referred to by the number of rows, columns, and subplot number, which is 1 based. So, for the $O \times I$ matrix of frequency responses, for $o \in 0 \ldots O-1$, $i \in 0 \ldots I-1$, `tmfr[o][i]` is plotted at subplot number $o \cdot I + i + 1$.

Lines 5–14 are obvious with a basic understanding of matplotlib. The frequencies from the frequency responses are extracted in GHz and the responses are extracted in dB. There are other options for plotting the phase, but these are not shown.

The frequency responses are plotted in Figures 9.5(a) and 9.5(b). Figure 9.5(a) shows the magnitude response for the filter that converts the waveforms $V_s$ and $V_n$ to $V_t$. They are identical because the voltage sources are in series. If the transfer matrix was $\mathbf{H}$, the

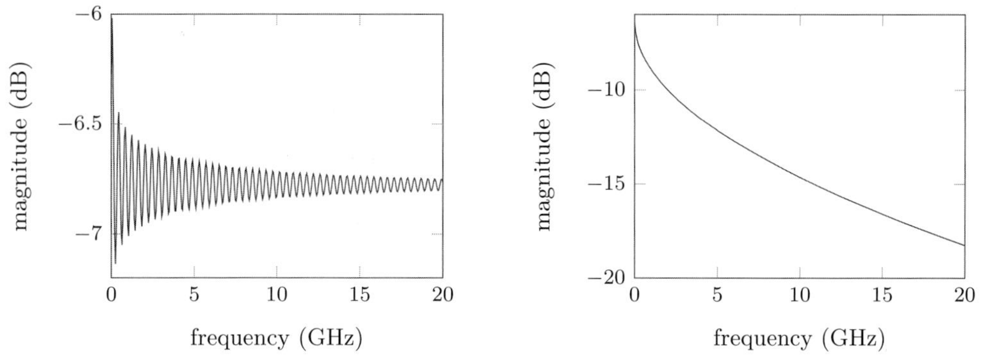


(a) `tmfr[0][0]`, `tmfr[0][1]` – $V_t$ due to $V_s$, $V_n$ (b) `tmfr[1][0]`, `tmfr[1][1]` – $V_r$ due to $V_s$, $V_n$

```
outnames=['Vt','Vr']
innames=['Vs','Vn']
tmfr=tm.FrequencyResponses()
import matplotlib.pyplot as plt
for i in range(len(innames)):
    for o in range(len(outnames)):
        plt.subplot(len(outnames),len(innames),o*len(innames)+i+1)
        plt.plot(tmfr[o][i].Frequencies('GHz'),tmfr[o][i].Response('dB'),
            label=outnames[o]+' due to '+innames[i],color='black')
        plt.legend(loc='upper right',labelspacing=0.1)
        plt.xlabel('frequency (GHz)');
        plt.ylabel('magnitude (dB)')
plt.show()
plt.cla()
```

(c) Python code to plot frequency response

**Figure 9.5** Transfer matrices frequency response

input waveforms are **vi** and the output waveforms are **vo**; then

$$\mathbf{vo} = \mathbf{H} \cdot \mathbf{vi}$$

or

$$\begin{pmatrix} V_t \\ V_r \end{pmatrix} = \begin{pmatrix} H_{00} & H_{01} \\ H_{10} & H_{11} \end{pmatrix} \cdot \begin{pmatrix} V_s \\ V_n \end{pmatrix},$$

and therefore

$$V_t = H_{00} \cdot V_s + H_{01} \cdot V_n,$$
$$V_r = H_{10} \cdot V_s + H_{11} \cdot V_n.$$

Since the responses $H_{00} = H_{01}$ and $H_{10} = H_{11}$, this can be written as follows:

$$V_t = H_{00} \cdot (V_s + V_n),$$
$$V_r = H_{11} \cdot (V_s + V_n).$$

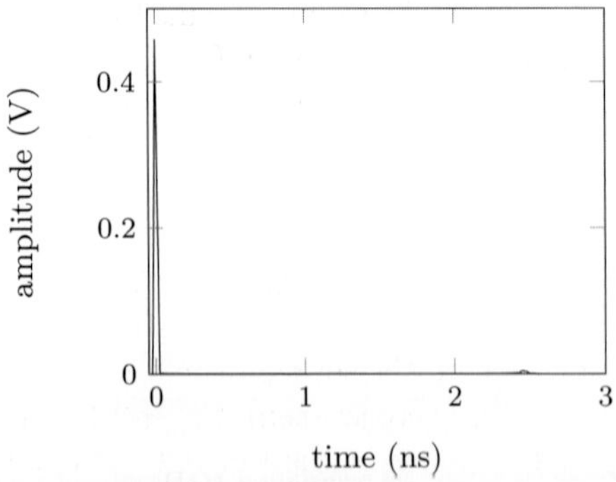


(a) `tmir[0][0]`, `tmir[0][1]` – $V_t$ due to $V_s$, $V_n$

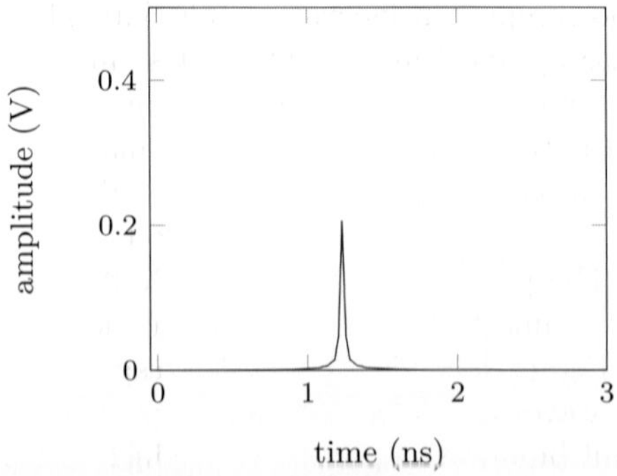


(b) `tmir[1][0]`, `tmir[1][1]` – $V_r$ due to $V_s$, $V_n$

```
tmir=tm.ImpulseResponses()
import matplotlib.pyplot as plt
for i in range(len(innames)):
    for o in range(len(outnames)):
        plt.subplot(len(outnames),len(innames),o*2+i+1)
        plt.plot(tmir[o][i].Times('ns'),tmir[o][i].Values(),
            label=outnames[o]+' due to '+innames[i],color='black')
        plt.legend(loc='upper right',labelspacing=0.1)
        plt.xlabel('time (ns)');
        plt.ylabel('amplitude (V)')
        plt.xlim(-0.05,3);
        plt.ylim(0,0.5)
plt.show()
plt.cla()
```

(c) Python code to plot impulse response

**Figure 9.6** Transfer matrices impulse response

The simulator solution was neither smart nor complicated enough to realize this. All of the simulation solutions are a matrix/vector multiplication (in the frequency domain).

In Figure 9.5(a), the response at the transmitter $V_t$ is seen to ripple and be down approximately 6 dB. The attenuation is due to the voltage divider formed by the source impedance and the characteristic impedance of the line, and would be exactly 6 dB if the impedances were equal. The ripples are due to the mismatch at each end of the transmission line.

In Figure 9.5(b), the response is mostly smooth (it is actually slightly ripply due to the small mismatch) and curves down with the skin-effect loss, eventually mostly leveling off due to the loss tangent. The intent is to transmit 5 Gb/s, so the response at 2.5 GHz is examined and seen to be down about 11 dB. Since there is about 6 dB attenuation initially, this means that alternately transitioning transmitted bits will appear half the size of the long runs of ones or zeros.

Users are encouraged to examine the impulse responses, which are shown in Figure 9.6. Originally, it was assumed that an impulse response length of 20 ns is sufficient. Since there is no comprehensive way to calculate what it should be, it ought to be examined mostly for causality and time aliasing issues, and for proper settling, as discussed in §12.2.1.

Figure 9.6(c) shows the Python code required to plot the impulse responses, again using matplotlib. This code is very similar to the code in Figure 9.5(c), except on line 1, where the impulse responses are extracted, and on line 6 they are plotted, using a time axis in nanoseconds and the values in volts.

The plots are shown in Figure 9.6(a) and Figure 9.6(b). Again, the impulse responses were found to be the same for the noise and signal waveforms.

Figure 9.6(a) is the impulse response to be convolved with the source waveforms to form the waveform at the transmitter. A large impulse at time zero is seen along with a very small bump at around 2.5 ns, which is what causes the ripples in the frequency response. Figure 9.6(b) is the impulse response that generates the waveform at the receiver. It is quite spread out due to the loss and dispersion in the channel.

Although not shown in these plots, the impulse response should be zoomed out and the step response should be examined and checked such that:

1. The step response does not start before time zero and settles completely at the end.
2. There are no time aliasing artifacts that cause portions of the impulse response to occur before time zero.

The input waveform is constructed in Figure 9.7 using methods not discussed until Chapter 13.

Referring to line 1 of Figure 9.7(c), the bit rate for the serial data waveform is set to 5 Gb/s and the length of one unit interval (UI) is set to the reciprocal.

On line 3, a raised cosine filter is constructed that has a half-width equal to 30 % of the unit interval.

On line 5, an upsample factor is specified for an interpolation filter that will be used at the end of the processing and on line 6 that filter is constructed.

On line 7, a fractional delay filter is constructed that will be used to add jitter to the waveform.

The reason why all of these filters are constructed ahead of time is to know their effect on the waveform, specifically the time axis of the waveform. Filters cause waveforms to shrink in time as the startup effects of the filter are removed after processing. Thus, a waveform tends to get smaller and smaller as it passes through filter stages. Because of this, it is desirable to reverse this effect by ensuring that the initial waveform is large enough so that the final result is properly sized.

On line 9, 300 UI are specified in the result, and on line 10 the amount of time that will have that many UI is calculated. On line 11, the time descriptor is constructed for the output waveform starting at zero time, containing a number of samples for the given length of time and a sample rate equal to the upsampled sample rate already specified as 40 GS/s upsampled by two to 80 GS/s.

As described in §13.1.3, when the time descriptor of a waveform is divided by a filter descriptor, it provides the time descriptor of a waveform that, when filtered, results in the numerator time descriptor. Thus, on line 12, the output waveform time descriptor is divided by the filter descriptor of the upsampling filter. On line 13, it is further divided by the filter descriptor of the transfer matrices. On line 14, it is further divided by the

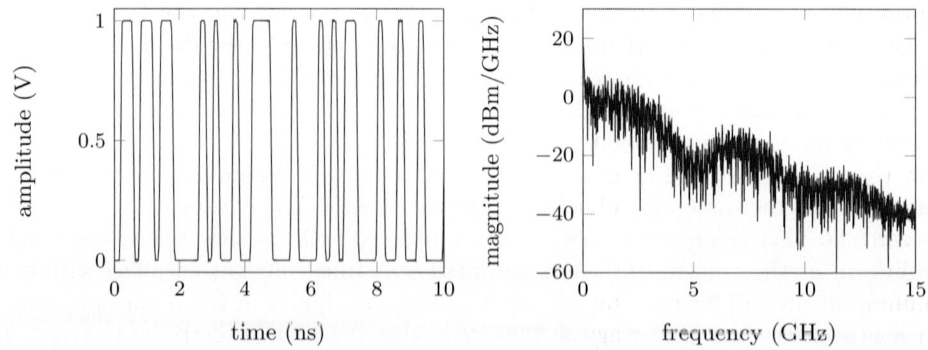


(a) Input waveform

(b) Frequency content of input waveform

```
bitRate=5e9; ui=1./bitRate

rcf=si.td.f.RaisedCosineFilter(int(0.3*ui/Ts))

uf=2
usf=si.td.f.InterpolatorSinX(uf)
fdf=si.td.f.FractionalDelayFilterSinX(0,False)

bitsResult=300
lengthResult=bitsResult*ui
tdr=si.td.wf.TimeDescriptor(0.,lengthResult/Ts*uf,Fs*uf)
tdi=((((tdr/usf.FilterDescriptor())/
        tmir[0][0].FirFilter().FilterDescriptor())/
            rcf.FilterDescriptor())/
                fdf.FilterDescriptor())
import random
bits=[1 if random.random()>0.5 else 0
    for _ in range(bitsResult+1)]; bits[0]=0
a=[(bits[k]-bits[k-1]) for k in range(1,bitsResult+1)]
import numpy
jitter=numpy.random.normal(0.,10.e-12,bitsResult).tolist()
wfi=sum([si.td.wf.StepWaveform(tdi,a[b],b*ui)*
         si.td.f.FractionalDelayFilterSinX(jitter[b]/Ts,False)
            for b in range(len(a))])
wfi=wfi*rcf
fci=wfi.FrequencyContent()

plt.plot(fci.Frequencies('GHz'),
    [v+90 for v in fci.Values('dBmPerHz')],color='black')
plt.xlabel('frequency (GHz)')
plt.ylabel('magnitude (dBm/GHz)')
plt.xlim(0,15)
plt.ylim(-60,30)
plt.show()
plt.cla()

wfn=si.td.wf.NoiseWaveform(tdi,20e-3)
```

(c) Python code to generate input waveform

**Figure 9.7** Input waveform

filter descriptor of the raised cosine filter, and on line 15 it is finally divided by the filter descriptor of the fractional delay filter. This is in the reverse order in which the filters will be applied. These are rough calculations because the fractional delay filter constructed in line 7 has a fractional delay of 0, and only one of the filters from the transfer matrices was used. On line 12, `tdi` is roughly the time descriptor of an input waveform that provides 300 UI of the serial data signal as the result.

Now that the time axis of the input waveform is known, the waveform is constructed. The random package imported on line 16 is used to build 300 bits of ones and zeros, starting with a zero bit.

On line 19, the amplitudes of the step waveforms are calculated; these waveforms will be summed together to create a random bit pattern.

On line 20 the *NumPy* package is imported, and the normal distribution function is used on line 21 to produce a normally distributed jitter times with a standard deviation of of 10 ps; theses times will be used to add jitter to the waveforms.

On lines 22–24, the input waveform is produced as the sum of the step waveforms with the given amplitudes at the times specified for a UI boundary fractionally delayed by the jitter amount. Note that the fractional delay is specified as a fraction of the sample period. Normally, fractional delay filters are employed to change the sample phase of a waveform (meaning that the time axis slides fractionally under the waveform without changing the timing of the waveform itself). This is the case if the second argument, which defaults to `False`, is not specified, but here it is set it to `True` actually to jitter the waveform.

On line 25, the final waveform is applied to the raised cosine filter. This waveform is shown (from 0 to 10 ns) in Figure 9.7(a).[1] On line 26, the frequency content of the waveform is calculated. In lines 28–35 the spectral density is plotted in dBm/GHz using matplotlib. Note that the values are extracted in dBm/Hz on line 29 and $10 \cdot \log\left(10^9\right) = 90$ added to put this in dBm/GHz. This plot is shown in Figure 9.7(b).

Finally, on line 37, the noise waveform is created with 20 mV of noise.

The output waveforms are produced in Figure 9.8. Figure 9.8(c) shows the Python code that produces the result.

On line 1, the transfer matrices processor is instantiated and the transfer matrices installed.

On line 2, the input waveforms (the transmit source and noise waveforms) are processed and then the upsampler is applied to each of these waveforms to produce the output waveform list.

In lines 4–14, the two resulting waveforms are plotted; they are shown in Figure 9.8(a). Here, the waveform at the transmitter $V_t$ with the added noise is seen to be roughly half the height of the input waveform in Figure 9.7(a). The waveform at the receiver $V_r$ is seen as a degraded version. On line 6, the electrical length of the transmission line was subtracted so that the bits can be plotted on top of each other.

On line 16, a crude eye diagram is generated for illustrative purposes by counting the times modulo three UI. On lines 19–27, these traces are separated into multiple traces that can be overlaid. On lines 29 and 30, these are plotted on top of each other. The resulting eye diagram is shown in Figure 9.8(b).

[1] In the interests of saving space, the function used to plot this waveform is not shown.

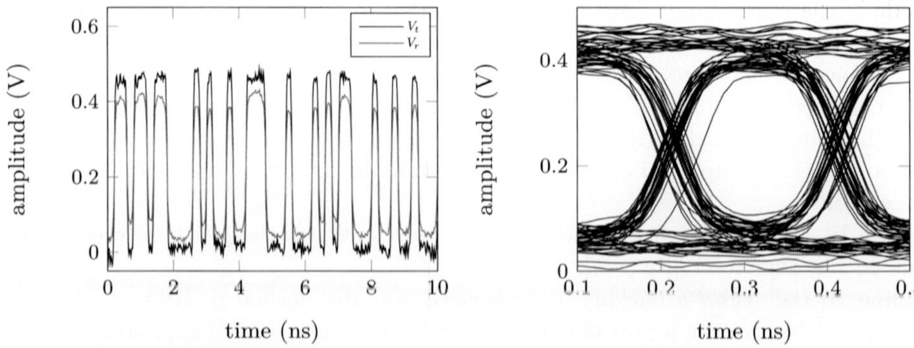


(a) Output waveforms (b) Eye diagram

```
tmp=si.td.f.TransferMatricesProcessor(tm)
wfolist=[wf*usf for wf in tmp.ProcessWaveforms([wfi,wfn])]

plt.plot(wfolist[0].Times('ns'),
         wfolist[0].Values(),label='Vt',color='black')
plt.plot([t-Td/1e-9 for t in wfolist[1].Times('ns')],
         wfolist[1].Values(),label='Vr',color='gray')
plt.legend(loc='upper right',labelspacing=0.1)
plt.xlim(0 ,10)
plt.ylim(-0.05,0.65)
plt.xlabel('time (ns)')
plt.ylabel('amplitude (V)')
plt.show()
plt.cla()

times=[(t-Td/1e-9)%(3*ui/1e-9) for t in wfolist[1].Times('ns')]
values = wfolist[1].Values()

pltt=[]; pltv=[]; tt=[]; vv=[]
for k in range(len(times)):
    if k==0:
        tt=[times[k]]; vv=[values[k]]
    elif times[k]>times[k-1]:
        tt.append(times[k]); vv.append(values[k])
    else:
        pltt.append(tt); pltv.append(vv)
        tt=[times[k]]; vv=[values[k]]

for e in range(len(pltt)):
    plt.plot(pltt[e],pltv[e],color='black')
plt.ylim(-0.00,0.5); plt.xlim(0.1,0.5)
plt.xlabel('time (ns)'); plt.ylabel('amplitude (V)')
plt.show()
plt.cla()
```

(c) Python code to generate output waveforms

**Figure 9.8** Output waveforms

One summarizing statement can be made about this simple example. The code involving the actual generation of the waveforms and the simulation involved very few lines. Most of the code in the example involved the plotting of the intermediate and final results.

# 10

# De-embedding

A PARTICULAR PRACTICAL DIFFICULTY that arises when dealing with s-parameters is that often they can only be obtained for systems consisting of both a device under test (DUT) and various fixtures, probes, and connectors used to interface the DUT to a measurement instrument. The problem is that usually the effects of these fixtures are not negligible and one really wants to obtain only the s-parameters of the DUT without the extraneous fixtures. The solution to the problem of removing unwanted fixtures from the measurement is classified as de-embedding. The concept of de-embedding is intermingled with the concept of *calibration* insofar as the calibration of network analysis instruments are concerned. This is because often calibration methods are utilized to perform the actual de-embedding. De-embedding is defined as follows:

> *Given a system of interconnected devices containing at least one device under test whose s-parameters are unknown and at least one other device whose s-parameters are known, with the system exposing ports for which the overall system s-parameters are known at these ports, de-embedding is the act of determining the unknown s-parameters.*

This definition unfortunately assumes the hard part of the practical de-embedding problem – that of knowing certain sets of s-parameters in the system which are often in difficult measurement arrangements. The practical de-embedding problem is really divided into two parts: devising arrangements for and determination of s-parameters systems of interconnected devices and the mathematical operation of determining the unknown s-parameters from system s-parameters and other known s-parameters.

De-embedding is such a common problem that many methods have been studied [21] and devised for dealing with it. Perhaps the most common method involves T-parameters [22, 23]. This was discussed briefly in §3.9. T-parameter methods are recommended only for very simple cases.

Another method commonly taught is through the use of network parameter conversions and manipulations using classic network parameter models [24] described in §1.1.2. This book avoids these techniques (and avoids the classic network parameters in general) because, while there is some elegance to these techniques, they are error prone and unnecessary.

This chapter will build from the simplest situations to the most complicated situations encountered, and in the end will rely on a single technique capable of performing all known de-embedding problems [25].

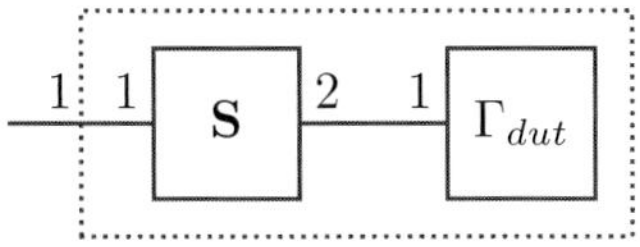


(a) Block diagram

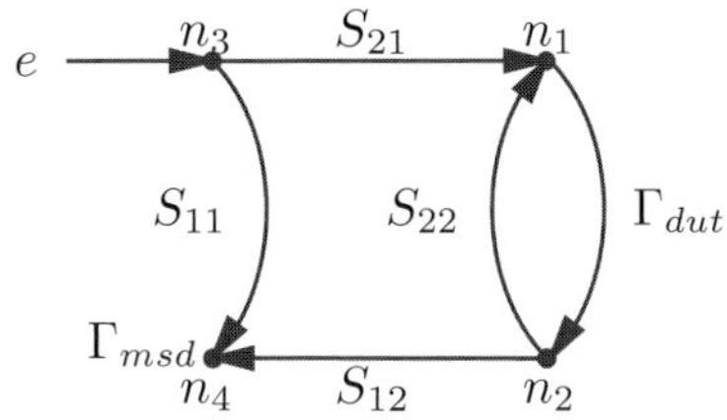


(b) Signal-flow diagram

**Figure 10.1** One-port de-embedding example

## 10.1 One-Port De-embedding

Consider a one-port de-embedding problem as shown in Figure 10.1. The block diagram for the problem is shown in Figure 10.1(a). Here there is a one-port DUT labeled $\Gamma_{dut}$ connected to a two-port element labeled **S**. There is a dotted-line box surrounding all of these elements with a single, exposed port on the left labeled 1. As a de-embedding exercise, it is presumed that the s-parameters of the device labeled **S** are known, as are the overall s-parameter measurements of the entire system within the dotted line. The de-embedding exercise is to determine the unknown s-parameters $\Gamma_{dut}$ from the known s-parameters of the system, the known s-parameters of **S**, and the interconnectedness of the system.

This particular system is not difficult and is solved in the prescribed manner of drawing the system signal-flow diagram and determining the set of equations pertaining to this system. The signal-flow diagram is shown in Figure 10.1(b).

The equations governing this system can be written as follows:

$$\begin{pmatrix} 1 & -S_{22} & -S_{21} & 0 \\ -\Gamma_{dut} & 1 & 0 & 0 \\ 0 & 0 & 1 & 0 \\ 0 & -S_{12} & -S_{11} & 1 \end{pmatrix} \cdot \begin{pmatrix} n_1 \\ n_2 \\ n_3 \\ n_4 \end{pmatrix} = \begin{pmatrix} 0 \\ 0 \\ e \\ 0 \end{pmatrix}.$$

The system is solved for node $n_4$, substituting 1 for $e$, and replacing $n_4$ with $\Gamma_{msd}$, which are now the measured system s-parameters:

$$n_4 = \Gamma_{msd} = \frac{S_{11} - |\mathbf{S}| \cdot \Gamma_{dut}}{1 - \Gamma_{dut} \cdot S_{22}}.$$

Solving for $\Gamma_{dut}$ yields

$$\Gamma_{dut} = \frac{\Gamma_{msd} - S_{11}}{\Gamma_{msd} \cdot S_{22} - |\mathbf{S}|}. \tag{10.1}$$

The de-embedding in this case is straightforward mathematically. Given a one-port measurement $\Gamma_{msd}$ of a system and the known s-parameters **S**, substituting these into (10.1) produces $\Gamma_{dut}$.

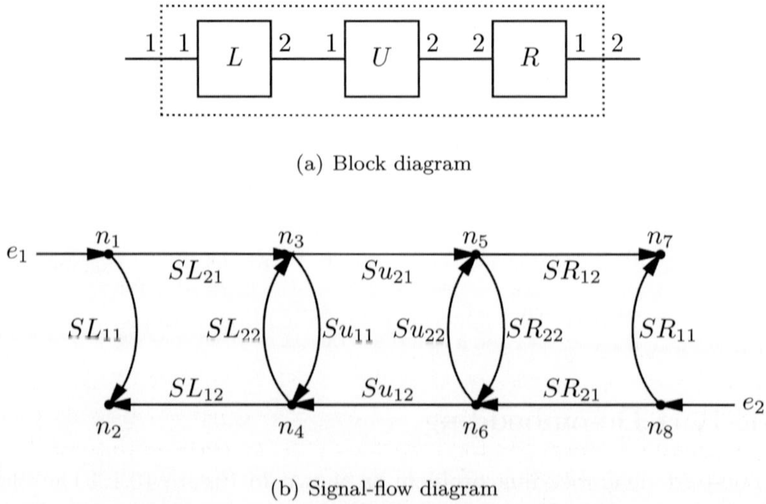


**Figure 10.2** Two-port de-embedding example

## 10.2 Two-Port De-embedding

The two-port example shown in Figure 10.2 introduces some new concepts not found in the simple one-port example. Consider the block diagram in Figure 10.2(a), which shows an unknown s-parameter device labeled $U$ bracketed by left and right devices labeled $L$ and $R$. As a de-embedding example, it presumes the s-parameters for $L$ and $R$ are known and the s-parameters for $U$ are unknown. It also presumes knowledge of the s-parameters for the entire block surrounded by the dotted lines.

A signal-flow diagram that represents this block diagram is shown in Figure 10.2(b). It has arbitrarily labeled nodes, known s-parameters (**SL** and **SR**) and unknown s-parameters (**Su**) inserted, and the inputs are driven with arbitrary signals $e_1$ and $e_2$. The equations governing this signal-flow diagram are shown in (10.2):

$$\begin{pmatrix} 1 & 0 & 0 & 0 & 0 & 0 & 0 & 0 \\ -SL_{11} & 1 & 0 & -SL_{12} & 0 & 0 & 0 & 0 \\ -SL_{21} & 0 & 1 & -SL_{22} & 0 & 0 & 0 & 0 \\ 0 & 0 & -Su_{11} & 1 & 0 & -Su_{12} & 0 & 0 \\ 0 & 0 & -Su_{21} & 0 & 1 & -Su_{22} & 0 & 0 \\ 0 & 0 & 0 & 0 & -SR_{22} & 1 & 0 & -SR_{21} \\ 0 & 0 & 0 & 0 & -SR_{12} & 0 & 1 & -SR_{11} \\ 0 & 0 & 0 & 0 & 0 & 0 & 0 & 1 \end{pmatrix} \cdot \begin{pmatrix} n_1 \\ n_2 \\ n_3 \\ n_4 \\ n_5 \\ n_6 \\ n_7 \\ n_8 \end{pmatrix} = \begin{pmatrix} e_1 \\ 0 \\ 0 \\ 0 \\ 0 \\ 0 \\ 0 \\ e_2 \end{pmatrix}. \tag{10.2}$$

Equation (10.2) is in a form with seemingly little hope of solving for the unknown s-parameters. Without any rationale provided as yet, observe that it is possible to group the nodes according to certain criteria:

- Nodes 1 and 8 are incident waves on the entire system and are therefore the measured incident waves. These are categorized as $a_{msd}$.
- Nodes 2 and 7 are reflected waves from the entire system and are therefore the measured reflected waves categorized as $b_{msd}$.
- Nodes 3 and 6 are the incident waves on the DUT categorized as $a$.
- Nodes 4 and 5 are the reflected waves from the DUT categorized as $b$.

It is desirable to rewrite (10.2) in a form that re-orders things according to a special grouping of nodes. Equations for systems are always written in a canonical form, meaning that the node ordering is always the same as the equation ordering. Here the nodes and equations are placed in order of categorization: $\begin{pmatrix} b_{msd} & a & a_{msd} & b \end{pmatrix}$. To do this, *permutation matrices* are utilized, as described in §4.2.2.

Assume that the current system of equations is in the form $\widehat{\mathbf{G}} \cdot \widehat{\mathbf{v}} = \widehat{\mathbf{e}}$, where $\widehat{\mathbf{G}}$ represents the system characteristics matrix, $\widehat{\mathbf{v}}$ represents an arbitrary node ordering (and equation ordering), and $\widehat{\mathbf{e}}$ represents the corresponding arbitrary stimulus ordering. A row permutation matrix $\mathbf{P}$ is developed such that $\mathbf{P} \cdot \widehat{\mathbf{v}} = \mathbf{v}$ has the new desired ordering. This must multiply $\mathbf{P}$ from the left of $\widehat{\mathbf{G}}$ to get the desired equation ordering. Less obviously, multiplying $\mathbf{P}^{-1} = \mathbf{P}^T$ from the right of $\widehat{\mathbf{G}}$ reorders the nodes. Thus,

$$\mathbf{P} \cdot \widehat{\mathbf{G}} \cdot \mathbf{P}^T \cdot \mathbf{P} \cdot \widehat{\mathbf{v}} = \mathbf{P} \cdot \widehat{\mathbf{e}}, \tag{10.3}$$

where $\mathbf{G} = \mathbf{P} \cdot \widehat{\mathbf{G}} \cdot \mathbf{P}^T$, $\mathbf{v} = \mathbf{P} \cdot \widehat{\mathbf{v}}$, $\mathbf{e} = \mathbf{P} \cdot \widehat{\mathbf{e}}$, and therefore $\mathbf{G} \cdot \mathbf{v} = \mathbf{e}$.

To group the nodes in order $\begin{pmatrix} b_{msd} & a & a_{msd} & b \end{pmatrix}^T$, the node order is written as $\begin{pmatrix} n_2 & n_7 & n_3 & n_6 & n_1 & n_8 & n_4 & n_5 \end{pmatrix}^T$. Equation (10.4) achieves such a reordering:

$$\begin{pmatrix} 0 & 1 & 0 & 0 & 0 & 0 & 0 & 0 \\ 0 & 0 & 0 & 0 & 0 & 0 & 1 & 0 \\ 0 & 0 & 1 & 0 & 0 & 0 & 0 & 0 \\ 0 & 0 & 0 & 0 & 0 & 1 & 0 & 0 \\ 1 & 0 & 0 & 0 & 0 & 0 & 0 & 0 \\ 0 & 0 & 0 & 0 & 0 & 0 & 0 & 1 \\ 0 & 0 & 0 & 1 & 0 & 0 & 0 & 0 \\ 0 & 0 & 0 & 0 & 1 & 0 & 0 & 0 \end{pmatrix} \cdot \begin{pmatrix} n_1 \\ n_2 \\ n_3 \\ n_4 \\ n_5 \\ n_6 \\ n_7 \\ n_8 \end{pmatrix} = \begin{pmatrix} n_2 \\ n_7 \\ n_3 \\ n_6 \\ n_1 \\ n_8 \\ n_4 \\ n_5 \end{pmatrix}. \tag{10.4}$$

So, $\mathbf{P}$ is defined as the left side of (10.4) and (10.2) is substituted into (10.3) to obtain

$$\begin{pmatrix} 1 & 0 & 0 & 0 & -SL_{11} & 0 & -SL_{12} & 0 \\ 0 & 1 & 0 & 0 & 0 & -SR_{11} & 0 & -SR_{12} \\ 0 & 0 & 1 & 0 & -SL_{21} & 0 & -SL_{22} & 0 \\ 0 & 0 & 0 & 1 & 0 & -SR_{21} & 0 & -SR_{22} \\ 0 & 0 & 0 & 0 & 1 & 0 & 0 & 0 \\ 0 & 0 & 0 & 0 & 0 & 1 & 0 & 0 \\ 0 & 0 & -Su_{11} & -Su_{21} & 0 & 0 & 1 & 0 \\ 0 & 0 & -Su_{12} & -Su_{22} & 0 & 0 & 0 & 1 \end{pmatrix} \cdot \begin{pmatrix} n_2 \\ n_7 \\ n_3 \\ n_6 \\ n_1 \\ n_8 \\ n_4 \\ n_5 \end{pmatrix} = \begin{pmatrix} 0 \\ 0 \\ 0 \\ 0 \\ e_1 \\ e_2 \\ 0 \\ 0 \end{pmatrix}. \tag{10.5}$$

Examining (10.5), the system can be driven somewhat arbitrarily, so it is chosen to drive the system first with $e_1 = 1$, then with $e_2 = 1$, and $\mathbf{G}$ can then be grouped into $2 \times 2$ element block matrices to obtain

$$\begin{pmatrix} \mathbb{I} & 0 & -\mathbf{F_{11}} & -\mathbf{F_{12}} \\ \mathbf{0} & \mathbb{I} & -\mathbf{F_{21}} & -\mathbf{F_{22}} \\ \mathbf{0} & \mathbf{0} & \mathbb{I} & \mathbf{0} \\ \mathbf{0} & -\mathbf{Su} & \mathbf{0} & \mathbb{I} \end{pmatrix} \cdot \begin{pmatrix} \mathbf{B_{msd}} \\ \mathbf{A} \\ \mathbf{A_{msd}} \\ \mathbf{B} \end{pmatrix} = \begin{pmatrix} \mathbf{0} \\ \mathbf{0} \\ \mathbb{I} \\ \mathbf{0} \end{pmatrix}; \tag{10.6}$$

all block matrices are $2 \times 2$. Expanding these equations yields

$$\mathbf{B_{msd}} - \mathbf{F_{11}} \cdot \mathbf{A_{msd}} - \mathbf{F_{12}} \cdot \mathbf{B} = \mathbf{0}, \tag{10.7}$$

$$\mathbf{A} - \mathbf{F_{21}} \cdot \mathbf{A_{msd}} - \mathbf{F_{22}} \cdot \mathbf{B} = \mathbf{0}, \tag{10.8}$$

$$\mathbf{A_{msd}} = \mathbb{I}, \tag{10.9}$$

$$-\mathbf{Su} \cdot \mathbf{A} = \mathbf{B}. \tag{10.10}$$

Equation (10.9) says that the measured incident waves are $\mathbb{I}$ as expected because the ports have been alternately driven with 1. Under these conditions, the measured reflected waves are equal to the measured overall system s-parameters, so $\mathbf{B_{msd}} = \mathbf{Sk}$, where $\mathbf{Sk}$ represents the known system s-parameters. Substituting (10.9) into (10.7) and solving for $\mathbf{B}$ yields

$$\mathbf{B} = \mathbf{F_{12}}^{-1} \cdot (\mathbf{Sk} - \mathbf{F_{11}}). \tag{10.11}$$

Substituting (10.11) and (10.9) into (10.8) and solving for $\mathbf{A}$:

$$\mathbf{A} = \mathbf{F_{21}} + \mathbf{F_{22}} \cdot \mathbf{B}. \tag{10.12}$$

Finally, (10.11) and (10.12) are substituted into (10.10) and solved for the unknown s-parameters $\mathbf{Su}$:

$$\mathbf{Su} = \mathbf{B} \cdot \mathbf{A}^{-1}. \tag{10.13}$$

Equations (10.11), (10.12), and (10.13) provide the recipe for de-embedding. Here is the full calculation for the two-port case.

$$\mathbf{F_{11}} = \begin{pmatrix} SL_{11} & 0 \\ 0 & SR_{11} \end{pmatrix}; \qquad \mathbf{F_{12}} = \begin{pmatrix} SL_{12} & 0 \\ 0 & SR_{12} \end{pmatrix};$$

$$\mathbf{F_{21}} = \begin{pmatrix} SL_{21} & 0 \\ 0 & SR_{21} \end{pmatrix}; \qquad \mathbf{F_{22}} = \begin{pmatrix} SL_{22} & 0 \\ 0 & SR_{22} \end{pmatrix};$$

$$\mathbf{B} = \begin{pmatrix} SL_{12} & 0 \\ 0 & SR_{12} \end{pmatrix}^{-1} \cdot \left[ \begin{pmatrix} Sk_{11} & Sk_{12} \\ Sk_{21} & Sk_{22} \end{pmatrix} - \begin{pmatrix} SL_{11} & 0 \\ 0 & SR_{11} \end{pmatrix} \right]$$

$$= \begin{pmatrix} \frac{Sk_{11} - SL_{11}}{SL_{12}} & \frac{Sk_{12}}{SL_{12}} \\ \frac{Sk_{21}}{SR_{12}} & \frac{Sk_{22} - SR_{11}}{SR_{12}} \end{pmatrix};$$

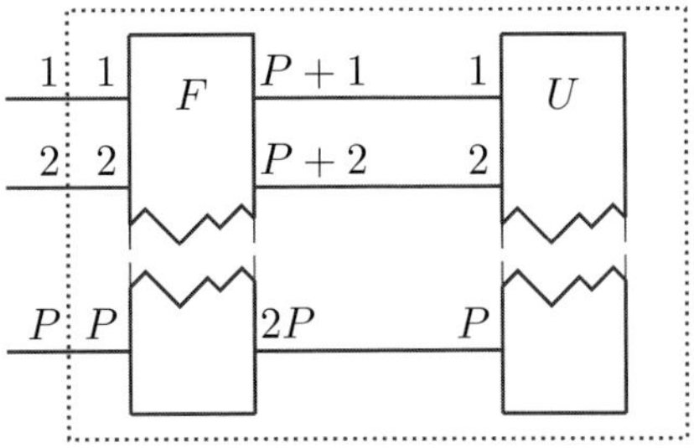


**Figure 10.3** Fixture de-embedding example

$$\mathbf{A} = \begin{pmatrix} SL_{21} & 0 \\ 0 & SR_{21} \end{pmatrix} + \begin{pmatrix} SL_{22} & 0 \\ 0 & SR_{22} \end{pmatrix} \cdot \begin{pmatrix} \frac{Sk_{11}-SL_{11}}{SL_{12}} & \frac{Sk_{12}}{SL_{12}} \\ \frac{Sk_{21}}{SR_{12}} & \frac{Sk_{22}-SR_{11}}{SR_{12}} \end{pmatrix}$$

$$= \begin{pmatrix} \frac{SL_{22}\cdot Sk_{11}-|\mathbf{SL}|}{SL_{12}} & \frac{SL_{22}\cdot Sk_{12}}{SL_{12}} \\ \frac{SR_{22}\cdot Sk_{21}}{SR_{12}} & \frac{SR_{22}\cdot Sk_{22}-|\mathbf{SR}|}{SR_{12}} \end{pmatrix};$$

$$\mathbf{Su} = \begin{pmatrix} \frac{Sk_{11}-SL_{11}}{SL_{12}} & \frac{Sk_{12}}{SL_{12}} \\ \frac{Sk_{21}}{SR_{12}} & \frac{Sk_{22}-SR_{11}}{SR_{12}} \end{pmatrix} \cdot \begin{pmatrix} \frac{SL_{22}\cdot Sk_{11}-|\mathbf{SL}|}{SL_{12}} & \frac{SL_{22}\cdot Sk_{12}}{SL_{12}} \\ \frac{SR_{22}\cdot Sk_{21}}{SR_{12}} & \frac{SR_{22}\cdot Sk_{22}-|\mathbf{SR}|}{SR_{12}} \end{pmatrix}^{-1}. \tag{10.14}$$

Equation (10.14) is the least messy form of the result.

It is interesting to reduce (10.14) for a system with only a single known two-port device. The situation where **SR** is not present is emulated by setting $\mathbf{SR} = \left(\begin{smallmatrix} 0 & 1 \\ 1 & 0 \end{smallmatrix}\right)$, thereby making it an ideal thru element. Substituting into (10.14):

$$\mathbf{Su} = \begin{pmatrix} \frac{Sk_{11}-SL_{11}}{SL_{12}} & \frac{Sk_{12}}{SL_{12}} \\ Sk_{21} & Sk_{22} \end{pmatrix} \cdot \begin{pmatrix} \frac{SL_{22}\cdot Sk_{11}-|\mathbf{SL}|}{SL_{12}} & \frac{SL_{22}\cdot Sk_{12}}{SL_{12}} \\ 0 & 1 \end{pmatrix}^{-1}$$

$$= \frac{1}{SL_{22}\cdot Sk_{11} - |\mathbf{SL}|} \begin{pmatrix} Sk_{11} - SL_{11} & SL_{21}\cdot Sk_{12} \\ SL_{12}\cdot Sk_{21} & SL_{22}\cdot|\mathbf{Sk}| - Sk_{22}\cdot|\mathbf{SL}| \end{pmatrix}. \tag{10.15}$$

Equation (10.15) is equivalent to (3.40), which utilized T-parameters to perform the de-embedding.

## 10.3 Fixture De-embedding

The fixture de-embedding problem can be established by considering Figure 10.3.

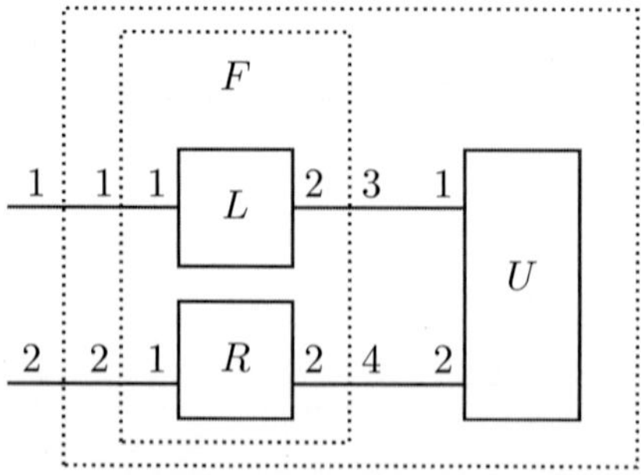


**Figure 10.4** Two-port fixture style equivalent block diagram

Figure 10.3 shows a system consisting of a device connected to both system ports on one side and the unknown DUT ports on the other. When the intervening device is connected in this way, it is called a *fixture*. This is a special case of a $P$-port DUT, $P$-port known system s-parameters, and a $2 \cdot P$-port intervening fixture. It is special in the sense that the number of system ports is exactly the same as the number of DUT ports and that the number of fixture ports is the sum or twice the number of the system or DUT ports.

The discussion in §10.2 was the perfect setup for the general fixture de-embedding case because it allows this work to be reused by simple inspection. Consider the fixture-style equivalent block diagram shown in Figure 10.4, which is equivalent to the block diagram shown in Figure 10.2(a) with some additional information. In Figure 10.4 there is exactly the same connection of the two-port elements $L$ and $R$ between the DUT and the system ports; in fact, it is the exact same de-embedding problem. Here, however, all of the system ports and the DUT ports are placed to the left. Furthermore, a box has been drawn around elements $L$ and $R$, treating them in aggregate as a four-port fixture device $F$. The ports on $F$ have been numbered exactly as specified for the fixture de-embedding arrangement as shown in Figure 10.3.

The s-parameters of the fixture as drawn are

$$\mathbf{F} = \begin{pmatrix} SL_{11} & 0 & SL_{12} & 0 \\ 0 & SR_{11} & 0 & SR_{12} \\ SL_{21} & 0 & SL_{22} & 0 \\ 0 & SR_{21} & 0 & SR_{22} \end{pmatrix}. \tag{10.16}$$

These s-parameters (or, more precisely, the negative of these s-parameters) fit neatly into the upper right quadrant of the system characteristics matrix shown in (10.5), and one can solve the general fixture de-embedding problem as a more general case of the two-port de-embedding case.

To de-embed a $2 \cdot P$-port fixture with s-parameters $\mathbf{F}$ from a $P$-port unknown DUT with s-parameters $\mathbf{Su}$, given known system s-parameters $\mathbf{Sk}$ for a fixture de-embedding arrangement as shown in Figure 10.3, the fixture s-parameters are partitioned in the block

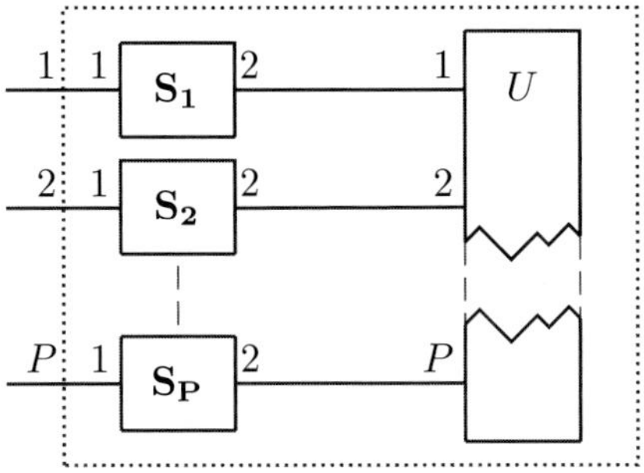


**Figure 10.5** Two-port tip de-embedding block diagram

matrices:

$$\underset{2\cdot P\times 2\cdot P}{\mathbf{F}} = \begin{pmatrix} \underset{P\times P}{\mathbf{F_{11}}} & \underset{P\times P}{\mathbf{F_{12}}} \\ \underset{P\times P}{\mathbf{F_{21}}} & \underset{P\times P}{\mathbf{F_{22}}} \end{pmatrix}.$$

Solving for the unknown DUT s-parameters using the solution already established in §10.2 results in

$$\mathbf{B} = \mathbf{F_{12}}^{-1} \cdot (\mathbf{Sk} - \mathbf{F_{11}}), \tag{10.17}$$

$$\mathbf{A} = \mathbf{F_{21}} + \mathbf{F_{22}} \cdot \mathbf{B}, \tag{10.18}$$

$$\mathbf{Su} = \mathbf{B} \cdot \mathbf{A}^{-1}. \tag{10.19}$$

## 10.4 Two-Port Tip De-embedding

A two-port de-embedding situation was considered in §10.2, and in §10.3 this was extended to a general fixture de-embedding situation. However, in (10.16), where the fixture equivalent of the two-port case was established, one can observe a certain pattern in the generation of $\mathbf{F}$ which contains the fixture s-parameters. This pattern extends into the block sub-matrices $\mathbf{F_{11}}$, $\mathbf{F_{12}}$, $\mathbf{F_{21}}$, and $\mathbf{F_{22}}$. The pattern is that, for a system containing two-port devices whose s-parameters are known at the tips of a device whose s-parameters are not known, the s-parameters of these two-port devices are located along the diagonal of the block sub-matrices. In fact, for a $P$-port device with unknown s-parameters, with each port $p \in 1 \ldots P$ connected to port 2 of a two-port device with s-parameters designated as $S_p$ and port 1 of the two-port device connected to system port $p$, the de-embedding problem is solved by setting each element of the, in this case diagonal, block matrices as

$$\mathbf{F_{rc}}_{pp} = S_{p_{rc}}, \tag{10.20}$$

where both $r \in 1, 2$ and $c \in 1, 2$.

Block matrix $\mathbf{F_{11}}$ contains all of the $S_{11}$ s-parameters of the two-port elements. Similarly, $\mathbf{F_{12}}$, $\mathbf{F_{21}}$, and $\mathbf{F_{22}}$ contain all of the $S_{12}$, $S_{21}$, and $S_{22}$ s-parameters, respectively, of the two-port devices. These s-parameters are located on the diagonal at a location corresponding to the port number of both the system and the DUT.

Therefore, given the known s-parameters at the system ports and the known s-parameters of the two-port devices connected to the tips of the DUT, as shown in Figure 10.5, the DUT is solved by applying (10.20) followed by (10.17), (10.18), and (10.19).

The method outlined here can be utilized even when only a subset of the ports of the DUT are connected to these two-port devices by setting the s-parameters of the known two-port device at a given port to those of an ideal thru element as was done in §10.2.

## 10.5 Extensions to the Fixture De-embedding Problem

The results of fixture de-embedding were provided in §10.3. These results provided a branch point for extension to the solution of other de-embedding varieties that can arise. Here, the concept is extended to the following situations:

- Systems containing multiple connected devices instead of a single fixture, which generates internal nodes.
- Systems containing numbers of system ports that differ from the number of unknown DUT ports.
- Systems containing system ports connected directly to the unknown DUT ports.
- Systems containing multiple unknown DUTs.
- Systems containing dense system characteristic matrices.

### 10.5.1 Systems With Internal Nodes

In this section the situation is addressed in which there is not a single fixture, but instead there is a larger circuit connected between the system ports and the DUT. Specifically, the calculation is extended to the case in which there are *internal nodes.* The solution to the fixture de-embedding problem results from grouping nodes in the system into four distinct categories:

- $A_{msd}$, nodes that are incident on the system;
- $B_{msd}$, nodes that are reflected from the system;
- $A$, nodes that are incident on the unknown DUT;
- $B$, nodes that are reflected from the unknown DUT.

For the system shown in Figure 10.3, all ports produce nodes that can be classified into these four categories. Even in the case shown in Figure 10.4, where there are two devices, one is able to use the fixture de-embedding method because when the two devices are grouped into a single fixture, all of the ports produce nodes that fall into these four categories. Problems come about when there are device connections such as those in Figure 10.6.

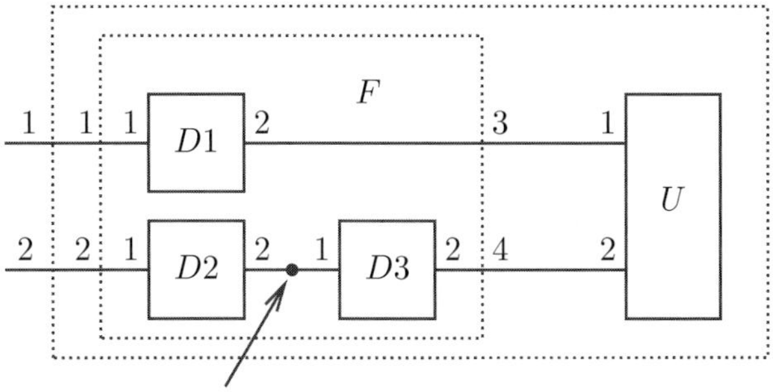


**Figure 10.6** De-embedding block diagram with internal port connection

Figure 10.6 shows devices connected in a manner such that there is an internal port, which produces internal nodes. Internal nodes are defined as nodes that are not at the system port interface and are not at the DUT interface. Internal nodes are produced by device port connections that represent neither a system port nor a DUT port.

Internal nodes expand the system characteristics equation, adding nodes that are not not classified into the four categories and are basically extraneous. When the system of equations governing a circuit with internal nodes is written out, a set of equations in block matrix form results as follows:

$$\begin{pmatrix} \underset{K\times K}{\mathbb{I}} & \underset{K\times U}{\mathbf{0}} & \underset{K\times L}{\mathbf{G_{13}}} & \underset{K\times K}{\mathbf{G_{14}}} & \underset{K\times U}{\mathbf{G_{15}}} \\ \underset{U\times K}{\mathbf{0}} & \underset{U\times U}{\mathbb{I}} & \underset{U\times L}{\mathbf{G_{23}}} & \underset{U\times K}{\mathbf{G_{24}}} & \underset{U\times U}{\mathbf{G_{25}}} \\ \underset{L\times K}{\mathbf{0}} & \underset{L\times U}{\mathbf{0}} & \underset{L\times L}{\mathbf{G_{33}}} & \underset{L\times K}{\mathbf{G_{34}}} & \underset{L\times U}{\mathbf{G_{35}}} \\ \underset{K\times K}{\mathbf{0}} & \underset{K\times U}{\mathbf{0}} & \underset{K\times L}{\mathbf{0}} & \underset{K\times K}{\mathbb{I}} & \underset{K\times U}{\mathbf{0}} \\ \underset{U\times K}{\mathbf{0}} & \underset{U\times U}{-\mathbf{Su}} & \underset{U\times L}{\mathbf{0}} & \underset{U\times K}{\mathbf{0}} & \underset{U\times U}{\mathbb{I}} \end{pmatrix} \cdot \begin{pmatrix} \underset{K\times K}{\mathbf{B_{msd}}} \\ \underset{U\times K}{\mathbf{A}} \\ \underset{L\times K}{\mathbf{U}} \\ \underset{K\times K}{\mathbf{A_{msd}}} \\ \underset{U\times K}{\mathbf{B}} \end{pmatrix} = \begin{pmatrix} \underset{K\times K}{\mathbf{0}} \\ \underset{U\times K}{\mathbf{0}} \\ \underset{L\times K}{\mathbf{0}} \\ \underset{K\times K}{\mathbb{I}} \\ \underset{U\times K}{\mathbf{0}} \end{pmatrix}, \tag{10.21}$$

where the sizes of the blocks are listed underneath each block, as they are not of equal size. Here, $K$ represents the number of system ports, $U$ represents the number of unknown DUT ports, and L represents the number of internal nodes in the system. For now,assume $K = U$.

Equation (10.21) is justified as follows:

- The first row contains the equations for the $\mathbf{B_{msd}}$ nodes, which are the waves reflected from the system ports. No $\mathbf{B_{msd}}$ node can have arrows coming from other $\mathbf{B_{msd}}$ nodes, hence the $\mathbb{I}$ in block $\mathbf{G_{11}}$. In addition, these nodes cannot have arrows coming from nodes representing incident waves on the DUT, hence the $\mathbf{0}$ for block $\mathbf{G_{12}}$. There may be arrows coming from the internal nodes $\mathbf{U}$, the waves incident on the system $\mathbf{A_{msd}}$ and the waves reflected from the unknown DUT $\mathbf{B}$, which accounts for the remainder of the first row.

- The second row is the equations for the waves incident on the DUT $\mathbf{A}$. No $\mathbf{A}$ node can have arrows coming from nodes incident on the DUT, hence the $\mathbf{0}$ in block $\mathbf{G_{21}}$. Furthermore, $\mathbf{A}$ nodes cannot have arrows coming from other $\mathbf{A}$ nodes, hence the $\mathbb{I}$ in block $\mathbf{G_{22}}$. There may be arrows coming from internal nodes $\mathbf{U}$, the waves incident on the system $\mathbf{A_{msd}}$ and the waves reflected from the unknown DUT $\mathbf{B}$, which accounts for the remainder of the second row.
- The third row contains the equations for the internal nodes. These nodes cannot have arrows coming into them from $\mathbf{B_{msd}}$ nodes representing waves reflected from the system nor from $\mathbf{A}$ nodes representing waves incident on the DUT, but they may have all other connections, hence the $\mathbf{0}$ blocks in blocks $\mathbf{G_{31}}$ and $\mathbf{G_{32}}$.
- The fourth row contains the equations for the waves incident on the system. These waves can only come from external stimuli, and therefore the entire row is $\mathbf{0}$ except for $\mathbb{I}$ in block $\mathbf{G_{44}}$. In order to generate the known s-parameters, these nodes are stimulated one at a time, which produces $\mathbb{I}$ in block $\mathbf{e_5}$.
- The fifth row contains the equations for the waves reflected from the DUT. These nodes can only have arrows coming from nodes incident on the unknown DUT, and these arrows must be the weights of the s-parameters of the DUT itself. There is an $-\mathbf{Su}$ at block $\mathbf{G_{52}}$ because it contains the negative of the actual DUT s-parameters, and $\mathbf{0}$ in all other blocks for the reasons given.

This same justification can also be looked at column wise. While the rows contain the weights of arrows coming into a node, the columns contain the weights of arrows going out of the node. In other words, the columns reflect the weights of arrows coming out of a node and the rows of the columns reflect the nodes where these arrows terminate. As an example, in the first column, representing the terminal nodes for nodes originating in $\mathbf{B_{msd}}$ nodes, no $\mathbf{B_{msd}}$ node representing waves reflected from the system can originate on a $\mathbf{B_{msd}}$ node and terminate on another, hence the $\mathbf{0}$ blocks for the entire first column except for block $\mathbf{G_{11}}$. As another example, the second column representing the weights of arrows originating from the $\mathbf{A}$ nodes incident on the DUT can only terminate in the $\mathbf{B}$ nodes reflected from the DUT, and their weights must be the unknown s-parameters contained in block $\mathbf{G_{25}}$.

It should now be apparent why (10.21) is in the form that it is. Examining (10.21), this problem can be solved in two ways. The first way is to solve the system of equations in (10.21) directly. This solution is followed by an alternative method which, while producing the same result, provides additional insight. The equations generated by (10.21) are listed as follows:

$$\mathbf{B_{msd}} + \mathbf{G_{13}} \cdot \mathbf{U} + \mathbf{G_{14}} \cdot \mathbf{A_{msd}} + \mathbf{G_{15}} \cdot \mathbf{B} = \mathbf{0}, \tag{10.22}$$

$$\mathbf{A} + \mathbf{G_{23}} \cdot \mathbf{U} + \mathbf{G_{24}} \cdot \mathbf{A_{msd}} + \mathbf{G_{25}} \cdot \mathbf{B} = \mathbf{0}, \tag{10.23}$$

$$\mathbf{G_{33}} \cdot \mathbf{U} + \mathbf{G_{34}} \cdot \mathbf{A_{msd}} + \mathbf{G_{35}} \cdot \mathbf{B} = \mathbf{0}, \tag{10.24}$$

$$\mathbf{A_{msd}} = \mathbb{I}, \tag{10.25}$$

$$-\mathbf{Su} \cdot \mathbf{A} + \mathbf{B} = 0. \tag{10.26}$$

Substituting (10.25) into (10.24) and solving for $\mathbf{U}$ in terms of $\mathbf{B}$:

$$\mathbf{U} = -\mathbf{G_{33}}^{-1} \cdot (\mathbf{G_{34}} + \mathbf{G_{35}} \cdot \mathbf{B}). \tag{10.27}$$

Substituting (10.25) and (10.27) into (10.22), and making use of the fact that when $\mathbf{A_{msd}} = \mathbb{I}$, $\mathbf{B_{msd}} = \mathbf{Sk}$, and solving for $\mathbf{B}$:

$$\mathbf{B} = \left(\mathbf{G_{13}} \cdot \mathbf{G_{33}}^{-1} \cdot \mathbf{G_{35}} - \mathbf{G_{15}}\right)^{-1} \cdot \left[\mathbf{Sk} - \left(\mathbf{G_{13}} \cdot \mathbf{G_{33}}^{-1} \cdot \mathbf{G_{34}} - \mathbf{G_{14}}\right)\right]. \tag{10.28}$$

Substituting (10.25) and (10.27) into (10.23) and solving for $\mathbf{A}$ in terms of $\mathbf{B}$:

$$\mathbf{A} = \left(\mathbf{G_{23}} \cdot \mathbf{G_{33}}^{-1} \cdot \mathbf{G_{34}} - \mathbf{G_{24}}\right) + \left(\mathbf{G_{23}} \cdot \mathbf{G_{33}}^{-1} \mathbf{G_{35}} - \mathbf{G_{25}}\right) \cdot \mathbf{B}. \tag{10.29}$$

Substituting (10.28) and (10.29) into (10.26) and solving for the unknown s-parameters $\mathbf{Su}$:

$$\mathbf{Su} = \mathbf{B} \cdot \mathbf{A}^{-1}. \tag{10.30}$$

Thus, a more general de-embedding case has been solved for the situation where there are internal nodes. Another way to look at this is by manipulating (10.21) into the form of (10.6) and continuing from there. What this really means for systems with internal nodes as in Figure 10.6 is that the circuit inside the boundary of $F$ must be solved, and then the s-parameters of the fixture represented by $F$ must be substituted into the equations in §10.3 to solve for the unknown s-parameters. In order to do this, one recognizes that solving for the circuit inside the boundary of $F$ is the same as removing the unknown DUT from the system, and that, in addition to driving and measuring the s-parameters at the system ports, one must also drive and measure the s-parameters at the DUT connection ports.

The equation for a system with internal nodes driven at the system ports and the DUT ports is

$$\begin{pmatrix} \underset{K\times K}{\mathbb{I}} & \underset{K\times U}{\mathbf{0}} & \underset{K\times L}{\mathbf{G_{13}}} & \underset{K\times K}{\mathbf{G_{14}}} & \underset{K\times U}{\mathbf{G_{15}}} \\ \underset{U\times K}{\mathbf{0}} & \underset{U\times U}{\mathbb{I}} & \underset{U\times L}{\mathbf{G_{23}}} & \underset{U\times K}{\mathbf{G_{24}}} & \underset{U\times U}{\mathbf{G_{25}}} \\ \underset{L\times K}{\mathbf{0}} & \underset{L\times U}{\mathbf{0}} & \underset{L\times L}{\mathbf{G_{33}}} & \underset{L\times K}{\mathbf{G_{34}}} & \underset{L\times U}{\mathbf{G_{35}}} \\ \underset{K\times K}{\mathbf{0}} & \underset{K\times U}{\mathbf{0}} & \underset{K\times L}{\mathbf{0}} & \underset{K\times K}{\mathbb{I}} & \underset{K\times U}{\mathbf{0}} \\ \underset{U\times K}{\mathbf{0}} & \underset{U\times U}{\mathbf{0}} & \underset{U\times L}{\mathbf{0}} & \underset{U\times K}{\mathbf{0}} & \underset{U\times U}{\mathbb{I}} \end{pmatrix} \cdot \begin{pmatrix} \underset{K\times K}{\mathbf{F_{11}}} & \underset{K\times U}{\mathbf{F_{12}}} \\ \underset{U\times K}{\mathbf{F_{21}}} & \underset{U\times U}{\mathbf{F_{22}}} \\ \underset{L\times K}{\mathbf{U_1}} & \underset{L\times U}{\mathbf{U_2}} \\ \underset{K\times K}{\mathbb{I}} & \underset{K\times U}{\mathbf{0}} \\ \underset{U\times K}{\mathbf{0}} & \underset{U\times U}{\mathbb{I}} \end{pmatrix} = \begin{pmatrix} \underset{K\times K}{\mathbf{0}} & \underset{K\times U}{\mathbf{0}} \\ \underset{U\times K}{\mathbf{0}} & \underset{U\times U}{\mathbf{0}} \\ \underset{L\times K}{\mathbf{0}} & \underset{L\times U}{\mathbf{0}} \\ \underset{K\times K}{\mathbb{I}} & \underset{K\times U}{\mathbf{0}} \\ \underset{U\times K}{\mathbf{0}} & \underset{U\times U}{\mathbb{I}} \end{pmatrix}. \tag{10.31}$$

In (10.31), there are two columns of block node equations and two columns of stimuli. The first column corresponds to driving the $\mathbf{A_{msd}}$ system ports with $\mathbb{I}$, and the second column corresponds to driving the $\mathbf{B}$ DUT ports with $\mathbb{I}$. Keep in mind that the $\mathbf{B}$ DUT ports, while generally representing reflections from the unknown DUT, also represent incident waves on the fixture. Note that the s-parameters of the unknown DUT have been removed because the DUT is not in the system for this calculation. The $\mathbf{F}$ block matrices have been installed in the upper two rows and columns of the node matrix to represent that these values are actually the fixture s-parameters. The last two pairs of equations are superfluous now. Examining the first three pairs of equations, from column 1:

$$\mathbf{F_{11}} + \mathbf{G_{13}} \cdot \mathbf{U_1} + \mathbf{G_{14}} = \mathbf{0}, \tag{10.32}$$

$$\mathbf{F_{21}} + \mathbf{G_{23}} \cdot \mathbf{U_1} + \mathbf{G_{24}} = \mathbf{0}, \tag{10.33}$$

$$\mathbf{G_{33}} \cdot \mathbf{U_1} + \mathbf{G_{34}} = \mathbf{0}. \tag{10.34}$$

From column 2:

$$\mathbf{F_{12}} + \mathbf{G_{13}} \cdot \mathbf{U_2} + \mathbf{G_{15}} = \mathbf{0}, \tag{10.35}$$

$$\mathbf{F_{22}} + \mathbf{G_{23}} \cdot \mathbf{U_2} + \mathbf{G_{25}} = \mathbf{0}, \tag{10.36}$$

$$\mathbf{G_{33}} \cdot \mathbf{U_2} + \mathbf{G_{35}} = \mathbf{0}. \tag{10.37}$$

Substituting $\mathbf{U_1} = -\mathbf{G_{33}}^{-1} \cdot \mathbf{G_{34}}$ from (10.34) into (10.32) and (10.33), and $\mathbf{U_2} = -\mathbf{G_{33}}^{-1} \cdot \mathbf{G_{35}}$ from (10.37) into (10.35) and (10.36), results in

$$\mathbf{F_{11}} = \mathbf{G_{13}} \cdot \mathbf{G_{33}}^{-1} \cdot \mathbf{G_{34}} - \mathbf{G_{14}}, \tag{10.38}$$

$$\mathbf{F_{12}} = \mathbf{G_{13}} \cdot \mathbf{G_{33}}^{-1} \cdot \mathbf{G_{35}} - \mathbf{G_{15}}, \tag{10.39}$$

$$\mathbf{F_{21}} = \mathbf{G_{23}} \cdot \mathbf{G_{33}}^{-1} \cdot \mathbf{G_{34}} - \mathbf{G_{24}}, \tag{10.40}$$

$$\mathbf{F_{22}} = \mathbf{G_{23}} \cdot \mathbf{G_{33}}^{-1} \cdot \mathbf{G_{35}} - \mathbf{G_{25}}. \tag{10.41}$$

Equations (10.38), (10.39), (10.40), and (10.41) can be substituted into (10.17) and (10.18) to solve the internal node de-embedding problem with the fixture de-embedding equations. By substitution, this provides the same result as (10.28), (10.29), and (10.30).

### 10.5.2 Different Number of System Ports and DUT Ports

The sizes of the block matrices in the systems of equations were labeled in (10.21), but there was no discussion of the sizes. The equations had $K = U$. In this section, the situation is addressed where $K \neq U$. This comes about when the number of known system ports differs from the number of unknown DUT ports. In the case of the system with internal nodes and $K \neq U$, the block sizes in (10.31) can be rationalized by understanding that the number of DUT ports forms half the number of the driven ports on the fixture and therefore sets the number of columns in the block matrices in the second column of the node and stimulus matrix in (10.31). From this rationalization, it becomes clear that the fixture s-parameter block matrices shown must be the sizes specified and that they can be determined from (10.38), (10.39), (10.40), and (10.41). These equations work out fine because matrix $\mathbf{G_{33}}$ is always square, so, although this matrix must be invertible, it produces no additional problems. That being said, the sizes of some of the $\mathbf{F}$ matrices are not necessarily square and therefore require special handling in the fixture de-embedding equations (10.17) and (10.19). These equations are copied here, with the addition of the block matrix sizes (with the inverse matrices placed back on the other sides):

$$\underset{K\times U}{\mathbf{F_{12}}} \cdot \underset{U\times K}{\mathbf{B}} = \underset{K\times K}{\mathbf{Sk}} - \underset{K\times K}{\mathbf{F_{11}}}, \tag{10.42}$$

$$\underset{U\times K}{\mathbf{A}} = \underset{U\times K}{\mathbf{F_{21}}} + \underset{U\times K}{\mathbf{F_{22}}} \cdot \underset{U\times K}{\mathbf{B}}, \tag{10.43}$$

$$\underset{U\times U}{\mathbf{Su}} \cdot \underset{U\times K}{\mathbf{A}} = \underset{U\times K}{\mathbf{B}}. \tag{10.44}$$

Equation (10.42) is the solution for $\mathbf{B}$. When $K > U$, $\mathbf{F_{12}}$ is a tall, skinny matrix indicating a least-squares solution for $\mathbf{B}$. Said differently, there are too many equations and too few unknowns. In this case, both sides of (10.42) are multiplied by $\mathbf{F_{12}}^H$ from the left, and $\mathbf{B}$ is solved as

$$\underset{U\times K}{\mathbf{B}} = \left(\underset{K\times U}{\mathbf{F_{12}}}^H \cdot \underset{K\times U}{\mathbf{F_{12}}}\right)^{-1} \cdot \underset{K\times U}{\mathbf{F_{12}}}^H \cdot \left(\underset{K\times K}{\mathbf{Sk}} - \underset{K\times K}{\mathbf{F_{11}}}\right). \tag{10.45}$$

This now makes $\mathbf{B}$ a short and fat matrix, and, when (10.45) is substituted into (10.43), one finds that $\mathbf{A}$ is also a short, fat matrix. Although the matrices in (10.44) are short and fat, they are multiplied from the right of the unknown $\mathbf{Su}$ and this is also a least-squares solution. Here, both sides of (10.44) are multiplied from the right by $\mathbf{A}^H$ and solved for $\mathbf{Su}$:

$$\underset{U\times U}{\mathbf{Su}} = \underset{U\times K}{\mathbf{B}} \cdot \underset{U\times K}{\mathbf{A}}^H \cdot \left(\underset{U\times K}{\mathbf{A}} \cdot \underset{U\times K}{\mathbf{A}}^H\right)^{-1}.$$

For solutions such as this, one typically writes

$$\mathbf{B} = \mathbf{F_{12}}^\dagger \cdot (\mathbf{Sk} - \mathbf{F_{11}}), \tag{10.46}$$

$$\mathbf{A} = \mathbf{F_{21}} + \mathbf{F_{22}} \cdot \mathbf{B}, \tag{10.47}$$

$$\mathbf{Su} = \mathbf{B} \cdot \mathbf{A}^\dagger. \tag{10.48}$$

The notation $\mathbf{X}^\dagger$ (pronounced X-dagger) refers to the pseudo-inverse of $\mathbf{X}$. Equations (10.46), (10.47), and (10.48) are the general form of the fixture de-embedding equations for $K \neq U$. Note that, for $K < U$, there are often viable solutions to these equations that are still written with a dagger, but special techniques must be utilized. Refer to Appendix C for more details about this.

### 10.5.3 Dense System Characteristics Matrix Solution

In §10.5.1, a certain symmetry of the system characteristics matrix was utilized. In this section, the de-embedding problem is solved without considering this symmetry. The dense reordered system is

$$\mathbf{G} \cdot \mathbf{v} = \mathbf{e},$$

or

$$\begin{pmatrix} \underset{K\times K}{\mathbf{G_{11}}} & \underset{K\times U}{\mathbf{G_{12}}} & \underset{K\times L}{\mathbf{G_{13}}} & \underset{K\times K}{\mathbf{G_{14}}} & \underset{K\times U}{\mathbf{G_{15}}} \\ \underset{U\times K}{\mathbf{G_{21}}} & \underset{U\times U}{\mathbf{G_{22}}} & \underset{U\times L}{\mathbf{G_{23}}} & \underset{U\times K}{\mathbf{G_{24}}} & \underset{U\times U}{\mathbf{G_{25}}} \\ \underset{L\times K}{\mathbf{G_{31}}} & \underset{L\times U}{\mathbf{G_{32}}} & \underset{L\times L}{\mathbf{G_{33}}} & \underset{L\times K}{\mathbf{G_{34}}} & \underset{L\times U}{\mathbf{G_{35}}} \\ \underset{K\times K}{\mathbf{G_{41}}} & \underset{K\times U}{\mathbf{G_{42}}} & \underset{K\times L}{\mathbf{G_{43}}} & \underset{K\times K}{\mathbf{G_{44}}} & \underset{K\times U}{\mathbf{G_{45}}} \\ \underset{U\times K}{\mathbf{G_{51}}} & \underset{U\times U}{-\mathbf{Su}} & \underset{U\times L}{\mathbf{G_{53}}} & \underset{U\times K}{\mathbf{G_{54}}} & \underset{U\times U}{\mathbf{G_{55}}} \end{pmatrix} \cdot \begin{pmatrix} \underset{K\times K}{\mathbf{B_{msd}}} \\ \underset{U\times K}{\mathbf{A}} \\ \underset{L\times K}{\mathbf{U}} \\ \underset{K\times K}{\mathbf{A_{msd}}} \\ \underset{U\times K}{\mathbf{B}} \end{pmatrix} = \begin{pmatrix} \underset{K\times K}{\mathbf{0}} \\ \underset{U\times K}{\mathbf{0}} \\ \underset{L\times K}{\mathbf{0}} \\ \underset{K\times K}{\mathbb{I}} \\ \underset{U\times K}{\mathbf{0}} \end{pmatrix}.$$

The only thing being relied on here is that, with the reordering of the nodes and equations, the negative of the unknown s-parameters has been moved to the second column in the last row.

$\mathbf{G}$ and $\mathbf{v}$ are separated into known and unknown portions:

$$\mathbf{E} = \begin{pmatrix} \mathbf{G_{11}} & \mathbf{G_{12}} & \mathbf{G_{13}} & \mathbf{G_{14}} & \mathbf{G_{15}} \\ \mathbf{G_{21}} & \mathbf{G_{22}} & \mathbf{G_{23}} & \mathbf{G_{24}} & \mathbf{G_{25}} \\ \mathbf{G_{31}} & \mathbf{G_{32}} & \mathbf{G_{33}} & \mathbf{G_{34}} & \mathbf{G_{35}} \\ \mathbf{G_{41}} & \mathbf{G_{42}} & \mathbf{G_{43}} & \mathbf{G_{44}} & \mathbf{G_{45}} \\ \mathbf{G_{51}} & \mathbf{0} & \mathbf{G_{53}} & \mathbf{G_{54}} & \mathbf{G_{55}} \end{pmatrix}, \qquad \mathbf{S} = \begin{pmatrix} \mathbf{0} & \mathbf{0} & \mathbf{0} & \mathbf{0} & \mathbf{0} \\ \mathbf{0} & \mathbf{0} & \mathbf{0} & \mathbf{0} & \mathbf{0} \\ \mathbf{0} & \mathbf{0} & \mathbf{0} & \mathbf{0} & \mathbf{0} \\ \mathbf{0} & \mathbf{0} & \mathbf{0} & \mathbf{0} & \mathbf{0} \\ \mathbf{0} & -\mathbf{Su} & \mathbf{0} & \mathbf{0} & \mathbf{0} \end{pmatrix},$$

$$\mathbf{x} = \begin{pmatrix} \mathbf{0} \\ \mathbf{A} \\ \mathbf{U} \\ \mathbf{A_{msd}} \\ \mathbf{B} \end{pmatrix}, \qquad \mathbf{b} = \begin{pmatrix} \mathbf{B_{msd}} \\ \mathbf{0} \\ \mathbf{0} \\ \mathbf{0} \\ \mathbf{0} \end{pmatrix},$$

such that

$$(-\mathbf{S} + \mathbf{E}) \cdot (\mathbf{x} + \mathbf{b}) = \mathbf{e}. \tag{10.49}$$

Equation (10.49) is expanded and multiplied from the left by $\mathbf{E}^{-1}$ (noting that $\mathbf{S} \cdot \mathbf{b} = \mathbf{0}$):

$$-\mathbf{E}^{-1} \cdot \mathbf{S} \cdot \mathbf{x} + \mathbf{x} + \mathbf{b} = \mathbf{E}^{-1} \cdot \mathbf{e}. \tag{10.50}$$

Note that

$$\mathbf{S} \cdot \mathbf{x} = \begin{pmatrix} \mathbf{0} \\ \mathbf{0} \\ \mathbf{0} \\ \mathbf{0} \\ \mathbf{Su} \cdot \mathbf{A} \end{pmatrix} = \begin{pmatrix} \mathbf{0} \\ \mathbf{0} \\ \mathbf{0} \\ \mathbf{0} \\ \mathbf{B} \end{pmatrix}.$$

Define $\mathbf{Ei} = \mathbf{E}^{-1}$ partitioned similarly as $\mathbf{E}$. When (10.50) is expanded, the following set of equations is obtained:

$$-\mathbf{Ei_{15}} \cdot \mathbf{B} + \mathbf{B_{msd}} = \mathbf{Ei_{14}}, \tag{10.51}$$

$$-\mathbf{Ei_{25}} \cdot \mathbf{B} + \mathbf{A} = \mathbf{Ei_{24}}, \tag{10.52}$$

$$-\mathbf{Ei_{35}} \cdot \mathbf{B} + \mathbf{U} = \mathbf{Ei_{34}},$$

$$-\mathbf{Ei_{45}} \cdot \mathbf{B} + \mathbf{A_{msd}} = \mathbf{Ei_{44}}, \tag{10.53}$$

$$-\mathbf{Ei_{55}} \cdot \mathbf{B} + \mathbf{B} = \mathbf{Ei_{54}}. \tag{10.54}$$

Equation (10.53) is solved for $\mathbf{A_{msd}}$, and both sides are multiplied from the left by $\mathbf{Sk}$:

$$\mathbf{Sk} \cdot \mathbf{A_{msd}} = \mathbf{B_{msd}} = \mathbf{Sk} \cdot (\mathbf{Ei_{44}} + \mathbf{Ei_{45}} \cdot \mathbf{B}). \tag{10.55}$$

Equation (10.51) is solved for $\mathbf{B_{msd}}$:

$$\mathbf{B_{msd}} = \mathbf{Ei_{14}} + \mathbf{Ei_{15}} \cdot \mathbf{B}. \tag{10.56}$$

Setting (10.55) equal to (10.56) and solving[1] for **B**:

$$\mathbf{B} = -\left(\mathbf{Sk} \cdot \mathbf{Ei}_{45} - \mathbf{Ei}_{15}\right)^{\dagger} \cdot \left(\mathbf{Sk} \cdot \mathbf{Ei}_{44} - \mathbf{Ei}_{14}\right). \tag{10.57}$$

Equation (10.52) is solved for **A** in terms of the now known **B**:

$$\mathbf{A} = \mathbf{Ei}_{24} + \mathbf{Ei}_{25} \cdot \mathbf{B}. \tag{10.58}$$

All that is left is to solve for the unknown s-parameters. If there is a single device **Su**, the solution is given by

$$\mathbf{Su} = \mathbf{B} \cdot \mathbf{A}^{\dagger}.$$

It is useful to check the results with the situation in §10.5.1, where, based on symmetry, one has

$$\mathbf{E} = \begin{pmatrix} \mathbb{I} & \mathbf{0} & \mathbf{G}_{13} & \mathbf{G}_{14} & \mathbf{G}_{15} \\ \mathbf{0} & \mathbb{I} & \mathbf{G}_{23} & \mathbf{G}_{24} & \mathbf{G}_{25} \\ \mathbf{0} & \mathbf{0} & \mathbf{G}_{33} & \mathbf{G}_{34} & \mathbf{G}_{35} \\ \mathbf{0} & \mathbf{0} & \mathbf{0} & \mathbb{I} & \mathbf{0} \\ \mathbf{0} & \mathbf{0} & \mathbf{0} & \mathbf{0} & \mathbb{I} \end{pmatrix}.$$

The inverse of **E** is described symbolically as follows:

$$\mathbf{Ei} = \mathbf{E}^{-1}$$
$$= \begin{pmatrix} \mathbb{I} & \mathbf{0} & -\mathbf{G}_{13} \cdot \mathbf{G}_{33}^{-1} & \mathbf{G}_{13} \cdot \mathbf{G}_{33}^{-1} \cdot \mathbf{G}_{34} - \mathbf{G}_{14} & \mathbf{G}_{13} \cdot \mathbf{G}_{33}^{-1} \cdot \mathbf{G}_{35} - \mathbf{G}_{15} \\ \mathbf{0} & \mathbb{I} & -\mathbf{G}_{23} \cdot \mathbf{G}_{33}^{-1} & \mathbf{G}_{23} \cdot \mathbf{G}_{33}^{-1} \cdot \mathbf{G}_{34} - \mathbf{G}_{24} & \mathbf{G}_{23} \cdot \mathbf{G}_{33}^{-1} \cdot \mathbf{G}_{35} - \mathbf{G}_{25} \\ \mathbf{0} & \mathbf{0} & \mathbf{G}_{33}^{-1} & -\mathbf{G}_{33}^{-1} \cdot \mathbf{G}_{34} & -\mathbf{G}_{33}^{-1} \cdot \mathbf{G}_{35} \\ \mathbf{0} & \mathbf{0} & \mathbf{0} & \mathbb{I} & \mathbf{0} \\ \mathbf{0} & \mathbf{0} & \mathbf{0} & \mathbf{0} & \mathbb{I} \end{pmatrix}.$$

Therefore, using (10.57),

$$\mathbf{B} = -\left(\mathbf{G}_{13} \cdot \mathbf{G}_{33}^{-1} \cdot \mathbf{G}_{35} - \mathbf{G}_{15}\right)^{\dagger} \cdot \left[\mathbf{Sk} - \left(\mathbf{G}_{13} \cdot \mathbf{G}_{33}^{-1} \cdot \mathbf{G}_{34} - \mathbf{G}_{14}\right)\right]. \tag{10.59}$$

Equation (10.59) matches (10.28) (with † substituted for the inverse). For **A** in terms of **B**, (10.58) is used:

$$\mathbf{A} = \left(\mathbf{G}_{23} \cdot \mathbf{G}_{33}^{-1} \cdot \mathbf{G}_{34} - \mathbf{G}_{24}\right) + \left(\mathbf{G}_{23} \cdot \mathbf{G}_{33}^{-1} \cdot \mathbf{G}_{35} - \mathbf{G}_{25}\right) \cdot \mathbf{B}. \tag{10.60}$$

Equation (10.60) matches (10.29). Thus it is shown that the de-embedding algorithm provided here, which makes no assumptions about the denseness of **G**, is in fact a superset of all of the methods shown elsewhere in this chapter.

### 10.5.4 System Ports Connected Directly to DUT Ports

Sometimes a particular problem arises that fortunately is easily mitigated. To understand this problem consider the two-port solution provided in §10.2. Originally, this was solved

[1] Note that (10.54) is not used to solve for **B** because usually $\mathbf{Ei}_{55}$ is unity, $\mathbf{Ei}_{54}$ is zero, and (10.54) is usually the equation $\mathbf{B} = \mathbf{B}$. It seems unreasonable anyway because if one could use this equation, then **B** would not be dependent on the known system s-parameters.

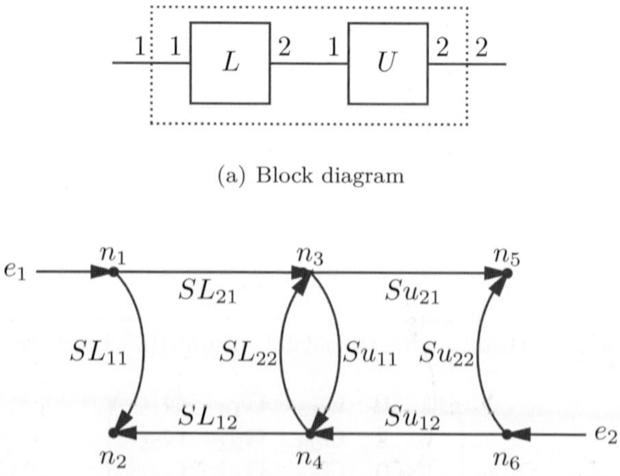


(a) Block diagram

(b) Signal-flow diagram

**Figure 10.7** Two-port de-embedding example with a single left element

for an unknown two-port device bracketed by two known two-port devices labeled $L$ and $R$. Then, at the end of the section, this was solved for a single left element labeled $L$. Here, an attempt is made to solve immediately for a single left element, as shown in Figure 10.7. The block diagram of such a system is shown in Figure 10.7(a).

The signal-flow diagram corresponding to Figure 10.7(a) is shown in Figure 10.7(b). The set of equations corresponding to Figure 10.7(b) is

$$\begin{pmatrix} 1 & 0 & 0 & 0 & 0 & 0 \\ -SL_{11} & 1 & 0 & -SL_{12} & 0 & 0 \\ -SL_{21} & 0 & 1 & -SL_{22} & 0 & 0 \\ 0 & 0 & -Su_{11} & 1 & 0 & -Su_{12} \\ 0 & 0 & -Su_{21} & 0 & 1 & -Su_{22} \\ 0 & 0 & 0 & 0 & 0 & 1 \end{pmatrix} \cdot \begin{pmatrix} n_1 \\ n_2 \\ n_3 \\ n_4 \\ n_5 \\ n_6 \end{pmatrix} = \begin{pmatrix} e_1 \\ 0 \\ 0 \\ 0 \\ 0 \\ e_2 \end{pmatrix}.$$

Following the same principle as in §10.2, the nodes are classified into the four (and including internal nodes, five) categories, and it is found that:

- Nodes 1 and 6 are incident waves on the entire system and are therefore the measured incident waves labeled $a_{msd}$.
- Nodes 2 and 5 are the reflected waves from the entire system and are therefore the measured reflected waves labeled $b_{msd}$.
- Nodes 3 and 6 are the incident waves on the DUT labeled $a$.
- Nodes 4 and 5 are the reflected waves from the DUT labeled $b$.

Herein lies the problem: nodes 5 and 6 are classified in two categories. With regard to node 5, it is classified as both a reflected wave from the system and a reflected wave from

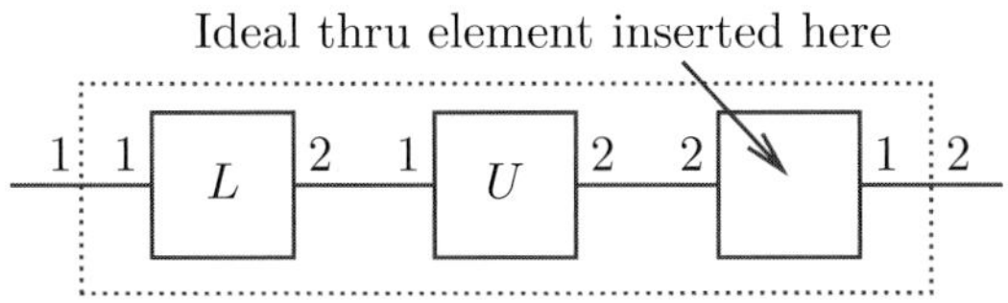


(a) Block diagram

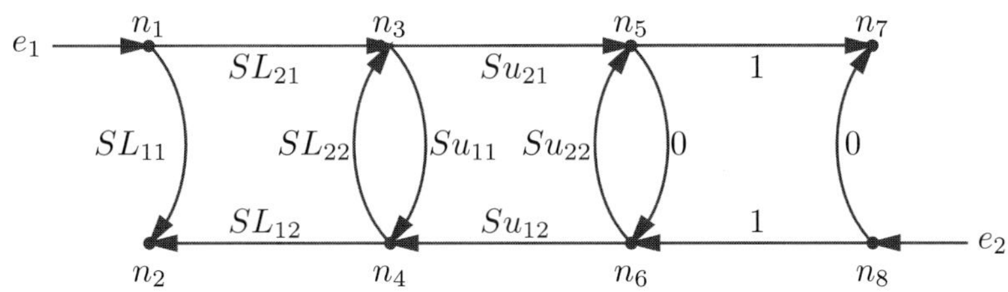


(b) Signal-flow diagram

**Figure 10.8** Two-port de-embedding example with a single left element with ideal thru

the DUT, while node 6 is classified as both an incident wave on the system and an incident wave on the DUT. Nodes 5 and 6 cannot be classified in both categories because the next step would be to generate a permutation matrix to generate the proper node ordering by classification, and this would not be possible. Fortunately, the solution to this problem is simple: anywhere a system port connects directly to an unknown DUT port, an ideal thru is inserted, as shown in Figure 10.8. The signal-flow diagram corresponding to Figure 10.8(a) is shown in Figure 10.8(b), and the system equations are

$$\begin{pmatrix} 1 & 0 & 0 & 0 & 0 & 0 & 0 & 0 \\ -SL_{11} & 1 & 0 & -SL_{12} & 0 & 0 & 0 & 0 \\ -SL_{21} & 0 & 1 & -SL_{22} & 0 & 0 & 0 & 0 \\ 0 & 0 & -Su_{11} & 1 & 0 & -Su_{12} & 0 & 0 \\ 0 & 0 & -Su_{21} & 0 & 1 & -Su_{22} & 0 & 0 \\ 0 & 0 & 0 & 0 & 0 & 1 & 0 & -1 \\ 0 & 0 & 0 & 0 & -1 & 0 & 1 & 0 \\ 0 & 0 & 0 & 0 & 0 & 0 & 0 & 1 \end{pmatrix} \cdot \begin{pmatrix} n_1 \\ n_2 \\ n_3 \\ n_4 \\ n_5 \\ n_6 \\ n_7 \\ n_8 \end{pmatrix} = \begin{pmatrix} e_1 \\ 0 \\ 0 \\ 0 \\ 0 \\ 0 \\ 0 \\ e_2 \end{pmatrix}. \tag{10.61}$$

It is left to the reader to compare (10.61) and (10.2), and to verify that, by the insertion of the thru (i) the system no longer has a node classification problem and (ii) the solution is given by (10.15).

In general, when this sort of de-embedding is worked out programmatically, one can always insert an ideal thru between a system port and the rest of the system simply to ensure that this case does not need special handling. It increases the work a little computationally,

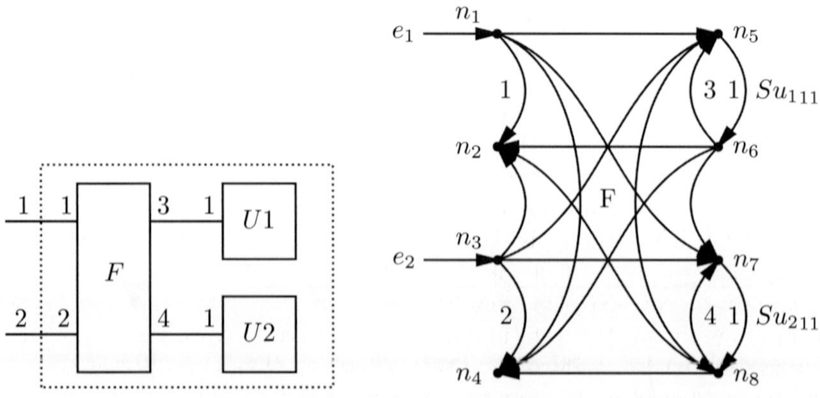


(a) Block diagram (b) Signal-flow diagram

**Figure 10.9** De-embedding example with two unknown devices

but simplifies things algorithmically. An even better alternative is to add the thru only if a system port is added that connects directly to an unknown device, as shown in Listing 8.3.

### 10.5.5 Multiple Unknown DUTs

Sometimes it becomes necessary to solve the de-embedding problem with multiple unknown DUTs. This requires some extra care, but does not deviate significantly from the methods presented thus far.

Consider the example in Figure 10.9. Figure 10.9(a) is a block diagram of a de-embedding example with two unknown DUTs. Here, there is a four-port fixture labeled $F$ and the two unknown DUTs labeled $U1$ and $U2$. The corresponding signal-flow diagram is shown in Figure 10.9(b), where, instead of labeling every arrow, only the ports are labeled so as not to clutter the diagram. The two unknown devices are shown with s-parameters $Su1_{11}$ and $Su2_{11}$. With some effort at expediency, the $b_{msd}$ nodes are identified as $n_2$ and $n_4$, $a$ nodes as $n_5$ and $n_7$, no internal nodes, $a_{msd}$ nodes as $n_1$ and $n_3$, and $b$ nodes as $n_6$ and $n_8$; the system of equations is drawn directly as follows:

$$\begin{pmatrix} 1 & 0 & 0 & 0 & -F_{11} & -F_{12} & -F_{13} & -F_{14} \\ 0 & 1 & 0 & 0 & -F_{21} & -F_{22} & -F_{23} & -F_{24} \\ 0 & 0 & 1 & 0 & -F_{31} & -F_{32} & -F_{33} & -F_{34} \\ 0 & 0 & 0 & 1 & -F_{41} & -F_{42} & -F_{43} & -F_{44} \\ 0 & 0 & 0 & 0 & 1 & 0 & 0 & 0 \\ 0 & 0 & 0 & 0 & 0 & 1 & 0 & 0 \\ 0 & 0 & -Su1_{11} & 0 & 0 & 0 & 1 & 0 \\ 0 & 0 & 0 & -Su2_{11} & 0 & 0 & 0 & 1 \end{pmatrix} \cdot \begin{pmatrix} n_2 \\ n_4 \\ n_5 \\ n_7 \\ n_1 \\ n_3 \\ n_6 \\ n_8 \end{pmatrix} = \begin{pmatrix} 0 \\ 0 \\ 0 \\ 0 \\ e_1 \\ e_2 \\ 0 \\ 0 \end{pmatrix}. \tag{10.62}$$

Examining (10.62), it becomes clear that these equations look identical to the standard fixture de-embedding equations except for one thing – the unknown s-parameters are labeled differently and they sit along the diagonal of what used to be the unknown s-parameter block. This is the standard feature of multiple unknown s-parameter elements. This is because, just as circuit elements between the system ports and the DUT ports can be grouped together into a fixture, the unknown s-parameter devices can be grouped into a single, larger unknown s-parameter device with the caveat that certain connections between the internal unknown s-parameter devices do not exist and that zeros are enforced where they are separated. The strategy is therefore to consider these unknown devices as grouped until the very end, at which point the individual s-parameters are solved for.

In block matrix form, a system of multiple s-parameter devices therefore looks exactly as in (10.6), where all of the separate unknown devices are grouped into one element labeled **Su**. For a system with a total of $U = U_1 + U_2 + \ldots$ unknown device ports (where $U_n$ represents the number of ports in unknown device $n$), **Su** is a $U \times U$ matrix as in (10.44), and, as such, the solution for **Su** utilizes all of the equations developed thus far except for the final solution. This is (10.19), for the fixture de-embedding case with no internal nodes, (10.30) for the more general case with internal nodes, and (10.48) for the most general case with both internal nodes and different numbers of known system ports and unknown DUT ports.

Before explaining the special handling required here, examine the solution for (10.62). Here, the fixture de-embedding case can be used, with (10.17) employed to solve for **A** and (10.18) to solve for **B**, which are both $2 \times 2$. Thus, $\mathbf{Su} \cdot \mathbf{A} = \mathbf{B}$ expands in this case to

$$\begin{pmatrix} Su_{111} & 0 \\ 0 & Su_{211} \end{pmatrix} \cdot \begin{pmatrix} A_{11} & A_{12} \\ A_{21} & A_{22} \end{pmatrix} = \begin{pmatrix} B_{11} & B_{12} \\ B_{21} & B_{22} \end{pmatrix}.$$

This provides four equations:

$$Su_{111} \cdot A_{11} = B_{11}, \quad Su_{111} \cdot A_{12} = B_{12}, \quad Su_{211} \cdot A_{21} = B_{21}, \quad Su_{211} \cdot A_{22} = B_{22},$$

with two unknowns. This is an overconstrained system that is usually solved in a least-squares sense, meaning the least-squared error between the solutions of all four solutions is minimized simultaneously. Therefore, the first and second pairs of equations become

$$Su_{111} \cdot \begin{pmatrix} A_{11} & A_{12} \end{pmatrix} = \begin{pmatrix} B_{11} & B_{12} \end{pmatrix},$$

$$Su_{211} \cdot \begin{pmatrix} A_{21} & A_{22} \end{pmatrix} = \begin{pmatrix} B_{21} & B_{22} \end{pmatrix}.$$

The solution uses the pseudo-inverse as

$$Su_{111} = \begin{pmatrix} B_{11} & B_{12} \end{pmatrix} \cdot \begin{pmatrix} A_{11} & A_{12} \end{pmatrix}^{\dagger}, \tag{10.63}$$

$$Su_{211} = \begin{pmatrix} B_{21} & B_{22} \end{pmatrix} \cdot \begin{pmatrix} A_{21} & A_{22} \end{pmatrix}^{\dagger}. \tag{10.64}$$

While these equations started out as two-port system s-parameters and two ports of unknowns, they have resulted in a least-squares solution because two of the unknown s-parameters (in **Su**) are constrained to be zero.

Following the pattern seen in this example, all of the steps for all of the methods provided so far are exactly the same with a single unknown device as for multiple unknown devices, up to the final step, which is explained more generally in the following.

Given a system with $N$ unknown devices, where each device $n \in 1 \dots N$ has unknown s-parameters $\mathbf{Su}_n$ with $U_n$ ports. All of the unknown devices are aggregated into an overall unknown device $\mathbf{Su}$ given by (10.65), where $U$ is the total number of ports in the unknown devices:

$$\underset{U\times U}{\mathbf{Su}} = \begin{pmatrix} \underset{U_1\times U_1}{\mathbf{Su_1}} & \mathbf{0} & \cdots & \mathbf{0} \\ \mathbf{0} & \underset{U_2\times U_2}{\mathbf{Su_2}} & \cdots & \mathbf{0} \\ \vdots & \vdots & \ddots & \vdots \\ \mathbf{0} & \mathbf{0} & \cdots & \underset{U_N\times U_N}{\mathbf{Su_N}} \end{pmatrix}. \tag{10.65}$$

To solve the systems, the $\mathbf{A}$ and $\mathbf{B}$ block matrices are solved using any of the aforementioned methods involving fixtures, internal nodes, etc. to arrive at (10.66), where $K$ is the total number of system ports at the interface:

$$\underset{U\times U}{\mathbf{Su}} \cdot \underset{U\times K}{\mathbf{A}} = \underset{U\times K}{\mathbf{B}}. \tag{10.66}$$

The $\mathbf{A}$ and $\mathbf{B}$ matrices are partitioned according to the unknown devices held in $\mathbf{Su}$ as follows:

$$\underset{U\times K}{\mathbf{A}} = \begin{pmatrix} \underset{U_1\times K}{\mathbf{A_1}} \\ \underset{U_2\times K}{\mathbf{A_2}} \\ \vdots \\ \underset{U_n\times K}{\mathbf{A_n}} \end{pmatrix}, \qquad \underset{U\times K}{\mathbf{B}} = \begin{pmatrix} \underset{U_1\times K}{\mathbf{B_1}} \\ \underset{U_2\times K}{\mathbf{B_2}} \\ \vdots \\ \underset{U_n\times K}{\mathbf{B_n}} \end{pmatrix}.$$

The s-parameters are finally solved for the individual unknown elements as, for $n \in 1 \dots N$:

$$\mathbf{Su_n} = \mathbf{B_n} \cdot \mathbf{A_n}^{\dagger}.$$

In order for this method to work, the $\mathbf{A}$ and $\mathbf{B}$ node ordering has some additional constraints. Previously, the node ordering for $\mathbf{A}$ and $\mathbf{B}$ was required to be in the same order as the port ordering of the unknown DUT. Here, the node ordering must also match the port ordering of the unknown DUTs and the order of the unknown DUTs.

## 10.6 The Deembedder Class

Many ways of dealing with issues concerning solutions to de-embedding problems have been provided in this chapter. For a programmatic solution, one overall procedure that considers all of the issues can be used. This procedure is a combination of the following:

- fixture de-embedding, as presented in §10.3;

**Listing 10.1 Deembedder** class

```
1  class Deembedder(SystemSParameters):
2      def __init__(self,sd=None):
3          SystemSParameters.__init__(self,sd)
4      def AddUnknown(self,Name,Ports):
5          self.AddDevice(Name,Ports,SParams=None,Type='unknown')
6      def DutANames(self):
7          return [p.A for d in self for p in d if d.Type=='unknown']
8      def DutBNames(self):
9          return [p.B for d in self for p in d if d.Type=='unknown']
10     def UnknownNames(self):
11         return [d.Name for d in self if d.Type=='unknown']
12     def UnknownPorts(self):
13         return [len(d) for d in self if d.Type=='unknown']
14     def Partition(self,A):#a list of arrays, one per unknown device
15         PL=self.UnknownPorts()
16         Result=[]
17         S=0
18         for d in range(len(PL)):
19             Result.append(A[S:S+PL[d],])
20             S=S+PL[d]
21         return Result
```

- fixtures consisting of multiple interconnected networks with potential internal nodes, as presented in §10.5.1;
- potentially different numbers of known system ports and unknown DUT ports, as presented in §10.5.2;
- potentially multiple unknown DUTs, as presented in §10.5.5.

The situation whereby a system port is connected directly to an unknown DUT port, as presented in §10.5.4, is unconcerning because the problem can be mitigated either by inserting an ideal thru device whenever a system port is connected directly to the unknown DUT port, or by indiscriminately adding an ideal thru whenever a system port is added for de-embedding. The dense matrix possibilities as outlined in §10.5.3 are not considered because these situations simply do not arise in the construction of the system description from systems of multiple interconnected network devices.

Listing 10.1 shows a Python listing of a new class called **Deembedder**. On line 1, it is seen to be derived from the **SystemSParameters** class shown in Listing 8.5, which is in turn derived from the **SystemDescription** class shown in Listing 8.3 and Listing 8.4. The **Deembedder** class is provided as a base class for derived classes that provide symbolic or numeric de-embedding solutions. On line 2, the `__init__()` function used to instantiate this class takes an optional argument `sd` which is an optional already created instance of a **SystemDescription**. The **Deembedder** class is derived from the **SystemSParameters** class because, in addition to the tools for building the system description, it needs to use some of the methods provided in the **SystemSParameters** class, most notably the `WeightsMatrix()` method.

On lines 6–21, the helper functions `DutANames()`, `DutBNames()`, `UnknownPorts()`, `UnknownNames()`, and `Partition()` are provided. Remember that in the solutions of system s-parameters the weights matrix was partitioned into nine sections based on three possible node possibilities: whether the node was incident on the system, reflected from the system, or something else. In the de-embedding solution, two more node possibilities are required: whether nodes are waves incident on the unknown DUT or reflected from the unknown DUT. `DutANames()` and `DutBNames()` provide for these added node identification requirements. The functions `UnknownPorts()`, `UnknownNames()`, and `Partition()` deal with multiple unknown DUTs. `UnknownNames()` provides a list of all of the unknown devices and `UnknownPorts()` provides a list of the numbers of ports in the corresponding unknown devices. The `Partition()` function uses these functions to extract the block matrices from the diagonals of **A** and **B** in solutions with multiple unknowns and to provide the correctly named unknown s-parameters.

## 10.7 Symbolic De-embedding Solutions

As in §8.3.2, where symbolic solutions to systems of interconnected devices were determined, it is not very difficult to provide symbolic de-embedding solutions. The main problem, after all, is the setup of the system, and the **SystemDescription** class solves this problem in a programmatic way.

Analogous to the symbolic solutions provided in §8.3.2, the class **DeembedderSymbolic** is shown in Listing 10.2; this derives from both the **Deembedder** and **Symbolic** classes. It can be constructed with a pre-built instance of a **SystemDescription** that is converted to a **Deembedder** during instantiation of a **DeembedderSymbolic**. This class has the main method `SymbolicSolution()`, which produces the text version of the symbolic solution, and a method called `Emit()` that prints it out.

The remainder of this chapter contains the symbolic solutions to most of the examples already solved by hand. This is to demonstrate that, while the math was straightforward once the equations were set up, the setting up of the equations is actually the most tedious part. These worked out examples show that, using the Python code provided, one needs only to describe the problem in a netlist form and that the production of the answer (even symbolically) is very simple.

### 10.7.1 One-Port Symbolic De-embedding Example

The example provided in Figure 10.10 solves the first de-embedding problem presented at the beginning of this chapter, in §10.1. Consider the code in Figure 10.10(a). The first step is to instantiate an empty **DeembedderSymbolic** class called `ds` on line 2, telling the class to produce symbolic equations in the LaTeX equation environment by default (i.e. bracketed with `'\['` and `'\]'`) and to use the *smallmatrix* when producing LaTeX matrices. The argument `known='\\Gamma_{msd}'` specifies the name of the known system s-parameters. On line 3, the known two-port device called `'S'` is added through a call to `AddDevice()`. On line 4, the one-port unknown device called `'\\Gamma_{dut}'` is added.

On line 5, port 2 of the device called `'S'` is connected to port 1 of the unknown device through a call to `ConnectDevicePort()`. On line 6, a system port 1 is added as port 1 of device `'S'` through a call to `AddPort()`. At this point, the description is complete and matches the diagram shown in Figure 10.1(a).

On line 7, the symbolic solution is generated through the call to `SymbolicSolution()`, and the solution is immediately printed through a subsequent call to `Emit()`.

**Listing 10.2 DeembedderSymbolic** class

```
class DeembedderSymbolic(Deembedder,Symbolic):
    def __init__(self,sd=None,**args):
        Deembedder.__init__(self,sd)
        Symbolic.__init__(self,**args)
        self.m_Sk = args['known'] if 'known' in args else 'Sk'
    def SymbolicSolution(self):
        Bmsd=self.PortANames(); Amsd=self.PortBNames()
        Adut=self.DutANames(); Bdut=self.DutBNames()
        Internals=self.OtherNames(Bmsd+Amsd+Adut+Bdut)
        sF11=snG14=self._LaTeXMatrix(self.WeightsMatrix(Bmsd,Amsd))
        sF12=snG15=self._LaTeXMatrix(self.WeightsMatrix(Bmsd,Bdut))
        sF21=snG24=self._LaTeXMatrix(self.WeightsMatrix(Adut,Amsd))
        sF22=snG25=self._LaTeXMatrix(self.WeightsMatrix(Adut,Bdut))
        if len(Internals)>0:# internal nodes
            snG13=self._LaTeXMatrix(self.WeightsMatrix(Bmsd,Internals))
            snG23=self._LaTeXMatrix(self.WeightsMatrix(Adut,Internals))
            snG34=self._LaTeXMatrix(self.WeightsMatrix(Internals,Amsd))
            snG35=self._LaTeXMatrix(self.WeightsMatrix(Internals,Bdut))
            snG33=self._LaTeXMatrix(self.WeightsMatrix(Internals,Internals))
            sGi33='\\left['+self._Identity()+'-'+snG33+'\\right]^{-1}'
            self._AddEq('\\mathbf{Gi_{33}} = '+sGi33)
            sGi33=' \\cdot\\mathbf{Gi_{33}}\\cdot '
            sF11=snG13+sGi33+snG34+((' + '+snG14) if snG14!='0' else '')
            sF12=snG13+sGi33+snG35+((' + '+snG15) if snG15!='0' else '')
            sF21=snG23+sGi33+snG34+((' + '+snG24) if snG24!='0' else '')
            sF22=snG23+sGi33+snG35+((' + '+snG25) if snG25!='0' else '')
        self._AddEq('\\mathbf{F_{11}} = '+sF11)
        self._AddEq('\\mathbf{F_{12}} = '+sF12)
        self._AddEq('\\mathbf{F_{21}} = '+sF21)
        self._AddEq('\\mathbf{F_{22}} = '+sF22)
        sF12='\\mathbf{F_{12}}'
        if len(Bmsd)!=len(Bdut): #if long and skinny F12 then
            sF12i='\\left[ '+sF12+'^H \\cdot '+sF12+\
             '\\right]^{-1} \\cdot' + sF12+'^H \\cdot'
        else: #square F12
            sF12i=sF12+'^{-1}\\cdot'
        BText=sF12i+'\\left[ \\mathbf{'+self.m_Sk+'} - \\mathbf{F_{11}} \\right] '
        self._AddEq('\\mathbf{B}='+BText)
        self._AddEq('\\mathbf{A}=\\mathbf{F_{21}}+ \\mathbf{F_{22}}\\cdot\\mathbf{B}')
        A=Device.SymbolicMatrix('A',len(Bdut),len(Bmsd))
        B=Device.SymbolicMatrix('B',len(Bdut),len(Bmsd))
        AL=self.Partition(matrix(A)); BL=self.Partition(matrix(B))
        AL=[AL[u].tolist() for u in range(len(AL))]
        BL=[BL[u].tolist() for u in range(len(BL))]
        un=self.UnknownNames(); up=self.UnknownPorts()
        if len(AL)==1: #only one unknown device
            if len(Bmsd)!=len(Bdut): #if short and fat A
                sAi='\\cdot \\mathbf{A}^H \\cdot\\left[ \\mathbf{A} \\cdot'+\
                '\\mathbf{A}^H \\right]^{-1}'
            else: #square A
                sAi='\\cdot \\mathbf{A}^{-1}'
            self._AddEq('\\mathbf{'+un[0]+'} = \\mathbf{B} '+sAi)
        else: #multiple unknown devices
            for u in range(len(AL)):
                AText=self._LaTeXMatrix(AL[u])
                BText=self._LaTeXMatrix(BL[u])
                if un[u]==len(Bmsd): # square A and B
                    SuText=BText+'\\cdot '+AText+'^{-1}'
                else: #short and fat A and B
                    SuText=BText+'\\cdot '+AText+'^H \\cdot \\left[ '+AText+\
                    '\\cdot'+AText+'^H \\right]^{-1}'
                self._AddEq('\\mathbf{'+un[u]+'}= '+SuText)
        return self
```

```
import SignalIntegrity.Lib as si
ds=si.sd.DeembedderSymbolic(size='small',known='\\Gamma_{msd}')
ds.AddDevice('S',2)
ds.AddUnknown('\\Gamma_{dut}',1)
ds.ConnectDevicePort('S',2,'\\Gamma_{dut}',1)
ds.AddPort('S',1,1)
ds.SymbolicSolution().Emit()
```

(a) Python code

$$\mathbf{F_{11}} = S_{11}$$

$$\mathbf{F_{12}} = S_{12}$$

$$\mathbf{F_{21}} = S_{21}$$

$$\mathbf{F_{22}} = S_{22}$$

$$\mathbf{B} = \mathbf{F_{12}}^{-1} \cdot [\mathbf{\Gamma_{msd}} - \mathbf{F_{11}}]$$

$$\mathbf{A} = \mathbf{F_{21}} + \mathbf{F_{22}} \cdot \mathbf{B}$$

$$\mathbf{\Gamma_{dut}} = \mathbf{B} \cdot \mathbf{A}^{-1}$$

(b) LaTeX processed equations

**Figure 10.10** One-port symbolic de-embedding example solution

The result of the `Emit()` call is raw LaTeX code and the typeset result is provided in Figure 10.10(b).

The result shown in Figure 10.10(b) is not in the ultimately simplified form and can be simplified either by hand or with a symbolic math processor like Mathcad or Maple. Using Mathcad, entering all of the equations provided in Figure 10.10(b) and simplifying, one obtains

$$\Gamma_{dut} = \frac{\Gamma_{msd} - S_{11}}{S_{21} \cdot S_{12} + S_{22} \cdot \Gamma_{msd} - S_{22} \cdot S_{11}}. \tag{10.67}$$

The equation in (10.67) is equivalent to the result provided in (10.1).

### 10.7.2 Cascaded Two-Port Symbolic De-embedding Example

Here the cascaded two-port device de-embedding example in §10.2 is repeated according to Figure 10.2(a). The example is shown in Figure 10.11, where the system is constructed and solved in Figure 10.11(a). Note that the **DeembedderParser** class used here is described later, in §10.8. Using Mathcad, entering all of the equations provided in Figure 10.11(b)

```
import SignalIntegrity.Lib as si
dp=si.p.DeembedderParser()
dp.AddLines(['device L 2','unknown U 2','device R 2',
    'port 1 L 1 2 R 1','connect L 2 U 1','connect R 2 U 2'])
ds = si.sd.DeembedderSymbolic(dp.SystemDescription(),size='small')
ds.SymbolicSolution().Emit()
```

(a) Python code

$$\mathbf{F_{11}} = \left(\begin{smallmatrix} L_{11} & 0 \\ 0 & R_{11} \end{smallmatrix}\right)$$

$$\mathbf{F_{12}} = \left(\begin{smallmatrix} L_{12} & 0 \\ 0 & R_{12} \end{smallmatrix}\right)$$

$$\mathbf{F_{21}} = \left(\begin{smallmatrix} L_{21} & 0 \\ 0 & R_{21} \end{smallmatrix}\right)$$

$$\mathbf{F_{22}} = \left(\begin{smallmatrix} L_{22} & 0 \\ 0 & R_{22} \end{smallmatrix}\right)$$

$$\mathbf{B} = \mathbf{F_{12}}^{-1} \cdot [\mathbf{Sk} - \mathbf{F_{11}}]$$

$$\mathbf{A} = \mathbf{F_{21}} + \mathbf{F_{22}} \cdot \mathbf{B}$$

$$\mathbf{U} = \mathbf{B} \cdot \mathbf{A}^{-1}$$

(b) LATEX processed equations

**Figure 10.11** Two-port symbolic de-embedding example solution

and simplifying, one obtains

$$\mathbf{U} = \begin{pmatrix} \frac{Sk_{11}-L_{11}}{L_{12}} & \frac{Sk_{12}}{L_{12}} \\ \frac{Sk_{21}}{R_{12}} & \frac{Sk_{22}-R_{11}}{R_{12}} \end{pmatrix} \cdot \begin{pmatrix} \frac{L_{21}\cdot L_{12}+L_{22}\cdot Sk_{11}-L_{22}\cdot L_{11}}{L_{12}} & \frac{L_{22}\cdot Sk_{12}}{L_{12}} \\ \frac{R_{22}\cdot Sk_{21}}{R_{12}} & \frac{R_{21}\cdot R_{12}+R_{22}\cdot Sk_{22}-R_{22}\cdot R_{11}}{R_{12}} \end{pmatrix}^{-1}. \tag{10.68}$$

Like the example in §10.2, the result could be simplified to the point just before the final matrix inverse. Simplifying further just creates a giant equation. The result in (10.68) is equivalent to the result provided in (10.14).

### 10.7.3 Multiple Unknowns De-embedding Example

Figure 10.12 demonstrates the solution for multiple unknown DUTs, as shown in Figure 10.9(a). In Figure 10.12(a), code is provided for constructing the netlist and the emitted equations are shown in Figure 10.12(b). These equations are equivalent to (10.63) and (10.64).

```
import SignalIntegrity.Lib as si
ds=si.sd.DeembedderSymbolic(size='small')
ds.AddDevice('F',4)
ds.AddUnknown('U_1',1)
ds.AddUnknown('U_2',1)
ds.AddPort('F',1,1)
ds.AddPort('F',2,2)
ds.ConnectDevicePort('F',3,'U_1',1)
ds.ConnectDevicePort('F',4,'U_2',1)
ds.SymbolicSolution().Emit()
```

(a) Python code

$$\mathbf{F_{11}} = \begin{pmatrix} F_{11} & F_{12} \\ F_{21} & F_{22} \end{pmatrix}$$

$$\mathbf{F_{12}} = \begin{pmatrix} F_{13} & F_{14} \\ F_{23} & F_{24} \end{pmatrix}$$

$$\mathbf{F_{21}} = \begin{pmatrix} F_{31} & F_{32} \\ F_{41} & F_{42} \end{pmatrix}$$

$$\mathbf{F_{22}} = \begin{pmatrix} F_{33} & F_{34} \\ F_{43} & F_{44} \end{pmatrix}$$

$$\mathbf{B} = \mathbf{F_{12}}^{-1} \cdot [\mathbf{Sk} - \mathbf{F_{11}}]$$

$$\mathbf{A} = \mathbf{F_{21}} + \mathbf{F_{22}} \cdot \mathbf{B}$$

$$\mathbf{U_1} = \begin{pmatrix} B_{11} & B_{12} \end{pmatrix} \cdot \begin{pmatrix} A_{11} & A_{12} \end{pmatrix}^H \cdot \left[ \begin{pmatrix} A_{11} & A_{12} \end{pmatrix} \cdot \begin{pmatrix} A_{11} & A_{12} \end{pmatrix}^H \right]^{-1}$$

$$\mathbf{U_2} = \begin{pmatrix} B_{21} & B_{22} \end{pmatrix} \cdot \begin{pmatrix} A_{21} & A_{22} \end{pmatrix}^H \cdot \left[ \begin{pmatrix} A_{21} & A_{22} \end{pmatrix} \cdot \begin{pmatrix} A_{21} & A_{22} \end{pmatrix}^H \right]^{-1}$$

(b) LaTeX processed equations

**Figure 10.12** Multiple unknowns symbolic de-embedding example solution

### 10.7.4 De-embedding Example with an Internal Node

In this example, the symbolic solution for the system shown in Figure 10.6 is solved. While this is not the most practical example, it highlights certain aspects of this type of solution.

The solution using the **DeembedderSymbolic** class is provided in Figure 10.13. In Figure 10.13(a) the schematic in Figure 10.6 has been redrawn to make the port numbering clearer. Shown in Figure 10.13(b) is the Python code used to assemble the schematic and emit the symbolic solutions, which are shown in Figure 10.13(c).

This particular example was chosen because it contains an internal node that is connected neither to a system port nor to a port of the unknown DUT. In Listing 10.2, there is a distinctly different, and more complicated, solution based on the number of internal nodes. This example exercises that case, hence the existence of the matrix $\mathbf{Gi_{33}}$ in the solution.

The solution shown in Figure 10.13(c) ought not to be simplified further. That being said, this particular problem could have been solved using T-parameters, using (3.32) and (3.34), keeping in mind the orientation of the devices in Figure 10.13(a) and the order of

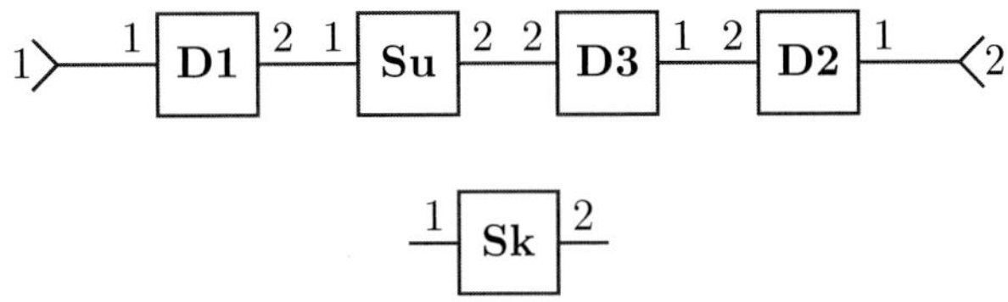


(a) Schematic

```
import SignalIntegrity.Lib as si
ds=si.sd.DeembedderSymbolic(size='small')
ds.AddDevice('D1',2)
ds.AddDevice('D2',2)
ds.AddDevice('D3',2)
ds.AddUnknown('Su',2)
ds.AddPort('D1',1,1)
ds.AddPort('D2',1,2)
ds.ConnectDevicePort('D1',2,'Su',1)
ds.ConnectDevicePort('D2',2,'D3',1)
ds.ConnectDevicePort('D3',2,'Su',2)
ds.SymbolicSolution().Emit()
```

(b) Python code

$$\mathbf{Gi_{33}} = \left[\mathbb{I} - \begin{pmatrix} 0 & D2_{22} \\ D3_{11} & 0 \end{pmatrix}\right]^{-1}$$

$$\mathbf{F_{11}} = \begin{pmatrix} 0 & 0 \\ 0 & D2_{12} \end{pmatrix} \cdot \mathbf{Gi_{33}} \cdot \begin{pmatrix} 0 & D2_{21} \\ 0 & 0 \end{pmatrix} + \begin{pmatrix} D1_{11} & 0 \\ 0 & D2_{11} \end{pmatrix}$$

$$\mathbf{F_{12}} = \begin{pmatrix} 0 & 0 \\ 0 & D2_{12} \end{pmatrix} \cdot \mathbf{Gi_{33}} \cdot \begin{pmatrix} 0 & 0 \\ 0 & D3_{12} \end{pmatrix} + \begin{pmatrix} D1_{12} & 0 \\ 0 & 0 \end{pmatrix}$$

$$\mathbf{F_{21}} = \begin{pmatrix} 0 & 0 \\ D3_{21} & 0 \end{pmatrix} \cdot \mathbf{Gi_{33}} \cdot \begin{pmatrix} 0 & D2_{21} \\ 0 & 0 \end{pmatrix} + \begin{pmatrix} D1_{21} & 0 \\ 0 & 0 \end{pmatrix}$$

$$\mathbf{F_{22}} = \begin{pmatrix} 0 & 0 \\ D3_{21} & 0 \end{pmatrix} \cdot \mathbf{Gi_{33}} \cdot \begin{pmatrix} 0 & 0 \\ 0 & D3_{12} \end{pmatrix} + \begin{pmatrix} D1_{22} & 0 \\ 0 & D3_{22} \end{pmatrix}$$

$$\mathbf{B} = \mathbf{F_{12}}^{-1} \cdot [\mathbf{Sk} - \mathbf{F_{11}}]$$

$$\mathbf{A} = \mathbf{F_{21}} + \mathbf{F_{22}} \cdot \mathbf{B}$$

$$\mathbf{Su} = \mathbf{B} \cdot \mathbf{A}^{-1}$$

(c) LaTeX processed equations

**Figure 10.13** Symbolic de-embedding example solution with an internal node

the multiplication of the inverse T-parameter matrices:

$$\mathbf{Tu} = \left[\frac{\begin{pmatrix} -|\mathbf{D1}| & D1_{11} \\ -D1_{22} & 1 \end{pmatrix}}{D1_{21}}\right]^{-1} \cdot \frac{\begin{pmatrix} -|\mathbf{Sk}| & Sk_{11} \\ -Sk_{22} & 1 \end{pmatrix}}{Sk_{21}} \cdot \left[\frac{\begin{pmatrix} -|\mathbf{D2}| & D2_{22} \\ -D2_{11} & 1 \end{pmatrix}}{D2_{12}}\right]^{-1} \cdot \left[\frac{\begin{pmatrix} -|\mathbf{D3}| & D3_{22} \\ -D3_{11} & 1 \end{pmatrix}}{D3_{12}}\right]^{-1}$$

**Listing 10.3 DeembedderParser** class

```
class DeembedderParser(SystemDescriptionParser):
    def __init__(self, f=None, args=None):
        SystemDescriptionParser.__init__(self, f, args)
    def _ProcessDeembedderLine(self,line):
        lineList=self.ReplaceArgs(line.split())
        if len(lineList) == 0: return
        if lineList[0] == 'system':
            dev=DeviceParser(self.m_f,None,lineList[1:])
            if not dev.m_spf is None:
                self.m_spc.append(('system',dev.m_spf))
        elif lineList[0] == 'unknown':
            self.m_sd.AddUnknown(lineList[1],int(lineList[2]))
        else: self.m_ul.append(line)
    def _ProcessLines(self):
        SystemDescriptionParser._ProcessLines(self,['connect','port'])
        self.m_sd = Deembedder(self.m_sd)
        lines=copy.deepcopy(self.m_ul); self.m_ul=[]
        for line in lines: self._ProcessDeembedderLine(line)
        lines=copy.deepcopy(self.m_ul); self.m_ul=[]
        for line in lines: SystemDescriptionParser._ProcessLine(self,line,[])
        return self
```

and

$$\mathbf{Su} = \frac{\begin{pmatrix} Tu_{12} & |\mathbf{Tu}| \\ 1 & -T_{21} \end{pmatrix}}{Tu_{22}}.$$

While the T-parameter solution is correct, and is just as good as the result in Figure 10.13(c), arrival at this solution required T-parameter and s-parameter conversions and attention to port orientations. There are situations that cannot be easily placed in a T-parameter solution format. The result shown in Figure 10.13(c) was arrived at without any analysis of the circuit topology and was generated by describing only the interconnections within the system.

## 10.8 The DeembedderParser Class

The **DeembedderParser** class shown in Listing 10.3 derives from the **Deembedder** class and adds netlist parsing capability. As shown, it implements the private member functions `_ProcessLines()` and `_ProcessDeembedderLine()`. The intent is that a user instantiates the derived **DeembedderNumericParser** class outlined in §10.9 for numeric solutions. For symbolic solutions, a user will instantiate this class and use it to build and extract an instance of the **SystemDescription** class, as will be shown in the next section.

Netlists parsed with instances of **DeembedderParser** or derived classes will recognize two new keywords needed for de-embedding solutions. These are:

- `'system'`, which allows the specification of the system s-parameters in the netlist that was specified in the `CalculateUnknown()` call in the **DeembedderNumeric** class;
- `'unknown'`, which allows the addition of an unknown to the system from a netlist that was specified in an `AddUnknown()` call provided by the **Deembedder** class.

```
device L 2
unknown Su 2
port 1 L 1 2 Su 2
connect L 2 Su 1
```

(a) Text netlist file

```
import SignalIntegrity.Lib as si
dp=si.p.DeembedderParser().File('SymbolicDeembedding3.txt')
ds = si.sd.DeembedderSymbolic(dp.SystemDescription(),size='small')
ds.SymbolicSolution().Emit()
```

(b) Python code

$$\mathbf{F_{11}} = \begin{pmatrix} L_{11} & 0 \\ 0 & 0 \end{pmatrix}$$

$$\mathbf{F_{12}} = \begin{pmatrix} L_{12} & 0 \\ 0 & 1 \end{pmatrix}$$

$$\mathbf{F_{21}} = \begin{pmatrix} L_{21} & 0 \\ 0 & 1 \end{pmatrix}$$

$$\mathbf{F_{22}} = \begin{pmatrix} L_{22} & 0 \\ 0 & 0 \end{pmatrix}$$

$$\mathbf{B} = \mathbf{F_{12}}^{-1} \cdot [\mathbf{Sk} - \mathbf{F_{11}}]$$

$$\mathbf{A} = \mathbf{F_{21}} + \mathbf{F_{22}} \cdot \mathbf{B}$$

$$\mathbf{Su} = \mathbf{B} \cdot \mathbf{A}^{-1}$$

(c) LaTeX processed equations

**Figure 10.14** Symbolic de-embedding parser example

## 10.8.1 DeembedderParser Symbolic Example

In the example provided in Figure 10.14, the **DeembedderParser** class is utilized in conjunction with the **DeembedderSymbolic** class to provide a solution to the example in Figure 10.7(a). This is a simple example of de-embedding one element on the left in a cascaded arrangement of this left element and an unknown element on the right.

In Figure 10.14(a), a netlist is provided that shows a two-port left element and a two-port unknown element connected together. Note the declaration of the unknown element with the `'unknown'` keyword.[2]

In Figure 10.14(b) this netlist is loaded from the file `'SymbolicDeembedding3.txt'` with an instance of the **DeembedderParser** class. The system description is extracted through a call to `SystemDescription()` and passed to an instance of the **DeembedderSymbolic** class, and the symbolic is solution emitted; this is shown in Figure 10.14(c).

[2]The system s-parameters are not declared in the netlist because they are implied and not required for a symbolic solution.

This solution can be expanded and simplified:

$$\mathbf{Su} = \frac{1}{L_{22} \cdot Sk_{11} - |\mathbf{L}|} \cdot \begin{pmatrix} Sk_{11} - L_{11} & Sk_{12} \cdot L_{21} \\ Sk_{21} \cdot L_{12} & |\mathbf{Sk}| \cdot L_{22} - Sk_{22} \cdot |\mathbf{L}| \end{pmatrix}.$$

This is the same result that would be obtained using T-parameters:

$$\mathbf{Tu} = \begin{pmatrix} Tu_{11} & Tu_{12} \\ Tu_{21} & Tu_{22} \end{pmatrix} = \left[ \frac{\begin{pmatrix} -|\mathbf{L}| & L_{11} \\ -L_{22} & 1 \end{pmatrix}}{L_{21}} \right]^{-1} \cdot \frac{\begin{pmatrix} -|\mathbf{Sk}| & Sk_{11} \\ -Sk_{22} & 1 \end{pmatrix}}{Sk_{21}}$$

and

$$\mathbf{Su} = \frac{\begin{pmatrix} Tu_{12} & |\mathbf{Tu}| \\ 1 & -T_{21} \end{pmatrix}}{Tu_{22}}.$$

The solution shown in Figure 10.14 benefits from simply specifying the interconnection of the two devices.

## 10.9 Numeric De-embedding Solutions

The numeric counterpart to the **DeembedderSymbolic** class is the **Deembedder-Numeric** class provided in Listing 10.4. This class provides a constructor on line 2 that takes an optional system description, with the caveat that the system description cannot have utilized the member function `AddUnknown()` in the **Deembedder** class in Listing 10.1; this call would be made post construction of the **DeembedderNumeric** class. The typical usage would be to construct this class without the system description and then make calls on the class to the member functions in the derivation hierarchy. Since the class derives from **Numeric** (which provides the `Dagger()` function) and the **Deembedder** class, all members in **Deembedder** and all classes it derives from are available (see Chapter 17 for more information on the class hierarchy).

After construction, a call to `CalculateUnknown()` on line 4, with the known s-parameters of the system provided in the argument `Sk`, produces the s-parameters of the unknown device `Su`. The result will be a list of matrices for the unknowns, if there are more than one, or a single s-parameter matrix. This s-parameter matrix is for a single frequency. To use this class, one loops over the frequencies, filling in the s-parameters of each device for each frequency prior to the call to `CalculateUnknown()`, thus obtaining the set of s-parameters for the unknown through repeated calls.

A more practical method of obtaining numerical solutions is to use the **Deembedder-NumericParser** class shown in Listing 10.5. Here, there is a constructor on line 2 that takes an optional list of frequencies. A netlist is added through calls to `AddLines()`, which is way up the hierarchy in the **SystemDescriptionParser** class shown in Listing 8.9. The special `'system'` and `'unknown'` keywords in the netlist are handled in the `_ProcessLines()` and `_ProcessDeembedderLine()` on lines 14 and 4 of the **DeembedderParser** class shown in Listing 10.3. Thus, an entire netlist is provided to the **DeembedderNumericParser** class prior to the call to `Deembed()` on line 5. Again, if there is a single unknown device, the result is an instance of the **SParameters** class (see Listing 3.16); otherwise a list of results is

**Listing 10.4 DeembedderNumeric** class

```
class DeembedderNumeric(Deembedder,Numeric):
    def __init__(self,sd=None):
        Deembedder.__init__(self,sd)
    def CalculateUnknown(self,Sk):
        Bmsd=self.PortANames()
        Amsd=self.PortBNames()
        Adut=self.DutANames()
        Bdut=self.DutBNames()
        Internals=self.OtherNames(Bmsd+Amsd+Adut+Bdut)
        G14=-matrix(self.WeightsMatrix(Bmsd,Amsd))
        G15=-matrix(self.WeightsMatrix(Bmsd,Bdut))
        G24=-matrix(self.WeightsMatrix(Adut,Amsd))
        G25=-matrix(self.WeightsMatrix(Adut,Bdut))
        if len(Internals)>0:# internal nodes
            G13=-matrix(self.WeightsMatrix(Bmsd,Internals))
            G23=-matrix(self.WeightsMatrix(Adut,Internals))
            G33=matrix(identity(len(Internals)))-\
                matrix(self.WeightsMatrix(Internals,Internals))
            G34=-matrix(self.WeightsMatrix(Internals,Amsd))
            G35=-matrix(self.WeightsMatrix(Internals,Bdut))
            F11=self.Dagger(G33,Left=G13,Right=G34,Mul=True)-G14
            F12=self.Dagger(G33,Left=G13,Right=G35,Mul=True)-G15
            F21=self.Dagger(G33,Left=G23,Right=G34,Mul=True)-G24
            F22=self.Dagger(G33,Left=G23,Right=G35,Mul=True)-G25
        else:# no internal nodes
            F11=-G14; F12=-G15; F21=-G24; F22=-G25
        #if long and skinny F12 then
        #F12.getI()=(F12.transpose()*F12).getI()*F12.transpose()
        #if short and fat F12, F12.getI() is wrong
        B=self.Dagger(F12,Right=(Sk-F11),Mul=True)
        A=F21+F22*B
        AL=self.Partition(A)# partition for multiple unknown devices
        BL=self.Partition(B)
        Su=[self.Dagger(AL[u],Left=BL[u],Mul=True).tolist() for u in range(len(AL))]
        if (len(Su)==1): return Su[0]# return the one result, not as a list
        return Su# return the list of results
```

**Listing 10.5 DeembedderNumericParser** class

```
class DeembedderNumericParser(DeembedderParser,CallBacker,LinesCache):
    def __init__(self, f=None, args=None, callback=None, cacheFileName=None):
        DeembedderParser.__init__(self, f, args)
        self.sf = None
    def Deembed(self,systemSParameters=None):
        self._ProcessLines()
        self.m_sd.CheckConnections()
        NumUnknowns=len(self.m_sd.UnknownNames())
        result=[[] for i in range(NumUnknowns)]
        for n in range(len(self.m_f)):
            system=None
            for d in range(len(self.m_spc)):
                if self.m_spc[d][0] == 'system': system=self.m_spc[d][1][n]
                else: self.m_sd.AssignSParameters(self.m_spc[d][0],self.m_spc[d][1][n])
            if not systemSParameters is None: system=systemSParameters[n]
            unl=DeembedderNumeric(self.m_sd).CalculateUnknown(system)
            if NumUnknowns == 1: unl=[unl]
            for u in range(NumUnknowns): result[u].append(unl[u])
        self.sf=[SParameters(self.m_f,r) for r in result]
        if len(self.sf)==1: self.sf=self.sf[0]
        return self.sf
```

produced. The `Deembed()` function performs the aforementioned looping over frequencies, assigning s-parameters of devices, and ends up using the **DeembedderNumeric** class to produce each frequency result.

## 10.10 Numeric De-embedding Example

A practical, yet uncomplicated, de-embedding example is provided in Figure 10.15. A de-embedding situation that occurs in measurements within the Teledyne LeCroy WavePulser 40iX is shown in Figure 10.15(a).[3] This instrument calibrates internally to a reference plane indicated by internal ports 1 and 2. It performs this calibration by measuring the known calibration standards labeled S1, O1, L1, S2, O2, L2, and Thru on each of the internal ports. It uses the techniques described in Chapter 15. Despite the fact that there exists internal cabling between the pulser/sampler (analogous to the source/receiver of the VNA), the measurements of the internal calibration standards are made directly at the internal port locations, and thus the calibration reference plane becomes internal ports 1 and 2. Calibration itself is a form of de-embedding by establishing the reference plane.

The s-parameters of the DUT are measured by taking a *raw* s-parameter measurement of the DUT through the microwave relay, an internal semi-rigid cable, a bulkhead connector, and a user cable. This measurement is called raw because it is simply a ratio of waves measured at ports 1 and 2 internally through the system. Calibration is then performed to arrive at the raw, calibrated s-parameters. These raw, calibrated s-parameters are the correct, calibrated s-parameters of the device connected between internal ports 1 and 2, with the understanding that the user only wants the s-parameters of the DUT, which will be calculated through de-embedding.

The de-embedding problem is shown schematically in Figure 10.15(b). Here, the internal relay path, semi-rigid cable, and bulkhead connector have been aggregated into a measurement made at the factory of this path. The path from internal port 1 to external port 1 has s-parameters in the file `'I1_O1.s2p'`. Similarly, the path from internal port 2 to external port 2 has s-parameters in the file `'I2_O2.s2p'`. Note the orientation of the ports of these devices.

The user cable s-parameter files `'Cable1.s2p'` and `'Cable2.s2p'` have also been previously measured and are placed into the schematic. The schematic shown in Figure 10.15(b) was generated in the *SignalIntegrity* GUI based application, which knows about system and unknown parts. It shows unknowns with a question mark in the middle, and it shows systems with an "S" in the middle. It does not allow connection of systems, as they are used only to describe the s-parameters of the system.

As explained, the system defines the measured s-parameters of the entire system, and the schematic portion with the known and unknown devices, and ports, defines the interconnections and the location of the desired, de-embedded device.

The *SignalIntegrity* GUI based application is mostly a shortcut for producing a netlist to provide the application through the **DeembedderNumericParser** class. The netlist corresponding to the schematic is shown in Figure 10.15(c). Here, the keyword `system`

[3]The WavePulser 40iX is a time-domain reflectometer (TDR) instrument manufactured by Teledyne LeCroy that measures s-parameters using time-domain techniques.

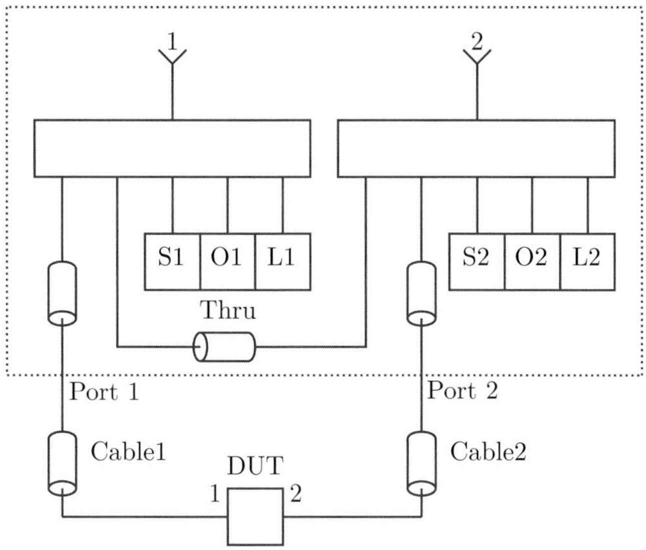


(a) De-embedding environment

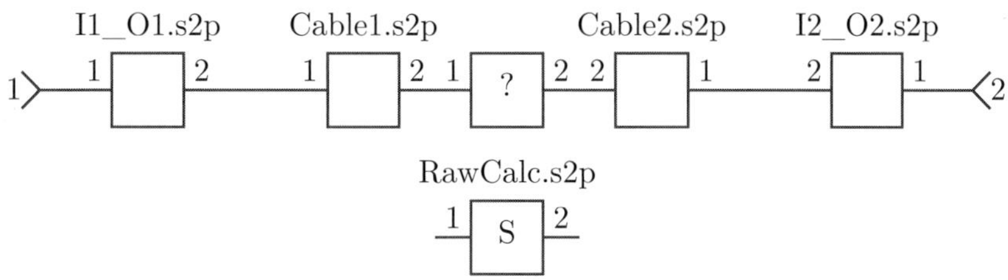


(b) De-embedding schematic

```
system file RawCalc.s2p
unknown U1 2
device D4 2 file Cable1.s2p
device D5 2 file Cable2.s2p
device D6 2 file I1_O1.s2p
device D7 2 file I2_O2.s2p
port 1 D6 1
port 2 D7 1
connect U1 1 D4 2
connect D5 2 U1 2
connect D6 2 D4 1
connect D5 1 D7 2
```

(c) De-embedding netlist

**Figure 10.15** Teledyne LeCroy WavePulser 40iX de-embedding example

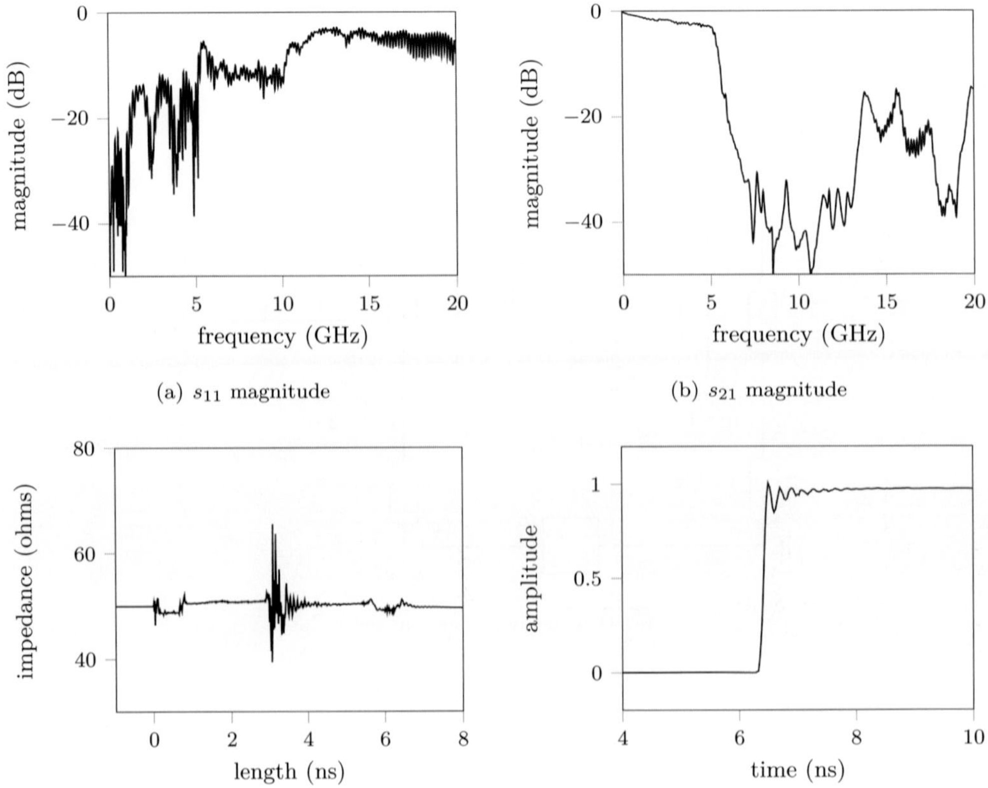


(a) $s_{11}$ magnitude

(b) $s_{21}$ magnitude

(c) Impedance

(d) Step response

**Figure 10.16** Numeric de-embedding example raw measurement

corresponds to the system component with s-parameters in the file `'RawCalc.s2p'`, and the keyword `'unknown'` corresponds to the unknown device.

A raw measurement of the low-pass (6 GHz) band of a triplexer is shown in Figure 10.16. Only the interesting measurements are shown.[4] The $s_{11}$ magnitude response is shown in Figure 10.16(a), where it is seen to be fuzzy due to the reflections in the long path between the ports of the measurement. The $s_{21}$ magnitude response is shown in Figure 10.16(b), where there is a slight amount of fuzziness, again due to the long path and some small reflections. The impedance profile is shown in Figure 10.16(c), where the reason for the fuzziness in Figure 10.16(a) is readily apparent. The DUT is seen to be in the middle of an approximately 6 ns total path. The slight mismatch in the relay, semi-rigid cable, and user

[4]Usually, the phase and impulse response would be shown, and for all of the s-parameters. Note that the impedance is somewhat equivalent to the step response of $s_{11}$ (see Chapter 14).

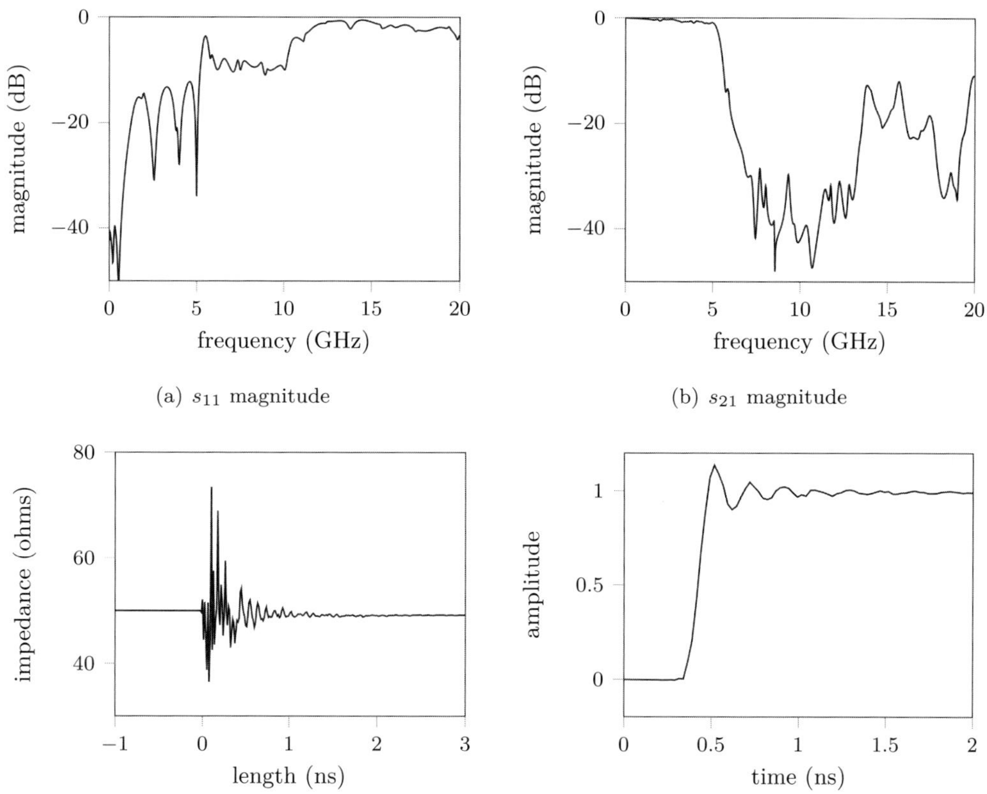


(a) $s_{11}$ magnitude

(b) $s_{21}$ magnitude

(c) Impedance

(d) Step response

**Figure 10.17** Numeric de-embedding example measurement result

cable can be seen. The $s_{21}$ step response is shown in Figure 10.16(d), and again the delay through the system is approximately 6–7 ns.

The result of the de-embedding calculation is shown in Figure 10.17. In Figure 10.17(a) and Figure 10.17(b), the de-embedded $s_{11}$ and $s_{22}$ magnitude responses are shown on the same scale as the raw measurement. The $s_{11}$ magnitude response shows a better match with the de-embedded structure removed at very low frequency, but the relatively non-ideal return loss of the triplexer dominates the measurement. The fuzziness in the measurement is, however, gone. The $s_{21}$ magnitude shows less loss than in the raw measurement. Here, with the loss of the structure de-embedded, it is important to check the result for passivity violations that can occur due to errors in the de-embedding; passivity can be enforced on the result, if desired (see §15.7.1). Similarly, one should compare $s_{21}$ with $s_{12}$ if reciprocity is expected (see §15.7.3).

The impedance profile of the de-embedded result is shown in Figure 10.17(c), zoomed from 0–2 ns. The impedance is slightly higher due to the de-embedding operation, as the impedance profile is approximated (see Chapter 14). It is important to look for causality violations, as discussed in §15.7.4. Small causality violations will invariably occur in practice, can be enforced, and most often occur in the $s_{11}$ and $s_{22}$ measurements, although WavePulser measurements should not exhibit these violations. Here, the impedance profile begins cleanly at time zero.

Finally, the $s_{21}$ step response of the result is shown in Figure 10.17(d), where the delay of the de-embedded filter is seen to be about 500 ps. The true step response of the filter exhibits more overshoot than the raw step response, but this is as expected.

# 11

# Virtual Probing

VIRTUAL PROBING IS A TECHNIQUE INVENTED AT LECROY CORPORATION [26, 27] for time-domain processing that enables a multitude of de-embedding and embedding possibilities.

About fifteen years ago, a trend arose in serial data transmission – that of equalization. Equalization is not something new [28], being integral in some of the first undersea cables [29]. It was in common use in telephone systems [30], but its presence in serial data communications on PCBs was suddenly ubiquitous; equalization was being relied on heavily to make the next speed step. In the world of oscilloscope design, this presented a huge problem because oscilloscopes had become the instrument of choice for analyzing and characterizing serial data systems and it had become clear that the analysis of closed eyes at the receiver required equalizer emulation in the oscilloscope software [31].[1] Equalizer emulation opened the eyes, but with a cost: most equalizers boost high frequency content relative to low frequency content in the signal, which boosts the noise in the oscilloscope measurement, sometimes to an intolerable level. The impetus for the creation of the virtual probing technique was to measure at the transmitter where the signal is strong, emulate the signal at the receiver with its closed eye, and open the eye with equalizer emulation, thus recovering the waveform at the receiver in a noise-neutral manner [33].

Later, the efficacy of the technique was shown in solving other problems of inaccessible probing points [34], and for validation of measurements [16]. To explain the concept, various cases are considered in this chapter.

## 11.1 A Simple Case of Virtual Probing

Consider a simple virtual probing example as shown in Figure 11.1. The block diagram in Figure 11.1(a) shows a simple system consisting of a transmitter (or source element), a channel, and a receiver (or terminating element). Two items are worth noting: there is a node *Vo* labeled as an output probe and a node *Vm* labeled as a measure probe. In virtual probing, nodes that are labeled as a measure are nodes where signals are *acquired.* Acquired signals are used because virtual probing is utilized with oscilloscopes and the measurement node represents a location where waveforms are acquired or digitized by the oscilloscope. Nodes that are labeled as output nodes are nodes for which the virtual probe system is being asked to produce waveforms. The problem statement in Figure 11.1(a) is: "Given a waveform measured at node *Vm*, produce the waveform that is appearing at node *Vo*." Generally a waveform is a voltage waveform, but it can also be a current waveform.

[1]The modern oscilloscope continues to evolve with ever more analysis capability [32].

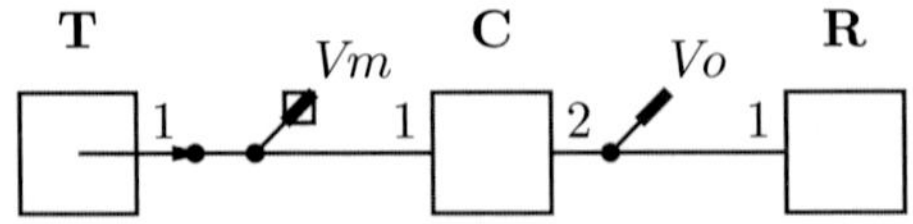


(a) Block diagram

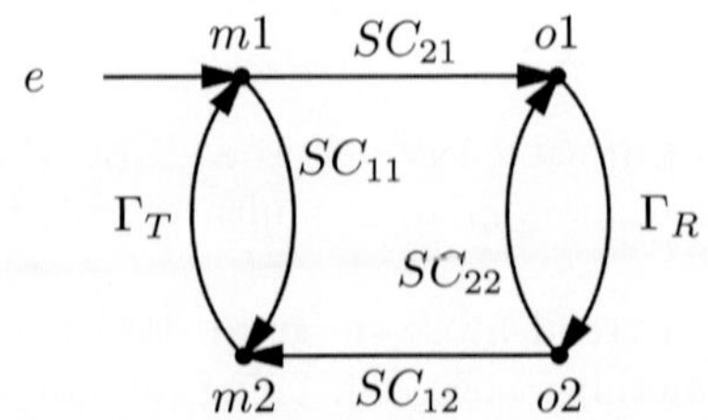


(b) Signal-flow diagram

**Figure 11.1** Simple virtual probing example

The signal-flow diagram representation of Figure 11.1(a) is shown in Figure 11.1(b). There is an arbitrarily labeled input stimulus $e$, the nodes corresponding to the voltage $Vm$ are labeled $m1$ and $m2$, and the nodes corresponding to the voltage $Vo$ are labeled $o1$ and $o2$. The equations governing this system are

$$\begin{pmatrix} 1 & -\Gamma_T & 0 & 0 \\ -SC_{11} & 1 & 0 & -SC_{12} \\ -SC_{21} & 0 & 1 & -SC_{22} \\ 0 & 0 & -\Gamma_R & 1 \end{pmatrix} \cdot \begin{pmatrix} m1 \\ m2 \\ o1 \\ o2 \end{pmatrix} = \begin{pmatrix} e \\ 0 \\ 0 \\ 0 \end{pmatrix}. \tag{11.1}$$

The following matrices and vectors are defined:

$$\mathbf{n} = \begin{pmatrix} m1 \\ m2 \\ o1 \\ o2 \end{pmatrix}, \quad \mathbf{m} = \begin{pmatrix} e \\ 0 \\ 0 \\ 0 \end{pmatrix}, \quad \mathbf{S} = \begin{pmatrix} 1 & -\Gamma_T & 0 & 0 \\ -SC_{11} & 1 & 0 & -SC_{12} \\ -SC_{21} & 0 & 1 & -SC_{22} \\ 0 & 0 & -\Gamma_R & 1 \end{pmatrix}. \tag{11.2}$$

Equation (11.1) corresponds to

$$\mathbf{S} \cdot \mathbf{n} = \mathbf{m},$$

and therefore

$$\mathbf{n} = \mathbf{S}^{-1} \cdot \mathbf{m}.$$

Define

$$\mathbf{Si} = \mathbf{S}^{-1} \tag{11.3}$$

therefore

$$\mathbf{n} = \mathbf{Si} \cdot \mathbf{m}.$$

Since the vector $\mathbf{m}$ contains a lot of zeros, a vector $\mathbf{m}'$ is defined that is the vector $\mathbf{m}$ with all rows with zeros removed. $\mathbf{Si}'$ is defined as the matrix $\mathbf{Si}$ with the corresponding columns removed. Therefore,

$$\mathbf{n} = \mathbf{Si}' \cdot \mathbf{m}' = \begin{pmatrix} m1 \\ m2 \\ o1 \\ o2 \end{pmatrix} = \begin{pmatrix} Si_{11} \\ Si_{21} \\ Si_{31} \\ Si_{41} \end{pmatrix} \cdot (e) . \tag{11.4}$$

A vector $\mathbf{v}$ is defined that contains all of the node voltages. In this case,

$$\mathbf{v} = \begin{pmatrix} Vm \\ Vo \end{pmatrix} .$$

A *voltage extraction matrix* is a matrix that, when multiplied by the node vector, produces the voltage vector $\mathbf{v}$. Recalling (2.8), a suitable voltage extraction matrix is

$$\mathbf{VE} = \begin{pmatrix} 1 & 1 & 0 & 0 \\ 0 & 0 & 1 & 1 \end{pmatrix} \cdot \sqrt{Z0},$$

and therefore

$$\mathbf{v} = \mathbf{VE} \cdot \mathbf{n} = \begin{pmatrix} Vm \\ Vo \end{pmatrix} = \begin{pmatrix} 1 & 1 & 0 & 0 \\ 0 & 0 & 1 & 1 \end{pmatrix} \cdot \sqrt{Z0} \cdot \begin{pmatrix} m1 \\ m2 \\ o1 \\ o2 \end{pmatrix} .$$

Utilizing (11.4),

$$\mathbf{v} = \begin{pmatrix} Vm \\ Vo \end{pmatrix} = \begin{pmatrix} 1 & 1 & 0 & 0 \\ 0 & 0 & 1 & 1 \end{pmatrix} \cdot \sqrt{Z0} \cdot \begin{pmatrix} Si_{11} \\ Si_{21} \\ Si_{31} \\ Si_{41} \end{pmatrix} \cdot (e) . \tag{11.5}$$

Two vectors are defined: $\mathbf{vm} = (Vm)$, which is a list of the voltage measurements (in this case, one voltage), and $\mathbf{vo} = (Vo)$, which is a list of the outputs (also one voltage). Examining (11.5), two key equations can be written:

$$\mathbf{vm} = \begin{pmatrix} Vm \end{pmatrix} = \begin{pmatrix} 1 & 1 & 0 & 0 \end{pmatrix} \cdot \sqrt{Z0} \cdot \begin{pmatrix} Si_{11} \\ Si_{21} \\ Si_{31} \\ Si_{41} \end{pmatrix} \cdot (e) , \tag{11.6}$$

$$\mathbf{vo} = \begin{pmatrix} Vo \end{pmatrix} = \begin{pmatrix} 0 & 0 & 1 & 1 \end{pmatrix} \cdot \sqrt{Z0} \cdot \begin{pmatrix} Si_{11} \\ Si_{21} \\ Si_{31} \\ Si_{41} \end{pmatrix} \cdot (e) . \tag{11.7}$$

Equations (11.6) and (11.7) are key because they have something important in common – they are both defined in terms of the single non-zero stimulus $\mathbf{m}' = (e)$, and therefore

$$\mathbf{m}' = (e) = \left[ \begin{pmatrix} 1 & 1 & 0 & 0 \end{pmatrix} \cdot \sqrt{Z0} \cdot \begin{pmatrix} Si_{11} \\ Si_{21} \\ Si_{31} \\ Si_{41} \end{pmatrix} \right]^{-1} \cdot \mathbf{vm}$$
$$= \left[ \begin{pmatrix} 0 & 0 & 1 & 1 \end{pmatrix} \cdot \sqrt{Z0} \cdot \begin{pmatrix} Si_{11} \\ Si_{21} \\ Si_{31} \\ Si_{41} \end{pmatrix} \right]^{-1} \cdot \mathbf{vo}. \tag{11.8}$$

Equation (11.8) allows **vo** to be written in terms of **vm**:

$$\mathbf{vo} = \left[ \begin{pmatrix} 0 & 0 & 1 & 1 \end{pmatrix} \cdot \sqrt{Z0} \cdot \begin{pmatrix} Si_{11} \\ Si_{21} \\ Si_{31} \\ Si_{41} \end{pmatrix} \right] \cdot \left[ \begin{pmatrix} 1 & 1 & 0 & 0 \end{pmatrix} \cdot \sqrt{Z0} \cdot \begin{pmatrix} Si_{11} \\ Si_{21} \\ Si_{31} \\ Si_{41} \end{pmatrix} \right]^{-1} \cdot \mathbf{vm}. \tag{11.9}$$

Equation (11.9) can be simplified to

$$\mathbf{vo} = \frac{Si_{31} + Si_{41}}{Si_{11} + Si_{21}} \cdot \mathbf{vm}.$$

Therefore,

$$\mathbf{vo} = \mathbf{H} \cdot \mathbf{vm},$$

where

$$\mathbf{H} = \frac{Si_{31} + Si_{41}}{Si_{11} + Si_{21}}. \tag{11.10}$$

Equation (11.10) is a transfer function that converts waveforms measured at **vm** into output waveforms at **vo**.

Returning to (11.2) and (11.3):

$$\mathbf{Si} = \mathbf{S}^{-1} = \frac{\begin{pmatrix} 1-SC_{22}\cdot\Gamma_R & \Gamma_T\cdot(1-SC_{22}\cdot\Gamma_R) & \Gamma_T\cdot SC_{12}\cdot\Gamma_R & \Gamma_T\cdot SC_{12} \\ SC_{11}-|\mathbf{SC}|\cdot\Gamma_R & 1-SC_{22}\cdot\Gamma_R & SC_{12}\cdot\Gamma_R & SC_{12} \\ SC_{12} & SC_{21}\cdot\Gamma_T & 1-SC_{11}\cdot\Gamma_T & SC_{22}-|\mathbf{SC}|\cdot\Gamma_T \\ SC_{12}\cdot\Gamma_R & SC_{21}\cdot\Gamma_T\cdot\Gamma_R & -SC_{11}\cdot\Gamma_T\cdot\Gamma_R+\Gamma_R & 1-SC_{11}\cdot\Gamma_T \end{pmatrix}}{[|\mathbf{SC}| \cdot \Gamma_T - SC_{22}] \cdot \Gamma_R + 1 - SC_{11} \cdot \Gamma_T}. \tag{11.11}$$

Substituting (11.11) in (11.10) and simplifying a little yields

$$\mathbf{H} = \frac{SC_{21} \cdot (1 + \Gamma_R)}{1 + SC_{11} - \Gamma_R \cdot (SC_{22} + |\mathbf{SC}|)}. \tag{11.12}$$

It is interesting to note that if $\Gamma_R = 0$, then

$$\underset{\Gamma_R=0}{\mathbf{H}} = \frac{SC_{21}}{1 + SC_{11}},$$

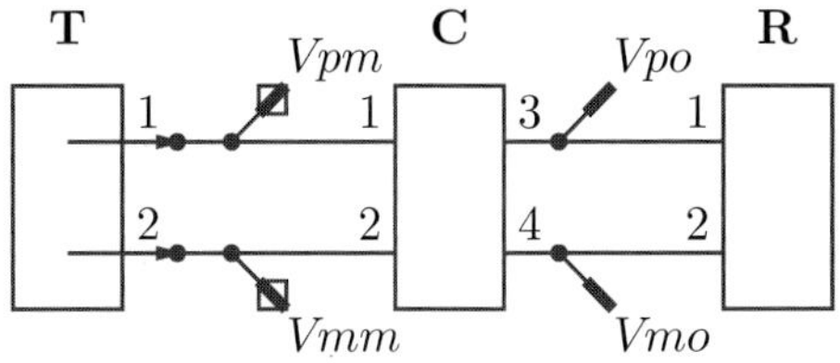


(a) Block diagram

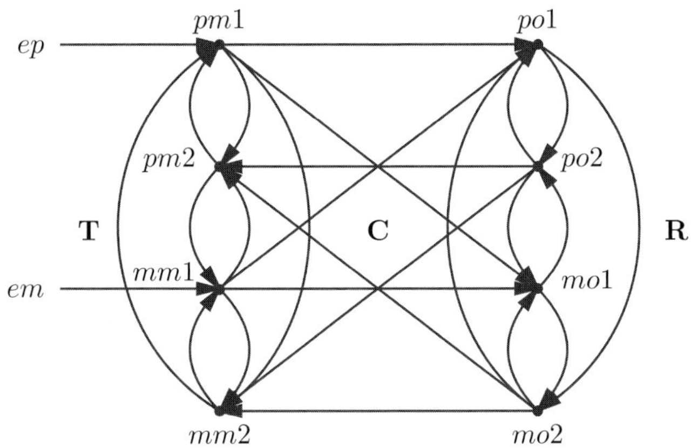


(b) Signal-flow diagram

**Figure 11.2** Virtual probing multiple input/output example

and that if the channel return loss is assumed small, then $\underset{\Gamma_R=0}{\mathbf{H}} \approx SC_{21}$. This is why oversimplified software for calculating a virtual probing transfer function tends to utilize $SC_{21}$. Here, however, a transfer function is calculated that takes care of the complete effects as produced in (11.12). Counterintuitively, (11.12) has no contribution from $\Gamma_T$. This frequency response can be converted to an impulse response, as discussed in §12.4.2, and this impulse response can be used as a filter to process waveforms, as discussed in Chapter 13.

## 11.2 A Multiple Input and Output Example

The previous section provided the simplest example of virtual probing. This concept is now expanded by considering an example with multiple measurement points and multiple output points, as shown in Figure 11.2. Figure 11.2(a) shows an example of a virtual probing arrangement with two measurement nodes *Vpm* and *Vmm* and two output nodes *Vpo* and *Vmo*. The intent of this arrangement is to produce output waveforms for both *Vpo* and

$Vmo$ as a function of measured waveforms $Vpm$ and $Vmm$. The signal-flow diagram for this system is shown in Figure 11.2(b), from which the system equation is formed:

$$\begin{pmatrix} 1 & -ST_{11} & 0 & -ST_{12} & 0 & 0 & 0 & 0 \\ -SC_{11} & 1 & -SC_{12} & 0 & 0 & -SC_{13} & 0 & -SC_{14} \\ 0 & -ST_{21} & 1 & -ST_{22} & 0 & 0 & 0 & 0 \\ -SC_{21} & 0 & -SC_{22} & 1 & 0 & -SC_{23} & 0 & -SC_{24} \\ -SC_{31} & 0 & -SC_{32} & 0 & 1 & -SC_{33} & 0 & -SC_{34} \\ 0 & 0 & 0 & 0 & -SR_{11} & 1 & -SR_{12} & 0 \\ -SC_{41} & 0 & -SC_{42} & 0 & 0 & -SC_{43} & 1 & -SC_{44} \\ 0 & 0 & 0 & 0 & -SR_{21} & 0 & -SR_{22} & 1 \end{pmatrix} \cdot \begin{pmatrix} pm1 \\ pm2 \\ mm1 \\ mm2 \\ po1 \\ po2 \\ mo1 \\ mo2 \end{pmatrix} = \begin{pmatrix} ep \\ 0 \\ em \\ 0 \\ 0 \\ 0 \\ 0 \\ 0 \end{pmatrix}. \tag{11.13}$$

As in the previous example, (11.13) is in the form

$$\mathbf{S} \cdot \mathbf{n} = \mathbf{m},$$

and therefore

$$\mathbf{n} = \mathbf{S}^{-1} \cdot \mathbf{m}.$$

Furthermore,

$$\mathbf{Si} = \mathbf{S}^{-1},$$

and therefore

$$\mathbf{n} = \mathbf{Si} \cdot \mathbf{m}.$$

Since the vector $\mathbf{m}$ contains a lot of zeros, $\mathbf{m}'$ is defined as the vector $\mathbf{m}$ with all rows with zeros removed, and further $\mathbf{Si}'$ is defined as the matrix $\mathbf{Si}$ with the corresponding columns removed:

$$\mathbf{n} = \mathbf{Si}' \cdot \mathbf{m}' = \begin{pmatrix} pm1 \\ pm2 \\ mm1 \\ mm2 \\ po1 \\ po2 \\ mo1 \\ mo2 \end{pmatrix} = \begin{pmatrix} Si_{11} & Si_{13} \\ Si_{21} & Si_{23} \\ Si_{31} & Si_{33} \\ Si_{41} & Si_{43} \\ Si_{51} & Si_{53} \\ Si_{61} & Si_{63} \\ Si_{71} & Si_{73} \\ Si_{81} & Si_{83} \end{pmatrix} \cdot \begin{pmatrix} ep \\ em \end{pmatrix}. \tag{11.14}$$

A vector $\mathbf{v}$ is defined that contains all of the node voltages. In this case,

$$\mathbf{v} = \begin{pmatrix} Vpm \\ Vmm \\ Vpo \\ Vmo \end{pmatrix}.$$

A suitable voltage extraction matrix is found to be

$$\mathbf{VE} = \begin{pmatrix} 1 & 1 & 0 & 0 & 0 & 0 & 0 & 0 \\ 0 & 0 & 1 & 1 & 0 & 0 & 0 & 0 \\ 0 & 0 & 0 & 0 & 1 & 1 & 0 & 0 \\ 0 & 0 & 0 & 0 & 0 & 0 & 1 & 1 \end{pmatrix} \cdot \sqrt{Z0},$$

therefore

$$\mathbf{v} = \mathbf{VE} \cdot \mathbf{n} = \begin{pmatrix} Vpm \\ Vmm \\ Vpo \\ Vmo \end{pmatrix} = \begin{pmatrix} 1 & 1 & 0 & 0 & 0 & 0 & 0 & 0 \\ 0 & 0 & 1 & 1 & 0 & 0 & 0 & 0 \\ 0 & 0 & 0 & 0 & 1 & 1 & 0 & 0 \\ 0 & 0 & 0 & 0 & 0 & 0 & 1 & 1 \end{pmatrix} \cdot \sqrt{Z0} \cdot \begin{pmatrix} pm1 \\ pm2 \\ mm1 \\ mm2 \\ po1 \\ po2 \\ mo1 \\ mo2 \end{pmatrix}.$$

Utilizing (11.14),

$$\mathbf{v} = \begin{pmatrix} Vpm \\ Vmm \\ Vpo \\ Vmo \end{pmatrix} = \begin{pmatrix} 1 & 1 & 0 & 0 & 0 & 0 & 0 & 0 \\ 0 & 0 & 1 & 1 & 0 & 0 & 0 & 0 \\ 0 & 0 & 0 & 0 & 1 & 1 & 0 & 0 \\ 0 & 0 & 0 & 0 & 0 & 0 & 1 & 1 \end{pmatrix} \cdot \sqrt{Z0} \cdot \begin{pmatrix} Si_{11} & Si_{13} \\ Si_{21} & Si_{23} \\ Si_{31} & Si_{33} \\ Si_{41} & Si_{43} \\ Si_{51} & Si_{53} \\ Si_{61} & Si_{63} \\ Si_{71} & Si_{73} \\ Si_{81} & Si_{83} \end{pmatrix} \cdot \begin{pmatrix} ep \\ em \end{pmatrix}. \tag{11.15}$$

Two vectors are defined: $\mathbf{vm} = \begin{pmatrix} Vpm & Vmm \end{pmatrix}^T$, which is a list of the voltage measurements (in this case two voltages), and $\mathbf{vo} = \begin{pmatrix} Vpo & Vmo \end{pmatrix}^T$, which is a list of the outputs (also two voltages). Examining (11.15), two key equations can be written:

$$\begin{aligned} \mathbf{vm} &= \begin{pmatrix} Vpm \\ Vmm \end{pmatrix} = \begin{pmatrix} 1 & 1 & 0 & 0 \\ 0 & 0 & 1 & 1 \end{pmatrix} \cdot \sqrt{Z0} \cdot \begin{pmatrix} Si_{11} & Si_{13} \\ Si_{21} & Si_{23} \\ Si_{31} & Si_{33} \\ Si_{41} & Si_{43} \end{pmatrix} \cdot \begin{pmatrix} ep \\ em \end{pmatrix} \\ &= \begin{pmatrix} Si_{11} + Si_{21} & Si_{13} + Si_{23} \\ Si_{31} + Si_{41} & Si_{33} + Si_{43} \end{pmatrix} \cdot \sqrt{Z0} \cdot \begin{pmatrix} ep \\ em \end{pmatrix}, \end{aligned} \tag{11.16}$$

$$\begin{aligned} \mathbf{vo} &= \begin{pmatrix} Vpo \\ Vmo \end{pmatrix} = \begin{pmatrix} 1 & 1 & 0 & 0 \\ 0 & 0 & 1 & 1 \end{pmatrix} \cdot \sqrt{Z0} \cdot \begin{pmatrix} Si_{51} & Si_{53} \\ Si_{61} & Si_{63} \\ Si_{71} & Si_{73} \\ Si_{81} & Si_{83} \end{pmatrix} \cdot \begin{pmatrix} ep \\ em \end{pmatrix} \\ &= \begin{pmatrix} Si_{51} + Si_{61} & Si_{53} + Si_{63} \\ Si_{71} + Si_{81} & Si_{73} + Si_{83} \end{pmatrix} \cdot \sqrt{Z0} \cdot \begin{pmatrix} ep \\ em \end{pmatrix}. \end{aligned} \tag{11.17}$$

In (11.16) and (11.17), many zero columns were removed from the voltage extraction matrices, as were the corresponding rows of **Si**. Both (11.16) and (11.17) are defined in terms of the single non-zero stimulus vector $\mathbf{m}' = \begin{pmatrix} ep & em \end{pmatrix}^T$, and therefore

$$\mathbf{m}' = \begin{pmatrix} ep \\ em \end{pmatrix} = \left[ \begin{pmatrix} 1 & 1 & 0 & 0 \\ 0 & 0 & 1 & 1 \end{pmatrix} \cdot \sqrt{Z0} \cdot \begin{pmatrix} Si_{11} & Si_{13} \\ Si_{21} & Si_{23} \\ Si_{31} & Si_{33} \\ Si_{41} & Si_{43} \end{pmatrix} \right]^{-1} \cdot \mathbf{vm}$$
$$= \left[ \begin{pmatrix} 1 & 1 & 0 & 0 \\ 0 & 0 & 1 & 1 \end{pmatrix} \cdot \sqrt{Z0} \cdot \begin{pmatrix} Si_{51} & Si_{53} \\ Si_{61} & Si_{63} \\ Si_{71} & Si_{73} \\ Si_{81} & Si_{83} \end{pmatrix} \right]^{-1} \cdot \mathbf{vo}.$$

This is simplified as

$$\mathbf{m}' = \begin{pmatrix} ep \\ em \end{pmatrix} = \begin{pmatrix} Si_{11} + Si_{21} & Si_{13} + Si_{23} \\ Si_{31} + Si_{41} & Si_{33} + Si_{43} \end{pmatrix}^{-1} \cdot \mathbf{vm}$$
$$= \begin{pmatrix} Si_{51} + Si_{61} & Si_{53} + Si_{63} \\ Si_{71} + Si_{81} & Si_{73} + Si_{83} \end{pmatrix}^{-1} \cdot \mathbf{vo}. \tag{11.18}$$

Equation (11.18) allows **vo** to be written in terms of **vm**:

$$\mathbf{vo} = \begin{pmatrix} Si_{51} + Si_{61} & Si_{53} + Si_{63} \\ Si_{71} + Si_{81} & Si_{73} + Si_{83} \end{pmatrix} \cdot \begin{pmatrix} Si_{11} + Si_{21} & Si_{13} + Si_{23} \\ Si_{31} + Si_{41} & Si_{33} + Si_{43} \end{pmatrix}^{-1} \cdot \mathbf{vm}, \tag{11.19}$$

and therefore

$$\mathbf{vo} = \mathbf{H} \cdot \mathbf{vm},$$

where

$$\mathbf{H} = \begin{pmatrix} Si_{51} + Si_{61} & Si_{53} + Si_{63} \\ Si_{71} + Si_{81} & Si_{73} + Si_{83} \end{pmatrix} \cdot \begin{pmatrix} Si_{11} + Si_{21} & Si_{13} + Si_{23} \\ Si_{31} + Si_{41} & Si_{33} + Si_{43} \end{pmatrix}^{-1}. \tag{11.20}$$

Equation (11.20) is a transfer function that converts waveforms measured at **vm** into output waveforms at **vo**.

The structure of **H** dictates that the transfer function, in this case, is a $2 \times 2$ matrix such that

$$\mathbf{H} = \begin{pmatrix} H_{11} & H_{12} \\ H_{21} & H_{22} \end{pmatrix},$$

$$\begin{pmatrix} Vpo \\ Vmo \end{pmatrix} = \begin{pmatrix} H_{11} & H_{12} \\ H_{21} & H_{22} \end{pmatrix} \cdot \begin{pmatrix} Vpm \\ Vmm \end{pmatrix}. \tag{11.21}$$

Equation (11.21) implies that each output voltage is generated by summing the result of applying two filters to each measured voltage. For example, $Vpo = H_{11} \cdot Vpm + H_{12} \cdot Vmm$ and $Vmo = H_{21} \cdot Vpm + H_{22} \cdot Vmm$. This processing is discussed in §13.5.

## 11.3 A Degree of Freedom Example

Suppose that in the example in §11.2, one were able to measure only the differential voltage $Vdm = Vpm - Vmm$. Equation (11.16) would be modified to generate **vm** as follows:

$$\begin{aligned}
\mathbf{vm} &= \begin{pmatrix} 1 & -1 \end{pmatrix} \cdot \begin{pmatrix} Vpm \\ Vmm \end{pmatrix} \\
&= \begin{pmatrix} 1 & -1 \end{pmatrix} \cdot \begin{pmatrix} 1 & 1 & 0 & 0 \\ 0 & 0 & 1 & 1 \end{pmatrix} \cdot \sqrt{Z0} \cdot \begin{pmatrix} Si_{11} & Si_{13} \\ Si_{21} & Si_{23} \\ Si_{31} & Si_{33} \\ Si_{41} & Si_{43} \end{pmatrix} \cdot \begin{pmatrix} ep \\ em \end{pmatrix} \\
&= \begin{pmatrix} 1 & 1 & -1 & -1 \end{pmatrix} \cdot \sqrt{Z0} \cdot \begin{pmatrix} Si_{11} & Si_{13} \\ Si_{21} & Si_{23} \\ Si_{31} & Si_{33} \\ Si_{41} & Si_{43} \end{pmatrix} \cdot \begin{pmatrix} ep \\ em \end{pmatrix} \\
&= \begin{pmatrix} Si_{11} + Si_{21} - Si_{31} - Si_{41} & Si_{13} + Si_{23} - Si_{33} - Si_{43} \end{pmatrix} \cdot \sqrt{Z0} \cdot \begin{pmatrix} ep \\ em \end{pmatrix} \\
&= \begin{pmatrix} Si_{11} + Si_{21} - Si_{31} - Si_{41} \\ Si_{13} + Si_{23} - Si_{33} - Si_{43} \end{pmatrix}^T \cdot \sqrt{Z0} \cdot \begin{pmatrix} ep \\ em \end{pmatrix}.
\end{aligned} \tag{11.22}$$

This means that (11.18) becomes

$$\begin{aligned}
\mathbf{m}' = \begin{pmatrix} ep \\ em \end{pmatrix} &= \begin{pmatrix} Si_{11} + Si_{21} - Si_{31} - Si_{41} & Si_{13} + Si_{23} - Si_{33} - Si_{43} \end{pmatrix}^{-1} \cdot \mathbf{vm} \\
&= \begin{pmatrix} Si_{51} + Si_{61} & Si_{53} + Si_{63} \\ Si_{71} + Si_{81} & Si_{73} + Si_{83} \end{pmatrix}^{-1} \cdot \mathbf{vo};
\end{aligned}$$

(11.19) becomes

$$\mathbf{vo} = \begin{pmatrix} Si_{51} + Si_{61} & Si_{53} + Si_{63} \\ Si_{71} + Si_{81} & Si_{73} + Si_{83} \end{pmatrix} \cdot \left[ \begin{pmatrix} Si_{11} + Si_{21} - Si_{31} - Si_{41} \\ Si_{13} + Si_{23} - Si_{33} - Si_{43} \end{pmatrix}^T \right]^{-1} \cdot \mathbf{vm};$$

and (11.20) becomes

$$\mathbf{H} = \begin{pmatrix} Si_{51} + Si_{61} & Si_{53} + Si_{63} \\ Si_{71} + Si_{81} & Si_{73} + Si_{83} \end{pmatrix} \cdot \left[ \begin{pmatrix} Si_{11} + Si_{21} - Si_{31} - Si_{41} \\ Si_{13} + Si_{23} - Si_{33} - Si_{43} \end{pmatrix}^T \right]^{-1}. \tag{11.23}$$

Equation (11.23) highlights a problem, however, in that the right side of the equation is not invertible. This example points out an issue with virtual probing that must sometimes be dealt with.

This situation arose because the matrix on the right side of (11.23) that needs to be inverted became a $1 \times 2$ matrix when it used to be $2 \times 2$. This happened when the measured voltages were changed from two voltages to one. It did not really matter that the voltage measured was changed to differential, only that it became one. Furthermore, nothing about the output voltages has any effect on what happened here. Although not discussed here, one can have as many output voltages as one wants – the problem is all about measured voltages.

The number of columns in the matrix multiplied by **vm** and **vo** will always match the number of stimuli specified in the system. In other words, the number of columns matches the number of non-zero rows of **m**, which is the number of rows in **m**′. The number of rows in the matrix that is multiplied by **vm**, however, depends on the number of measurement nodes in the system. Therefore, in order to evaluate (11.23), one ideally has the same number of measurement nodes as stimuli.

This problem arises frequently, especially when other uses are considered for virtual probing beyond the simple examples shown here. So far, because of the way the equations have been written, the number of stimuli in the system represents the number of *degrees of freedom* in the system. There must always be at least as many measurement nodes as there are degrees of freedom. One way to resolve a degrees of freedom problem is by adding a constraint, such as an assumption that the system is driven in a balanced fashion. Depending on the problem, this may or may not be a valid assumption. Fortunately, there are ways of testing the sensitivity to these assumptions. For example, one might assume that only one stimulus was driven, then solve the equations in this manner and compare results to see how the assumption affects the situation. But, for now, an assumption of balanced drive is made. This means that a new stimulus can be defined, $\widehat{m} = (e)$, for example such that

$$\mathbf{m}' = \begin{pmatrix} ep \\ em \end{pmatrix} = \begin{pmatrix} 1 \\ -1 \end{pmatrix} \cdot \widehat{\mathbf{m}}.$$

Substituting for $\mathbf{m}'$ in (11.22):

$$\begin{aligned}\mathbf{vm} &= \begin{pmatrix} Si_{11}+Si_{21}-Si_{31}-Si_{41} \\ Si_{13}+Si_{23}-Si_{33}-Si_{43} \end{pmatrix}^T \cdot \sqrt{Z0} \cdot \begin{pmatrix} 1 \\ -1 \end{pmatrix} \cdot (e) \\ &= (Si_{11}+Si_{21}-Si_{31}-Si_{41}-Si_{13}-Si_{23}+Si_{33}+Si_{43}) \cdot \sqrt{Z0} \cdot (e). \end{aligned} \tag{11.24}$$

Substituting for $\mathbf{m}'$ in (11.17):

$$\begin{aligned}\mathbf{vo} &= \begin{pmatrix} Si_{51}+Si_{61} & Si_{53}+Si_{63} \\ Si_{71}+Si_{81} & Si_{73}+Si_{83} \end{pmatrix} \cdot \sqrt{Z0} \cdot \begin{pmatrix} 1 \\ -1 \end{pmatrix} \cdot (e) \\ &= \begin{pmatrix} Si_{51}+Si_{61}-Si_{53}-Si_{63} \\ Si_{71}+Si_{81}-Si_{73}-Si_{83} \end{pmatrix} \cdot \sqrt{Z0} \cdot (e). \end{aligned} \tag{11.25}$$

One relies on the fact that both (11.24) and (11.25) are defined in terms of $\widehat{\mathbf{m}}$ to solve for **vo** in terms of **vm** and writes

$$\mathbf{H} = \frac{\begin{pmatrix} Si_{51}+Si_{61}-Si_{53}-Si_{63} \\ Si_{71}+Si_{81}-Si_{73}-Si_{83} \end{pmatrix}}{(Si_{11}+Si_{21}-Si_{31}-Si_{41}-Si_{13}-Si_{23}+Si_{33}+Si_{43})}. \tag{11.26}$$

In this case, $\mathbf{H}$ is a $2 \times 1$ matrix such that

$$\mathbf{H} = \begin{pmatrix} H_{11} \\ H_{21} \end{pmatrix},$$

$$\begin{pmatrix} Vpo \\ Vmo \end{pmatrix} = \begin{pmatrix} H_{12} \\ H_{22} \end{pmatrix} \cdot Vdm. \tag{11.27}$$

Originally, the number of measurements did not match the number of stimuli in the system and therefore there were too many degrees of freedom to determine a solution. The stimuli were constrained in a manner that reduced the number of degrees of freedom. This extra constraint adds assumptions to the virtual probing results, and possibly reduces the accuracy and validity of the virtual probing solution, but allows for a viable solution.

In this example, a constraint of balanced drive was enforced, meaning that the stimulus *ep* is the inverse of stimulus *em*; this constraint caused neither the values *Vpm* and *Vmm* nor the outputs to be balanced, since no such enforcement is seen in (11.27).

As a note on the nomenclature, if the stimuli must be constrained, the original stimuli (the ones that emanate directly from a device port) are called *dependent* stimuli, and the new stimuli that these dependent stimuli depend on are called *independent* stimuli. The relationship between the independent and dependent stimuli is called a *stimdef*, which always comes in the form of a matrix, which, when multiplied by the independent stimuli, produces the dependent stimuli.

## 11.4 The Virtual Probe General Case Equations

In the previous examples, specific, simple virtual probing problems were solved just to illustrate the concept. Here, a general method for virtual probing is developed.

Consider a system consisting of devices interconnected at nodes; it is preferable that the voltages are labeled at these nodes (because virtual probing preferably uses voltages), but currents can be labeled at these nodes as well. The system under consideration further comprises identified sources of stimuli, and these stimuli are labeled. Finally, the system has specific nodes labeled as measurement nodes and output nodes, remembering that a measurement node is an input to the virtual probing system and that output nodes are outputs from the virtual probing system.

This system can be described by a system characteristics matrix $\mathbf{S}$, a node vector $\mathbf{n}$, and a stimulus vector $\mathbf{m}$ such that

$$\mathbf{S} \cdot \mathbf{n} = \mathbf{m}. \tag{11.28}$$

The formation of this set of equations is not yet considered.

Note that $\mathbf{Si} = \mathbf{S}^{-1}$, therefore

$$\mathbf{n} = \mathbf{Si} \cdot \mathbf{m}.$$

As $\mathbf{m}$ contains a certain number of non-zero elements, corresponding to the number of stimuli in the system, a new vector $\mathbf{m}'$ can be generated that contains only these non-zero elements; in other words, the stimuli in the system. For each zero row of $\mathbf{m}$ removed to form vector $\mathbf{m}'$, the corresponding column is removed from $\mathbf{Si}$ to form $\mathbf{Si}'$, resulting in

$$\mathbf{n} = \mathbf{Si}' \cdot \mathbf{m}'.$$

A constraints matrix $\mathbf{D}$ is identified (the *stimdef*), which constrains the values in $\mathbf{m}'$ (the dependent stimuli) based on a new list of stimuli $\widehat{\mathbf{m}}$ (the independent stimuli) such that

$$\mathbf{m}' = \mathbf{D} \cdot \widehat{\mathbf{m}},$$

$$\mathbf{n} = \mathbf{Si}' \cdot \mathbf{D} \cdot \widehat{\mathbf{m}}.$$

Considering $\mathbf{v}$ to be a list of all voltage nodes in the system, there exists a voltage extraction matrix that, when multiplied by the vector of node values, generates the voltages in $\mathbf{v}$:

$$\mathbf{VE} \cdot \mathbf{n} = \mathbf{v}. \tag{11.29}$$

Given a list of measurement node and output node names, these voltages are a subset of the voltages in $\mathbf{v}$. Therefore, there are voltage extraction matrices $\mathbf{VE_m}$ and $\mathbf{VE_o}$ that consist of rows of $\mathbf{VE}$ that extract voltages from $\mathbf{n}$ to form measurement and output node voltage vectors $\mathbf{vm}$ and $\mathbf{vo}$ such that

$$\mathbf{VE_m} \cdot \mathbf{n} = \mathbf{vm} = \mathbf{VE_m} \cdot \mathbf{Si}' \cdot \mathbf{D} \cdot \widehat{\mathbf{m}},$$

$$\mathbf{VE_o} \cdot \mathbf{n} = \mathbf{vo} = \mathbf{VE_o} \cdot \mathbf{Si}' \cdot \mathbf{D} \cdot \widehat{\mathbf{m}}.$$

Since all of these equations are in terms of $\widehat{\mathbf{m}}$,

$$\left[\mathbf{VE_m} \cdot \mathbf{Si}' \cdot \mathbf{D}\right]^{-1} \cdot \mathbf{vm} = \widehat{\mathbf{m}} = \left[\mathbf{VE_o} \cdot \mathbf{Si}' \cdot \mathbf{D}\right]^{-1} \cdot \mathbf{vo}, \tag{11.30}$$

$$\mathbf{vo} = \left[\mathbf{VE_o} \cdot \mathbf{Si}' \cdot \mathbf{D}\right] \cdot \left[\mathbf{VE_m} \cdot \mathbf{Si}' \cdot \mathbf{D}\right]^{-1} \cdot \mathbf{vm}.$$

And since a transfer function is desired,

$$\mathbf{vo} = \mathbf{H} \cdot \mathbf{vm}, \tag{11.31}$$

$$\mathbf{H} = \left[\mathbf{VE_o} \cdot \mathbf{Si}' \cdot \mathbf{D}\right] \cdot \left[\mathbf{VE_m} \cdot \mathbf{Si}' \cdot \mathbf{D}\right]^{-1}. \tag{11.32}$$

It is useful to review the dimensions of the matrices in (11.32). To begin, the dimensions in (11.28) are examined. A value *Nodes* is defined as the number of nodes in the system; $\mathbf{n}$ and $\mathbf{m}$ are both *Nodes* element vectors and both $\mathbf{S}$ and $\mathbf{Si}$ are square $Nodes \times Nodes$ matrices. A value *Stims* is defined as the number of stimuli in the system; $\mathbf{m}'$ is a *Stims* element vector, and therefore $\mathbf{Si}'$ becomes a $Nodes \times Stims$ matrix. A value *Degrees* is defined as the number of degrees of freedom in the system, and therefore the number of elements in $\widehat{\mathbf{m}}$. Since $\mathbf{m}'$ is a *Stims* element vector, $\mathbf{D}$ is a $Stims \times Degrees$ matrix. A value $Voltages$ is defined as the number of voltages in the system, which is incidentally half the value of *Nodes*. Examining (11.29) shows that $\mathbf{VE}$ must be a $Voltages \times Nodes$ matrix. A value *Meas* is defined as the number of measurement voltages, and *Outputs* is defined as the number of output voltages. The measurement and output voltages are a subset of the total number of voltages in the system, and therefore $\mathbf{VE_o}$ and $\mathbf{VE_m}$ have the same number of columns as $\mathbf{VE}$. The number of rows in $\mathbf{VE_m}$ is the number of measurement voltages *Meas*, and the number of rows in $\mathbf{VE_o}$ is the number of output voltages *Outputs*. Therefore, $\mathbf{VE_m}$ is $Meas \times Nodes$ and $\mathbf{VE_o}$ is $Outputs \times Nodes$.

With all of this, (11.32) can be written with dimensions as follows:

$$\begin{aligned}\underset{Outputs\times Meas}{\mathbf{H}} &= \left[\left(\underset{Outputs\times Nodes}{\mathbf{VE_o}}\right) \cdot \left(\underset{Nodes\times Stims}{\mathbf{Si}'}\right) \cdot \left(\underset{Stims\times Degrees}{\mathbf{D}}\right)\right] \\ &\quad \cdot \left[\left(\underset{Meas\times Nodes}{\mathbf{VE_m}}\right) \cdot \left(\underset{Nodes\times Stims}{\mathbf{Si}'}\right) \cdot \left(\underset{Stims\times Degrees}{\mathbf{D}}\right)\right]^{-1}.\end{aligned} \tag{11.33}$$

Equation (11.33) implies that *Meas* must equal *Degrees* because the matrix on the right that needs to be inverted is ideally square, and it is shown to be $Meas \times Degrees$. Also, $\mathbf{H}$ must be $Outputs \times Meas$ in order to be used in (11.31). This means that if $Stims > Meas$, then there are two ways to mitigate this. One is to increase the number of measurement nodes and therefore *Meas*. The other is to reduce the degrees of freedom in the system by constraining the stimuli such that $Degrees \leq Meas$. If $Meas > Degrees$, there are two options. One is to reduce the constraints on the stimuli, if possible, or to increase the value of *Degrees*. The other is to solve for $\mathbf{H}$ in a least-squares sense:

$$\begin{aligned}\mathbf{H} &= \left[\mathbf{VE_o} \cdot \mathbf{Si'} \cdot \mathbf{D}\right] \cdot \left(\left[\mathbf{VE_m} \cdot \mathbf{Si'} \cdot \mathbf{D}\right]^H \cdot \left[\mathbf{VE_m} \cdot \mathbf{Si'} \cdot \mathbf{D}\right]\right)^{-1} \cdot \left[\mathbf{VE_m} \cdot \mathbf{Si'} \cdot \mathbf{D}\right]^H \\ &= \left[\mathbf{VE_o} \cdot \mathbf{Si'} \cdot \mathbf{D}\right] \cdot \left[\mathbf{VE_m} \cdot \mathbf{Si'} \cdot \mathbf{D}\right]^{\dagger}. \end{aligned} \tag{11.34}$$

It is left to the reader to verify that the matrix to invert is $Degrees \times Degrees$, and that the final result $\mathbf{H}$ is indeed $Outputs \times Meas$.

## 11.5 Virtually Probing a Virtual Circuit

The use of virtual probing involving the measurement and production of waveforms in a circuit at various output nodes was demonstrated in the simple examples discussed in §11.1 and §11.2. In that model, virtual probing was utilized to produce waveforms at potentially inaccessible probing locations based on measurements made at presumably accessible locations. It turns out that this is only part of the power of virtual probing. This section will demonstrate ways of producing waveforms not only in inaccessible locations, but also in circuits that do not physically exist and are not being physically probed. This application, it turns out, is the primary usage of virtual probing where de-embedding is concerned, because in de-embedding one is sometimes not interested in probing alternative locations, but instead is interested in producing waveforms as if various circuit elements were removed from the circuit. This application is also the primary usage where embedding is concerned. In an embedding application, one wants to see waveforms that would be present if additional circuit elements were added to the circuit.

These applications of de-embedding and embedding utilize the same paradigm: that of a measurement circuit and an output circuit. These circuits, representing different probing circumstances, are related in the same manner as previously shown. In fact, all of the math provided in §11.4 is applicable.

Figure 11.3 shows a probe de-embedding problem where the goal is to remove the probe loading effects. A real probe loading effect example may be seen in [34]. Probe loading is the effect of the probe parasitics when the tips are connected to the circuit. Basically, it is the effect that this new circuit has on the measurement result as a result of being in the circuit. Probing effects in lower frequency applications (below 5 GHz) are uncompensated, but, in higher frequency applications, digital oscilloscopes generally compensate to some extent. The type of compensation depends on the probe and the oscilloscope manufacturer in the sense that some manufacturers try to provide compensation that shows the waveform present with the probe connected. Other manufacturers try to provide compensation that

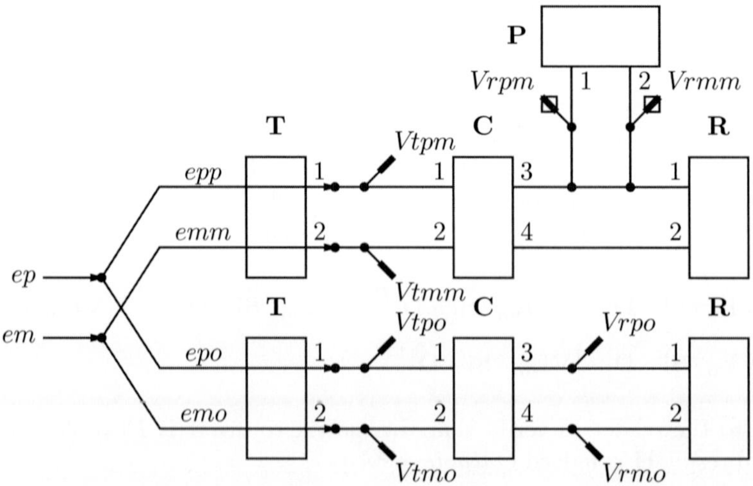


**Figure 11.3** Probe de-embedding example block diagram

shows the waveform present with the probe unconnected. The compensation strategy needs to be obtained from the scope and probe manufacturer.

Figure 11.3 depicts a strategy for removing the probe loading effect from a measurement made with a probe that presumably does not compensate for this effect. The most interesting difference between this and previous diagrams is the existence of two parallel circuits. In this arrangement, the top circuit is called the *measurement condition* and the bottom circuit is called the *output condition*. Basically, the measurement condition reflects the condition under which waveforms are acquired from the circuit. The measurement nodes *Vrpm* and *Vrmm* are in this circuit. The output condition reflects a *different circuit condition* under which the desired output waveforms are shown at output nodes *Vrpo* and *Vrmo*. Note here that the difference between the two circuits is the existence and connection of the element labeled **P**, which is representative of the probe and is placed in the measurement condition circuit but is not present in the output condition circuit. Another interesting observation is the connection of the stimuli. This connection is intended to show the constraints placed on the stimuli, whereby the dependent stimuli *epm* and *epo* are set equal to the independent stimulus *ep* and the dependent stimuli *emm* and *emo* are set equal to the independent stimulus *em*. In this arrangement, *ep* and *em* are called the *independent stimuli* because these reflect the number of degrees of freedom in the system. Likewise, *epm*, *emm*, *epo*, and *emo* are referred to as *dependent stimuli* because they depend on the independent stimuli. So, the dependent stimulus vector can be $\mathbf{m}' = \begin{pmatrix} epm & emm & epo & emo \end{pmatrix}^T$, the independent stimulus vector can be $\widehat{\mathbf{m}} = \begin{pmatrix} ep & em \end{pmatrix}^T$, with the matrix that determines the relationship between the two given by $\mathbf{D} = \begin{pmatrix} 1 & 0 \\ 0 & 1 \\ 1 & 0 \\ 0 & 1 \end{pmatrix}$. Take a moment to understand why $\mathbf{D}$ is what it is based on the definitions of $\mathbf{m}'$ and $\widehat{\mathbf{m}}$: $\mathbf{D}$ is the *stimdef*.

**Listing 11.1 VirtualProbe** class

```
class VirtualProbe(Simulator,object):
    def __init__(self,sd):
        Simulator.__init__(self,sd)
        self.m_ml = sd.m_ml if hasattr(sd, 'm_ml') else None
        self.m_D = sd.m_D if hasattr(sd, 'm_D') else None
...
    def pMeasurementList(self):
        return self.m_ml
    @pMeasurementList.setter
    def pMeasurementList(self,ml=None):
        if not ml is None: self.m_ml = ml
        return self
    @property
    def pStimDef(self):
        return self.m_D
    @pStimDef.setter
    def pStimDef(self,D=None):
        if not D is None: self.m_D = D
        return self
```

There's really nothing dramatic in the solution of this arrangement. The result is produced using the general case method provided in §11.4. The result is that the transfer function **H** produced converts measurement waveforms *Vrpm* and *Vrmm* into output waveforms *Vrpo* and *Vrmo* and that the output waveforms are the waveforms that would be at the terminals of the receiver labeled **R** if the probe were not in the circuit. The only real requirement for this to work, aside from the ability to invert matrices and, for now, ignoring numerical problems, is the existence of the independent stimuli in both the measurement and output conditions. It is the fact that the independent stimuli are exactly the same in both conditions that enable the solution. Note that the relationship between the independent and dependent stimuli does not even have to be the same for this to work. This conclusion can be drawn by simply examining the ramifications of (11.30) in §11.4.

Note also that, while it doesn't really make any sense other than for the measurement nodes to exist in only one of the circuits shown, the output nodes are not restricted to either the measurement or output condition. In fact, all nodes could be selected as output nodes in whichever condition the node exists. For example, one might want to compare the waveforms at the terminals of the transmitter labeled **T** when the probe is in or out of the circuit based on the measured waveforms. This is certainly possible by adding *Vtpm*, *Vtmm*, *Vtpo*, and *Vtmo* to the list of output voltages **vo**.

There is no restriction on having only one output condition. It is possible to add as many virtual circuits as desired as output conditions as long as the dependent stimuli are all constrained by the independent stimuli *ep* and *em*. Reference [16], for example, provides an example with one measurement condition and three output conditions.

**Listing 11.2 VirtualProbeSymbolic** class

```
class VirtualProbeSymbolic(SystemSParametersSymbolic, VirtualProbe):
    def __init__(self,sd=None,**args):
        SystemSParametersSymbolic.__init__(self,sd,**args)
        VirtualProbe.__init__(self, sd)
    def LaTeXTransferMatrix(self):
        self.Check()
        self._LaTeXSi()
        numMeas=len(self.pMeasurementList)
        vemsi = MatrixMultiply(
            self.VoltageExtractionMatrix(self.pMeasurementList), self.SIPrime(True))
        oneElementVemsi = False
        if len(vemsi) == 1:
            if len(vemsi[0]) == 1: oneElementVemsi = True
        vemsi = Matrix2LaTeX(vemsi, self._SmallMatrix())
        veosi = MatrixMultiply(
            self.VoltageExtractionMatrix(self.pOutputList), self.SIPrime(True))
        oneElementVeosi = False
        if len(veosi) == 1:
            if len(veosi[0]) == 1: oneElementVeosi = True
        veosi = Matrix2LaTeX(veosi, self._SmallMatrix())
        if self.pStimDef is None:
            numDeg=len(self.SIPrime(True)[0])
            inverse='^{-1}' if numDeg==numMeas else '^\\dagger'
            if oneElementVemsi and oneElementVeosi:
                line = lfrac(' '+veosi+' ',' '+vemsi+' ')+' '
            elif oneElementVemsi and not oneElementVeosi:
                line = veosi+' '+lfrac('1',' '+vemsi+' ')+' '
            elif not oneElementVemsi and oneElementVeosi:
                line = '\\left( '+veosi+'\\right)\\cdot '+vemsi+inverse
            elif not oneElementVemsi and not oneElementVeosi:
                line = veosi+'\\cdot '+vemsi+inverse
        else:
            numDeg=len(self.pStimDef[0])
            inverse='^{-1}' if numDeg==numMeas else '^\\dagger'
            D = Matrix2LaTeX(self.pStimDef, self._SmallMatrix())
            fveosi='\\left( '+veosi+'\\right)' if oneElementVeosi else veosi
            fvemsi='\\left( '+vemsi+'\\right)' if oneElementVemsi else vemsi
            line = '\\left[ '+fveosi+'\\cdot '+D+'\\right]\\cdot'+\
                '\\left[ '+fvemsi+'\\cdot '+D+'\\right]'+inverse
        if len(self.pMeasurementList) == 1 and len(self.pOutputList) == 1: H = 'H'
        else: H = '\\mathbf{H}'
        return self._AddEq(H+' = '+line)
    def LaTeXEquations(self):
        self.LaTeXSystemEquation()
        self.LaTeXTransferMatrix()
        return self
```

## 11.6 Programmatic Methods

Programmatically, the virtual probe application most closely resembles the simulation application, which is why the first class, the **VirtualProbe** class shown in Listing 11.1, is seen to derive from the **Simulator** class provided in Listing 9.1 and Listing 9.2.

The constructor of the **VirtualProbe** class, the `__init__()` member function shown on line 2 of Listing 11.1, takes a system description, but augments the system description information with a measurement list and a *stimdef*. Note that if, for some reason, the system description has these (because it is actually an instance of **VirtualProbe**, which derives eventually from the **SystemDescription** base class), it includes these attributes which might already be present. The **Simulator** class that it derives from itself deals with the output list, and thus there exist all of the additions to the system description required to define a virtual probing application.

The properties `pMeasurementList` and `pStimDef` are used to deal with these extra definitions. So, the basic strategy in supplying a virtual probe definition programmatically is either to instantiate a **SystemDescription** and fill in all of the devices and device connectivity, instantiating a **VirtualProbe** with this system description and filling in the remaining information through the added properties, or simply to instantiate a **VirtualProbe** class and fill in all the information for both the system description and the virtual probe.

The **VirtualProbe** class on its own is useful only for defining the problem. For producing symbolic results, the **VirtualProbeSymbolic** class shown in Listing 11.2 is utilized; this can be seen to derive from the **VirtualProbe** class along with the **SystemSParametersSymbolic** class in Listing 8.8. The derivation from **SystemSParametersSymbolic** gives it access to the internal function `_LaTeXSi()` on line 4 of Listing 8.8, which is part of the symbolic solution, along with the other general symbolic functions provided in the base classes **SystemDescriptionSymbolic** and **Symbolic** previously discussed in Chapter 8.

To generate symbolic problem solutions, one instantiates a **VirtualProbeSymbolic** class, defines the system, defines the stimuli, output list, and measurement list, and defines the *stimdef*. The stimuli are supplied with calls to `AssignM()` with arguments of the device name, device port, and stimulus name. The output list and measurement list are supplied by accessing the properties `pOutputList` and `pMeasurementList`, each as a list of tuples containing device/port pairs. The *stimdef*, which is optional, defines the relationship between independent and dependent stimuli in the system and must be specified manually as a matrix (in Python list form) such that, when multiplied by a vector of independent stimuli, it produces the dependent stimuli. This is supplied through the property `pStimDef`. If no stimuli depend on an independent stimulus, then no stimdef is provided. After defining the problem, either a call to `LaTeXTransferMatrix()` on line 5 of Listing 11.2 produces the solution, or a call to `LaTeXEquations()` on line 43 produces the system equation along with the solution.

### 11.6.1 Symbolic Examples

The operation of the symbolic solutions is most easily seen by examining a few examples.

The first example is shown in Figure 11.4. It is a repeat of the worked example in §11.1, whose block diagram is shown in Figure 11.1(a). The Python code for this example is provided in Figure 11.4(a).

On line 1, the *SignalIntegrity* package is imported as `si`. On line 2, a **VirtualProbeSymbolic** class is instantiated with the argument `size='small'` to indicate that the symbolic result is to be typeset with small matrices.

Lines 3–7 add the three devices `'T'`, `'C'`, and `'R'` and connect them as shown in the block diagram. Line 8 defines the single stimulus `'m1'` to emanate from port 1 of device `'T'`. Line 9 defines the measurement list as a list with a single tuple that defines port 1 of `'T'` as the measurement point. Line 10 similarly defines the output list as a list with a

```
1  import SignalIntegrity.Lib as si
2  vps=si.sd.VirtualProbeSymbolic(size='small')
3  vps.AddDevice('\\Gamma_T',1)
4  vps.AddDevice('C',2)
5  vps.AddDevice('\\Gamma_R',1)
6  vps.ConnectDevicePort('\\Gamma_T',1,'C',1)
7  vps.ConnectDevicePort('C',2,'\\Gamma_R',1)
8  vps.AssignM('\\Gamma_T',1,'m1')
9  vps.pMeasurementList = [('\\Gamma_T',1)]
10 vps.pOutputList = [('\\Gamma_R',1)]
11 vps.LaTeXEquations().Emit()
```

(a) Python code

$$\left[\mathbb{I}-\begin{pmatrix} 0 & \Gamma_T & 0 & 0 \\ C_{11} & 0 & 0 & C_{12} \\ C_{21} & 0 & 0 & C_{22} \\ 0 & 0 & \Gamma_R & 0 \end{pmatrix}\right]\cdot\begin{pmatrix} n_1 \\ n_2 \\ n_3 \\ n_4 \end{pmatrix}=\begin{pmatrix} m_1 \\ 0 \\ 0 \\ 0 \end{pmatrix}$$

$$\mathbf{Si}=\left[\mathbb{I}-\begin{pmatrix} 0 & \Gamma_T & 0 & 0 \\ C_{11} & 0 & 0 & C_{12} \\ C_{21} & 0 & 0 & C_{22} \\ 0 & 0 & \Gamma_R & 0 \end{pmatrix}\right]^{-1}$$

$$H=\frac{Si_{31}+Si_{41}}{Si_{11}+Si_{21}}$$

(b) LaTeX processed equations

**Figure 11.4** Symbolic simple virtual probing example

single tuple that defines port 1 of 'R' as the output point. With the problem completely defined, line 11 makes a call to `LaTeXEquations()` and subsequently emits the result.

The resulting output is in LaTeX as discussed in §8.3, and the typeset LaTeX result is provided in Figure 11.4(b). This is the same result as that provided in (11.10), and, if simplified through a symbolic processor, the result provided in (11.12) is obtained (after matching up the variable naming).

The second example is shown in Figure 11.5, which is a repeat of the worked example in §11.2, whose block diagram is shown in Figure 11.2(a). The Python code for this example is provided in Figure 11.5(a).

On line 1, the *SignalIntegrity* package is imported as `si`. On line 2, a **VirtualProbeSymbolic** class is instantiated with the argument `size='small'` to indicate that the symbolic result is to be typeset with small matrices.

Lines 3–9 add the three devices 'T', 'C', and 'R' and connect them as shown in the block diagram. These are two-, four-, and two-port devices. Lines 10 and 11 define the two stimuli: 'm1', which emanates from port 1 of device 'T'; and 'm2', which emanates from port 2 of device 'T'. Line 12 defines the measurement list as a list with two tuples that define ports 1 and 2 of 'T' as the measurement points. Line 13 similarly defines the output list as a list with two tuples that define ports 1 and 2 of 'R' as the output points. With the problem completely defined, line 14 makes a call to `LaTeXEquations()` and subsequently emits the result.

```
import SignalIntegrity.Lib as si
vps=si.sd.VirtualProbeSymbolic(size='small')
vps.AddDevice('T',2)
vps.AddDevice('C',4)
vps.AddDevice('R',2)
vps.ConnectDevicePort('T',1,'C',1)
vps.ConnectDevicePort('T',2,'C',2)
vps.ConnectDevicePort('C',3,'R',1)
vps.ConnectDevicePort('C',4,'R',2)
vps.AssignM('T',1,'m1')
vps.AssignM('T',2,'m2')
vps.pMeasurementList = [('T',1),('T',2)]
vps.pOutputList = [('R',1),('R',2)]
vps.LaTeXEquations().Emit()
```

(a) Python code

$$\left[\mathbb{I}-\begin{pmatrix} 0 & 0 & T_{11} & T_{12} & 0 & 0 & 0 & 0 \\ 0 & 0 & T_{21} & T_{22} & 0 & 0 & 0 & 0 \\ C_{11} & C_{12} & 0 & 0 & 0 & 0 & C_{13} & C_{14} \\ C_{21} & C_{22} & 0 & 0 & 0 & 0 & C_{23} & C_{24} \\ C_{31} & C_{32} & 0 & 0 & 0 & 0 & C_{33} & C_{34} \\ C_{41} & C_{42} & 0 & 0 & 0 & 0 & C_{43} & C_{44} \\ 0 & 0 & 0 & 0 & R_{11} & R_{12} & 0 & 0 \\ 0 & 0 & 0 & 0 & R_{21} & R_{22} & 0 & 0 \end{pmatrix}\right]\cdot\begin{pmatrix} n_1 \\ n_3 \\ n_2 \\ n_4 \\ n_5 \\ n_7 \\ n_6 \\ n_8 \end{pmatrix}=\begin{pmatrix} m_1 \\ m_2 \\ 0 \\ 0 \\ 0 \\ 0 \\ 0 \\ 0 \end{pmatrix}$$

$$\mathbf{Si}=\left[\mathbb{I}-\begin{pmatrix} 0 & 0 & T_{11} & T_{12} & 0 & 0 & 0 & 0 \\ 0 & 0 & T_{21} & T_{22} & 0 & 0 & 0 & 0 \\ C_{11} & C_{12} & 0 & 0 & 0 & 0 & C_{13} & C_{14} \\ C_{21} & C_{22} & 0 & 0 & 0 & 0 & C_{23} & C_{24} \\ C_{31} & C_{32} & 0 & 0 & 0 & 0 & C_{33} & C_{34} \\ C_{41} & C_{42} & 0 & 0 & 0 & 0 & C_{43} & C_{44} \\ 0 & 0 & 0 & 0 & R_{11} & R_{12} & 0 & 0 \\ 0 & 0 & 0 & 0 & R_{21} & R_{22} & 0 & 0 \end{pmatrix}\right]^{-1}$$

$$\mathbf{H}=\begin{pmatrix} Si_{51}+Si_{71} & Si_{52}+Si_{72} \\ Si_{61}+Si_{81} & Si_{62}+Si_{82} \end{pmatrix}\cdot\begin{pmatrix} Si_{11}+Si_{31} & Si_{12}+Si_{32} \\ Si_{21}+Si_{41} & Si_{22}+Si_{42} \end{pmatrix}^{-1}$$

(b) LaTeX processed equations

**Figure 11.5** Symbolic multiple input/output virtual probing example

The resulting output, when typeset by a LaTeX processor, produces the result provided in Figure 11.5(b). This is technically the same result as that provided in (11.20). It does not have the same index numbers on the inverse **Si** because the system equation is in a different permuted, but canonical, form than that provided in (11.13). It is still the correct answer. In other words, if numbers were put into these equations, they would produce the same numeric results.

### 11.6.2 Virtual Probe Parser

As with all of the applications, a parser class is provided to aid in the description of the problem. The **VirtualProbeParser** class is shown in Listing 11.3. It derives from the **SystemDescriptionParser** class shown in Listing 8.9. When constructed through the `__init__()` member function on line 2, it initializes its base **SystemDescriptionParser** class.

**Listing 11.3 VirtualProbeParser** class

```
1  class VirtualProbeParser(SystemDescriptionParser):
2      def __init__(self, f=None, args=None):
3          SystemDescriptionParser.__init__(self, f, args)
4      def _ProcessVirtualProbeLine(self,line):
5          lineList=self.ReplaceArgs(line.split())
6          if len(lineList) == 0: return
7          if lineList[0] == 'meas':
8              if self.m_sd.pMeasurementList is None: self.m_sd.pMeasurementList = []
9              for i in range(1,len(lineList),2):
10                 self.m_sd.pMeasurementList.append((lineList[i],int(lineList[i+1])))
11         elif lineList[0] == 'output':
12             if self.m_sd.pOutputList is None: self.m_sd.pOutputList = []
13             for i in range(1,len(lineList),2):
14                 self.m_sd.pOutputList.append((lineList[i],int(lineList[i+1])))
15         elif lineList[0] == 'stim':
16             for i in range((len(lineList)-1)//3):
17                 self.m_sd.AssignM(lineList[i*3+2],int(lineList[i*3+3]),lineList[i*3+1])
18         elif lineList[0] == 'stimdef':
19             self.m_sd.pStimDef = [[float(e) for e in r] for r in [s.split(',')
20                 for s in ''.join(lineList[1:]).strip(' ').strip('[[').
21                     strip(']]').split('],[') ]]
22         else: self.m_ul.append(line)
23     def _ProcessLines(self):
24         SystemDescriptionParser._ProcessLines(self)
25         self.m_sd = VirtualProbe(self.m_sd)
26         lines=copy.deepcopy(self.m_ul); self.m_ul=[]
27         for line in lines: self._ProcessVirtualProbeLine(line)
28         return self
```

It works by overriding the base `_ProcessLines()` member function on line 23, thus intercepting these calls, and calls the `_ProcessLines()` base class member function on **SystemDescriptionParser** first. The base class keeps track of all unprocessed lines (actually unrecognized lines), and these lines are subsequently processed through calls to `_ProcessVirtualProbeLine()` on line 4 of Listing 11.3.

The internal function `_ProcessVirtualProbeLine()` splits a line of text into space separated tokens and handles token lists whose first token is one of four keywords: `'stim'`, `'meas'`, `'output'`, and `'stimdef'`. These work as follows:

- `'stim arg1 arg2 arg3 ...'` adds a stimulus named `arg1` as emanating from port `arg3` of a device named `arg2`. If there are more tokens, it can also add a stimulus named `arg4` as emanating from port `arg6` of a device named `arg5`, and so on. The tokens must come in triplets. It provides the same result as the `AssignM()` function on **SystemDescription** class.
- `'meas arg1 arg2 ...'` appends a tuple containing a device named `arg1` and a port of that device `arg2` to a list of measures. If more than one measure is provided, it adds them in groups of two. The measurement list formed in this way looks just like that specified in Figure 11.5(a). Thus, a command such as `'meas T 1 T 2'` to the **VirtualProbeParser** class looks just like a call to `pMeasurementList=[('T',1),('T',2)]` on the **VirtualProbe** class.
- `'output arg1 arg2 ...'` appends a tuple containing a device named `arg1` and a port of that device `arg2` to a list of outputs. If more than one output

is provided, it adds them in groups of two. The output list formed in this way looks just like that specified in Figure 11.5(a). Thus, a command such as `'output R 1 R 2'` to the **VirtualProbeParser** class looks just like a call to `pOutputList=[('R',1),('R',2)]` on the **VirtualProbe** class.

- `'stimdef arg1'` assigns a stimdef with a Python matrix, in list form as `arg1`. If one had a set of $S$ dependent stimuli, $\mathbf{m}$, and a list of $I$ independent stimuli, $\mathbf{i}$ , the matrix would be $\mathbf{D}$, such that

$$\begin{pmatrix} D_{11} & D_{12} & \cdots & D_{1I} \\ D_{21} & D_{22} & \cdots & D_{2I} \\ \vdots & \vdots & \ddots & \vdots \\ D_{S1} & D_{S2} & \cdots & D_{SI} \end{pmatrix} \cdot \begin{pmatrix} i_1 \\ i_2 \\ \vdots \\ i_I \end{pmatrix} = \begin{pmatrix} m_1 \\ m_2 \\ \vdots \\ m_S \end{pmatrix},$$

and the equivalent Python matrix describing this would look like

$$[[D_{11}, D_{12}, \ldots, D_{1I}], [D_{21}, D_{22}, \ldots, D_{2I}], \ldots, [D_{S1}, D_{S2}, \ldots, D_{SI}]].$$

### 11.6.3 Symbolic Parser Examples

In the symbolic virtual probing example shown in Figure 11.6, the second example provided in §11.6.1 is repeated, specifically the example in Figure 11.5 with a small difference. The block diagram in Figure 11.6(a) resembles the diagram in Figure 11.2(a), except for the two stimuli emanating from ports 1 and 2 of $\mathbf{T}$ that are now dependent on a single independent stimulus $m$.

In Figure 11.6(b), line 1, the *SignalIntegrity* package is imported as `si`, and on line 2 a **VirtualProbeParser** is instantiated as `vpp`. On lines 3–5, a list of lines is added. The first three lines have the `'device'` keyword and specify the three devices `'T'`, `'C'`, and `'R'` that have two, four, and two ports. The next four lines have the `'connect'` keyword and specify the device connections, as in the diagram in Figure 11.6(a). The next two lines have the `'stim'` keyword and specify the two stimuli in the system emanating from ports 1 and 2 of device `'T'`. The next line has the `'meas'` keyword and specifies the two measurement points as ports 1 and 2 of `'T'`. The next line has the output keyword and specifies the two output points as ports 1 and 2 of `'R'`. At this point, the internal system description would resemble that produced by Figure 11.5(a), but the final line provides the `'stimdef'` keyword and provides a matrix in the Python format $[[1], [-1]]$. This says that the two stimuli depend on a single independent stimulus. This is known because, according to (11.33), $\mathbf{D}$, the stimdef, is a $Stims \times Degrees$ matrix. There are two rows for two stims $m_1$ and $m_2$ and one column for one degree. The name of the independent stimulus is irrelevant.

On line 6, the **VirtualProbeSymbolic** class is instantiated with the instance of the system description provided by a call to `SystemDescription()` on `vpp`. This means that the **VirtualProbeParser** is used to parse the netlist and then to extract an instance of **SystemDescription** used to instantiate **VirtualProbeSymbolic**.

On line 7, the symbolic result is produced through a call to `LaTeXEquations()` and subsequently emitted.

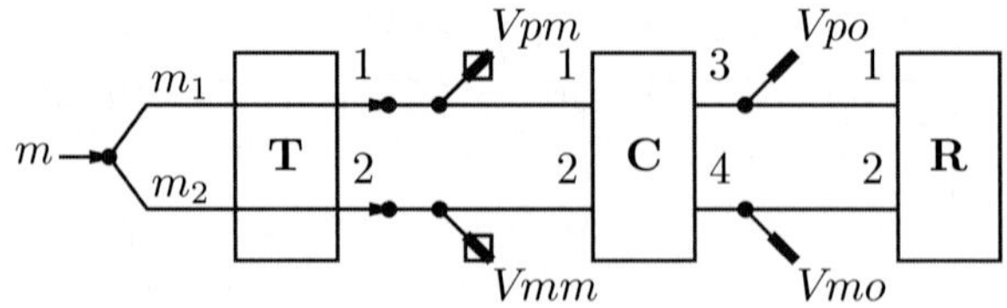


(a) Block diagram

```
import SignalIntegrity.Lib as si
vpp=si.p.VirtualProbeParser()
vpp.AddLines(['device T 2','device C 4','device R 2','connect T 1 C 1',
    'connect T 2 C 2','connect C 3 R 1','connect C 4 R 2','stim m1 T 1',
    'stim m2 T 2','meas T 1 T 2','output R 1 R 2','stimdef [[1],[-1]]'])
vps=si.sd.VirtualProbeSymbolic(vpp.SystemDescription(),size='small')
vps.LaTeXEquations().Emit()
```

(b) Python code

$$\left[\mathbb{I}-\begin{pmatrix} 0 & 0 & T_{11} & T_{12} & 0 & 0 & 0 & 0\\ 0 & 0 & T_{21} & T_{22} & 0 & 0 & 0 & 0\\ C_{11} & C_{12} & 0 & 0 & 0 & 0 & C_{13} & C_{14}\\ C_{21} & C_{22} & 0 & 0 & 0 & 0 & C_{23} & C_{24}\\ C_{31} & C_{32} & 0 & 0 & 0 & 0 & C_{33} & C_{34}\\ C_{41} & C_{42} & 0 & 0 & 0 & 0 & C_{43} & C_{44}\\ 0 & 0 & 0 & 0 & R_{11} & R_{12} & 0 & 0\\ 0 & 0 & 0 & 0 & R_{21} & R_{22} & 0 & 0 \end{pmatrix}\right]\cdot\begin{pmatrix} n_1\\ n_3\\ n_2\\ n_4\\ n_5\\ n_7\\ n_6\\ n_8 \end{pmatrix}=\begin{pmatrix} m_1\\ m_2\\ 0\\ 0\\ 0\\ 0\\ 0\\ 0 \end{pmatrix}$$

$$\mathbf{Si}=\left[\mathbb{I}-\begin{pmatrix} 0 & 0 & T_{11} & T_{12} & 0 & 0 & 0 & 0\\ 0 & 0 & T_{21} & T_{22} & 0 & 0 & 0 & 0\\ C_{11} & C_{12} & 0 & 0 & 0 & 0 & C_{13} & C_{14}\\ C_{21} & C_{22} & 0 & 0 & 0 & 0 & C_{23} & C_{24}\\ C_{31} & C_{32} & 0 & 0 & 0 & 0 & C_{33} & C_{34}\\ C_{41} & C_{42} & 0 & 0 & 0 & 0 & C_{43} & C_{44}\\ 0 & 0 & 0 & 0 & R_{11} & R_{12} & 0 & 0\\ 0 & 0 & 0 & 0 & R_{21} & R_{22} & 0 & 0 \end{pmatrix}\right]^{-1}$$

$$\mathbf{H}=\left[\begin{pmatrix} Si_{51}+Si_{71} & Si_{52}+Si_{72}\\ Si_{61}+Si_{81} & Si_{62}+Si_{82} \end{pmatrix}\cdot\begin{pmatrix} 1\\ -1 \end{pmatrix}\right]\cdot\left[\begin{pmatrix} Si_{11}+Si_{31} & Si_{12}+Si_{32}\\ Si_{21}+Si_{41} & Si_{22}+Si_{42} \end{pmatrix}\cdot\begin{pmatrix} 1\\ -1 \end{pmatrix}\right]^{\dagger}$$

(c) LaTeX processed equations

**Figure 11.6** Symbolic overconstrained virtual probing example

The result is typeset in Figure 11.6(c). The system equation and the matrix **Si** look exactly as in Figure 11.5(b). After all, they are the same system aside from the *stimdef.* But the final result for **H** is different. The matrix to the right is the product of a $2 \times 2$ matrix and the $2 \times 1$ matrix **D** to form a $2 \times 1$ matrix to be inverted. But, since the matrix is not square, the † operator is used to indicate that this is an overconstrained result and the pseudo-inverse is used (see Appendix C, §C.2). The final result, when simplified, will take on a solution of the form shown in (11.34).

Another example, in Figure 11.7, demonstrates the usage of the **VirtualProbeParser**, but uses a file input. This example duplicates the problem discussed in §11.3, where only the differential voltage could be probed. This will also show how differential voltage probing

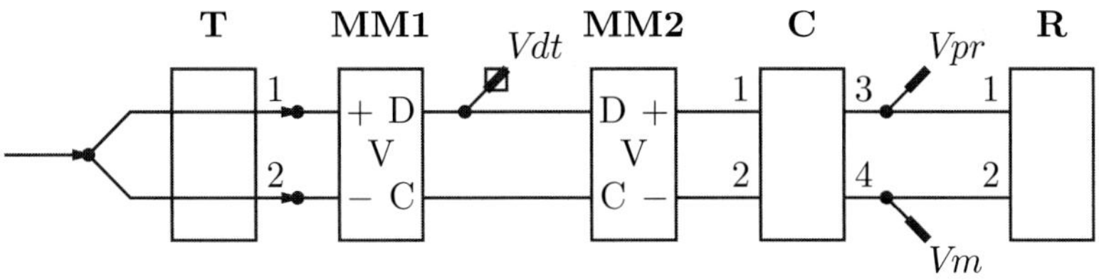


(a) Block diagram

```
device T 2
device MM1 4 mixedmode voltage
device MM2 4 mixedmode voltage
device C 4
device R 2
connect T 1 MM1 1
connect T 2 MM1 2
connect MM1 3 MM2 3
connect MM1 4 MM2 4
connect MM2 1 C 1
connect MM2 2 C 2
connect C 3 R 1
connect C 4 R 2
stim m1 T 1 m2 T 2
meas MM1 3
output R 1 R 2
stimdef [[1],[-1]]
```

(b) Text netlist file

```
import SignalIntegrity.Lib as si
vpp=si.p.VirtualProbeParser().File('VirtualProbe4.txt')
vps=si.sd.VirtualProbeSymbolic(vpp.SystemDescription(),size='small')
vps.LaTeXTransferMatrix().Emit()
```

(c) Python code

$$
\mathbf{Si} = \left[ \mathbb{I} - \begin{pmatrix}
0 & 0 & T_{11} & T_{12} & 0 & 0 & 0 & 0 & 0 & 0 & 0 & 0 & 0 & 0 & 0 & 0 \\
0 & 0 & T_{21} & T_{22} & 0 & 0 & 0 & 0 & 0 & 0 & 0 & 0 & 0 & 0 & 0 & 0 \\
0 & 0 & 0 & 0 & 0 & 0 & 0 & 0 & \frac{1}{2} & 1 & 0 & 0 & 0 & 0 & 0 & 0 \\
0 & 0 & 0 & 0 & 0 & 0 & 0 & 0 & -\frac{1}{2} & 1 & 0 & 0 & 0 & 0 & 0 & 0 \\
1 & -1 & 0 & 0 & 0 & 0 & 0 & 0 & 0 & 0 & 0 & 0 & 0 & 0 & 0 & 0 \\
\frac{1}{2} & \frac{1}{2} & 0 & 0 & 0 & 0 & 0 & 0 & 0 & 0 & 0 & 0 & 0 & 0 & 0 & 0 \\
0 & 0 & 0 & 0 & \frac{1}{2} & 1 & 0 & 0 & 0 & 0 & 0 & 0 & 0 & 0 & 0 & 0 \\
0 & 0 & 0 & 0 & -\frac{1}{2} & 1 & 0 & 0 & 0 & 0 & 0 & 0 & 0 & 0 & 0 & 0 \\
0 & 0 & 0 & 0 & 0 & 0 & 0 & 0 & 0 & 0 & 1 & -1 & 0 & 0 & 0 & 0 \\
0 & 0 & 0 & 0 & 0 & 0 & 0 & 0 & 0 & 0 & \frac{1}{2} & \frac{1}{2} & 0 & 0 & 0 & 0 \\
0 & 0 & 0 & 0 & 0 & 0 & C_{11} & C_{12} & 0 & 0 & 0 & 0 & 0 & 0 & C_{13} & C_{14} \\
0 & 0 & 0 & 0 & 0 & 0 & C_{21} & C_{22} & 0 & 0 & 0 & 0 & 0 & 0 & C_{23} & C_{24} \\
0 & 0 & 0 & 0 & 0 & 0 & C_{31} & C_{32} & 0 & 0 & 0 & 0 & 0 & 0 & C_{33} & C_{34} \\
0 & 0 & 0 & 0 & 0 & 0 & C_{41} & C_{42} & 0 & 0 & 0 & 0 & 0 & 0 & C_{43} & C_{44} \\
0 & 0 & 0 & 0 & 0 & 0 & 0 & 0 & 0 & 0 & 0 & 0 & R_{11} & R_{12} & 0 & 0 \\
0 & 0 & 0 & 0 & 0 & 0 & 0 & 0 & 0 & 0 & 0 & 0 & R_{21} & R_{22} & 0 & 0
\end{pmatrix} \right]^{-1}
$$

$$
\mathbf{H} = \left[ \begin{pmatrix} Si_{13,1}+Si_{15,1} & Si_{13,2}+Si_{15,2} \\ Si_{14,1}+Si_{16,1} & Si_{14,2}+Si_{16,2} \end{pmatrix} \cdot \begin{pmatrix} 1 \\ -1 \end{pmatrix} \right] \cdot \left[ \begin{pmatrix} Si_{51}+Si_{91} & Si_{52}+Si_{92} \end{pmatrix} \cdot \begin{pmatrix} 1 \\ -1 \end{pmatrix} \right]^{-1}
$$

(d) LaTeX processed equations

**Figure 11.7** Symbolic differential virtual probing example

is handled. The file is shown in Figure 11.7(b), which is a text file stored on the disk called `'VirtualProbe4.txt'`. This file contains lines of a netlist in the form used in the preceding example.

The Python code for processing this netlist is provided in Figure 11.7(c). On line 1, the *SignalIntegrity* package is imported, and on line 2, a **VirtualProbeParser** is instantiated and the netlist file is immediately read in. On line 3, a call is made to `SystemDescription()`, which causes the lines in the file to be processed to produce the system description used to instantiate the **VirtualProbeSymbolic** class. On line 4, the symbolic transfer matrix is produced through a call to `LaTeXTransferMatrix()`, and the result is emitted.

In the netlist file in Figure 11.7(c) there are five devices: the two-, four-, and two-port devices `'T'`, `'C'`, and `'R'`, along with two four-port devices called `'MM1'` and `'MM2'`. These are specified as devices called `'mixedmode'` with an argument `'voltage'`. These are voltage mixed-mode converters, as shown in Figure 7.6, placed back-to-back in the circuit to expose the differential and common modes. When used in this way, typically the differential- and common-mode ports are connected together and probed, and the single-ended positive and negative ports are connected to the rest of the circuit. Because of this connection, they have no impact on the circuit whatsoever, except to provide access to the mixed-mode voltages. In this usage, the common definitions for differential- and common-mode voltages are used:

$$V_D = V_p - V_m, \qquad V_C = \frac{V_p + V_m}{2}.$$

This is the behavior of the mixed-mode converter when voltage is specified. These two back-to-back mixed-mode converters are shown in Figure 11.7(a), where the measurement probe *Vdt* was placed in between the mixed-mode converters.

The LaTeX result is typeset in Figure 11.7(d); the result is similar to that in (11.26). It does not have the same index numbers on the inverse **Si** because the system equation is in a different permuted, but canonical, form than that provided in (11.26). It is still the correct answer.

### 11.6.4 Numeric Solutions

The **VirtualProbeNumeric** class is shown in Listing 11.4. This is the analogous numeric class to the symbolic **VirtualProbeSymbolic** class provided in Listing 11.2. This class inherits all of the capability of the **VirtualProbe** class in Listing 11.1.

There are only two member functions in this class. The constructor on line 2 takes an optional instance of a **VirtualProbe** class (or a **SystemDescription** class), which could have been constructed previously using the member functions in **VirtualProbe**, **Simulator**, and **SystemDescription**. Otherwise, the class can be instantiated without a system description, and the system can be defined similarly post construction.

As in the simulation described in Chapter 9, the **VirtualProbeNumeric** provides a solution in the form of a transfer matrix through a call to the `TransferMatrix()` function shown on line 4 in Listing 11.4. The idea is to loop over all of the desired frequency points to solve for each transfer matrix for each frequency.

**Listing 11.4 VirtualProbeNumeric** class

```
class VirtualProbeNumeric(VirtualProbe,Numeric):
    def __init__(self,sd=None):
        VirtualProbe.__init__(self,sd)
    def TransferMatrix(self):
        if self.m_D is None:
            D=matrix(identity(len(self.StimsPrime())))
        else:
            D=self.m_D
        VE_m=matrix(self.VoltageExtractionMatrix(self.m_ml))
        VE_o=matrix(self.VoltageExtractionMatrix(self.m_ol))
        SIPrime_m=matrix(self.SIPrime(Left=VE_m,Right=D))
        SIPrime_o=matrix(self.SIPrime(Left=VE_o,Right=D))
        Left=(VE_o*SIPrime_m*matrix(D))
        Result=(Left*self.Dagger(VE_m*SIPrime_o*matrix(D),Left=Left)).tolist()
        return Result
```

**Listing 11.5 VirtualProbeNumericParser** class

```
1  class VirtualProbeNumericParser(VirtualProbeParser,CallBacker,LinesCache):
2      def __init__(self, f=None, args=None, callback=None, cacheFileName=None):
3          VirtualProbeParser.__init__(self, f, args)
4          self.transferMatrices = None
5          self.m_tm=None
6      def TransferMatrices(self):
7          self.SystemDescription()
8          self.m_sd.CheckConnections()
9          spc=self.m_spc
10         result=[]
11         for n in range(len(self.m_f)):
12             for d in range(len(self.m_spc)):
13                 self.m_sd.AssignSParameters(spc[d][0],spc[d][1][n])
14             tm=VirtualProbeNumeric(self.m_sd).TransferMatrix()
15             result.append(tm)
16         self.transferMatrices=TransferMatrices(self.m_f,result)
17         return self.transferMatrices
```

This class is difficult to use. More commonly, one utilizes the **VirtualProbeNumericParser** class shown in Listing 11.5. Again, there is a constructor on line 2 and the `TransferMatrices()` member function on line 6. The **VirtualProbeNumericParser** is constructed ideally with a list of frequencies, and the call to `TransferMatrices()` provides a list of transfer matrices, one matrix for each frequency.

The netlist for a **VirtualProbeNumericParser** is the same as the netlist for system descriptions along with the additional keywords parsed in the **VirtualProbeParser** class in Listing 11.3. These additional keywords, as previously described, define the measure probe locations (where known, previously measured waveforms are supplied), output probe locations (the desired output waveforms), and the stimulus definitions provided by stims and the stimdef statement.

Each transfer matrix has a row corresponding to an output measurement probe, where the row index is in the order that they were supplied in the netlist, and a column corresponding to a measurement probe, where the column index is in the order that the measurement probe occurs in the netlist.

The call to `TransferMatrices()` provides an instance of the class **TransferMatrices**, as shown in Listing 9.7. The `TransferMatrices()` member function builds the system description by processing the lines of the netlist and then checks the netlist for validity. If an exception is not raised, then it proceeds to loop over all of the frequencies in the list provided and fills in the s-parameters of the devices in the system description with the s-parameters stored during processing. This system description is then passed to a new instance of **VirtualProbeNumeric**, and the transfer matrix for a single frequency is calculated and added to the result. Once the transfer matrices have been produced, the result is returned. As far as the **VirtualProbeNumericParser** class is concerned, the job is complete once the transfer matrices have been calculated; there's really nothing more for it to do. Measured waveforms are processed through a call to `ProcessWaveforms()` on the **TransferMatrices** class shown in Figure 13.9(b), with a list of measured waveforms provided. This processing is discussed in §13.5. The transfer matrices can be stored by obtaining them as s-parameters, as shown in Listing 9.7 and writing them to a file. Thus they can be stored and used over and over again with different waveforms.

## 11.7 Virtual Probing Numeric Example

A numeric example of virtual probing is provided in this section that is produced by the *SignalIntegrity* application software described in Chapter 18. This software was used to produce the schematic example shown in Figure 11.8. An advantage of this software, in addition to providing a GUI interface for drawing the schematic, is that the interaction with the *SignalIntegrity* library is entirely through a netlist generated by the application and passed to the **VirtualProbeNumericParser** class. The netlist for the schematic in Figure 11.8 is shown in Figure 11.9. The generation of the virtual probing solution could have been scripted entirely using the netlist in Figure 11.9 without any reference to the GUI based *SignalIntegrity* application; the application largely automated the process of converting the schematic to a netlist.

The virtual probing problem being solved in Figure 11.8 shows a solution to a wide variety of problems, including the main problem that virtual probing was intended to solve in the first place (see the Introduction to this chapter). It solves these problems with a probe model in the top subcircuit, a channel model with the probe connected in the middle subcircuit, and the channel model alone in the bottom subcircuit. The problems addressed here are:

1. probe compensation;
2. receiver waveform generation, while probing at the transmitter;
3. differential waveform generation from a single-ended probing measurement;
4. recovery of waveforms that would be measured at the transmitter and receiver without the probe loading the circuit.
5. While not shown here, a discussion is provided for how the situation could be modified to avoid physically having and probing with the actual channel hardware (here, the SPARQ demo board).

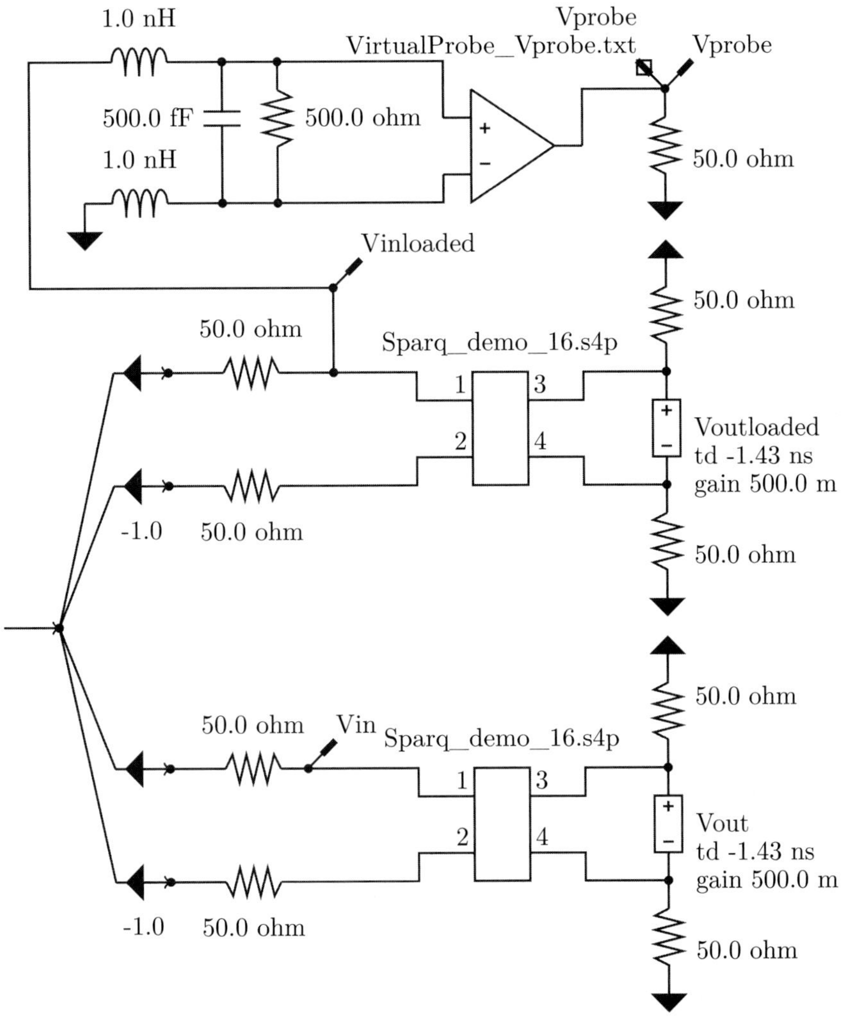


**Figure 11.8** Virtual probing numeric example

Before discussing the problem solutions, it is worthwhile pointing out the underlying assumptions that are being made and that must hold true to produce a valid outcome. The first is that of wave sources in the system. On the left of the circuit shown in Figure 11.8, there is a single arrow that connects to four other arrows at the location where four grounds connect to four source termination resistors. The single arrow to the left is an independent stim, and the four arrows connecting to the ground are dependent stims. Two of the dependent stims that are at the bottom of each subcircuit have a weight of $-1$, indicating differential drive. Note also that the dependent stims connect to the ground symbol connection point, which has been offset from the connection to the resistor. In other

```
device L1 2 L 1e-09
device L2 2 L 1e-09
device C1 2 C 5e-13 esr 0.0 df 0.0
device R1 2 R 500.0
device G1 1 ground
device D1 4 file Sparq_demo_16.s4p
device R2 2 R 50.0
device R3 2 R 50.0
device G2 1 ground
device R4 1 R 50.0
device R5 1 R 50.0
device G3 1 ground
device D2 4 file Sparq_demo_16.s4p
device Vout 4 voltagecontrolledvoltagesource 1.0
device Voutloaded 4 voltagecontrolledvoltagesource 1.0
device R8 2 R 50.0
device R9 2 R 50.0
device G5 1 ground
device G6 1 ground
device R6 1 R 50.0
device R7 1 R 50.0
device R10 1 R 50.0
device D3 3 opamp gain 1.0 zi 100000000.0 zo 0.0
output R2 2
connect R2 2 L1 1 D1 1
connect L1 2 R1 2 C1 2 D3 2
connect L2 1 G1 1
connect R1 1 C1 1 D3 1 L2 2
connect D1 2 R3 2
connect R5 1 Voutloaded 2 D1 3
connect R4 1 D1 4 Voutloaded 1
connect G2 1 R2 1
stim m1 G2 1
connect G3 1 R3 1
stim m2 G3 1
output R8 2
connect R8 2 D2 1
connect R9 2 D2 2
connect Vout 2 D2 3 R7 1
connect Vout 1 D2 4 R6 1
meas R10 1
output R10 1
connect R10 1 D3 3
connect G5 1 R8 1
stim m3 G5 1
connect G6 1 R9 1
stim m4 G6 1
stimdef [[1.0], [-1.0], [1.0], [-1.0]]
device Vout_2 1 ground
device Vout_3 1 open
connect Vout 3 Vout_2 1
connect Vout 4 Vout_3 1
output Vout 4
device Voutloaded_2 1 ground
device Voutloaded_3 1 open
connect Voutloaded 3 Voutloaded_2 1
connect Voutloaded 4 Voutloaded_3 1
output Voutloaded 4
```

**Figure 11.9** Virtual probing numeric example netlist

words, the resistor is connected to the ground through a small wire (not directly abutting it), and the dependent stim is connected to the port on the ground. This is a necessary condition because the connection implies the port from which the waves indicated by the dependent stim emanate; if the ground and resistor abutted, there would be no way to distinguish whether the waves emanate from the ground or the resistor.

Two important assumptions here are that waves enter the system only from the dependent stims and that there are no other sources of waves entering the system. These assumptions are absolutely crucial to the virtual probing solution. While all of the calculations shown previously prove mathematically that this method works, in layman's terms the reasons why it works are as follows:

1. All sources of entry of waves in the system are known.
2. The measured waveforms represent measurements made *for all time*.

While there is no way with a finite length waveform to represent measurements made for all time, the second statement means that the measurement has been made over a long enough time for waves entering the system prior to the beginning of the measured waveform to have died down. The validity of this assumption will be discussed later when studying the time-domain transfer function generated.

Previously it was shown that the solution depends on there being an equal number of measurements and stims. Here, there is a single-ended probe with a loading model (shown at the top of Figure 11.8) with a single waveform measurement $V_{probe}$. There is one measurement being made and there are four stims. Therefore, a constraint is added to reduce these four stims to one: the single independent stim on the left. In the middle and bottom subcircuits, there are two dependent stims. One assumption is that the waves entering each of these two subcircuits are identical. This is not problematic because the assumption is embodied in the problem definition: that of determining what the waveforms would look like in the loaded and unloaded circuits. Because the probe is single ended and the system is differential, an assumption is made that the waves entering the positive leg of the differential line are equal and opposite to the waves entering the negative leg, hence the weight of $-1$. This assumption must be verified for the validity of the virtual probing to hold.[2] This assumption can be avoided by using a differential probe. In the netlist in Figure 11.9, one can see the stim statements corresponding to the four dependent stims labeled $m_1$, $m_2$, $m_3$, and $m_4$, along with the *stimdef* that defines a $4 \times 1$ vector, defining

$$\mathbf{m}' = \begin{pmatrix} m_1 \\ m_2 \\ m_3 \\ m_4 \end{pmatrix} = \begin{pmatrix} 1 \\ -1 \\ 1 \\ -1 \end{pmatrix} \cdot \hat{\mathbf{m}}.$$

The stims and the *stimdef* were automatically generated by the *SignalIntegrity* application.

In the example shown, the subcircuit at the top of Figure 11.8 is assumed to be an accurate model of the probe. In this example, a loading network model is shown at the front

[2] This assumption does not enforce balanced voltage on each leg. Nor does it enforce any knowledge of the wave entering at the transmitter, only the similarity between the waves on each side of the transmitter.

of a differential-to-single-ended amplifier driving the 50 Ω oscilloscope channel termination. Probe loading models can usually be obtained for high frequency probes and can be fairly accurate. In this example, there is some inductance due to the probe leads, some internal capacitance between them, and the desired DC resistance of 500 Ω. This is a contrived example and bears no similarity to any known probe, with respect to the magnitude of the values in the circuit, although the topology of the loading model is typical [35]. This type of probe, with its relatively high DC resistance, would be meant to pick off a voltage, as opposed to a probe that is meant to terminate a line in 50 Ω. A note of caution when doing this in practice is to be aware of the internal probe compensation methodology. Most high-speed active probes designed today have some form of digital compensation applied as digital signal processing (DSP) within the oscilloscope. This means that the probe is calibrated as a system and it might not be possible to understand the probe performance from a circuit model alone. One must understand the system in which the probe is measured for compensation and whether the compensation is intended to remove the probe loading effects or to provide accurate measurements of the loaded circuit, with the understanding that probe loading will, to some extent, affect the measurement [36].

Finally, it is assumed here that the transmitter and the receiver are perfect 50 Ω terminations. If this assumption is not valid, one could replace the 50 Ω resistors with two-port networks from measurements.

In the example, there is a single measurement $V_{probe}$ along with four additional output waveforms: $V_{probe}$ represents a waveform acquired through the probe, presumably with a digital storage oscilloscope (DSO). The other four output waveforms are:

- $V_{inloaded}$, the single-ended input voltage present on the positive leg at the output of the transmitter while the probe is attached to the circuit.
- $V_{in}$, the single-ended input voltage present on the positive leg at the output of the transmitter while the probe is not attached to the circuit.
- $V_{outloaded}$, the differential output voltage present at the receiver while the probe is attached to the circuit.
- $V_{out}$, the differential output voltage present at the receiver while the probe is not attached to the circuit.

The construction of the differential probes in the netlist should be pointed out. There is no such circuit element as a differential probe within the *SignalIntegrity* application. Instead, the application automatically creates a voltage controlled voltage source, as provided in Figure 6.11(a), with ports 1 and 2 connected to the intended minus and plus terminals of the circuit, port 3 grounded, port 4 open, and the output probe connected to port 4. This can be seen in the netlist in Figure 11.9, with these connections made for the differential probes at the very end.[3]

Thus, the virtual probing problem has been constructed to produce these four output voltages, given the measured voltage waveform $V_{probe}$. Therefore, the virtual probe problem, given the measured waveform $V_{probe}$, is to answer the following questions:

[3]There is a similar situation for current probes. If current probes are used, a current controlled voltage source is connected in this same manner.

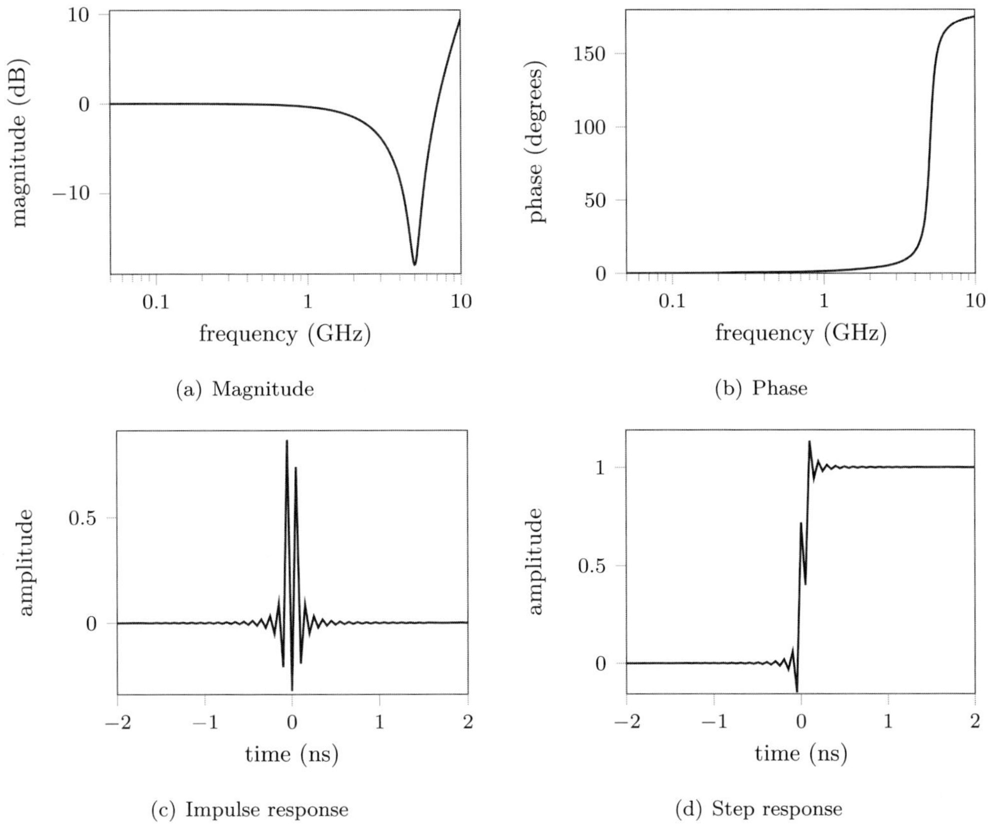


**Figure 11.10** Virtual probe example $V_{in}$ (loaded) due to $V_{probe}$ transfer function

- $V_{inloaded}$ Given the fact that the probe modifies the probed waveform, what does the single-ended waveform actually look like at the transmitter?
- $V_{in}$ Given the fact that the probe not only modifies the probed waveform, but also loads the circuit at the probing point, what would the single-ended waveform actually look like at the transmitter if the probe were not loading the circuit?
- $V_{outloaded}$ Despite the fact that the probe is placed at the transmitter and modifies the probed waveform, what does the actual differential waveform look like at the receiver?
- $V_{out}$ Given the fact that the probe is placed at the transmitter, modifies the probed waveform, and loads the circuit, thus affecting the measurement, what would the differential waveform actually look like at the receiver if the probe were not loading the circuit?

The $V_{in}$ and $V_{out}$ waveforms are the usual, real end goal in analysis (although $V_{inloaded}$ and $V_{outloaded}$ are useful in debugging situations when the circuit changes behavior when

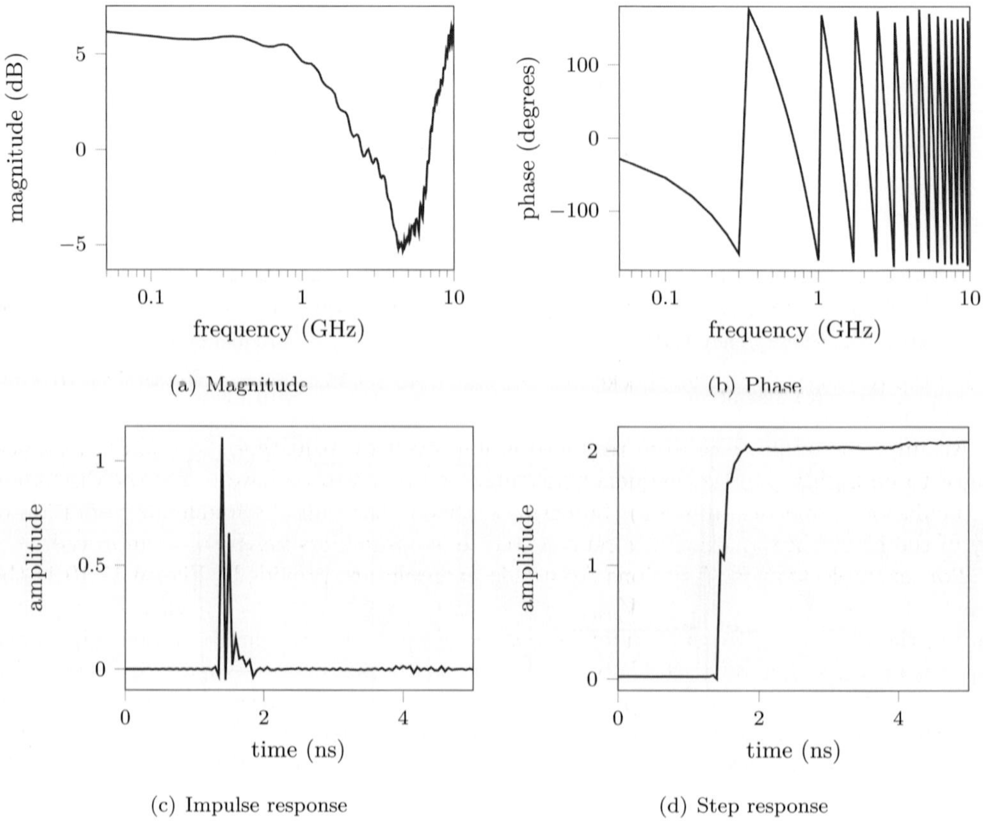


**Figure 11.11** Virtual probe example $V_{out}$ due to $V_{probe}$ transfer function

probed). Note that if one really could develop good models, one does not even need to physically have the channel circuitry to develop $V_{in}$ and $V_{out}$. To do this, the channel is removed and measurements are made of the terminated waveforms at the transmitter.

In this manner, transmitter compliance testing can be performed simply using models of the channel, or measurements of a golden channel. This makes these compliance measurements less cumbersome, cheaper, and more repeatable from test setup to test setup.

As mentioned earlier, the first step in solving a virtual probing problem is the generation of the transfer matrices. Remember, the transfer matrices are lists containing one matrix per frequency, where each (not necessarily square) matrix element is the frequency response of a filter at a given frequency point. Therefore, if each element of a given row and column is extracted for each frequency and placed in a vector, the filter frequency response is obtained. This can and ought to be viewed in both the frequency and time domains for the following characteristics:

1. Magnitude response should be examined for boost. Boost in response means boost in noise.
2. Impulse response should be examined for die down on each side of the response and for a proper time relationship (i.e. for time aliasing, as discussed in §12.2.1). Note that the proper time relationship may be non-causal. Die down of the impulse response (which can be directly observed) without time aliasing (which cannot be proven simply by inspection) satisfies the assumption stated above of the measured waveform existing for all time. This is because the impulse response length will be removed from the processed waveforms. In oscilloscope measurements using virtual probing (and any filtering), it is customary to acquire more samples to account for the waveform loss due to filtering.
3. Step response should be checked to see that it starts from zero and stabilizes to some final value.

An improper impulse or step response is improved by adjusting the impulse response length to be higher (i.e. the frequency spacing is adjusted to be lower – see the discussion of calculation properties in §18.3). Sometimes, this means that the frequency resolution of any of the blocks in the schematic represented by s-parameters needs to be improved.

Two example transfer functions from this example are provided. Figure 11.10 is the transfer function for $V_{in}$ due to $V_{probe}$ (i.e. it converts $V_{probe}$ into $V_{in}$) when the probe is loading the circuit. The magnitude response in Figure 11.10(a) shows a large dip in the response at 5 GHz and a boost of about 10 dB at 10 GHz. This can boost noise, as alluded to earlier. If the noise boost is problematic, and/or the useful frequency of the probe is lower than 10 GHz, this filter should be low-pass filtered. Here, the 10 dB boost is not too bad and is therefore ignored. The phase response shown in Figure 11.10(b) tends upwards, indicating non-causal behavior. This is seen also in Figure 11.10(c), where the impulse response has been zoomed to ±2 ns (the total impulse response, because of the frequency spacing of 50 MHz to 10 GHz, is ±10 ns). Here a non-causality of about 75 ps is seen. As stated earlier, non-causality is not an issue because $V_{probe}$ arrives later than $V_{in}$ because of the length of the probe.[4] One can see that the impulse response in Figure 11.10(c) is well settled to zero on each side of the main response. The step response shown in Figure 11.10(d) is also zoomed to ±2 ns and is also seen to be well settled.

Figure 11.11 is the transfer function for $V_{out}$ due to $V_{probe}$ (i.e. it converts $V_{probe}$ into $V_{out}$). This transfer function allows the waveform at the receiver with the probe detached from, and not loading, the circuit to be generated from $V_{probe}$. The magnitude response in Figure 11.11(a) shows an overall 6 dB boost, due to the conversion of a single-ended probe measurement into a differential signal, but, aside from that and the dip at 5 GHz, no overall additional boost is provided. The 6 dB overall boost does not change the signal-to-noise ratio (SNR) of the measurement. Often, the goal in a virtual probing measurement is to arrive at only $V_{out}$ and one can see that the boost previously seen in the conversion of $V_{probe}$ to $V_{in}$ is compensated by the loss in the channel. This means that by actually probing at the transmitter, one obtains the waveform at the receiver without any boost, again aside from the 6 dB already explained. And, if the probe had higher bandwidth and/or the channel was

[4] The length of the cable in the probe is not even considered here because, in this configuration, it would only cause more delay to deal with anyway. It is, however, common to include this cable length.

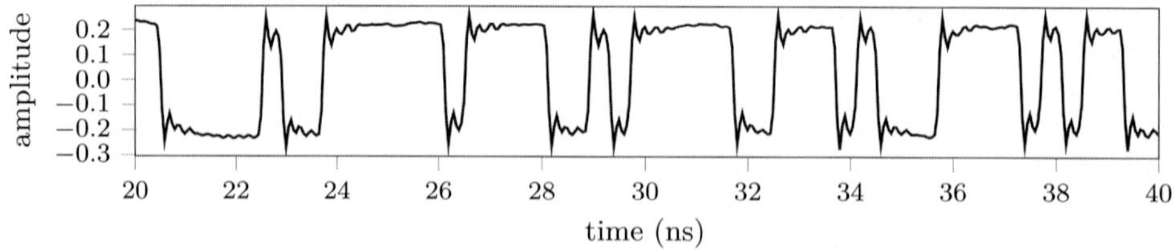


(a) $V_{probe}$

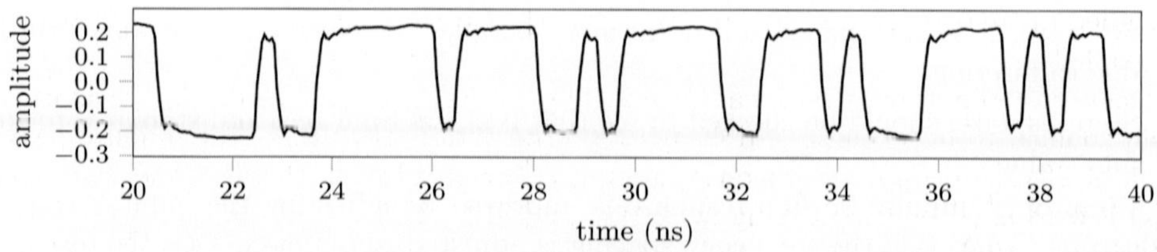


(b) $V_{in}$ (loaded)

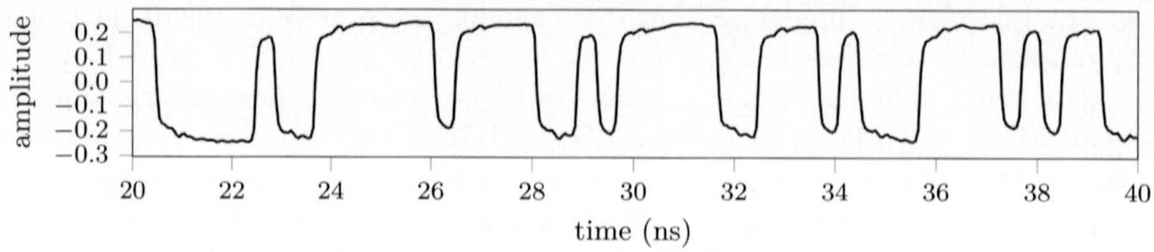


(c) $V_{in}$ (unloaded)

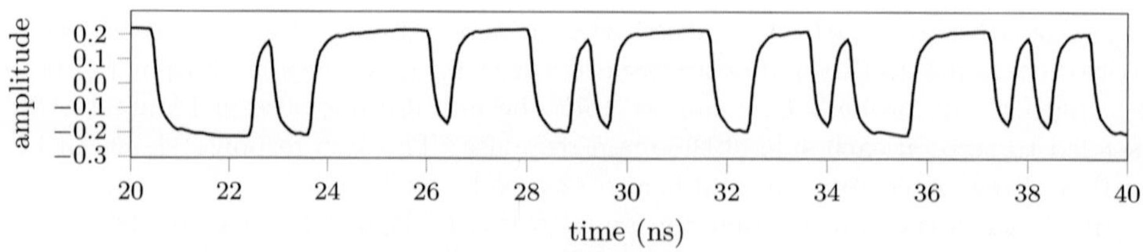


(d) $V_{out}$ (loaded)

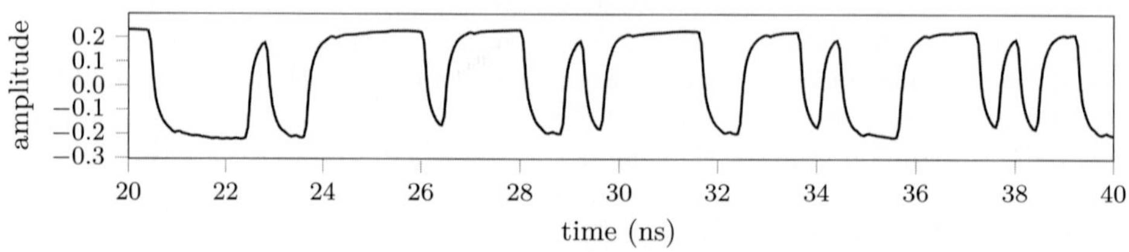


(e) $V_{out}$ (unloaded)

**Figure 11.12** Virtual probe example waveforms

lossier, Figure 11.11(a) would show loss at 10 GHz, mitigating any noise amplification that might occur through equalizer emulation. The phase response shown in Figure 11.11(b) doesn't show much useful information, aside from the fact that the transfer function delays the waveform. This is observed in Figure 11.11(c), and seen even better in Figure 11.11(d), where both the impulse and step response are zoomed to between 0 and 5 ns. The delay is seen to be about 1.4 ns. (The delay is the difference between the delay of the channel and the delay through the probe.) This delay has been removed in the differential probes shown in Figure 11.8, where a delay of −1.43 ns has been applied to the two differential probes measuring at the receiver. This causes the waveform comparisons in Figure 11.12 to be aligned in time at the receiver and transmitter. Both the impulse and step response are well settled.

The acquired and processed waveforms are provided in Figure 11.12. All waveforms are shown zoomed to between 20 and 40 ns to see the detail. Because of the impulse response length of 20 ns (an impulse response ranging between ±10 ns), the first and last 10 ns of $V_{probe}$ is consumed by the filtering operation (see §13.1.3).

Figure 11.12(a) shows $V_{probe}$ to be exhibiting some overshoot and ringing. This is expected based on Figure 11.10(a), which is similar to the inverse of the probe response. Figure 11.12(b) depicts $V_{inloaded}$, the processed waveform showing the voltage waveform at the transmitter with the probe loading the circuit. As expected, the ringing and overshoot have been removed and a cleaner waveform is provided. Figure 11.12(c) shows $V_{in}$, which is the input waveform at the transmitter without probe loading. There are small differences between the loaded and unloaded waveforms at the input, and usually the unloaded waveform is the desired waveform. It is useful during measurement to compare $V_{in}$ to $V_{inloaded}$, just to see the probe loading effects, which are both probe and signal dependent.

$V_{outloaded}$ and $V_{out}$ are shown in Figure 11.12(d) and Figure 11.12(e). Without overlaying them, the waveform at the receiver appears to be very similar, regardless of whether the probe is loading the circuit or not.

# Part III

## Signal Processing and Measurement

### Introduction

THE MICROWAVE TREATMENT of s-parameters deals only with the frequency domain. In signal integrity, the real action occurs in the time domain, and the engineer must deal with the interaction between sampled, time-domain waveforms and sampled, frequency-domain s-parameters. In Chapter 9 and Chapter 11, where these interactions appeared, the simulation and virtual probing solutions ended with the determination of a filter (or matrix of filters). The first two chapters of this part are related to the interaction of the time domain and the frequency domain. Chapter 12 discusses the interpretation of discrete-frequency s-parameters and discrete-time waveforms. Chapter 13 discusses how to process waveforms with filters to complete the simulation and virtual probing solutions.

The three remaining chapters in Part III deal with three important topics in signal integrity. Chapter 14 describes the impedance profile, which is essentially the time-domain view of return loss and is an important tool. Chapter 15 covers the topic of measurement of s-parameters and the calibration of s-parameter measurement instruments such as the VNA and TDR. Finally, in Chapter 16, methods are provided for fitting models to s-parameter data.

# 12

# Frequency Responses, Impulse Responses, and Convolution

THE PRIMARY DOMAIN FOR SIGNAL INTEGRITY is the time domain, and the analysis is primarily of waveforms. Waveforms in a system are assumed to be either an input or source waveform or an output waveform with some *effect* applied to the input waveform by the system.

Effects are most often measured or described as frequency-domain behavior, and the input and output waveforms are most often described or acquired in the time domain. In signal integrity, the most common instrument for measuring the effect of a system is the vector network analyzer (VNA), which directly measures s-parameters. The most common instrument for acquiring time-domain waveforms is the oscilloscope. Both the oscilloscope and the VNA create sampled measurements. Thus, the oscilloscope is utilized to acquire both input and output sampled, time-domain waveforms in a system and the VNA is utilized to measure the sampled, frequency-domain effect of a system.

In simulation, a common tool for measuring the effect of a system in signal integrity is the electromagnetic field solver. Field solvers also provide the s-parameters of a system. Simulation tools for providing time-domain waveforms abound, with the most common being SPICE and its derivatives. SPICE is a *transient* simulator that handles nonlinearities. These occur in transistors, field-effect transistors (FETs), and other active devices in circuits, or in other similarly constructed devices, such as diodes. Transient simulations performed properly are the most accurate simulations, but their computational requirements are generally large. This book is concerned instead with *linear* simulations. Linear simulations can be utilized when the system consists entirely of linear electromagnetic passive elements such as transmission lines, coaxial cables, printed circuit board traces, etc., or with devices that are sufficiently linear.

In a linear system, the input, output, and effects rely on the mathematics of convolution. The equation that describes convolution is

$$y(t) = (x * h)(t) = \int_{\lambda=-\infty}^{\infty} x(\lambda) \cdot h(t-\lambda) \cdot \mathrm{d}\lambda, \tag{12.1}$$

where $x(t)$ is a continuous-time input waveform, $h(t)$ is a continuous-time *response,* and $y(t)$ is a continuous-time output waveform. Note that $h(t)$ is referred to as the impulse response of the system and is the waveform produced by applying a unit impulse to a system.

As stated, the effect of a system is most often described, measured, and simulated in the frequency domain, therefore there is generally at least one conversion required between the frequency domain and the time domain. The most common transformations that convert between time and frequency domains are the Laplace transform, the Fourier transform, and the z-transform. Each of these transforms relates to various ways of looking at systems.

The Laplace transform relies on the fact that, in a linear system, an input waveform can be decomposed into a train of impulses and an output can be formed through convolution with each impulse and summed at the end. Alternatively, it can be viewed as the decomposition of the response into a train of impulses and the convolution of each of these impulses with the continuous-time signal, again with summation of the results after convolution. Finally, both the input signal and the response can be decomposed into individual impulses. The properties of superposition and time invariance are relied upon.

The Fourier transform relies on a different decomposition: the fact that an input signal and a response can be decomposed into an infinite set of exponential waves (or cosine waves). In the Fourier transform, the wave response at each frequency is applied to each wave in the input at that same frequency and the resulting waves are summed at the end to form the output waveform. In linear systems, the only effect that a system can have on an input wave at a given frequency is to modify the amplitude and/or phase of the wave at that frequency. Thus, the frequency response of a system defines the amplitude and phase effects in both the magnitude response and the phase response of a system.

The z-transform is the discrete time and frequency version of these transformations.

The main reason for these transformations is the fact that multiplication in one domain of the transformation is the same as convolution in the other domain. Thus, the convolution integral with regard to the Fourier transform can be written as follows:[1]

$$y(t) = (x * h)(t) = \mathcal{F}^{-1}\{\mathcal{F}[x(t)] \cdot \mathcal{F}[h(t)]\},$$

where $X(f) = \mathcal{F}[x(t)]$ is the frequency content of $x(t)$ and $H(f) = \mathcal{F}[h(t)]$ is the frequency response of the system. If $Y(f) = \mathcal{F}[y(t)]$ is the frequency content of the output waveform $y(t)$, then

$$Y(f) = X(f) \cdot H(f).$$

Generally (but not always) one considers the effect of a system as a frequency response and the input and output of the system as time-domain waveforms.

## 12.1 Discrete-Time Waveforms

The most common instruments used in the analysis of systems in signal integrity are the VNA and the oscilloscope. The VNA produces frequency responses of a system in the form of s-parameters at individual *discrete* frequencies. These discrete-frequency responses are both sampled and limited; limited in the sense that they start and end at some frequency and sampled in the sense that between the beginning and the end frequency a sequence of frequencies is chosen. The oscilloscope (at least the modern digital oscilloscope) produces waveforms at discrete times. Again, these discrete-time time-domain waveforms are both sampled and limited; limited in the time extent of the waveform, meaning they begin and

[1] In this book, uppercase letters are used to express the frequency domain and lower case letters to express the time domain.

end at some time, and sampled in the sense that, during the acquisition time window the sequence consists of samples taken at discrete times. The sampling and limiting in both the VNA and oscilloscope results force us to use discrete-time and discrete-frequency behavior, while always trying to approximate continuous-time and continuous-frequency behavior. Despite the fact that instruments are used that measure limited and discrete frequency and time responses and content, one is always hoping to achieve an understanding of the continuous time and frequency nature of the system. Therefore, it is important to understand how the approximations involving discreteness affect the understanding of the continuous time and frequency.

When talking about input and output discrete-time waveforms, the input waveform can be described as

$$x[k] = x(t)\big|_{t=k\cdot T}\,,$$

for which the following shorthand notation is used:

$$x[k] = x(k \cdot T)\,.$$

The output waveform can be described as

$$y[k] = y(k \cdot T)\,.$$

When the input and output waveforms in the system have been acquired by an oscilloscope, the output waveform is truly a sampled version of the continuous-time output waveform with the exception that the oscilloscope has some response of its own. In other words,

$$y[k] = \{[x(t) * h(t)] * h_o(t)\}(k \cdot T)\,.$$

Despite the presence of a discrete version of the input waveform $x[k]$, the discrete version of the output waveform $y[k]$ is a sampled version of the convolution of the continuous-time responses of both the input and the system. This is desirable as one only needs to deal with the sampled nature of the time-domain waveforms.

Assuming that the response of the oscilloscope can be approximated only by its bandwidth limiting effect on the frequency response, the relationship between the ideal continuous-time waveform and the sampled waveform can be described as follows:

$$\hat{y}(t) = y(t) * \mathrm{sinc}(t \cdot 2 \cdot F_{bw}) * \Delta_{KT}\,(t) \cdot \Delta_T\,(t)\,.$$

The way to understand this equation is by using Fourier transforms. The first operator on $y(t)$ is the sinc function, which is defined by the follwing Fourier transform pair:

$$\Pi\,(f \cdot T) = \mathrm{rect}(f \cdot T) = \begin{cases} 0 & \text{if } |f \cdot T| > \frac{1}{2}, \\ \frac{1}{2} & \text{if } |f \cdot T| = \frac{1}{2}, \\ 1 & \text{if } |f \cdot T| < \frac{1}{2}; \end{cases} \tag{12.2}$$

$$\mathcal{F}^{-1}\{\Pi\,(f \cdot T)\} = \mathrm{sinc}\left(\frac{t}{T}\right) = \frac{\sin\left(\pi \cdot \frac{t}{T}\right)}{\pi \cdot \frac{t}{T}}.$$

Thus, the following relationship holds:

$$y(t) * \mathrm{sinc}(t \cdot 2 \cdot F_{bw}) = \mathcal{F}^{-1}\left\{Y(f) \cdot \Pi\left(\frac{f}{2 \cdot F_{bw}}\right)\right\}. \tag{12.3}$$

This means that convolution with the sinc function is equivalent to bandwidth limiting the signal to $F_{bw}$. This is one known effect of the oscilloscope. Although the sinc function dies down far from $t = 0$, it has infinite length. This is an effect of strict band limiting. The convolved signal therefore has infinite length, even if $y(t)$ is of limited duration.

The second operation is that of convolution with the sampling function $\Delta_{KT}(t)$. The sampling function is defined such that

$$\Delta_T(t) = \sum_{k=-\infty}^{+\infty} \delta(t - kT) = \mathcal{F}^{-1}\left\{\Delta_{\frac{1}{T}}(f)\right\}, \tag{12.4}$$

$$\Delta_{\frac{1}{T}}(f) = \frac{1}{T} \cdot \sum_{n=-\infty}^{+\infty} \delta\left(f - \frac{n}{T}\right) = \mathcal{F}\{\Delta_T(t)\}. \tag{12.5}$$

In (12.4) and (12.5), multiplication by the sampling function in one domain (i.e. sampling the signal) is the same as convolution with a similar function in the other domain, which causes the signal to repeat. This is a very confusing situation because it states that the bandwidth limited waveform is expected to repeat with a period of $K \cdot T$, where $K$ is the number of sample points in the waveform and $T$ is the sample period. This is where one might argue that, in fact, the boxcar function $\Pi([t - 1/2] \cdot K \cdot T)$ should be applied to limit the acquired waveform to a duration of time equal to $K \cdot T$ (i.e. to make it zero outside this extent). Boxcar limiting is not a realistic choice because of the desire to use the discrete Fourier transform. Actually, repetitiveness of the waveform is assumed through convolution with $\Delta_{KT}(t)$. Incidentally, convolution with the sampling function in the time domain causes the frequency content of the resulting signal to become discrete in the frequency domain; $K$ can be taken to be arbitrarily large, so causing the frequency content to become, in the limit, continuous.

Finally, the result is multiplied by the sampling function $\Delta_T(t)$, which samples the waveform, making the result band limited, repetitive, and discrete time. From the previous discussion, this means that the resulting frequency content now repeats in frequency.

These three operations on the continuous-time waveform cause serious changes that affect the usefulness of the resulting discrete-time waveform. To summarize, these effects are:

1. bandwidth limiting the waveform;
2. assuming the waveform repeats outside the extent of the waveform points in the sequence;
3. sampling the waveform.

### 12.1.1 Frequency Aliasing

Sampling of a signal can introduce a deleterious effect called frequency aliasing, which is related to bandwidth limiting. Most readers will be familiar with this concept, but how it appears mathematically will be highlighted. First, bandwidth limiting causes the frequency content to be limited to $\pm F_{bw}$. All real-valued signals come with positive and negative frequency content. This means that the useful frequency content is limited to positive frequencies between zero and $F_{bw}$. Second, sampling causes the resulting frequency content to repeat with a period of $1/T$. More important than the repetition, however, is that the sampling function indicates that the result is the sum of frequency images that are $2 \cdot F_{bw}$ wide (containing the frequency content $Y(F)$ restricted to $\pm F_{bw}$), where each image is centered at multiples of $1/T$. Thus, if $F_{bw} < 1/(2 \cdot T)$, these images don't overlap each other and the result simply yields repeating images of the frequency content $Y(F)$ limited to bandwidth $F_{bw}$. If, however, $F_{bw} > 1/(2 \cdot T)$, these images do overlap and the resulting frequency content is not repeating images of the frequency content $Y(F)$ limited to bandwidth $F_{bw}$ (even though it still repeats with period $1/T$). This effect is called aliasing because, when examining the frequency content, one finds that some of the original content appears at another location in the spectrum as if it actually belonged there. Once aliasing has occurred, it cannot be undone if two frequencies that were in the original frequency spectrum end up summed. In other words, only the sum is obtained, and it is impossible to figure out which frequency or combinations of frequencies it was. To avoid aliasing, a restriction is imposed that the sample rate of the system must be higher than twice the frequency content of the signal, or, said differently, half the sample rate of the system must be higher than the frequency content in the signal. Half of the sample rate is called the Nyquist rate (the name is associated with Harry Nyquist, the famous Bell Labs engineer). Thus, it is bad enough that the bandwidth limiting effect may have an effect on the discrete-time approximation if this results in truncation of frequency content, but in attempts to utilize higher bandwidth, the sample rate requirements are driven higher as well.

### 12.1.2 Periodicity and the Discrete-Frequency Nature

The interrelationship between bandwidth and sample rate were examined in §12.1.1. If one were able to acquire a signal with enough bandwidth to contain all of the frequency content of the signal and with a high enough sample rate so that no aliasing occurred, the remaining issue is only the assumption of periodicity of the waveform. Again, this means that the $K$ sample points that occur between times zero and $K \cdot T$ are assumed to repeat with a period of $K \cdot T$. This assumption involves convolution with the sampling function in the time domain, which means sampling of the frequency content of the signal in the frequency domain. This means that this assumed periodic signal is described by discrete-frequency content. This should not be surprising if one considers the Fourier series. The Fourier series contains coefficients applied to discrete, harmonically related frequencies (i.e. all frequencies occur at multiples of the first, fundamental, frequency), which contain all of the frequency content that make up a periodic signal. It has already been seen that the sampling of the waveform causes the frequency content to become periodic, but this frequency content is repeating discrete content already contained in the discrete-frequency points between $\pm 1/(2 \cdot T)$. If the periodicity is ignored and the frequency content at these discrete frequencies limited to $\pm 1/(2 \cdot T)$ are summed, one finds that, by summing a finite number of discrete-frequency cosine waves, a continuous version of the discrete-time waveform is obtained. Even though the number of cosine waves being summed is finite and discrete, each cosine wave is continuous, and therefore their sums are continuous. Furthermore, this frequency content can be found directly from the time-limited, discrete-time waveform, meaning that having only these sampled values allows for full reconstruction of the continuous-time waveform.

### 12.1.3 The Relationship Between Discrete-Time and Continuous-Time Waveforms

The $K$ points acquired by sampling a continuous-time waveform, sample period $T$, provide all of the information needed to construct fully the continuous-time waveform if:

1. the frequency content of the continuous-time signal is limited to $1/(2 \cdot T)$;
2. the continuous-time signal is periodic with period $K \cdot T$.

If these conditions are met, then the frequency content is fully described by $K$ elements of discrete-frequency content. Items 1 and 2 are interrelated. A strictly band limited waveform must have existed for all time. Periodic signals could have an infinite number of discrete, harmonically related frequencies, but some periodic signals can be constructed from a finite number of such frequencies. After all, a cosine wave is one such signal, and the sums of a finite number of cosine waves are still bandwidth limited as long as their frequencies fall within the stated bandwidth limit. The periodicity requirement, therefore, not only discretizes the frequency content, but also bandwidth limits the content. A non-periodic signal that is band limited requires an infinite amount of time to describe. This is understood by realizing that the sinc function goes on forever, so, even if the time extent of a waveform is finite, convolution with the sinc function causes it to become of infinite length.

In signal integrity applications, these assumptions of periodicity are overcome in many ways:

1. by analyzing truly periodic signals (many communications signals can be made to be periodic);
2. by not analyzing the frequency content of signals;
3. by approximating the truncation of the frequency content to the discrete frequencies between $-1/(2 \cdot T)$ and $+1/(2 \cdot T)$;
4. by analyzing signals that can be assumed to be zero outside the acquisition window.

Item 1 goes without saying, and allows for the use of circular convolution techniques.

Item 2 needs some explaining, but is an important point insofar as oscilloscope waveforms are concerned. In understanding the effect of acquisition time limiting and sampling, there is general comfort with the frequency content limiting but discomfort with the periodicity requirement. However, the periodicity requirement is needed only when examining the effect on the frequency content, which is needed to reconstruct the continuous-time waveform perfectly. Let's say for the moment that the examination of the frequency content can be forestalled and that, in the reconstruction of the continuous-time waveform, it might never be considered. After all, in many cases in signal integrity the acquired oscilloscope waveform, assuming the bandwidth limitations and sample rate are not impactive, can be assumed to be exact sample points on a time-domain waveform. There is no need necessarily to look in between the sample points – they are used as is.

Assuming that the choices made in items 1 and 2 are unpalatable, item 3 remains. Item 3 states that the truncation of the frequency content should be approximated. The word "approximate" is used because, if the frequency content were truly truncated, the waveform is periodic, and therefore any attempt to look between the discrete-time points at the continuous-time content by summing cosine waves results in a periodic waveform. The effects of periodicity are most prominent at the beginning and end of the waveform. This can be understood by simply appending the repeating sequences to each other endlessly and examining the sequence. If the sequence is highly discontinuous at the seams between the end of the waveform (the end of the period) and the beginning of the waveform (the beginning of the period), then the continuous-time waveform must make large gyrations to make the waveform periodic, and these gyrations will occur mostly at the seam.

It was seen in (12.3), however, that truncation of the frequency response is equivalent to convolution with the sinc function. While the sinc function goes on forever, the sinc function itself can be truncated. The Fourier transform of the truncated sinc function therefore becomes an approximation of the strict bandwidth limiting. Its finite length, however, means that this truncated sinc function can be convolved with the sampled waveform, and, since the sinc function is continuous, the convolved waveform becomes continuous. The longer the sinc function, the closer it is an approximation to a hard bandwidth limit, and, because of the finite length of the truncated sinc function, the effect on the convolution is limited to a finite number of points in the discrete-time waveform. After convolution is performed, one usually throws away the beginning and end points anyway such that the remaining points in the discrete waveform are unaffected. This discussion of convolution with the sinc function comes under the category of interpolation, and there are many ways to perform interpolation. For example, if the sample rate is such that the Nyquist rate is much higher than the frequency content of a signal, then the underlying continuous-time waveform can be approximated simply by drawing lines between the sample points in the discrete-time waveform.

A final note on item 3 is that one should not assume that the approximation employed can solve any aliasing problem. The continuous-time waveform that is sampled must contain frequency content below $1/(2 \cdot T)$ prior to sampling. Once sampled, item 3 approximates the truncation of the content, but the content already includes the aliasing, which will cause errors in the relationship between the continuous-time and discrete-time waveforms.

Item 4 can be used in conjunction with item 3. Certain types of waveforms, such as impulse responses, can assume that the waveform is zero outside the acquisition window, if the acquisition window contains a waveform that has died down sufficiently. This means that the time of application of a signal prior to the acquisition window is so far in the past that the waveform has reached zero, and that both the response and the signal applied are sufficiently short with respect to the acquisition window. This allows convolution with a truncated sinc function that is arbitrarily long, where the waveform is extended by adding zeros so that it encompasses the truncated sinc function length. In this way, convolution can be performed without discarding any points from the waveform. Item 4 is important when working with the response of a system.

## 12.2 Discrete-Frequency Responses

The analysis of discrete-frequency responses arises when systems are measured and described by s-parameters. While the approximations made in forming time-domain waveforms are generally well understood, discrete-frequency responses, even though they are completely analogous, are less well understood.

Within certain limitations, discrete-time waveforms contain information that allows the determination of the continuous-time waveform. Similarly, discrete-frequency responses, like s-parameters, describe continuous-frequency behavior.

When talking about discrete-frequency responses, the response can be described as

$$H[n] = H(f)|_{f=\frac{n}{2 \cdot T}} ,$$

where the following shorthand notation is used:

$$H[n] = H\left(\frac{n}{2 \cdot T}\right) .$$

When the response has been generated in the form of s-parameters, the measurement is a sampled version of the continuous-frequency response. In other words,

$$\hat{H}(f) = H(f) \cdot \Pi(f \cdot T) \cdot \Delta_{\frac{1}{KT}}(f) * \Delta_{\frac{1}{T}}(f) . \tag{12.6}$$

This equation is understood by using Fourier transforms. The first operator on $H(f)$ is the boxcar function, which is defined by the Fourier transform pair in (12.2). The boxcar function limits the extent of the frequency measurement to $\pm 1/(2 \cdot T)$. The next step is to sample the frequency response with the sampling function $\Delta_{\frac{1}{KT}}(f)$, which is defined by the Fourier transform pair in (12.4) and (12.5). The final step is to convolve the discrete-frequency response with the sampling function $\Delta_{\frac{1}{T}}(f)$. Convolution with the sampling function causes the response to repeat over and over again with a spacing of $1/T = Fs$. The frequency-domain implications of (12.6) are shown graphically in Figure 12.1.

In the examination of the relationship between discrete-time and continuous-time waveforms, it was found that the frequency-domain implications are the most important. Conversely, in examining the relationship between discrete-frequency and continuous-frequency responses, it is the time-domain implications that are most important. The time-domain implications of (12.6) are shown graphically in Figure 12.2.

It can be seen that the three operations on the continuous-frequency waveform cause serious changes that affect the usefulness of the resulting discrete-frequency response. These effects are:

1. bandwidth limiting the response;
2. sampling the frequency response;
3. repeating the frequency response points in the sequence.

### 12.2.1 Time Aliasing

Sampling the frequencies of a frequency response can introduce errors in the form of *time aliasing*. Most readers will be familiar with the frequency aliasing effect of time sampling,

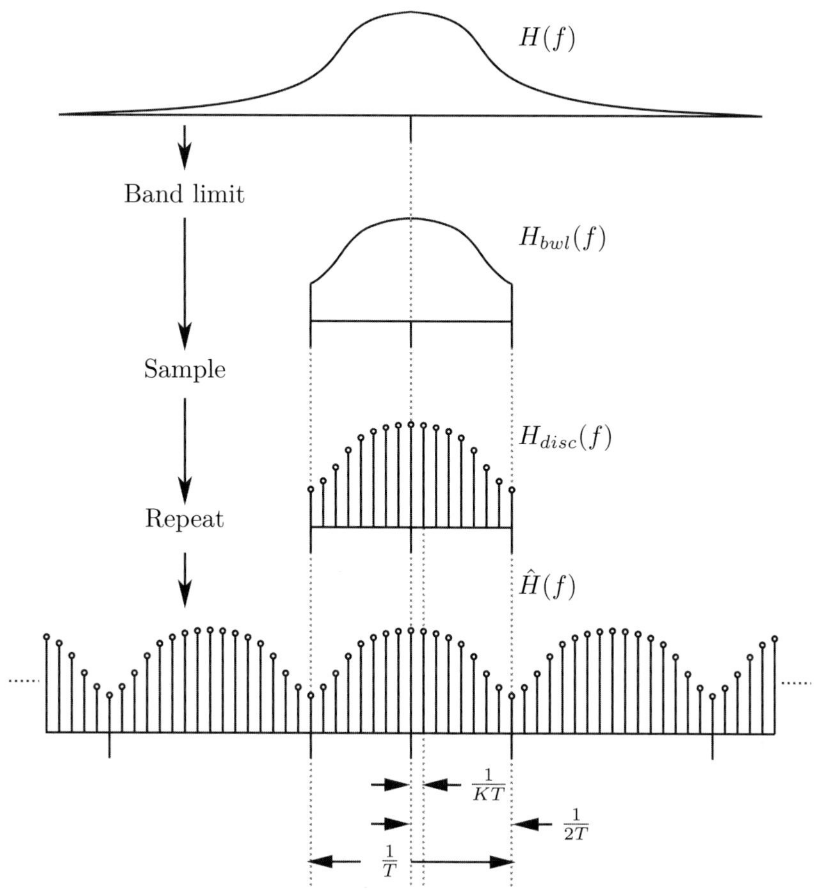


**Figure 12.1** Continuous- and discrete-frequency response frequency-domain relationship

but will be less familiar with the analogous frequency-domain effect. The sampling of frequencies causes the resulting time-domain content to repeat with a period $K \cdot T$. More importantly than the repetition, however, is that the sampling function indicates that the result is the sum of response images that are possibly infinitely long. Thus, sampling turns the bandwidth limited impulse response into what is more accurately called the *impulse train response*. If the response dies down sufficiently in between each pulse in the train separated by $K \cdot T$, then these responses do not overlap each other and the result simply yields repeating images of the impulse response. If, however, the response time is larger than $K \cdot T$, then these responses do overlap, and the resulting time-domain response is not repeating images of the time-domain impulse response (even though it still repeats with period $K \cdot T$), as illustrated in Figure 12.3. This effect is called time aliasing because some elements of the response occur at times other than when they actually occurred. This can

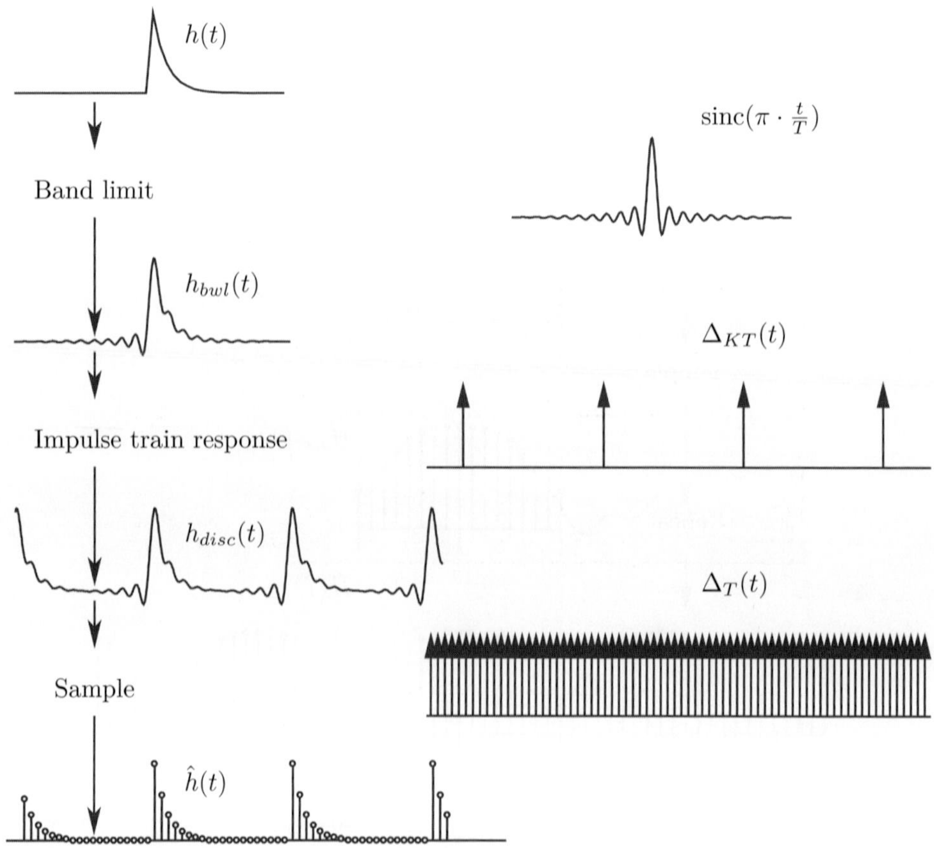


**Figure 12.2** Continuous- and discrete-frequency response time-domain relationship

lead to non-causality issues (i.e. the response occurs before the actual impulse response) and misunderstandings about time relationships. Once time aliasing has occurred, it cannot be undone if two elements of the impulse response at different times end up summed together at a given time. In other words, only the sum is obtained, and one cannot figure out which time it really was. To avoid aliasing, a restriction is imposed that the reciprocal of the frequency spacing must be higher than twice the impulse response length. Twice the length is used assuming that the middle of the impulse response is time zero in the response, so half of the points in the impulse response occur prior to the impulse. This is shown in Figure 12.4.

The frequency spacing is directly dictated by the length of the impulse response. In the time domain, time spacing (the sample period) is directly dictated by the highest frequency (the bandwidth) of the system and the waveform. In the frequency domain, the frequency spacing is directly dictated by the highest time during which the impulse response has not died down to zero.

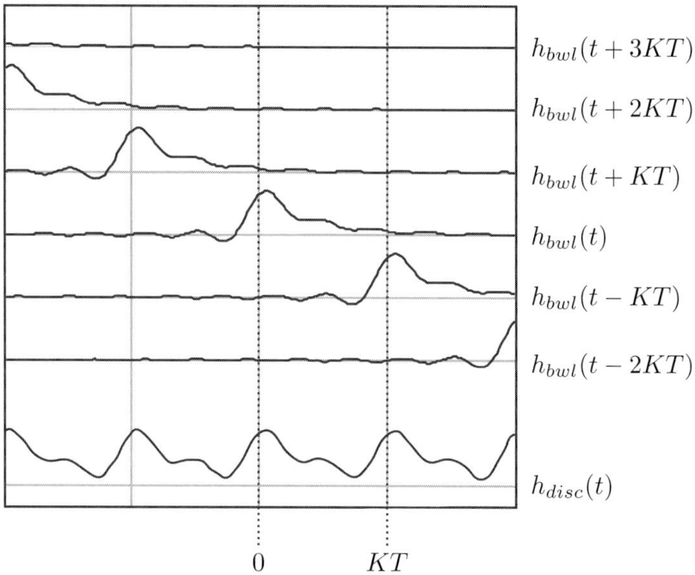


**Figure 12.3** The impulse train response

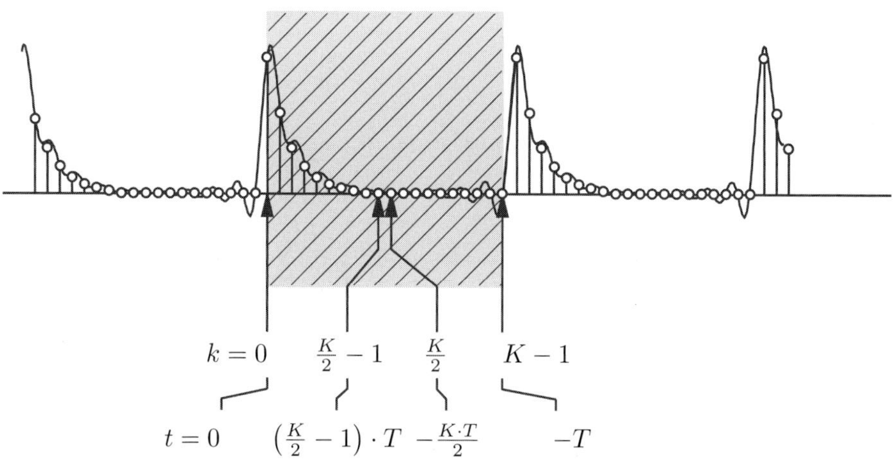


**Figure 12.4** Impulse train response time sense

In the time domain, even though aliasing occurs through improper relationships between the sample rate and signal content, each point sampled is still assumed to be an exact sample of the waveform. This means that the aliasing only affects the analysis of the frequency content and the inter-point behavior of the underlying continuous-time waveform. The analogous situation holds in the frequency domain. Even with improper choice of frequency spacing, the frequency response measured is assumed to be an exact sample of the frequency response. It is only the analysis of the impulse response and the interpolation of inter-point behavior of the frequency response that suffer.

Frequency spacing of s-parameter measurements is extremely important. The lack of understanding of this concept is a leading cause of problems in the usage of s-parameters in time-domain simulations.

### 12.2.2 Periodicity and Discrete-Time Nature

The effect of frequency spacing was discussed in §12.2.1. A frequency response to a high enough frequency to contain all of the frequency response of the system, and measured with fine enough resolution such that no time aliasing occurs, leaves only the assumption of periodicity of the response as a potential issue. Periodicity means that the $K/2$ frequency samples that occur between zero frequency and $K/(2 \cdot T)$ are assumed to repeat with a period of $1/(K \cdot T)$ (with negative frequencies implied). Since convolution with the sampling function in the frequency domain is involved, this is the same as sampling the impulse response of the signal in the time domain. This means that this assumed periodic frequency response is described by a discrete impulse response. This may be surprising, and will be addressed further in discussions of the discrete Fourier transform (see §12.3.6). The Fourier transform properties that pertain to this situation are

$$\mathcal{F}\{\delta\left(t\right)\} = 1$$

and

$$\mathcal{F}\{x\left(t-a\right)\} = \mathcal{F}\{x(t)\} \cdot e^{-j \cdot 2 \cdot \pi \cdot f \cdot a}$$

Thus, one can compute the Fourier transform of the assumed repetitive $K$ point impulse response sequence $h\left[k\right]$:

$$H(f) = \mathcal{F}\{\mathbf{h}\} = \mathcal{F}\left\{\sum_{k=0}^{K-1} h\left[k\right] \cdot \delta(t - k \cdot T)\right\} = \sum_{k=0}^{K-1} h[k] \cdot e^{-j \cdot 2 \cdot \pi \cdot f \cdot k \cdot T}.$$

This result is continuous in frequency and thus describes a continuous-frequency response, but this response is periodic in the frequency domain with a period of $1/(2 \cdot T)$.

## 12.3 The Discrete Fourier Transform

Until this point, the relationships between discrete and continuous time and frequency signals were discussed using Fourier analysis. Now, the discrete Fourier transform, the tool of choice for dealing with discrete times and frequencies, is discussed.

The discrete Fourier transform (DFT) is defined for a sequence of $K$ points $x[k]$, $k \in 0 \dots K-1$, as the z-transform of the sequence evaluated at specific equiangularly spaced points on the unit circle. The definition of the z-transform is

$$X(z) = \sum_{k=-\infty}^{\infty} x[k] \cdot z^{-k},$$

where

$$z = e^{j \cdot 2\pi \cdot f \cdot T}$$

and $T$ is the sample period of the sequence. Since the sequence is limited to $K$ points, this is

$$X(z) = \sum_{k=0}^{K-1} x[k] \cdot z^{-k}.$$

The unit circle is defined as $|z| = 1$, and a point on the unit circle at a given angle $\theta$ is $e^{j \cdot \theta}$. Selecting specific angles $\theta_n = \frac{n}{K} \cdot 2\pi$ for $n \in 0 \dots K-1$ produces

$$X[n] = X\left(z = e^{j \cdot 2\pi \cdot \frac{n}{K}}\right) = \sum_{k=0}^{K-1} x[k] \cdot e^{-j \cdot 2\pi \cdot \frac{k \cdot n}{K}}.$$

The inverse discrete Fourier transform (IDFT) is defined as the inverse of this transform:

$$x[k] = \frac{1}{K} \cdot \sum_{n=0}^{K-1} X[n] \cdot e^{j \cdot 2\pi \cdot \frac{k \cdot n}{K}}.$$

### 12.3.1 The Frequency Response Effect View of the DFT

One way of looking at the DFT is to view it as the frequency-domain *effect* that a given impulse response has on another waveform. In other words, as the discrete sequence $h[k]$ analogous with the continuous-time impulse response. To understand this effect, the unit impulse response defined for a $M$ point sequence, $m \in 0 \dots M-1$, is used:

$$h[m] = \begin{cases} 1 & \text{if } m = 0, \\ 0 & \text{otherwise.} \end{cases}$$

From the definition of the DFT,

$$H[n] = 1.$$

Similarly, a DFT given as $H[n] = 1$ provides the unit impulse sequence as the IDFT. This means that the definition provided scales the DFT to provide the frequency response of a sequence.

### 12.3.2 The Frequency Content View of the DFT

One can also view the DFT as the frequency-domain *content* of a signal. The IDFT formula shows that the DFT provides the weights of $K-1$ complex exponentials, but these weights are scaled by $K$ (hence the $1/K$ in the IDFT). Complex exponentials are confusing with regard to frequency response and it is easier to look at the frequency content in terms of cosine waves.

One can obtain an alternative equation for the IDFT, applicable for real-valued sequences $\mathbf{x}$, that provides an insight into what the DFT is all about:

$$x[k] = \frac{1}{K} \cdot \sum_{n=0}^{K/2} \begin{cases} 2 & \text{if } 0 < n < \frac{K}{2} \\ 1 & \text{otherwise} \end{cases} \cdot |X[n]| \cdot \cos\left(2\pi \cdot \frac{n \cdot k}{K} + \arg\left(X[n]\right)\right).$$

In other words, although the DFT provides $K$ frequency points, it actually provides $K/2+1$ amplitudes and phases for $K/2+1$ cosine waves (assuming $K$ to be even). Within this book, the first choice for $K$ is as an even number, and $N = K/2$ is defined with $n \in 0 \ldots N$. Thus, if $K$ is even, a $K$ element real-valued time-domain sequence produces an $N+1$ element set of weights for cosine waves. The times associated with the time-domain sequence are $t[k] = k \cdot T$, and therefore, examining the cosine function, there are frequencies at $f[n] = (n/N) \cdot 1/(2 \cdot T)$. Defining the sample rate as $Fs = 1/T$, there are equally spaced frequencies at $f[n] = (n/N) \cdot Fs/2$. Furthermore, the weighting of the cosine waves is different for two elements corresponding to $n = 0$ and $n = N$, which correspond to zero frequency and the Nyquist rate at $Fs/2$.

### 12.3.3 DFT Interpretation and Mechanics, and Even and Odd Points

The DFT is a powerful tool for analyzing and manipulating data and is the mechanism for moving between the time and frequency domains. That being said, the details can be confusing, especially when trying to analyze sampled data representative of continuous time and frequency systems, as is often the case in signal integrity.

To help with understanding and to ensure proper mechanics, interpretation, and conversions, Table 12.1 contains formulas to be used when dealing with the DFT. The mechanics are quite particular with regards to whether the time-domain waveform contains an even or odd number of points.

Assuming some familiarity with the DFT, the DFT of a time-domain sequence of $K$ points $\mathbf{x}$ technically yields a $K$ point frequency-domain sequence $\mathbf{X}$. However, nearly half of the frequency-domain points pertain to negative frequencies, which are only mathematically interesting. The $K$ element DFT contains only $N+1 = \lfloor K/2 \rfloor + 1$ points of interest; the points $X_{[n]}$ for $n \in 0 \ldots N$.

Two columns appear in Table 12.1, depending on whether $K$ is even or odd, but the table has been carefully crafted such that all formulas are common to each case (except for variable *Keven*, which might determine some outcome).

When possible, even $K$ is preferred because, in this situation, the last point $Fe = f_{[N]} = Fs/2$, which makes things convenient when working with waveforms; the delta frequency is also often a round number. Otherwise, the last frequency point is a strange value, not the

**Table 12.1** DFT odd and even point handling and conversions[a]

| Value | Even $K$ | Odd $K$ |
|---|---|---|
| Frequency-domain points | $N = \lfloor K/2 \rfloor$ | |
| Time-domain points | $K = 2 \cdot N + \begin{cases} 0 & \text{if } Keven \\ 1 & \text{otherwise} \end{cases}$ | |
| End frequency | $Fe = \frac{N}{K} \cdot Fs = N \cdot \Delta f$ | |
| Sample rate | $Fs = \frac{K}{N} \cdot Fe = K \cdot \Delta f$ | |
| Delta frequency | $\Delta f = Fs/K = Fe/N$ | |
| Frequency $n \in 0 \ldots N$ | $f_{[n]} = n \cdot \Delta f = \frac{n}{K} \cdot Fs = \frac{n}{N} \cdot Fe$ | |
| Half to full | $X_{[N+\sigma]} = X^*_{[N-\sigma]}$ for $\sigma \in 1 \ldots N-1$ | $X_{[N+\sigma]} = X^*_{[N-\sigma+1]}$ for $\sigma \in 1 \ldots N$ |
| Amplitude from DFT | $A_{[n]} = \frac{X_{[n]}}{K} \cdot \begin{cases} 1 & \text{if } n = 0 \\ 1 & \text{if } (n = N) \wedge Keven \\ 2 & \text{otherwise} \end{cases}$ | |
| Phase from DFT | $\theta_{[n]} = \frac{180}{\pi} \cdot \arg\left(X_{[n]}\right)$ | |
| rms[b] from amplitude | $rms_{[n]} = A_{[n]} / \begin{cases} 1 & \text{if } n = 0 \\ 1 & \text{if } (n = N) \wedge Keven \\ \sqrt{2} & \text{otherwise} \end{cases}$ | |
| dBm from rms | $dBm_{[n]} = 20 \cdot \log\left(rms_{[n]}\right) + 13.010$ | |
| Spectral density from rms | $\rho_{rms_{[n]}} = rms_{[n]} / \sqrt{\Delta f} \cdot \begin{cases} \sqrt{2} & \text{if } n = 0 \\ \sqrt{2} & \text{if } (n = N) \wedge Keven \\ 1 & \text{otherwise} \end{cases}$ | |
| Spectral density from dBm | $\rho_{dBm_{[n]}} = dBm_{[n]} - 10 \cdot \log(\Delta f) + \begin{cases} 3 & \text{if } n = 0 \\ 3 & \text{if } (n = N) \wedge Keven \\ 0 & \text{otherwise} \end{cases}$ | |
| Total spectral content in rms | $\sigma = \sqrt{\sum_n rms^2_{[n]}}$ | |
| Total spectral content in dBm | $20 \cdot \log(\sigma) + 13.010 = 10 \cdot \log\left(\sum_n 10^{\frac{dBm_{[n]}}{10}}\right)$ | |

[a] $Keven = \lfloor K/2 \rfloor \cdot 2 == K$.
[b] Root mean squared.

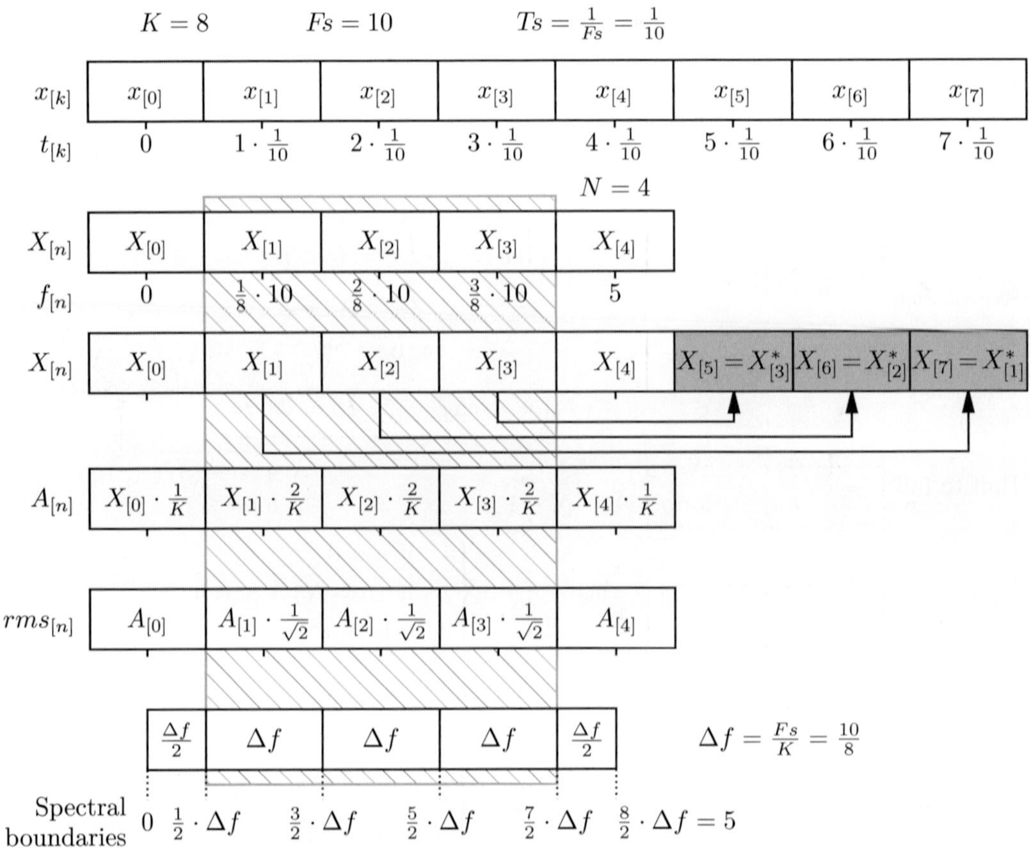

**Figure 12.5** DFT mechanics and interpretation for even $K$

Nyquist rate point, and the frequency spacing is unusual. Of course, whether $K$ is even or odd can only be controlled when presented first with frequency-domain data. Then, one is free to decide whether the $N+1$ frequency points represent a $K = 2 \cdot N$ or $K = 2 \cdot N + 1$ point time sequence.

Figures 12.5 and 12.6 better explain the rules and reasoning behind dealing with even or odd $K$.

In Figure 12.5, a $K = 8$ element waveform $\mathbf{x}$ is provided such that for $k \in 0 \ldots K-1$ there are waveform points $x_{[k]}$ corresponding to a voltage sampled at times $t_{[k]}$. In this example, the sample rate $Fs = 10$. Computing the forward DFT, there are $N + 1 = \lfloor K/2 \rfloor + 1$ frequency points produced, where the frequency of each point corresponds to $f_{[n]}$, for $n \in 0 \ldots N$. Since $K$ is even, the last point $f_{[N]} = Fe = Fs/2$.

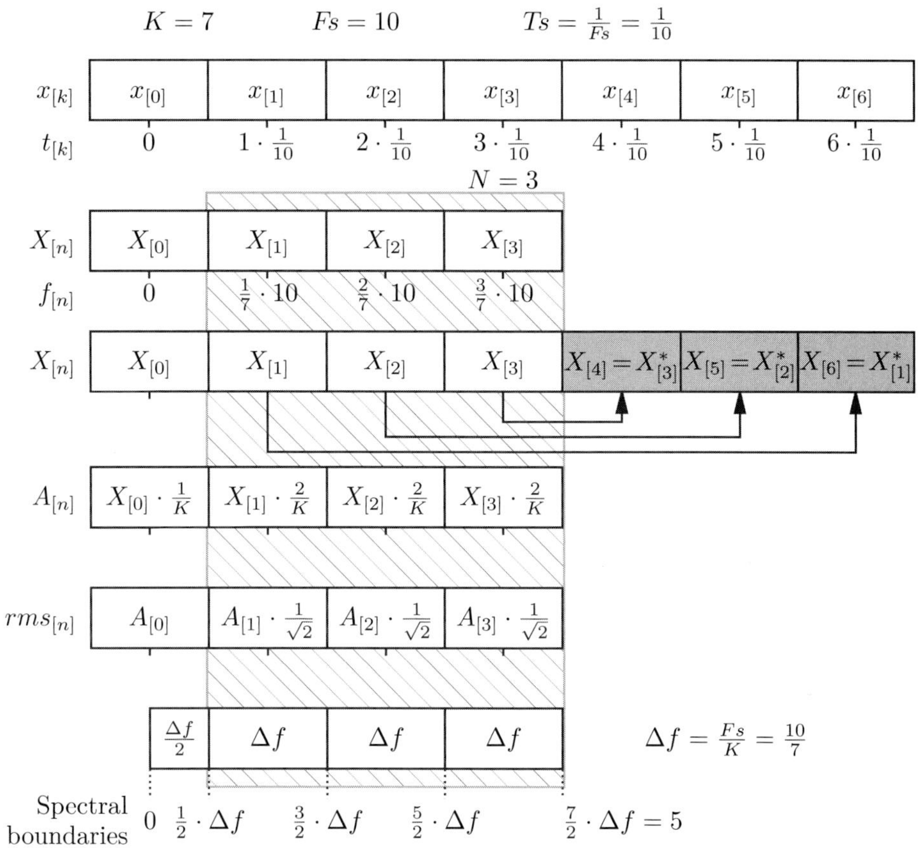


**Figure 12.6** DFT mechanics and interpretation for odd $K$

The middle three points of the DFT are shaded, because these are the points where images in the actual DFT have negative frequencies. Thus, given a frequency content in $\mathbf{X}$, containing values $X_{[n]}$ for $n \in 0 \ldots N$, the actual DFT has the three negative frequencies filled in as shown, making it a $K$ element sequence suitable for conversion back to the time domain with the IDFT. To interpret the frequency content, the complex vector $\mathbf{A}$ is provided, where $\left|A_{[n]}\right|$ contains the amplitude of a cosine wave and $\theta_{[n]} = 180/\pi \cdot \arg\left(A_{[n]}\right)$ contains the phase in degrees. This interpretation dictates that the time-domain waveform can be computed from this vector as follows:

$$x_{[k]} = \sum_{n}^{N} \left|A_{[n]}\right| \cdot \cos\left(2\pi \cdot f_{[n]} \cdot t_{[k]} + \theta_{[n]}\right) . \tag{12.7}$$

Thus, (12.7) is seen as a strict definition of the correspondence between the time and frequency domains. Note the scaling of the DFT values $X_{[n]}$ in the computation of each $A_{[n]}$. The points in the shaded band in Figure 12.5 are scaled by $2/K$ while the points at DC and the Nyquist rate are scaled by $1/K$. This is related to the existence of the negative frequency images and the definition of the cosine using Euler's equation

$$\cos(\theta) = \frac{e^{j\cdot\theta} + e^{-j\cdot\theta}}{2}.$$

Therefore, the images in the DFT are half size and this scaling accounts for that.

The root mean squared (rms) values in each bin are also scaled differently. The rms value of a DC voltage is the voltage itself, and the rms value of a sinusoid is its amplitude divided by the square root of 2. The proper scaling on the Nyquist point is unity. This ensures that rms voltages computed in the time and frequency domains meet the following equality condition:[2]

$$\sqrt{\frac{1}{K}\cdot\sum_{k=0}^{K-1}\left(x_{[k]}\right)^2} = \sigma = \sqrt{\sum_{n=0}^{N}\left(rms_{[n]}\right)^2}. \tag{12.8}$$

In interpretations of spectral density, extra care must be taken with regard to the width of each frequency bin. Up to this point in the discussion, bin-centered components have been considered (i.e. components with frequencies associated with $f_{[n]}$). With regard to spectral density, one must consider both the boundaries and the width of each bin. The normal spacing for a bin is $\Delta f = Fs/K$, but the first and last bin containing the DC and Nyquist rate point have a width of only $\Delta f/2$, and the entire spectral content fits between DC and $Fs/2$. This accounts for the scaling of the spectral density in $\mathrm{V}/\sqrt{\mathrm{Hz}}$ by $\sqrt{2}$ at DC and the Nyquist rate in Table 12.1 for even $K$.

Consider now the odd $K = 7$ example provided in Figure 12.6. Here, there is one less point in the time-domain waveform $\mathbf{x}$. In the DFT obtained, there are still three negative frequency image points, but the Nyquist rate point is no longer included. The last frequency point is at the strange frequency $Fe = N/K \cdot Fs$. The full $K$ element DFT is created by filling in these three negative frequency points, as shown (shaded). The lack of the Nyquist rate point accounts for most of the differences in the remaining equations.

Since all of the non-DC frequencies contain negative frequency images, all of the non-DC elements are scaled by $2/K$ to form the amplitude vector $\mathbf{A}$. This vector also corresponds to the strict definition between the time and frequency domains using (12.7). Similarly, the rms values are calculated by scaling all of the non-DC amplitudes by $1/\sqrt{2}$, and (12.8) also holds.

Regarding spectral densities, the first bin is of size $\Delta f/2$; this bin, in conjunction with the remaining $\Delta f$ size bins and the strange end frequency, produce, yet again, a spectrum that covers DC to the Nyquist rate $Fs/2$. All of this assumes no scaling on the forward DFT and a scaling of $1/K$ on the IDFT.

Python software for dealing with frequency content is provided in the **Frequency-Content** class shown in Listing 12.1, which derives from the **FrequencyDomain** class

[2] This assumes a total lack of correlation between points in the time and frequency domains for ideal noise; in practice, the actual results can vary slightly.

**Listing 12.1 FrequencyContent class**

```
class FrequencyContent(FrequencyDomain):
    def __init__(self,wf,fd=None):
        td=wf.td
        if fd is None:
            X=fft.fft(wf.Values())
            K=int(td.K)
            Keven=(K//2)*2 == K
            fd=td.FrequencyList()
        else:
            K=fd.N*2
            Keven=True
            X=CZT(wf.Values(),td.Fs,0,fd.Fe,fd.N,True)
            td=TimeDescriptor(td.H,fd.N*2,fd.Fe*2.)
        FrequencyDomain.__init__(self,fd,[X[n]/K*\
            (1. if (n==0 or ((n==fd.N) and Keven)) else 2.)*\
            cmath.exp(-1j*2.*math.pi*fd[n]*td.H) for n in range(fd.N+1)])
        self.td=td
    def Values(self,unit=None):
        if unit=='rms':
            Keven=(self.td.K/2)*2==self.td.K
            A=FrequencyDomain.Values(self,'mag')
            return [A[n]/(1 if (n==0 or ((n==self.m_f.N) and Keven))
                    else math.sqrt(2)) for n in range(len(A))]
        elif unit=='dBm':
            return [-3000. if r < 1e-15 else 20.*math.log10(r)-self.LogRP10
                        for r in self.Values('rms')]
        elif unit=='dBmPerHz':
            Keven=(self.td.K/2)*2==self.td.K
            Deltaf=self.m_f.Fe/self.m_f.N
            adder=-10*math.log10(Deltaf)
            dBm=self.Values('dBm')
            return [dBm[n]+adder+
                    (self.dB3 if (n==0 or ((n==self.m_f.N) and Keven))
                    else 0) for n in range(len(dBm))]
        else: return FrequencyDomain.Values(self,unit)
    def Waveform(self,td=None):
        Keven=(self.td.K//2)*2==self.td.K
        X=self.Values()
        X=[X[n]*self.td.K*\
            (1. if (n==0 or ((n==self.m_f.N) and Keven)) else 0.5)*\
            cmath.exp(1j*2.*math.pi*self.m_f[n]*self.td.H)
            for n in range(self.m_f.N+1)]
        if Keven:
            X2=[X[self.m_f.N-n].conjugate() for n in range(1,self.m_f.N)]
        else:
            X2=[X[self.m_f.N-n+1].conjugate() for n in range(1,self.m_f.N+1)]
        X.extend(X2)
        x=[xk.real for xk in fft.ifft(X).tolist()]
        wf=Waveform(self.td,x)
        if not td is None:
            wf=wf.Adapt(td)
        return wf
    def WaveformFromDefinition(self,td=None):
        absX=self.Values('mag')
        theta=self.Values('deg')
        wf=Waveform(self.td)
        for n in range(self.m_f.N+1):
            wf=wf+SineWaveform(self.td,Frequency=self.m_f[n],
                Amplitude=absX[n],Phase=theta[n]+90)
        if not td is None:
            wf=wf.Adapt(td)
        return wf
```

**Listing 12.2 FrequencyDomain class**

```
class FrequencyDomain(list):
    def __init__(self,f=None,resp=None):
        self.m_f=FrequencyList(f)
        if not resp is None:
            list.__init__(self,resp)
    def FrequencyList(self):
        return self.m_f
    def Frequencies(self,unit=None):
        return self.m_f.Frequencies(unit)
    def Values(self,unit=None):
        if unit==None:
            return list(self)
        elif unit =='dB':
            return [-3000. if (abs(self[n]) < 1e-15) else
                    20.*math.log10(abs(self[n]))
                        for n in range(len(self.m_f))]
        elif unit == 'mag':
            return [abs(self[n]) for n in range(len(self.m_f))]
        elif unit == 'rad':
            return [cmath.phase(self[n]) for n in range(len(self.m_f))]
        elif unit == 'deg':
            return [cmath.phase(self[n])*180./math.pi
                        for n in range(len(self.m_f))]
        elif unit == 'real':
            return [self[n].real for n in range(len(self.m_f))]
        elif unit == 'imag':
            return [self[n].imag for n in range(len(self.m_f))]
    def ReadFromFile(self,fileName):
        with open(fileName,'rU' if sys.version_info.major < 3 else 'r') as f:
            data=f.readlines()
        if data[0].strip('\n')!='UnevenlySpaced':
            N = int(str(data[0]))
            Fe = float(str(data[1]))
            frl=[line.split(' ') for line in data[2:]]
            resp=[float(fr[0])+1j*float(fr[1]) for fr in frl]
            self.m_f=EvenlySpacedFrequencyList(Fe,N)
            list.__init__(self,resp)
        else:
            frl=[line.split(' ') for line in data[1:]]
            f=[float(fr[0]) for fr in frl]
            resp=[float(fr[1])+1j*float(fr[2]) for fr in frl]
            self.m_f=GenericFrequencyList(f)
            list.__init__(self,resp)
        return self
    def WriteToFile(self,fileName):
        fl=self.FrequencyList()
        with open(fileName,"w") as f:
            if fl.CheckEvenlySpaced():
                f.write(str(fl.N)+'\n')
                f.write(str(fl.Fe)+'\n')
                for v in self.Values():
                    f.write(str(v.real)+' '+str(v.imag)+'\n')
            else:
                f.write('UnevenlySpaced\n')
                for n in range(len(fl)):
                    f.write(str(fl[n])+' '+str(self.Values()[n].real)+' '+
                    str(self.Values()[n].imag)+'\n')
        return self
...
```

provided in Listing 12.2. These classes corresponds to the equations provided in Table 12.1 and the previous discussion.

During construction, the **FrequencyDomain** class is provided with an instance of a **Waveform** class (see Listing 13.4) and converts the **Waveform** instance internally to the frequency domain. The values are stored internally in the form of the previously described amplitude vector **A**, such that values directly read out correspond to the amplitude and phase of the element. In addition to the DFT, a mechanism is provided for converting waveforms to specific frequencies using the chirp z-transform (CZT), as described in §12.5.2, if a **FrequencyList** (see Listing 12.4) is provided in `fd`. Otherwise, note that, in addition to what has been previously described, a mechanism for dealing with the time of the first point is provided, as waveforms do not need to start at time zero.

The `Values()` method extracts the values described as either `'rms'`, `'dBm'`, or `'dBmPerHz'`. If none of these units is specified, it defers to the **FrequencyDomain** base class, where `'mag'` and `'deg'` extract the amplitude and phase.

The `Waveform()` method converts the frequency content back into a waveform. Optionally, a **TimeDescriptor** (see Listing 13.1) is provided in `td` that allows the method to resample the waveform onto a desired time axis.

A very slow method, `WaveformFromDefinition()`, is provided for educational purposes; this reconstructs the waveform according to the strict definition as provided in (12.7). This is used mostly for testing the software and as a form of documentation.

### 12.3.4 Convolution and the DFT

This chapter began by explaining the motivation for waveform processing involving the convolution integral, which is continuous. Convolution for discrete-time sequences is defined as

$$y\,[k] = \sum_{\sigma=0}^{M-1} h[\sigma] \cdot x[k-\sigma]\,,$$

where $\mathbf{x}$ is a $K$ element discrete-time sequence defined for $k \in 0 \ldots K-1$ and $\mathbf{h}$ is an $M$ element discrete-time sequence defined for $m \in 0 \ldots M-1$.

Problems arise in this definition when $k - \sigma < 0$; various options are possible. One option is to assume that $x[k-\sigma] = 0$ for $k - \sigma < 0$. This is the *filter* view, whereby it is assumed that the sequence $\mathbf{h}$ is a FIR filter. In this view, the first $M-1$ points in the result are generally discarded. Another option is to pad both sequences so that both sequences are $K$ elements long and to assume the periodicity of both waveforms:

$$y[k] = \sum_{\sigma=0}^{K-1} h[\sigma] \cdot x\left[\begin{cases} k-\sigma & \text{if } k-\sigma \geq 0 \\ K-(\sigma-k) & \text{if } k-\sigma < 0 \end{cases}\right].$$

**Table 12.2** Frequency response, content, and convolution depending on DFT scaling

| Operation | Scaling employed | |
|---|---|---|
| | Unity on DFT, $1/K$ on IDFT | $1/K$ on DFT, unity on IDFT |
| DFT view | frequency response | frequency content |
| Filling in frequency responses | $H[n] = R[n]$ | $H[n] = \frac{1}{K} \cdot R[n]$ |
| Filling in frequency content | $X[n] = C[n] \cdot K \cdot \begin{cases} \frac{1}{2} & \text{if } 0 < n < N \\ 1 & \text{otherwise} \end{cases}$ | $X[n] = C[n] \cdot \begin{cases} \frac{1}{2} & \text{if } 0 < n < N \\ 1 & \text{otherwise} \end{cases}$ |
| Convolution | $\text{IDFT}\left[\overrightarrow{\text{DFT}(\mathbf{x}) \cdot \text{DFT}(\mathbf{h})}\right]$ | $K \cdot \text{IDFT}\left[\overrightarrow{\text{DFT}(\mathbf{x}) \cdot \text{DFT}(\mathbf{h})}\right]$ |

This is the circular convolution view; when using this view, one computes the z-transform of each side and finds that

$$\begin{aligned} Y(z) &= \sum_{k=0}^{K-1} \left\{ \sum_{\sigma=0}^{K-1} h[\sigma] \cdot x[k-\sigma] \right\} \cdot z^{-k} = \sum_{\sigma=0}^{K-1} h[\sigma] \cdot \sum_{k=0}^{K-1} x[k-\sigma] \cdot z^{-k} \\ &= \sum_{\sigma=0}^{K-1} h[\sigma] \cdot X(z) \cdot z^{-\sigma} = X(z) \cdot \sum_{\sigma=0}^{K-1} h[\sigma] \cdot z^{-\sigma} = H(z) \cdot X(z), \end{aligned}$$

which allows the following to be written:

$$\mathbf{y} = \text{IDFT}\left[\overrightarrow{\text{DFT}(\mathbf{x}) \cdot \text{DFT}(\mathbf{h})}\right].$$

### 12.3.5 Scaling Factors and the DFT

The DFT has been defined such that it is scaled properly for response and convolution. A scaling factor of $1/K$ appears in the IDFT in order to satisfy the property IDFT $[\text{DFT}(\mathbf{x})] = \mathbf{x}$. These are generally accepted definitions, but sometimes other scaling factors are used. These include the scaling $1/K$ on the DFT, with unity scaling on the IDFT, and a less commonly used factor of $1/\sqrt{K}$ on both the DFT and IDFT, which makes it symmetric. The main reason for the two different scaling factors of $1/K$ on either the DFT or IDFT is to simplify the understanding of either the response effects or the frequency content. These scaling rules are shown in Table 12.2.

In the case of the definition used in this book, the choice has been made to simplify the view of the response and convolution. The DFT view is the frequency response, shown in Table 12.2. This ensures that the DFT of the unit impulse sequence is all ones and that convolution using the DFT requires no scaling. When looking at the frequency content, the

DFT of a sequence of all ones is $K$ at all DFT locations, and the DFT of a unity amplitude cosine wave has a magnitude of $K/2$ at the bin corresponding to the frequency of the cosine wave (again, except for the Nyquist rate, where it is $K$).

In the case of the $1/K$ scaling on the DFT, this would be used to simplify the view of the frequency content. Thus, the DFT of a sequence of all ones becomes unity at the first point and zero elsewhere, and the DFT of a sequence representing a unity amplitude cosine wave has a magnitude of $1/2$ at the frequency bin containing the cosine wave frequency (unless it is at the Nyquist rate, where it is unity). However, the DFT of the unit impulse sequence is $1/K$ at all locations.

This scaling becomes important for filling in DFT values that came from an external source (like a frequency response of a system, or the frequency content of a waveform) and when performing convolution. Note that the scaling for the frequency response of a system or the frequency content of a waveform must be handled differently.

Consider the frequency content in a vector $\mathbf{C}$, whose magnitude represents the amplitude and whose argument represents the phase of cosine waves, and a frequency response vector $\mathbf{R}$, whose values represent the complex frequency response of a system. Table 12.2 shows the two commonly used scaling factors on various operations with respect to $\mathbf{C}$ and $\mathbf{R}$, and the DFTs $\mathbf{X}$ and $\mathbf{H}$, for even $K$.

If the scaling rules for convolution are not clear, consider that there is a Fourier matrix $\mathbf{F}$ defined, for $n, k \in 0 \ldots K-1$ as

$$F_{n,k} = e^{-j \cdot 2\pi \cdot \frac{n \cdot k}{K}}$$

such that the DFT of a vector $\mathbf{x}$ is defined as

$$\mathbf{X} = \mathrm{DFT}\left(\mathbf{x}\right) = \mathbf{F} \cdot \mathbf{x}.$$

When the Fourier matrix is defined in this way, the inverse matrix becomes

$$\left(\mathbf{F}^{-1}\right)_{n,k} = \frac{1}{K} \cdot e^{j \cdot 2\pi \cdot \frac{n \cdot k}{K}}$$

such that

$$\mathbf{x} = \mathrm{IDFT}\left(\mathbf{X}\right) = \mathbf{F}^{-1} \cdot \mathbf{X}.$$

The convolution is such that

$$\mathbf{x} * \mathbf{h} = \mathrm{IDFT}\left[\overrightarrow{\mathrm{DFT}(\mathbf{x}) \cdot \mathrm{DFT}(\mathbf{h})}\right] = \mathbf{F}^{-1} \cdot \mathrm{diag}(\mathbf{F} \cdot \mathbf{h}) \cdot \mathbf{F} \cdot \mathbf{x}.$$

This is the frequency response view and uses the scaling employed in this book. If the scaling on the forward DFT had a factor $K$ in the denominator,

$$\left(\frac{\mathbf{F}}{K}\right)^{-1} \cdot \mathrm{diag}\left(\frac{\mathbf{F}}{K} \cdot \mathbf{h}\right) \cdot \frac{\mathbf{F}}{K} \cdot \mathbf{x} = \frac{1}{K} \cdot \mathbf{F}^{-1} \cdot \mathrm{diag}(\mathbf{F} \cdot \mathbf{h}) \cdot \mathbf{F} \cdot \mathbf{x},$$

and it is seen that this would require scaling by $K$ to produce the correct result.

### 12.3.6 Periodicity and Discreteness Implied by the DFT

A lot has been discussed about periodicity and discreteness with regard to signals prior to discussing the DFT. Here, this topic is explained further with regard to the DFT itself. Again, this involves the implicit assumption that discreteness in the domain implies periodicity in the other domain with regard to the time-domain sequence $\mathbf{x}$ and the frequency-domain sequence $\mathbf{X}$.

It is clear that a time-domain sequence is discrete (or, more properly, discrete time) as the sequence is defined only at the points $x[k]$ for $k \in 0 \ldots K-1$. A question (and potential misunderstanding) arises as to what is being assumed by the DFT when interpretations are made about the behavior of the sequence between sample points, and before $k = 0$ and after $k = K - 1$. The implied behavior of the sequence outside the specified $K$ points is tackled first.

The IDFT formula, or at least the alternative form provided in §12.3.2, includes a sum of cosine waves. The IDFT defines $K/2 + 1$ such cosine waves summed together to form a waveform that is sampled to produce the original discrete-time points in the sequence. Stop for a moment to consider the meaning of this. The input signal is expected to be formed by a sum of cosine waves. Furthermore, the cosine wave frequencies are harmonically related: the fundamental frequency (the first frequency after DC) is for $n = 1$ and is at $f[n] = (1/N) \cdot Fs/2$, and all other subsequent frequencies are multiples, or harmonics, of this frequency. This means that $\mathbf{X}$ provides a list of complex Fourier series coefficients. Any waveform described by a sum of harmonically related frequencies must be repetitive. Furthermore, the repetition frequency is the period of the fundamental, and is therefore $2 \cdot N/Fs$, which is, in fact the duration of the time-domain sequence. This means that the time-domain sequence is definitely assumed to repeat.

This is surprising because often the intent is that the sequence starts at time zero and ends at time $K/Fs$. This built-in assumption of repetition causes numerous problems when using the DFT for many purposes.

This repetition in the time domain stems from the discreteness of the frequencies in the frequency domain, and the same should be expected in reverse. This is understood by realizing that the definition of the DFT is of the z-transform computed at equiangularly spaced points on the unit circle, but only $K$ such points were computed. It is not hard to see that if the value of the z-transform is computed at point $K + n$, one obtains

$$
\begin{aligned}
X[K+n] = X\left(z = e^{j \cdot 2 \cdot \pi \cdot \frac{K+n}{K}}\right) &= \sum_{k=0}^{K-1} x[k] \cdot e^{-j \cdot 2 \cdot \pi \cdot \frac{k \cdot (K+n)}{K}} \\
&= \sum_{k=0}^{K-1} x[k] \cdot e^{-j \cdot 2\pi \cdot k} \cdot e^{-j \cdot 2\pi \cdot \frac{k \cdot n}{K}} = X\left[n\right].
\end{aligned}
$$

Therefore, the DFT provided only the information necessary to construct the repetitive frequency information, but the implication is that the sequence repeats. This implied repetition of the Fourier series coefficients in the frequency domain is what enforces the discreteness of the time-domain waveform. Although shown for the frequency domain, this same implication exists for the time domain.

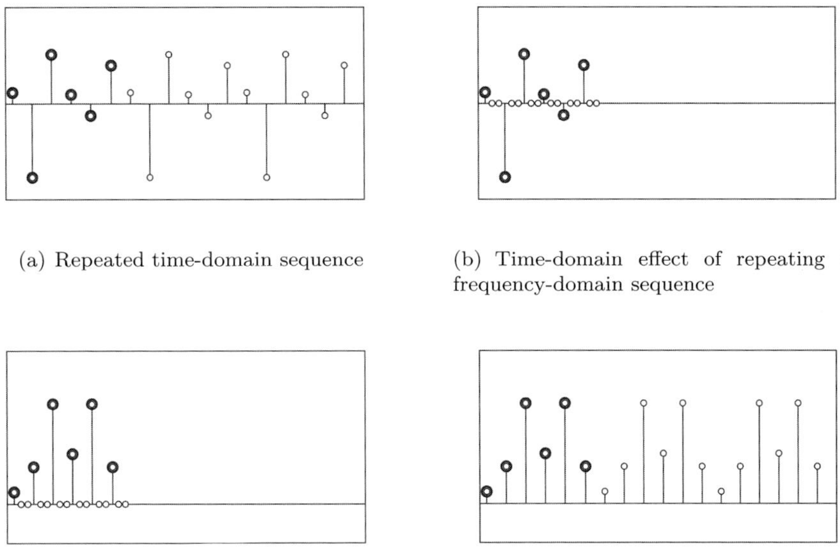

(a) Repeated time-domain sequence

(b) Time-domain effect of repeating the frequency-domain sequence

(c) Frequency-domain effect of repeating the time-domain sequence

(d) Repeated frequency-domain sequence

**Figure 12.7** Effects of repetition in the time and frequency domains

The implied discreteness and repetitiveness relationships can be seen graphically through two simple examples provided in Figure 12.7. Figure 12.7(a) shows a time-domain sequence along with that same sequence repeated three times. The effect of the repetition of the time-domain signal on the DFT is shown in Figure 12.7(c), in which the index for the DFT of the repeated sequence has been normalized so that the two overlaying plots lie on the same frequency axis. The effect of repeating the time-domain sequence three times is to fill in two zeros between each element in the frequency-domain sequence. Thus, if one were to repeat endlessly the time-domain signal, one would find that there is nothing between the points in the frequency domain.

Next, in Figure 12.7(d), the frequency-domain sequence is repeated three times. The result of this frequency-domain repetition is shown in the time domain through the IDFT in Figure 12.7(b). The similar effect shown is that the repetition of the frequency-domain sequence three times has the effect of filling two zeros after each time-domain element.

From this simple exercise it can be seen that the implication of the DFT and the IDFT is that both the time-domain and frequency-domain sequences repeat forever, and that the samples are zero in between both the time- and frequency-domain points.

These assumptions of repetitiveness and discreteness violate one's intentions when working with s-parameters in the analysis of continuous time and frequency systems. Fortunately, these assumptions can be worked around to some degree using methods provided in the following section.

### 12.3.7 Padding and Interpolation

When working with continuous time and frequency systems, the discrete time and frequency nature and the periodicity become inconveniences. This is because, despite the fact that the actual systems are discrete in neither time nor frequency, nor are they periodic, the assumptions built into the DFT can be worked around to some extent. The workaround involves zero padding in both the time and frequency domains.

Generally, the intent is that the system, whether described in the time or frequency domain, has content only within the extent provided. Certainly, for the time domain, at least insofar as a time-domain system *response* is concerned, the intent is generally that the response is zero before some time, usually time zero, and that it ends at some point. Responses generally do not die down completely to zero, but they can usually be described for a sufficient time such that the residual amount at the end of this time is substantially zero. In this situation, it is desirable to assume that, for any time outside the given sample points, the sample points of the impulse response of the system are zero.

In the frequency domain, however, things become more complicated. For the frequency response of a filter, describing the input to output relationship of a system, it is true to say that there is a similar situation to that in the time domain when attempting to describe the system out to a frequency at which the magnitude of the response is very small. When this can be attained, one would like to assume that, for any frequencies above the end frequency, the frequency response is zero. This works for many filters, and most of the time for port–port relationships held in s-parameters (meaning the off-diagonal elements), but oftentimes the s-parameters on the diagonal actually reach a steady-state value of unity not zero at high frequency. In these situations, one might want to employ some extrapolations other than the simple assumption that the response goes to zero. Similarly, with instruments like the VNA, it is not possible to take low frequency measurements, thus low frequency extrapolation is required. These are subjects for another section. In this section, the assumption is that the intent is that the time-domain sequence is zero beyond the points provided, and that the frequency-domain sequence is similarly provided from zero frequency to some end frequency, where beyond the end frequency the response is assumed to be zero. This situation leads to the concept of zero padding of time- and frequency-domain data.

To avoid the math, the effects of zero padding in each domain are shown graphically in Figure 12.8. An analogous situation to Figure 12.7(a) is seen in Figure 12.8(a), except here, instead of repeating the sequence three times, two sequences of the same length as the original are appended that contain all zeros. The effect of this on the DFT of the sequence is shown in Figure 12.8(c). Here, instead of zeros being inserted into the sequence, as in Figure 12.7(c), the frequency content is *interpolated*. In other words, the zero padding of the time-domain sequence has allowed the generation of the frequency-domain sequence with higher resolution. The frequency-domain plots have been normalized to utilize the same frequency scale. One can conclude that, with the assumption that the time-domain sequence can be padded endlessly with zeros, the frequency-domain sequence can be interpolated to an arbitrary degree and can be assumed to be continuous-frequency content.

Similarly, zero padding the frequency response as shown in Figure 12.8(d) is analogous to repeating the frequency response, as shown previously in Figure 12.7(d). When zero padding, the original frequency-domain sequence is decreased to maintain the division by

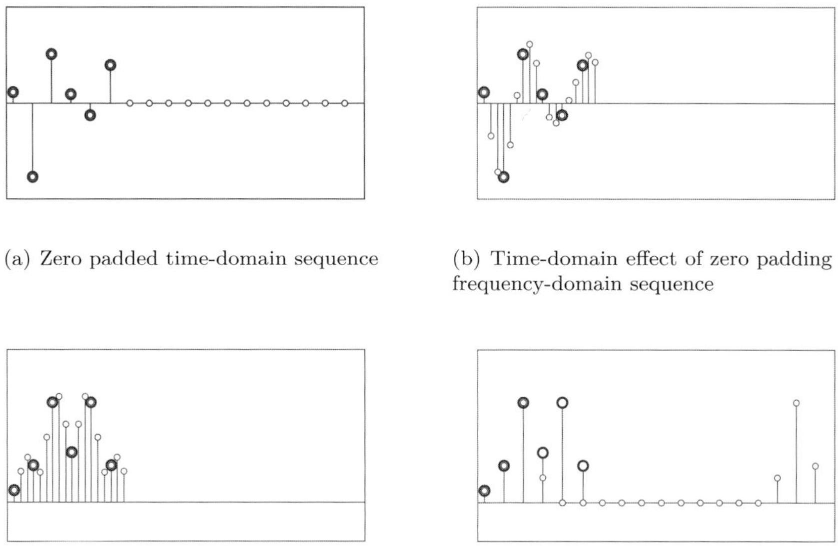

(a) Zero padded time-domain sequence

(b) Time-domain effect of zero padding the frequency-domain sequence

(c) Frequency-domain effect of zero padding the time-domain sequence

(d) Zero padded frequency-domain sequence

**Figure 12.8** Effects of zero padding in the time and frequency domains

$K$ in the DFT formula; this content is shown multiplied again by a factor of three for comparison. This zero padding has the time-domain effect in the IDFT shown in Figure 12.8(b), where the time-domain sequence is seen to be interpolated.

The zero padding of the frequency sequence shown in Figure 12.8(d) is not completely straightforward. The concept is that the sequence elements between elements $K/2$ and $K-1$ inclusively are placed near the end and that the response at $K/2$ is halved and duplicated. This is a consequence of the intent of this zero padding. In the frequency domain, the intent is that the response of the system is low-pass filtered at the supplied end frequency, and these mechanics are consistent with that intent. In subsequent sections it will be shown that, by doing things a certain way, one needn't worry about this detail.

The zero padding effects allow the assumption that the discrete-time and frequency-domain signals are representative of continuous-time and frequency-domain signals. This was seen in §12.2.2, where it was found that the, albeit periodic, continuous-frequency response could be found from the discrete-time time-domain sequence:

$$H(f) = \sum_{k=0}^{K-1} h[k] \cdot e^{-j \cdot 2 \cdot \pi \cdot f \cdot k \cdot T}.$$

When using the DFT, this is equivalent to zero padding of the time-domain sequence prior to computing the DFT.

**Listing 12.3 FrequencyResponse** class

```
class FrequencyResponse(FrequencyDomain):
    def __init__(self,f=None,resp=None):
        FrequencyDomain.__init__(self,f,resp)
    def Response(self,unit=None):
        return self.Values(unit)
...
```

Similarly, it is expected that

$$h(t) = \frac{1}{K} \cdot \sum_{n=0}^{K-1} H[n] \cdot e^{j \cdot 2\pi \cdot \frac{n \cdot t}{K \cdot T}},$$

which is simply the alternative form of the sum of cosine waves equation put forth in §12.3.2, where here it is for an arbitrary value of $t$. Interpolation is accomplished by zero padding the DFT prior to computing the IDFT.

## 12.4 Frequency Responses and Impulse Responses

### 12.4.1 Frequency Responses

The general form of a frequency response is a vector of complex values representing the response. This can come from s-parameters, for example. Associated with this vector is a frequency list that contains information for ideally evenly spaced frequency values starting from zero and ending at an end frequency $Fe$, along with the number of frequency elements $N + 1$ (i.e. $N$ is one less than the number of elements). Thus, the implication is that $f[n] = (n/N) \cdot Fe$ for $n \in 0 \ldots N$, and that the response vector $H[n]$ contains the complex values corresponding to these frequencies. The class **FrequencyResponse** provided in Listing 12.3 serves as a container for such responses. The **FrequencyResponse** class takes a **FrequencyList** providing the frequency information along with a list of complex numbers representing the response. If the frequency information is not provided as a **FrequencyList**, then it is converted into one.

S-parameters are the likely source of frequency response information, and they cause certain problems in that the frequency information is completely free-form. The frequencies might be a list of any arbitrary frequencies without regard to spacings. To deal with this, the **FrequencyList** class is shown in Listing 12.4. This class can be instantiated directly only with another instance of **FrequencyList** or a list of frequencies. This is only for compatibility with actual lists of frequencies so that, if one is holding a list of frequencies and the class wants a **FrequencyList**, the caller doesn't need to think about it. There are two preferred derived classes shown in Listing 12.5 and Listing 12.6. The class **EvenlySpacedFrequencyList** is designed to specify directly the kind of frequencies required, that is $N + 1$ points to an end frequency $Fe$. The class **GenericFrequencyList** provides an explicit **FrequencyList** just as if a list of frequencies were provided. The **SParameters** class shown in Listing 3.15 always provides a **GenericFrequencyList**, as

### Listing 12.4 FrequencyList class

```
class FrequencyList(list):
...
    def SetEvenlySpaced(self,Fe,N):
        self.Fe=Fe
        self.N=int(N)
        list.__init__(self,[Fe/N*n for n in range(self.N+1)])
        self.m_EvenlySpaced=True
        return self
    def SetList(self,fl):
        list.__init__(self,fl)
        self.N=len(fl)-1
        self.Fe=fl[-1]
        self.m_EvenlySpaced=False
        return self
    def EvenlySpaced(self): return self.m_EvenlySpaced
    def Frequencies(self,unit=None):
        if unit == None: return list(self)
        elif isinstance(unit,float): return (self/unit).Frequencies()
        elif unit == 'GHz': return (self/1.e9).Frequencies()
        elif unit == 'MHz': return (self/1.e6).Frequencies()
        elif unit == 'kHz': return (self/1.e3).Frequencies()
    def CheckEvenlySpaced(self,epsilon=0.01):
        if self.m_EvenlySpaced: return True
        for n in range(self.N+1):
            if abs(self[n]-self.Fe/self.N*n) > epsilon:
                self.m_EvenlySpaced=False
                return False
        self.SetEvenlySpaced(self.Fe,self.N)
        return True
    def __div__(self,d):
        return self.__truediv__(d)
    def __truediv__(self,d):
        if self.EvenlySpaced(): return EvenlySpacedFrequencyList(self.Fe/d,self.N)
        else: return GenericFrequencyList([v/d for v in self])
    def __mul__(self,d):
        if self.EvenlySpaced(): return EvenlySpacedFrequencyList(self.Fe*d,self.N)
        else: return GenericFrequencyList([v*d for v in self])
    def TimeDescriptor(self,Keven=True):
        N=self.N
        K=2*N
        if not Keven: K=K+1
        Fs=self.Fe*K/N
        return TimeDescriptor(-K/2./Fs,K,Fs)
...
```

### Listing 12.5 EvenlySpacedFrequencyList class

```
class EvenlySpacedFrequencyList(FrequencyList):
    def __init__(self,Fe,Np):
        FrequencyList.__init__(self)
        self.SetEvenlySpaced(Fe,Np)
```

### Listing 12.6 GenericFrequencyList class

```
class GenericFrequencyList(FrequencyList):
    def __init__(self,fl):
        FrequencyList.__init__(self)
        self.SetList(fl)
```

**Listing 12.7 FrequencyResponse** class `ImpulseResponse()` method

```
class FrequencyResponse(FrequencyDomain):
...
    def ImpulseResponse(self,td=None,adjustDelay=True):
        fd = self.FrequencyList()
        if isinstance(td,float) or isinstance(td,int):
            Fs=float(td)
            td=fd.TimeDescriptor()
            td = TimeDescriptor(0.,2*int(math.ceil(Fs*td.K/2./td.Fs)),Fs)
        evenlySpaced = fd.CheckEvenlySpaced()
        if not evenlySpaced and td is None: return None
        if not evenlySpaced and not td is None:
            newfd = td.FrequencyList()
            oldfd = fd
            Poly=Spline(oldfd,self.Response())
            newresp=[Poly.Evaluate(f) if f <= oldfd[-1] else 0.0001 for f in newfd]
            newfr=FrequencyResponse(newfd,newresp)
            return newfr.ImpulseResponse(None,adjustDelay)
        if evenlySpaced and td is None and not adjustDelay:
            yfp=self.Response()
            ynp=[yfp[fd.N-nn].conjugate() for nn in range(1,fd.N)]
            y=yfp+ynp
            y[0]=y[0].real
            y[fd.N]=y[fd.N].real
            Y=fft.ifft(y)
            td=fd.TimeDescriptor()
            tp=[Y[k].real for k in range(td.K//2)]
            tn=[Y[k].real for k in range(td.K//2,td.K)]
            Y=tn+tp
            return ImpulseResponse(td,Y)
        if evenlySpaced and td is None and adjustDelay:
            TD=self._FractionalDelayTime()
            return self._DelayBy(-TD).ImpulseResponse(None,False).DelayBy(TD)
        if evenlySpaced and not td is None:
            # if td is a float and not a time descriptor, it's assumed to be a
            # sample rate.  In this case, the number of points in a
            # time descriptor are filled in representing the time content of self
            return self.Resample(td.FrequencyList()).ImpulseResponse()
...
```

there is no general control over this. Frequency responses in this chapter and in Chapter 13 will only consider evenly spaced points, as these are preferred.

### 12.4.2 Frequency Response and Impulse Response Duality

The two classes **FrequencyResponse** and **ImpulseResponse** are highly interrelated, in the sense that, given an instance of a **FrequencyResponse**, an **ImpulseResponse** can be generated, and vice versa. The general intent is that, given an instance of the class **FrequencyResponse** as `fr`,

$$\texttt{fr.ImpulseResponse().FrequencyResponse()} == \texttt{fr}$$

Furthermore, given an instance of the class **ImpulseResponse** as `ir`,

$$\texttt{ir.FrequencyResponse().ImpulseResponse()} == \texttt{ir}$$

**Listing 12.8 ImpulseResponse** class `FrequencyResponse()` method

```
class ImpulseResponse(Waveform):
...
    def FrequencyResponse(self,fd=None,adjustLength=True):
        if not fd and not adjustLength:
            X=fft.fft(self.Values())
            fd=self.td.FrequencyList()
            return FrequencyResponse(fd,[X[n] for n in range(fd.N+1)]).\
                _DelayBy(self.td.H)
        if not fd and adjustLength:
            return self._AdjustLength().FrequencyResponse(None,adjustLength=False)
        if fd:
            return self.FrequencyResponse().Resample(fd)
    def _AdjustLength(self):
        td = self.td
        PositivePoints = int(max(0,math.floor(td.H*td.Fs+td.K+0.5)))
        NegativePoints = int(max(0,math.floor(-td.H*td.Fs+0.5)))
        P=max(PositivePoints,NegativePoints)*2
        return self._Pad(P)
...
```

There are a few practical details to work through in this regard. The code for generating an impulse response from a frequency response is shown in Listing 12.7. This method covers a lot of situations. For the moment, the situation is considered where the points are evenly spaced, no time descriptor has been specified, and `adjustDelay` is `False`.

To convert a frequency response to an impulse response, the DFT is formed by first filling in the conjugate pairs associated with the frequency response. The response vector is appended to by filling in $H[N+\sigma] = H[N-\sigma]^*$, for $\sigma \in 1 \ldots N-1$. Now there is a $K = 2 \cdot N$ element vector containing the DFT. Note that $K$ is even in this case, which is the motivation for defining $K$ to be even under all circumstances.

The IDFT of this sequence is the impulse response $\mathbf{h}$ defined for $h[k]$, for $k \in 0 \ldots K-1$. With the frequency descriptor containing the number of frequency points and the end frequency, a time descriptor is required to associate with the sequence. Figure 12.4 showed the implied time sense of the time-domain response. Although the IDFT provides a time sequence that looks like it starts from time zero and ends at time $(K-1) \cdot T$, this time sequence is not actually the impulse response, but is in fact the impulse train response. Since all sense of time is lost, it has been chosen to assume that the first $K/2$ points are for times from zero to $(K/2-1) \cdot T$ and that the last $K/2$ points are for times from $-K/2 \cdot T$ to $-T$. Thus, the first half of the IDFT points are swapped with the second half and a set of times $t[k] = -K/2 \cdot T + k \cdot T$ is assumed that goes with the swapped vector. As will be seen when waveform time descriptors are discussed in §13.1, this waveform has a horizontal offset (time of the first point) $H = -K/2 \cdot T$, a number of points $K$, and a sample rate $Fs = 1/T = 2 \cdot Fe$.

Thus, given an $N+1$ point frequency response $X[n]$ with end frequency $Fe$, where, for $n \in 0 \ldots N$ frequencies of $f[n] = n/N \cdot Fe$ are assumed, there is a corresponding $K$ point impulse response $x[k]$ with a sample rate $Fs = 2 \cdot Fe$, a sample period $T = 1/Fs$, a horizontal offset $H = -K/2 \cdot T$, and an associated time vector $t[k] = H + k \cdot T$.

The code for going back to the frequency domain is given in Listing 12.8, in which, given an **ImpulseResponse** instance, one invokes `FrequencyResponse()` to provide an

**Listing 12.9 FrequencyResponse** class delay methods

```
class FrequencyResponse(FrequencyDomain):
...
    def _DelayBy(self,TD):
        fd=self.FrequencyList()
        return FrequencyResponse(fd,
        [self[n]*cmath.exp(-1j*2.*math.pi*fd[n]*TD)
            for n in range(fd.N+1)])
...
    def _FractionalDelayTime(self):
        ir = self.ImpulseResponse(None,adjustDelay=False)
        idx = ir.Values('abs').index(max(ir.Values('abs')))
        TD = ir.td[idx] # the time of the main peak
        # calculate the frequency response with this delay taken out
        # the fractional delay is based on the minimum adjustment to the phase of
        # the last point to make that point real
        theta = self._DelayBy(-TD).Response('rad')[self.FrequencyList().N]
        # do not make this adjustment if the phase is of a tiny magnitude!
        if self.Response('dB')[self.FrequencyList().N]<-90: theta=0.
        if theta < -math.pi/2.: theta=-(math.pi+theta)
        elif theta > math.pi/2.: theta = math.pi-theta
        else: theta = -theta
        # calculate the fractional delay
        TD=theta/2./math.pi/self.FrequencyList()[-1]
        return TD
...
```

equivalent **FrequencyResponse** instance. The DFT of this time sequence is computed and the end frequency is calculated as $Fe = Fs/2$; there are $N+1$ points such that $N = K/2$. Previously, the time-domain elements were rearranged prior to conversion to the frequency domain. Here, they are not rearranged prior to computing the DFT. Instead, the DFT is computed directly on the impulse response, and then the DFT elements are delayed by the horizontal offset $H$, as shown in Listing 12.8. Remember that $H$ is negative, so it amounts to an advance of the DFT. This delay (advance) is accomplished by multiplying each element $X[n]$ by $e^{-j\cdot 2\pi\cdot f[n]\cdot H}$. Thus, the equivalence between the frequency response and the impulse response has been achieved. Listing 12.8 contains other code for dealing with other situations. For the moment, the case has been discussed in which there is no frequency descriptor specified and `adjustLength` is `False`.

### 12.4.3 Delay and the Nyquist Rate Point

The basic equivalence between the impulse response and the frequency response was described in §12.4.2; however, some important details were not considered. The first detail is that of the requirement that the Nyquist rate point be a real number. In other words, given an $N+1$ element DFT $\mathbf{X}$, there is an assumption that $X[N]$ is real (i.e. has an argument of zero or $\pi$). This is a requirement (and is enforced) when $\mathbf{X}$ is computed from a real-valued time sequence, but often the frequency response has been generated from, for example, a set of s-parameters, which do not meet this constraint. Thus, in Listing 12.7 an argument called `adjustDelay` is added to the `ImpulseResponse()` function. This argument defaults to `True`, so the normal way of converting a frequency response to an impulse response is to manage the delay as in this section.

**Listing 12.10 ImpulseResponse** class delay methods

```
class ImpulseResponse(Waveform):
...
    def DelayBy(self,d):
        return ImpulseResponse(self.td.DelayBy(d),self.Values())
...
    def _FractionalDelayTime(self):
        td=self.td
        TD=-(-td.H*td.Fs-math.floor(-td.H*td.Fs+0.5))/td.Fs
        return TD
...
```

The constraint that the Nyquist rate point being real is best handled by making $X[N]$ real by delaying or advancing $\mathbf{X}$ in the frequency domain, then undoing the delay or advance in the time domain; two functions that address delay are shown in Listing 12.9 (for frequency responses) and Listing 12.10 (for impulse responses). The idea here is to delay the frequency response using Listing 12.9 and to advance the impulse response using Listing 12.10 to undo the effect.

The amount by which to delay or advance is determined by evaluating $\theta = \arg(X[N])$ in conjunction with the frequency $f[N] = Fe$. The argument $\theta$ can fall into three categories and is adjusted depending on the minimum delay or advance that would be performed. The adjustment is given by

$$\theta = \begin{cases} \theta - \pi & \text{if } \theta < -\pi/2, \\ \pi - \theta & \text{if } \theta > \pi/2, \\ -\theta & \text{if } -\pi/2 \leq \theta \leq \pi/2. \end{cases} \tag{12.9}$$

The delay is calculated as

$$TD = \frac{\theta}{2\pi} \cdot \frac{1}{Fe}. \tag{12.10}$$

It is incorrect to make this delay adjustment when the argument $\theta$ is of a tiny complex number, so a check is made on the magnitude to see if $\theta$ is valid. This check is essential.[3]

The new frequency response is then calculated by advancing the frequency response by the time delay value by multiplying $X[n]$ by $e^{j \cdot 2\pi \cdot f[n] \cdot TD}$ prior to computing the IDFT. The effect on the last frequency point can be seen by recognizing that, for $-\pi/2 \leq \theta \leq \pi/2$,

$$X[N] \cdot e^{-j \cdot 2\pi \cdot Fe \cdot \frac{\arg(X[N])}{2\pi} \cdot \frac{1}{Fe}} \in \mathcal{R}.$$

Once the impulse response has been calculated, the impulse response itself is delayed by $TD$ to counteract the effect in the frequency domain caused by advancing by $TD$. This is performed simply by adding $TD$ to the horizontal offset.

[3] A good example of this is the frequency response of a terminated transmission line that is a single sample time in electrical length. By varying termination values, one finds that the last point of the frequency response for non-zero termination resistances is real, meaning no delay is required. But, in the limit as the termination resistance approaches zero, the magnitude goes to zero as the phase goes to $-\pi/2$. If this check is not made, an incorrect delay adjustment of one-half a sample point is made.

The value of $TD$ calculated is restricted such that $-\pi/2 < \theta < \pi/2$, which means that

$$-\frac{1}{2}\cdot T \leq TD \leq \frac{1}{2}\cdot T.$$

An important feature of this calculation is that it enables fractional delay. Fractional delay, or non-bin centering of the time-domain response, is a leading cause of non-causality issues in the impulse response. A frequency response calculated from an impulse response will always have the phase tending to zero or $\pm 180°$ in the phase response if the horizontal offset is a multiple of the sample period. Actual frequency responses generated from s-parameters do not have such tendencies. The fractional delay $TD$ takes care of this.

In practice, one wants the delay or advance employed in the form of $TD$ to be the best fractional delay (i.e. fraction of a sample point) to do the job. Therefore, prior to the computation of this delay, the waveform is delayed or advanced by an amount that takes out the principal delay in the waveform. This is achieved by computing the IDFT, finding the largest magnitude element in the time-domain sequence, calculating its time, advancing the waveform by the time amount, and then calculating $TD$.

A very important point to consider here is the ability to delay or advance the signal in either domain. Again, to apply delay $TD$ to a frequency response DFT given by $\mathbf{X}$, $X[n]\cdot e^{-j\cdot 2\pi\cdot f[n]\cdot TD}$ is computed for $n \in 0\ldots N$. To apply delay to the time-domain impulse response, $TD$ is simply added to the horizontal offset.

Figure 12.9 shows an example of how all of this is done. Figure 12.9(b) shows the original frequency response which is wrapping about $\pm 180°$. Its impulse response (by forcing $f[N]$ to be real) is shown in Figure 12.9(a). The principal delay comprises three samples. The impulse response with this principal delay removed is shown in Figure 12.9(a), and Figure 12.9(b) shows that the phase response has been unwrapped, but is not zero at the last frequency point. The fractional delay is calculated to be 0.2 samples and is removed. This brings the phase response to zero at the last frequency point and the clean IDFT is shown in Figure 12.9(a). The final step is to delay the impulse response to put it back into the correct location, as shown in the last impulse response in Figure 12.9(a). The phase response of this impulse response is not exactly the original phase response until the fractional delay contained in the horizontal offset is applied.

### 12.4.4 Unbalanced Positive and Negative Time in the Impulse Response

Section §12.4.3 handled the details for producing impulse responses from frequency responses. The methods used ensured that fractional delay was properly considered and that a frequency response calculated again from the generated impulse response would match the original frequency response.

This section discusses some practicalities of frequency responses generated directly from impulse responses, specifically the timing associated with the impulse response and its length. An argument is introduced to the `FrequencyResponse()` method of the **ImpulseResponse** class called `adjustLength`, which defaults to `True` when converting impulse responses into frequency responses. Just as delay is adjusted when converting frequency

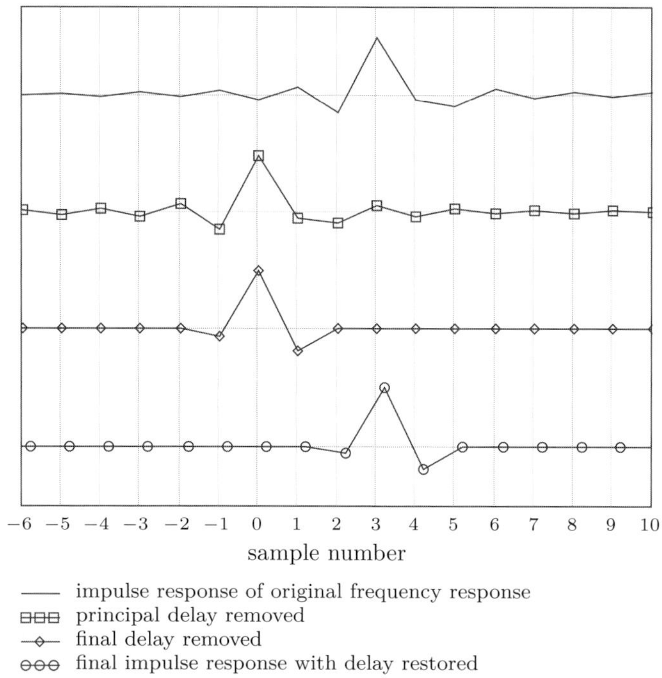


(a) Impulse responses with various delays removed

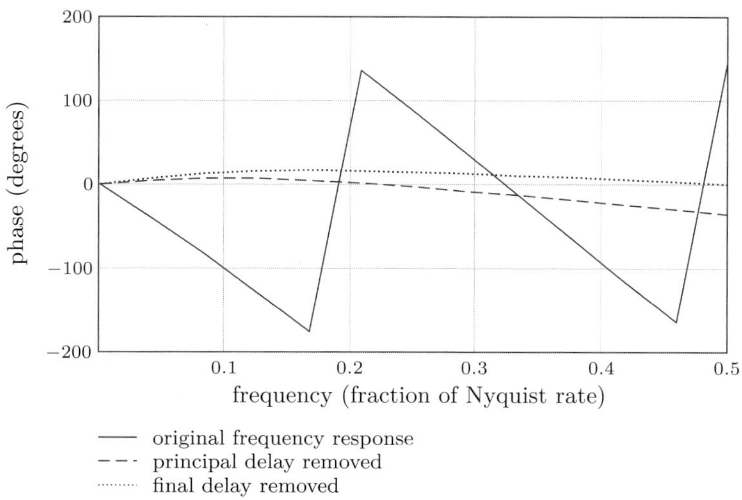


(b) Phase responses with various delays removed

**Figure 12.9** Effects of delay on impulse and phase responses

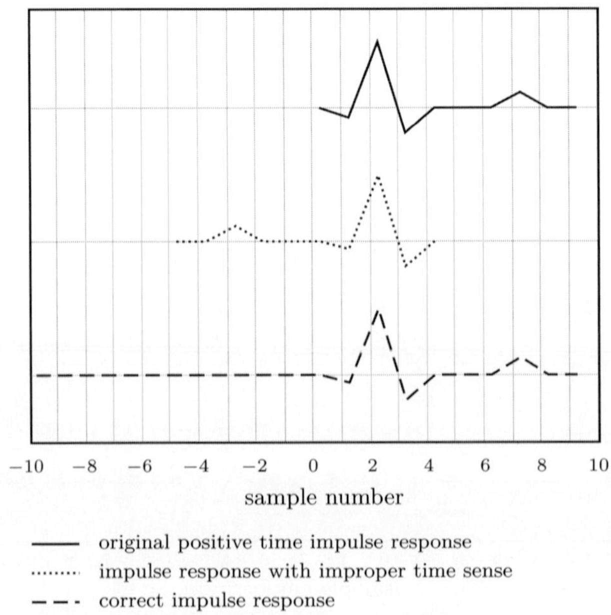


**Figure 12.10** Impulse response length adjustment

responses to impulse responses, length is adjusted when converting impulse responses to frequency responses. This was shown in Listing 12.8.

In the definition of the impulse response, half of the points occur prior to time zero and half occur after time zero. That being said, it is customary to provide impulse responses that start from time zero when generating frequency responses. This would be okay if the times that go with the impulse response were remembered by the frequency response. This problem can be seen when computing the frequency response of an impulse response generated with positive times only and then computing the impulse response again from the calculated frequency response. Ideally, the same result is produced. Unless handled properly, the time sense of the newly computed impulse response will be incorrect.

This is shown graphically in Figure 12.10. The top waveform depicts a short impulse response that has a small bump around sample time 7.2. This impulse consists of ten points. When the frequency response is computed and converted back to the impulse response, the impulse response shown in the middle waveform is produced. Five points of the impulse response are in the correct place, but another five points placed prior to time zero are in the wrong place. This occurs because of the rules imposed on the impulse response when it is created from the frequency response. The main problem is the number of points in the frequency response, which dictate where the points in the impulse response must end up.

The way to mitigate this is to deal with the imbalance between positive and negative time points prior to going to the frequency response. The basic idea is to calculate the number of positive and negative time points and then pad the response so that the impulse

**Listing 12.11 FrequencyResponse** class `_Pad()` method

```
class FrequencyResponse(FrequencyDomain):
...
    def _Pad(self,P):
        fd=self.FrequencyList()
        if P == fd.N: X=self.Response()
        elif P < fd.N: X=[self.Response()[n] for n in range(P+1)]
        else: X=self.Response()+[0 for n in range(P-fd.N)]
        return FrequencyResponse(EvenlySpacedFrequencyList(P*fd.Fe/fd.N,P),X)
...
```

response has the same number of each. In this case, the impulse response is padded to twenty points by introducing ten points prior to time zero in the impulse response. Then, when the frequency response is calculated and converted back to the impulse response, these extra ten points are retained in the twenty point impulse response shown as the lower waveform.

In this way, the frequency response remembers the time sense of the original impulse response and pads the response properly. When this is done, the correct impulse response is recovered. The impulse response returned has ten extra zero points (it needed to have these ten points anyway to follow the rules). These zero points can always be trimmed from the impulse response prior to convolution, so they don't really affect anything.

Since impulse responses may contain horizontal offsets that are not integer multiples of the sample rate, the number of positive and negative time points must account for any fractional delay. This is accounted for by calculating these numbers of points as

$$K_p = \lfloor \max\left(0, \lfloor H/T + K \rceil\right) \rfloor,$$

$$K_n = \lfloor \max\left(0, \lfloor -H/T \rceil\right) \rfloor,$$

with the final number of points $K$ calculated as

$$K = \max\left(K_p, K_n\right) \cdot 2.$$

The padding of impulse responses is covered in the following section.

### 12.4.5 Padding Impulse and Frequency Responses

The padding of impulse and frequency responses is generally described as extending through the addition of zeros a $K$ element impulse response, or an $N+1$ element frequency response, to a new value of $K$, or $N$, specified as $P$. The Python code for padding frequency and impulse responses is provided in Listing 12.11 and Listing 12.12.

For an impulse response, there are three possible scenarios for $P$. The simplest scenario is $P = K$, where the padded response is the response itself. The next scenario is $P < K$. In this scenario, $x[k]$ is returned for $k \in \frac{K-P}{2} \ldots K - \frac{K-P}{2} - 1$. The final scenario is $P > K$, where $(P - K)/2$ zeros are placed before and after the impulse response. Just as there is the requirement that $K$ is an even number, $P$ must also be an even number. The time descriptor that accompanies the impulse response must also be adjusted. As mentioned

**Listing 12.12 ImpulseResponse** class `_Pad()` method

```
class ImpulseResponse(Waveform):
...
    def _Pad(self,P):
        K=len(self)
        if P==K: x = self.Values()
        elif P<K: x=[self[k] for k in range((K-P)//2,K-(K-P)//2)]
        else:
            x=[0 for p in range((P-K)//2)]
            x=x+self.Values()+x
        td = self.td
        return ImpulseResponse(TimeDescriptor(td.H-(P-K)/2./td.Fs,P,td.Fs),x)
...
```

previously, the time descriptor consists of the horizontal offset $H$, the number of points $K$, and the sample rate $Fs$. The sample rate stays the same, the number of points becomes $P$, and the horizontal offset is adjusted to be $H - (P - K)/2/Fs$. This adjustment should not change the time associated with any of the original impulse response points and it makes the time associated with any zeros added to the response correct.

Frequency response padding is simpler. Again there are three scenarios. The scenario where $P = N$ means that the original frequency response is returned. The scenario where $P < N$ means that $x[n]$ is returned for $n \in 0 \ldots P$. Finally, the scenario where $P > N$ means that $P - N$ zeros are appended to the response.

The frequency descriptor associated with the frequency response containing the number of points (minus one) $N$ and end frequency $Fe$ are adjusted such that the new number of points (minus one) is simply $P$ and the new end frequency is $(P/N) \cdot Fe$.

### 12.4.6 Trimming the Impulse Response

The impulse response can always be trimmed by removing either points that are zero or points whose absolute value falls below some threshold. This can be used to shorten the impulse response for two reasons. One simple reason is to reduce the computation required for convolution. Another is to reduce the resolution of a frequency response.

The algorithm employed for trimming an impulse response is provided in Listing 12.13. Here, the point with the maximum absolute value is found and a threshold value is generated that is some fraction of this value. Then, a walk is performed from both negative and positive time inwards. The first point from either direction that exceeds the threshold defines the time extent of the impulse response.

Trimming can be utilized to remove the zeros that were added to adjust the length of the impulse response during frequency response determination prior to using the impulse response for convolution, but converting the trimmed response back to the frequency domain and back to the time domain will restore these zeros.

**Listing 12.13 ImpulseResponse** class `TrimToThreshold()` method

```
class ImpulseResponse(Waveform):
...
    def TrimToThreshold(self,threshold):
        x=self.Values()
        td=self.td
        maxabsx=max(self.Values('abs'))
        minv=maxabsx*threshold
        for k in range(len(x)):
            if abs(x[k]) >= minv:
                startidx = k
                break
        for k in range(len(x)):
            ki = len(x)-1-k
            if abs(x[ki]) >= minv:
                endidx = ki
                break
        if (endidx-startidx+1)//2*2 != endidx-startidx+1:
            # the result would not have an even number of points
            if endidx < len(x)-1:
                # keep a point at the end if possible
                endidx = endidx + 1
            elif startidx > 0:
                # keep a point at the beginning if possible
                startidx = startidx - 1
            else:
                # append a zero to the end and calculate number of
                # points with endidx+1
                return ImpulseResponse(TimeDescriptor(td[startidx],
                    (endidx+1)-startidx+1,td.Fs),
                    [x[k] for k in range(startidx,endidx+1)]+[0.])
        return ImpulseResponse(TimeDescriptor(td[startidx],
            endidx-startidx+1,td.Fs),
            [x[k] for k in range(startidx,endidx+1)])
...
```

## 12.5 Resampling

Resampling frequency responses is tricky. Consider a frequency response $\mathbf{H}$ consisting of $N+1$ points and an end frequency $Fe$ such that, for $n \in 0 \ldots N$, $H[n]$ is the complex frequency response for frequencies $f[n] = (n/N) \cdot Fe$. For the resampling, a new end frequency $Fe'$ and a new number of points (minus one) $N'$ is provided. The desired frequency response $\mathbf{H}'$ is for $N'+1$ frequency points such that, for $n' \in 0 \ldots N'$, $f'[n'] = (n'/N') \cdot Fe'$.

The first way to think about this is to determine a uniform frequency spacing $\Delta f'$ that would contain all of the desired points, albeit with a possible decimation factor, and all of the original points, also with a possible decimation factor. In other words, with this new spacing and a decimation factor $D_1$ applied to the original points, $f[n \cdot \Delta f' \cdot D_1] = f[n]$, which states that $\Delta f' \cdot D_1 = Fe/N$. Also, with this new spacing and a decimation factor $D_2$ applied to these new points, $f'[n' \cdot \Delta f' \cdot D_2] = f'[n']$, which states that $\Delta f' \cdot D_2 = Fe'/N'$. This produces the following equality:

$$\frac{Fe}{D_1 \cdot N} = \frac{Fe'}{D_2 \cdot N'},$$

**Listing 12.14 FrequencyResponse** class `Resample()` method

```
class FrequencyResponse(FrequencyDomain):
...
    def Resample(self,fdp):
        fd=self.FrequencyList()
        evenlySpaced = fd.CheckEvenlySpaced() and fdp.CheckEvenlySpaced()
        if not evenlySpaced: return self._SplineResample(fdp)
        R=Rat(fd.Fe/fdp.Fe*fdp.N); ND1=R[0]; D2=R[1]
        if ND1 < fd.N: R=Rat(fd.Fe/fdp.Fe*fdp.N/fd.N); ND1=R[0]*fd.N; D2=R[1]
        if  ND1 > 50000: return self.ResampleCZT(fdp)
        if ND1 == fd.N: fr=self
        else: fr=self.ImpulseResponse()._Pad(2*ND1).FrequencyResponse(None,False)
        if D2*fdp.N != ND1: fr=fr._Pad(D2*fdp.N)
        if D2==1: return fr
        else: return fr._Decimate(D2)
...
```

or

$$\frac{Fe}{Fe'} \cdot \frac{N'}{N} = \frac{D_1}{D_2}.$$

Thus, the requirement is to find the integers $D_1$ and $D_2$.

This means that the condition for the resampled specification can be met if the number of points in the original response is changed from $N$ to $D_1 \cdot N$, the response is padded to $D_2 \cdot N'$, and finally is decimated by $D_2$.

This specification means that the intermediate resampled response, having $D_1 \cdot N$ points, contains all of the points in the original response, and all of the points in the desired response out to the original end frequency $Fe$. It may contain many superfluous points that are removed when the final response is decimated by $D_2$. This specification is therefore potentially overly restrictive; it is therefore useful to change the specification on $D_1$ and $D_2$ and solve

$$\frac{Fe}{Fe'} \cdot N' = \frac{D_1 \cdot N}{D_2}. \tag{12.11}$$

This requirement tends to lower $D_2$ by removing the common factors of both $D_1 \cdot N$ and $D_2$, but it means that all of the points in the original response are no longer in the intermediate response.

If this newly calculated $D_1 \cdot N$ is less than $N$, it should not be used because it would involve trimming points from the impulse response associated with the original frequency response, and that should be avoided. In this case, the original equation is used to calculate $D_1 \geq 1$, which means that $D_1 \cdot N \geq N$. This is shown in Listing 12.14.

Remember, $D_1$, $N$, and $D_2$ are all integers, so the first step is to find the two integers $P = D_1 \cdot N$ and $D_2$ such that their ratio is equal to the left-hand side of (12.11). This is performed using the `Rat()` function, provided in Listing 12.15, that determines these integers. Given a number $x$ and a goal to determine the integer ratio, the integer $I_0$ and the residual $R_0$ are found that satisfy $x = I_0 + R_0$. This is of course satisfied by $I_0 = \lfloor x \rfloor$ and $R_0 = x - I_0$. Then, the best integer $I_1$ and residual $R_1$ are found that satisfy $R_0 = 1/(I_1 + R_1)$; these are $I_1 = \lfloor 1/R_0 \rfloor$ and $R_1 = 1/R_0 - I_1$. Next, the best integer $I_2$ and residual $R_2$ are found that satisfy $R_1 = 1/(I_2 + R_2)$, and so on. Eventually, the

**Listing 12.15** `Rat()` function

```
def Rat(R,tol=0.0001):
    N=10
    D=[]
    for n in range(N+1):
        D.append(int(math.floor(R)))
        B=R-D[n]
        if B < tol:
            break
        R = 1./B
    N = len(D)-1
    R=(1,0)
    for n in range(N+1):
        R=(R[0]*D[N-n]+R[1],R[0])
    return R
```

residual goes to zero (or becomes sufficiently small) and the calculation stops. Thus, an approximation is formed:

$$x = I_0 + \cfrac{1}{I_1 + \cfrac{1}{I_2 + \cfrac{1}{I_3 + \dots}}}$$

All of the values $I_0$, $I_1$, $I_2$, $I_3$, etc. are integers, so the result can be algebraically rearranged into a ratio of two integers.

For the moment, the only requirement is for the numerator in this ratio, $D_1 \cdot N$, which forms the desired intermediate points in the frequency response prior to padding and decimation. In order to accomplish this new number of points, it is realized that, if an impulse response with sample rate $Fs = 2 \cdot Fe$ and number of points $K = 2 \cdot D_1 \cdot N$ were provided, the frequency response of such an impulse response would meet the criteria. Therefore, the impulse response is created from the original frequency response, padded to $P_1 = 2 \cdot D_1 \cdot N$ points, and the frequency response is determined from this impulse response.

At this point, the frequency response has been resampled to only part of the criteria required. The frequency spacing is now correct, but the extent of the frequency response might not be correct. Therefore, the frequency response is padded to a new number of points, $P_2 = D_2 \cdot N'$.

It may turn out that this new padded frequency response is exactly what was desired; if so, here is where it stops. This is the case when $D_2 = 1$. It is more likely that, if everything went well, the new padded frequency response contains a number of points that are an integer multiple of the desired frequency response. This occurs when $D_2 > 1$. The final response is generated by decimating the padded response by $D_2$. Observe that one could also decimate by $D_2$ prior to padding and then just pad to $N'$.

This resampling was accomplished using the discrete Fourier transform by simply padding impulse responses (to change frequency resolution) or padding frequency responses (to change frequency extent); all that is needed is to determine integer amounts of padding to use.

**Listing 12.16 ImpulseResponse** class `Resample()` method

```
class ImpulseResponse(Waveform):
...
    def Resample(self,td):
        fr=self.FrequencyResponse()
        return fr.ImpulseResponse(td)
...
```

Sometimes unusual requirements mean that the factors $D_1 \cdot N$ and $D_2$ become huge. When this occurs, it is impractical or impossible to pad responses and compute huge DFTs. In this case, the chirp z-transform is used. This is described in §12.5.2.

In the resampling of frequency responses, there is often a need to specify both the number of points and the end frequency in order to resample s-parameters onto common frequency points prior to computations. One must always remember that this resampling affects the impulse response length. Generally, the end frequency is chosen to be equal to or higher than the largest end frequency in the s-parameters, and the number of points is chosen in conjunction with this end frequency, based on estimates of impulse response length.

With regard to the resampling of impulse responses, the frequency response is computed corresponding to the impulse response, and that frequency response is resampled onto a number of points and end frequency corresponding to the desired number of points and sample rate of the resampled impulse response. This is shown in Listing 12.16.

The need to resample impulse responses most often arises when convolving with waveforms to match the sample rates. In these cases, the length of the impulse response is not generally specified. The horizontal offset is never changed specifically (it may change due to sample rate and length changes). In this case, given an impulse response with a number of points $K$, sample rate $Fs$, a new specification of $K'$ points, and a sample rate of $Fs'$, the frequency response is computed first that will have $N+1$ points, where $N = K/2$, and an end frequency of $Fe = Fs/2$. Then the frequency response is resampled to $N' = K'/2$ and $Fe' = Fs/2$, and converted back to an impulse response. In the usual case, where one is given a new sample rate only, the desired $K'$ is computed from the frequency response as $K' = 2 \cdot \lceil Fs'/Fs \cdot N \rceil$. This causes the duration of the impulse response to be unaffected in the resampling operation.

The resampling of s-parameters using the methods described here is provided in Listing 12.17.

### 12.5.1 Resampling Example

A simple resampling exampling is provided in Figure 12.11. A frequency response is provided whose response is a set of eleven ($N = 10$) points from zero to the end frequency $Fe = 20.0$ (presumably GHz, but that doesn't matter). The goal is to resample this response to $N' = 21$ and $Fe' = 24.0$. These odd points and frequencies were contrived to construct a good example.

The first step is to find a frequency spacing that would accommodate both points from the input and the desired points. This spacing is calculated by using the `Rat()` function to

**Listing 12.17 SParameters** class `Resample()` method

```
class SParameters(SParameterManipulation):
    def Resample(self,fl):
        if self.m_d is None:
            self.m_f=fl
            copy.deepcopy(self)
        fl=FrequencyList(fl)
        f=FrequencyList(self.f()); f.CheckEvenlySpaced()
        SR=[empty((self.m_P,self.m_P)).tolist() for n in range(fl.N+1)]
        for o in range(self.m_P):
            for i in range(self.m_P):
                res = FrequencyResponse(f,self.Response(o+1,i+1)).Resample(fl)
                for n in range(len(fl)):
                    SR[n][o][i]=res[n]
        return SParameters(fl,SR,self.m_Z0)
...
```

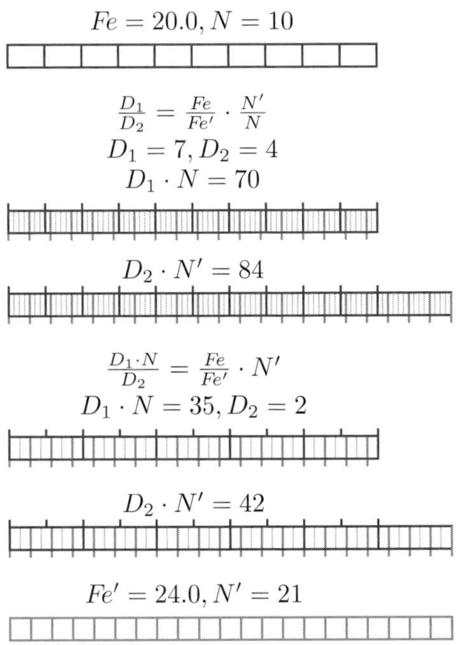


**Figure 12.11** Resampling example

find the following ratio:

$$\frac{D_1}{D_2} = \frac{Fe}{Fe'} \cdot \frac{N'}{N} = \frac{20.0}{24.0} \cdot \frac{21}{10} = \frac{7}{4}.$$

On examining the input frequency response spacing shown at the top of Figure 12.11, the goal is to refine the frequency spacing by a factor of $D_1 = 7$, which means resampling the response to $D_1 \cdot N = 70$ (71 points) with the same end frequency. This is performed by converting the frequency response to an impulse response, which has $K = 2 \cdot N = 20$ points, and padding this response with zeros to $2 \cdot D_1 \cdot N = 140$ points. Converting back to the frequency domain provides the desired points. This is shown in Figure 12.11, in which all of the input frequency points are still on the grid, as expected. Although many points are superfluous, the output points (the output is shown at the bottom of Figure 12.11) are also in this new response.

Since the response only goes to an end frequency of $Fe = 20.0$, it is calculated that $D_2 \cdot N' = 4 \cdot 21 = 84$ is greater than $D_1 \cdot N = 70$, and therefore the frequency response is padded to 84 points, noting that, because of the frequency spacing calculated, the remaining points for the desired end frequency of $Fe' = 24.0$ were provided. Since $D_2 = 4$, there are four times the number of points needed, so it is decimated by $D_2 = 4$ to obtain the $N' = 21$ points desired. This is the simplest way of looking at the resampling problem.

A small improvement can be made by noting that the statement of the problem required more points than needed to be calculated. This is because it was specified that all points generated in the first resampling (by padding the impulse response) provided points in the original response. However, the problem can be restated by desiring that only the output points are in this response out to the original end frequency. When the problem is described in this way,

$$\frac{D_1 \cdot N}{D_2} = \frac{Fe}{Fe'} \cdot N' = \frac{20.0}{24.0} \cdot 21 = \frac{35}{2}.$$

This suggests that one should first resample the response to $D_1 \cdot N = 35$ (36 points) with the same end frequency, which is now half the number of points in the first part of the example. As stated previously, if the new value of $D_1 \cdot N < N$, then one needs to revert to the previous manner and use those calculated values of $D_1$ and $D_2$.

The frequency response is converted to an impulse response which is padded with zeros to $2 \cdot D_1 \cdot N = 70$ points. When this is converted back to the frequency domain, not all of the input frequency points in Figure 12.11 are still on the grid. The end frequency is, however, as specified.

Since $D_2 \cdot N' = 2 \cdot 21 = 42$ is greater than $D_1 \cdot N = 35$, the frequency response is padded to 42 points, and since $D_2 = 2$, there are twice the number of points as required and the padded frequency response is decimated by $D_2 = 2$ to obtain the $N' = 21$ points desired.

In this example, there was still a decimation by $D_2 = 2$ in the end. The problem could have been improved by not requiring that the original end frequency appear in the points, which is not really a requirement, but this would overly complicate the problem. Besides, all of this is handled by using the chirp z-transform if desired, which is discussed in the following section. This example showed that, given a discrete-frequency response, it is possible to obtain any other response by simply padding impulse responses (to change frequency resolution) or padding frequency responses (to change frequency extent), and that all that is needed is the determination of the integer amounts of padding to use. Furthermore, all of this was accomplished using the discrete Fourier transform.

### 12.5.2 The Chirp Z-Transform

In §12.3, the discrete Fourier transform was defined as the z-transform evaluated at specific equiangularly spaced points on the unit circle. Specifically, for a $K$ point sequence, the locations on the unit circle were $e^{j\cdot 2\pi\cdot \frac{n\cdot k}{K}}$ for $n \in 0 \ldots K-1$. The chirp z-transform is a variation in that, instead of equiangularly spaced points on the unit circle, an arc in the z plane is used, as shown in Figure 12.12.

The CZT is a transform that provides for $M+1$ elements, $m \in 0 \ldots M$, a set of points $\zeta$ to consist of points on an equiangularly spaced spiral or arc. The location of these points is described such that

$$\zeta_m = A \cdot W^m, \quad A = A_0 \cdot e^{j\cdot 2\pi\cdot \theta_0}, \quad W = W_0 \cdot e^{j\cdot 2\pi \cdot \phi_0}.$$

The chirp z-transform is defined as

$$X[m] = H_{CZT}(\zeta_m) = \sum_{k=0}^{K-1} x[k] \cdot \zeta_m^{-k} = \sum_{k=0}^{K-1} x[k] \cdot (A \cdot W^m)^{-k} = \sum_{k=0}^{K-1} x[k] \cdot A^{-k} \cdot W^{-m\cdot k},$$

where $A_0$, $\theta_0$, $W_0$, and $\phi_0$ are arbitrary constants that define the arc. To see how the arc is defined, consider that when $m=0$ and $\zeta_0 = A$ the beginning of the arc is at radius $A_0$ and at a beginning angle of $2\pi \cdot \theta_0$, and that when $m = M$, $\zeta_M = A_0 \cdot W_0^M \cdot e^{j\cdot 2\pi\cdot(\theta_0 + M\cdot\phi_0)}$, meaning that the end of the arc is at radius $A_0 \cdot W_0^M$ at an angle $2\pi \cdot (\theta_0 + M \cdot \phi_0)$, and therefore the arc spans the angle $2\pi \cdot M \cdot \phi_0$. Thus, for each $m \in 0 \ldots M$:

$$\zeta_m = A_0 \cdot W_0^m \cdot e^{j\cdot 2\pi\cdot(\theta_0 + m\cdot\phi_0)}.$$

The angle between each point on the arc is $2\pi \cdot \phi_0$. This is all shown graphically in Figure 12.12(a).

The CZT has many uses and has a fast implementation [37]. The concern here is only the ability to resample onto a desired equally spaced frequency scale. For these purposes, given a specification of $M+1$ frequency points with an end frequency *Fe*, $A_0 = 1$, $W_0 = 1$, $\theta_0 = 0$, and $\phi_0 = Fe/(Fs \cdot M)$, so, for $M+1$ frequency points,

$$X[m] = \sum_{k=0}^{K-1} x[k] \cdot \zeta_m^{-k} = \sum_{k=0}^{K-1} x[k] \cdot e^{-j\cdot 2\pi\cdot \frac{k\cdot m}{M}\cdot\frac{Fe}{Fs}}. \tag{12.12}$$

Examining (12.12) closely, one sees that, for this application, it is the same as the DFT with a different choice of points and end frequency. If $M = K$ and $Fe = Fs$, it would be the same as the DFT.[4] In other words, in this application, the z-transform of the input sequence is evaluated at chosen frequencies.

[4] Another way to look at this, more closely aligned with how it is used, is $M = N = K/2$ and $Fe = Fs/2$, whereby the algorithm extracts the positive frequency points of the DFT.

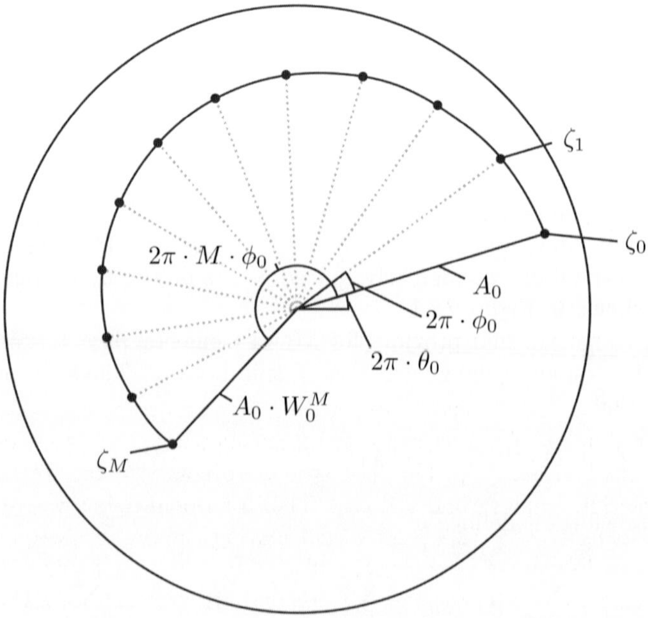


(a) Graphical interpretation

```
def CZT(x,Fs,fs,fe,M,highSpeed=True):
    M=int(M); K=len(x); fs=float(fs); Fs=float(Fs); fe=float(fe)
    theta0=fs/Fs; phi0=(fe-fs)/Fs*1./M; A0=1.; W0=1.
    A=A0*cmath.exp(1j*2.*math.pi*theta0)
    W=W0*cmath.exp(1j*2.*math.pi*phi0)
    if highSpeed:
        L=pow(2,int(math.ceil(math.log(float(M+K),2.))))*2
        v=[0 for l in range(L)]
        for k in range(M+1): v[k]=pow(W,float(k*k)/2.)
        for k in range(M+1,L-K+1): v[k]=0.
        for k in range(L-K+1,L): v[k]=pow(W,(L-k)*(L-k)/2.)
        V=fft.fft(v)
        y = [0 for l in range(L)]
        for k in range(K): y[k]=x[k]*pow(A,-k)*pow(W,-(k*k)/2.)
        for k in range(K,L): y[k]=0
        Y=fft.fft(y)
        P=[Y[l]*V[l] for l in range(L)]
        P=fft.ifft(P)
        X = [P[m]/v[m] for m in range(M+1)]
    else:
        Zeta=[A*pow(W,m) for m in range(M+1)]
        X=[]
        for m in range(M+1):
            Xm=0.
            for k in range(K): Xm=Xm+x[k]*pow(Zeta[m],-k)
            X.append(Xm)
    return X
```

(b) Python code

**Figure 12.12** The chirp z-transform

**Listing 12.18 FrequencyResponse** class `ResampleCZT()` method

```
class FrequencyResponse(FrequencyDomain):
...
    def ResampleCZT(self,fdp,speedy=True):
        fd=self.FrequencyList()
        evenlySpaced = fd.CheckEvenlySpaced() and fdp.CheckEvenlySpaced()
        if not evenlySpaced: return self._SplineResample(fdp)
        ir=self.ImpulseResponse()
        TD=ir._FractionalDelayTime()
        Ni=int(min(math.floor(fd.Fe*fdp.N/fdp.Fe),fdp.N))
        Fei=Ni*fdp.Fe/fdp.N
        return FrequencyResponse(EvenlySpacedFrequencyList(Fei,Ni),
            CZT(ir.DelayBy(-TD).Values(),ir.td.Fs,0,Fei,Ni,speedy)).\
            _Pad(fdp.N)._DelayBy(-fd.N/2./fd.Fe+TD)
```

Listing 12.18 shows the Python code utilized to perform resampling using the CZT. There are quite a number of practical items to consider here. The first involves fractional delay time, discussed in §12.4.3. The first step is to compute the impulse response and remember the fractional delay time utilized. This fractional delay was applied to the frequency response using Listing 12.9 prior to calculating the impulse response using Listing 12.7, and is extracted from the resulting impulse response using Listing 12.10.

Next, the extent to which to resample must be calculated as an intermediate $N_i$ and $Fe_i$. These describe the number of points that are actually contained in the original response.

Finally, the CZT in Figure 12.12(b) resamples the values of the impulse response with the fractional delay removed to these new intermediate points $N_i$ and $Fe_i$, pads the result to $N'$ points, and restores the delay taken out using Listing 12.9.

The CZT provides the same result as the previous methods discussed, albeit in a more direct, brute-force way. It is often slower than using the simpler DFT based method, but it can always be used regardless of unusual resampling requirements. This is why in Listing 12.14 it is only utilized when the padding requirements become too onerous.

# 13

# Waveforms and Filters

THE PRECEDING CHAPTER introduced the concept of frequency responses, impulse responses, and convolution. This chapter discusses the practicalities of performing the processing of waveforms with filters. As such, this is exactly the convolution of two waveforms, with one waveform being the input waveform and the other being an impulse response. In this chapter, the waveform serving as the impulse response is called a filter; filters are applied to input waveforms to produce output waveforms.

In Chapter 12, the mechanics of convolution showed that, given $\mathbf{x}$, which is a $K$ element discrete-time sequence defined for $k \in 0 \ldots K-1$, and $\mathbf{h}$, which is an $M$ element discrete-time impulse response sequence defined for $m \in 0 \ldots M-1$, the output waveform values $\mathbf{y}$ are determined as follows:

$$y[k] = \sum_{\sigma=0}^{M-1} h[\sigma] \cdot \begin{cases} x[k-\sigma] & \text{if } k-\sigma \geq 0, \\ 0 & \text{otherwise.} \end{cases}$$

This is the filter view of convolution. Generally, the first $M-1$ points are discarded, and for such input and impulse response sequences, there is a $K-(M-1)$ point output sequence defined for $k \in 0 \ldots K-(M-1)-1$:

$$y[k] = \sum_{\sigma=0}^{M-1} h[\sigma] \cdot x[k+M-1-\sigma] \, .$$

This equation is the exact definition of how convolution is performed when applying filters to waveforms, and one finds that it is easy to describe the values of the output sequence. The trouble occurs not in calculating the values, but instead in formulating the time axis that goes underneath the result. Keeping track of times when waveforms and filters are involved is a complicated issue and is the main subject of this chapter.

## 13.1 Convolution and Time

In order to determine the time effects of applying an impulse response to a waveform, the definition of convolution in the continuous-time domain is reexamined. The convolution integral provided in (12.1) is repeated below:

$$y(t) = (x * h)(t) = \int_{-\infty}^{+\infty} x(\lambda) \cdot h(t-\lambda) \cdot \mathrm{d}\lambda.$$

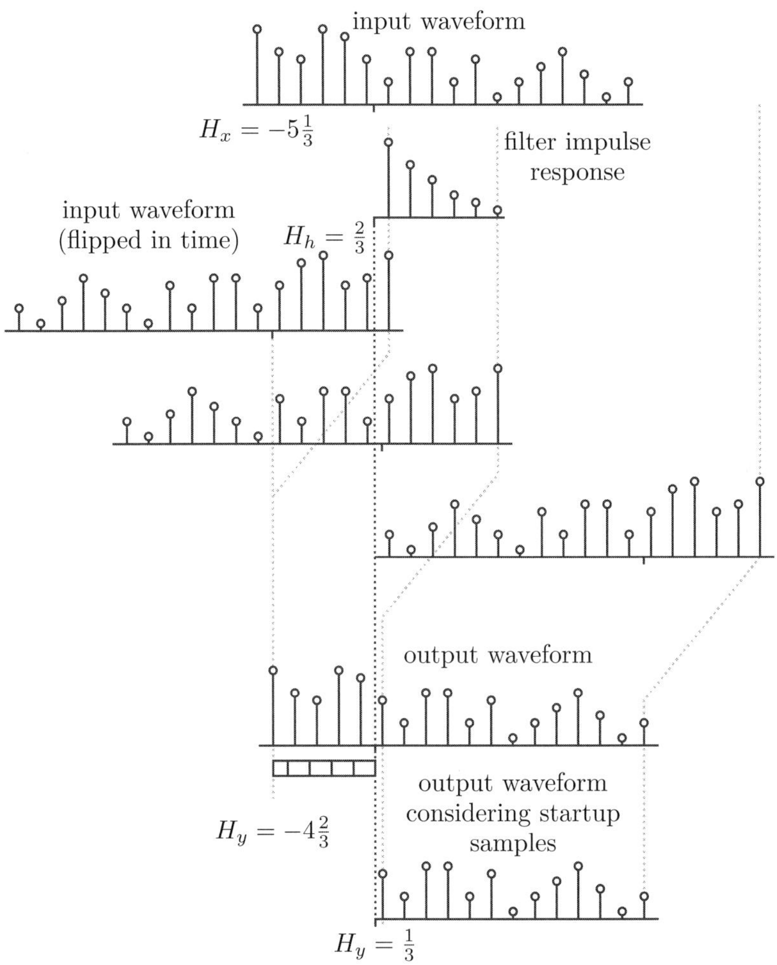


**Figure 13.1** Convolution example

For a given discrete-time waveform containing $K_x$ points, where $x[k]$, $k \in 0 \ldots K_x - 1$, is defined for times $H_x + k/Fs$, and a discrete-time filter containing $K_h$ points, where $h[m]$, $m \in 0 \ldots K_h - 1$, is defined for times $H_h + m/Fs$, the continuous-time versions of **x** and **h** are

$$x(t) = \sum_{k=0}^{K_x-1} x[k] \cdot \delta(t - [H_x + k/Fs])$$

and

$$h(t) = \sum_{m=0}^{K_h-1} h[m] \cdot \delta(t - [H_h + m/Fs])\,.$$

The convolution integral therefore becomes

$$y(t) = \int_{-\infty}^{+\infty} \left\{ \sum_{k=0}^{K_x-1} x[k] \cdot \delta\left(\lambda - [H_x + k/Fs]\right) \right\} \cdot \left\{ \sum_{m=0}^{K_h-1} h[m] \cdot \delta(t - [H_h + m/Fs] - \lambda) \right\} \cdot \mathrm{d}\lambda.$$

Because of the structure of this equation, the result will be a sequence of impulses. These impulses will be located only where the delta functions evaluate to one. This is for $\lambda = [H_x + k/Fs]$ and therefore for $t = [H_h + m/Fs] + [H_x + k/Fs]$. The first point of the waveform result $\mathbf{y}$ is located at $t = H_x + H_h$ and the delta functions evaluate to one only when $k = 0$ and $m = 0$, and so are calculated as $x[0] \cdot y[0]$. The second point is located at $t = H_x + H_h + 1/Fs$. At this point, the delta functions evaluate to one for $(k, m) = (0, 1)$ and $(1, 0)$, and thus are calculated as $x[1] \cdot h[0] + x[0] \cdot h[1]$. The third point is at $t = H_x + H_h + 2/Fs$ and it is not too hard to see that the result is $x[2] \cdot h[0] + x[1] \cdot h[1] + x[0] \cdot h[2]$.

Because both the waveform and filter impulse response are finite in length, it is customary and advantageous to remove the beginning portion of the waveform where the full filter length is not being considered. If the waveform were truly finite, and all other portions of the waveform for $k < 0$ were zero, this would not be necessary, but, in general, this is not the situation. The waveform sequence is usually assumed to be a portion of a longer waveform in which the points are simply unknown for $k < 0$. Because of this thinking, the first $K_h - 1$ points of the result are discarded and the resulting sequence is therefore of length $K_x - (K_h - 1)$. The first point in the resulting sequence is at $t = H_x + H_h + (K_h - 1)/Fs$. Thus, the resulting waveform has a horizontal offset of $H_x + H_h + (K_h - 1)/Fs$, a number of points equal to $K_x - (K_h - 1)$, and the original sample rate $Fs$.

In order to visualize the process of convolution, consider the example shown in Figure 13.1, which depicts an input waveform and the impulse response of a filter. Convolution can be visualized by reversing one of the waveforms and sliding it underneath the other, as commonly shown in linear systems textbooks. In most textbooks, both the filter impulse response and the waveform start at time zero, so the time effects are not interesting. Here, the input waveform has a horizontal offset (the time of the first point) shown as $-5\frac{1}{3}$ samples and it has a unity sample rate, for convenience. The time of the first point of the filter impulse response is shown as $\frac{2}{3}$ samples. As the flipped input waveform is slid forward under the filter impulse response, the first output point is encountered when the first point of the input waveform lines up with the first point of the filter. The time of this first point is at the sum of the horizontal offsets of the filter impulse response and the input waveform, which is $-5\frac{1}{3} + \frac{2}{3} = -4\frac{2}{3}$. This location can be seen as the faint line dropped from the zero time tick mark for the flipped input waveform. As the flipped input waveform is slid forward, the first output waveform point is encountered that is a function of all points in the filter impulse response and all points of the input waveform. For the filter length $K_h = 6$, this is at a time $-4\frac{2}{3} + 5 = \frac{1}{3}$, which is the time of the first actual sample in the output waveform. Finally, continuing to slide the flipped input waveform forward, the last point is reached that considers all points of the filter impulse response and all points of the input waveform. This is the last sample in the output waveform.

Not considering startup samples, the resulting waveform length would be the input waveform length, $K_x = 18$, but, after considering startup samples, it is $K_y = K_x - (K_h - 1) = 18 - 5 = 13$ samples. From the example, the output waveform length is seen to be determined by considering only the samples produced when the input waveform and filter impulse response overlap completely.

Although the time considerations were calculated in this example, they were not entirely explained. Consider an input waveform convolved with a filter impulse response, where both waveforms start at zero; in this case, the output waveform, not considering startup samples, starts at time zero. Next, consider that delaying the filter impulse response is equivalent to making the horizontal offset of the filter impulse response negative by an amount equal to the delay, and that this causes the output waveform to be delayed, taking on the filter's horizontal offset. Furthermore, delaying the input waveform would have the same effect; therefore, delaying both waveforms causes the output waveform, not considering startup samples, to take on a horizontal offset equal to the sum of the filter and input waveform's horizontal offsets. The startup samples increase the horizontal offset by an amount equal to the number of startup samples multiplied by the sample period.

### 13.1.1 Waveform Time Descriptors

In order to formalize the effects of convolution, or, more specifically, filtering operations on waveforms, the notion of a waveform time descriptor, or simply a time descriptor, is introduced.

This type of descriptor has actually been discussed all along; it is the information that comes with a waveform that provides the time of each point. The time descriptor $\left\{\begin{smallmatrix} H \\ K \\ Fs \end{smallmatrix}\right\}$ contains the horizontal offset $H$, the number of points $K$, and the sample rate $Fs$, such that, for $k \in 0 \ldots K-1$, the time of each point in the waveform is given as $t[k] = H + k/Fs$.

A waveform $w$ has a time descriptor $\{w\}$,which has horizontal offset, number of points, and sample rate given as $H_w$, $K_w$, and $Fs_w$. The time descriptor $\{w\}$ is written as $\left\{\begin{smallmatrix} H_w \\ K_w \\ Fs_w \end{smallmatrix}\right\}$.

Time descriptors are an integral part of a waveform and carry the time information. The **TimeDescriptor** class is shown in Listing 13.1.

### 13.1.2 Filter Descriptors

Unlike time descriptors, filter descriptors have not yet been encountered. A filter descriptor describes how a filter affects the time information in a waveform. A filter, by definition, has three characteristics that become important when dealing with time:

- Upsample factor: the factor that states the change in sample rate. It is the ratio of two integers where the numerator represents the sample rate increase and the denominator is the decimation factor.
- Delay samples: the number of samples by which the filter delays the waveform. It is important to note that this is not an actual delay, but rather an apparent delay that is removed by accounting for the filter delay. This will hopefully become apparent shortly. It is a delay in samples, not time.

**Listing 13.1 TimeDescriptor class**

```
class TimeDescriptor(object):
    def __init__(self,HorOffset,NumPts,SampleRate):
        self.H = HorOffset
        self.K = int(NumPts)
        self.Fs=SampleRate
    def __len__(self):
        return self.K
    def __getitem__(self,item):
        return item/self.Fs+self.H
...
    def Times(self,unit=None):
        if unit==None:
            return [self[k] for k in range(len(self))]
        elif isinstance(unit,float):
            return [self[k]/unit for k in range(len(self))]
        elif unit=='ps':
            return [self[k]/1.e-12 for k in range(len(self))]
        elif unit=='ns':
            return [self[k]/1.e-9 for k in range(len(self))]
        elif unit =='us':
            return [self[k]/1.e-6 for k in range(len(self))]
        elif unit == 'ms':
            return [self[k]/1.e-3 for k in range(len(self))]
    def ApplyFilter(self,F):
        return TimeDescriptor(
            HorOffset=self.H+(F.S-F.D)/self.Fs,
            NumPts=int(max(0,(self.K-F.S)*F.U)),
            SampleRate=self.Fs*F.U)
    def __mul__(self,F):
        return self.ApplyFilter(F)
    def __div__(self,other):
        return self.__truediv__(other)
    def __truediv__(self,other):
        if isinstance(other,FilterDescriptor):
            return TimeDescriptor(
                HorOffset=self.H+other.U/self.Fs*(other.D-other.S),
                NumPts=self.K/other.U+other.S,
                SampleRate=self.Fs/other.U)
        elif isinstance(other,TimeDescriptor):
            UpsampleFactor=self.Fs/other.Fs
            return FilterDescriptor(
                UpsampleFactor,
                DelaySamples=other.K-self.K/UpsampleFactor-
                    (self.H-other.H)*other.Fs,
                StartupSamples=other.K-self.K/UpsampleFactor)
    def DelayBy(self,D):
        return TimeDescriptor(self.H+D,self.K,self.Fs)
    def FrequencyList(self):
        K=int(self.K)
        N=K//2
        Fe=float(self.Fs)*N/K
        return EvenlySpacedFrequencyList(Fe,N)
...
    def TimeOfPoint(self,k):
        return self.H+float(k)/self.Fs
...
```

**Listing 13.2 FilterDescriptor** class

```
class FilterDescriptor(object):
    def __init__(self,UpsampleFactor,DelaySamples,StartupSamples):
        self.U = UpsampleFactor
        self.D = DelaySamples
        self.S = StartupSamples
    def __eq__(self,other):
        if abs(self.U - other.U) > 1e-15: return False
        if abs(self.D - other.D) > 1e-15: return False
        if abs(self.S - other.S) > 1e-15: return False
        return True
    def __mul__(self,other):
        if isinstance(other,FilterDescriptor):
            return FilterDescriptor(
                UpsampleFactor=self.U*other.U,
                DelaySamples=float(self.U*self.D+other.D)/self.U,
                StartupSamples=float(self.U*self.S+other.S)/self.U)
...
```

- Startup samples: the number of samples that it takes for the filter to start up and also the number of points removed from the waveform as a result of the filtering operation in order to remove the filter startup effects.

Thus, the nomenclature for a filter descriptor is $\begin{bmatrix} U \\ D \\ S \end{bmatrix}$, where $U$ is the upsample factor, $D$ the delay samples, and $S$ the startup samples. $[F]$ refers to the filter descriptor of filter $F$, and $U_F$, $D_F$, and $S_F$ refer to the upsample factor, delay samples, and startup samples of filter $F$. Filter $F$ has the filter descriptor $\begin{bmatrix} U_F \\ D_F \\ S_F \end{bmatrix}$.

The **FilterDescriptor** class is provided in Listing 13.2.

### 13.1.3 Time Descriptor and Filter Descriptor Multiplication and Division

To deal with the application of filters to waveforms, the abstract concept of time and filter descriptor multiplication is introduced. This involves the time effect of applying a filter to a waveform. Thus, given a filter and a waveform, the filter is applied to the waveform to determine the new waveform points, and then the waveform time descriptor is multiplied by the filter descriptor to form a new waveform time descriptor. The following equation defines the basis for all waveform processing as far as time effects are concerned:

$$\left\{ \begin{array}{c} H_i \\ K_i \\ Fs_i \end{array} \right\} \cdot \begin{bmatrix} U \\ D \\ S \end{bmatrix} \equiv \left\{ \begin{array}{c} H_o = H_i + \frac{S-D}{Fs_i} \\ K_o = (K_i - S) \cdot U \\ Fs_o = Fs_i \cdot U \end{array} \right\}. \tag{13.1}$$

This is an input waveform time descriptor multiplied by a filter descriptor, and the result is the output waveform time descriptor. This is a handy equation and allows for the simple calculation of the resulting waveform descriptor that goes with a convolution result. The equivalent code for this is shown in the `ApplyFilter()` member function in the **TimeDescriptor** class shown in Listing 13.1. Here, the multiplication operation (defined in Python by providing a `__mul__()` function to a class) treats multiplication with a filter in the same way as applying the filter.

A related equation is that of time descriptor division. As such, it is the solution for the filter descriptor between two waveform time descriptor results. Thus, provided time descriptors of an input waveform $\{w_i\}$, and an output waveform $\{w_o\}$, the unknown filter descriptor $[F]$ for the filter between them is the solution of $[F] = \{w_i\}^{-1} \cdot \{w_o\}$.

This is solved by looking at the waveform and filter multiplication problem having the following three equations:

$$H_i + \frac{S - D}{Fs} = H_o, \qquad (K_i - S) \cdot U = K_o, \qquad Fs_i \cdot U = Fs_o.$$

The first step in the solution utilizes the last equation, rearranged as $U = Fs_o/Fs_i$. Substituting this result into the second equation leads to $S = K_i - K_o/\left(Fs_o/Fs_i\right)$. Finally, substituting $S$ and $U$ into the first equation allows the solution for $D = K_i - K_o/\left(Fs_o/Fs_i\right) - (H_o - H_i) \cdot Fs_i$. Thus,

$$\left\{ \begin{array}{c} H_i \\ K_i \\ Fs_i \end{array} \right\}^{-1} \cdot \left\{ \begin{array}{c} H_o \\ K_o \\ Fs_o \end{array} \right\} = \left[ \begin{array}{c} \frac{Fs_o}{Fs_i} \\ K_i - \frac{K_o}{\frac{Fs_o}{Fs_i}} - (H_o - H_i) \cdot Fs_i \\ K_i - \frac{K_o}{\frac{Fs_o}{Fs_i}} \end{array} \right]. \tag{13.2}$$

Therefore, given two waveforms, the filter descriptor for the filter between the two waveforms can be found. The equivalent code for this is shown in the `__div__()` member function in the **TimeDescriptor** class shown in Listing 13.1 for the case where the argument supplied is an instance of **TimeDescriptor**.

Finally, time descriptors can be divided by filter descriptors. Given an output waveform time descriptor and a filter descriptor, this is equivalent to solving for the time descriptor of the input waveform that would produce that of the output waveform. This can be written as

$$\left\{ \begin{array}{c} H_o \\ K_o \\ Fs_o \end{array} \right\} \cdot \left[ \begin{array}{c} U \\ S \\ D \end{array} \right]^{-1} = \left\{ \begin{array}{c} H_i = H_o + \frac{U}{Fs_o} \cdot (D - S) \\ K_i = \frac{K_o}{U} + S \\ Fs_i = \frac{Fs_o}{U} \end{array} \right\}. \tag{13.3}$$

This can be proved through induction by substituting the right-hand side of (13.3) into (13.1) and solving for the output waveform descriptor.

The division of waveform time descriptors by filter descriptors is shown in the `__div__()` member function in the **TimeDescriptor** class shown in Listing 13.1 for the case where the argument supplied is an instance of **FilterDescriptor**.

Filter descriptor multiplication is achieved by employing (13.1) first for a left filter and again for a right filter, and employing (13.2) to determine the filter between the initial waveform and the final waveform calculated by employing two filters in cascade. It is defined as

$$\left[ \begin{array}{c} U_L \\ D_L \\ S_L \end{array} \right] \cdot \left[ \begin{array}{c} U_R \\ D_R \\ S_R \end{array} \right] = \left[ \begin{array}{c} U_L \cdot U_R \\ \frac{U_L \cdot D_L + D_R}{U_L} \\ \frac{U_L \cdot S_L + S_R}{U_L} \end{array} \right]. \tag{13.4}$$

**Listing 13.3 ImpulseResponse** class `FirFilter()` method

```
class ImpulseResponse(Waveform):
...
    def FirFilter(self):
        td=self.td
        return FirFilter(FilterDescriptor(1,-td.H*td.Fs,td.K-1),self.Values())
```

Note that filter descriptors do not necessarily commute, but they do when $U_L = U_R = 1$, as will be seen in §13.4.6. Filter descriptor multiplication is shown in the `__mul__()` member function in the **FilterDescriptor** class shown in Listing 13.2, and is only defined when the argument supplied is an instance of **FilterDescriptor**.

### 13.1.4 Filter Descriptor of the Impulse Response

Returning to the result found in discussing convolution, the time descriptor of the output waveform can be written as follows:

$$\{y\} = \left\{ \begin{array}{c} H_x + H_h + (K_h - 1)/Fs_x \\ K_x - (K_h - 1) \\ Fs_x \end{array} \right\},$$

where the time descriptor of the input waveform is given by

$$\{x\} = \left\{ \begin{array}{c} H_x \\ K_x \\ Fs_x \end{array} \right\},$$

with the unknown filter descriptor for the impulse response $\{x\} \cdot [F] = \{y\}$, or $[F] = \{x\}^{-1} \cdot \{y\}$. Using (13.2),

$$\begin{aligned} [F] = \left[ \begin{array}{c} U_F \\ D_F \\ S_F \end{array} \right] &= \left\{ \begin{array}{c} H_x \\ K_x \\ Fs_x \end{array} \right\}^{-1} \cdot \left\{ \begin{array}{c} H_y = H_x + H_h + (K_h - 1)/Fs_x \\ K_y = K_x - (K_h - 1) \\ Fs_y = Fs_x \end{array} \right\} \\ &= \left[ \begin{array}{c} \frac{Fs_y}{Fs_x} \\ K_x - \frac{K_y}{\frac{Fs_y}{Fs_x}} - (H_y - H_x) \cdot Fs_x \\ K_x - \frac{K_y}{\frac{Fs_y}{Fs_x}} \end{array} \right] = \left[ \begin{array}{c} 1 \\ K_x - K_y - (H_y - H_x) \cdot Fs_x \\ K_x - K_y \end{array} \right] \\ &= \left[ \begin{array}{c} 1 \\ -H_h \cdot Fs \\ K_h - 1 \end{array} \right]. \end{aligned}$$

Thus, the filter descriptor corresponding to an impulse response is given by

$$F_{impulse} = \left[ \begin{array}{c} 1 \\ -H_h \cdot Fs \\ K_h - 1 \end{array} \right]. \tag{13.5}$$

**Listing 13.4 Waveform** class

```
class Waveform(list):
    def __init__(self,x=None,y=None):
        if isinstance(x,Waveform):
            self.td=x.td
            list.__init__(self,x)
        elif isinstance(x,TimeDescriptor):
            self.td=x
            if isinstance(y,list):
                list.__init__(self,y)
            elif isinstance(y,(float,int,complex)):
                list.__init__(self,[y.real for _ in range(x.K)])
            else:
                list.__init__(self,[0 for _ in range(x.K)])
        else:
            self.td=None
            list.__init__(self,[])
    def Times(self,unit=None):
        return self.td.Times(unit)
    def TimeDescriptor(self):
        return self.td
    def Values(self,unit=None):
        if unit==None:
            return list(self)
        elif unit =='abs':
            return [abs(y) for y in self]
...
```

Given an instance of **ImpulseResponse** (which is derived from the not yet discussed **Waveform** class), an instance of **FirFilter** is obtained by calling `FirFilter()` as shown in Listing 13.3, where the result is formed simply by defining the filter descriptor according to (13.5) along with the impulse response waveform values.

### 13.1.5 Waveform and Filter Convolution

Now that time and filter descriptors have been defined and their interaction with the waveform time axis discussed, the convolution of waveforms with filters can be summarized; more exactly, the convolution is summarized from the perspective of time.

As shown in Listing 13.4, a waveform consists internally of a list of sample points and a time descriptor associated with these samples (i.e. an instance of a **TimeDescriptor** class as shown in Listing 13.1). Waveforms can be read and written as files, as shown in Listing 13.5, where the file format for a waveform is a horizontal offset (time of first point), a number of points, a sample rate, and the list of values associated with each sample point. A waveform file comprises text with each number on one line.

There are a number of classes deriving from **Waveform**. One of the most important is the **ImpulseResponse** class (all impulse responses are waveforms). There are also other generic classes for creating special purpose waveforms, including the **StepWaveform** class shown in Listing 13.6, the **PulseWaveform** class shown in Listing 13.7, and the **SineWaveform** class shown in Listing 13.8. These classes offer simple ways to instantiate waveforms designed for different purposes.

**Listing 13.5 Waveform** class file handling methods

```
class Waveform(list):
    def ReadFromFile(self,fileName):
        with open(fileName,'rU' if sys.version_info.major < 3 else 'r') as f:
            data=f.readlines()
            HorOffset=float(data[0])
            NumPts=int(float(data[1])+0.5)
            SampleRate=float(data[2])
            Values=[float(data[k+3]) for k in range(NumPts)]
        self.td=TimeDescriptor(HorOffset,NumPts,SampleRate)
        list.__init__(self,Values)
        return self
    def WriteToFile(self,fileName):
        with open(fileName,"w") as f:
            td=self.td
            f.write(str(td.H)+'\n')
            f.write(str(int(td.K))+'\n')
            f.write(str(td.Fs)+'\n')
            for v in self:
                f.write(str(v)+'\n')
        return self
...
```

**Listing 13.6 StepWaveform** class

```
class StepWaveform(Waveform):
    def __init__(self,td,Amplitude=1.,StartTime=0.,risetime=0.):
        x=[0 if t < StartTime else Amplitude for t in td.Times()]
        T=risetime/self.rtvsT
        rcStart=max(0,td.IndexOfTime(StartTime-T/2.))
        if td.TimeOfPoint(rcStart)<StartTime-T/2: rcStart=min(rcStart+1,len(td)-1)
        rcEnd=min(len(td)-1,td.IndexOfTime(StartTime+T/2.))
        if td.TimeOfPoint(rcEnd)>StartTime+T/2: rcEnd=max(rcEnd-1,0)
        for i in range(rcStart,rcEnd+1):
            try:
                x[i]=Amplitude*\
                    (math.sin((td.TimeOfPoint(i)-StartTime)/T*math.pi)+1.)/2.
            except ZeroDivisionError:
                pass
        Waveform.__init__(self,td,x)
```

**Listing 13.7 PulseWaveform** class

```
class PulseWaveform(Waveform):
    def __init__(self,td,Amplitude=1.,StartTime=0.,PulseWidth=0.,Risetime=0.):
        StopTime=StartTime+PulseWidth
        stepup=StepWaveform(td,Amplitude,StartTime,Risetime)
        stepdown=StepWaveform(td,Amplitude,StopTime,Risetime)
        Waveform.__init__(self,td,[stepup[k]-stepdown[k] for k in range(len(stepup))])
```

**Listing 13.8 SineWaveform** class

```
class SineWaveform(Waveform):
    def __init__(self,td,Amplitude=1.,Frequency=1e6,Phase=0.):
        x=[Amplitude*math.sin(2.*math.pi*Frequency*t+Phase/180.*math.pi)
                for t in td.Times()]
        Waveform.__init__(self,td,x)
```

**Listing 13.9 FirFilter** class

```
1  class FirFilter(WaveformProcessor):
2      def __init__(self,fd,ft):
3          self.m_fd = fd
4          self.m_ft=ft
5      def ProcessWaveform(self, wf):
6          return self.FilterWaveform(wf)
7      def FilterTaps(self):
8          return self.m_ft
9      def FilterDescriptor(self):
10         return self.m_fd
11     def FilterWaveform(self,wf):
12         td = wf.td*self.FilterDescriptor()
13         filteredwf=convolve(wf.Values(),self.FilterTaps(),'valid').tolist()
14         return Waveform(td,filteredwf)
15     def Print(self):
16         self.FilterDescriptor().Print()
17         print(str(self.FilterTaps()))
```

**Listing 13.10 Waveform** class multiplication method

```
class Waveform(list):
    def __mul__(self,other):
        if isinstance(other,WaveformProcessor):
            return other.ProcessWaveform(self)
        elif isinstance(other,(float,int,complex)):
            return Waveform(self.td,[v*other.real for v in self])
...
```

Although an **ImpulseResponse** class is a **Waveform**, it can be seen from Listing 13.3 that it can be easily converted into a **FirFilter** (see Listing 13.9). Thus, a **FirFilter** is simply an **ImpulseResponse** (which is a **Waveform**) that has a **FilterDescriptor** instead of a **TimeDescriptor** associated with the data.

To filter a waveform, `FilterWaveform()` is called on an instance of **FirFilter** with a **Waveform** as the argument, as shown on line 11 of Listing 13.9. Alternatively, the multiplication operator of **Waveform** is overloaded to deal with convolution as multiplication of a **Waveform** and **FirFilter**, as shown in Listing 13.10. Either way, convolution is performed by first multiplying the waveform time descriptor by the filter descriptor of the FIR filter to compute the new time descriptor and then convolving the waveform values by the filter taps. The convolution function in NumPy[1] has an argument called `'valid'` that tells the function to remove the startup samples, according to the time descriptor calculated.

## 13.2 Upsampling and Interpolation

Upsampling is frequency employed in the processing of waveforms to obtain higher sample point resolution. The DSP definition of upsampling involves zero insertion; upsampling by a factor $U$ means inserting $U-1$ samples of zeros in between every point in a waveform. Interpolation involves the addition of other samples in between points in a waveform.

[1]see `https://www.numpy.org/`.

The DSP definition of upsampling and interpolation are linked topics, as interpolation can be viewed as zero insertion followed by a filtering operation. Usually (and everywhere in this book), the filter operation following the zero insertion involves low-pass filtering, usually up to the Nyquist rate. An interpolating filter can be produced very easily: simply define a $K$ point impulse response containing a single impulse in the center, convert this impulse response to a frequency response, pad the frequency response to $N' = U \cdot N$ frequency points (where $N$ is the number frequency points corresponding to the $K$ point impulse response, and $U$ is the desired rate change, or upsample factor), and convert it back to an impulse response. The padding operation has the same outcome as repeating copies of the frequency response (zero insertion in the time domain) and then filtering them out with a brick-wall low-pass filter (the boxcar filter), which is the same as convolution with the sinc function in the time domain. Thus the interpolating filter is basically a sinc function with zeros occurring every $U$ points, and is unity at time zero. The points between the zeros serve to interpolate the waveform. Note that the resulting impulse response forming the sinc function is at $U$ times the original sample rate. The interpolation is therefore carried out by upsampling the input waveform by inserting $U - 1$ zeros between every sample and convolving with the sinc function impulse response obtained.

Whether zero insertion or interpolation has been utilized to effect a sample rate change, one says that the waveform has been upsampled by the upsample factor $U$ representing the factor by which the sample rate has been changed.

### 13.2.1 Upsampling

Upsampling, in order to be fully useful, requires two arguments: the upsample factor $U$ and the *upsample phase* $\Phi$. Upsampling is strictly defined to always produce $U$ times the number of input points. For a $K$ point input waveform $\mathbf{x}$ defined with points at $x[k]$ for $k \in 0 \ldots K - 1$, the new upsampled waveform $\mathbf{x}'$ is $K \cdot U$ points long such that $x'[U \cdot k + \Phi] = x[k]$ and all other points are zero. In other words, $K' = K \cdot U$ and, for $k' \in 0 \ldots K' - 1$,

$$x'[k'] = \begin{cases} x\left[\frac{k'-\Phi}{U}\right] & \text{if } \left\lfloor\frac{k'-\Phi}{U}\right\rfloor \cdot U + \Phi = k', \\ 0 & \text{otherwise.} \end{cases}$$

This definition allows for *time interleaving* of multiple data streams. Thus, if data were received from two input streams $x_0$ and $x_1$, each stream alternately sampling a waveform, such that $x_0$ is sampled at phase 0 and $x_1$ is sampled at phase 1, $x_0$ could be upsampled to $x'_0$ by upsampling with an upsample factor of two and an upsample phase of 0, and $x_1$ could be upsampled to $x'_1$ with the same upsample factor of two and an upsample phase of 1. The two upsampled streams could be added together to form a time interleaved waveform at twice the sample rate of $x_0$ and $x_1$. This is in fact how modern oscilloscopes operate.

If one upsamples with an upsample phase of 0, the horizontal offset (i.e. the time of the first point) remains unchanged, the number of points increases from $K$ to $K \cdot U$, and the sample rate changes by a factor of $U$. If one upsamples with an upsample phase of $\Phi$, then the number of points and the sample rate are calculated similarly, but the horizontal offset becomes the original horizontal offset minus $\Phi/(U \cdot Fs)$. Thus, an input waveform with a waveform descriptor $\left\{\begin{smallmatrix} H \\ K \\ Fs \end{smallmatrix}\right\}$ produces an output waveform with descriptor $\left\{\begin{smallmatrix} H-\Phi/(U\cdot Fs) \\ K\cdot U \\ U\cdot Fs \end{smallmatrix}\right\}$. Using (13.2), the filter descriptor defining the upsampler is given by

$$\left\{\begin{matrix} H \\ K \\ Fs \end{matrix}\right\}^{-1} \cdot \left\{\begin{matrix} H - \frac{\Phi}{U\cdot Fs} \\ K \cdot U \\ U \cdot Fs \end{matrix}\right\} = \left[\begin{matrix} \frac{U\cdot Fs}{Fs} \\ K - \frac{K\cdot U}{\frac{U\cdot Fs}{Fs}} - \left(H - \frac{\Phi}{U\cdot Fs} - H\right) \cdot Fs \\ K - \frac{K\cdot U}{\frac{U\cdot Fs}{Fs}} \end{matrix}\right] = \left[\begin{matrix} U \\ \frac{\Phi}{U} \\ 0 \end{matrix}\right].$$

Upsamplers defined in this way go together with downsamplers, as will be explained. A downsampler is defined by a decimation factor $D$ and a decimation phase $\Phi$. Downsampling, or decimation, is also strictly defined. For a $K$ point input waveform $\mathbf{x}$ defined with points at $x[k]$ for $k \in 0 \ldots K-1$, the new downsampled waveform $\mathbf{x}'$ is $\lceil (K-\Phi)/D \rceil$ points long. For $K' = \lceil (K-\Phi)/D \rceil$ and for $k' \in 0 \ldots K'-1$:

$$x'[k'] = x[k' \cdot D + \Phi].$$

If one downsamples with a decimation phase of zero, the horizontal offset (i.e. the time of the first point) remains unchanged, the number of points decreases from $K$ to $\lceil K/D \rceil$, and the sample rate changes by a factor of $1/D$. If one downsamples with a decimation phase of $\Phi$, then $\lceil (K-\Phi)/D \rceil$ is obtained for the number of points, the sample rate changes by $1/D$, and the horizontal offset becomes the original horizontal offset plus $\Phi/Fs$. Thus, an input waveform with a waveform descriptor $\left\{\begin{smallmatrix} H \\ K \\ Fs \end{smallmatrix}\right\}$ produces an output waveform with descriptor $\left\{\begin{smallmatrix} H+\Phi/Fs \\ \lceil (K-\Phi)/D \rceil \\ Fs/D \end{smallmatrix}\right\}$. Using (13.2), the filter descriptor defining the downsampler is calculated as follows:

$$\begin{aligned} \left\{\begin{matrix} H \\ K \\ Fs \end{matrix}\right\}^{-1} \cdot \left\{\begin{matrix} H + \frac{\Phi}{Fs} \\ \lceil (K-\Phi)/D \rceil \\ Fs/D \end{matrix}\right\} &= \left[\begin{matrix} \frac{Fs}{D\cdot Fs} \\ K - \frac{\lceil (K-\Phi)/D \rceil}{\frac{Fs}{D\cdot Fs}} - \left(H + \frac{\Phi}{Fs} - H\right) \cdot Fs \\ K - \frac{\lceil (K-\Phi)/D \rceil}{\frac{Fs}{D\cdot Fs}} \end{matrix}\right] \\ &= \left[\begin{matrix} 1/D \\ K - \left\lceil \frac{(K-\Phi)}{D} \right\rceil \cdot D - \Phi \\ K - \left\lceil \frac{(K-\Phi)}{D} \right\rceil \cdot D \end{matrix}\right]. \end{aligned} \tag{13.6}$$

Equation (13.6) is awkward because the filter descriptor depends on the input waveform. It is preferable to use downsamplers with waveforms whose length is an integer multiple of the decimation factor. When this is the case, $\left\lceil \frac{(K-\Phi)}{D} \right\rceil = K/D$, and the downsampler becomes

$$[d] = \left[\begin{matrix} 1/D \\ -\Phi \\ 0 \end{matrix}\right].$$

Furthermore, if an upsampler is cascaded with a downsampler with identical decimation and upsample factors, $D = U$, and the same sample phase $\Phi$, the result of the upsampled waveform will always contain a number of points that is a multiple of the upsample factor, and therefore of the decimation factor in this case. Then, using (13.4):

$$\begin{bmatrix} U \\ \frac{\Phi}{U} \\ 0 \end{bmatrix} \cdot \begin{bmatrix} 1/U \\ -\Phi \\ 0 \end{bmatrix} = \begin{bmatrix} U \cdot \frac{1}{U} \\ \frac{U \cdot \frac{\Phi}{U} - \Phi}{U} \\ \frac{U \cdot 0 + 0}{U} \end{bmatrix} = \begin{bmatrix} 1 \\ 0 \\ 0 \end{bmatrix}. \tag{13.7}$$

When using an upsampler followed by an interpolation filter and a sample phase of $\Phi = 0$, one could include the zeros before the actual sample taken in the input waveform. In other words, one might insert $U - 1$ samples after each waveform point in the input waveform *plus* $U - 1$ *samples before the first waveform point.* This would produce an upsampled waveform with a descriptor $\left\{ \begin{smallmatrix} H-(U-1)/(U \cdot Fs) \\ K \cdot U + U - 1 \\ U \cdot Fs \end{smallmatrix} \right\}$ and a filter descriptor of

$$\begin{bmatrix} \frac{U \cdot Fs}{Fs} \\ K - \frac{K \cdot U + U - 1}{U \cdot Fs / Fs} - \left(H - \frac{U-1}{U \cdot Fs} - H\right) \cdot Fs \\ K - \frac{K \cdot U + U - 1}{U \cdot Fs / Fs} \end{bmatrix} = \begin{bmatrix} U \\ 0 \\ -\frac{U-1}{U} \end{bmatrix}.$$

Note that the filter descriptor has negative startup samples (i.e. adds points to the waveform). This can save points that were removed during interpolation, but it does not describe the implementation of an upsampler in the strictest sense because the result of cascading an upsampler with a downsampler with the same decimation phase and $U = D$ will not be a filter that has no effect, as calculated in (13.7).

### 13.2.2 Interpolation

Sinc interpolation can be used to interpolate between points in the time domain, as shown graphically for an impulse in Figure 13.2. Since it is shown interpolating the unit impulse, it is therefore the impulse response of a filter used to perform interpolation. Mathematically, the sinc function is infinite in length to accomplish the hard bandwidth limiting effect of the padding. Furthermore, the interpolating function obtained from padding the frequency response of the unit impulse is often not the best function to use because it is basically a filter created by the frequency response sampling method. Filters created using this method have the exact response at the points specified, but often have wild inter-point behavior. There are many variations of interpolating filters.

The interpolation filter length is preferably defined by an integer number of *base side samples* $S_b$ such that the filter length is $K = 1 + 2 \cdot S_b \cdot U$. The reason for the specification of side samples is to preserve the symmetry of the filter. Thus, it makes the impulse response length odd, with the middle point being time zero. The filter length is specified as a base number (i.e. independent of upsample factor) because the performance of the filter insofar as artifacts and imperfections are concerned is not a function of the upsample factor.

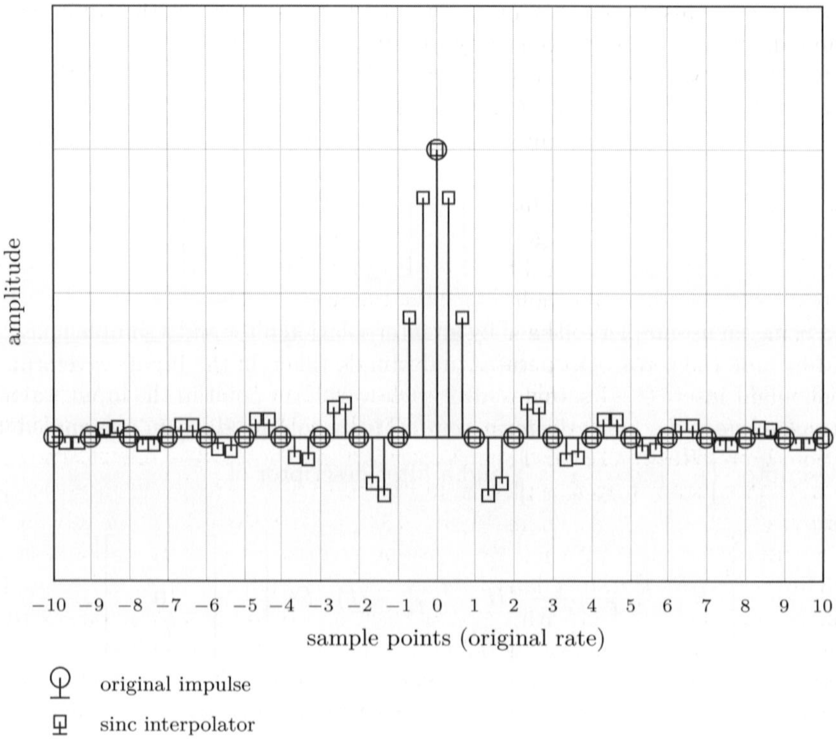


**Figure 13.2** Sinc interpolator

In other words, the base filter length determines the underlying continuous-time waveform assumed and the upsample factor determines only how many more samples are taken from it. Note that $t[0] = -t[K-1]$ is required and therefore the horizontal offset is given as $H_i = -(1/2) \cdot (K-1)/Fs = -S_b \cdot U/Fs$, where $Fs$ is the upsampled waveform sample rate and the argument to the sinc function is zero for $k = (K-1)/2 = S_b \cdot U$. The argument to the sinc function therefore becomes $(k - S_b \cdot U)/U = k/U - S_b$. Therefore, the sinc interpolator is defined, for $k \in 0 \ldots 2 \cdot S_b \cdot U$, as

$$h[k] = \begin{cases} 1 & \text{if } k/U - S_b = 0, \\ \frac{\sin(\pi \cdot (k/U - S_b))}{\pi \cdot (k/U - S_b)} \cdot \left[\frac{1}{2} + \frac{1}{2} \cdot \cos\left(\pi \cdot \frac{k/U - S_b}{S_b}\right)\right] & \text{otherwise.} \end{cases} \tag{13.8}$$

There are many variations on waveform interpolation; most are based on whether the interpolating filter really needs to be a brick-wall filter at the Nyquist rate. If one were to accept degradation of the filter response at lower than the Nyquist rate (usually due to the lack of substantial frequency content of the waveform in regions close to the Nyquist rate), then the interpolating filter can begin its roll-off earlier and reach its frequency of desired attenuation later. This serves to shorten the impulse response length requirement and to reduce time-domain artifacts produced by interpolation.

Another simple variation is to use a triangular filter for the interpolating filter; this is known as linear interpolation. A triangular filter applied to an upsampled waveform produces linearly interpolated points. Linear interpolation uncovers none of the artifacts that sinc interpolation creates, and is only applicable when the frequency content is very low relative to the Nyquist rate. This is why oscilloscopes prefer 10× oversampling: so that linear interpolation of points is a good approximation.

Some alternative interpolating filters are shown in Figure 13.3. The plots shown are for 3× interpolators. The sinc interpolating filter impulse response is given as a reference in Figure 13.3(b), where 72 taps of the filter are shown. In reality, many more than 72 taps are required to provide the correct operation; for example, the *SignalIntegrity* software would use $128 \cdot 3 = 384$ tap filters for this purpose. As mentioned, the sinc interpolation filter is specified as a brick-wall filter in the frequency domain, with a mostly equivalent time-domain specification as provided in (13.8). The brick-wall filter aspect of the sinc interpolating filter is shown in the frequency response plotted in Figure 13.3(a). The linear interpolation filter, whose impulse response is shown in Figure 13.3(c), is specified completely in the time domain, requiring only five taps for 3× interpolation. Its frequency response is shown in Figure 13.3(a), where it does not look ideal at all. Remember that the filter applied after upsampling is filtering a signal with zeros inserted and therefore the frequency response is repeating. The linear interpolation filter has a large effect on the magnitude response of the signal and retains a lot of the repeating frequency spectrum. That being said, if the system is highly oversampled, meaning that the highest amount of frequency content is constrained to about 10 to 20 % of the Nyquist rate, linear interpolation works just fine. Finally, what is referred to as a *reduced range* interpolation filter is provided. This filter is specified as flat out to some frequency of interest (one-half the Nyquist rate in the example provided in Figure 13.3). At frequencies above one minus the frequency of interest and beyond, it is specified as zero, and in the transition region, in this case, it is specified as a raised cosine:

$$H(f) = \begin{cases} 1 & \text{if } f < f_{max}, \\ 0 & \text{if } f > 1 - f_{max}, \\ \frac{1}{2} + \frac{1}{2}\cos\left(\pi \cdot \frac{f - f_{max}}{1 - 2 \cdot f_{max}}\right) & \text{otherwise.} \end{cases}$$

Remember, the frequency response repeats in a particular manner. The signal content of the original waveform was concentrated in this first Nyquist band. The signal content from the Nyquist rate out to the sample rate is this same frequency content, but reversed (and conjugated). The signal content from the sample rate to 1.5 times the sample rate is a repeat of the signal content from zero frequency to 0.5 times the sample rate. This means that if there is signal content only between zero frequency and $f_{max}$, the repeated image content will be between $1 - f_{max}$ and $1 + f_{max}$. The filter specified would filter this frequency content out entirely, and it would be a very good choice of interpolation filter under these circumstances.[2] The frequency response of this filter is shown in Figure 13.3(a) and the

[2]Technically, this filter could be allowed to rise after $1 + f_{max}$ if this would shorten the impulse response because no image is presumed to be present there.

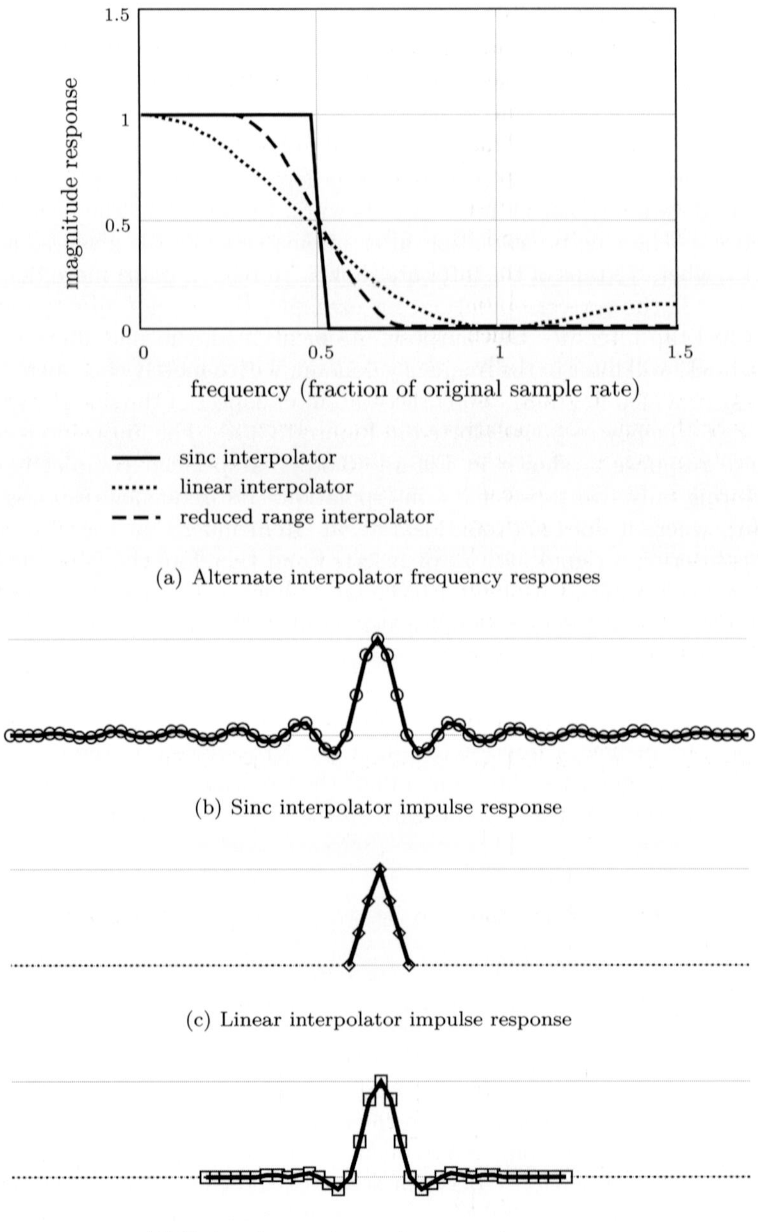


(a) Alternate interpolator frequency responses

(b) Sinc interpolator impulse response

(c) Linear interpolator impulse response

(d) Reduced range interpolator impulse response

**Figure 13.3** Alternative interpolating filters

impulse response is shown in Figure 13.3(d), where it is seen to be very short (36 taps); the amount of ringing is significantly reduced when compared with the sinc interpolating filter impulse response in Figure 13.3(b).

Some examples of how these interpolators perform are provided in Figure 13.4. Figures 13.4(a) and 13.4(b) are two example pulses with pulse widths and time constants for a simple single-pole response (e.g. an $RC$ network). In Figure 13.4(a) there is a 5 sample pulse with a time constant of 1/2 sample. This would exemplify a high frequency content signal. In Figure 13.4(b) there is a 20 sample pulse with a time constant of 3 samples. This would exemplify a low frequency content signal. Also shown are the interpolation effects: linear interpolation in Figure 13.4(c) and Figure 13.4(d), reduced range interpolation in Figure 13.4(e) and Figure 13.4(f), and sinc interpolation in Figure 13.4(g) and Figure 13.4(h). For the high frequency content signal, one can see that the linear interpolation preserves lines in between the sample points, which might be considered unacceptable, whereas the sinc interpolation shows the theoretically correct ringing. The reduced range interpolation falls somewhere in between. For the low frequency content plots on the right-hand side, it's hard to argue with linear interpolation as a good strategy. When looking at these plots, one might be tempted to reject the sinc interpolator because of all of the ringing, but remember that it is the theoretically correct interpolator for hard band limited content. One might also not like the anticipatory behavior exhibited by the reduced range and sinc interpolation highlighted by Figure 13.4(e) and Figure 13.4(g), but if the signal content of the input waveforms has a reasonable frequency content and the impulse responses of the filters are limited in a reasonable manner, there will be group delay effects that will work to improve the behavior of these interpolators.

The key takeaway point is that there are options other than sinc interpolation.

The filter descriptor for an upsampler with an upsample phase of zero is $[u] = \begin{bmatrix} U \\ 0 \\ 0 \end{bmatrix}$. Thus, for a waveform $\{x\}$ applied to an upsampler, using (13.1):

$$\{x\} \cdot [u] = \left\{ \begin{array}{c} H_x \\ K_x \cdot U \\ Fs_x \cdot U \end{array} \right\}.$$

Regarding the interpolator, the impulse response has a waveform time descriptor $\{h\} = \left\{ \begin{array}{c} -S_b \cdot U/Fs \\ 1+2 \cdot S \cdot U \\ Fs \end{array} \right\}$, where $Fs$ refers to the sample rate of the upsampled waveform (i.e. the waveform in front of the interpolator), and, using (13.5), a filter descriptor

$$[h] = \begin{bmatrix} 1 \\ S_b \cdot U \\ 2 \cdot S_b \cdot U \end{bmatrix}.$$

The filter descriptor of the entire interpolator (i.e. upsampler followed by interpolation filter) is therefore calculated using (13.4) as[3]

$$[i] = [u] \cdot [h] = \begin{bmatrix} U \\ S_b \\ 2 \cdot S_b \end{bmatrix}. \tag{13.9}$$

[3]The startup and delay samples are with respect to the input waveform.

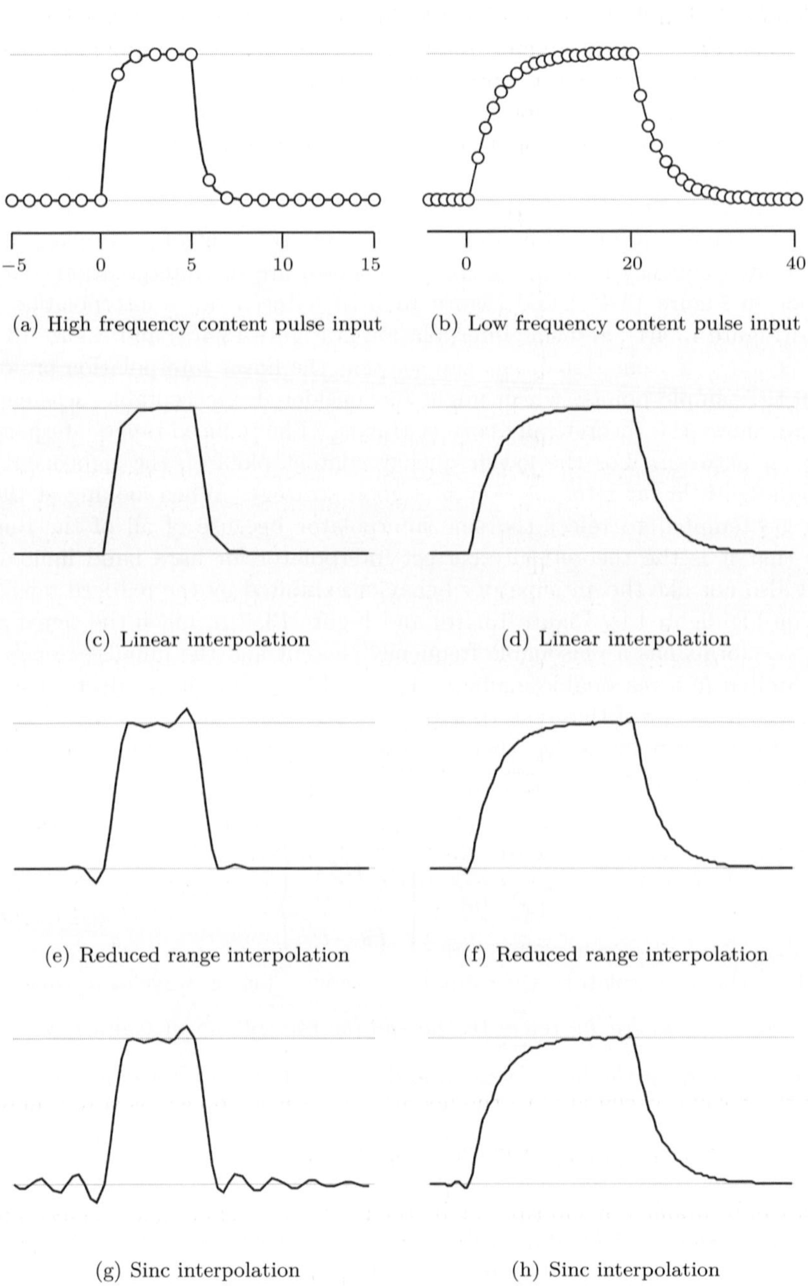


(a) High frequency content pulse input

(b) Low frequency content pulse input

(c) Linear interpolation

(d) Linear interpolation

(e) Reduced range interpolation

(f) Reduced range interpolation

(g) Sinc interpolation

(h) Sinc interpolation

**Figure 13.4** Alternative interpolator pulse responses. Left-hand plots represent high frequency content pulses with a pulse width of 5 samples and a time constant of 1/2 sample. Right-hand plots represent low frequency content pulses with a pulse width of 20 samples and a time constant of 3 samples.

Equation (13.9) is a direct result of employing an upsampler as described in §13.2.1 with the requirements defined as producing exactly $K \cdot U$ points from a $K$ point input waveform, and with a symmetric interpolator containing an odd number of points, as defined in (13.8). While alternative definitions of upsamplers and interpolators can be used, (13.9) is valid only for the symmetric interpolators that have been described here.

## 13.3 Fractional Delay Filters

Upon examining the relationship between discrete time-domain impulse responses and frequency-domain frequency responses, an interesting enforcement of the handling of the concept of fractional delay emerges. Fractional delay is defined as delay that is built into the impulse response and is a fraction of a sample in time. This concept arose from the fact that frequency responses may have a phase response that does not meet the criterion of being either 0 or $\pm 180°$ at the last frequency point, which is the Nyquist rate of the associated impulse response *unless the frequency response was generated directly from an impulse response*. This said, by the definition of impulse responses, this will not be true unless the horizontal offset is a multiple of the sample period. When generating impulse responses, this constraint is usually met; in fact, most impulse responses start exactly at time zero and each sample point of the impulse is for a sample taken at a multiple of the sample period.

How fractional delay is handled can be seen most clearly by examining a discrete-time fractional delay filter. One way to imagine the generation of a fractional delay filter is in a similar manner to that of upsampling, which was discussed in §13.2. Simply define a $K$ point impulse response $\mathbf{h}$ at a sample rate $Fs$ with a time descriptor defined with a horizontal offset of $-K/\left(2 \cdot Fs\right)$. Each sample $k \in 0 \ldots K-1$ has an associated time $t[k] = (1/Fs) \cdot (k - K/2)$ and thus has time zero at $k = K/2$. All of the samples in the impulse response $h[k]$ are set to zero except the point at time zero, which is set to $h[K/2] = 1$.

When this impulse response is converted to a frequency response, $N + 1$ frequency points are obtained in $\mathbf{H}$, where, for $n \in 0 \ldots N$, $h\left[n\right] = 1$. The associated frequencies are $f[n] = (n/N) \cdot Fs/2$. When one computes the DFT, these values will actually be as follows:

$$X[n] = \sum_{k=0}^{K-1} x[k] \cdot e^{-j \cdot 2 \cdot \pi \cdot \frac{k \cdot n}{K}} = e^{-j \cdot \pi \cdot n},$$

but, when the delay associated with the horizontal offset is applied as stated in §12.4,

$$H[n] = X[n] \cdot e^{-j \cdot 2\pi \cdot f[n] \cdot H} = e^{-j \cdot \pi \cdot n} \cdot e^{j \cdot 2\pi \cdot \frac{n}{N} \cdot \frac{Fs}{2} \cdot \frac{K}{2 \cdot Fs}} = 1.$$

A delay can be applied by multiplying by the frequency response, as discussed in §12.4.3. The response is multiplied by some fractional sample delay $F \cdot T = F/Fs$, and this new response is given by

$$H_f[n] = H[n] \cdot e^{-j \cdot 2\pi \cdot f[n] \cdot \frac{F}{Fs}} = e^{-j \cdot 2\pi \cdot \frac{n}{N} \cdot \frac{Fs}{2} \cdot \frac{F}{Fs}} = e^{-j \cdot \pi \cdot \frac{n}{N} \cdot F}.$$

When this response is converted back to an impulse response, the first step after finding the principal delay is to evaluate the phase at the Nyquist rate, $n = N$, which is $\theta = -\pi \cdot F$.

If the fractional delay is limited to between plus or minus one half sample, or $|F| < 1/2$, then applying (12.9) produces $\theta = \pi \cdot F$; applying (12.10), the time delay calculated is $F/Fs$, which is the delay that was introduced. Therefore the frequency response is advanced by this amount:

$$H'_f[n] = H_f[n] \cdot e^{j \cdot 2\pi \cdot f[n] \cdot \frac{F}{Fs}} = e^{-j \cdot \pi \cdot \frac{n}{N} \cdot F} \cdot e^{j \cdot 2\pi \cdot \frac{n}{N} \cdot \frac{Fs}{2} \cdot \frac{F}{Fs}} = 1.$$

Both $\mathbf{H}'_f$ and its corresponding IDFT are the same as the original response, but the delay is applied to the impulse response by adding the delay $F/Fs$ to the horizontal offset of the impulse response. The result is the original impulse in $\mathbf{h}$ insofar as the values are concerned, except that the horizontal offset is now $-(K/2 + F)/Fs$. Thus, the values in the impulse response have not been modified at all, but the waveform has been delayed by changing the horizontal offset. As pointed out in §12.4.3, introducing delay in the frequency domain (through multiplication of the frequency response by complex values representing the delay) or by increasing the value of the horizontal offset has the same effect on the impulse response.

This handling of fractional delay works fine on individual waveforms; the fractional delay filter is simply an impulse and its horizontal offset is all that is affected. Furthermore, convolution with this filter to achieve fractional delay will have no effect on the waveform values; only the horizontal offset of the waveform is changed. This can be seen by imagining the fractional delay filter as a single impulse point (i.e. $K = 1$) with a horizontal offset of $H_h = F$. In §13.1, it was seen that the new waveform has a horizontal offset of $H_x + H_h + (K_h - 1)/Fs$, which is simply $H_x + F$. This means that a fractional delay filter simply modifies the horizontal offset of a waveform by a fraction of a sample time. This works well on individual waveforms because, if there is only one waveform passing through one filter, the results that are processed and plotted are all correct, although the *sample phase* (the locations of the samples on the new waveform relative to the original) changes. A waveform that originally had samples at times that were multiples of the sample period will not have times at multiples of the sample period after being processed in this manner by a fractional delay filter. This causes problems when processing involves two or more waveforms. The simplest example of this is when two waveforms are added or subtracted. Since the samples are at different times, the waveform values cannot be directly added together.

These problems are resolved by waveform *adaption*: by adapting one or all waveforms to a common sample phase. In order to adapt a waveform to the sample phase of a reference waveform, one needs to fractionally delay or advance the waveform with a fractional delay filter *that does not simply change the horizontal offset*; it must affect the waveform points. One way to accomplish this is to convert the fractionally delayed frequency response back to an impulse response without the accounting for delay. In other words, a fractional delay filter is generated in the frequency domain, the Nyquist rate phase response is set to either zero or $\pi$, and is then immediately converted back to the impulse response. A fractional delay filter generated in such a manner will have the correct behavior at low frequency and over most of the frequency range, but will need to make wild changes near the Nyquist rate to get to zero or $\pi$. Therefore, much like with interpolators in §13.2, one prefers to modify the sinc interpolator in (13.8) so that it can function as both an interpolator and a

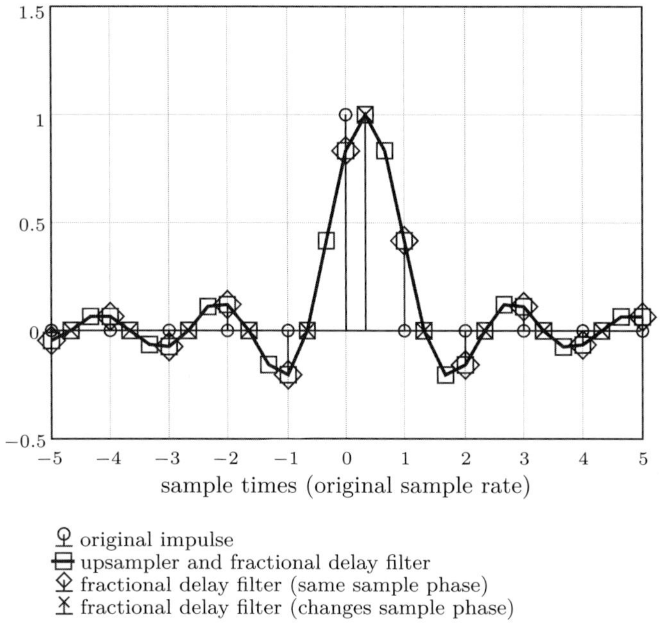


**Figure 13.5** Fractional delay filters

fractional delay filter:

$$h[k] = \begin{cases} 1 & \text{if } \frac{k}{U} - F - S_b = 0, \\ \frac{\sin\left(\pi\cdot\left(\frac{k}{U}-F-S_b\right)\right)}{\pi\cdot\left(\frac{k}{U}-F-S_b\right)} \cdot \left[\frac{1}{2} + \frac{1}{2}\cdot\cos\left(\pi\cdot\frac{\frac{k}{U}-S_b}{S_b}\right)\right] & \text{otherwise.} \end{cases} \tag{13.10}$$

In (13.10), it is preferable to introduce the fractional delay into the sinc function, but not into the raised cosine portion of the function.

Figure 13.5 is a graphical representation of what has been discussed in this section. The original impulse has a sample equal to one at time zero and zero everywhere else. Imposing a fractional delay on the impulse changes its horizontal axis. A delay of one-third of a sample is shown in Figure 13.5.

When a fractional delay filter that does not change the horizontal axis is constructed, it is formed from samples of the sinc function as shown. This is particularly obvious, for example, when one constructs an interpolator with an upsample factor of three and a fractional delay (referred to the input signal sample rate) of one-third.

The same considerations that apply to fractional delay filters are the same as those for interpolators, since the two concepts are linked. Both interpolators and fractional delay filters assume an underlying continuous-time waveform associated with a discrete-time waveform. The interpolator explicitly adds points on the underlying waveform, while the fractional delay filter exposes the assumed underlying waveform in a more subtle manner.

**Table 13.1** Filter delay rules

| When the goal is to... | Then the process is... |
|---|---|
| Properly account for and counteract the effects of filter delay | to specify the exact amount of delay provided by the filter in the filter descriptor. |
| Modify the sample phase of a waveform | to introduce a fractional delay filtering element and properly specify the amount of fractional delay provided by the filtering element in the filter descriptor |
| Delay in time physically the points of a given waveform | to introduce a filter element that provides delay and specify the amount of delay as zero in the filter descriptor<br>*or*<br>to do nothing physically to the waveform, but to specify the amount of delay as the negative of the amount of physical delay desired |

The filter descriptor of a fractional delay filter whose fractional delay is not taken into account is given by

$$[f] = \begin{bmatrix} 1 \\ S_b \\ 2 \cdot S_b \end{bmatrix}.$$

The filter descriptor of a fractional delay filter whose fractional delay is taken into account is given by

$$[f] = \begin{bmatrix} 1 \\ S_b + F \\ 2 \cdot S_b \end{bmatrix}.$$

The usage of these filter descriptors that do or do not account for the fractional delay introduced can be summarized in Table 13.1. Insofar as this book is concerned, the filter descriptors provided always counteract the effects of filters so that waveforms processed by filters end up on the correct time scale. One is not generally concerned with physically delaying waveforms. In other words, whenever a frequency response needs to be delayed or advanced to make its phase suitable for the DFT, the waveform is "physically" delayed, but then the opposite of this delay is introduced in the impulse response descriptor, thus counteracting the delay added. So, although physically delaying waveforms is sometimes useful, and the methods provide for that possibility, that is not the usual requirement. And again, if a waveform needs to be physically delayed, one usually opts to alter directly the descriptor of the waveform that leads to a sample phase change. The primary usage of fractional delay filters is to change the sample phase of a waveform, as discussed in §13.4.

```
def SinX(S,U,F):
    sl=[1. if float(k)/U-F-S==0 else
        math.sin(math.pi*(float(k)/U-F-S))/(math.pi*(float(k)/U-F-S))*\
        (1./2.+1./2.*math.cos(math.pi*(float(k)/U-S)/S))
        for k in range(2*U*S+1)]
    s=sum(sl)/U
    sl=[sle/s for sle in sl]
    return sl
```

(a) `SinX()`

```
class FractionalDelayFilterSinX(FirFilter):
    def __init__(self,F,accountForDelay=True):
        U=1
        FirFilter.__init__(self,
            FilterDescriptor(U,self.S+F if accountForDelay else self.S,2*self.S),
                SinX(self.S,U,F))
```

(b) **FractionalDelayFilterSinX** class

```
class InterpolatorSinX(FirFilter):
    def __init__(self,U):
        F=0.
        FirFilter.__init__(self,FilterDescriptor(U,self.S+F,2*self.S),SinX(self.S,U,F))
    def FilterWaveform(self,wf):
        fd=self.FilterDescriptor()
        us=[0. for k in range(len(wf)*fd.U)]
        for k in range(len(wf)):
            us[k*fd.U]=wf[k]
        return FirFilter.FilterWaveform(self,Waveform(wf.td,us))
```

(c) **InterpolatorSinX** class

```
class InterpolatorFractionalDelayFilterSinX(WaveformProcessor):
    def __init__(self,U,F,accountForDelay=True):
        self.fdf = FractionalDelayFilterSinX(F,accountForDelay)
        self.usf = InterpolatorSinX(U)
    def ProcessWaveform(self, wf):
        return self.FilterWaveform(wf)
    def FilterWaveform(self,wf):
        return self.usf.FilterWaveform(self.fdf.FilterWaveform(wf))
```

(d) **InterpolatorFractionalDelayFilterSinX** class

**Figure 13.6** Sinc processing Python code

The Python code for the interpolation and fractional delay discussed thus far is provided in Figure 13.6. In Figure 13.6(a), as a practical matter, the truncated sinc function is windowed, as shown in (13.10), for better performance. The two main classes, **FractionalDelayFilterSinX** shown in Figure 13.6(b) and **InterpolatorSinX** shown in Figure 13.6(c), can be used independently or combined in the class **Interpolator-FractionalDelayFilterSinX** shown in Figure 13.6(d). The former two derive from **Fir-Filter** and overload the `FilterWaveform()` member function, so they can be applied to

```
class FractionalDelayFilterLinear(FirFilter):
    def __init__(self,F,accountForDelay=True):
        FirFilter.__init__(self,FilterDescriptor(1,
            (F if F >= 0 else 1+F) if accountForDelay else 0,1),
            [1-F,F] if F >= 0 else [-F,1+F])
```

(a) **FractionalDelayFilterLinear** class

```
class InterpolatorLinear(FirFilter):
    def __init__(self,U):
        FirFilter.__init__(self,
            FilterDescriptor(U,(U-1.)/float(U),2*(U-1.)/float(U)),
            [float(u+1)/float(U) for u in range(U)]+
            [1-float(u+1)/float(U) for u in range(U-1)])
    def FilterWaveform(self,wf):
        fd=self.FilterDescriptor()
        us=[0. for k in range(len(wf)*fd.U)]
        for k in range(len(wf)):
            us[k*fd.U]=wf[k]
        return FirFilter.FilterWaveform(self,Waveform(wf.td,us))
```

(b) **InterpolatorLinear** class

```
class InterpolatorFractionalDelayFilterLinear(WaveformProcessor):
    def __init__(self,U,F,accountForDelay=True):
        self.fdf = FractionalDelayFilterLinear(F,accountForDelay)
        self.usf = InterpolatorLinear(U)
    def ProcessWaveform(self, wf):
        return self.FilterWaveform(wf)
    def FilterWaveform(self,wf):
        return self.usf.FilterWaveform(self.fdf.FilterWaveform(wf))
```

(c) **InterpolatorFractionalDelayFilterLinear** class

**Figure 13.7** Linear processing Python code

waveforms just like any other filter. The length of the interpolation filter is hard coded to $64 \cdot 2 \cdot U$ samples, which provides somewhat expensive, but high, performance.

### 13.3.1 Linear Fractional Delay

The linear fractional delay filter is a simple and useful fractional delay filter. As with interpolation, this filter derives from the linear interpolation filter.

The impulse response of the linear fractional delay filter is simple. Given a fractional delay $F$, it has two taps, $h[0] = 1-F$ and $h[1] = F$. The time axis for this impulse response is denoted by $t[0] = -F$ and $t[1] = 1 - F$. The filter has one startup sample, and delay samples are specified as either zero, if not accounting for the delay, or $F$ if accounting for delay.

The filter descriptor of a linear fractional delay filter whose fractional delay is not taken into account is given by

$$[f] = \begin{bmatrix} 1 \\ 0 \\ 1 \end{bmatrix}.$$

The filter descriptor of a linear fractional delay filter whose fractional delay is taken into account is given by

$$[f] = \begin{bmatrix} 1 \\ F \\ 1 \end{bmatrix}.$$

The Python code for the linear interpolation and fractional delay is provided in Figure 13.7. The two main classes, **FractionalDelayFilterLinear** shown in Figure 13.7(a) and **InterpolatorLinear** shown in Figure 13.7(b), can be used independently or combined in the class **InterpolatorFractionalDelayFilterLinear** shown in Figure 13.7(c). The former two derive from **FirFilter** and overload the `FilterWaveform()` member function, so they can be applied to waveforms just like any other filter.

## 13.4 Waveform Adaption

When there is the need to process two waveforms together, both waveforms must be arranged to have the same sample phase. Also, the overlapping portions of the two waveforms must be determined because they cannot be processed together in regions where either of the waveforms do not exist. The process of arranging waveforms in this way is called *adaption.*

Returning to the concepts of filter and time descriptors, the adaption problem is solved by considering two waveforms with time descriptors $\{a\}$ and $\{b\}$ that need to be processed together. One of these waveforms is specified as the reference waveform, whose sample phase will be retained. It is customary to use the first waveform $\{a\}$. In the adaption process, points may be trimmed from waveform $\{a\}$, while waveform $\{b\}$ might be fractionally delayed and trimmed to form the overlapping portions.

### 13.4.1 Determination of Fractional Delay

The fractional delay on waveform $\{b\}$ is calculated by finding a sample point in the same sample phase as $\{a\}$ that would be closest to a sample point in $\{b\}$, calculating the time of that sample point in $\{a\}$, and subtracting from it the time of the chosen sample point in $\{b\}$. The time of the first sample point in $\{b\}$ is $H_b$. The integer $k$ is determined that satisfies $|H_a + k/Fs - H_b| \cdot Fs \leq 1/2$. The solution to this is $\lfloor (H_b - H_a) \cdot Fs \rceil$. The time of this point in $\{a\}$ is $H_a + \lfloor (H_b - H_a) \cdot Fs \rceil / Fs$, and therefore the amount to fractionally delay $\{b\}$, to get it onto the same sample phase as $\{a\}$, is

$$F = (H_b - H_a) \cdot Fs - \lfloor (H_b - H_a) \cdot Fs \rceil . \tag{13.11}$$

The fractional delay amount is in samples and obeys the inequality $-1/2 \leq F \leq 1/2$.

Fractional delay filters first delay the waveform (on the same sample phase as the original) and then slide the waveform backward in time. Thus, if the desired sample phase of a point in a waveform is backward, the fractional delay will be positive.

### 13.4.2 Determination of Overlapping Waveform Portion

If the amount of required fractional delay is $F \neq 0$, the effect of the fractional delay filter must be anticipated. If $F = 0$, the fractionally delayed waveform $\{c\} = \{b\}$, otherwise $\{c\} = \{b\} \cdot [h]$, where $[h] = \begin{bmatrix} 1 \\ S_b + F \\ 2 \cdot S_b \end{bmatrix}$. The fractional delay filter effect is anticipated by accounting for its fractional delay in the filter descriptor. This causes the application of the fractional delay filter to adjust the sample phase of the waveform, as stated in Table 13.1.

Now that $\{c\}$ reflects a waveform that is on the same sample phase as $\{a\}$, the minimum time in the overlapping waveform can be calculated (i.e. the horizontal offset of the overlapping waveform):

$$t_{min} = H_0 = \max\left(H_a, H_c\right).$$

The maximum time is calculated as

$$t_{max} = \min\left[H_a + \frac{K_a - 1}{Fs}, H_c + \frac{K_c - 1}{Fs}\right],$$

and the number of points in the overlapping waveform is

$$K_o = \max\left[0, \left(t_{max} - H_o\right) \cdot Fs + 1\right].$$

Thus, the waveform descriptor for the overlapping portion is found to be

$$\{o\} = \left\{ \begin{matrix} \max\left(H_a, H_c\right) \\ \max\left[0, \left(\min\left[H_a + \frac{K_a-1}{Fs}, H_c + \frac{K_c-1}{Fs}\right] - \max\left(H_a, H_c\right)\right) \cdot Fs + 1\right] \\ Fs \end{matrix} \right\}. \quad (13.12)$$

The waveform descriptor for the overlapping waveform portion is the time descriptor of the desired result. This time descriptor $\{o\}$ is for a waveform that simply has points removed from the left and/or right side of waveforms $\{a\}$ and $\{c\}$. This means that points are lopped off each side of $\{a\}$, and $\{b\}$ is processed through the fractional delay filter with points lopped off each side of the result $\{c\}$.

### 13.4.3 Waveform Trimming

Equation (13.2) is utilized to determine the filter descriptors for filters that could be used to remove points from waveforms. In other words, $[t_a] = \{a\}^{-1} \cdot \{o\}$ is the filter descriptor of a filter *that converts the time axis of waveform* $\{a\}$ *to the time axis of* $\{o\}$. Similarly, $[t_c] = \{c\}^{-1} \{o\}$. These filters are not actually applied (although they could be), but instead the descriptor information is utilized to determine what is required.

To utilize the descriptor information for this purpose, the *trimmer / filter duality* must be understood.

**Listing 13.11 FilterDescriptor** class trimming methods

```
class FilterDescriptor(object):
...
    def TrimLeft(self):
        return self.S-self.D
    def TrimRight(self):
        return self.D
    def TrimTotal(self):
        return self.S
...
```

**Listing 13.12 WaveformTrimmer** class

```
class WaveformTrimmer(FilterDescriptor,WaveformProcessor):
    def __init__(self,TrimLeft,TrimRight):
        FilterDescriptor.__init__(self,1,TrimRight,TrimLeft+TrimRight)
    def ProcessWaveform(self, wf):
        return self.TrimWaveform(wf)
    def TrimWaveform(self,wf):
        K=wf.td.K
        TL=self.TrimLeft()
        TT=self.TrimTotal()
        return Waveform(wf.td*self,
            [wf[k+TL] if 0 <= k+TL < K else 0. for k in range(K-TT)])
```

Consider a filter produced from an impulse response with a single element of one at time zero, and with unity sample rate. Applying this filter (i.e. convolving this impulse response) with a waveform has no effect. It has no delay samples because it starts from time zero, and it has no startup samples because it has one element. Now consider a filter produced from this same impulse response with a single zero added to the right (i.e at sample number 1 corresponding to time 1); this is now a two element filter. This filter still has no delay samples, but now it has one startup sample. This startup sample is trimmed from the left of the waveform result, as expected for convolution with an impulse response that starts from time zero. Furthermore, if $L$ zero points are placed to the right of the unit impulse occurring at the first sample point at time zero, $L$ points are trimmed from the left of the waveform.

Consider another impulse response, but this time with $R$ zero samples preceding a single unit impulse at the end, and take the time of the first zero sample to be $-R$; the result of convolution with this filter is to delay the input waveform, thereby pushing samples off the end, and thus $R$ samples are trimmed from the right of the waveform. Now a filter is formed by cascading two filters. The first filter $[r]$ is formed with $R$ zeros followed by a one, where the first sample occurs at time $-R$ (assuming a sample rate of one). The second filter $[l]$ is formed with a one followed by $L$ zeros, which starts at time zero. Using (13.5), the first filter has a filter descriptor of $\begin{bmatrix} 1 \\ R \\ R \end{bmatrix}$. The second filter has a filter descriptor of $\begin{bmatrix} 1 \\ 0 \\ L \end{bmatrix}$. Using (13.4) to cascade the two filters results in $\begin{bmatrix} 1 \\ R \\ L+R \end{bmatrix}$. This filter trims $R$ points from the right of the waveform and a total of $L + R$ points from the waveform, and thus $L$ points are trimmed from the left of the waveform.

Since filter descriptors are in terms of delay and startup samples, if a filter descriptor $[f]$ is found for the filter between two waveforms (i.e. formed by the waveform division equation in (13.2)), then the number of points trimmed from the right equals the delay samples $D$, the total number of points trimmed equals the startup samples $S$, and the number of points trimmed from the left is given by $S - D$.

When using filter descriptors to define point trimming, the following nomenclature is used for the number of points trimmed:

$$\langle L|R\rangle = \begin{bmatrix} 1 \\ R \\ L+R \end{bmatrix} ; \qquad \begin{bmatrix} 1 \\ D \\ S \end{bmatrix} = \langle S-D|D\rangle \, . \tag{13.13}$$

In summary, given an output waveform $\{o\}$ and an input waveform $\{i\}$, the filter that transforms the time axis of $\{i\}$ into the time axis of $\{o\}$ by using (13.2) is $[f] = \{i\}^{-1}\{o\}$. When viewed as a point trimmer, the number of points to trim is found from the delay and startup samples of the filter descriptor.

Listing 13.12 shows the **WaveformTrimmer** class, which derives from **FilterDescriptor**. In the constructor on line 2 it takes the number of points to trim from each side and constructs a **FilterDescriptor** according to (13.13). The `TrimWaveform()` member function on line 6 trims the waveform by multiplying the waveform time descriptor by itself and removing the data values from the waveform.

In the **FilterDescriptor** class shown in Listing 13.11 there are three member functions, `TrimLeft()` on line 3, `TrimRight()` on line 5, and `TrimTotal()` on line 7, that supply the trimming information from a **FilterDescriptor**. This reinforces the fact that any filter descriptor can be viewed as supplying waveform trimming information, regardless of the type of filter employed.

### 13.4.4 Waveform Adaption Summary

Given two waveforms with time descriptors $\{a\}$ and $\{b\}$ that need to be adapted to each other, and considering $\{a\}$ as the reference waveform whose sample phase is to be retained, the steps of adaption are as follows:

1. Determine a fractional delay required to put $\{b\}$ on the same sample phase as $\{a\}$.
2. Apply the effect of a fractional delay filter applied to $\{b\}$ to form a resulting waveform descriptor $\{c\}$.
3. Determine the overlapping portion of $\{a\}$ and $\{c\}$ defined as $\{o\}$.
4. Determine the filter descriptors that transform the waveform descriptors $\{a\}$ and $\{c\}$ into $\{o\}$.
5. Interpret the filter descriptors as waveform trimmers and determine the number of points to trim from $\{a\}$ and $\{c\}$ to obtain the adapted results.

The fractional delay filter can then be applied to $\{b\}$ to produce $\{c\}$, and the appropriate points can be trimmed from $\{a\}$ and $\{c\}$ to form the adapted result. The adapted result will be two waveforms having the same length and horizontal offset, and, assuming the sample rate is the same, they will have identical time descriptors.

**Listing 13.13 Waveform** class `Adapt()` method

```
class Waveform(list):
    def Adapt(self,td):
        wf=self
        (upsampleFactor,decimationFactor)=Rat(td.Fs/wf.td.Fs)
        if upsampleFactor>1:
            wf=wf*(InterpolatorSinX(upsampleFactor) if wf.adaptionStrategy=='SinX'
                else InterpolatorLinear(upsampleFactor))
        ad=td/wf.td
        f=ad.D-int(ad.D)
        if not f==0.0:
            wf=wf*(FractionalDelayFilterSinX(f,True) if wf.adaptionStrategy=='SinX'
                else FractionalDelayFilterLinear(f,True))
            ad=td/wf.td
        if decimationFactor>1:
            decimationPhase=int(round(ad.TrimLeft())) % decimationFactor
            wf=wf*WaveformDecimator(decimationFactor,decimationPhase)
            ad=td/wf.td
        tr=WaveformTrimmer(max(0,int(round(ad.TrimLeft()))),
                           max(0,int(round(ad.TrimRight()))))
        wf=wf*tr
        return wf
...
```

Listing 13.13 is the **Waveform** class `Adapt()` member function used to adapt one waveform to another as outlined. Waveform multiplication by trimmers is also defined in Listing 13.10, and that is employed at the end of Listing 13.13.

### 13.4.5 Waveform Adaption Example

Figure 13.8 shows two waveforms that are to be adapted. The time descriptors are $\{a\} = \left\{\begin{smallmatrix} -3.205 \\ 12 \\ 1 \end{smallmatrix}\right\}$ and $\{b\} = \left\{\begin{smallmatrix} 0.451 \\ 12 \\ 1 \end{smallmatrix}\right\}$. The first step is to determine the amount of fractional delay required to get waveform $\{b\}$ onto the same sample phase as $\{a\}$. This is calculated according to (13.11) as follows:

$$\begin{aligned} F &= (H_b - H_a) \cdot Fs - \lfloor (H_b - H_a) \cdot Fs \rceil \\ &= (0.451 - -3.205) - \lfloor (0.451 - -3.205) \rceil \\ &= -0.344. \end{aligned}$$

Remembering that positive fractional delay means to slide the sample phase backward and negative fractional delay means to slide the sample phase forward, this result makes sense, because sliding the sample phase of $\{b\}$ forward by 0.344 looks right.

To accomplish the fractional delay, a linear phase fractional delay filter will be applied. This has a filter descriptor of $[f] = \left[\begin{smallmatrix} 1 \\ F \\ 1 \end{smallmatrix}\right]$, so, applying this to $\{b\}$ and using (13.1),

$$\{c\} = \{b\} \cdot [f] = \left\{ \begin{array}{c} H + \frac{S-D}{Fs} \\ (K - S) \cdot U \\ Fs \cdot U \end{array} \right\} = \left\{ \begin{array}{c} 0.451 + \frac{1 - -0.344}{1} \\ (12 - 1) \cdot 1 \\ 1 \cdot 1 \end{array} \right\} = \left\{ \begin{array}{c} 1.795 \\ 11 \\ 1 \end{array} \right\}.$$

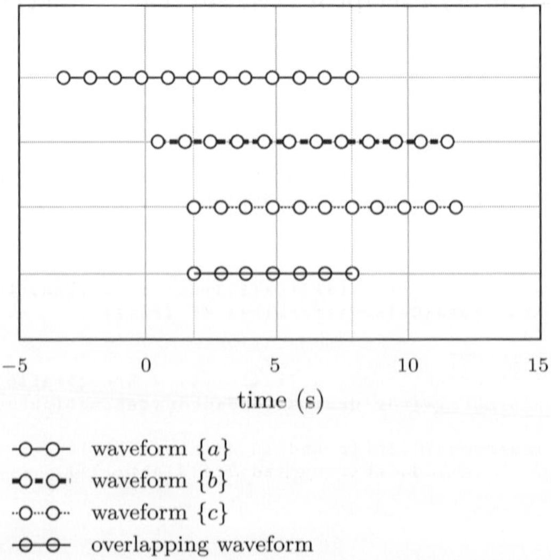


**Figure 13.8** Waveform adaption example

Checking, one finds that $1.795 - -3.205 = 5$ is indeed a multiple of the sample period $1/Fs = 1$.

Now the overlapping waveform portion of $\{a\}$ and $\{c\}$ is determined. Applying (13.12):

$$\{o\} = \left\{ \begin{array}{c} \max\left(H_a, H_c\right) \\ \max\left[0, \left(\min\left[H_a + \frac{K_a-1}{Fs}, H_c + \frac{K_c-1}{Fs}\right] - \max\left(H_a, H_c\right)\right) \cdot Fs + 1\right] \\ Fs \end{array} \right\}$$

$$= \left\{ \begin{array}{c} \max\left(-3.205, 1.795\right) \\ \max\left[0, \left(\min\left[-3.205 + \frac{12-1}{1}, 1.795 + \frac{11-1}{1}\right] - \max\left(-3.205, 1.795\right)\right) + 1\right] \\ 1 \end{array} \right\}$$

$$= \left\{ \begin{array}{c} 1.795 \\ 7 \\ 1 \end{array} \right\};$$

this is the time descriptor of the final result.

Next, the filter descriptor of the filter that transforms $\{a\}$ into $\{o\}$ is found. Using (13.2):

$$[f_{ao}] = \left[ \begin{array}{c} \frac{Fs_o}{Fs_i} \\ K_i - \frac{K_o}{\frac{Fs_o}{Fs_i}} - (H_o - H_i) \cdot Fs_i \\ K_i - \frac{K_o}{\frac{Fs_o}{Fs_i}} \end{array} \right] = \left[ \begin{array}{c} 1 \\ 12 - \frac{7}{1} - (1.795 - -3.205) \cdot 1 \\ 12 - \frac{7}{1} \end{array} \right] = \left[ \begin{array}{c} 1 \\ 0 \\ 5 \end{array} \right].$$

**Listing 13.14 FilterDescriptor** class `Before()` and `After()` methods

```
class FilterDescriptor(object):
...
    def Before(self,other):
        return FilterDescriptor(
                UpsampleFactor=self.U,
                DelaySamples=(other.D*other.U+self.D)/other.U-other.D/self.U,
                StartupSamples=(other.U*other.S+self.S)/other.U-other.S/self.U)
    def After(self,other):
        return FilterDescriptor(
                UpsampleFactor=self.U,
                DelaySamples=(self.D*self.U+other.D)*other.U/self.U-other.D*other.U,
                StartupSamples=(self.S*self.U+other.S)*other.U/self.U-other.U*other.S)
...
```

The filter descriptor of the filter that transforms $\{c\}$ into $\{o\}$ is

$$[f_{co}] = \begin{bmatrix} \frac{Fs_o}{Fs_i} \\ K_i - \frac{K_o}{\frac{Fs_o}{Fs_i}} - (H_o - H_i) \cdot Fs_i \\ K_i - \frac{K_o}{\frac{Fs_o}{Fs_i}} \end{bmatrix} = \begin{bmatrix} 1 \\ 11 - \frac{7}{1} - (1.795 - 1.795) \cdot 1 \\ 11 - \frac{7}{1} \end{bmatrix} = \begin{bmatrix} 1 \\ 4 \\ 4 \end{bmatrix}.$$

Using (13.13) to convert the filter descriptors to waveform trimmers yields $[f_{ao}] = \langle 5|0\rangle$ and $[f_{co}] = \langle 0|4\rangle$.

Therefore, to adapt the two waveforms $\{a\}$ and $\{b\}$, five points are trimmed from the left of $\{a\}$. A linear fractional delay filter is applied to $\{b\}$ with a fractional delay of 0.344 and four points are trimmed from the right of the result. The adapted waveforms have a common waveform descriptor of $\{o\} = \left\{ \begin{smallmatrix} 1.795 \\ 7 \\ 1 \end{smallmatrix} \right\}$. The results are shown graphically in Figure 13.8.

### 13.4.6 Changing Filter Order in Cascade Arrangements

In §13.4.4, waveform adaption was seen to involve the trimming of points from a waveform. Since this waveform could be a processed result, this seems wasteful, and one wonders whether the points could be trimmed prior to processing. This question involves the determination of whether filter order *from a time axis standpoint* is valid and how to handle it. Filter order might be important for reasons other than the time axis, but in general filter orders can be modified; this section discusses the time axis implications of filter order.

In response to this question, two filters in cascade are considered. Consider a left filter $[L]$, a right filter $[R]$, and an object filter $[F]$. Given, for example, a cascade arrangement $[L] \cdot [F]$, a question might be: What should the filter descriptor of $[L]$ become if it is moved after $[F]$ to become a right filter $[F] \cdot [R]$? Similarly, given a filter descriptor $[R]$ in the cascade arrangement $[F] \cdot [R]$, a question might be: What should the filter descriptor of $[R]$ become if it is moved to the left of $[F]$ in an arrangement $[L] \cdot [F]$?

The solution to these questions is to consider the filter cascade equation (13.4) and solve $[L] \cdot [F] = [F] \cdot [R]$. Solving for both $[L]$ and $[R]$ in terms of the other unknowns determines the Before and After operators for filters:

$$[L] = \ [R] \quad \text{Before} \quad [F] = \begin{bmatrix} U_R \\ \frac{U_F \cdot D_F + D_R}{U_F} - \frac{D_F}{U_R} \\ \frac{U_F \cdot S_F + S_R}{U_F} - \frac{S_F}{U_R} \end{bmatrix}; \tag{13.14}$$

$$[R] = \ [L] \quad \text{After} \quad [F] = \begin{bmatrix} U_L \\ (U_L \cdot D_L + D_F) \cdot \frac{U_F}{U_L} - U_F \cdot D_F \\ (U_L \cdot S_L + S_F) \cdot \frac{U_F}{U_L} - U_F \cdot S_F \end{bmatrix}. \tag{13.15}$$

These equations are read as follows: in (13.14), the right filter becomes the left filter when moved before a given filter; and, in (13.15), the left filter becomes the right filter when moved after a given filter.

When the upsample factors of the filters equal one, i.e. $U_L = U_R = U_F = 1$,

$$[L] = \ [R] \quad \text{Before} \quad [F] = \begin{bmatrix} 1 \\ D_R \\ S_R \end{bmatrix} = [R]$$

and

$$[R] = \ [L] \quad \text{After} \quad [F] = \begin{bmatrix} 1 \\ D_L \\ S_L \end{bmatrix} = [L] \ .$$

In other words, when the upsample factors of the filters equal unity, filter order is unimportant. This is expected.

This means that during waveform adaption it is possible to trim waveform points *prior* to fractional delay filtering. If the sequence of waveform processing is known, it may even be possible to trim the input waveform prior to processing to save on computation requirements.

Listing 13.14 shows the member functions `Before()` (line 3) implemented according to (13.14) and `After()` (line 8) implemented according to (13.15).

## 13.5 Transfer Matrices Processing

In time-domain applications such as simulation (see Chapter 9) and virtual probing (see Chapter 11), the problem solution involves *transfer matrices*. For example, the **Simulator-Numeric** class shown in Listing 9.5 and the **VirtualProbeNumeric** class shown in Listing 11.4 each provide a transfer matrix, and the **SimulatorNumericParser** class shown in Listing 9.6 and the **VirtualProbeNumericParser** class shown in Listing 11.5 each provide transfer matrices as their solutions. From a frequency-domain perspective, these are the solutions, but they are primarily intended for time-domain processing.

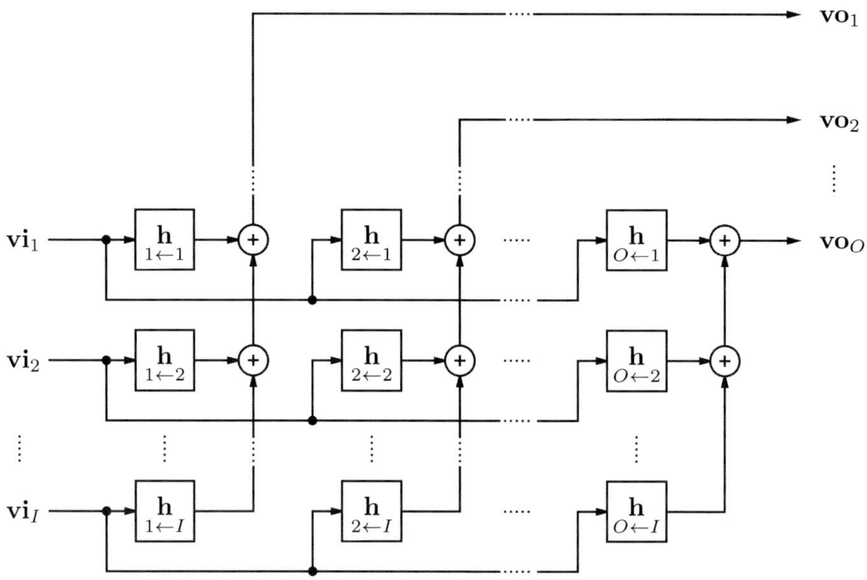


(a) Graphical representation

```
class TransferMatricesProcessor(CallBacker):
    def __init__(self,transferMatrices,callback=None):
        self.TransferMatrices=transferMatrices
    def ProcessWaveforms(self,wfl,td=None):
        if td is None:
            td = [wflm.td.Fs for wflm in wfl]
        ir = self.TransferMatrices.ImpulseResponses(td)
        result=[]
        for o in range(len(ir)):
            acc=[]
            for i in range(len(ir[o])):
                acc.append(ir[o][i].FirFilter().FilterWaveform(wfl[i]))
            result.append(sum(acc))
        return result
```

(b) **TransferMatricesProcessor** class

**Figure 13.9** Transfer matrices processor

**Listing 13.15 TransferMatrices** class

```
1  class TransferMatrices(list):
2      def __init__(self,f,d):
3          self.f=FrequencyList(f)
4          list.__init__(self,d)
5          self.Inputs=len(d[0][0])
6          self.Outputs=len(d[0])
7  ...
8      def FrequencyResponse(self,o,i):
9          return FrequencyResponse(self.f,[Matrix[o-1][i-1]
10             for Matrix in self])
11     def FrequencyResponses(self):
12         return [[self.FrequencyResponse(o+1,s+1)
13             for s in range(self.Inputs)] for o in range(self.Outputs)]
14     def ImpulseResponses(self,td=None):
15         fr = self.FrequencyResponses()
16         if td is None or isinstance(td,float) or isinstance(td,int):
17             td = [td for m in range(len(fr[0]))]
18         return [[fro[m].ImpulseResponse(td[m]) for m in range(len(fro))]
19             for fro in fr]
```

The **TransferMatrices** class was provided in Listing 9.7 for frequency-domain purposes only, but the remainder of the member functions utilized primarily for time-domain processing are shown in Listing 13.15. An instance of **TransferMatrices** is used to initialize a **TransferMatricesProcessor** through the `__init__()` function on line 2.

The transfer matrices processor is shown in Figure 13.9. Waveforms are processed by passing a list of instances of the **Waveform** class provided in Listing 13.4 to the `ProcessWaveforms()` member function on line 4 of the **TransferMatricesProcessor** class in Figure 13.9(b). Here, the transfer matrices are converted into a matrix of impulse responses that are applied to the waveforms as FIR filters. The function returns a list of output waveforms that are the sum of the filtered input waveforms.

For simulation and virtual probing, at each frequency there is a transfer matrix that converts an array of input waveforms to an array of output waveforms. This can be described mathematically as follows. Given $I$ input waveforms in a vector **VI** and $O$ output waveforms in a vector **VO**, for a set of frequencies, $n \in 0 \ldots N$, $f[n] = \frac{n}{N} \cdot \frac{Fs}{2}$,

$$\begin{pmatrix} \mathbf{vo}[n]_{[1]} \\ \mathbf{vo}[n]_{[2]} \\ \vdots \\ \mathbf{vo}[n]_{[O]} \end{pmatrix} = \begin{pmatrix} \mathbf{H}[n]_{[1][1]} & \mathbf{H}[n]_{[1][2]} & \cdots & \mathbf{H}[n]_{[1][I]} \\ \mathbf{H}[n]_{[2][1]} & \mathbf{H}[n]_{[2][2]} & \cdots & \mathbf{H}[n]_{[2][I]} \\ \vdots & \vdots & \ddots & \vdots \\ \mathbf{H}[n]_{[O][1]} & \mathbf{H}[n]_{[O][2]} & \cdots & \mathbf{H}[n]_{[O][I]} \end{pmatrix} \cdot \begin{pmatrix} \mathbf{vi}[n]_{[1]} \\ \mathbf{vi}[n]_{[2]} \\ \vdots \\ \mathbf{vi}[n]_{[I]} \end{pmatrix},$$

or, for $o \in 1 \ldots O$ , $i \in 1 \ldots I$,

$$\mathbf{vo}[n]_{[o]} = \sum_{i=1}^{I} \mathbf{H}[n]_{[o][i]} \cdot \mathbf{vi}[n]_{[i]} .$$

Since $\mathbf{H}[n]_{[o][i]}$ is a complex number at a given frequency, the data can be reorganized as follows:

$$\underset{o\leftarrow i}{H}[n] = \mathbf{H}[n]_{[o][i]}, \qquad VO_o[n] = \mathbf{vo}[n][o], \qquad VI_i[n] = \mathbf{vi}[n][i].$$

In other words, where previously there was a list of matrices and vectors per frequency, now there is a matrix of frequency responses and vectors of frequency content. The frequency responses of the outputs due to the inputs can be described as follows:

$$VO_o[n] = \sum_{i=1}^{I} \underset{o\leftarrow i}{\mathbf{H}}[n] \cdot VI_i[n],$$

where $VO_o$ and $VI_i$ are $N+1$ element vectors describing the frequency content of an output and input, respectively, and $\underset{o\leftarrow i}{\mathbf{H}}$ is an $N+1$ element frequency response characteristic that supplies the part of the response of an output due to an input. Each frequency response is converted to an impulse response:

$$\underset{o\leftarrow i}{\mathbf{h}} = \text{IDFT}\left(\underset{o\leftarrow i}{\mathbf{H}}\right),$$

and the result is the sum of multiple convolutions:

$$vo_o = \sum_{i=1}^{I} \underset{o\leftarrow i}{\mathbf{h}} * vi_i,$$

where $vo_o$ and $vi_i$ are output and input time-domain waveforms. This leads to the filter structure shown in Figure 13.9(a) that describes all linear, time-domain simulations.

# 14

# The Impedance Profile

S-PARAMETERS, WHEN VIEWED IN THE FREQUENCY DOMAIN, do not offer much insight into signal integrity applications. Signal integrity analysis is best carried out in the time domain, where one has a sense of time and distance. This is especially true when transmission lines are involved, which is almost always.

S-parameters, or more specifically the s-parameters on the diagonal of the s-parameter matrix, have a very useful time-domain transformation. This transformation is the impedance profile and is the subject of this chapter.

The impedance profile is a plot whose x axis is time (or distance if the propagation velocity is known) and whose y axis is either the instantaneous reflection coefficient ($\rho$) or impedance ($z$).[1] It is generated by assuming that a system can be considered to consist of multiple, cascaded, usually small and constant in length, transmission line sections. This, of course, is an abstraction, but this abstraction is very useful in understanding what is happening along the transmission line. Furthermore, this model has two other enabling features:

- If loss is neglected, or known per unit time and frequency, a measurement of $s_{11}$ (i.e. a one-port measurement) can be used to generate a two-port model of a system. This is particularly useful when looking into the port of a connector, fixture, or probe, or when using measurements where only one port could be connected to a network analysis instrument.
- Since the impedance profile is vs. time, and essentially vs. length, s-parameters can be de-embedded by cutting the impedance profile.

This chapter will demonstrate these features.

## 14.1 Impedance and Time-Domain Reflectometry

The time-domain reflectometer (TDR) has been a common tool for signal integrity engineers. It utilizes a step with a very fast risetime to stimulate a circuit. By observing the step response, one can measure impedances [38]. Figure 14.1 illustrates how this works. A circuit simulation is employed in Figure 14.1(a), and a VNA measurement is made in Figure 14.1(b). In Figure 14.1(a), a step is applied with a step generator through a 50 Ω resistor to a 50 Ω, 1 ns cable to three possible terminations: open, 50 Ω load, and a short to ground. The step responses generated by this circuit are shown in Figure 14.1(c), where the step is launched at time zero and remains unchanged at 0.5 V for 2 ns. At 2 ns, it rises to 1 V for the

[1] This is not to be confused with z-parameters which are something entirely different.

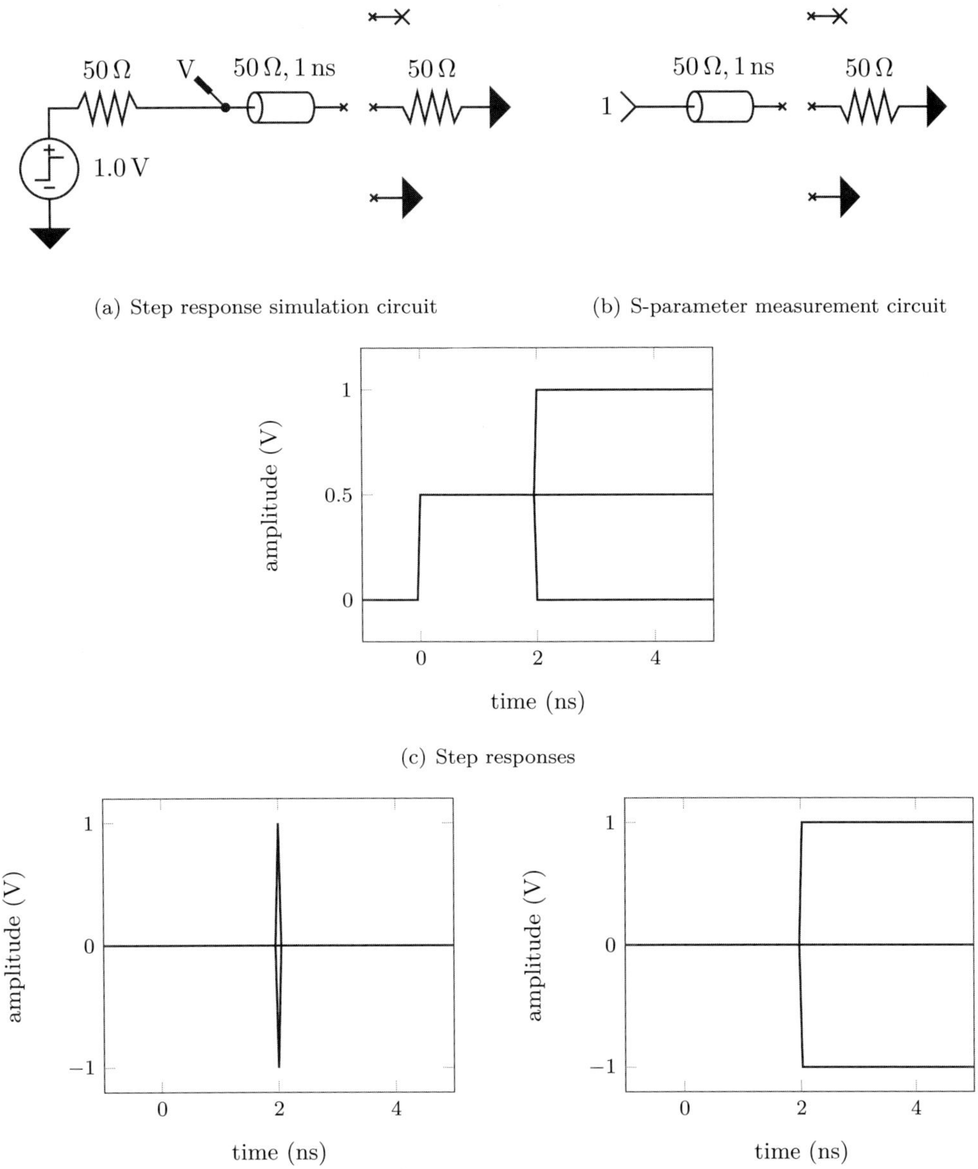


(a) Step response simulation circuit

(b) S-parameter measurement circuit

(c) Step responses

(d) $s_{11}$ impulse response

(e) Integrated $s_{11}$ impulse response

**Figure 14.1** Measurement of single-port impedance

open, continues at 0.5 V in the case of the 50 Ω load, or goes to zero in the case of the short to ground. Using a TDR, one can therefore measure the time location of an impedance as one-half the time measured to the impedance (1 ns in this case), and the impedance of the termination.

The impedance of the termination is measured very simply. For a circuit with an internal step driving voltage of $V$ and a source impedance of $Z_s$, the final value of the voltage $v$ across the impedance being measured $Z$ is

$$v = \frac{V}{Z + Z_s} \cdot Z,$$

and thus $Z$ is measured as:

$$Z = \frac{v}{V - v} \cdot Z_s. \tag{14.1}$$

If $V$ and $Z_s$ are unknown, the system can be calibrated by applying an open-circuit impedance, in which case the final voltage is measured as $V$, and by applying a calibration standard with a known impedance $Z_{std}$:

$$v_{std} = \frac{V}{Z_{std} + Z_s} \cdot Z_{std},$$

and therefore

$$Z_s = Z_{std} \cdot \frac{V - v_{std}}{v_{std}}.$$

The reflection coefficient Γ is given by (3.41) as

$$\Gamma = \frac{Z - Z0}{Z + Z0} = \frac{\frac{v}{V-v} \cdot Z_s - Z0}{\frac{v}{V-v} \cdot Z_s + Z0} = \frac{\left(\frac{Z_s}{Z0} + 1\right) \cdot v - V}{\left(\frac{Z_s}{Z0} - 1\right) \cdot v - V}. \tag{14.2}$$

In the case where $Z_s = Z0$,

$$\Gamma = \frac{2 \cdot v - V}{V} = \frac{v}{V/2} - 1 = \frac{v - V/2}{V/2}. \tag{14.3}$$

This means that for the three cases of a short, an open, and an impedance $Z_s$, Γ equals −1, 1, and 0, respectively.

Since the TDR is often used in a somewhat crude fashion, it is sometimes assumed that $Z_s = Z0$ and that (14.3) is the assumed equation. Even so, it should be understood that this measured Γ can be converted to a Γ′ in the reference impedance $Z0$ using (5.13) as follows:

$$\Gamma' = \frac{\Gamma - \frac{Z0 - Z_s}{Z + Z_s}}{1 - \Gamma \cdot \frac{Z0 - Z_s}{Z + Z_s}}; \tag{14.4}$$

simplifying (14.4), one arrives at exactly (14.2) for Γ in the reference impedance $Z0$.

If Γ is calculated in a given reference impedance $Z0$, then this Γ is converted to an impedance by solving for $Z$ in (3.41):

$$Z = Z0 \cdot \frac{1 + \Gamma}{1 - \Gamma}.$$

In other words, the TDR is measuring in a reference impedance $Z_s$, and, if $Z_s$ is known, the measurement can be converted to any other reference impedance, for example $50\,\Omega$.

Figure 14.1(d) shows the impulse response of the single-port s-parameters measured by the VNA. When the impulse response is integrated to form the step response, Figure 14.1(e) is obtained, which is similar to what is measured by the TDR. This similarity will be discussed in §14.3.

## 14.2 Impedance Profile Approximation with the Step Response

In §14.1, a single impedance was measured using the TDR. In this section the *impedance profile* is measured. The measurement of an impedance profile assumes that the system being measured consists essentially of cascaded transmission lines, each with a characteristic impedance. A simulation of such a system is shown in Figure 14.2. In Figure 14.2(a) a circuit is shown with a step generator providing a step through a $50\,\Omega$ source impedance to five cascaded transmission line sections, each with 100 ps delay, but with various characteristic impedances, all terminated finally in $50\,\Omega$. A voltage probe is placed at the input to the system being measured, after the $50\,\Omega$ source resistance.

The simulated voltage waveform is shown in Figure 14.2(b), where the applied step rises at time zero to a voltage of 1 V and bounces around for about 1 ns, presumably due to the impedance discontinuities, before mostly settling out to the final 1 V steady-state value.

In subsequent sections, rigor is applied to the calculation, but for now, without full justification, the measurement of the impedance is made utilizing (14.1) for each sample point taken. It is seen in Figure 14.2(a) that the total electrical length of the lines is 500 ps and the extent of the impedance discontinuities in Figure 14.2(b) is 1 ns. Each sample taken is assumed to be the impedance of a transmission line section and the electrical length of the section is half the sample period, considering the fact that any wave launched into the system must go to the discontinuity and back again during this period. Although $\Gamma$ was used to represent the reflection coefficient of the lumped impedance in §14.1, here the vector $\rho$ is used, where $\rho_{[k]}$ represents the reflection coefficient of a transmission line section with propagation constant $\gamma = j \cdot 2\pi \cdot f \cdot Ts/2$, as shown in (7.16).

Thus, given the step-like waveform $v_{[k]}$, due to a step of amplitude $V$ driven by a source through a source impedance $Z0$, the reflection coefficients are calculated as

$$\rho_{[k]} = \frac{v_{[k]} - V/2}{V/2}. \tag{14.5}$$

Given the vector $\boldsymbol{\rho}$, the impedance of each section of transmission line is calculated as

$$Z_{[k]} = Z0 \cdot \frac{1 + \rho_{[k]}}{1 - \rho_{[k]}}. \tag{14.6}$$

This is an estimate of the impedance of each transmission line section. (The reason it is an estimate will be explained in the following sections.) Taking a series expansion about $\rho_{[k]}$, a further approximation is calculated:

$$Z_{[k]} \approx Z0 + 2 \cdot Z0 \cdot \rho_{[k]} + \mathcal{O}\left(\rho_{[k]}^2\right). \tag{14.7}$$

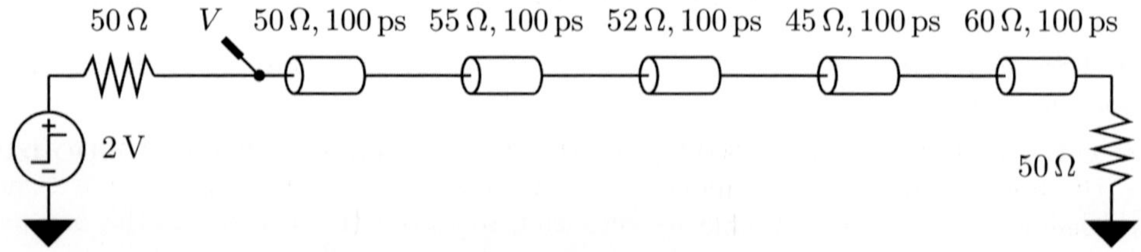


(a) Circuit

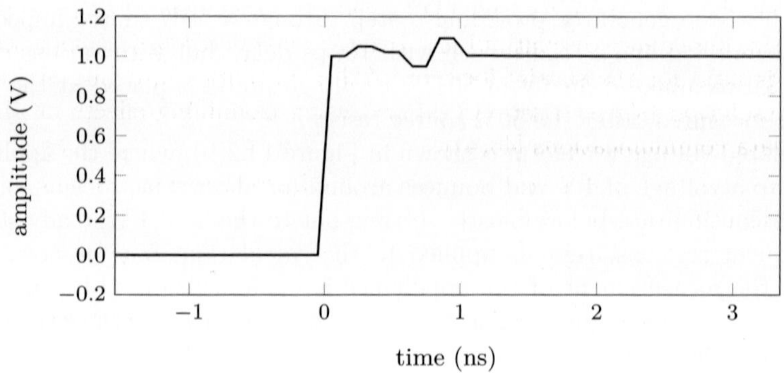


(b) Step response

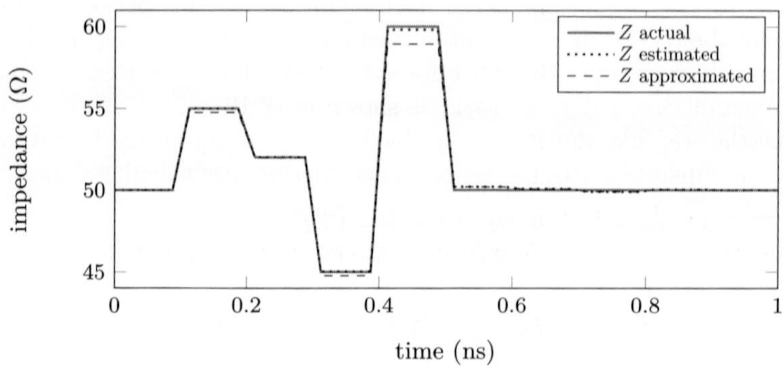


(c) Impedance

**Figure 14.2** Impedance profile simulation experiment

The actual transmission line impedances are compared to the estimated, and further approximated, impedance measurements in Figure 14.2(c).

## 14.3 Impedance Profile Approximation from S-Parameters

It has been increasingly popular to measure the impedance profile from s-parameters for the following reasons:

- The VNA is a commonly used instrument in signal integrity, with the caveat that the DC point needs to be extrapolated or determined in some other manner as network analyzers are typically unable to measure this point.[2]
- Network analyzers can produce impedance profile measurements that are calibrated and are in the desired reference impedance.
- Using simulation techniques, the TDR risetime can be controlled to emphasize or de-emphasize impedance discontinuities. Usually, the full resolution [39] of the impedance profile is used for the spatial location of discontinuities in debug and analysis, and lower resolution (slower risetime) is used when examining effects on things such as serial data communications [40, 41].

In the simulation of the step response, there are the following source stimulus and source reflection coefficient from Figure 6.7:

$$m = \frac{1}{\sqrt{Z0}} \cdot \frac{Z_s}{Z_s + Z0} \cdot V(z), \qquad \Gamma_s = \frac{Z_s - Z0}{Z_s + Z0}.$$

If $Z_s = Z0$,

$$m = \frac{1}{2 \cdot \sqrt{Z0}} \cdot V(z), \qquad \Gamma_s = 0.$$

Thus, the incident and reflected waves are given by

$$a = \frac{1}{2 \cdot \sqrt{Z0}} \cdot V(z), \qquad b = \frac{1}{2 \cdot \sqrt{Z0}} \cdot \Gamma(z) \cdot V(z).$$

(Of course, the ratio $b/a$ is $\Gamma$, the $s_{11}$ of the DUT.)

The voltage measured at the interface is defined from (2.8) as follows:

$$v(z) = \sqrt{Z0} \cdot (a + b) = \frac{1}{2} \cdot [1 + \Gamma(z)] \cdot V(z).$$

If the source supplies an impulse $A \cdot \delta(k)$, then, in the z domain and the time domain,

$$v(z) = \frac{1}{2} \cdot A + \frac{1}{2} \cdot A \cdot \Gamma(z), \qquad v_{[k]} = \frac{1}{2} \cdot A \cdot \delta(k) + \frac{1}{2} \cdot A \cdot \Gamma_{[k]},$$

where $\Gamma_{[k]}$ is the inverse z-transform of the frequency-domain s-parameters $\Gamma(z)$.

[2]TDRs that produce s-parameter measurements are able to provide the DC point measurement.

If the source supplies a step $A \cdot u(k)$, then, in the z domain and the time domain,

$$v(z) = \frac{1}{2} \cdot A \cdot \frac{1}{1 - z^{-1}} + \frac{1}{2} \cdot A \cdot \frac{1}{1 - z^{-1}} \cdot \Gamma(z), \quad v_{[k]} = \frac{1}{2} \cdot A \cdot u(k) + \frac{1}{2} \cdot A \cdot \sum_{\sigma=0}^{k} \Gamma_{[k]}.$$

This looks a bit confusing, and it is good to consider what $\Gamma_{[k]}$ actually is. Using (14.5), it becomes clear that the estimation of the reflection coefficients is given by

$$\rho_{[k]} = \frac{v_{[k]} - V/2}{V/2} = \sum_{\sigma=0}^{k} \Gamma_{[k]}.$$

In other words, the reflection coefficient of the transmission line section $k$ is the integral (the sum) of the impulse response of $\mathbf{\Gamma}$.

In order to rationalize this and to understand better the impulse response $\Gamma_{[k]}$, the signal-flow model of cascaded transmission line sections is provided in Figure 14.3. The basic model is shown in Figure 14.3(a), where

$$\rho_{[k]} = \frac{Z_{[k]} - Z0}{Z_{[k]} + Z0};$$

here, $Z_{[k]}$ is the characteristic impedance of the $k$th lossless transmission line and $z^{-\frac{1}{2}}$ is one-half of a unit delay. These transmission line sections correspond to the model provided in Figure 7.3. As the interface between each transmission line section is essentially two cascaded two-port devices, it can be simplified as shown in Figure 14.3(b). In Figure 14.3(a), the reflection coefficients $\rho_{[k]}$ were referenced to the reference impedance $Z0$, and the forward and reverse propagating waves $a_{[k]}$ and $b_{[k]}$ are the waves at the transmission line interfaces. In Figure 14.3(b), $\rho_{[k]}$, $a_{[k]}$, and $b_{[k]}$ are replaced with primed variables to represent that $\rho'_{[k]}$ represents the impedance mismatch between the two interfaces, and $a'_{[k]}$ and $b'_{[k]}$ no longer represent waves exactly at the transmission line interfaces. Figure 14.3(a) and Figure 14.3(b) are, however, equivalent models as viewed at the system interface:

$$\rho'_{[k]} = \begin{cases} \frac{Z_{[k]} - Z0}{Z_{[k]} + Z0} & \text{if } k = 0, \\ \frac{Z_{[k]} - Z_{[k-1]}}{Z_{[k]} + Z_{[k-1]}} & \text{otherwise.} \end{cases}$$

Figure 14.3(c) depicts an approximation of Figure 14.3(b). This approximation is justified if the mismatch between each section is relatively small, such that the products of any $\rho'_{[k]}$ are negligibly small. In the first section of Figure 14.3(b), the loop around the following path is approximated as follows:

$$a'_0 \to a'_1 \to b'_1 \to b'_0 = (1 + \rho'_0) \cdot z^{-\frac{1}{2}} \cdot \rho'_1 \cdot z^{-\frac{1}{2}} \cdot (1 - \rho'_0) = z^{-1} \cdot \rho'_1 - z^{-1} \cdot \rho'_1 \cdot (\rho'_0)^2 \approx z^{-1} \cdot \rho'_1.$$

With the approximation in Figure 14.3(c), an incident wave of $\delta(k)$ results in

$$a'_0(z) = 1, \qquad b'_0(z) = \rho'_0 + \rho'_1 \cdot z^{-1} + \rho'_2 \cdot z^{-2} + = \sum_{k=0}^{K-1} \rho'_k \cdot z^{-k},$$

$$(a'_0)_{[k]} = \delta(k), \qquad (b'_0)_{[k]} = \rho'_{[k]}.$$

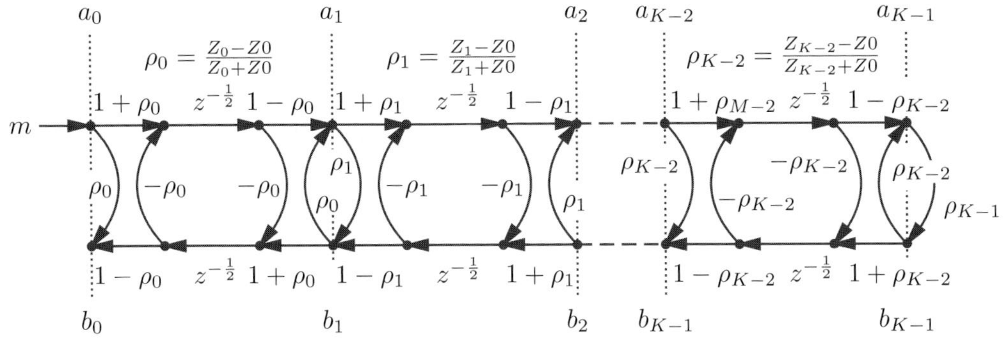


(a) Model

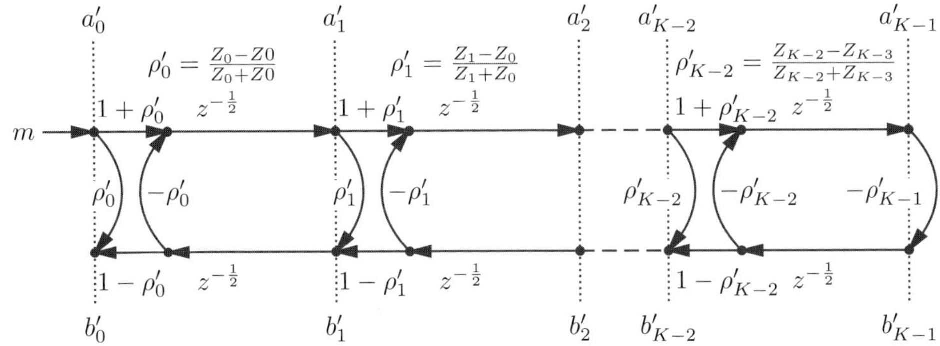


(b) Simplified

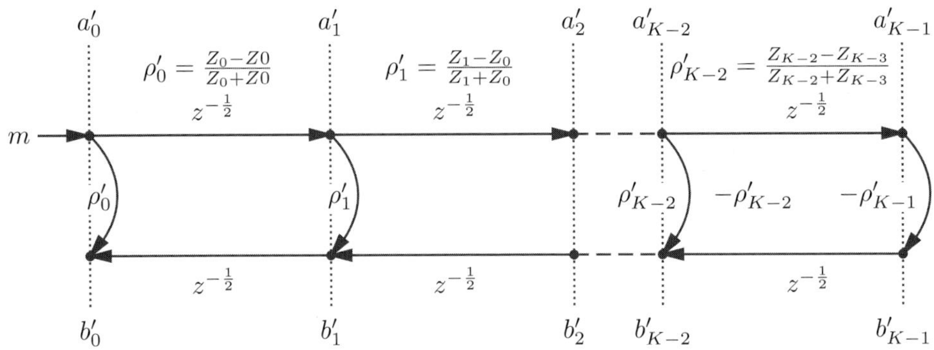


(c) Approximate

**Figure 14.3** Impedance profile cascaded transmission line model signal-flow diagram

It is therefore clear that, according to the approximation, the impulse response of the single-port s-parameters $\mathbf{\Gamma}$ is such that $\Gamma_{[k]} \approx \rho'_{[k]}$ and that the reflection coefficients are calculated as follows:

$$\rho_{[k]} \approx \sum_{\sigma=0}^{k} \rho'_{[\sigma]}.$$

The calculation is made by computing the IDFT of the $s_{11}$ (i.e. the impulse response), integrating this to form the step response, and then applying (14.6). This is shown in Figure 14.4, where the circuit from Figure 14.2(a) is duplicated in Figure 14.4(a) with a port instead of a step generator. S-parameters are calculated from DC to 10 GHz at 200 MHz spacing, which is equivalent to a 5 ns total impulse response length. The simulated magnitude and phase are shown in Figure 14.4(b) and Figure 14.4(c). The impulse and step response are shown in Figure 14.4(d) and Figure 14.4(e), and the final computation using (14.6) is shown in Figure 14.4(f).

## 14.4 Impedance Profile Calculation Using Peeling

In the preceding sections, the impedance profile has been approximated using a TDR by considering the step response from a simulation using s-parameters, and directly from the s-parameters themselves. These methods for calculation actually work very well, despite the fact that an approximation is made. There are methods for producing theoretically exact measurements of the impedance profile through techniques called *peeling*.

Peeling is a method wherein the impedance profile is calculated in an iterative fashion, meaning that each reflection coefficient is calculated from the previous value. A simple way to view this is first to examine the approximate model of the cascaded transmission line sections in Figure 14.3(c). (It is approximate because various possible loops traversed by the waves are ignored.) In other words, the wave incident on the system passes unmolested through the interfaces, returning small reflections as the wave hits each interface. So, reverse going waves never reflect from interfaces, to become forward going waves.

This being said, the very first calculated reflection coefficient is not an approximation; $\rho'_{[0]} = \rho_{[0]}$ is the reflection coefficient at the very front of the system, where the incident wave enters.

Because the delay of the system is shown as $z^{-\frac{1}{2}}$, the first sample of the IDFT of the s-parameters $\Gamma$ is presumed to be the reflection coefficient of a transmission line section of electrical length $Ts/2$, one-half the sample period. If the s-parameters consist of uniformly spaced frequency points starting from zero and ending at a frequency $Fe$, then the sample rate of the system is assumed to be $Fs = 2 \cdot Fe$ and the sample period is assumed to be $Ts = 1/Fs$.

Therefore, given a reflection coefficient for the first section $\rho_{[0]}$ and an electrical length $Ts/2$, a transmission line model as shown in Figure 7.3 is assumed for this section. Since $z = e^{s \cdot Ts}$, $\gamma = j \cdot 2\pi \cdot f \cdot Ts/2$, and the model for the first section $k = 0$ is written using

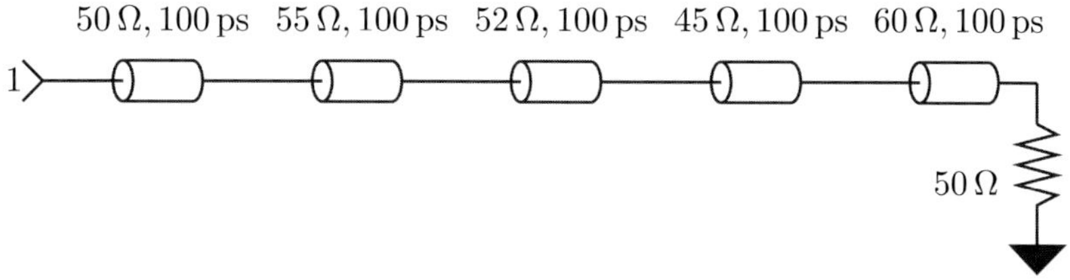


(a) S-parameters circuit

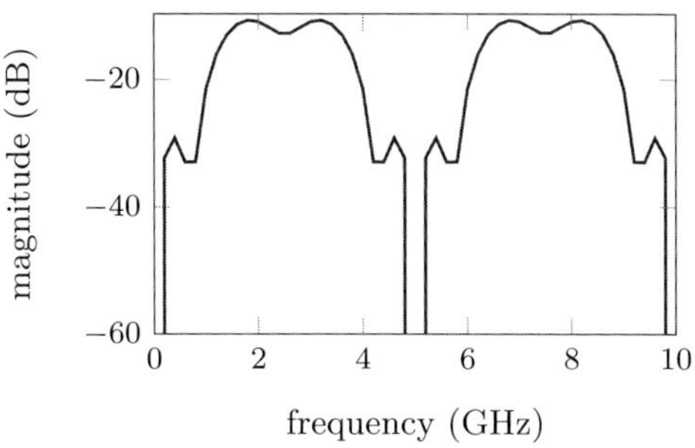


(b) Magnitude

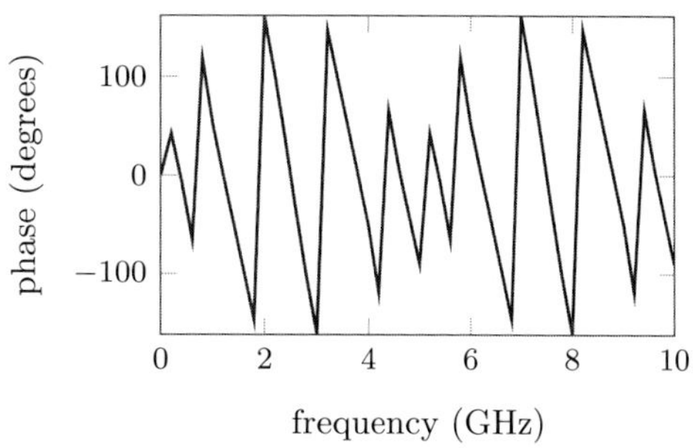


(c) Phase

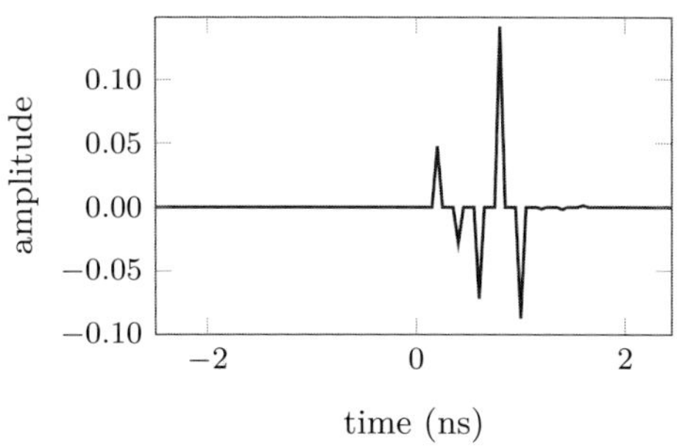


(d) Impulse response

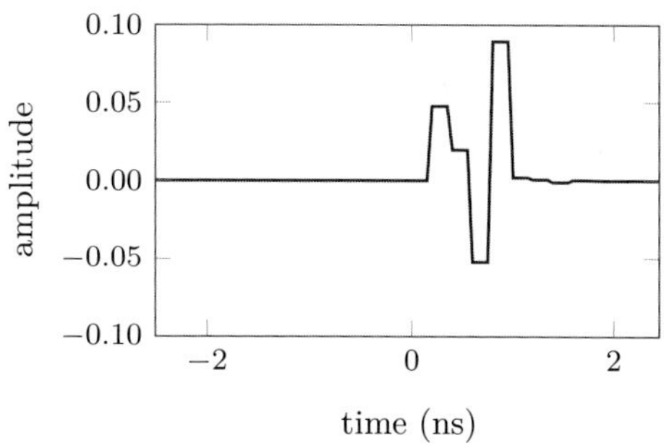


(e) Step response

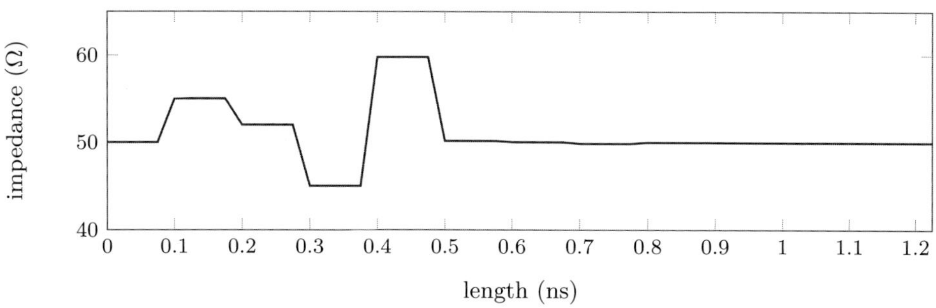


(f) Calculated impedance

**Figure 14.4** Impedance profile from s-parameters

(7.16) as follows:

$$\mathbf{S}_{[k]} = \begin{pmatrix} \dfrac{\rho_{[k]} \cdot \left(1 - z^{-1}\right)}{1 - \rho_{[k]}^2 \cdot z^{-1}} & \dfrac{\left(1 - \rho_{[k]}^2\right) \cdot z^{-\frac{1}{2}}}{1 - \rho_{[k]}^2 \cdot z^{-1}} \\ \dfrac{\left(1 - \rho_{[k]}^2\right) \cdot z^{-\frac{1}{2}}}{1 - \rho_{[k]}^2 \cdot z^{-1}} & \dfrac{\rho_{[k]} \cdot \left(1 - z^{-1}\right)}{1 - \rho_{[k]}^2 \cdot z^{-1}} \end{pmatrix}. \tag{14.8}$$

As previously explained, only the very first reflection coefficient calculated at the front of the cascaded transmission line model is theoretically accurate and the remaining reflection coefficients are approximations, but, if this section is de-embedded from the remainder of the system, then the second transmission line section becomes the first in the model and its reflection coefficient can be calculated accurately.

Thus, a peeling strategy becomes that of calculating a single reflection coefficient, building a transmission line model using that reflection coefficient, de-embedding that transmission line model from the system, and repeating. The de-embedding can be calculated using T-parameters, or by using (10.1). Given the model of the first transmission line section $\mathbf{S}$ as in (14.8), and the s-parameters of the system $\mathbf{\Gamma}$, the remainder of the system $\Gamma'$ with this section de-embedded is calculated at a given frequency as follows:

$$\Gamma' = \frac{\Gamma - S_{11}}{\Gamma \cdot S_{22} - |\mathbf{S}|} = \frac{\Gamma \cdot \rho_{[k]}^2 \cdot z^{-1} - \Gamma + \rho_{[k]} - \rho_{[k]} \cdot z^{-1}}{\rho_{[k]}^2 + \Gamma \cdot \rho_{[k]} \cdot z^{-1} - \Gamma \cdot \rho_{[k]} - z^{-1}}.$$

Thus, the impedance peeling algorithm is calculated, given $\mathbf{\Gamma}$ as the $N+1$ element $s_{11}$ of the system ending with frequency $Fe$, $z_{[n]} = e^{j \cdot \pi \cdot \frac{n}{N}}$, and starting with $k = 0$:

1. Calculate the reflection coefficient $\rho_{[k]}$:

$$\rho_{[k]} = \text{IDFT}\left(\mathbf{\Gamma}\right)_{[0]} = \frac{1}{2 \cdot N} \cdot \left[\Gamma_{[0]} + \text{Re}\left(\Gamma_{[N]}\right) + 2 \cdot \sum_{n=1}^{N-1} \text{Re}\left(\Gamma_{[n]}\right)\right].$$

For $n \in 0 \ldots N$, calculate the new $\mathbf{\Gamma}$ as

$$\Gamma_{[n]} = \frac{\Gamma_{[n]} \cdot \rho_{[k]}^2 \cdot z_{[n]}^{-1} - \Gamma + \rho_{[k]} - \rho_{[k]} \cdot z_{[n]}^{-1}}{\rho_{[k]}^2 + \Gamma_{[n]} \cdot \rho_{[k]} \cdot z_{[n]}^{-1} - \Gamma_{[n]} \cdot \rho_{[k]} - z_{[n]}^{-1}}.$$

2. Increment $k$ and go to step 1 or stop when a sufficient number of sections have been calculated.

## 14.5 Python Impedance Profile Software

The impedance profile peeling algorithm is shown in Listing 14.1 as the **ImpedanceProfile** class. The class derives from **list** and as such is a list of reflection coefficients. The constructor is declared in the `__init__()` function on line 2, where s-parameters are supplied, along with the number of sections desired and the port of the s-parameters from which

**Listing 14.1 ImpedanceProfile** class

```
1  class ImpedanceProfile(list):
2      def __init__(self,sp,sections,port):
3          list.__init__(self,[])
4          N = len(sp)-1
5          self.m_Td = 1./(4.*sp.f()[N])
6          self.m_Z0 = sp.m_Z0
7          fr=sp.FrequencyResponse(port,port)
8          self.m_fracD=fr._FractionalDelayTime()
9          fr=fr._DelayBy(-self.m_fracD)
10         S11 = fr.Values()
11         zn1 = [cmath.exp(-1j*math.pi*n/N) for n in range(N+1)]
12         finished=False
13         rho=0.0
14         for _ in range(sections):
15             if finished:
16                 self.append(rho)
17                 continue
18             rho = 1/(2.*N)*(S11[0].real + S11[N].real +
19                  sum([2.*S11[n].real for n in range(1,N)]))
20             rho=max(-self.rhoLimit,min(rho,self.rhoLimit))
21             self.append(rho)
22             if abs(rho)==self.rhoLimit:
23                 finished=True
24                 continue
25             rho2=rho*rho
26             S11=[(-S11[n]+S11[n]*rho2*zn1[n]-rho*zn1[n]+rho)/
27                 (rho2+S11[n]*rho*zn1[n]-S11[n]*rho-zn1[n])
28                 for n in range(N+1)]
29     def Z(self):
30         return [max(0.,min(self.m_Z0*(1+rho)/(1-rho),self.ZLimit))
31             for rho in self]
32     def DelaySection(self):
33         return self.m_Td
34     def SParameters(self,f):
35         N = len(f)-1
36         Gsp=[None for n in range(N+1)]
37         gamma=[1j*2.*math.pi*f[n]*self.m_Td for n in range(N+1)]
38         for n in range(N+1):
39             tacc=matrix([[1.,0.],[0.,1.]])
40             for m in range(len(self)):
41                 tacc=tacc*matrix(S2T(IdealTransmissionLine(self[m],gamma[n])))
42             Gsp[n]=T2S(tacc.tolist())
43         sp = SParameters(f,Gsp,self.m_Z0)
44         return sp
```

peeling takes place. After computation, the impedances can be fetched using the function `Z()` on line 29, and the time length of each section can be fetched using the function `DelaySection()` on line 32, allowing the plotting of the impedance profile in time.

The impedance profile so generated can be converted back into s-parameters, if desired, using the function `SParameters()` on line 34.

The **ImpedanceProfileWaveform** class is shown in Listing 14.2. The class derives from **Waveform** and consists of only a constructor; it constructs an impedance profile waveform from the arguments provided, including the s-parameters and the port used to construct the impedance profile, along with the arguments that supply the method for calculation, the alignment of the waveform, whether to include the port impedance, and whether to adjust for delay.

**Listing 14.2 ImpedanceProfileWaveform** class

```
class ImpedanceProfileWaveform(Waveform):
    def __init__(self,sp,port=1,method='exact',align='middle',includePortZ=True,
                 adjustForDelay=True):
        tdsp=sp.m_f.TimeDescriptor()
        # assumes middle and no portZ
        tdip=TimeDescriptor(1./(tdsp.Fs*4),tdsp.K/2,tdsp.Fs*2)
        if not align == 'middle':
            tdip.H=tdip.H-1./(tdsp.Fs*4)
        if method == 'exact':
            ip=ImpedanceProfile(sp,tdip.K,port)
            Z=ip.Z()
            delayAdjust=ip.m_fracD
        elif method == 'estimated' or method == 'approximate':
            fr=sp.FrequencyResponse(port,port)
            rho=fr.ImpulseResponse().Integral(addPoint=True,scale=False)
            delayAdjust=fr._FractionalDelayTime()
            finished=False
            for m in range(len(rho)):
                if finished:
                    rho[m]=rho[m-1]
                    continue
                rho[m]=max(-self.rhoLimit,min(self.rhoLimit,rho[m]))
                if abs(rho[m])==self.rhoLimit:
                    finished=True
            if method == 'estimated':
                Z=[max(0.,min(sp.m_Z0*(1+rho[tdsp.K//2+1+k])/
                    (1-rho[tdsp.K//2+1+k]),self.ZLimit)) for k in range(tdip.K)]
            else:
                Z=[max(0.,min(sp.m_Z0+2*sp.m_Z0*rho[tdsp.K//2+1+k],self.ZLimit))
                    for k in range(tdip.K)]
        if includePortZ:
            tdip.H=tdip.H-1./(tdsp.Fs*2)
            tdip.K=tdip.K+1
            Z=[sp.m_Z0]+Z
        if adjustForDelay: tdip.H=tdip.H+delayAdjust/2
        Waveform.__init__(self,tdip,Z)
```

The calculation methods that can be specified include:

- `'exact'` (default), which uses the **ImpedanceProfile** class provided in Listing 14.1;
- `'estimated'`, which uses the estimate of the impedance profile outlined in §14.3;
- `'approximate'`, which uses the estimate of the reflection coefficients outlined in §14.3, but uses the approximate formula in (14.7).

If the alignment is specified as `'middle'` (the default), then the waveform is generated with the impedance corresponding to the time of the middle of the transmission line section; otherwise, the time is that of the left-hand side of the transmission line interface.

If the inclusion of the port impedance is specified (default `True`), then the first point in the waveform is the port impedance $Z0$ obtained from the s-parameters.

Whether the adjustment for delay (default `True`) is specified determines whether the waveform should be shifted to account for any fractional delay calculated when converting the s-parameters into the time domain. This is explained in §12.4.3.

The impedance profile is particularly useful for de-embedding small launch structures. If the launch structure is very small and can be assumed to be essentially lossless, the

**Listing 14.3 PeeledPortSParameters** class

```
class PeeledPortSParameters(SParameters):
    def __init__(self,sp,port,timelen,method='estimated'):
        ip=ImpedanceProfileWaveform(sp,port,method,includePortZ=False)
        Ts=1./ip.td.Fs; sections=int(math.floor(timelen/Ts+0.5))
        tp1=[identity(2) for n in range(len(sp.f()))]
        for k in range(sections):
            tp1=[tp1[n]*matrix(S2T(TLineTwoPortLossless(ip[k],Ts,sp.m_f[n])))
                for n in range(len(sp.m_f))]
        SParameters.__init__(self,sp.m_f,[T2S(tp.tolist()) for tp in tp1])
```

**Listing 14.4 PeeledLaunches** class

```
class PeeledLaunches(SParameters):
    def __init__(self,sp,timelen,method='estimated'):
        spp=[PeeledPortSParameters(sp,p+1,timelen[p],method) for p in range(sp.m_P)]
        sddp=DeembedderParser().AddLine('unknown S '+str(sp.m_P))
        for ps in [str(p+1) for p in range(sp.m_P)]:
            sddp.AddLines(['device D'+ps+' 2','connect D'+ps+' 2 S '+ps,
                           'port '+ps+' D'+ps+' 1'])
        sddn=DeembedderNumeric(sddp.SystemDescription()); spd=[]
        for n in range(len(sp)):
            for p in range(sp.m_P): sddn.AssignSParameters('D'+str(p+1),spp[p][n])
            spd.append(sddn.CalculateUnknown(sp[n]))
        SParameters.__init__(self,sp.m_f,spd)
```

impedance profile can be calculated for a small number of sections, and the aggregate s-parameters corresponding to the small structure can be obtained and de-embedded from the system [42, 43]. In Listing 14.3, the class **PeeledPortSParameters** provides a class that is a constructor only and derives from the class **SParameters** (see Listing 3.16). This class generates the impedance profile at a given port of a given set of s-parameters for a certain length of time. The methods that are possible are the same as those previously described. This impedance profile is then aggregated into a set of s-parameters that presumably contains the s-parameters of the launch into that port.

The class **PeeledLaunches** in Listing 14.4 utilizes the class **PeeledPortSParameters** to de-embed the launches from one or more ports of a given set of s-parameters. Here, the `timelen` argument contains a list of times, one time for each port, which determines the length of time to peel from the s-parameter set.

A typical example of the usage of the **PeeledLaunches** class is provided in Figure 14.5. Figure 14.5(a) shows the impedance profile of a cable that is just over 2 ns in electrical length. The characteristic impedance of the cable throughout its length is approximately 51 Ω. The boot of the cable shows a very short spike in impedance, up to about 53 Ω for about 65 ps. A zoom of this area on port 1 is shown in Figure 14.5(b). This discontinuity is present in the cable, so it belongs with the cable model, but there are many reasons why one might want to de-embed this area. Two reasons come immediately to mind:

1. One might want to know the effect that this discontinuity has on the s-parameters of the cable or on signals passed through this cable. Therefore, a comparison might be desired between a simulation with and without this discontinuity.

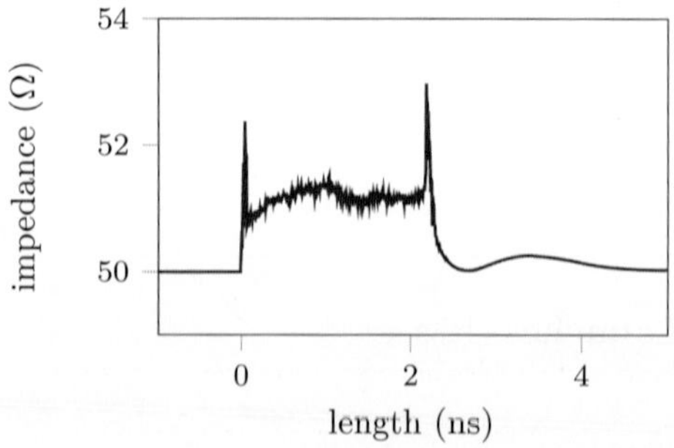


(a) Cable impedance profile

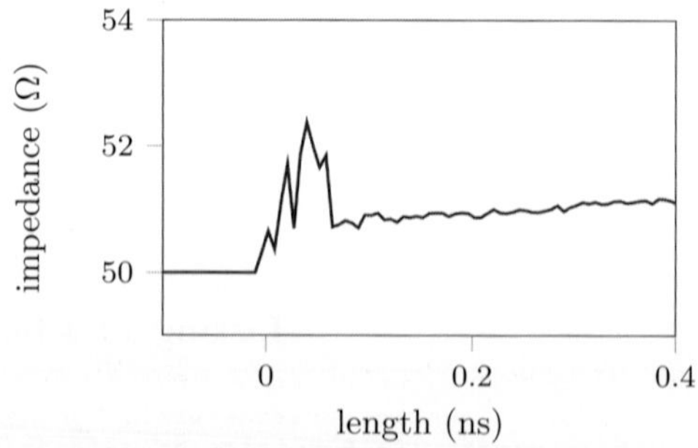


(b) Cable impedance profile launch, zoomed

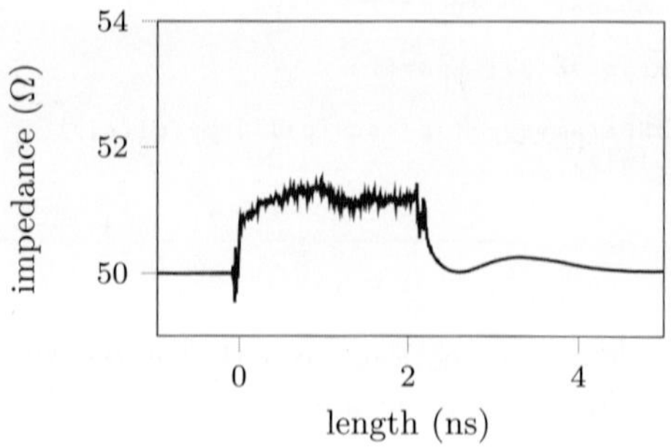


(c) Cable impedance profile with launches de-embedded

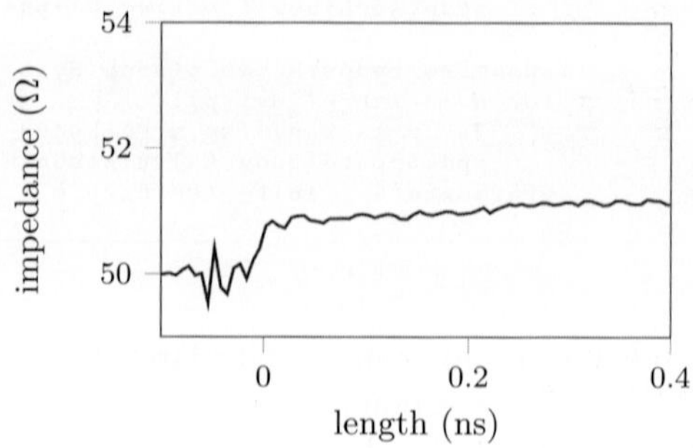


(d) Cable impedance profile with launches de-embedded, zoomed

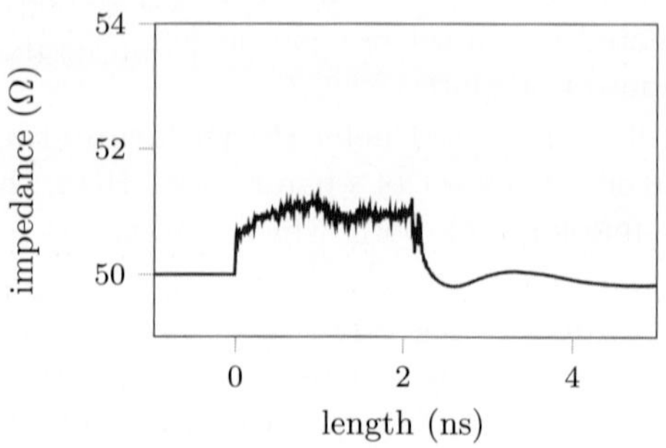


(e) Cable impedance profile with launches de-embedded with causality enforced

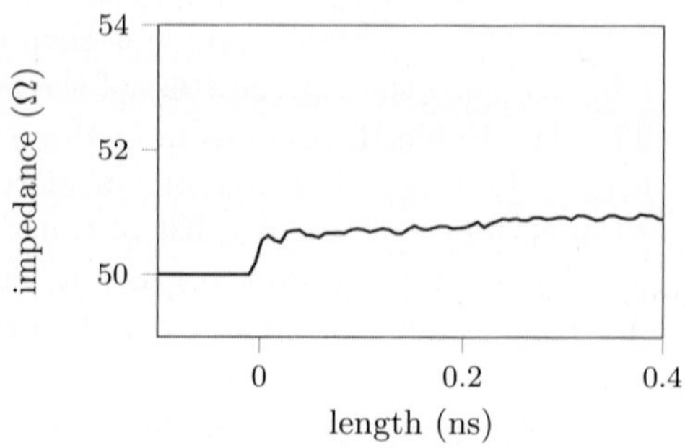


(f) Cable impedance profile with launches de-embedded with causality enforced, zoomed

**Figure 14.5** Cable launch de-embedding

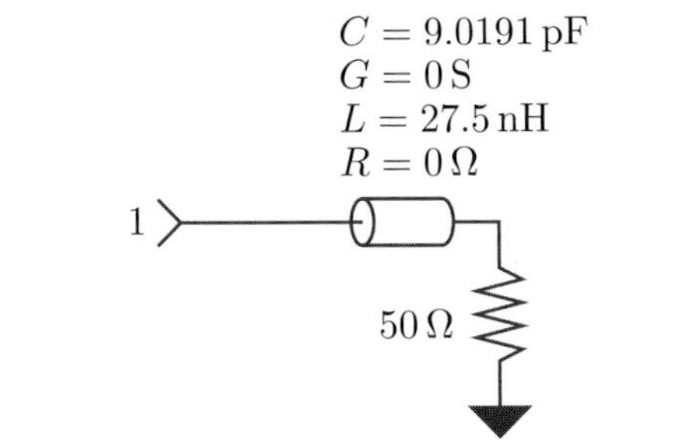


(a) Transmission line without series resistance

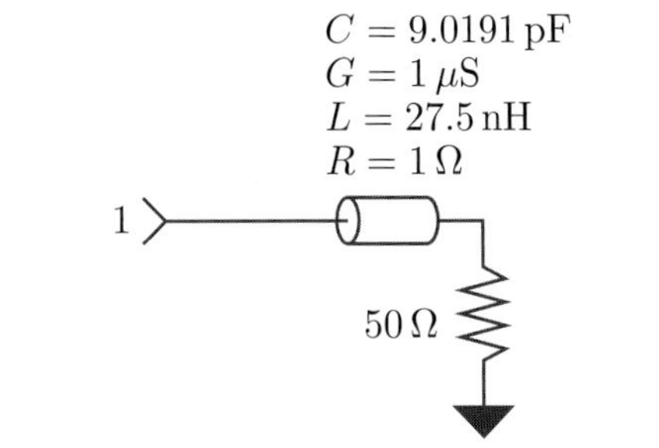


(b) Transmission line with series resistance

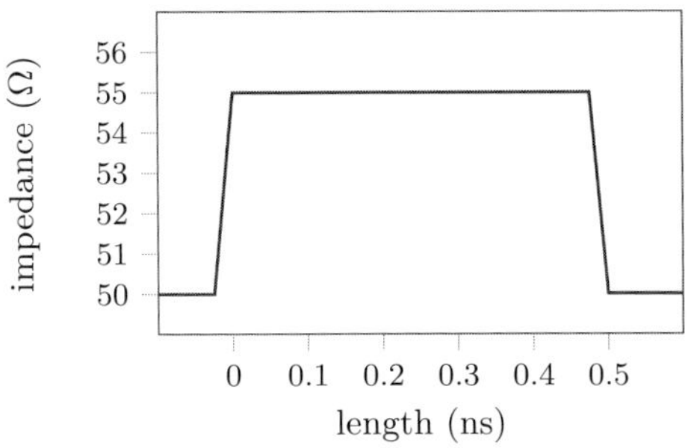


(c) Impedance profile of transmission line without series resistance

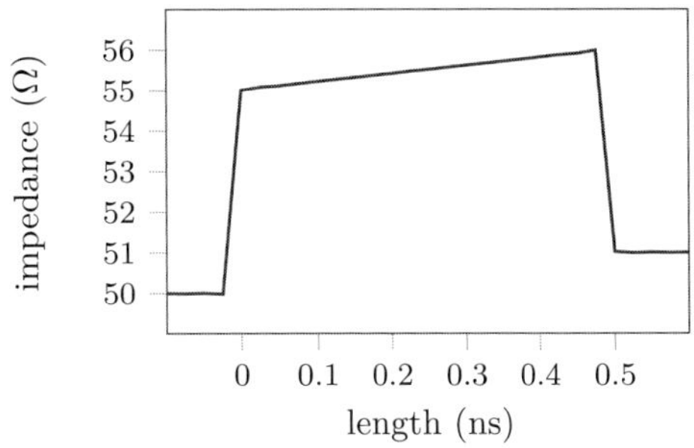


(d) Impedance profile of transmission line with series resistance

**Figure 14.6** Comparison of transmission line with and without series distributed resistance

2. It might be desirable to extract the characteristics of the internal part of the cable using the methods provided in Chapter 16, Model Extraction, and these discontinuities might confound the extraction.

Either way, in this example, 65 ps of launch is de-embedded from each port of the cable, and the result is shown in Figure 14.5(c).

Since this method of de-embedding is not perfect, one observes some non-causality in the result, especially in Figure 14.5(d), which is a zoom of Figure 14.5(c). The non-causality results from the fact that the de-embedded portion is close to, but not exactly, the s-parameters of the launch.

While not a panacea, causality enforcement (see §15.7.4) improves the situation somewhat by nulling the impulse response prior to time zero. The anticipated effect is seen in the impedance profile in Figure 14.5(e), and especially in the zoomed in portion in Figure 14.5(f). Causality enforcement is not an ideal solution because it takes care of only the portion prior to time zero, and does nothing about the portion around 2 ns in either the $s_{11}$ or $s_{22}$ of the cable with the launch de-embedded. As usual in these kinds of manipulations, it is up to the engineer to decide whether the results are good enough for the desired analysis.

## 14.6 Problems with the Impedance Profile

There are three particular problems that arise in the computation of the impedance profile that are worth mentioning. The first is the instability of the peeling algorithm. Since the algorithm computes a reflection coefficient and de-embeds a section to compute subsequent ones, any mistakes propagate and create further errors in the computation of future reflection coefficients. For this reason, when computing the impedance profile of long lines, when the impedances are all relatively close to the reference impedance, the estimated method is the most reliable.

The second problem is that the calculated impedances are always real. In Chapter 7, it was clear that transmission lines will have mostly real, but often slightly complex, characteristic impedances. This is seen in Chapter 16, where the models certainly have a slightly complex impedance, especially due to the skin effect.

The final problem is related to the fact that the impedance profile calculated from a line with small amounts of distributed series resistance will be wrong in that it will tend to be tilted due to the accumulation of this resistance [44]. This effect is seen in Figure 14.6. The simulation of a transmission line without distributed series resistance in Figure 14.6(a) produces an impedance profile that is flat (Figure 14.6(c)); compare this with the simulation of a transmission line with distributed series resistance in Figure 14.6(b), which produces an impedance profile that slopes upwards (Figure 14.6(d)).

# 15

# Measurement

IDEALLY, THE MEASUREMENT OF S-PARAMETERS is performed by driving a DUT with incident waves and measuring both the incident and reflected waves from the DUT, as shown in Figure 15.1, in which measurements are shown for both the VNA in Figure 15.1(a) and the TDR in Figure 15.1(b). The voltage measurements are shown as simple probes on the schematic, whereas in reality the measurements of these voltages are performed through complex circuitry involving samplers and analog-to-digital converters (ADCs), and, in the case of the VNA, additional filters and downconverters.

In Figure 15.1(a), the DUT is stimulated by standing waves produced by the generators. The theory behind the VNA is that, at a given frequency, a standing wave is produced and the forward and reverse going waves are measured through a device called a *directional coupler*, which ideally picks off the waves going only in certain directions. In practice, directional couplers have many imperfections in that they can have impedance mismatch, imperfect directionality, and passive devices must attenuate either the picked off wave or the wave flowing through the coupler. An unavoidable imperfection is that these devices cannot go down to DC.

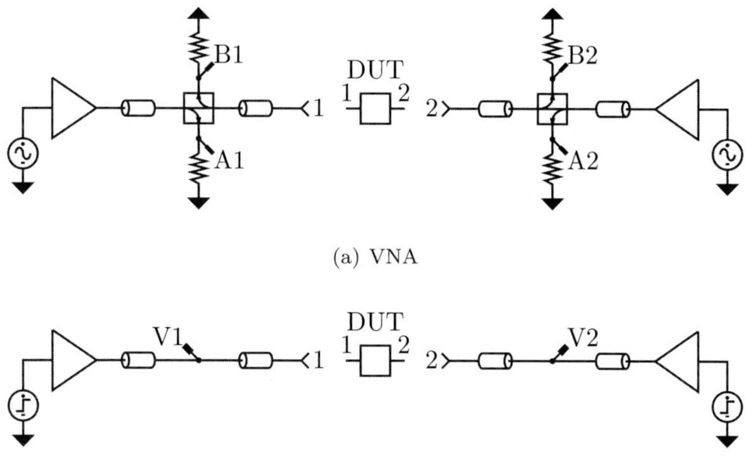


(a) VNA



(b) TDR

**Figure 15.1** S-parameter measurement methods

In Figure 15.1(b), the DUT is stimulated by a step waveform. The theory behind the TDR is that all frequencies are generated in a wave front and that only a voltage is measured. In the wave definition equations provided in Table 2.1, voltage is seen to be proportional to the sum of the forward and reverse propagating waves, and, in the TDR, the waves are separated in time, with the implication that, ideally, the incident wave is produced in the rising edge of the step and the reflected waves come afterwards. This will be discussed in detail in this chapter.

All measurements are performed under a number of given *measurement conditions*. These measurement conditions are not completely arbitrary; they must be sufficient for the measurement of the s-parameters. For example, given a DUT with s-parameters $\mathbf{S}$, if the DUT is driven with incident waves $\mathbf{a}$ and reflected waves $\mathbf{b}$, the measurement obeys the following equation:

$$\mathbf{b} = \mathbf{S} \cdot \mathbf{a}.$$

If a number of such measurements $M$ were taken, where for $m \in 1 \ldots M$ incident waves $\mathbf{a}_m$ and reflected waves $\mathbf{b}_m$ are measured, then the following matrices can be formed:

$$\mathbf{A} = \begin{pmatrix} \mathbf{a}_1 & \mathbf{a}_2 & \ldots & \mathbf{a}_M \end{pmatrix},$$
$$\mathbf{B} = \begin{pmatrix} \mathbf{b}_1 & \mathbf{b}_2 & \ldots & \mathbf{b}_M \end{pmatrix},$$

and the s-parameters of the DUT can be determined as

$$\mathbf{S} = \mathbf{B} \cdot \mathbf{A}^{\dagger}.$$

Keep in mind that each vector of the measured incident and reflected waves $\mathbf{a}_m$ and $\mathbf{b}_m$ is ideally $P$ elements long, where $P$ is the number of ports in the measurement instrument and in the DUT, and therefore the augmented matrices $\mathbf{A}$ and $\mathbf{B}$ are $P \times M$. If $M = P$, both $\mathbf{A}$ and $\mathbf{B}$ are square and $\mathbf{A}^{\dagger} = \mathbf{A}^{-1}$. This is covered in Appendix C, §C.2. The measurement conditions must be chosen such that $\mathbf{A}$ is invertible – ideally, $\mathbf{A}$ is orthonormal, such as, for example, the identity matrix.

In practice, it is not possible to drive the DUT either arbitrarily or directly. This is accounted for by introducing a *fixture* between the measurement instrument and the DUT and introducing the measurements $\hat{\mathbf{a}}_m$ and $\hat{\mathbf{b}}_m$, which are the measurements of the incident and reflected waves internal to the measurement instrument; they are different from the actual incident and reflected waves on the DUT, $\mathbf{a}_m$ and $\mathbf{b}_m$, as shown in Figure 15.2. This might lead one to think that this becomes a *fixture de-embedding* problem as provided in Chapter 10, but it tends to be more complicated than that because, traditionally, the fixture is different for each measurement condition, as indicated by denoting the fixture s-parameters by $\mathbf{F}_m$.

It should be pointed out that, in a sense, the fixture $\mathbf{F}_m$ is fictitious because, although it is expressed as s-parameters, it is not completely known and is constructed to account for many impairments in the measurement and in a manner only to take accurate measurements of a DUT. It is not an actual device that is either directly measured or measured completely.

In terminology applicable to s-parameter measurements, each fixture $\mathbf{F}_m$ is said to contain *error terms*. These error terms are computed during a *calibration* process. After calibration, the error terms are known, and therefore each $\mathbf{F}_m$ is known, and the remaining challenge is to use measurements of a DUT in combination with each $\mathbf{F}_m$ to determine the actual s-parameters of the DUT.

## 15.1 The Twelve-Term Error Model

The twelve-term error model is an abstraction that allows for the solution of s-parameters that accounts for three main effects:

- The frequency response characteristics of the source and receiver in the network analyzer.
- The fixturing effects of the path between the source/receiver and the DUT.
- The fact that not all the information about frequency responses and fixturing effects can be known.

The last item is particularly important as, during the calibration (explained in §15.2), usually certain products and ratios of the first two items can be found. This leads to an error model as shown in Figure 15.2. This is for a two-port system and uses two fixture models, one for the case when port 1 is driven (Figure 15.2(a)) and another for the case when port 2 is driven (Figure 15.2(b)). During calibration, the twelve error terms are determined, and during measurement, this model is used to calculate the DUT.

For the twelve-term error model, there is an error term matrix **ET** preferably structured as follows:

$$\mathbf{ET} = \begin{pmatrix} \begin{pmatrix} E_{D_1} \\ E_{R_1} \\ E_{S_1} \end{pmatrix} & \begin{pmatrix} E_{X_{12}} \\ E_{T_{12}} \\ E_{L_{12}} \end{pmatrix} & \begin{pmatrix} E_{X_{13}} \\ E_{T_{13}} \\ E_{L_{13}} \end{pmatrix} & \begin{pmatrix} E_{X_{14}} \\ E_{T_{14}} \\ E_{L_{14}} \end{pmatrix} \\ \begin{pmatrix} E_{X_{21}} \\ E_{T_{21}} \\ E_{L_{21}} \end{pmatrix} & \begin{pmatrix} E_{D_2} \\ E_{R_2} \\ E_{S_2} \end{pmatrix} & \begin{pmatrix} E_{X_{23}} \\ E_{T_{23}} \\ E_{L_{23}} \end{pmatrix} & \begin{pmatrix} E_{X_{24}} \\ E_{T_{24}} \\ E_{L_{24}} \end{pmatrix} \\ \begin{pmatrix} E_{X_{31}} \\ E_{T_{31}} \\ E_{L_{31}} \end{pmatrix} & \begin{pmatrix} E_{X_{32}} \\ E_{T_{32}} \\ E_{L_{32}} \end{pmatrix} & \begin{pmatrix} E_{D_3} \\ E_{R_3} \\ E_{S_3} \end{pmatrix} & \begin{pmatrix} E_{X_{34}} \\ E_{T_{34}} \\ E_{L_{34}} \end{pmatrix} \\ \begin{pmatrix} E_{X_{41}} \\ E_{T_{41}} \\ E_{L_{41}} \end{pmatrix} & \begin{pmatrix} E_{X_{42}} \\ E_{T_{42}} \\ E_{L} \end{pmatrix} & \begin{pmatrix} E_{X_{43}} \\ E_{T_{43}} \\ E_{L_{43}} \end{pmatrix} & \begin{pmatrix} E_{D_4} \\ E_{R_4} \\ E_{S_4} \end{pmatrix} \end{pmatrix}.$$

The error terms for the driven port $m$ and the measured port $n$ are the reflect error terms when $m = n$ and the thru error terms when $m \neq n$:

$$\mathbf{ET}_{nm} = \begin{cases} \begin{pmatrix} E_{D_n} \\ E_{R_n} \\ E_{S_n} \end{pmatrix} & \text{if } n = m, \\ \begin{pmatrix} E_{X_{nm}} \\ E_{T_{nm}} \\ E_{L_{nm}} \end{pmatrix} & \text{if } n \neq m. \end{cases}$$

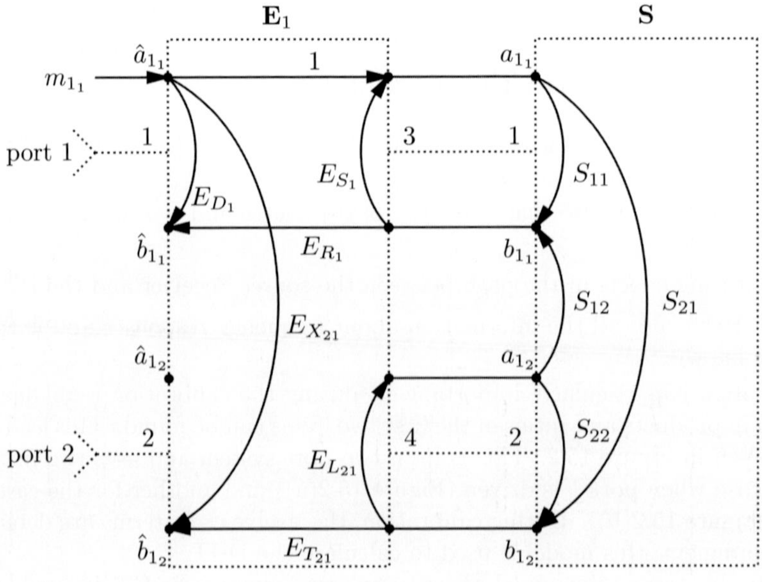


(a) Port 1 driven

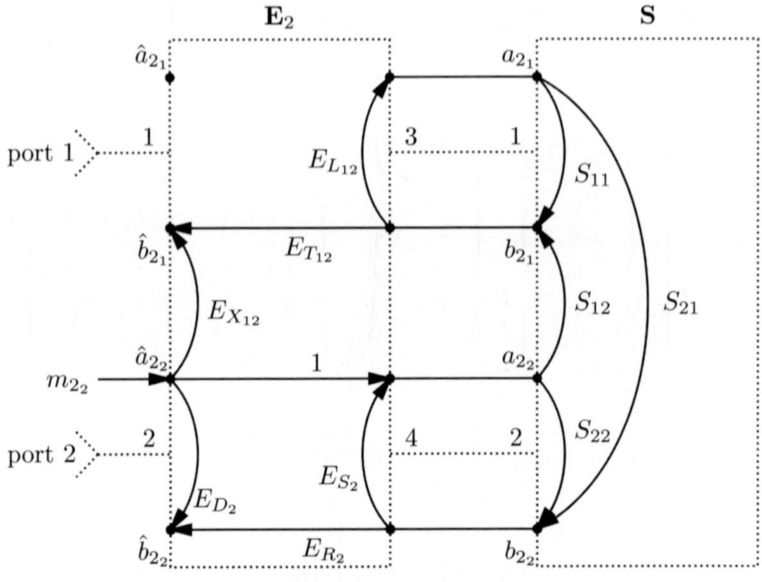


(b) Port 2 driven

**Figure 15.2** Twelve-term error model

The grouping of these error terms will become apparent. The fixtures shown in Figure 15.2 and provided in Listing 15.1 contain these error terms and are constructed to be partitioned into four pieces for the solution. To understand the general structure of the partitioned matrices, consider the four-port situation:

$$\mathbf{E}_1 = \begin{pmatrix} \begin{pmatrix} E_{D_1} & 0 & 0 & 0 \\ E_{X_{21}} & 0 & 0 & 0 \\ E_{X_{31}} & 0 & 0 & 0 \\ E_{X_{41}} & 0 & 0 & 0 \end{pmatrix} & \begin{pmatrix} E_{R_1} & 0 & 0 & 0 \\ 0 & E_{T_{21}} & 0 & 0 \\ 0 & 0 & E_{T_{31}} & 0 \\ 0 & 0 & 0 & E_{T_{41}} \end{pmatrix} \\ \begin{pmatrix} 1 & 0 & 0 & 0 \\ 0 & 0 & 0 & 0 \\ 0 & 0 & 0 & 0 \\ 0 & 0 & 0 & 0 \end{pmatrix} & \begin{pmatrix} E_{S_1} & 0 & 0 & 0 \\ 0 & E_{L_{21}} & 0 & 0 \\ 0 & 0 & E_{L_{31}} & 0 \\ 0 & 0 & 0 & E_{L_{41}} \end{pmatrix} \end{pmatrix};$$

$$\mathbf{E}_2 = \begin{pmatrix} \begin{pmatrix} 0 & E_{X_{12}} & 0 & 0 \\ 0 & E_{D_2} & 0 & 0 \\ 0 & E_{X_{32}} & 0 & 0 \\ 0 & E_{X_{42}} & 0 & 0 \end{pmatrix} & \begin{pmatrix} E_{T_{12}} & 0 & 0 & 0 \\ 0 & E_{R_2} & 0 & 0 \\ 0 & 0 & E_{T_{32}} & 0 \\ 0 & 0 & 0 & E_{T_{42}} \end{pmatrix} \\ \begin{pmatrix} 0 & 0 & 0 & 0 \\ 0 & 1 & 0 & 0 \\ 0 & 0 & 0 & 0 \\ 0 & 0 & 0 & 0 \end{pmatrix} & \begin{pmatrix} E_{L_{12}} & 0 & 0 & 0 \\ 0 & E_{S_2} & 0 & 0 \\ 0 & 0 & E_{L_{32}} & 0 \\ 0 & 0 & 0 & E_{L_{42}} \end{pmatrix} \end{pmatrix};$$

$$\mathbf{E}_3 = \begin{pmatrix} \begin{pmatrix} 0 & 0 & E_{X_{13}} & 0 \\ 0 & 0 & E_{X_{23}} & 0 \\ 0 & 0 & E_{D_3} & 0 \\ 0 & 0 & E_{X_{43}} & 0 \end{pmatrix} & \begin{pmatrix} E_{T_{13}} & 0 & 0 & 0 \\ 0 & E_{T_{23}} & 0 & 0 \\ 0 & 0 & E_{R_3} & 0 \\ 0 & 0 & 0 & E_{T_{43}} \end{pmatrix} \\ \begin{pmatrix} 0 & 0 & 0 & 0 \\ 0 & 0 & 0 & 0 \\ 0 & 0 & 1 & 0 \\ 0 & 0 & 0 & 0 \end{pmatrix} & \begin{pmatrix} E_{L_{13}} & 0 & 0 & 0 \\ 0 & E_{L_{23}} & 0 & 0 \\ 0 & 0 & E_{S_3} & 0 \\ 0 & 0 & 0 & E_{L_{43}} \end{pmatrix} \end{pmatrix};$$

$$\mathbf{E}_4 = \begin{pmatrix} \begin{pmatrix} 0 & 0 & 0 & E_{X_{14}} \\ 0 & 0 & 0 & E_{X_{24}} \\ 0 & 0 & 0 & E_{X_{34}} \\ 0 & 0 & 0 & E_{D_4} \end{pmatrix} & \begin{pmatrix} E_{T_{14}} & 0 & 0 & 0 \\ 0 & E_{T_{24}} & 0 & 0 \\ 0 & 0 & E_{T_{34}} & 0 \\ 0 & 0 & 0 & E_{R_4} \end{pmatrix} \\ \begin{pmatrix} 0 & 0 & 0 & 0 \\ 0 & 0 & 0 & 0 \\ 0 & 0 & 0 & 0 \\ 0 & 0 & 0 & 1 \end{pmatrix} & \begin{pmatrix} E_{L_{14}} & 0 & 0 & 0 \\ 0 & E_{L_{24}} & 0 & 0 \\ 0 & 0 & E_{L_{34}} & 0 \\ 0 & 0 & 0 & E_{S_4} \end{pmatrix} \end{pmatrix}.$$

By inspection, it becomes clear that a general strategy can be developed for the construction of these matrices for a given driven port $m$ of an $M$-port system by setting

$$\mathbf{E}_m = \begin{pmatrix} \underset{M \times M}{\mathbf{0}} & \underset{M \times M}{\mathbf{0}} \\ \underset{M \times M}{\mathbf{0}} & \underset{M \times M}{\mathbf{0}} \end{pmatrix}$$

**Listing 15.1 ErrorTerms** class `Fixture()` method

```
class ErrorTerms(object):
...
    def Fixture(self,m,pl=None):
        if pl is None: pl = [p for p in range(self.numPorts)]
        numPorts=len(pl)
        E=[[zeros((numPorts,numPorts),complex).tolist(),
            zeros((numPorts,numPorts),complex).tolist()],
           [zeros((numPorts,numPorts),complex).tolist(),
            zeros((numPorts,numPorts),complex).tolist()]]
        for n in range(numPorts):
            ETn=self[pl[n]][pl[m]]
            E[0][0][n][m]=ETn[0]
            E[0][1][n][n]=ETn[1]
            E[1][1][n][n]=ETn[2]
        E[1][0][m][m]=1.
        return E
...
```

and then setting $\mathbf{E}_{m21_{[m][m]}} = 1$ and, for $n \in 1 \ldots M$,

$$\mathbf{E}_{m11_{[n][m]}} = \mathbf{ET}_{[n][m]_1}, \qquad \mathbf{E}_{m12_{[n][n]}} = \mathbf{ET}_{[n][m]_2}, \qquad \mathbf{E}_{m22_{[n][n]}} = \mathbf{ET}_{[n][m]_3}.$$

This strategy accounts for the choice of the structure of **ET**, whereby the terms appearing in a given partitioned matrix are placed in the same row in an **ET** column vector. This is shown in Listing 15.1, using zero-based matrix indices.

## 15.2 Calibration

Figure 15.3 shows the signal-flow diagram of an s-parameter measurement system model corresponding to Figure 15.1(a). The characteristics of this system must be dealt with practically in order to take s-parameter measurements. The theory here is that a DUT is connected in the middle of this system, stimulated from the ports of the instrument, and measurements taken at the instrument ports. These measurements are of the forward and reverse going waves at each port $\hat{a}_1$, $\hat{a}_2$, $\hat{b}_1$, and $\hat{b}_2$. Note that the actual forward and reverse going waves are $a'_1$, $a'_2$, $b'_1$, and $b'_2$, and the measured waves are scaled by $\alpha_1$, $\alpha_2$, $\beta_1$, and $\beta_2$. The transmitting portion injects waves as $m_1$ and $m_2$, usually alternating between driven ports in two separate acquisitions. The internal instrument connections are shown with s-parameters $\mathbf{F_1}$ and $\mathbf{F_2}$. Note the orientation of these internal s-parameters: port 1 of both connections is where the waves are injected and measured and port 2 connects to the DUT. The DUT is shown with its ports oriented such that port 1 connects to port 1 of the instrument and port 2 connects to port 2 of the instrument. The instrument ports are shown with non-zero reflection coefficients $\Gamma_1$ and $\Gamma_2$, and there is crosstalk between the ports shown as $X_{21}$ and $X_{12}$.

The system of equations that governs the behavior of this system is provided in (15.1):

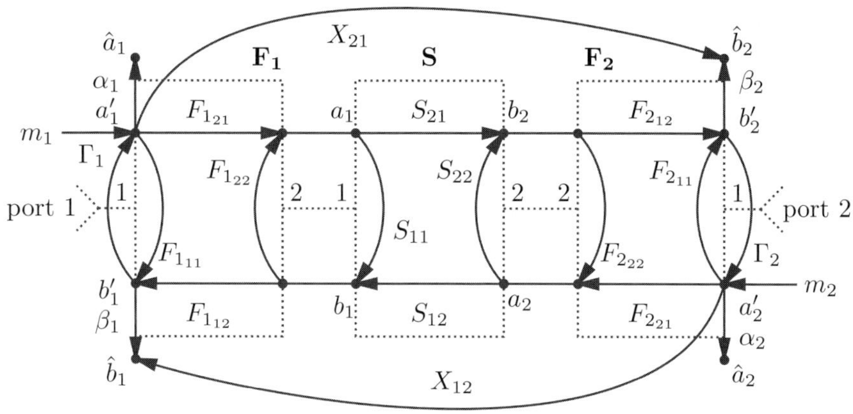


**Figure 15.3** An s-parameter measurement system model

$$
\left[\mathbb{I}-\begin{pmatrix}
0&0&0&0&\alpha_1&0&0&0&0&0&0&0\\
0&0&0&0&0&\alpha_2&0&0&0&0&0&0\\
0&0&0&0&0&X_{12}&\beta_1&0&0&0&0&0\\
0&0&0&0&X_{21}&0&0&\beta_2&0&0&0&0\\
0&0&0&0&0&0&\Gamma_1&0&0&0&0&0\\
0&0&0&0&0&0&0&\Gamma_2&0&0&0&0\\
0&0&0&0&F_{1_{11}}&0&0&0&0&0&F_{1_{12}}&0\\
0&0&0&0&0&F_{2_{11}}&0&0&0&0&0&F_{2_{12}}\\
0&0&0&0&F_{1_{21}}&0&0&0&0&0&F_{1_{22}}&0\\
0&0&0&0&0&F_{2_{21}}&0&0&0&0&0&F_{2_{22}}\\
0&0&0&0&0&0&0&0&S_{11}&S_{12}&0&0\\
0&0&0&0&0&0&0&0&S_{21}&S_{22}&0&0
\end{pmatrix}\right]\cdot\begin{pmatrix}\hat{a}_1\\ \hat{a}_2\\ \hat{b}_1\\ \hat{b}_2\\ a'_1\\ a'_2\\ b'_1\\ b'_2\\ a_1\\ a_2\\ b_1\\ b_2\end{pmatrix}=\begin{pmatrix}0\\0\\0\\0\\m_1\\m_2\\0\\0\\0\\0\\0\\0\end{pmatrix}. \tag{15.1}
$$

Equation (15.1) can be written in block form as follows:

$$
\left[\mathbb{I}-\begin{pmatrix}
\mathbf{0}&\mathbf{0}&\boldsymbol{\alpha}&\mathbf{0}&\mathbf{0}&\mathbf{0}\\
\mathbf{0}&\mathbf{0}&\mathbf{X}&\boldsymbol{\beta}&\mathbf{0}&\mathbf{0}\\
\mathbf{0}&\mathbf{0}&\mathbf{0}&\boldsymbol{\Gamma}&\mathbf{0}&\mathbf{0}\\
\mathbf{0}&\mathbf{0}&\mathbf{F_{11}}&\mathbf{0}&\mathbf{0}&\mathbf{F_{12}}\\
\mathbf{0}&\mathbf{0}&\mathbf{F_{21}}&\mathbf{0}&\mathbf{0}&\mathbf{F_{22}}\\
\mathbf{0}&\mathbf{0}&\mathbf{0}&\mathbf{0}&\mathbf{S}&\mathbf{0}
\end{pmatrix}\right]\cdot\begin{pmatrix}\hat{\mathbf{A}}\\ \hat{\mathbf{B}}\\ \mathbf{A}'\\ \mathbf{B}'\\ \mathbf{A}\\ \mathbf{B}\end{pmatrix}=\begin{pmatrix}\mathbf{0}\\ \mathbf{0}\\ \mathbf{M}\\ \mathbf{0}\\ \mathbf{0}\\ \mathbf{0}\end{pmatrix},
$$

which provides the following six equations for the system:

$$\hat{\mathbf{A}} = \boldsymbol{\alpha} \cdot \mathbf{A}', \tag{15.2}$$

$$\hat{\mathbf{B}} = \boldsymbol{\beta} \cdot \mathbf{B}' + \mathbf{X} \cdot \mathbf{A}', \tag{15.3}$$

$$\mathbf{A}' = \boldsymbol{\Gamma} \cdot \mathbf{B}' + \mathbf{M},$$

$$\mathbf{B}' = \mathbf{F_{11}} \cdot \mathbf{A}' + \mathbf{F_{12}} \cdot \mathbf{B}, \tag{15.4}$$

$$\mathbf{A} = \mathbf{F_{21}} \cdot \mathbf{A}' + \mathbf{F_{22}} \cdot \mathbf{B}, \tag{15.5}$$

$$\mathbf{B} = \mathbf{S} \cdot \mathbf{A}. \tag{15.6}$$

To start, one wants to eliminate the internal $\mathbf{A}'$ and $\mathbf{B}'$, so the first two equations, (15.2) and (15.3), are rewritten in terms of $\mathbf{A}'$ and $\mathbf{B}'$:

$$\mathbf{A}' = \boldsymbol{\alpha}^{-1} \cdot \hat{\mathbf{A}}, \tag{15.7}$$

$$\mathbf{B}' = \boldsymbol{\beta}^{-1} \cdot \hat{\mathbf{B}} - \boldsymbol{\beta}^{-1} \cdot \mathbf{X} \cdot \boldsymbol{\alpha}^{-1} \cdot \hat{\mathbf{A}}. \tag{15.8}$$

Equations (15.7) and (15.8) are substituted into (15.4):

$$\boldsymbol{\beta}^{-1} \cdot \hat{\mathbf{B}} - \boldsymbol{\beta}^{-1} \cdot \mathbf{X} \cdot \boldsymbol{\alpha}^{-1} \cdot \hat{\mathbf{A}} = \mathbf{F_{11}} \cdot \boldsymbol{\alpha}^{-1} \cdot \hat{\mathbf{A}} + \mathbf{F_{12}} \cdot \mathbf{B}, \tag{15.9}$$

and terms collected to yield

$$\hat{\mathbf{B}} = \left(\boldsymbol{\beta} \cdot \mathbf{F_{11}} \cdot \boldsymbol{\alpha}^{-1} + \mathbf{X} \cdot \boldsymbol{\alpha}^{-1}\right) \cdot \hat{\mathbf{A}} + \boldsymbol{\beta} \cdot \mathbf{F_{12}} \cdot \mathbf{B}.$$

Equations (15.6) and (15.7) are then substituted into (15.5):

$$\mathbf{S}^{-1} \cdot \mathbf{B} = \mathbf{F_{21}} \cdot \boldsymbol{\alpha}^{-1} \cdot \hat{\mathbf{A}} + \mathbf{F_{22}} \cdot \mathbf{B},$$

and $\mathbf{B}$ is solved for in terms of $\hat{\mathbf{A}}$ to give

$$\mathbf{B} = (\mathbb{I} - \mathbf{S} \cdot \mathbf{F_{22}})^{-1} \cdot \mathbf{S} \cdot \mathbf{F_{21}} \cdot \boldsymbol{\alpha}^{-1} \cdot \hat{\mathbf{A}}. \tag{15.10}$$

Substituting (15.10) into (15.9):

$$\boldsymbol{\beta}^{-1} \cdot \hat{\mathbf{B}} - \boldsymbol{\beta}^{-1} \cdot \mathbf{X} \cdot \boldsymbol{\alpha}^{-1} \cdot \hat{\mathbf{A}} = \mathbf{F_{11}} \cdot \boldsymbol{\alpha}^{-1} \cdot \hat{\mathbf{A}} + \mathbf{F_{12}} \cdot (\mathbb{I} - \mathbf{S} \cdot \mathbf{F_{22}})^{-1} \cdot \mathbf{S} \cdot \mathbf{F_{21}} \cdot \boldsymbol{\alpha}^{-1} \cdot \hat{\mathbf{A}},$$

and solving for $\hat{\mathbf{B}}$ in terms of $\hat{\mathbf{A}}$ yields

$$\hat{\mathbf{B}} = \left(\boldsymbol{\beta} \cdot \mathbf{F_{11}} \cdot \boldsymbol{\alpha}^{-1} + \mathbf{X} \cdot \boldsymbol{\alpha}^{-1} + \boldsymbol{\beta} \cdot \mathbf{F_{12}} \cdot (\mathbb{I} - \mathbf{S} \cdot \mathbf{F_{22}})^{-1} \cdot \mathbf{S} \cdot \mathbf{F_{21}} \cdot \boldsymbol{\alpha}^{-1}\right) \cdot \hat{\mathbf{A}}.$$

The raw, measured s-parameters are defined as

$$\hat{\mathbf{S}} = \hat{\mathbf{B}} \cdot \hat{\mathbf{A}}^{-1},$$

and the raw, measured s-parameters are written in terms of the s-parameters of the DUT are

$$\hat{\mathbf{S}} = \boldsymbol{\beta} \cdot \mathbf{F_{11}} \cdot \boldsymbol{\alpha}^{-1} + \mathbf{X} \cdot \boldsymbol{\alpha}^{-1} + \boldsymbol{\beta} \cdot \mathbf{F_{12}} \cdot (\mathbb{I} - \mathbf{S} \cdot \mathbf{F_{22}})^{-1} \cdot \mathbf{S} \cdot \mathbf{F_{21}} \cdot \boldsymbol{\alpha}^{-1}.$$

**Listing 15.2 ErrorTerms** class crosstalk calibration method

```
class ErrorTerms(object):
...
    def ExCalibration(self,b2a1,n,m):
        [_,Et,El]=self[n][m]
        Ex=b2a1
        self[n][m]=[Ex,Et,El]
        return self
...
```

If all of the parameters of this model were known, one would be able to take a raw measurement and recover the DUT by solving for **S**. There is no need to write that equation, because all of these parameters cannot be known uniquely. What can be discovered about the model is found through a process called *calibration.* In these types of measurements, calibration involves comparing raw measured s-parameters of known *standards* to the knowledge of the standards themselves.

To start, the application of a *reflect* standard to the system is considered. A reflect standard is one that makes no connection between the ports and simply terminates them. Here, the standard is considered as a two-port standard, with s-parameters

$$\mathbf{S}_s = \begin{pmatrix} \Gamma_s & 0 \\ 0 & \Gamma_s \end{pmatrix}.$$

This makes the s-parameters completely diagonal. Since all of the other block matrices are diagonal, except for **X**, this means that the raw measured s-parameters of this system are

$$\hat{\mathbf{S}}_s = \begin{pmatrix} \frac{\beta_1}{\alpha_1} \cdot \left( F_{1_{11}} + \frac{F_{1_{12}} \cdot F_{1_{21}} \cdot \Gamma_s}{1 - \Gamma_s \cdot F_{1_{22}}} \right) & \frac{X_{12}}{\alpha_2} \\ \frac{X_{21}}{\alpha_1} & \frac{\beta_2}{\alpha_2} \cdot \left( F_{2_{11}} + \frac{F_{2_{12}} \cdot F_{2_{21}} \cdot \Gamma_s}{1 - \Gamma_s \cdot F_{2_{22}}} \right) \end{pmatrix}.$$

The first piece of information gained from this measurement concerns the crosstalk terms:

$$\mathbf{E_X} = \mathbf{X} \cdot \alpha^{-1} = \begin{pmatrix} 0 & \hat{S}_{s_{12}} \\ \hat{S}_{s_{21}} & 0 \end{pmatrix} = \begin{pmatrix} 0 & E_{X_{12}} \\ E_{X_{21}} & 0 \end{pmatrix};$$

the code for this calculation is provided in Listing 15.2.

Otherwise, an equation for port $p$ can be written as

$$\hat{\Gamma}_{s_p} = \hat{S}_{s_{pp}} = \frac{\beta_p}{\alpha_p} \cdot \left( F_{p_{11}} + \frac{F_{p_{12}} \cdot F_{p_{21}} \cdot \Gamma_s}{1 - \Gamma_s \cdot F_{p_{22}}} \right).$$

This can be rearranged as

$$\hat{\Gamma}_{s_p} = \frac{\beta_p}{\alpha_p} \cdot F_{p_{11}} + \hat{\Gamma}_{s_p} \cdot \Gamma_s \cdot F_{p_{22}} - \Gamma_s \cdot \frac{\beta_p}{\alpha_p} \cdot |\mathbf{F_p}|, \tag{15.11}$$

**Listing 15.3 ErrorTerms** class reflect calibration method

```
class ErrorTerms(object):
...
    def ReflectCalibration(self,hatGamma,Gamma,m):
        A=[[1.,Gamma[r]*hatGamma[r],-Gamma[r]] for r in range(len(Gamma))]
        B=[[hatGamma[r]] for r in range(len(Gamma))]
        EdEsDeltaS=(matrix(A).getI()*matrix(B)).tolist()
        Ed=EdEsDeltaS[0][0]
        Es=EdEsDeltaS[1][0]
        DeltaS=EdEsDeltaS[2][0]
        Er=Ed*Es-DeltaS
        self[m][m]=[Ed,Er,Es]
        return self
...
```

where multiple measurements of multiple standards are taken:

$$\begin{pmatrix} \hat{\Gamma}_{s_{p1}} \\ \hat{\Gamma}_{s_{p2}} \\ \vdots \\ \hat{\Gamma}_{s_{pM}} \end{pmatrix} = \begin{pmatrix} 1 & \hat{\Gamma}_{s_{p1}} \cdot \Gamma_{s1} & -\Gamma_{s1} \\ 1 & \hat{\Gamma}_{s_{p2}} \cdot \Gamma_{s2} & -\Gamma_{s2} \\ \vdots & \vdots & \vdots \\ 1 & \hat{\Gamma}_{s_{pM}} \cdot \Gamma_{sM} & -\Gamma_{sM} \end{pmatrix} \cdot \begin{pmatrix} \frac{\beta_p}{\alpha_p} \cdot F_{p11} \\ F_{p22} \\ \frac{\beta_p}{\alpha_p} \cdot |\mathbf{F_p}| \end{pmatrix}.$$

Since the s-parameters of the standards and the measurements are known,

$$\begin{pmatrix} \frac{\beta_p}{\alpha_p} \cdot F_{p11} \\ F_{p22} \\ \frac{\beta_p}{\alpha_p} \cdot |\mathbf{F_p}| \end{pmatrix} = \begin{pmatrix} 1 & \hat{\Gamma}_{s_{p1}} \cdot \Gamma_{s1} & -\Gamma_{s1} \\ 1 & \hat{\Gamma}_{s_{p2}} \cdot \Gamma_{s2} & -\Gamma_{s2} \\ \vdots & \vdots & \vdots \\ 1 & \hat{\Gamma}_{s_{pM}} \cdot \Gamma_{sM} & -\Gamma_{sM} \end{pmatrix}^{\dagger} \cdot \begin{pmatrix} \hat{\Gamma}_{s_{p1}} \\ \hat{\Gamma}_{s_{p2}} \\ \vdots \\ \hat{\Gamma}_{s_{pM}} \end{pmatrix}. \tag{15.12}$$

The first term is called the *directivity* term for port $p$, defined as

$$E_{D_p} = \frac{\beta_p}{\alpha_p} \cdot F_{p11}.$$

The second term is called the *source-match* term for port $p$, defined as

$$E_{S_p} = F_{p22}.$$

The *reverse-transmission* term for port $p$ is defined as

$$E_{R_p} = \frac{\beta_p}{\alpha_p} \cdot F_{p12} \cdot F_{p21} = E_{D_p} \cdot E_{S_p} - \frac{\beta_p}{\alpha_p} \cdot |\mathbf{F_p}|.$$

The code for performing this calculation is provided in Listing 15.3.

Thus, (15.11) can be rewritten for an arbitrary measurement on port $p$ as

$$\hat{\Gamma}_p = E_{D_p} + \hat{\Gamma}_p \cdot \Gamma \cdot E_{S_p} - \Gamma \cdot \left(E_{D_p} \cdot E_{S_p} - E_{R_p}\right)$$

and solved for $\Gamma$ as

$$\Gamma = \frac{\hat{\Gamma}_p - E_{D_p}}{\hat{\Gamma}_p \cdot E_{S_p} - \left(E_{D_p} \cdot E_{S_p} - E_{R_p}\right)}. \tag{15.13}$$

Equation (15.13) shows that port $p$ is now calibrated as a single-port network analyzer capable of determining an unknown $\Gamma$ from a measurement $\hat{\Gamma}_p$ and knowledge of these error terms.

Defining the matrices

$$\mathbf{E_D} = \begin{pmatrix} E_{D_1} & 0 \\ 0 & E_{D_2} \end{pmatrix}, \mathbf{E_S} = \begin{pmatrix} E_{S_1} & 0 \\ 0 & E_{S_2} \end{pmatrix}, \mathbf{E_R} = \begin{pmatrix} E_{R_1} & 0 \\ 0 & E_{R_2} \end{pmatrix}$$

allows

$$\hat{\mathbf{S}} = \mathbf{E_D} + \mathbf{E_X} + \boldsymbol{\beta} \cdot \mathbf{F_{12}} \cdot \left(\mathbb{I} - \mathbf{S} \cdot \mathbf{E_S}\right)^{-1} \cdot \mathbf{S} \cdot \mathbf{F_{21}} \cdot \boldsymbol{\alpha}^{-1}. \tag{15.14}$$

To continue the calibration, a known *thru* standard is connected between ports 1 and 2 and a measurement is taken. The s-parameters of the thru standard are given as

$$\mathbf{S_t} = \begin{pmatrix} S_{t_{11}} & S_{t_{12}} \\ S_{t_{21}} & S_{t_{22}} \end{pmatrix}.$$

The inner product of (15.14) is known:

$$\mathbf{C_t} = \left(\mathbb{I} - \mathbf{S_t} \cdot \mathbf{E_S}\right)^{-1} \cdot \mathbf{S_t}.$$

Other knowns are gathered from the measurement as

$$\mathbf{K_t} = \hat{\mathbf{S}}_\mathbf{t} - \mathbf{E_D} - \mathbf{E_X};$$

therefore,

$$\mathbf{K_t} = \boldsymbol{\beta} \cdot \mathbf{F_{12}} \cdot \mathbf{C_t} \cdot \mathbf{F_{21}} \cdot \boldsymbol{\alpha}^{-1}$$

and

$$\begin{pmatrix} K_{t_{11}} & K_{t_{12}} \\ K_{t_{21}} & K_{t_{22}} \end{pmatrix} = \begin{pmatrix} \frac{\beta_1}{\alpha_1} \cdot F_{1_{12}} \cdot F_{1_{21}} \cdot C_{t_{11}} & \frac{\beta_1}{\alpha_2} \cdot F_{1_{12}} \cdot F_{2_{21}} \cdot C_{t_{12}} \\ \frac{\beta_2}{\alpha_1} \cdot F_{2_{12}} \cdot F_{1_{21}} \cdot C_{t_{21}} & \frac{\beta_2}{\alpha_2} \cdot F_{2_{12}} \cdot F_{2_{21}} \cdot C_{t_{22}} \end{pmatrix}.$$

The reverse-transmission terms are on the diagonal. The off-diagonal elements contain the *forward-transmission* terms, where it is defined that[1]

$$\begin{pmatrix} E_{R_1} & E_{T_{12}} \\ E_{T_{21}} & E_{R_2} \end{pmatrix} = \begin{pmatrix} \frac{\beta_1}{\alpha_1} \cdot F_{1_{12}} \cdot F_{1_{21}} & \frac{\beta_1}{\alpha_2} \cdot F_{1_{12}} \cdot F_{2_{21}} \\ \frac{\beta_2}{\alpha_1} \cdot F_{2_{12}} \cdot F_{1_{21}} & \frac{\beta_2}{\alpha_2} \cdot F_{2_{12}} \cdot F_{2_{21}} \end{pmatrix} = \begin{pmatrix} \frac{K_{t_{11}}}{C_{t_{11}}} & \frac{K_{t_{12}}}{C_{t_{12}}} \\ \frac{K_{t_{21}}}{C_{t_{21}}} & \frac{K_{t_{22}}}{C_{t_{22}}} \end{pmatrix}.$$

[1]The $E_{R_1}$ and $E_{R_2}$ terms are not actually calculated here as $K_{11}$, $K_{22}$, $C_{11}$, and $C_{22}$ tend to be very small values.

Given these error terms, $\mathbf{K_t}$ and $\mathbf{C_t}$ can be substituted for and, for general s-parameters in $\mathbf{S}$,

$$\hat{\mathbf{S}} = \begin{pmatrix} E_{R_1} \cdot \left[ \left( \mathbb{I} - \mathbf{S} \cdot \mathbf{E_S} \right)^{-1} \cdot \mathbf{S} \right]_{11} & E_{T_{12}} \cdot \left[ \left( \mathbb{I} - \mathbf{S} \cdot \mathbf{E_S} \right)^{-1} \cdot \mathbf{S} \right]_{12} \\ E_{T_{21}} \cdot \left[ \left( \mathbb{I} - \mathbf{S} \cdot \mathbf{E_S} \right)^{-1} \cdot \mathbf{S} \right]_{21} & E_{R_2} \cdot \left[ \left( \mathbb{I} - \mathbf{S} \cdot \mathbf{E_S} \right)^{-1} \cdot \mathbf{S} \right]_{22} \end{pmatrix} + \mathbf{E_D} + \mathbf{E_X}$$

$$= \frac{\begin{pmatrix} (S_{11} - |\mathbf{S}| \cdot E_{S_2}) \cdot E_{R1} & S_{12} \cdot E_{T_{12}} \\ S_{21} \cdot E_{T_{21}} & (S_{22} - |\mathbf{S}| \cdot E_{S_1}) \cdot E_{R_2} \end{pmatrix}}{(|\mathbf{S}| \cdot E_{S_2} - S_{11}) \cdot E_{S_1} - S_{22} \cdot E_{S_2} + 1} + \begin{pmatrix} E_{D_1} & E_{X_{12}} \\ E_{X_{21}} & E_{D_2} \end{pmatrix}. \quad (15.15)$$

This allows the forward-transmission term $E_{T_{12}}$ to be written in a direct form:

$$E_{T_{12}} = \frac{\hat{S}_{t_{12}} - E_{X_{12}}}{S_{t_{12}}} \cdot \left[ \left( |\mathbf{S_t}| \cdot E_{S_2} - S_{t_{11}} \right) \cdot E_{S_1} - S_{t_{22}} \cdot E_{S_2} + 1 \right],$$

and, correspondingly,

$$E_{T_{21}} = \frac{\hat{S}_{t_{21}} - E_{X_{21}}}{S_{t_{21}}} \cdot \left[ \left( |\mathbf{S_t}| \cdot E_{S_2} - S_{t_{11}} \right) \cdot E_{S_1} - S_{t_{22}} \cdot E_{S_2} + 1 \right].$$

Generally, when a thru is connected between ports $p$ and $o$, with port 1 of the standard at port $p$,

$$E_{T_{op}} = \frac{\hat{S}_{t_{21}} - E_{X_{op}}}{S_{t_{21}}} \cdot \left[ \left( |\mathbf{S_t}| \cdot E_{S_o} - S_{t_{11}} \right) \cdot E_{S_p} - S_{t_{22}} \cdot E_{S_o} + 1 \right] \quad (15.16)$$

and

$$E_{T_{po}} = \frac{\hat{S}_{t_{12}} - E_{X_{po}}}{S_{t_{12}}} \cdot \left[ \left( |\mathbf{S_t}| \cdot E_{S_o} - S_{t_{11}} \right) \cdot E_{S_p} - S_{t_{22}} \cdot E_{S_o} + 1 \right]. \quad (15.17)$$

On a side note, each port has been calibrated with reflect standards and therefore $E_{D_p}$, $E_{R_p}$, and particularly $E_{S_p}$, which is equal to $F_{p_{22}}$, are known. However, sometimes the model does not permit knowledge of the $S_{22}$ of a port through reflect standard measurements. Because of this, when a thru is connected between ports $p$ and $o$, and port $p$ has been calibrated with reflect standards, thus knowing $E_{D_p}$, $E_{R_p}$, and $E_{S_p}$, it is customary to measure $E_{S_o}$ from the thru standard measurement. The easiest way to see how this is performed is by using port $p$ as a calibrated reflectometer. Examining Figure 15.3, this means that the raw measurement $\hat{S}_{11}$ with the thru connected can be converted using (15.13) into a calibrated measurement of an $S'_{11}$, defined as the $S_{11}$ of the thru standard connected to $\mathbf{F_2}$. Since the s-parameters of the standard $\mathbf{S_t}$ are known, the thru standard can be de-embedded from this measurement to form $F_{2_{22}}$. The one-port equation in (10.1) is used to perform this de-embedding step.

This means that, with the thru connected between ports $p$ and $o$, and port 1 of the standard at port $p$, using (15.13) at port $p$ yields

$$S'_{pp} = \frac{\hat{S}_{pp} - E_{D_p}}{\hat{S}_{pp} \cdot E_{S_p} - \left( E_{D_p} \cdot E_{S_p} - E_{R_p} \right)};$$

using (10.1), solved for the *load-match* term at port $o$ due to port $p$,

$$E_{L_{op}} = \frac{S'_{pp} - S_{t_{11}}}{S'_{pp} \cdot S_{t_{22}} - |\mathbf{S_t}|}.$$

Combining all of this:[2]

$$E_{L_{op}} = E_{S_o} = F_{o22} = \frac{E_{R_p} \cdot S_{t_{11}} - \left(\hat{S}_{t_{pp}} - E_{D_p}\right) \cdot \left(1 - E_{S_p} \cdot S_{t_{11}}\right)}{\left(\hat{S}_{t_{pp}} - E_{D_p}\right) \cdot \left(E_{S_p} \cdot |\mathbf{S_t}| - S_{t_{22}}\right) + E_{R_p} \cdot |\mathbf{S_t}|}. \tag{15.18}$$

If, with the thru connected in the same manner, the measurement is reversed, measuring at port $o$, but assuming the standard connected in the same orientation,[3]

$$E_{L_{po}} = E_{S_p} = F_{p22} = \frac{E_{R_o} \cdot S_{t_{22}} - \left(\hat{S}_{t_{oo}} - E_{D_o}\right) \cdot \left(1 - E_{S_o} \cdot S_{t_{22}}\right)}{\left(\hat{S}_{t_{oo}} - E_{D_o}\right) \cdot \left(E_{S_o} \cdot |\mathbf{S_t}| - S_{t_{11}}\right) + E_{R_o} \cdot |\mathbf{S_t}|}.$$

This $E_{L_{po}}$ and $E_{L_{op}}$ are therefore substituted into (15.16) and (15.17) for the assumed unknown $E_{s_o}$ and $E_{S_p}$, respectively, where $E_{T_{op}}$ and $E_{T_{po}}$ are calculated after the $E_L$ calculations:

$$E_{T_{op}} = \frac{\hat{S}_{t_{21}} - E_{X_{op}}}{S_{t_{21}}} \cdot \left[\left(|\mathbf{S_t}| \cdot E_{L_{op}} - S_{t_{11}}\right) \cdot E_{S_p} - S_{t_{22}} \cdot E_{L_{op}} + 1\right] \tag{15.19}$$

and

$$E_{T_{po}} = \frac{\hat{S}_{t_{12}} - E_{X_{po}}}{S_{t_{12}}} \cdot \left[\left(|\mathbf{S_t}| \cdot E_{S_o} - S_{t_{11}}\right) \cdot E_{L_{po}} - S_{t_{22}} \cdot E_{S_o} + 1\right].$$

In the DUT calculation in §15.3, $E_{L_{op}}$ and $E_{L_{po}}$ will be interchanged with $E_{S_o}$ and $E_{S_p}$, respectively, depending on which port is driven, but the equality of these two terms will be relied upon in §15.2.3.

Before moving on, an element of possible confusion should be addressed. When measuring the load-match term, it was stated that a fully calibrated single-port reflectometer was used, but it was not used strictly in a single-port measurement. If it were used in that manner, and Figure 15.3 is examined, one sees that a calibrated measurement would have been made of a system containing the thru standard (as $\mathbf{S}$) connected to the portion of the analyzer with s-parameters $\mathbf{F_2}$ terminated in $\Gamma_2$. This measurement would not be of $F_{2_{22}}$, and in fact $\Gamma_2$ would be in the result. But a strict one-port measurement was not taken; a two-port measurement in $\hat{\mathbf{S}} = \hat{\mathbf{B}} \cdot \hat{\mathbf{A}}^{-1}$ was taken with $\hat{\mathbf{S}}_{11}$ then taken out. This way of measuring removes $\Gamma_2$ along with the whole loop containing $F_{2_{21}}$, $F_{2_{12}}$, and $F_{2_{11}}$ from the measurement. If this is not understood, it can be confirmed by calculating, for example, $E_{L_{12}}$ from (15.15) by writing the equation for the $\hat{S}_{t_{11}}$ term and solving for the unknown

[2]The subscripts for the ports of the instrument are $o$ and $p$, whereas the subscripts for the ports of the standard remain as 1 and 2.

[3]When this is performed in software that takes the thru measurements and the s-parameter model of the thru standard, and when measuring the other port (i.e. driving into port 2 of the standard), the s-parameters of the thru standard are flipped. Fortunately, most thru standard models are symmetric.

$E_{S_2} = E_{L_{21}}$ and by writing the equation for the $\hat{S}_{t_{22}}$ term and solving for the unknown $E_{S_1} = E_{L_{12}}$.

This completes the basic short-open-load-thru (SOLT) calibration.

### 15.2.1 Overconstrained Thru Measurements

The reflect error terms were shown to be measured with three or more standards in (15.12). However, the thru error terms were shown for only one thru connection. It is sometimes advantageous to take multiple measurements of multiple thrus in order to make a least-squares set of error terms. To make overconstrained thru measurements, (15.18) and (15.19) are rearranged:

$$\left[\left(\hat{S}_{t_{11}} - E_{D_p}\right) \cdot \left(E_{S_p} \cdot |\mathbf{S_t}| - S_{t_{22}}\right) + E_{R_p} \cdot |\mathbf{S_t}|\right] \cdot E_{L_{op}} = E_{R_p} \cdot S_{t_{11}} - \left(\hat{S}_{t_{11}} - E_{D_p}\right) \cdot \left(1 - E_{S_p} \cdot S_{t_{11}}\right),$$

$$\frac{\hat{S}_{t_{21}} - E_{X_{op}}}{S_{t_{21}}} \cdot \left(|\mathbf{S_t}| \cdot E_{S_p} - S_{t_{22}}\right) \cdot E_{L_{op}} - E_{T_{op}} = \frac{\hat{S}_{t_{21}} - E_{X_{op}}}{S_{t_{21}}} \cdot \left(S_{t_{11}} \cdot E_{S_p} - 1\right).$$

For $M$ measurements, with $m \in 1 \ldots M$, on each measurement a two-port thru is connected between ports $p$ and $o$, with port $p$ connected to port 1 of the thru and port $o$ connected to port 2 of the thru, the s-parameters of the standard are denoted $\mathbf{S_{t_m}}$, and the two-port measurement of the standard denoted $\hat{\mathbf{S}}_{\mathbf{t}_m}$, then the stack of block matrices $\mathbf{A}$ and $\mathbf{B}$ for the multiple measurements are

$$\mathbf{A}_m = \begin{pmatrix} \left(\hat{S}_{t_{m_{11}}} - E_{D_p}\right) \cdot \left(E_{S_p} \cdot |\mathbf{S_{t_m}}| - S_{t_{m_{22}}}\right) + E_{R_p} \cdot |\mathbf{S_{t_m}}| & 0 \\ \frac{\hat{S}_{t_{m21}} - E_{X_{op}}}{S_{t_{m21}}} \cdot \left(|\mathbf{S_{t_m}}| \cdot E_{S_p} - S_{t_{m_{22}}}\right) & -1 \end{pmatrix}$$

and

$$\mathbf{B}_m = \begin{pmatrix} E_{R_p} \cdot S_{t_{m_{11}}} - \left(\hat{S}_{t_{m_{11}}} - E_{D_p}\right) \cdot \left(1 - E_{S_p} \cdot S_{t_{m_{11}}}\right) \\ \frac{\hat{S}_{t_{m21}} - E_{X_{op}}}{S_{t_{m21}}} \cdot \left(S_{t_{m_{11}}} \cdot E_{S_p} - 1\right) \end{pmatrix},$$

and the overconstrained calculation of the thru error terms is given by

$$\begin{pmatrix} E_{L_{op}} \\ E_{T_{op}} \end{pmatrix} = \mathbf{A}^{\dagger} \cdot \mathbf{B}.$$

To construct the calculations for $E_{L_{po}}$ and $E_{T_{po}}$, the $p$ and $o$ subscripts are swapped on the error term and the 1 and 2 subscripts are swapped on the thru standards and the thru standard measurements.

The overconstrained thru calculation code is provided in Listing 15.4.

**Listing 15.4 ErrorTerms** class thru calibration method

```
class ErrorTerms(object):
...
    def ThruCalibration(self,b1a1,b2a1,S,n,m):
        [Ed,Er,Es]=self[m][m]
        Ex=self[n][m][0]
        A=zeros((2*len(b1a1),2)).tolist()
        B=zeros((2*len(b1a1),1)).tolist()
        for i in range(len(b1a1)):
            Sm=S[i]
            detS=det(matrix(Sm))
            A[2*i][0]=(Es*detS-Sm[1][1])*(Ed-b1a1[i])-Er*detS
            A[2*i][1]=0.
            A[2*i+1][0]=(Es*detS-Sm[1][1])*(Ex-b2a1[i])
            A[2*i+1][1]=Sm[1][0]
            B[2*i][0]=(1.-Es*Sm[0][0])*(b1a1[i]-Ed)-Er*Sm[0][0]
            B[2*i+1][0]=(1-Es*Sm[0][0])*(b2a1[i]-Ex)
        ElEt=(matrix(A).getI()*matrix(B)).tolist()
        (El,Et)=(ElEt[0][0],ElEt[1][0])
        self[n][m]=[Ex,Et,El]
        return self
...
```

### 15.2.2 Method of the Unknown Thru

Sometimes it is not possible to know the thru standard accurately enough to perform a good thru calibration. In such cases, specifying the thru incorrectly will lead to potentially large errors in the calibration. A remeasurement of the thru with the system will appear to be fine, but measurements of other thru elements are badly affected. A method has been determined that resolves this problem by utilizing certain mathematical facts about the thru calibration, not only to calibrate the load-match and forward-transmission terms without knowledge of the thru, but also to measure the thru in the process [45]. This method is called short-open-load-reciprocal (SOLR).

When performing the calibration with the thru standard between an initial port $p$ and another port $o$, one obtains the raw measured s-parameters of the thru as

$$\hat{\mathbf{S}}_{\mathbf{t}} = \begin{pmatrix} \hat{S}_{t_{pp}} & \hat{S}_{t_{po}} \\ \hat{S}_{t_{op}} & \hat{S}_{t_{oo}} \end{pmatrix}.$$

Ferrero and Pisani [45] found that, if the thru standard and the fixtures $\mathbf{F_p}$ and $\mathbf{F_o}$ are assumed reciprocal,

$$\frac{\hat{S}_{t_{po}} - E_{X_{po}}}{\hat{S}_{t_{op}} - E_{X_{op}}} = \frac{\beta_p \cdot \alpha_p}{\beta_o \cdot \alpha_o} = p \tag{15.20}$$

for any thru length, loss, or match. As such, $p$ encapsulates the frequency responses of the measurements. This can be exploited by recognizing that

$$E_{R_p} \cdot E_{R_o} = \frac{\beta_p}{\alpha_p} \cdot F_{p12} \cdot F_{p21} \cdot \frac{\beta_o}{\alpha_o} \cdot F_{o12} \cdot F_{o21}$$

and

$$E_{T_{op}}^2 \cdot p = \frac{\beta_o}{\alpha_p} \cdot F_{o12} \cdot F_{p21} \cdot \frac{\beta_o}{\alpha_p} \cdot F_{o12} \cdot F_{p21} \cdot \frac{\beta_p \cdot \alpha_p}{\beta_o \cdot \alpha_o} = \frac{\beta_p}{\alpha_p} \cdot F_{p21} \cdot F_{p21} \cdot \frac{\beta_o}{\alpha_o} \cdot F_{o12} \cdot F_{o12} \cdot$$

**Listing 15.5 ErrorTerms** class unknown thru calibration method

```
class ErrorTerms(object):
...
    def UnknownThruCalibration(self,Sm,Sest,firstPort,secondPort):
        [ED1,ER1,ES1]=self[firstPort][firstPort]
        [ED2,ER2,ES2]=self[secondPort][secondPort]
        [EX12,ET12,EL12]=self[firstPort][secondPort]
        [EX21,ET21,EL21]=self[secondPort][firstPort]
        p=cmath.sqrt((Sm[0][1]-EX12)/(Sm[1][0]-EX21))
        [EX12,ET12,EL12]=[EX12,cmath.sqrt(ER1)*cmath.sqrt(ER2)*p,ES1]
        [EX21,ET21,EL21]=[EX21,cmath.sqrt(ER1)*cmath.sqrt(ER2)/p,ES2]
        DutCalc1=ErrorTerms([[[ED1,ER1,ES1],[EX12,ET12,EL12]],
                             [[EX21,ET21,EL21],[ED2,ER2,ES2]]]).DutCalculation(Sm)
        DutCalc2=ErrorTerms([[[ED1,ER1,ES1],[EX12,-ET12,EL12]],
                             [[EX21,-ET21,EL21],[ED2,ER2,ES2]]]).DutCalculation(Sm)
        if norm(matrix(DutCalc1)-matrix(Sest)) < norm(matrix(DutCalc2)-matrix(Sest)):
            return DutCalc1
        else:
            return DutCalc2
...
```

Therefore, assuming reciprocity of the fixture,[4]

$$E_{R_p} \cdot E_{R_o} = E_{T_{op}}^2 \cdot p$$

and thus

$$E_{T_{op}} = \frac{\sqrt{E_{R_p}} \cdot \sqrt{E_{R_o}}}{p}.$$

Although the math is not shown for the other port, it is found that

$$E_{T_{po}} = \sqrt{E_{R_p}} \cdot \sqrt{E_{R_o}} \cdot p,$$

where, in this calibration, $E_{L_{op}} = E_{S_o}$ and $E_{L_{po}} = E_{S_p}$ are utilized.

There is one detail in this calculation that has not been considered, and this is regarding which branch of the square root to use. Fortunately, there are only two choices possible, which involve using either both positive or both negative values of the $E_T$ terms calculated. Usually there is at least an estimate of the thru standard available. If so, the DUT calculation in (15.22) (see §15.3) is utilized to calculate the DUT and the number closest to the thru standard estimate is used. If no thru standard estimate is available, and the measurement is made to DC, then an ideal thru of zero length is estimated for the zero frequency point for the first calculation. When the thru is recovered, it is used as the estimate for the next calculation. If the frequency steps are fine enough, the correct sign will be chosen.

The code for the calculation of the unknown thru is provided in Listing 15.5, where one sees that the result of the calibration is the recovery of the thru, not the calculation of the thru error terms. This is provided for use with the overconstrained thru measurement in §15.2.1. In this way, multiple SOLR thru calibrations can be made and an overconstrained solution can be calculated using these now known thru standards.

[4]It is not only the reciprocity of the fixture being assumed here, because (15.20) is true only if both the fixtures and the thru are reciprocal.

**Listing 15.6 ErrorTerms** class `TransferThruCalibration()` method

```
class ErrorTerms(object):
...
    def TransferThruCalibration(self):
        didOne=True
        while didOne:
            didOne=False
            for otherPort in range(self.numPorts):
                for drivenPort in range(self.numPorts):
                    if (otherPort == drivenPort):
                        continue
                    if ((self[otherPort][drivenPort][1]==0) and
                        (self[otherPort][drivenPort][2]==0)):
                        for mid in range(self.numPorts):
                            if ((mid != otherPort) and
                                (mid != drivenPort) and
                                ((self[otherPort][mid][1]!=0) or
                                 (self[otherPort][mid][2]!=0)) and
                                ((self[mid][drivenPort][1]!=0) or
                                 (self[mid][drivenPort][2]!=0))):
                                (_,Etl,_)=self[otherPort][mid]
                                (_,Etr,_)=self[mid][drivenPort]
                                (_,Erm,_)=self[mid][mid]
                                (_,_,Eso)=self[otherPort][otherPort]
                                (Ex,Et,El)=self[otherPort][drivenPort]
                                Et=Etl*Etr/Erm
                                El=Eso
                                self[otherPort][drivenPort]=[Ex,Et,El]
                                didOne=True
                                continue
        return self
...
```

### 15.2.3 Transfer Thru Calibration

Until now, a system has been calibrated by measuring at least three reflect standards on each port and measuring a thru standard between every combination of two port connections. As an example, for a four-port system the requirement is a thru standard measured for the 1–2, 1–3, 1–4, 2–3, 2–4, and 3–4 port connections.

Although at least three reflect measurements are required on every port, using the transfer thru method the requirement on thru standard connections is only that it is possible to chain together actual port connections to form the desired thru connection. As an example, for a four-port system, if a thru is measured between ports 1–2, 2–3, and 3–4, the 1–3 connection can be formed by chaining 1–2 and 2–3, the 2–4 connection by chaining 2–3 and 3–4, and the 1–4 connection by chaining the 1–2, 2–3, and 3–4 connections. In practice, this calibration is performed after all of the reflect and crosstalk terms are measured and the desired thru connections have been made and measured. Terms are then filled in that are the chains of two others. By continuing to do this until no other chains of two exist, the error terms are filled in. In the two-port example, this means that the 1–3 and 2–4 connections are found first, and then the 1–4 connection is found as a chain of either 1–2 and 2–4 or 1–3 and 3–4.

If the thru connection $o$–$d$ has not been made, then one looks for a port $m$ where the $o$–$m$ and $m$–$d$ connection exists. In this case, from the previous discussion,

$$E_{T_{om}} = \frac{\beta_o}{\alpha_m} \cdot F_{o_{12}} \cdot F_{m_{21}}$$

and

$$E_{T_{md}} = \frac{\beta_m}{\alpha_d} \cdot F_{m_{12}} \cdot F_{d_{21}},$$

and, from the reflect calibration on port $m$,

$$E_{R_m} = \frac{\beta_m}{\alpha_m} \cdot F_{m_{12}} \cdot F_{m_{21}}.$$

Therefore, the desired error term can be computed as follows:

$$E_{T_{od}} = \frac{E_{T_{om}} \cdot E_{T_{md}}}{E_{R_m}} = \frac{\left(\frac{\beta_o}{\alpha_m} \cdot F_{o_{12}} \cdot F_{m_{21}}\right) \cdot \left(\frac{\beta_m}{\alpha_d} \cdot F_{m_{12}} \cdot F_{d_{21}}\right)}{\frac{\beta_m}{\alpha_m} \cdot F_{m_{12}} \cdot F_{m_{21}}} = \frac{\beta_o}{\alpha_d} \cdot F_{o_{12}} \cdot F_{d_{21}}.$$

It has already been seen that the load-match term is equal to the source-match term in the following way:

$$E_{L_{od}} = E_{S_o},$$

and thus two thru connections can be chained to form a missing connection.

The algorithm is shown in Listing 15.6.

## 15.3 Calculation of the Device Under Test

During the calibration calculations, it was determined that, given an unknown DUT with s-parameters $\mathbf{S}$ and raw measured s-parameters $\hat{\mathbf{S}}$,

$$\mathbf{K} = \hat{\mathbf{S}} - \mathbf{E_D} - \mathbf{E_X}$$

and

$$\mathbf{C} = \begin{pmatrix} \frac{K_{11}}{E_{R_1}} & \frac{K_{12}}{E_{T_{12}}} \\ \frac{K_{21}}{E_{T_{21}}} & \frac{K_{22}}{E_{R_2}} \end{pmatrix};$$

then,

$$\mathbf{C} = (\mathbb{I} - \mathbf{S} \cdot \mathbf{E_S})^{-1} \cdot \mathbf{S}$$

and the unknown DUT is solved as follows:

$$\mathbf{S} = \mathbf{C} \cdot (\mathbb{I} + \mathbf{E_S} \cdot \mathbf{C})^{-1}.$$

Expanding this completely to show all of the measurements and error terms:[5]

$$\begin{pmatrix} S_{11} & S_{12} \\ S_{21} & S_{22} \end{pmatrix} = \begin{pmatrix} \frac{\hat{S}_{11}-E_{D_1}}{E_{R_1}} & \frac{\hat{S}_{12}-E_{X_{12}}}{E_{T_{12}}} \\ \frac{\hat{S}_{21}-E_{X_{21}}}{E_{T_{21}}} & \frac{\hat{S}_{22}-E_{D_2}}{E_{R_2}} \end{pmatrix} \cdot \begin{pmatrix} 1+E_{S_1}\cdot\frac{\hat{S}_{11}-E_{D_1}}{E_{R_1}} & E_{L_{12}}\cdot\frac{\hat{S}_{12}-E_{X_{12}}}{E_{T_{12}}} \\ E_{L_{21}}\cdot\frac{\hat{S}_{21}-E_{X_{21}}}{E_{T_{21}}} & 1+E_{S_2}\cdot\frac{\hat{S}_{22}-E_{D_2}}{E_{R_2}} \end{pmatrix}^{-1}. \tag{15.21}$$

(Equation (15.21) provides the result for the two-port solutions put forth in [46] with different nomenclature.)

The solution to a single-port measurement in (15.13) is expressed here as

$$\Gamma = S_{11} = \left(\frac{\hat{S}_{11}-E_{D_1}}{E_{R_1}}\right)\cdot\left(1+E_{S_1}\cdot\frac{\hat{S}_{11}-E_{D_1}}{E_{R_1}}\right)^{-1}.$$

The general expression for $P$-port s-parameters is given by

$$\begin{pmatrix} S_{11} & S_{12} & \cdots & S_{1P} \\ S_{21} & S_{22} & \cdots & S_{2P} \\ \vdots & \vdots & \ddots & \vdots \\ S_{P1} & S_{P2} & \cdots & S_{PP} \end{pmatrix} = \begin{pmatrix} \frac{\hat{S}_{11}-E_{D_1}}{E_{R1}} & \frac{\hat{S}_{12}-E_{X_{12}}}{E_{T_{12}}} & \cdots & \frac{\hat{S}_{1P}-E_{X_{1P}}}{E_{T_{1P}}} \\ \frac{\hat{S}_{21}-E_{X_{21}}}{E_{T_{21}}} & \frac{\hat{S}_{22}-E_{D_2}}{E_{R2}} & \cdots & \frac{\hat{S}_{2P}-E_{X_{2P}}}{E_{T_{2P}}} \\ \vdots & \vdots & \ddots & \vdots \\ \frac{\hat{S}_{P1}-E_{X_{P1}}}{E_{T_{P1}}} & \frac{\hat{S}_{P2}-E_{X_{P2}}}{E_{T_{P2}}} & \cdots & \frac{\hat{S}_{PP}-E_{D_P}}{E_{RP}} \end{pmatrix}$$

$$\cdot \begin{pmatrix} 1+E_{S_1}\frac{\hat{S}_{11}-E_{D_1}}{E_{R1}} & E_{L_{12}}\cdot\frac{\hat{S}_{12}-E_{X_{12}}}{E_{T_{12}}} & \cdots & E_{L_{1P}}\cdot\frac{\hat{S}_{1P}-E_{X_{1P}}}{E_{T_{1P}}} \\ E_{L_{21}}\cdot\frac{\hat{S}_{21}-E_{X_{21}}}{E_{T_{21}}} & 1+E_{S_2}\frac{\hat{S}_{22}-E_{D_2}}{E_{R2}} & \cdots & E_{L_{2P}}\cdot\frac{\hat{S}_{2P}-E_{X_{2P}}}{E_{T_{2P}}} \\ \vdots & \vdots & \ddots & \vdots \\ E_{L_{P1}}\cdot\frac{\hat{S}_{P1}-E_{X_{P1}}}{E_{T_{P1}}} & E_{L_{P2}}\cdot\frac{\hat{S}_{P2}-E_{X_{P2}}}{E_{T_{P2}}} & \cdots & 1+E_{S_P}\frac{\hat{S}_{PP}-E_{D_P}}{E_{RP}} \end{pmatrix}^{-1}. \tag{15.22}$$

The code for this solution is provided in Listing 15.7, line 3. This solution is a simplified version of that provided by [47]. Line 13 of Listing 15.7 provides the calculation in the reverse direction, which is useful for calibration algorithm development.

## 15.4 Calibration and Measurement Summary

The error terms for the twelve-term model are summarized in Table 15.1. The SOLT, SOLR, and transfer thru methods have been shown but there are many more calibration methods in use including improvements to the SOLT algorithms provided here [48]. Furthermore, in addition to the twelve-term model, there are many other models in use, and they are all related in some manner [49–51], with some not making the assumption that there are two-port only connections per calibration [52].

To perform all of these calculations, a **Calibration** class is provided in Listing 15.8. The class is constructed on line 2 with the number of ports, the list of frequencies, and a

[5]The load-match terms are substituted for the source-match terms in the off-diagonal elements.

**Listing 15.7 ErrorTerms** class DUT calculation method

```
class ErrorTerms(object):
...
    def DutCalculation(self,sRaw,pl=None,reciprocal=False):
        P=len(sRaw); Pr=range(P)
        if pl is None: pl = [p for p in Pr]
        B=[[(sRaw[r][c]-self[pl[r]][pl[c]][0])/self[pl[r]][pl[c]][1]
            for c in Pr] for r in  Pr]
        A=[[B[r][c]*self[pl[r]][pl[c]][2]+(1 if r==c else 0) for c in Pr] for r in Pr]
        if not reciprocal: S=(matrix(B)*matrix(A).getI()).tolist()
        else: S=self._EnforceReciprocity(A,B)
        return S
...
    def DutUnCalculation(self,S,pl=None):
        if pl is None: pl = [p for p in range(len(S))]
        Sp=[[None for c in range(len(S))] for r in range(len(S))]
        Si=matrix(S).getI()
        for c in range(len(S)):
            E=self.Fixture(c,pl)
            Em=[[matrix(E[0][0]),matrix(E[0][1])],[matrix(E[1][0]),matrix(E[1][1])]]
            col=(Em[0][0]*Em[1][0]+Em[0][1]*(Si-Em[1][1]).getI()*Em[1][0]).tolist()
            for r in range(len(S)): Sp[r][c]=col[r][c]
        return Sp
```

**Table 15.1** Summary of error terms

| Term | Formula | Name | Notes |
|---|---|---|---|
| $E_{D_p}$ | $\frac{\beta_p}{\alpha_p} \cdot F_{p_{11}}$ | directivity term for port $p$ | |
| $E_{S_p}$ | $F_{p_{22}}$ | source-match term for port $p$ | |
| $E_{R_p}$ | $\frac{\beta_p}{\alpha_p} \cdot F_{p_{12}} \cdot F_{p_{21}}$ | reverse-transmission term for port $p$ | assumed that $F_{p_{12}} = F_{p_{21}}$ in SOLR |
| $E_{X_{op}}$ | $\frac{1}{\alpha_p} \cdot X_{op}$ | crosstalk term for port $o$ when port $p$ driven | |
| $E_{T_{op}}$ | $\frac{\beta_o}{\alpha_p} \cdot F_{o_{12}} \cdot F_{p_{21}}$ | forward-transmission term for port $o$ when port $p$ driven | |
| $E_{L_{op}}$ | $F_{o_{22}}$ | load-match term for port $o$ when port $p$ driven | assumed equal to $E_{S_o}$ in transfer thru and SOLR |

list of measurements. While the structure of the measurements is not shown, the measurements are self-explanatory, providing the raw measurements, the standard, and the port connections. Measurements come in four types, namely `'reflect'`, `'xtalk'`, `'thru'`, and `'reciprocal'`. The calibration class holds a list of the **ErrorTerms** class for each frequency point. As measurements are added on line 10, they are organized in a calibration matrix, based on the port connections.

**Listing 15.8 Calibration class**

```
1  class Calibration(object):
2      def __init__(self,ports,f,calibrationList=[]):
3          self.ports=ports
4          self.f=f
5          self.ET=None
6          self.calibrationMatrix=[[[] for _ in range(self.ports)]
7                                  for _ in range(self.ports)]
8          self.AddMeasurements(calibrationList)
9  ...
10     def AddMeasurements(self,calibrationList=[]):
11         self.ET=None
12         for calibrationMeasurement in calibrationList:
13             if calibrationMeasurement.type=='reflect':
14                 portDriven=calibrationMeasurement.port
15                 otherPort=portDriven
16                 self.calibrationMatrix[otherPort][portDriven].\
17                     append(calibrationMeasurement)
18             elif (calibrationMeasurement.type=='thru') or\
19                 (calibrationMeasurement.type=='xtalk') or\
20                 (calibrationMeasurement.type=='reciprocal'):
21                 portDriven=calibrationMeasurement.portDriven
22                 otherPort=calibrationMeasurement.otherPort
23                 self.calibrationMatrix[otherPort][portDriven].\
24                     append(calibrationMeasurement)
25         return self
26 ...
27     def CalculateErrorTerms(self,force=False):
28         if (not self.ET is None) and (not force): return self
29         self.ET=[ErrorTerms().Initialize(self.ports) for _ in range(len(self))]
30         measurements=copy.deepcopy(self.calibrationMatrix)
31         self._CalculateReflectErrorTerms(measurements)
32         self._CalculateXtalkErrorTerms(measurements)
33         self._CalculateUnknownThruErrorTerms(measurements)
34         self._CalculateThruErrorTerms(measurements)
35         self._CalculateTransferThruErrorTerms()
36         return self
37 ...
38     def DutCalculation(self,sRaw,portList=None,reciprocal=False):
39         self.CalculateErrorTerms()
40         return SParameters(self.f,[self[n].DutCalculation(sRaw[n],portList,reciprocal)
41                                    for n in range(len(self))])
42 ...
```

The error terms calculation is shown on line 27, which mostly calls private methods shown in Listing 15.9. They are calculated in a specific order, as many depend on error terms that have already been calculated. The reflect error terms are calculated first, the details of which are shown on line 2 of Listing 15.9. Then the crosstalk error terms are calculated (see line 9 of Listing 15.9).

The unknown thru calculation is next. As seen on line 21 of Listing 15.9, this calculation depends on the existence of the crosstalk and reflect error terms. Also, one can see the detail that the unknown thru calculation simply recovers the thru standard and converts the measurement into a known thru measurement, then adds these new measurements to the measurement list. This has the advantage of providing the capability of overconstrained thru measurements. Another advantage of this technique is that it allows the limiting of the impulse response length of the thru standard measurement. Typically, but not always,

**Listing 15.9 Calibration** class error terms calculation details

```
class Calibration(object):
    def _CalculateReflectErrorTerms(self,measurements):
        for port in range(self.ports):
            measurementList=measurements[port][port]
            for n in range(len(self)):
                hatGamma=[meas.gamma[n] for meas in measurementList]
                Gamma=[meas.Gamma[n][0][0] for meas in measurementList]
                self[n].ReflectCalibration(hatGamma,Gamma,port)
    def _CalculateXtalkErrorTerms(self,measurements):
        for other in range(self.ports):
            for driven in range(self.ports):
                if (other != driven):
                    measurementList=measurements[other][driven]
                    xtalkMeasurementList=[]
                    for meas in measurementList:
                        if meas.type=='xtalk': xtalkMeasurementList.append(meas)
                    if len(xtalkMeasurementList)!=0:
                        for n in range(len(self.f)):
                            self[n].ExCalibration(
                                xtalkMeasurementList[0].b2a1[n],other,driven)
    def _CalculateUnknownThruErrorTerms(self,measurements):
        for other in range(self.ports):
            for driven in range(self.ports):
                if (other != driven):
                    for meas in measurements[other][driven]:
                        if meas.type=='reciprocal':
                            Sestsp= [s for s in meas.S] if not (meas.S is None)\
                                else [Thru() for _ in range(len(self.f))]
                            for n in range(len(self.f)):
                                Sestsp[n]=self[n].UnknownThruCalibration(
                                    meas.Smeasured[n],
                                    Sestsp[n] if not meas.S is None
                                    else Sestsp[max(n-1,0)],driven,other)
                            Sest=SParameters(self.f,Sestsp)
                            Sest=Sest.LimitImpulseResponseLength(meas.limit)
                            measurements[other][driven].append(
                                ThruCalibrationMeasurement(
                                meas.Smeasured.FrequencyResponse(1,1),
                                meas.Smeasured.FrequencyResponse(2,1),
                                Sest,other,driven))
                            measurements[driven][other].append(
                                ThruCalibrationMeasurement(
                                meas.Smeasured.FrequencyResponse(2,2),
                                meas.Smeasured.FrequencyResponse(1,2),
                                Sest.PortReorder([1,0]),driven,other))
    def _CalculateThruErrorTerms(self,measurements):
        for other in range(self.ports):
            for driven in range(self.ports):
                if (other != driven):
                    measurementList=measurements[other][driven]
                    thruMeasurementList=[]
                    for meas in measurementList:
                        if meas.type=='thru': thruMeasurementList.append(meas)
                    if len(thruMeasurementList)!=0:
                        for n in range(len(self.f)):
                            b1a1=[meas.b1a1[n] for meas in thruMeasurementList]
                            b2a1=[meas.b2a1[n] for meas in thruMeasurementList]
                            S=[meas.S[n] for meas in thruMeasurementList]
                            self[n].ThruCalibration(b1a1,b2a1,S,other,driven)
    def _CalculateTransferThruErrorTerms(self):
        if Calibration.FillInTransferThru:
            for n in range(len(self.f)): self[n].TransferThruCalibration()
...
```

the thru standard is very short. Any errors in the measurement of the thru can cause errors in the other measurements. Limiting the impulse response length of the thru standard measurment smooths the response, which, for a good short thru, could not physically create rapid swings in magnitude and phase.

Next, the thru measurements are processed, as shown on line 46 of Listing 15.9. These calculations rely on all of the previously made error term measurements.

Finally, the transfer thru calculations are made, the details of which are provided on line 60 of Listing 15.9. This comprises the final step because until this point it was unknown how many thru connections might be missing, and the calculations rely on the thru connections that have actually been made. In order to work properly, this step must loop over all of the ports connecting thru measurements until none are missing.

The DUT calculation is shown on line 38 of Listing 15.8, which simply calls the `DutCalculation()` member of **ErrorTerms** on line 3 in Listing 15.7 for each frequency point.

## 15.5 Calibration Standards

The standards for SOLT calibration are provided by instrument manufacturers in the form of a calibration kit, or cal kit. A cal kit contains at least one of each standard, sometimes with female and male versions, and sometimes with a male–female version of the thru standard.

The calibration standards are not perfect in the sense that a short is not a perfect short, an open is not a perfect open, etc. In the calibration of an instrument (i.e. the measurement of the error terms), the s-parameters of the standards are provided. The s-parameters can be provided through direct measurement, or more generally as a set of models of each standard.

A cal kit has four types of standards, each with three coefficients pertaining to the offset (explained in §15.5.1) along with four more coefficients for each of the short and the open and a single coefficient for the load. This is a total of 21 coefficients. These coefficients and the code to read them in from a file are shown in Listing 15.10. The model for each standard is provided in the following sections.

### 15.5.1 Offset (and Thru Standard)

For all of the standards, the concept of an *offset* is employed. An offset is a length of transmission line and is characterized by three parameters:

- *offsetZ0*, $Z_c$, the real characteristic impedance of the offset, in Ω;
- *offsetDelay*, $T_d$, the electrical length of the offset, in time (usually provided in picoseconds of delay);
- *offsetLoss*, $R_0$, the frequency dependent loss in ohms per second (usually provided in GΩ/s at a frequency $f_0$).

**Listing 15.10 CalibrationConstants class**

```python
class CalibrationConstants(object):
...
    def ReadFromFile(self,filename):
        with open(filename) as f:
            lines=f.readlines()
        actualLines=[]
        for line in lines:
            if line.strip()[0]!='%':
                actualLines.append(line.strip())
        # % C0 (fF) - OPEN
        self.openC0=float(actualLines[0])*1e-15
        # % C1 (1e-27 F/Hz) - OPEN
        self.openC1=float(actualLines[1])*1e-27
        # % C2 (1e-36 F/Hz^2) - OPEN
        self.openC2=float(actualLines[2])*1e-36
        # % C3 (1e-45 F/Hz^3) - OPEN
        self.openC3=float(actualLines[3])*1e-45
        # % offset delay (pS) - OPEN
        self.openOffsetDelay=float(actualLines[4])*1e-12
        # % real(Zo) of offset length - OPEN
        self.openOffsetZ0=float(actualLines[5])
        # % offset loss (Gohm/s) - OPEN
        self.openOffsetLoss=float(actualLines[6])*1e9
        # % L0 (pH) - SHORT
        self.shortL0=float(actualLines[7])*1e-12
        # % L1 (1e-24 H/Hz) - SHORT
        self.shortL1=float(actualLines[8])*1e-24
        # % L2 (1e-33 H/Hz^2) - SHORT
        self.shortL2=float(actualLines[9])*1e-33
        # % L3 (1e-42 H/Hz^3) - SHORT
        self.shortL3=float(actualLines[10])*1e-42
        # % offset delay (pS) - SHORT
        self.shortOffsetDelay=float(actualLines[11])*1e-12
        # % real(Zo) of offset length - SHORT
        self.shortOffsetZ0=float(actualLines[12])
        # % offset loss (Gohm/s) - SHORT
        self.shortOffsetLoss=float(actualLines[13])*1e9
        # % load resistance (ohm) - LOAD
        self.loadZ=float(actualLines[14])
        # % offset delay (pS) - LOAD
        self.loadOffsetDelay=float(actualLines[15])*1e-12
        # % real(Zo) of offset length - LOAD
        self.loadOffsetZ0=float(actualLines[16])
        # % offset loss (Gohm/s) - LOAD
        self.loadOffsetLoss=float(actualLines[17])*1e9
        # % offset delay (pS) - THRU
        self.thruOffsetDelay=float(actualLines[18])*1e-12
        # % real(Zo) of offset length - THRU
        self.thruOffsetZ0=float(actualLines[19])
        # % offset loss (Gohm/s) - THRU
        self.thruOffsetLoss=float(actualLines[20])*1e9
        return self
...
```

**Listing 15.11 Offset** class

```
class Offset(SParameters):
    def __init__(self,fList,offsetDelay,offsetZ0,offsetLoss,f0=1e9):
        data=[]
        Z0=50.
        Zc=offsetZ0
        Td=offsetDelay
        R0=offsetLoss
        L=Td*Zc
        C=Td/Zc
        G=0
        for f in fList:
            R=R0*Td*cmath.sqrt(f/f0)
            Z=R+1j*2*cmath.pi*f*L
            Y=G+1j*2*cmath.pi*f*C
            y=cmath.sqrt(Z*Y)
            rho=(Zc-Z0)/(Zc+Z0)
            D=(1-rho*rho*cmath.exp(-2*y))
            S11=rho*(1-cmath.exp(-2*y))/D
            S21=(1-rho*rho)*cmath.exp(-y)/D
            data.append([[S11,S21],[S21,S11]])
        SParameters.__init__(self,fList,data)
```

Thus, given the calibration constants *offsetZ0*, *offsetDelay*, and *offsetLoss* from a cal kit, the model of the offset is defined as

$$Z_c = \mathit{offsetZ0}, \qquad T_d = \mathit{offsetDelay} \cdot 10^{-12}, \qquad R_0 = \mathit{offsetLoss} \cdot 10^{9},$$
$$L = T_d \cdot Z_c, \qquad C = T_d / Z_c, \qquad G = 0.$$

For each frequency, assume the following definitions:

- resistance $R(f) = R_0 \cdot T_d \cdot \sqrt{f/f_0}$;
- series impedance $Z(f) = R(f) + j \cdot 2 \cdot \pi \cdot f \cdot L$;
- shunt admittance $Y(f) = G(f) + j \cdot 2 \cdot \pi \cdot f \cdot C$;
- propagation constant $\gamma = \sqrt{Z(f) \cdot Y(f)}$;
- reflection coefficient[6] $\rho = (Z_c - Z0) / (Z_c + Z0)$; and finally
- the s-parameters of the offset,

$$\mathbf{S} = \frac{1}{1 - \rho^2 \cdot e^{-2 \cdot \gamma}} \cdot \begin{pmatrix} \rho \cdot \left(1 - e^{-2 \cdot \gamma}\right) & \left(1 - \rho^2\right) \cdot e^{-\gamma} \\ \left(1 - \rho^2\right) \cdot e^{-\gamma} & \rho \cdot \left(1 - e^{-2 \cdot \gamma}\right) \end{pmatrix}.$$

The code for the offset calculation is provided in Listing 15.11. This offset contains the complete model of the thru standard in the calibration kit.

[6] At the time of writing, it is unclear whether to use the real characteristic impedance $Z_c$ provided for the calculation of $\rho$ or to recalculate the characteristic impedance as $Z_c = \sqrt{Z(f)/Y(f)}$ with the understanding that if a divide by zero occurs the $Z_c$ provided should be used. The Python code uses the calculation shown here, with an option to calculate $Z_c$.

**Listing 15.12 ThruStandard** class

```
class ThruStandard(Offset):
    def __init__(self,f,offsetDelay=0.0,offsetZ0=50.0,offsetLoss=0.0,f0=1e9):
        Offset.__init__(self,f,offsetDelay,offsetZ0,offsetLoss,f0)
```

**Listing 15.13 LoadStandard** class

```
class LoadStandard(SParameters):
    def __init__(self,f,offsetDelay=0.0,offsetZ0=50.0,offsetLoss=0.0,
                 terminationZ0=50.0):
        sspn=SystemSParametersNumeric(SystemDescriptionParser().AddLines(
            ['device offset 2','device R 1','port 1 offset 1','connect offset 2 R 1']
            ).SystemDescription())
        offsetSParameters=Offset(f,offsetDelay,offsetZ0,offsetLoss)
        terminationSParameters=TerminationZ(terminationZ0)
        sp=[]
        for n in range(len(f)):
            sspn.AssignSParameters('offset',offsetSParameters[n])
            sspn.AssignSParameters('R',terminationSParameters)
            sp.append(sspn.SParameters())
        SParameters.__init__(self,f,sp)
```

### 15.5.2 Load Standard

The model for the load standard is an offset that is terminated:

- *loadResistance*, $R_{load}$, the load resistance of the termination (in Ω).

The s-parameters can be determined using (4.32). The code for the load standard is provided in Listing 15.13.

### 15.5.3 Short Standard

The short standard employs a termination inductance to model the short. Ideally, the termination inductance would be 0 H, and the termination would be a perfect short. Instead, four coefficients are supplied:

- *L0*, $L_0$, inductance in H (usually specified in pH);
- *L1*, $L_1$, inductance in H/Hz (usually specified in $10^{-24}$ H/Hz);
- *L2*, $L_2$, inductance in H/Hz$^2$ (usually specified in $10^{-33}$ H/Hz$^2$);
- *L3*, $L_3$, inductance in H/Hz$^3$ (usually specified in $10^{-42}$ H/Hz$^3$).

Thus, given the coefficients from the calibration kit, the model of the termination is the offset combined with a terminating impedance calculated as follows:

$$L_0 = \mathit{L0} \cdot 10^{-12},\ L_1 = \mathit{L1} \cdot 10^{-24},\ L_2 = \mathit{L2} \cdot 10^{-33},\ L_3 = \mathit{L3} \cdot 10^{-42},$$

and finally

$$L(f) = ((L_3 \cdot f + L_2) \cdot f + L_1) \cdot f + L_0.$$

The s-parameters can be determined using $Z = j \cdot 2\pi \cdot f \cdot L(f)$ with (3.41) and applying (4.32). The code for the short standard is shown in Listing 15.14.

**Listing 15.14 ShortStandard** class

```
class ShortStandard(SParameters):
    def __init__(self,f,offsetDelay=0.0,offsetZ0=50.0,offsetLoss=0.0,
                 L0=0.0,L1=0.0,L2=0.0,L3=0.0):
        sspn=SystemSParametersNumeric(SystemDescriptionParser().AddLines(
            ['device offset 2','device L 1','port 1 offset 1','connect offset 2 L 1']
            ).SystemDescription())
        offsetSParameters=Offset(f,offsetDelay,offsetZ0,offsetLoss)
        terminationSParameters=TerminationLPolynomial(f,L0,L1,L2,L3)
        sp=[]
        for n in range(len(f)):
            sspn.AssignSParameters('offset',offsetSParameters[n])
            sspn.AssignSParameters('L',terminationSParameters[n])
            sp.append(sspn.SParameters())
        SParameters.__init__(self,f,sp)
```

**Listing 15.15 OpenStandard** class

```
class OpenStandard(SParameters):
    def __init__(self,f,offsetDelay=0.0,offsetZ0=50.0,offsetLoss=0.0,
                 C0=0.0,C1=0.0,C2=0.0,C3=0.0):
        sspn=SystemSParametersNumeric(SystemDescriptionParser().AddLines(
            ['device offset 2','device C 1','port 1 offset 1','connect offset 2 C 1']
            ).SystemDescription())
        offsetSParameters=Offset(f,offsetDelay,offsetZ0,offsetLoss)
        terminationSParameters=TerminationCPolynomial(f,C0,C1,C2,C3)
        sp=[]
        for n in range(len(f)):
            sspn.AssignSParameters('offset',offsetSParameters[n])
            sspn.AssignSParameters('C',terminationSParameters[n])
            sp.append(sspn.SParameters())
        SParameters.__init__(self,f,sp)
```

### 15.5.4 Open Standard

Similar to the short standard, the open standard employs a termination capacitance to model the open. Ideally, the termination capacitance would be 0 F, and the termination would be a perfect open. Instead, four coefficients are supplied:

- *C0*, $C_0$, capacitance in F (usually specified in fF);
- *C1*, $C_1$, capacitance in F/Hz (usually specified in $10^{-27}$ F/Hz);
- *C2*, $C_2$, capacitance in F/Hz$^2$ (usually specified in $10^{-36}$ F/Hz$^2$);
- *C3*, $C_3$, capacitance in F/Hz$^3$ (usually specified in $10^{-45}$ F/Hz$^3$).

Thus, given the coefficients from the calibration kit, the model of the termination is the offset combined with a terminating impedance calculated as:

$$C_0 = C0 \cdot 10^{-15},\ C_1 = C1 \cdot 10^{-27},\ C_2 = C2 \cdot 10^{-36},\ C_3 = C3 \cdot 10^{-45},$$

and finally

$$C(f) = ((C_3 \cdot f + C_2) \cdot f + C_1) \cdot f + C_0.$$

The s-parameters can be determined using $Z = 1/(j \cdot 2\pi \cdot f \cdot C(f))$ with (3.41) and applying (4.32). The code for the open standard is shown in Listing 15.15.

### 15.5.5 Offset Loss

Offset loss can also be specified as a loss, in decibels, $dB_0$ found by measuring the magnitude of $S_{21}$ at the frequency $f_0$:

$$dB_0 = -20 \cdot \log\left(|S_{21}|\right)\big|_{f=f_0}.$$

To convert this to $R_0$, the formula for the transmission line given in (7.16) is used, where the loss is measured in the characteristic impedance of the line, meaning $\rho = 0$:

$$dB_0 = -20 \cdot \log\left(\left|e^{-\gamma}\right|\right) = -20 \cdot \log\left(e^{-\mathrm{Re}(\gamma)}\right) = -20 \cdot \log\left(e^{-\mathrm{Re}\left(\sqrt{\mathrm{Re}(\gamma^2)+j\cdot\mathrm{Im}(\gamma^2)}\right)}\right). \tag{15.23}$$

The real part of the square root of a complex number $z = a + j \cdot b$ can be written, according to [53], as

$$\mathrm{Re}\left(\sqrt{z}\right) = \sqrt{\frac{a + \sqrt{a^2 + b^2}}{2}}.$$

Therefore, (15.23) can be written as

$$\frac{\ln(10)}{20} \cdot dB_0 = \sqrt{\frac{\mathrm{Re}\left(\gamma^2\right) + \sqrt{\mathrm{Re}\left(\gamma^2\right)^2 + \mathrm{Im}\left(\gamma^2\right)^2}}{2}}. \tag{15.24}$$

Solving (15.24) for $\mathrm{Im}\left(\gamma^2\right)$ yields

$$\mathrm{Im}\left(\gamma^2\right) = \frac{\ln(10)}{10} \cdot dB_0 \cdot \sqrt{\left(\frac{\ln(10)}{20} \cdot dB_0\right)^2 - \mathrm{Re}\left(\gamma^2\right)} \approx \frac{\ln(10)}{10} \cdot dB_0 \cdot \sqrt{-\mathrm{Re}\left(\gamma^2\right)}. \tag{15.25}$$

Using $L = T_d \cdot Z_c$ for the inductance of the line and $C = T_d/Z_c$ for the capacitance, where $T_d$ and $Z_c$ have been defined for the offset, and setting the conductance $G = 0$:

$$\mathrm{Re}\left(\gamma^2\right) = -4 \cdot \pi^2 \cdot f^2 \cdot T_d^2, \qquad \mathrm{Im}\left(\gamma^2\right) = 2 \cdot R \cdot \pi \cdot f \cdot \frac{T_d}{Z_c}. \tag{15.26}$$

Substituting (15.26) into (15.25) and solving for $R$, remembering that this is the total series resistance at $f = f_0$, yields

$$R = \frac{Z_c \cdot dB_0 \cdot \ln(10)}{10} = R_0 \cdot T_d.$$

Solving this for $R_0$,

$$R_0 = \frac{\ln(10) \cdot dB_0 \cdot Z_c}{10 \cdot T_d}.$$

Usually, this offset loss is specified as *offsetLoss* $= R_0/10^9$ GΩ/s.

```
% DESCRIPTION Rosenberger RPC-2.92 Cal Kit
%       Model 02K30R-MSOTS3
%       Serial Number: 21A45
-4.3     % C0 (fF) - OPEN
431.00   % C1 (1e-27 F/Hz) - OPEN
-11.50   % C2 (1e-36 F/Hz^2) - OPEN
0.12     % C3 (1e-45 F/Hz^3) - OPEN
28.353   % offset delay (pS) - OPEN
50.0     % real(Zo) of offset length - OPEN
2.4      % offset loss (Gohm/s) - OPEN
0.0      % L0 (pH) - SHORT
0.0      % L1 (1e-24 H/Hz) - SHORT
0.0      % L2 (1e-33 H/Hz^2) - SHORT
0.0      % L3 (1e-42 H/Hz^3) - SHORT
28.353   % offset delay (pS) - SHORT
50       % real(Zo) of offset length - SHORT
2.4      % offset loss (Gohm/s) - SHORT
50.0     % load resistance (ohm) - LOAD
0.0      % offset delay (pS) - LOAD
50.0     % real(Zo) of offset length - LOAD
0.0      % offset loss (Gohm/s) - LOAD
65.712   % offset delay (pS) - THRU
50       % real(Zo) of offset length - THRU
2.7      % offset loss (Gohm/s) - THRU
```

**Figure 15.4** Rosenberger 2.92 mm calibration kit constants

### 15.5.6 Calibration Kit Example

An actual calibration kit file is shown in Figure 15.4. This file is used in conjunction with the **CalibrationKit** class shown in Listing 15.16.

## 15.6 Time-Domain Reflectometry

There is so much literature created around the VNA that it is unnecessary to provide any more detail on this instrument in this book. However, the literature is sparse regarding the TDR [54–56] especially when considering s-parameter calculation, aside from [57]. The author has much more knowledge of the TDR and, since it is an arcane subject, it is worth further discussion.

Table 15.2 shows a comparison between the VNA and the TDR. The VNA has traditionally been a microwave measurement instrument and as such is used inherently by microwave engineers. As the field of signal integrity has become much more involved with electromagnetic characteristics, mainly of transmission lines, the VNA, along with many microwave engineers, have migrated into this field. Although the TDR has been used in a variety of applications not related to electronics, it has been a mainstay electronic instrument for signal integrity engineers for some time. Part of the reason for this is the fact that the TDR is inherently time domain and, despite increasing frequency-domain analysis, the main domain of signal integrity remains the time domain. The use of the TDR for signal integrity is much more prevalent among engineers whose expertise is not with, or did not

**Listing 15.16 CalibrationKit** class

```
class CalibrationKit(object):
    def __init__(self,filename=None,f=None):
        self.Constants=CalibrationConstants()
        self.m_f=None
        if not filename is None:
            self.ReadFromFile(filename)
        if not f is None:
            self.InitializeFrequency(f)
    def InitializeFrequency(self,f):
        self.m_f=f
        self.openStandard=OpenStandard(self.m_f,self.Constants.openOffsetDelay,
            self.Constants.openOffsetZ0,self.Constants.openOffsetLoss,
            self.Constants.openC0,self.Constants.openC1,self.Constants.openC2,
            self.Constants.openC3)
        self.shortStandard=ShortStandard(self.m_f,self.Constants.shortOffsetDelay,
            self.Constants.shortOffsetZ0,self.Constants.shortOffsetLoss,
            self.Constants.shortL0,self.Constants.shortL1,self.Constants.shortL2,
            self.Constants.shortL3)
        self.loadStandard=LoadStandard(self.m_f,self.Constants.loadOffsetDelay,
            self.Constants.loadOffsetZ0,self.Constants.loadOffsetLoss,
            self.Constants.loadZ)
        self.thruStandard=ThruStandard(self.m_f,self.Constants.thruOffsetDelay,
            self.Constants.thruOffsetZ0,self.Constants.thruOffsetLoss)
        return self
    def ReadFromFile(self,filename):
        self.Constants=CalibrationConstants().ReadFromFile(filename)
        return self
    def WriteToFile(self,filename,calkitname=None):
        self.Constants.WriteToFile(filename, calkitname)
        return self
    def WriteStandardsToFiles(self,filenamePrefix=''):
        self.shortStandard.WriteToFile(filenamePrefix+'Short')
        self.openStandard.WriteToFile(filenamePrefix+'Open')
        self.loadStandard.WriteToFile(filenamePrefix+'Load')
        self.thruStandard.WriteToFile(filenamePrefix+'Thru')
        return self
    def ReadStandardsFromFiles(self,filenamePrefix):
        self.shortStandard=SParameterFile(filenamePrefix+'Short.s1p')
        self.openStandard=SParameterFile(filenamePrefix+'Open.s1p')
        self.loadStandard=SParameterFile(filenamePrefix+'Load.s1p')
        self.thruStandard=SParameterFile(filenamePrefix+'Thru.s2p')
        return self
```

originate in, microwave measurements. In general, the TDR is a less precise, less expensive, and easier to use instrument.

One of the better attributes of the TDR is that it dynamically updates in the time domain and therefore it is possible to see nearly instantaneous updates of the step response of the system. When the step response is viewed at a driven port, there is a good correlation between what is seen in the step response and the time-localized impedance of the system. This can be contrasted with the VNA's better attributes in the frequency domain, which shows nearly instantaneous sweeps of frequency response of the system. The VNA is much better and quicker at showing resonances in the frequency domain.

The VNA has a greater dynamic range, which allows it to see much smaller effects. In signal integrity, this can be useful for seeing tiny effects such as far-end crosstalk, but otherwise the huge dynamic range offered by the VNA is really not required in signal

**Table 15.2** Comparison between the VNA and the TDR

| Attribute | VNA | TDR |
|---|---|---|
| Cost[a] | very high (\$100 k–\$150 k) | much lower (\$35 k–\$55 k) |
| Complexity | very high | lower |
| Best for | frequency localization | time localization |
| Measurement speed | fast (seconds) | slower (seconds to minutes) |
| Dynamic range | very high (> 90 dB) | low to medium (30–70 dB) |
| Ease of use | very difficult | easier |
| Low frequency | no DC (100 kHz–1 MHz) | measures to DC |
| Calibration | difficult, time consuming | automatic, internal on some models |
| Sensitivity to electrostatic discharge (ESD) | less sensitive | very sensitive |

[a] Estimates are for four-port, 40 GHz instruments, with and without electronic calibration.

integrity. Plus, for measurements such as crosstalk, techniques such as wavelet denoising (see §15.7.5) are useful in bringing out these tiny effects, albeit while measuring these effects in the time domain.

Regarding the capability of measuring s-parameters for time-domain simulations, the VNA offers higher precision and accuracy, but is missing the DC measurement point. Although some VNAs can reach relatively low frequencies, the dynamic range at low frequency is poor due to the directional coupler; often, attempts to measure or extrapolate DC lead to s-parameters that do not perform well in the time domain. The TDR performs better in this area, and the time-domain analysis tools in the TDR are generally better performing than those in the VNA.

### 15.6.1 Raw Measured S-Parameters in the TDR

In the TDR, all frequencies for a driven port are sent in the form of an impulse or step, and only voltages are measured at each port. The raw s-parameters are measured by separating the incident voltage waveform from the reflected voltage waveform and computing ratios of the incident to the reflected in the frequency domain [58, 59].

An example of such a system is shown in Figure 15.5, where the impulse from a Teledyne LeCroy WavePulser 40iX is shown driving a short, open, and load standard.[7] Here, the impulse is about 10 ps wide and has a frequency content that is basically flat to 40 GHz.

[7]The impulse is actually inverted, but is shown reversed for better insight.

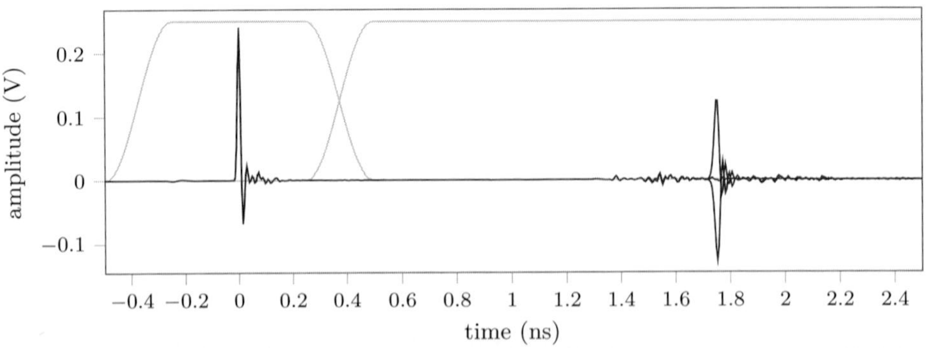


**Figure 15.5** Time-domain reflectometry s-parameter measurement method

The acquisition system acquires with an equivalent-time sample rate of 204.8 GS/s, so there is no frequency aliasing possible [60, 61]. The incident extraction window is set to encompass 250 ps on either side of the incident impulse, and the reflected extraction window is the remainder of the waveform. For measurements at the undriven port in multi-port measurements, the entire waveform is used as the incident impulse is not present.

If the reader is disoriented by the waveforms in Figure 15.5, it should be pointed out that the positive going reflection is the open, the negative going reflection is the short, and the load continues through. Any difficulty in conceptualizing this can be managed by mentally integrating the waveforms, in which case the more common and intuitive step-like time-domain reflectometry waveforms are formed.

This measurement method depends on there being some time separation between the pulser/sampler and the DUT, otherwise there is no way to separate the incident from the reflected waves. Some time-domain reflectometry measurement systems do not bother to extract the incident wave, assuming that it is an impulse of unity. These systems assume that the impulse is always the same and does not change from acquisition to acquisition, and that the actual size of the impulse gets taken up in the error terms. The method shown here is the preferred method; fortunately, it is easy to impose some amount of electrical length between the pulser/sampler and the DUT.

For waveforms that are steps, or are step-like, the traditional way of generating the raw measured s-parameters is to use DFT methods that deal with discontinuous waveforms [62,63]. These methods are needlessly complicated and not preferred. Instead, the derivative of the waveform is taken prior to windowing and separation of the incident and reflected portions. A common erroneous belief is that taking the derivative amplifies the noise in the waveform. Although in theory this is true, it amplifies the signal equally. In fact, the drop in signal content of 20 dB/decade of the step waveform is normalized and the noise floor then rises at 20 dB/decade. The distance between the signal and the noise does not change, however, and the benefit is that the raw measured s-parameters, the measurements of the frequency response of the pulser/sampler, and the error terms no longer contain the 20 dB/decade drop. Instead, they are basically flat, depending on the flatness of the pulser/sampler response, and can be compared to the VNA, as the frequency

**Listing 15.17 TDRWaveformToSParameterConverter** class

```
class TDRWaveformToSParameterConverter(object):
    def __init__(self,
                 WindowForwardHalfWidthTime=0,
                 WindowReverseHalfWidthTime=None,
                 WindowRaisedCosineDuration=0,
                 Step=True,
                 Length=0,
                 Denoise=False,
                 DenoisePercent=30.,
                 Inverted=False,
                 fd=None
                 ):
        self.wfhwt=WindowForwardHalfWidthTime
        self.wrhwt = self.wfhwt if WindowReverseHalfWidthTime is None\
            else WindowReverseHalfWidthTime
        self.wrcdr=WindowRaisedCosineDuration
        self.step=Step
        self.length=Length
        self.denoise=Denoise
        self.denoisePercent=DenoisePercent
        self.inverted=Inverted
        self.fd=fd
    def RawMeasuredSParameters(self,wfList):
        ports=len(wfList)
        S=[[None for _ in range(ports)] for _ in range(ports)]
        for d in range(ports):
            fc=self.Convert(wfList[d],d)
            for o in range(ports):
                S[o][d]=fc[o]
        f=S[0][0].Frequencies()
        return SParameters(f,
            [[[S[r][c][n] for c in range(ports)] for r in range(ports)]
            for n in range(len(f))])
...
```

content is similar to the flat sweeping frequencies of the VNA. As a final note on this topic, the derivative also enables *wavelet denoising*, discussed in §15.7.5, as the bumps and wiggles of impedance discontinuities relate better to wavelets and better facilitate wavelet decomposition when the waveform is impulse-like (i.e. derivative of step-like) as opposed to step-like.

Listing 15.17 shows the **TDRWaveformToSParameterConverter** class used to convert waveforms measured in the TDR to raw s-parameters. It takes parameters that dictate the following aspects: the time window around the main impulse in the driven port; the amount of time used to taper the window with a raised cosine window; whether the waveforms are steps (as opposed to impulses); the length to which to trim the final waveforms; whether to denoise the waveforms, and if so, the percentage of the end of the waveform used to estimate the noise; whether the waveforms are inverted; and, finally, the optional frequency descriptor used to describe the desired frequency content.

Once this class has been initialized with its options, raw measured s-parameters are generated by providing a list of waveforms to the `RawMeasuredSParameters()` method in Listing 15.17. This is actually a list of lists and should be square such that, for a given number of ports $P$ in an s-parameter measurement, for $r, c \in 0 \ldots P-1$, `wfList[r][c]` provides the waveform acquired at port $r$ when port $c$ is driven. In other words, the incident

**Listing 15.18 TDRWaveformToSParameterConverter** class `Convert()` method

```
class TDRWaveformToSParameterConverter(object):
    def Convert(self,wfListProvided,incidentIndex=0):
        if self.step:
            wfList=[wf.Derivative(removePoint=False,scale=False)
                    for wf in wfList]
        if self.inverted:
            wfList=[wf*-1. for wf in wfList]
        if self.denoise:
            wfList=[WaveletDenoiser.DenoisedWaveform(
                wf,isDerivative=self.step,
                mult=self.sigmaMultiple,
                pct=self.denoisePercent)
                    for wf in wfList]
        if self.length!=0:
            lengthSamples=int(self.length*
                wfList[incidentIndex].td.Fs+0.5)
            wfList=[wf*WaveformTrimmer(0,wf.td.K-lengthSamples)
                for wf in wfList]
        incwf=copy.deepcopy(wfList[incidentIndex])
        maxValueIndex=0
        maxValue=incwf[0]
        for k in range(1,len(incwf)):
            if incwf[k]>maxValue:
                maxValue=incwf[k]
                maxValueIndex=k
        forwardSamples=int(self.wfhwt*incwf.td.Fs)
        reverseSamples=int(self.wrhwt*incwf.td.Fs)
        raisedCosineSamples=int(self.wrcdr*incwf.td.Fs)
        (incidentExtractionWindow,reflectExtractionWindow)=self._ExtractionWindows(
            incwf.td,forwardSamples,reverseSamples,raisedCosineSamples,maxValueIndex)
        incwf=Waveform(incwf.td,[x*w
            for (x,w) in zip(incwf.Values(),incidentExtractionWindow.Values())])
        wfList=[Waveform(wf.td,[x*w
            for (x,w) in zip(wf.Values(),reflectExtractionWindow.Values())])
                for wf in wfList]
        #wfList[incidentIndex]=wfList[incidentIndex]-incwf
        incwffc=incwf.FrequencyContent(self.fd)
        res=[wf.FrequencyContent(self.fd) for wf in wfList]
        for fc in res:
            for n in range(len(fc)):
                fc[n]=fc[n]/incwffc[n]
        return res
...
```

waveform will be extracted from port $c$ and the reflected waveform will be extracted from port $r$. The raw measured s-parameters are the ratios of the reflected frequency content to the incident frequency content. These frequency content ratios are produced by making calls to the `Convert()` method, which does the actual work.

The `Convert()` method shown in Listing 15.18 takes a list of waveforms along with an index for the waveform in the list containing the incident waveform. The returned result will be the ratio of the frequency content of the reflected portion of all waveforms to the incident portion of the specified waveform. It proceeds by first taking the derivative of all of the waveforms if a step waveform is specified. It then inverts the waveforms if specified. It then denoises the waveforms if specified using wavelet denoising (see §15.7.5). The waveforms are then trimmed to the desired length.

The incident impulse is found by searching for the largest spike in the incident waveform. Once found, an extraction window is formed, which runs from the left edge of the waveform to a half-width of the window beyond the location of the incident impulse and then tapers down as a raised cosine. The incident waveform is then extracted by multiplication by this window. The reflected waveform in the waveform containing the incident portion is found by subtracting the incident portion from the waveform. The reflected portion of the waveforms not containing the incident waveforms are simply the waveforms themselves.

There are some improvements that can be made here. One is that all waveforms should be tapered at the left and right edges with at least a raised cosine window. This is to improve the performance when there are slight discontinuities at the waveform edges.

In any case, the final result is found by calculating the frequency content of all of the reflected waveforms and dividing these by the frequency content of the incident waveform. If a frequency descriptor was specified during initialization of the class, then this descriptor is used to dictate the frequency content generated, otherwise the frequency content is dictated by the time descriptor of the trimmed waveforms.

When this frequency content is returned to the `RawMeasuredSParameters()` method in Listing 15.17, all that remains is to reassemble the frequency content into s-parameter matrices. This result returns the raw measured s-parameters, so called because they look like s-parameters as they are ratios of reflected to incident waves (or a suitable ratio of frequency content) and they can be converted to calibrated measured s-parameters using the aforementioned calibration and correction techniques.

As the VNA operates by stimulating a DUT at specified frequencies, at any given frequency the forward and reverse going standing waves are measured under various driving conditions. For example, for a $P$-port DUT, one generally drives each port $p \in 1 \ldots P$ and acquires $\hat{\mathbf{A}}_{*p}$ and $\hat{\mathbf{B}}_{*p}$ to calculate $\hat{\mathbf{S}} = \hat{\mathbf{B}} \cdot \hat{\mathbf{A}}^{-1}$. In the TDR, only voltage waveforms are measured at ports. Due to the aforementioned processing, the equation for the raw measured s-parameters takes the following form:

$$
\begin{aligned}
\hat{\mathbf{S}} &= \left[ \begin{pmatrix} \hat{V}_{11} & \hat{V}_{12} & \cdots & \hat{V}_{1P} \\ \hat{V}_{21} & \hat{V}_{22} & \cdots & \hat{V}_{2P} \\ \vdots & \vdots & \ddots & \vdots \\ \hat{V}_{P1} & \hat{V}_{P2} & \cdots & \hat{V}_{PP} \end{pmatrix} - \begin{pmatrix} \hat{v}_1 & 0 & \cdots & 0 \\ 0 & \hat{v}_2 & \cdots & 0 \\ \vdots & \vdots & \ddots & \vdots \\ 0 & 0 & \cdots & \hat{v}_P \end{pmatrix} \right] \cdot \begin{pmatrix} \hat{v}_1 & 0 & \cdots & 0 \\ 0 & \hat{v}_2 & \cdots & 0 \\ \vdots & \vdots & \ddots & \vdots \\ 0 & 0 & \cdots & \hat{v}_P \end{pmatrix}^{-1} \\
&= \begin{pmatrix} \frac{\hat{V}_{11} - \hat{v}_1}{\hat{v}_1} & \frac{\hat{V}_{12}}{\hat{v}_2} & \cdots & \frac{\hat{V}_{1P}}{\hat{v}_2} \\ \frac{\hat{V}_{21}}{\hat{v}_1} & \frac{\hat{V}_{22} - \hat{v}_2}{\hat{v}_2} & \cdots & \frac{\hat{V}_{2P}}{\hat{v}_2} \\ \vdots & \vdots & \ddots & \vdots \\ \frac{\hat{V}_{P1}}{\hat{v}_1} & \frac{\hat{V}_{P2}}{\hat{v}_2} & \cdots & \frac{\hat{V}_{PP}}{\hat{v}_2} \end{pmatrix}.
\end{aligned}
\tag{15.27}
$$

In (15.27), $\hat{V}_{od}$ is the voltage measured on a port $o$ when port $d$ is driven in the frequency domain. The time-domain voltage is intended to be *impulsive*, meaning it represents the impulse response. If a step is used, as is usually the case, the derivative of the waveform is used.

The waveform $\hat{v}_d$ is the incident voltage waveform on the driven port $d$ in the frequency domain. This is measured by locating the rising edge of the step and windowing the time-domain derivative waveform about the incident impulse. The time-domain waveforms are converted to the frequency domain through the DFT.

### 15.6.2 TDR Calibration

Although the TDR operates fundamentally differently than the VNA with respect to measurement, given raw measured s-parameters as provided in (15.27) it is calibrated similarly using the methods provided in §15.2. The error terms calculated and the same final DUT calculation shown in (15.22) are used; however, the values of the error terms themselves will be very different.

For the TDR, (15.1) is altered in that the $\alpha$ and $\beta$ matrices are set to unity, and replaced by a similar gain matrix $\mathbf{G}$ with all of the gain terms of the voltage measurement devices on the diagonal, representing a non-ideal frequency response of the waveform sampling system employed:

$$\hat{\mathbf{V}} = \sqrt{Z0} \cdot \mathbf{G} \cdot \left( \hat{\mathbf{A}} + \hat{\mathbf{B}} \right).$$

The voltage measurement is calculated as a simulation, as explained in Chapter 9. For a single-port measurement with a reflect standard connected to the system, the voltage measured is given by

$$\hat{V}_p = \frac{\left(F_{p22} + |\mathbf{F_p}|\right) \cdot S_{11} - 1 - F_{p11}}{\left(F_{p22} - |\mathbf{F_p}| \cdot \Gamma_p\right) \cdot S_{11} - 1 + F_{p11} \cdot \Gamma_p} \cdot \sqrt{Z0} \cdot m_p \cdot g_p.$$

The voltage source is shown in Figure 6.7, which shows voltage in terms of the series impedance $Z$, where $\Gamma_p = (Z - Z0)/(Z + Z0)$ and therefore $Z = Z0 \cdot (1 + \Gamma_p)/(1 + \Gamma_p)$. Therefore, the value of the wave launched is given by

$$m_p = \frac{1}{\sqrt{Z0}} \cdot \frac{Z_p}{Z_p + Z0} \cdot V_p = \frac{(1 - \Gamma_p)}{2 \cdot \sqrt{Z0}} \cdot \frac{V_p}{N},$$

where $N$ is the number of frequency points and accounts for the fact that the impulsive voltage delivered has its energy spread over the $N$ frequency points in the spectrum. Despite this scaling, the spectral *density* remains constant.[8]

To find $\hat{v}_p$, a small trick must be employed. This value is the frequency-domain representation of the impulse incident on the system. In order to find its value, one must temporarily treat the fixture $\mathbf{F_p}$ as a transmission line. Considering the transmission line formula in (7.14), the voltage measured can be expressed as follows:

$$\begin{aligned}\hat{V}_p =& \sqrt{Z0} \cdot m_p \cdot g_p \\ &\cdot \frac{\left([1 + \rho_p] \cdot e^{-2 \cdot \gamma_p} - \rho_p \cdot [1 + \rho_p]\right) \cdot S_{11} + 1 + \rho_p - \rho_p \cdot [1 + \rho_p] \cdot e^{-2 \cdot \gamma_p}}{\left([\rho_p - \Gamma_p] \cdot e^{-2 \cdot \gamma_p} + \Gamma_p \cdot \rho_p^2 - \rho_p\right) \cdot S_{11} + \rho_p \cdot (\Gamma_p - \rho_p) \cdot e^{-2 \cdot \gamma_p} + 1 - \rho_p \cdot \Gamma_p}.\end{aligned}$$

[8]Technically, the content at DC must be halved, and, if the number of points in the waveform is even, the content at the last frequency point must also be halved if using DFT methods. This does not mean that the actual DFT elements are halved, only the amplitudes of the cosine waves they represent. See §12.3.2 for a discussion of frequency content.

Thus, $\hat{v}_p$ is found by taking the limit by making the transmission line infinitely long:

$$\hat{v}_p = \lim_{\gamma_p \to \infty} \hat{V}_p = \sqrt{Z0} \cdot m_p \cdot g_p \cdot \frac{1 + \rho_p}{1 - \rho_p \cdot \Gamma_p}.$$

It doesn't really matter whether the fixture $\mathbf{F_p}$ can be considered to be expressed as a transmission line in the manner used, only that its *interface* can be treated as such, noting that only $\rho_p$ survives the limit. Also, if $\mathbf{F_p}$ is perfectly matched at the interface (i.e. $\rho_p = 0$), then $\hat{v}_p$ is truly the forward going wave launched into the system, but mathematically it doesn't matter if it is perfectly matched and whether $\hat{v}_p$ truly represents the forward going wave. In any case, $\hat{V}_p - \hat{v}_p$ represents a combination of forward going waves and returning waves reflecting off its interface.

The raw measured s-parameter $\hat{S}_{11}$ is computed as follows:

$$\begin{aligned} \hat{S}_{pp} &= \frac{\hat{V}_p - \hat{v}_p}{\hat{v}_p} \\ &= \frac{([\rho_p \cdot F_{p22} - |\mathbf{F_p}|] \cdot \Gamma_p + \rho_p \cdot F_{p22} - |\mathbf{F_p}|) \cdot S_{pp} + (F_{p11} - \rho_p) \cdot \Gamma_p + F_{p11} - \rho_p}{(\rho_p + 1) \cdot ([|\mathbf{F_p}| \cdot \Gamma_p - F_{p22}] \cdot S_{pp} - \Gamma_p \cdot F_{p11} + 1)}. \end{aligned} \tag{15.28}$$

The raw measured s-parameter, because of the operations taken to compute it, no longer contains the absolute voltage incident on the system or the gain term. Equally important in practice is that raw s-parameter measurement is relatively insensitive to any amplitude variation or jitter on the waveform or gain variation because both the incident and reflected waveforms are scaled vertically or jittered horizontally together.

To compute the reflected error terms, (15.12) is repeated for at least three of the standards, and one computes

$$\begin{pmatrix} E_{D_p} \\ E_{S_p} \\ \Delta E_p \end{pmatrix} = \begin{pmatrix} 1 & \hat{\Gamma}_{s_{p1}} \cdot \Gamma_{s1} & -\Gamma_{s1} \\ 1 & \hat{\Gamma}_{s_{p2}} \cdot \Gamma_{s2} & -\Gamma_{s2} \\ \vdots & \vdots & \vdots \\ 1 & \hat{\Gamma}_{s_{pM}} \cdot \Gamma_{sM} & -\Gamma_{sM} \end{pmatrix}^{\dagger} \cdot \begin{pmatrix} \hat{\Gamma}_{s_{p1}} \\ \hat{\Gamma}_{s_{p2}} \\ \vdots \\ \hat{\Gamma}_{s_{pM}} \end{pmatrix},$$

$$E_{D_p} = \frac{\rho_p - F_{p11} - F_{p11} \cdot \Gamma_p + \rho_p \cdot \Gamma_p}{(1 + \rho_p) \cdot (F_{p11} \cdot \Gamma_p - 1)},$$

$$E_{S_p} = \frac{\Gamma_p \cdot |\mathbf{F_p}| - F_{p22}}{F_{p11} \cdot \Gamma_p - 1},$$

$$\Delta E_p = \frac{F_{p22} \cdot (1 + \Gamma_p) \cdot \rho_p - (1 + \Gamma_p) \cdot |\mathbf{F_p}|}{(1 + \rho_p) \cdot (\Gamma_p \cdot F_{p11} - 1)},$$

$$E_{R_p} = E_{D_p} \cdot E_{S_p} - \Delta E_p = F_{p21} \cdot F_{p12} \cdot \frac{1 + (1 - \rho_p) \cdot \Gamma_p - \rho_p \cdot \Gamma_p^2}{(1 + \rho_p) \cdot (F_{p11} \cdot \Gamma_p - 1)^2}.$$

The error terms are computed with non-ideal standards, but they are mathematically independent of the standards used. So, to compute the thru terms, an ideal thru is applied. This is because the actual expression for the voltages when non-ideal s-parameters are applied to the system is extremely long and complex. Instead, to verify the raw measured s-parameters, the error terms are computed here ideally, and the DUT `UnCalculation()` formula on line 13 of Listing 15.7 is used to calculate what is actually measured.

The voltage measured at the other port with the ideal thru standard connected is

$$\hat{V}_{t_o} = \frac{g_o \cdot F_{o_{12}} \cdot F_{p_{21}} \cdot (1 + \Gamma_o) \cdot m_p \cdot \sqrt{Z0}}{\begin{array}{c}[|\mathbf{F_p}| \cdot F_{o_{22}} - F_{p_{11}} + (F_{o_{11}} \cdot F_{p_{11}} - |\mathbf{F_p}| \cdot |\mathbf{F_o}|) \cdot \Gamma_o] \cdot \Gamma_p \\ + (F_{p_{22}} \cdot |\mathbf{F_o}| - F_{o_{11}}) \cdot \Gamma_o - F_{p_{22}} \cdot F_{o_{22}} + 1\end{array}}$$

and the raw measured s-parameter $\hat{S}_{t_{op}}$ is

$$\hat{S}_{t_{op}} = \frac{\hat{V}_{t_o}}{\hat{v}_p} = \frac{g_o}{g_p} \cdot \frac{1 - \rho_p \cdot \Gamma_p}{1 + \rho_p} \cdot \frac{F_{o_{12}} \cdot F_{p_{21}} \cdot (1 + \Gamma_o)}{\begin{array}{c}[|\mathbf{F_p}| \cdot F_{o_{22}} - F_{p_{11}} + (F_{o_{11}} \cdot F_{p_{11}} - |\mathbf{F_p}| \cdot |\mathbf{F_o}|) \cdot \Gamma_o] \cdot \Gamma_p \\ + (F_{p_{22}} \cdot |\mathbf{F_o}| - F_{o_{11}}) \cdot \Gamma_o - F_{p_{22}} \cdot F_{o_{22}} + 1\end{array}}.$$

When (15.28) is applied with the ideal thru standard, $\hat{S}_{pp}$ becomes the s-parameters of the aggregated fixture $\mathbf{F_o}$ and $\Gamma_0$, seen looking back into the other port $o$, and the load-match term is predictably verified as

$$E_{L_{op}} = E_{S_o} = \frac{\Gamma_o \cdot |\mathbf{F_o}| - F_{o_{22}}}{F_{o_{11}} \cdot \Gamma_o - 1}.$$

Using the DUT calculation equation in (15.22), one finds the following relationship when an ideal thru is used:

$$E_{T_{op}} = \left(1 - E_{L_{op}} \cdot E_{S_p}\right) \cdot \hat{S}_{t_{op}}.$$

Therefore, the forward-transmission error term is calculated as

$$E_{T_{op}} = \frac{g_o}{g_p} \cdot \frac{1 - \rho_p \cdot \Gamma_p}{1 + \rho_p} \cdot \frac{F_{o_{12}} \cdot F_{p_{21}} \cdot (1 + \Gamma_o)}{(F_{o_{11}} \cdot \Gamma_o - 1) \cdot (F_{p_{11}} \cdot \Gamma_p - 1)}.$$

All of the error terms for the TDR are summarized in Table 15.3. Although they differ significantly from the error terms computed for the VNA, as provided in Table 15.1, their method of computation is identical. And, despite the difference in how the raw measured s-parameters are computed in (15.27), the same DUT calculation method is used in (15.22).

### 15.6.3 TDR Dynamic Range

In [57], there is a complete derivation of dynamic range for the TDR employing step-like stimulus (systems for which the stimulus is essentially a step and the incident waveform is contained in the edge). Contrast this with impulse-like systems, where the waveform is essentially an impulse. Here, the dynamic range is derived for an impulse-like system and then compared to the dynamic range calculation in [57].

**Table 15.3** Summary of error terms for the TDR

| Term | Formula | Name | Notes |
|---|---|---|---|
| $E_{D_p}$ | $\frac{\rho_p - F_{p_{11}} - F_{p_{11}} \cdot \Gamma_p + \rho_p \cdot \Gamma_p}{(1+\rho_p)\cdot\left(F_{p_{11}}\cdot\Gamma_p - 1\right)}$ | directivity term for port $p$ | |
| $E_{S_p}$ | $\frac{\Gamma_p \cdot \lvert\mathbf{F_p}\rvert - F_{p_{22}}}{F_{p_{11}}\cdot\Gamma_p - 1}$ | source-match term for port $p$ | |
| $E_{R_p}$ | $F_{p_{21}} \cdot F_{p_{12}} \cdot \frac{1+(1-\rho_p)\cdot\Gamma_p - \rho_p\cdot\Gamma_p^2}{(1+\rho_p)\cdot\left(F_{p_{11}}\cdot\Gamma_p - 1\right)^2}$ | reverse-transmission term for port $p$ | assumed that $F_{p_{12}} = F_{p_{21}}$ in SOLR |
| $E_{X_{op}}$ | 0 | crosstalk term for port $o$ when port $p$ driven | ignored for the TDR |
| $E_{T_{op}}$ | $\frac{\frac{g_o}{g_p} \cdot \frac{1-\rho_p\cdot\Gamma_p}{1+\rho_p} \cdot F_{o_{12}}\cdot F_{p_{21}}\cdot(1+\Gamma_o)}{\left(F_{o_{11}}\cdot\Gamma_o - 1\right)\cdot\left(F_{p_{11}}\cdot\Gamma_p - 1\right)}$ | forward-transmission term for port $o$ when port $p$ driven | |
| $E_{L_{op}}$ | $\frac{\Gamma_o \cdot \lvert\mathbf{F_o}\rvert - F_{o_{22}}}{F_{o_{11}}\cdot\Gamma_o - 1}$ | load-match term for port $o$ when port $p$ driven | assumed equal to $E_{S_o}$ in transfer thru and SOLR |

Consider an impulsive waveform consisting of $K$ points of signal $\mathbf{x}$ plus noise $\boldsymbol{\varepsilon}$ sampled at an *equivalent-time* sample rate[9] $F_{se}$ (or a sample period $T_{se} = 1/F_{se}$) such that, for $k \in 0 \ldots K-1$,

$$w[k] = x[k] + \varepsilon[k]\,.$$

The dynamic range is the signal-to-noise ratio at each frequency, where the frequency vector consists of $N = K/2$ points and, for $n \in 0 \ldots N$,

$$f[n] = \frac{n}{N} \cdot \frac{F_{se}}{2}.$$

Starting with the noise $\boldsymbol{\varepsilon}$, with a rms noise value of $\sigma$, in each bin of the DFT the expected value of the noise is given by

$$\overline{E} = \frac{\sigma}{\sqrt{N}} \cdot \frac{1}{\sqrt{\frac{f_{bw}}{\frac{F_{se}}{2}}}}, \tag{15.29}$$

where $f_{bw}$ is the bandwidth limitation on how the noise is determined. Because the frequency domain is involved, one is not concerned with noise above an end frequency for the

[9]The equivalent-time sample rate is used because, usually, TDRs sample at an odd, slower sample rate relative to the pulse repetition rate and a waveform is built up from the resulting acquisition. These waveforms are called equivalent-time waveforms.

measurement. If $\sigma$ represents the total noise in the waveform, then $f_{bw} = F_{se}/2$; otherwise, if it is specified as band limited (or it is actually band limited), then $f_{bw}$ will be lower.

The actual signal $\mathbf{x}$ contains an impulsive waveform. The term "impulsive" here does not mean an actual impulse, just something impulse-like. This waveform is integrated to obtain the amplitude of a resulting step waveform $A$, which is in units of V · s, or webers (Wb). Dividing this by the sample period produces the height of an actual impulse $A/T_{se}$. For an impulse, the rms value in each DFT bin is

$$\hat{X}[n] = S = \frac{A}{T_{se} \cdot K} \cdot \frac{2}{\sqrt{2}}. \tag{15.30}$$

The ratio of (15.30) to (15.29) is the SNR:

$$\begin{aligned} SNR_{impulse} &= 20 \cdot \log\left(\frac{S}{E}\right) = 20 \cdot \log\left(\frac{A \cdot 2 \cdot \sqrt{N} \cdot \sqrt{2} \cdot \sqrt{f_{bw}}}{T_{se} \cdot K \cdot \sqrt{2} \cdot \sigma \cdot \sqrt{F_{se}}}\right) \\ &= 20 \cdot \log\left(\frac{A \cdot 2 \cdot \sqrt{N} \cdot \sqrt{f_{bw}}}{T_{se} \cdot K \cdot \sigma \cdot \sqrt{F_{se}}}\right). \end{aligned}$$

Recognizing that the acquisition length is $T_d = T_{se} \cdot K$, this can be simplified:

$$SNR_{impulse} = 20 \cdot \log\left(\frac{A \cdot \sqrt{2} \cdot \sqrt{f_{bw}}}{\sqrt{T_d} \cdot \sigma}\right) = 10 \cdot \log\left(\frac{A^2 \cdot 2 \cdot f_{bw}}{T_d \cdot \sigma^2}\right).$$

This is the SNR for a true impulse, but the signal is not an actual impulse, so the dynamic range is adjusted for the frequency response by computing the magnitude of the DFT of the signal $\mathbf{x}$ and dividing all values by the DC component in the first bin to obtain a shape. This shape, in dB, is added to the SNR:

$$SNR_{impulse}(f) = 10 \cdot \log\left(\frac{A^2 \cdot 2 \cdot f_{bw}}{T_d \cdot \sigma^2}\right) + P(f). \tag{15.31}$$

The noise is preferably expressed in dBm:[10]

$$\mathcal{N} = 20 \cdot \log(\sigma) - 10 \cdot \log\left(50 \cdot 10^{-3}\right)$$

or

$$\sigma = \sqrt{\frac{10^{\frac{\mathcal{N}}{10}}}{20}}. \tag{15.32}$$

Substituting (15.32) into (15.31):

$$\begin{aligned} SNR_{impulse}(f) &= 10 \cdot \log\left(\frac{A^2 \cdot 2 \cdot f_{bw}}{T_d \cdot \frac{10^{\frac{\mathcal{N}}{10}}}{20}}\right) + P(f) = 10 \cdot \log\left(\frac{A^2 \cdot f_{bw} \cdot 2 \cdot 20}{10^{\frac{\mathcal{N}}{10}} \cdot T_d}\right) + P(f) \\ &= 10 \cdot \log\left(\frac{A^2 \cdot f_{bw} \cdot 2}{T_d}\right) + 13 - \mathcal{N} + P(f). \end{aligned}$$

[10] The dBm is defined such that the rms voltage $v$ that delivers a given power $P$ to a load resistance $R$ corresponds to 0 dBm, or $dBm = 20 \cdot \log(v) - 10 \cdot \log(R \cdot P)$, where $R = 50\,\Omega$ and $P = 1\,\text{mW}$.

Averaging of the waveform achieves a 3 dB reduction in noise for every doubling of the number of averages, or $20 \cdot \log\left(\sqrt{avg}\right) = 10 \cdot \log(avg)$, so this can be inserted directly into the numerator:

$$SNR_{impulse}(f) = 10 \cdot \log\left(\frac{A^2 \cdot f_{bw} \cdot 2 \cdot avg}{T_d}\right) + 13 - \mathcal{N} + P(f). \tag{15.33}$$

The number of averages is not as important as the amount of time to wait for the acquisition. The number of averages taken in an amount of time $T_w$ is given by

$$avg = \frac{F_{sa}}{T_d \cdot F_{se}} \cdot T_w,$$

where $F_{sa}$ is the actual sample rate of the system. Substituting this into (15.33) yields

$$SNR_{impulse}(f) = 10 \cdot \log\left(\frac{A^2 \cdot f_{bw} \cdot 2 \cdot F_{sa} \cdot T_w}{T_d^2 \cdot F_{se}}\right) + 13 - \mathcal{N} + P(f).$$

Finally, two items are considered – the fractional amount of the waveform containing useful reflections and the cable losses:

$$SNR_{impulse}(f) = 10 \cdot \log\left(\frac{2 \cdot A^2 \cdot f_{bw} \cdot F_{sa} \cdot T_w}{T_d^2 \cdot F_{se} \cdot frac}\right) + 13 - \mathcal{N} + P(f) + 2 \cdot F(f).$$

This is for an impulsive system. The equation given in [57] for a step-like system is[11]

$$SNR_{step}(f) = 10 \cdot \log\left(\frac{2 \cdot A^2 \cdot f_{bw} \cdot F_{sa} \cdot T_w}{T_d^2 \cdot F_{se} \cdot frac \cdot f^2}\right) - 3 - \mathcal{N} + P(f) + 2 \cdot F(f).$$

The difference between the two equations is the $f^2$ in the denominator and an extra 16 dB in $SNR_{impulse}$. This extra 16 dB arises from the fact that a term $(2 \cdot \pi)^2$ was removed from the denominator and placed outside. The overall equation for the SNR is

$$\begin{aligned} SNR(f) = 10 \cdot log\left(\frac{f_{bw}}{F_{se/2}}\right) - \mathcal{N} \\ &+ \begin{cases} 20 \cdot \log\left(\frac{A}{f}\right) - 3 & \text{step} \\ 20 \cdot \log(A) + 13 & \text{impulse} \end{cases} \\ &+ 10 \cdot \log\left(\frac{F_{sa} \cdot T_w}{T_d^2}\right) \\ &- 10 \cdot \log(frac) \\ &+ P(f) + 2 \cdot F(f). \end{aligned}$$

The first term deals with the definition of the overall noise only; the second term deals with steps vs. impulses; the third term the acquisition speed; the fourth with the effects of denoising; and the last two terms deal with the response of the pulser/sampler and the fixturing effects.

[11] A 3 dB error in that equation has been corrected due to (15.29).

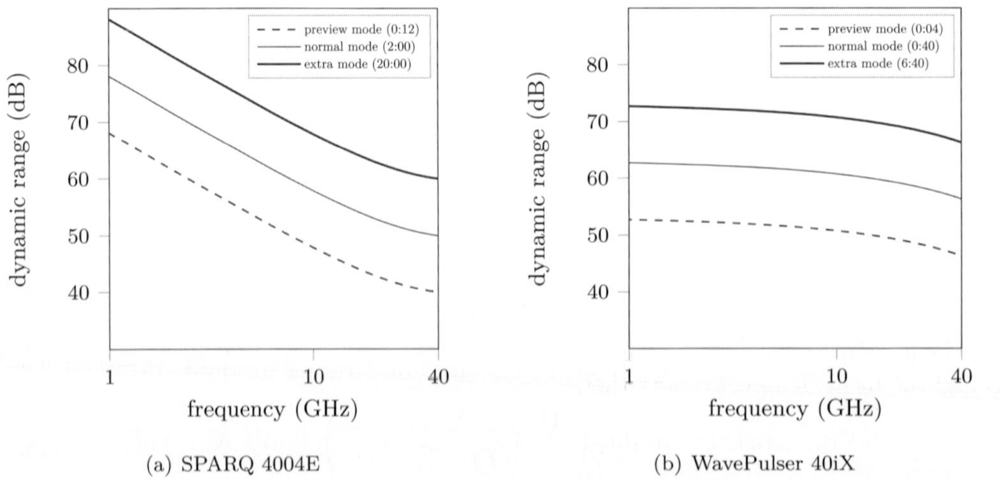


**Figure 15.6** Dynamic range of Teledyne LeCroy TDR instruments. Dynamic range is shown for a four-port measurement in preview, normal, and extra (dynamic range) measurement modes, along with the times required for the measurements, in (mm:ss).

Two high dynamic range TDRs are shown in Figure 15.6; both are Teledyne LeCroy instruments. Figure 15.6(a) shows the SPARQ 4004E, which generates a step-like stimulus, exhibiting the 20 dB/decade roll-off in dynamic range. Figure 15.6(b) shows the WavePulser 40iX, which generates impulses and has flatter dynamic range.[12]

## 15.7 S-Parameter Checking and Conditioning

This section will discuss certain characteristics that ought to exist in s-parameters that are measured or simulated. These characteristics can be checked in order to detect measurement or simulation errors and can often be corrected if small errors exist.

### 15.7.1 Passivity

Passive devices are devices that cannot generate power. When measuring s-parameters, sometimes small errors cause passive devices to be measured as non-passive. There are some instances where a non-passive network can cause loss of stability in a simulation [64], but even when harmless it is undesirable simply from a physicality aspect to have measured s-parameters of passive devices that can be shown to generate power, even minute amounts due to measurement or calculation error.

As outlined in §2.3, one can determine the amount of power generated or dissipated based on wave measurements only. Under the conditions outlined in §2.3, with positive,

[12]The SPARQ 4004E dynamic range tends upward near 40 GHz due to the extra 12 dB signal content, while the WavePulser 40iX tends slightly downward due to internal fixture losses.

real $\mathbf{Z0} = Z0 \cdot \mathbb{I}$, and $\sqrt{\mathbf{Z0}} = \sqrt{Z0}$, if $\mathbf{b}$ is a vector of reflected waves and $\mathbf{a}$ is a vector of incident waves at the ports of a device, the net real power delivered to a device with $P$ ports is measured, for $p \in 1 \dots P$, as

$$\mathrm{Re}\,(p) = \sum_p |a_p|^2 - \sum_p |b_p|^2 = \mathbf{a}^H \cdot \mathbf{a} - \mathbf{b}^H \cdot \mathbf{b},$$

or as the difference between the squares of the Euclidian norms of vectors $\mathbf{a}$ and $\mathbf{b}$:

$$\mathrm{Re}\,(p) = \|\mathbf{a}\|_2{}^2 - \|\mathbf{b}\|_2{}^2. \tag{15.34}$$

Having s-parameters $\mathbf{S}$ for a given device, the net power delivered is a function only of $\mathbf{S}$ and the incident waves $\mathbf{a}$, and therefore

$$\mathrm{Re}\,(p) = \|\mathbf{a}\|_2{}^2 - \|\mathbf{S} \cdot \mathbf{a}\|_2{}^2. \tag{15.35}$$

A passive device can never have negative net power delivered under any circumstances. Since the value of power delivered in (15.35) will depend on the value of $\mathbf{a}$, the test for passivity is that the worst case value of $\mathbf{a}$ delivers no negative power.

For this, we want the maximum Euclidian norm possible for $\mathbf{S} \cdot \mathbf{a}$ for a value of $\mathbf{a}$ whose Euclidian norm is unity. This is called the induced 2-norm of $\mathbf{S}$ and is defined according to [65] as follows:

$$\|\mathbf{S}\|_2 = \max_{\|\mathbf{a}\|_2 = 1} \left(\|\mathbf{S} \cdot \mathbf{a}\|_2\right) = \sqrt{\lambda_{max}},$$

where $\lambda_{max}$ is the largest value of $\lambda$ such that $\mathbf{S}^H \cdot \mathbf{S} - \lambda \cdot \mathbb{I}$ is singular. The value $\lambda_{max}$ is the largest eigenvalue of $\mathbf{S}^H \cdot \mathbf{S}$. Thus, a test for passivity is to compute the eigenvalues for $\mathbf{S}^H \cdot \mathbf{S}$, and, if any is larger than unity, the device is not passive. In other words, if the induced 2-norm of $\mathbf{S}$ is less than or equal to unity, the device is passive; otherwise, it is not. Using only the eigenvalues, an adjustment to $\mathbf{S}$ can be made to enforce passivity.

Despite the fact that eigenvalues are usually mentioned in conjunction with passivity, according to [66], the singular value decomposition (SVD) is more useful if there is a desire to enforce passivity by making presumed small changes to the s-parameters. The SVD for a matrix $\mathbf{S}$ is given by

$$\mathbf{S} = \mathbf{U} \cdot \mathbf{\Sigma} \cdot \mathbf{V}^H,$$

where $\mathbf{U}$ and $\mathbf{V}$ are unitary matrices and $\mathbf{\Sigma}$ is a diagonal matrix containing the singular values in descending order of magnitude. Also:[13]

$$\mathbf{S}^H \cdot \mathbf{S} = \left(\mathbf{U} \cdot \mathbf{\Sigma} \cdot \mathbf{V}^H\right)^H \cdot \mathbf{U} \cdot \mathbf{\Sigma} \cdot \mathbf{V}^H = \mathbf{V} \cdot \mathbf{\Sigma}^2 \cdot \mathbf{V}^H.$$

The eigenvalue decomposition of $\mathbf{S}^H \cdot \mathbf{S}$ is

$$\mathbf{S}^H \cdot \mathbf{S} = \mathbf{V} \cdot \mathbf{\Lambda} \cdot \mathbf{V}^H,$$

[13]This makes use of the identities $(\mathbf{A} \cdot \mathbf{B})^H = \mathbf{B}^H \cdot \mathbf{A}^H$, and for unitary matrices $\mathbf{U}^H \cdot \mathbf{U} = \mathbb{I}$.

**Listing 15.19 SParameterManipulation** class `EnforcePassivity()` method

```
class SParameterManipulation(object):
    def _LargestSingularValues(self):
        return [linalg.svd(m,full_matrices=False,compute_uv=False)[0]
            for m in self.m_d]
    def EnforcePassivity(self,maxSingularValue=1.):
        for n in range(len(self.m_d)):
            (u,s,vh)=linalg.svd(self.m_d[n],full_matrices=1,compute_uv=1)
            for si in range(len(s)): s[si]=min(maxSingularValue,s[si])
            self.m_d[n]=dot(u,dot(diag(s),vh)).tolist()
        return self
...
```

and therefore the eigenvalues are related as follows:[14]

$$\mathbf{\Lambda} = \mathbf{\Sigma}^2.$$

Therefore, an alternative method of testing for passivity is that the largest singular value of $\mathbf{S}$ (which is $\Sigma_{11}$) is less than or equal to unity. Furthermore, if the s-parameters are found to be non-passive, they can be made passive with minimal changes by calculating

$$\Sigma_{new_{r,r}} = \begin{cases} 1 & \text{if } \sqrt{\Sigma_{\mathrm{r,r}}} > 1, \\ \mathbf{\Sigma}_{r,r} & \text{otherwise,} \end{cases}$$

and

$$\mathbf{\Delta\Sigma} = \mathbf{\Sigma}_{\boldsymbol{new}} - \mathbf{\Sigma}.$$

Thus, the changes made to the s-parameters can be given as

$$\mathbf{\Delta S} = \mathbf{U} \cdot \mathbf{\Delta\Sigma} \cdot \mathbf{V}^H.$$

The new s-parameters are therefore $\mathbf{S} = \mathbf{S}+\mathbf{\Delta S}$. The changes made are thought to be the smallest changes that enforce passivity. The software for performing passivity enforcement is provided in Listing 15.19.

At the beginning of this section, it was noted that this passivity check, and the subsequent passivity enforcement, is valid only when the power calculation in (15.34) is valid. In cases where this calculation is not valid, a reference impedance and/or normalization factor change is required to perform this passivity check and enforcement. Although a reference impedance change might be required, it is important to remember that reference impedance is arbitrary and that passivity is a condition that does not depend on the reference impedance. In other words, if a device is passive in one reference impedance, it is passive in any other.

---

[14] This relationship holds for $\mathbf{S}^H \cdot \mathbf{S}$ only and not in general (where the eigenvalue decomposition is $\mathbf{S} = \mathbf{V} \cdot \mathbf{\Lambda} \cdot \mathbf{V}^{-1}$ and $\mathbf{V}$ is different from the matrix in the SVD and not necessarily unitary). Also, unlike the singular values, the eigenvalues have no preferred ordering, so the relationship holds only when $\mathbf{\Lambda}$ is ordered in the same way as $\mathbf{\Sigma}$.

**Listing 15.20 SParameterManipulation** class `EnforceReciprocity()` method

```
class SParameterManipulation(object):
    def EnforceReciprocity(self):
        for n in range(len(self.m_d)):
            for r in range(len(self.m_d[n])):
                for c in range(r,len(self.m_d[n])):
                    if c>r:
                        self.m_d[n][r][c]=(self.m_d[n][r][c]+self.m_d[n][c][r])/2.
                        self.m_d[n][c][r]=self.m_d[n][r][c]
        return self
```

### 15.7.2 Symmetry

A device is symmetric if its ports are completely interchangeable. Let $\mathbf{P}$ be a permutation matrix that reorders a vector of port numbers $\mathbf{v}$ to produce a new, reordered vector of port numbers. A device is symmetric if, for all such combinations of reordered ports, all of the s-parameter matrices obey

$$\mathbf{P} \cdot \mathbf{S} \cdot \mathbf{P}^T = \mathbf{S}.$$

All elemental impedance elements are symmetric, as can be seen from (3.42) and (3.43). Examining (3.35), one can show that parallel arrangements of symmetric elements retain their symmetry, but most series combinations of elements will not be symmetric unless the two symmetric devices are identical. Furthermore, using the tip embedding equation in (4.26), in conjunction with the symmetric s-parameters of a tee or circuit node provided in (3.15), shows that symmetric networks connected together at a circuit node produce a resulting symmetric network only if each of the interconnected symmetric networks are identical. In summary, symmetry is somewhat unusual. Even in the measurement of symmetric devices, lack of symmetry will often result from differing launch characteristics.

Methods for imposing symmetry are similar to methods for enforcing reciprocity, but when the symmetry in the measurement is lost due to launch characteristics, the best technique to improve symmetry is by de-embedding the launches, with the most successful technique involving the use of peeling, as outlined in §14.4, to determine the structure to de-embed.

### 15.7.3 Reciprocity

In circuit analysis, a reciprocal network [67] implies a mathematical definition of reciprocity with regard to s-parameters. That is, given each s-parameter matrix $\mathbf{S}$ of a device, the device is reciprocal[15] if $\mathbf{S}^T = \mathbf{S}$. Therefore, a test for reciprocity is simply a test of the validity of this equation.

Many systems are reciprocal. In (3.42) and (3.43), it was shown that all series and shunt impedance elements are reciprocal. Examining (3.35), one can see that any cascade of

[15] This causes some confusion between electrical engineering and mathematics terms because networks having the electrical engineering property of reciprocity have s-parameter matrices that have the mathematical property of symmetry, but electrically speaking reciprocal devices are not symmetric. To add to the confusion, in mathematics, a reciprocal matrix might be the matrix inverse!

reciprocal devices is reciprocal, meaning shunt reciprocal networks in parallel are reciprocal along with series combinations of reciprocal networks. Finally, while the math is not shown, using the tip embedding equation in (4.26), in conjunction with the reciprocal s-parameters of a tee or circuit node provided in (3.15), shows that reciprocal networks connected together produce aggregate reciprocal networks. Thus, when a network consists of interconnected reciprocal elements, there is an expectation of reciprocity in the measured s-parameters of the network.

Unfortunately, given an already calculated s-parameter file, all that is really possible for reciprocity enforcement post measurement is to average the off-diagonal elements. This means that, for a $P$-port device with $r, c \in 1 \ldots P$,

$$S_{r,c} = \begin{cases} \frac{S_{r,c}+S_{c,r}}{2} & \text{if } c \neq r, \\ S_{r,c} & \text{otherwise;} \end{cases}$$

this is shown in Listing 15.20.

In the event that one has access to the matrices in the last steps of the measurement calculation, as in (15.22) or (15.21), one is presented with the equation in the form $\mathbf{S} \cdot \mathbf{A} = \mathbf{B}$ and the reciprocal s-parameters can be calculated optimally [66]. Either way, enforcing reciprocity can cause causality and passivity violations, requiring multiple passes over the s-parameter enforcements.

For $c, r, p \in 0 \ldots P-1$, in that order,[16]

$$M_{r,c} = \begin{cases} M_{c,r} & \text{if } r < c, \\ r - c + \begin{cases} 0 & \text{if } c = 0, \\ M_{P-1,c-1} + 1 & \text{otherwise,} \end{cases} & \text{otherwise,} \end{cases}$$

$$L_{p \cdot P + r, M_{p,c}} = A_{c,r},$$

$$b_{r \cdot P + c} = B_{r,c},$$

$$\mathbf{s} = \mathbf{L}^{\dagger} \cdot \mathbf{b},$$

$$\mathbf{S}'_{r,c} = s_{M_{r,c}}.$$

The code that performs this is shown in Listing 15.21 and it is referenced in the calculation of the DUT shown in Listing 15.7.

This algorithm can be understood by first realizing that the equation $\mathbf{S} \cdot \mathbf{A} = \mathbf{B}$ can be rewritten in a vectorized form as follows:

$$\left( \underset{P}{\mathbb{I}} \otimes \mathbf{A}^T \right) \cdot \text{vec}\left( \mathbf{S}^T \right) = \text{vec}\left( \mathbf{B}^T \right).$$

The Kronecker product of the identity matrix and $\mathbf{A}^T$ form a block matrix containing copies of $\mathbf{A}^T$ on its diagonal. To solve the problem, $\text{vec}\left( \mathbf{S}^T \right)$ is replaced with a vector $\mathbf{s}$ containing only one of the two corresponding elements in $\mathbf{S}$ that has $\mathbf{S}_{r,c} = \mathbf{S}_{c,r}$. The length

[16]The indices in this algorithm are zero based.

**Listing 15.21 ErrorTerms** class `_EnforceReciprocity()` method

```
class ErrorTerms(object):
...
    def _EnforceReciprocity(self,A,B):
        P=len(A); Pr=range(P); M=[[None for _ in Pr] for _ in Pr]
        for c in Pr:
            for r in Pr:
                M[r][c]=M[c][r] if r < c else r-c+(0 if c==0 else M[P-1][c-1]+1)
        L=[[0. for c in range(P*(P+1)//2)] for r in range(P*P)]
        b=[None for r in range(P*P)]
        for r in Pr:
            for c in Pr:
                b[r*P+c]=[B[r][c]]
                for p in Pr: L[p*P+r][M[p][c]]=A[c][r]
        sv=(matrix(L).getI()*matrix(b)).tolist()
        S=[[sv[M[r][c]][0] for c in Pr] for r in Pr]
        return S
...
```

of this matrix is $T_P$, defined by the triangular number series.[17] Then, in the Kronecker product, for each of the blocks, and for any element that multiplies by an element $S_{r,c}$ that is not found in **s**, the column of **A** is moved to multiply the element that is multiplied by $S_{c,r}$, and that column in $\left(\underset{P}{\mathbb{I}} \otimes \mathbf{A}^T\right)$ is removed. This is **L**:

$$\underset{P \times T_P}{\mathbf{L}} \cdot \underset{T_P}{\mathbf{s}} = \underset{P}{\mathbf{b}}.$$

Matrix **L** was formed using a mapping matrix **M**, which contains the zero-based index of the row in **s** that corresponds to an element in **S** based on reciprocity.

An example of reciprocity enforcement for a two-port device is

$$\begin{pmatrix} S_{11} & S_{12} \\ S_{21} & S_{22} \end{pmatrix} \cdot \begin{pmatrix} A_{11} & A_{12} \\ A_{21} & A_{22} \end{pmatrix} = \begin{pmatrix} B_{11} & B_{12} \\ B_{21} & B_{22} \end{pmatrix}$$

or

$$\begin{pmatrix} A_{11} & A_{21} & 0 & 0 \\ A_{12} & A_{22} & 0 & 0 \\ 0 & 0 & A_{11} & A_{21} \\ 0 & 0 & A_{12} & A_{22} \end{pmatrix} \cdot \begin{pmatrix} S_{11} \\ S_{12} \\ S_{21} \\ S_{22} \end{pmatrix} = \begin{pmatrix} B_{11} \\ B_{12} \\ B_{21} \\ B_{22} \end{pmatrix}.$$

If the matrix **S** is reciprocal, this can be written as follows:

$$\begin{pmatrix} A_{11} & A_{21} & 0 & 0 \\ A_{12} & A_{22} & 0 & 0 \\ 0 & A_{11} & 0 & A_{21} \\ 0 & A_{12} & 0 & A_{22} \end{pmatrix} \cdot \begin{pmatrix} S_{11} \\ S_{12} \\ S_{21} \\ S_{22} \end{pmatrix} = \begin{pmatrix} B_{11} \\ B_{12} \\ B_{21} \\ B_{22} \end{pmatrix},$$

[17] $T_n$ is the triangular number series defined as $T_n = \sum_{k=1}^{n} k = \frac{n \cdot (n+1)}{2} = \binom{n+1}{2}$.

and further as

$$\mathbf{L}\cdot\mathbf{s}=\mathbf{b}=\begin{pmatrix} A_{11} & A_{21} & 0 \\ A_{12} & A_{22} & 0 \\ 0 & A_{11} & A_{21} \\ 0 & A_{12} & A_{22} \end{pmatrix}\cdot\begin{pmatrix} S_{11} \\ S_{12} \\ S_{22} \end{pmatrix}=\begin{pmatrix} B_{11} \\ B_{12} \\ B_{21} \\ B_{22} \end{pmatrix}.$$

The remainder of the solution becomes obvious at this point.

The matrix $\mathbf{M}$ serves to create $\mathbf{L}$ and to remap the s-parameters in $\mathbf{s}$ into $\mathbf{S}'$; in this case it is given by

$$\mathbf{M}=\begin{pmatrix} 0 & 1 \\ 1 & 2 \end{pmatrix}.$$

Incidentally, if $\mathbf{A}=\mathbb{I}$, we have the situation where the s-parameters are provided after calculation in $\mathbf{B}$. In this case, it is seen that

$$\mathbf{L}=\begin{pmatrix} 1 & 0 & 0 \\ 0 & 1 & 0 \\ 0 & 1 & 0 \\ 0 & 0 & 1 \end{pmatrix}$$

and

$$\mathbf{s}=\mathbf{L}^{\dagger}\cdot\mathbf{b}=\begin{pmatrix} 1 & 0 & 0 & 0 \\ 0 & \frac{1}{2} & \frac{1}{2} & 0 \\ 0 & 0 & 0 & 1 \end{pmatrix}\cdot\begin{pmatrix} S_{11} \\ S_{12} \\ S_{21} \\ S_{22} \end{pmatrix}.$$

This leads to

$$\mathbf{S}'=\begin{pmatrix} S_{11} & \frac{S_{12}+S_{21}}{2} \\ \frac{S_{12}+S_{21}}{2} & S_{22} \end{pmatrix}$$

and proves the earlier statement about the best that can be done with calculated s-parameters.

### 15.7.4 Causality

There are two types of causality that concern engineers when using s-parameters:

1. Circuits should not respond prior to time zero before a stimulus is applied.
2. Waves, voltages, and currents should not arrive at locations in a circuit before they could reasonably be expected to.

Although both types of causality violations are important, usually the first condition is the only one that can be checked without further knowledge of the circuit or device. While causality violations can cause havoc with simulators, the simulation techniques employed in this book are impervious to any causality violations from the standpoint of stability.[18]

[18] In fact, virtual probing (as discussed in Chapter 11) will often provide correct transfer functions that are non-causal depending on how the problem is posed. That said, causality violations due to insufficient frequency sampling of s-parameters will produce incorrect results.

**Listing 15.22 SParameterManipulation** class `EnforceCausality()` method

```
class SParameterManipulation(object):
    def EnforceCausality(self):
        for toPort in range(self.m_P):
            for fromPort in range(self.m_P):
                fr=self.FrequencyResponse(toPort+1,fromPort+1)
                ir=fr.ImpulseResponse()
                if ir is not None:
                    t=ir.td; Ts=1./ir.td.Fs
                    for k in range(len(t)):
                        if t[k]<=-Ts: ir[k]=0.
                    fr=ir.FrequencyResponse()
                    frv=fr.Response()
                    for n in range(len(frv)): self.m_d[n][toPort][fromPort]=frv[n]
        return self
...
```

Causality is easy to determine. One simply computes the impulse response of each s-parameter and verifies that nothing appears prior to zero time. The duality between the frequency response of an s-parameter and its impulse response, where the relationship between them is through the DFT and IDFT, was discussed in §12.4.2. Causality enforcement is equally easy; simply zero out the impulse response prior to zero time. This enforcement method is shown in Listing 15.22.

While the s-parameters of a device comprise an array of matrices with one matrix per frequency, when one refers to the frequency response of an s-parameter, what is meant is a view of the s-parameters as a matrix of arrays with one array per element in the s-parameter matrix, where each array has one element per frequency. When working with s-parameters in signal integrity applications, it is important that each impulse response is examined per s-parameter element for causality violations or other gross problems.

There are several major causes of causality violations:

1. Interpolation phenomena. These are due to truncation of the frequency response and non-bin centering of the time-domain response requiring some form of fractional delay and therefore some form of interpolation. This is covered in §13.2. Bin centering of time-domain data can be enforced only when one starts with direct measurements of time-domain data, as made with a digital oscilloscope. When s-parameters are measured or simulated, this bin centering cannot be enforced without processing, which sometimes creates artifacts before time zero (and throughout the waveform). Usually, these kinds of artifacts can be reduced by performing a measurement that goes high enough in frequency.

2. Extrapolation phenomena. These are due to the extrapolation of the DC point. Since the VNA is incapable of measuring DC, the DC and low frequency points must be extrapolated. In TDR based s-parameter measurements, the DC point is measured and these instruments produce s-parameters without these low frequency extrapolation effects. A common error made by engineers in attempts to reduce these effects is to measure as low in frequency as possible on the VNA. This is a mistake because the VNA is very noisy at low frequencies. A better approach is to recognize that the frequency resolution required is the reciprocal of twice the impulse response length

in time and use the frequency resolution (and the lowest point) that comes from this calculation. Also, in many cases the extrapolation algorithm should be questioned.

3. Time aliasing. If the s-parameter measurements are not made with sufficient resolution as stated in item 2, the impulse response will have elements that are not at the correct time. This is explained in §12.2.1. If time aliasing occurs due to insufficient frequency sampling of the s-parameters, nothing can really be done to correct them, and, if they are used in time-domain simulations, absolutely incorrect results will be obtained. There have been cases of s-parameter measurements involving complicated logistics where the s-parameter measurements were made, the measurement arrangement dismantled, and the s-parameters distributed to various groups, only to find that the insufficient sampling made them completely useless. The time aliasing effect can only be seen by examination of the impulse (and step) response; this should be checked at the time of the measurement.
4. Measurement noise and inaccuracies. Because s-parameter measurements are made with instruments that require calibration and that essentially triangulate the measurement, there will always be measurement errors; one can only try to minimize them. These measurement errors can manifest themselves as causality violations. If these errors are small enough, the quality of the s-parameters can be improved by applying a window of some form to the impulse responses used to evaluate causality violations and then converting these windowed impulses back to frequency responses, and these back to s-parameters.

Figure 15.7 provides an example of two potential causality issues. Figure 15.7(a) portrays a single-ended microstrip transmission line structure, called a *meander* because it wanders around the top of the substrate. A microstrip transmission line tends to be tightly coupled to the ground plane under the line. The line has 500 ps of electrical length on each side, and the meandering portion in the middle has 3 ns of transmission line. A schematic that simulates this line is shown in Figure 15.7(b). The meander is depicted as three four-port RLGC transmission lines. The inductance is 25 nH and the capacitance is 10 pF for a characteristic impedance of $Z_c = \sqrt{L/C} = 50\,\Omega$ and a propagation delay of $T_d = \sqrt{L \cdot C} = 500\,\text{ps}$. There is a small mutual inductance of 10 pH and a small mutual capacitance of 10 fF to simulate very weak coupling between the lines.

The line is first simulated with 200 points at 100 MHz/point, which is equivalent to a 10 ns impulse response length. The results are shown in Figure 15.8. Transmitted waves arrive at port 2, 4 ns after they enter port 1. Figure 15.8(a) and Figure 15.8(b) show the impulse and step response zoomed so that the vertical axis ranges between $\pm 0.01$, where small causality violations are observed. There are voltage bumps at $-4$ ns and $-2$ ns that occur before time zero, and waves arriving at 3 ns, which is 1 ns too early since the line is 4 ns long. Also, the step response seems to start below zero.

The first two causality violations are due to insufficient sampling of the s-parameters in the frequency domain.

Figure 15.9 provides the simulation run again with 400 points at 50 MHz/point, which is equivalent to a 20 ns impulse response length. Again, transmitted waves arrive at port 2, 4 ns after they enter port 1. Figure 15.9(a) and Figure 15.9(b) show the impulse and step response zoomed so that the vertical axis ranges between $\pm 0.01$. Here, the causality

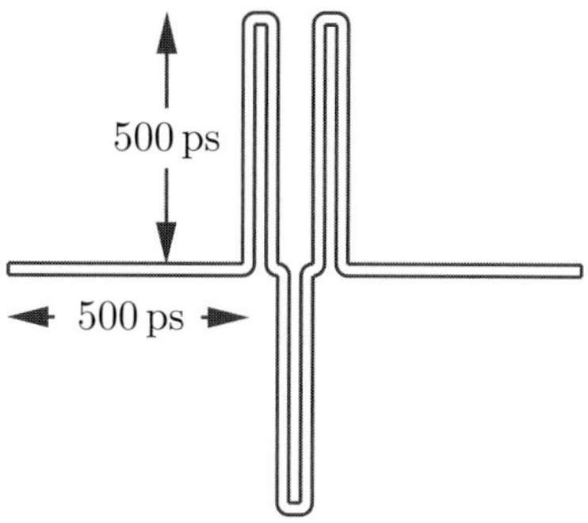


(a) Meandering transmission line

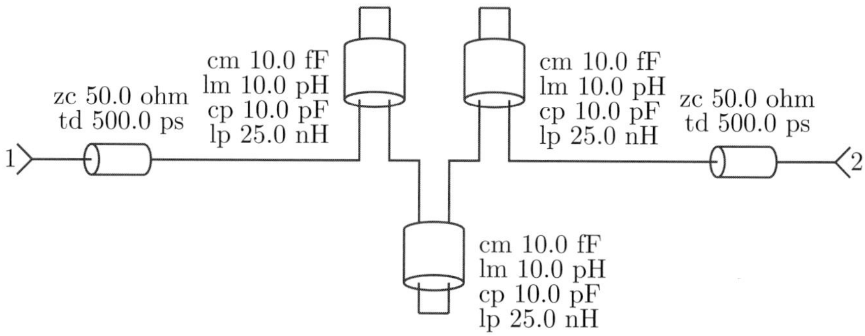


(b) Simulation schematic

**Figure 15.7** Meandering transmission line example

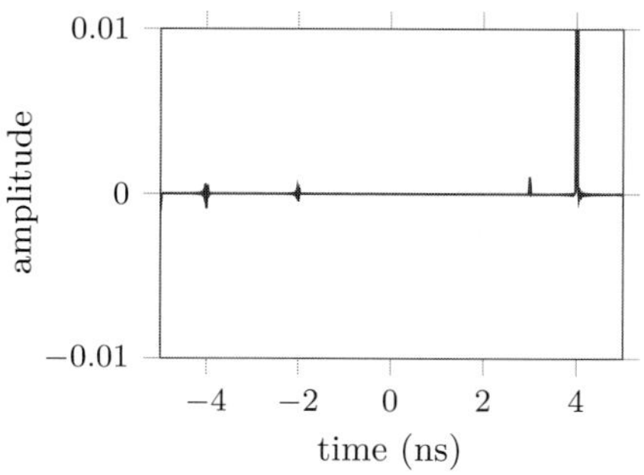


(a) Impulse response (zoomed)

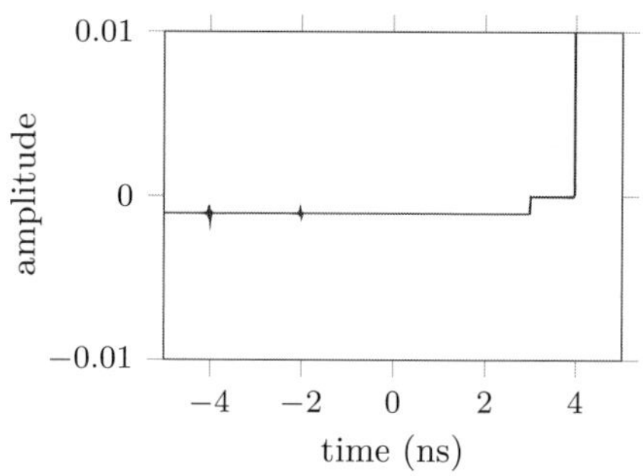


(b) Step response (zoomed)

**Figure 15.8** Meandering transmission line $S_{21}$ (200 points, 100 MHz/point)

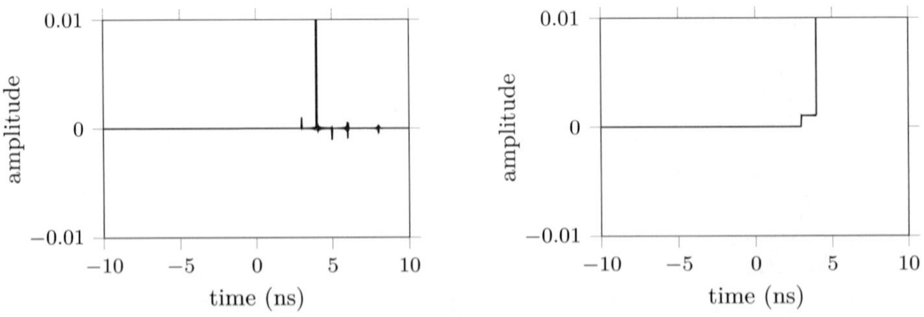


(a) Impulse response (zoomed) (b) Step response (zoomed)

**Figure 15.9** Meandering transmission line $S_{21}$ (400 points, 50 MHz/point)

violations have disappeared, and it is confirmed that the problem was in fact the frequency resolution. Once again, a small portion of the signal arrives at 3 ns, 1 ns too early. This early arrival is not a violation of any causality expectation as it is due to the weak coupling of the meandering lines. Despite the fact that the example specifically utilized weakly coupled meandering lines, in actual systems, even if this coupling is not expected, it may occur, and waves are free to travel to some extent in a direct path between ports. Remember, even though the line is long and meandering, the ground plane has a much shorter potential path. Here, that path is approximately 1 ns and one might now wonder why no signal arrives at port 2, 1 ns after its application at port 1.[19]

### 15.7.5 Wavelet Denoising

Wavelets [68, 69] and wavelet denoising in time-domain reflectometry [70] solve a basic problem that is seen when examining the dynamic range calculations (discussed in §15.6.3). In the TDR, the incident waveform is very short and the reflected waveform is very long. The length of the reflected waveform should be such that the impulse response dies to zero, but the TDR user does not know how long that is prior to measurement. As the waveform is lengthened beyond the end of the impulse response, the only effect is to add noise to the system. Furthermore, as outlined in [70], if the user is measuring a long cable, all of the portion of the waveform measured while waiting for the reflection is simply noise. As the incident and reflected waveforms are broadband, filtering and other techniques cannot be employed to remove noise.

In wavelet denoising, the waveform is decomposed into its wavelet coefficients and the absolute size of each coefficient is compared against a threshold developed based on the

[19] In fact, a very, very tiny amount does arrive 1 ns after being applied to port 1. It is so small that it cannot be seen in this simulation with any amount of zooming. Only the amount that is coupled at the first, middle, and last bend that jumps 1 ns total length is seen. This amount is less than one-thousandth of the signal and arrives 1 ns too early. The amount that jumps two loops is less than one-millionth, and the amount that jumps all three loops is less than one-billionth of the original signal size.

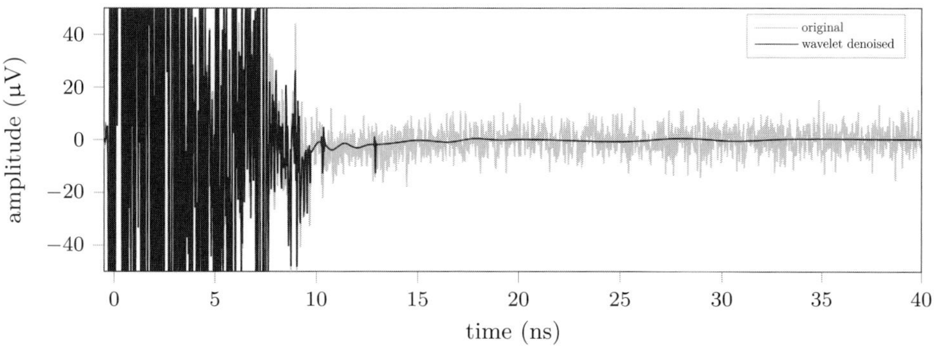


(a) Extreme vertical zoom

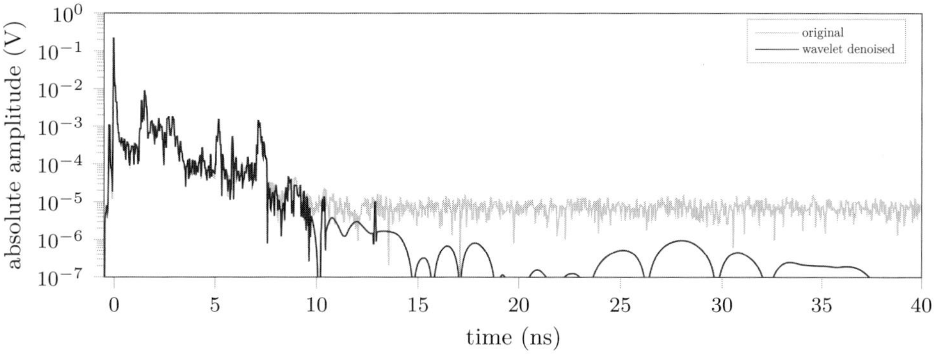


(b) Absolute voltage

**Figure 15.10** Wavelet denoised waveform

statistics of the noise.[20] Only wavelets deemed statistically valid by being higher than this threshold are retained; when the waveform is reassembled, it has large, sometimes huge, amounts of noise removed.

An example of wavelet denoising is shown in Figure 15.10 for a load measurement using a Teledyne LeCroy WavePulser 40iX. Here, the noise is about $10\,\mu V_{rms}$ (or $-87\,dBm$), and is heavily averaged. In Figure 15.10(a), the original and denoised waveforms are zoomed extremely vertically, where it is seen that, after about 10 ns, the denoised waveform becomes mostly a line with no noise. In Figure 15.10(b), the absolute value of the waveform is plotted semi-log versus time. Again, the reflections die out around 10 ns, where the waveform sinks below the noise floor, diminishing the noise by several orders of magnitude in this region. In any region containing actual reflections (or wavelet coefficients higher than the noise statistics support), the time-domain waveform is essentially unchanged. The most

[20]These statistics can be gathered from the portion of the waveform containing no reflections, or during a calibration phase.

**Listing 15.23 Wavelet** class transform and inverse transform methods

```
1  class Wavelet(object):
2      def __init__(self,h):
3          self.h=h
4          self.L=len(self.h)
5          self.g=[pow(-1.,l)*h[self.L-1-l] for l in range(self.L)]
6      def DWT(self,xi):
7          x=list(xi); N=len(x)
8          B=self.intlog2(N)-self.intlog2(self.L)+1
9          for i in range(B):
10             X=list(x)
11             for k in range(N//2):
12                 X[k]=sum([x[(2*k+l)%N]*self.h[l]
13                         for l in range(self.L)])
14                 X[k+N//2]=sum([x[(2*k+l+N-2)%N]*self.g[l]
15                         for l in range(self.L)])
16             x=list(X); N=N//2
17         return X
18     def IDWT(self,XI):
19         X=list(XI); N=len(X)
20         B=self.intlog2(N)-self.intlog2(self.L)+1
21         N=self.L
22         for i in range(B):
23             x=list(X)
24             for k in range(N//2):
25                 x[2*k]=sum([self.h[2*l]*X[(k-l+(N//2))%(N//2)]+
26                     self.g[2*l]*X[(k+1-l+(N//2))%(N//2)+(N//2)]
27                         for l in range(self.L//2)])
28                 x[2*k+1]=sum([self.h[2*l+1]*X[(k-l+(N//2))%(N//2)]+
29                     self.g[2*l+1]*X[(k+1-l+(N//2))%(N//2)+(N//2)]
30                         for l in range(self.L//2)])
31             X=list(x); N=2*N
32         return x
33     @staticmethod
34     def intlog2(x):
35         r=math.log(float(x))/math.log(2.)
36         return int(round(r))
```

**Listing 15.24 WaveletDaubechies4** class initialization

```
class WaveletDaubechies4(Wavelet):
    def __init__(self):
        Wavelet.__init__(self,[h*math.sqrt(2.)/2
            for h in [0.6830127,1.1830127,0.3169873,-0.1830127]])
```

important aspect of wavelet denoising is that the user doesn't need to worry about the length of the acquisition and the noise impact of lengthening it.

The code for implementing wavelet denoising is provided in Listings 15.23, 15.24, and 15.25. Listing 15.23 provides a constructor on line 2 that initializes the wavelet coefficients supplied by a derived class that specifies a particular wavelet, such as the Daubechies four-tap wavelet as shown in Listing 15.24. Lines 6 and 18 provide the forward and inverse wavelet transform functions.

In Listing 15.25, the wavelet denoising operation is shown. In `DenoisedWaveform()`, a waveform is provided for denoising, along with the percentage of the end of the waveform to use for noise statistics, a multiplier to use on the standard deviation as the threshold, and

**Listing 15.25 WaveletDenoiser** class `DenoisedWaveform()` method

```
class WaveletDenoiser(object):
    def DenoisedWaveform(wf,pct=30.,mult=5.,isDerivative=True):
        w=WaveletDenoiser.wavelet
        Ki=wf.td.K; Kf=int(pow(2,math.ceil(log2(wf.td.K))))
        PadLeft=Kf-Ki; pct=pct*Ki/Kf; pwf=wf*WaveformTrimmer(-PadLeft,0)
        X=w.DWT(pwf.Values())
        T=WaveletDenoiser.DerivativeThresholdCalc(X,w.h,pct,isDerivative)
        dwf=Waveform(pwf.td,w.IDWT(
            [0 if abs(x) < t*mult else x for (x,t) in zip(X,T)]))
        return dwf*WaveformTrimmer(PadLeft,0)
...
```

whether the waveform was formed from a step or impulsive TDR. The waveform is zero padded to the next higher power of two and the forward transform applied. A threshold calculation (not shown) calculates the noise statistics and the threshold. For step waveforms, this is quite complicated [71], but for impulse waveforms it is simply the multiplier on the standard deviation. Then, the waveform is processed, zeroing any wavelet coefficients that fall under the threshold. After applying the inverse transform, the waveform is trimmed to its original size and returned.

# 16

# Model Extraction

MOST OF THIS BOOK has been concerned with the reduction of circuit models to solutions. In many cases, network parameters are computed where the underlying model is a complete circuit with known components. This chapter deals with the opposite situation: the determination of the individual circuit parameters given s-parameters of a system that have been provided either through simulation or actual measurement.

This problem is sometimes called system identification or model fitting. The most difficult aspect of this problem is the determination of the model itself. This model is called the *topology*. The second part of the problem, the determination of values of parameters in the topological model, involves lots of math and computation that this chapter will explain.

The reasons for performing model extraction are numerous, but here are a few:

1. There are large drawbacks with network parameters as generally provided: as a list of matrices where each matrix represents the port–port relationship at a given frequency. The drawbacks are due to their sampled nature in that they end at some point, at an end frequency, and there are a finite number of frequency points that dictate the impulse response length. The measurement or simulation goal would be to produce network parameters to sufficient frequency with sufficient resolution, a situation allowing extrapolation and interpolation. This is sometimes not possible. A network parameter model is therefore desirable, which is a matrix of continuous functions that can be sampled in any way desired.
2. Network parameters do not always simulate well. In §15.7 it was seen that sometimes s-parameter measurements have minor or major issues that cause them to appear non-physical in simulation. Having an extracted model solves this problem to some extent because, generally, the extracted model cannot suffer from non-physicality or sampling issues and can be constrained to behave properly.
3. Network parameters supply no underlying insight. While network parameters can be useful in simulations and computations, they are basically black boxes used to produce more numerical results. They are difficult to adjust during the design phase, and it is difficult to make what-if or sensitivity analyses. For example, having the s-parameter measurement of a transmission line on a printed circuit board, how would one determine the performance of the transmission line if the traces were widened or on a different substrate? There's no real way to do that without first returning to a model of the line.

The focus here is on solutions that assume that the underlying circuit topology is known or can be selected, and methods will be provided for fitting the circuit element parameters to a model. Usually, this is done in a *least-squares sense*, meaning once the topology has been determined, the parameters are found that make the fit as good as possible. If a good fit cannot be made, the engineer is forced to assume that either the model topology is incorrect or incomplete, or that the measurements being fitted are in error. Along these lines, the reader is warned that the methods put forth in this chapter focus mainly on the method of fitting a model, not on the model itself. While the models used here are certainly reasonable, there is no implied recommendation of any particular model, and it is up to the engineer using these algorithms to determine the right model for a given situation.

Sometimes a model can be fitted using linear techniques. This means that an equation can be constructed and solved in one step, without an initial guess at the answer and without iteration. This is usually not possible, though, and nonlinear techniques are required. Nonlinear techniques involve making a guess at what the parameters ought to be and then adjusting them so that the fit to the data is improved. Typically, many steps or iterations are needed to arrive at the final answer. In order to determine the direction and amounts by which to change the model elements, the derivatives are used of the fitted result with respect to each parameter being modified. Using these numerical or analytic derivatives, one can either take small steps in the direction of improvement, called a gradient walk, or take large steps as supplied by Newton's method.

In this chapter, both linear and nonlinear techniques will be provided, ending in a powerful fitting algorithm called the Levenberg–Marquardt [72] algorithm.

## 16.1 Linear Equations

Consider a function $f(\mathbf{x}, \mathbf{a})$ and a vector $\mathbf{y}$. The $K$ elements of $\mathbf{x}$ and $\mathbf{y}$ are known and the $M$ variables in $\mathbf{a}$ are unknown. For linear equations, each element of the vector $\mathbf{x}$ can be used to form a row in a matrix $\mathbf{X}$ that has $K$ rows and $M$ columns, according to the function. There are $K$ equations with $M$ unknowns. The desire is to fit the equations by finding the value of the vector $\mathbf{a}$ such that, for all of the sets of $K$ equations, $f(\mathbf{X}_{k*}, \mathbf{a})$ is equal to $y_k$, or the values of $\mathbf{a}$ are at least the best values possible; what is meant by this will be explained subsequently, but, for now, a simple linear example is considered.

Given $K$ coordinates $\{x_k, y_k\}$ one has a desire to find the best fitting line to these coordinates. The equation for a line is $f(\mathbf{x}, \mathbf{a}) = a_{[0]} + a_{[1]} \cdot x_{[k][0]}$, and the desire is for this equation to equal $y_{[k]}$. If exactly two points are supplied, $\{x_0, y_0\}$ and $\{x_1, y_1\}$, then $\mathbf{X} = \left(\begin{smallmatrix} 1 & x_0 \\ 1 & x_1 \end{smallmatrix}\right)$, $\mathbf{y} = \left(\begin{smallmatrix} y_0 \\ y_1 \end{smallmatrix}\right)$ , and $f(\mathbf{X}, \mathbf{a}) = \mathbf{X} \cdot \mathbf{a}$. The solution to this equation for $\mathbf{a}$ is the solution to $\mathbf{X} \cdot \mathbf{a} = \mathbf{y}$, and therefore $\mathbf{a} = \mathbf{X}^{-1} \cdot \mathbf{y}$. This solution is as follows:

$$\mathbf{a} = \begin{pmatrix} b \\ m \end{pmatrix} = \begin{pmatrix} 1 & x_0 \\ 1 & x_1 \end{pmatrix}^{-1} \cdot \begin{pmatrix} y_0 \\ y_1 \end{pmatrix} = \begin{pmatrix} y_0 - \frac{y_1 - y_0}{x_1 - x_0} \cdot x_0 \\ \frac{y_1 - y_0}{x_1 - x_0} \end{pmatrix}. \tag{16.1}$$

The equation in (16.1) is written in a form so as to highlight what is most familiar. The second element in the solution is simply the slope of the line, which is the ratio of the differences between the $y$ and $x$ values, and the first element is simply the y-intercept found by plugging the slope into the equation of the line and supplying one of the coordinates.

When more than two coordinates are supplied, this problem is considered a *linear regression* problem. This type of solution is overconstrained and has no exact solution to the equation. That being said, it is often a very desirable situation, and what is desired is an exact solution to the *best fit possible*. The desirability of this situation is based on the fact that often data are provided from noisy measurements, and finding the best fit possible of a known function to a noisy data set actually provides the best solution, with the goodness of the solution increasing with the number of samples in the data.

The best fit possible criteria are often specified as being the best fit in a least mean-squared error (LMSE) sense. This means that for $K$ coordinates, where $k \in 0 \ldots K-1$, the coordinates are $\{x_k, y_k\}$. The best fit is found when the following function is minimized:

$$\sigma^2 = \frac{1}{K} \cdot \sum_k \left(f\left(x_k, \mathbf{a}\right) - y_k\right)^2, \tag{16.2}$$

where $\sigma^2$ is the variance, or mean-squared error, and $\sigma$ is the standard deviation of the absolute error. The goal is to find the value of $\mathbf{a}$ that minimizes this.

To solve this type of problem, the minimum is defined as the vector $\mathbf{a}$ such that all of the partial derivatives of the mean-squared error are zero:

$$\frac{\partial}{\partial a_m}\sigma^2 = \frac{\partial}{\partial a_m}\frac{1}{K} \cdot \sum_k \left(f\left(x_k, \mathbf{a}\right) - y_k\right)^2 = 0, \tag{16.3}$$

or where the residual error $r_k = f(x_k, \mathbf{a}) - y_k$:

$$\sum_k r_k \cdot \frac{\partial}{\partial a_m} f(x_k, a) = 0.$$

In the linear regression problem,

$$\mathbf{X} = \begin{pmatrix} 1 & x_0 \\ 1 & x_1 \\ \vdots & \vdots \\ 1 & x_{K-1} \end{pmatrix}, \quad \mathbf{a} = \begin{pmatrix} b \\ m \end{pmatrix}, \quad \mathbf{x} = \begin{pmatrix} x_0 \\ x_1 \\ \vdots \\ x_{K-1} \end{pmatrix}, \quad \mathbf{y} = \begin{pmatrix} y_0 \\ y_1 \\ \vdots \\ y_{K-1} \end{pmatrix},$$

$$\mathbf{r} = \mathbf{X} \cdot \mathbf{a} - \mathbf{y}, \qquad f(\mathbf{x}, \mathbf{a}) = \mathbf{X} \cdot \mathbf{a}, \qquad \frac{d}{d\mathbf{a}} f(\mathbf{x}, \mathbf{a}) = \mathbf{X}.$$

Therefore, the solution is written as

$$\mathbf{r}^H \cdot \mathbf{X} = \left(\mathbf{X} \cdot \mathbf{a} - \mathbf{y}\right)^H \cdot \mathbf{X} = \mathbf{X} \cdot \left(\mathbf{X} \cdot \mathbf{a} - \mathbf{y}\right) = \mathbf{X}^H \cdot \mathbf{X} \cdot \mathbf{a} - \mathbf{X}^H \cdot \mathbf{y} = 0.$$

Solving for $\mathbf{a}$:

$$\mathbf{a} = \begin{pmatrix} b \\ m \end{pmatrix} = \left(\mathbf{X}^H \cdot \mathbf{X}\right)^{-1} \cdot \mathbf{X}^H \cdot \mathbf{y} = \mathbf{X}^\dagger \cdot \mathbf{y}.$$

(See Appendix C, §C.3 for a discussion of $\mathbf{X}^\dagger$.)

The variance in the fit is given by

$$\sigma^2 = \frac{1}{K} \cdot \mathbf{r}^H \cdot \mathbf{r}.$$

This is a linear LMSE solution and can be used for any situation where, for a given set of variables $\mathbf{a}$, a suitable $\mathbf{X}$ can be formulated such that $f(\mathbf{x},\mathbf{a}) = \mathbf{X}\cdot\mathbf{a}$. Sometimes the problems don't look linear, but simply need to be posed in a linear manner.

Some examples are:

- $f(x,a) = a\cdot x^2$ is linear because one can write $\mathbf{X}_{[k][0]} = x_{[k]}^2$ and solve $a = \mathbf{X}^\dagger \cdot \mathbf{y}$;
- $f(x,a) = a^2\cdot x$ can be solved in a linear manner as one can write $g = \mathbf{X}^\dagger\cdot\mathbf{y}$ and solve $a = \sqrt{g}$, presuming one knows which branch to select for the answer;
- $f(x,a) = \ln(a\cdot x)$ looks nonlinear, but can be linearized by using $y_{[k][0]} = e^{y_{[k]}}$ and solving for $a$;
- $f(\mathbf{x},\mathbf{a}) = \ln\big(a_{[0]}\cdot x_{[0]}\big) + a_{[1]}\cdot x_{[1]}$ is nonlinear.

These examples illustrate some of the changes that sometimes need to be made to the problem to put it into linear form.

### 16.1.1 Linear Model Extraction Example

Given the magnitude $S_{21}$ of a cable, it is suspected to have a frequency dependent loss due to resistance, dielectric loss, conductor roughness, and skin effect. Assume this loss is modeled as the function $h(f,\mathbf{a}) = a_0 + a_1\cdot f + a_2\cdot\sqrt{f}$ and that one is provided with $K$ coordinates of frequency and gain, where, for $k \in 0\ldots K-1$, $\{f_k, y_k\}$. This is written in linear form as follows:[1]

$$\mathbf{X} = \begin{pmatrix} 1 & f_0 & \sqrt{f_0} \\ 1 & f_1 & \sqrt{f_1} \\ \vdots & \vdots & \vdots \\ 1 & f_{K-1} & \sqrt{f_{K-1}} \end{pmatrix}, \quad \mathbf{a} = \begin{pmatrix} a_0 \\ a_1 \\ a_2 \end{pmatrix}, \quad \mathbf{x} = \begin{pmatrix} f_0 \\ f_1 \\ \vdots \\ f_{K-1} \end{pmatrix}, \quad \mathbf{y} = \begin{pmatrix} y_0 \\ y_1 \\ \vdots \\ y_{K-1} \end{pmatrix},$$

$$h(\mathbf{x},\mathbf{a}) = \mathbf{X}\cdot\mathbf{a}, \qquad \mathbf{r} = \mathbf{X}\cdot\mathbf{a} - \mathbf{y}, \qquad \frac{d}{d\mathbf{a}}h(\mathbf{x},\mathbf{a}) = \mathbf{X}.$$

The solution is therefore

$$\mathbf{a} = \mathbf{X}^\dagger\cdot\mathbf{y} = \begin{pmatrix} 1 & f_0 & \sqrt{f_0} \\ 1 & f_1 & \sqrt{f_1} \\ \vdots & \vdots & \vdots \\ 1 & f_{K-1} & \sqrt{f_{K-1}} \end{pmatrix}^\dagger \cdot \begin{pmatrix} y_0 \\ y_1 \\ \vdots \\ y_{K-1} \end{pmatrix}.$$

The variance of the fit is given by

$$\sigma^2 = \frac{1}{K}\cdot\mathbf{r}^H\cdot\mathbf{r}.$$

[1] This type of fit should only be performed with the reference impedance as the characteristic impedance of the cable.

```
import math
from numpy import matrix
import SignalIntegrity.Lib as si
sp=si.sp.SParameterFile('cable.s2p')
s21=sp.FrequencyResponse(2,1)
f=s21.Frequencies('GHz')
mS21=s21.Values('mag')
K=len(f)
X=[[1,x,math.sqrt(x)] for x in f]
a=(matrix(X).getI()*[[y] for y in mS21]).tolist()
yf=(matrix(X)*matrix(a)).tolist()
r=(matrix(yf)-matrix([[y] for y in mS21])).tolist()
sigma=math.sqrt(((matrix(r).H*matrix(r)).tolist()[0][0])/K)
print('\\[a_0 = \\text{'+"{:10.4e}".format(a[0][0])+'}\\]')
print('\\[a_1 = \\text{'+"{:10.4e}".format(a[1][0])+'/GHz}\\]')
print('\\[a_2 = \\text{'+"{:10.4e}".format(a[2][0])+'}/\\sqrt{\\text{GHz}}\\]')
print('\\[\\sigma = \\text{'+"{:10.4e}".format(sigma)+'}\\]')
```

(a) Python code

$$a_0 = \text{1.0021e+00}$$

$$a_1 = \text{-9.1463e-06/GHz}$$

$$a_2 = \text{-2.6606e-02}/\sqrt{\text{GHz}}$$

$$\sigma = \text{5.8492e-04}$$

(b) Output

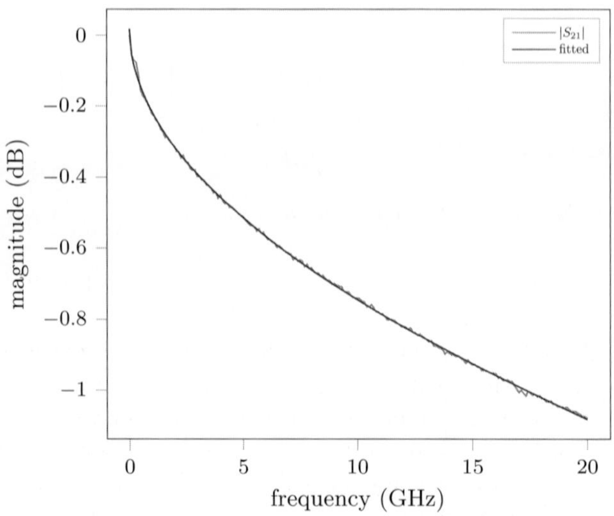


(c) Comparison to $S_{21}$ measured and fit

**Figure 16.1** Cable linear fit

An example of this type of fit is provided in Figure 16.1, with the Python code provided in Figure 16.1(a). The output of the code is shown in Figure 16.1(b), where it is seen that the dominant effect is presumably skin-effect loss, as indicated in the $a_2$ term. The fit has very small rms error. A plot comparing the raw data to the fit is shown in Figure 16.1(c).

Note that, from the $a_0$ term, the fit produces an $S_{21} > 1.0$, and this is nonsensical. If this is bothersome, the first term can be set to unity, but it would be better to fit the other terms with this constraint in place by setting

$$\mathbf{X} = \begin{pmatrix} f_0 & \sqrt{f_0} \\ f_1 & \sqrt{f_1} \\ \vdots & \vdots \\ f_{K-1} & \sqrt{f_{K-1}} \end{pmatrix}, \quad \mathbf{a} = \begin{pmatrix} a_1 \\ a_2 \end{pmatrix}, \quad \mathbf{x} = \begin{pmatrix} f_0 \\ f_1 \\ \vdots \\ f_{K-1} \end{pmatrix}, \quad \mathbf{y} = \begin{pmatrix} y_0 - 1 \\ y_1 - 1 \\ \vdots \\ y_{K-1} - 1 \end{pmatrix}.$$

## 16.2 Newton's Method

Newton's method is a very powerful iterative method used to solve equations and systems of equations when the given function is nonlinear.

For a given guess at the answer to a function $f(x, a) = y$, with $f(x, a)$ being the correct answer and $a + \Delta a$ being the guess supplied, in the vicinity of the correct answer the slope of the line can be approximated as

$$\frac{d}{da} f(x, a) \approx \frac{\Delta y}{\Delta x} = \frac{f(x, a + \Delta a) - f(x, a)}{(a + \Delta a) - a} = \frac{f(x, a + \Delta a) - f(x, a)}{\Delta a}.$$

Solving for $\Delta a$:

$$\Delta a = \frac{f(x, a + \Delta a) - f(x, a)}{\frac{d}{da} f(x, a)}.$$

This is arrived at similarly by writing the series expansion of $f(x, a)$ about $a$:

$$f(x, a + \Delta a) = f(x, a) + \Delta a \cdot \frac{d}{da} f(x, a) + \mathcal{O}\left(\Delta a^2\right).$$

Newton's method can be utilized to solve nonlinear functions. Given a problem $f(x, a) = y$, a guess at the answer is provided as $a + \Delta a$, and on each iteration $\Delta a$ is calculated and subtracted from the guess, in the hope that the correct answer will be eventually achieved.

A common usage for Newton's method is the calculation of functions within a computer. For example, when the square root key is used on a calculator, Newton's method is commonly utilized. Using the square root as an example, the goal is to solve $f(a) = y$, where $f(a) = a^2$; the value of $a$ that satisfies this equation is the square root of $y$. Thus, we have $f(a) = a^2$ and $\frac{d}{dx} f(a) = 2 \cdot a$. Given a guess $a$, the amount to adjust this guess by is

$$\Delta a = \frac{a^2 - y}{2 \cdot a} = \frac{1}{2} \cdot \left(a - \frac{y}{a}\right).$$

```
def newtonSquareRoot(Y):
    if Y<=0.0: raise ValueError('math domain error')
    if Y<=1e-32: return 0.0
    # in practice, exponent is directly extracted from fp number
    E=int(math.ceil(math.log(Y,2.)))
    Eeven=E//2*2==E
    y=Y/pow(2.0,E) # in practice, is the mantissa of the fp number
    seed=[0.72,0.737,0.76,0.78,0.8,0.82,0.84,0.856,0.876,
          0.892,0.91,0.927,0.943,0.961,0.975,0.993]
    # in practice, seed index taken from upper nybble of mantissa
    si=int(math.floor(y*32))-16
    x=seed[si]
    for _ in range(3): x=(x+y/x)/2.0
    x=x*pow(2.0,E//2)*(1.4142135623730951 if not Eeven else 1.0)
    return x
```

(a) Python code

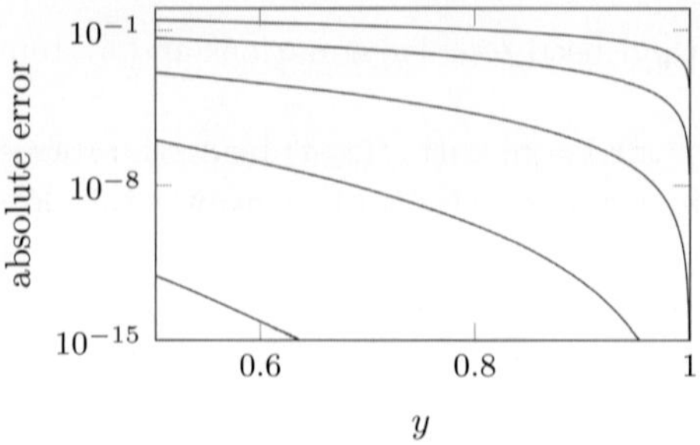


(b) Error vs. iterations with seed of 1.0

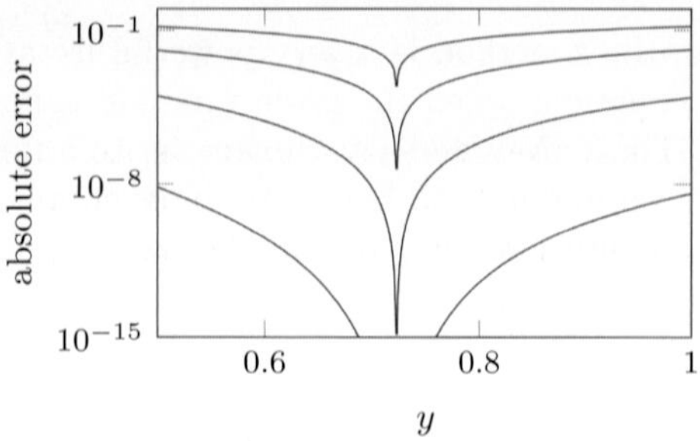


(c) Error vs. iterations with seed of 0.85

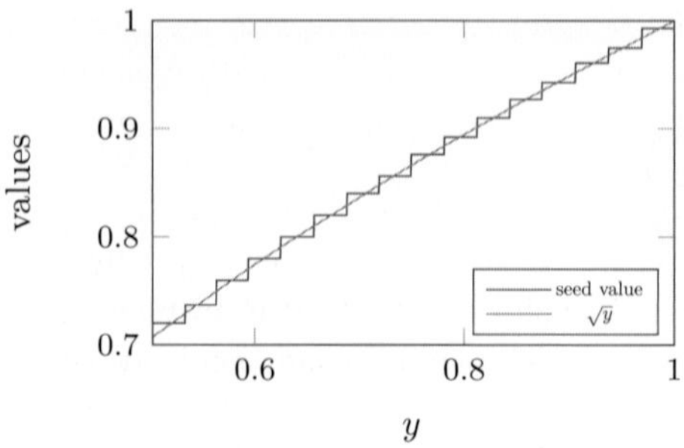


(d) Seed values

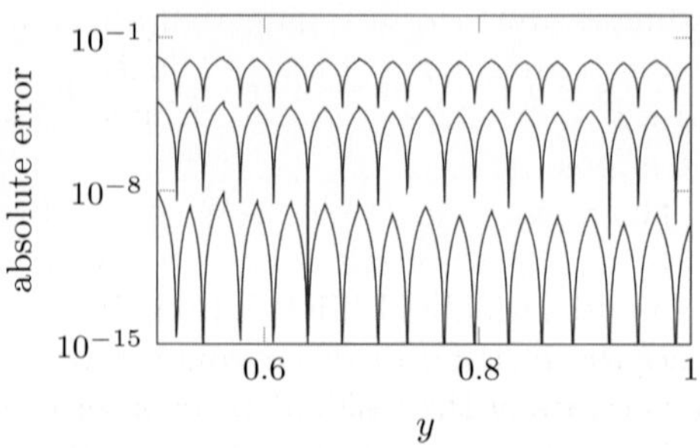


(e) Error vs. iterations with variable seeds

**Figure 16.2** Newton's method for square root calculation

An example of using Newton's method, in this case for the computation of a square root, is provided in Figure 16.2. In Figure 16.2(b) the convergence of the algorithm is shown on each iteration for an initial guess of 1.0 in the computation of square roots of numbers between 0.5 and 1.0. The plots show the absolute error between the square root calculated and the correct value on each iteration. Obviously, if the square root is 1.0 and the guess of 1.0 is provided, the answer is immediately correct, but for a square root of 0.5, five iterations are required before the answer is correct to better than $10^{-16}$.

While Newton's method has the possibility to diverge, if it is convergent, the convergence is *quadratic*. This means that the exponent of the error halves on each iteration, as can be seen in Figure 16.2(b). This is very rapid convergence.

The reason why the square roots of numbers between 0.5 and 1.0 are used is because the range of the numbers requiring a square root calculation can be restricted through a technique called *range reduction*. Floating point numbers represented within a computer are represented by a mantissa between 0.5 and 1.0 raised to a power of two given by an exponent. The square root of a number represented in this manner is calculated as follows:[2]

$$\sqrt{M \cdot 2^E} = \sqrt{M} \cdot 2^{\frac{E}{2}} \cdot \begin{cases} 1 & \text{if } E \text{ is even,} \\ \sqrt{2} & \text{otherwise.} \end{cases}$$

Thus, the integer exponent is extracted from the argument and divided by two. If the exponent is odd, the result is multiplied by the constant $\sqrt{2}$. The mantissa in this representation is restricted to between 0.5 and 1.0.

In Figure 16.2(c), the number of iterations required is reduced from five to four by using a guess of 0.85, but a commonly used method is to use a seed as the guess, where the seed is essentially a closer value to the answer found in a lookup table as a function of the mantissa. An example of 32 seed values is shown in Figure 16.2(d), and the convergence is shown in Figure 16.2(e), where the number of iterations required has been reduced to three. The software for the Newton's method based square root calculation using range reduction and seeding is provided in Figure 16.2(a). One note about this software is that the exponent is usually taken directly from the floating point number, not by using a logarithm as shown, and the seed values are usually found using the upper nybble of the mantissa as an index into a lookup table.

The previous example provides an insight into how Newton's method can be used, but the primary goal here is to use Newton's method for model extraction. This generally leads to a multi-variable LMSE solution. It was seen that the solution of a LMSE problem is one where the variance of the residual error is minimized, which occurs when the derivative of the variance of the residual error is zero.

Thus, given a function of $M$ variables $a_m$ for $m \in 0 \ldots M-1$, and $K$ coordinates $\{x_k, y_k\}$ for $k \in 0 \ldots K-1$, there is a function $f(\mathbf{a}, \mathbf{x})$ where the goal is to find $\mathbf{a}$ such that the variance of the residual error is minimized in a LMSE sense, where the variance is provided in (16.2) and the partial derivatives of the variance are provided in (16.3) as in the linear solution case. Here, however, the equation cannot be put in the form $f(\mathbf{a}, \mathbf{x}) = \mathbf{X} \cdot \mathbf{a}$, and no linear solution exists.

[2]The division $E/2$ is integer division.

In this case, Newton's method is used on the partial derivatives of the variance, and, through iteration, the solution to the derivative of the variance with respect to the variables is found where it equals zero. Therefore, the equation is

$$F(\mathbf{a},\mathbf{x}) = \frac{1}{K} \cdot \frac{d}{d\mathbf{a}} \sum_k (f(\mathbf{a},x_k) - y_k)^2 = 2 \cdot \frac{1}{K} \cdot \sum_k (f(\mathbf{a},x_k) - y_k) \cdot \frac{d}{d\mathbf{a}} f(\mathbf{a},x_k),$$

and the solution is found when

$$\sum_k (f(\mathbf{a},x_k) - y_k) \cdot \frac{\partial}{\partial a_m} f(\mathbf{a},x_k) = 0.$$

To use Newton's method to solve this, we need the derivative of this function, which for $n \in 0 \ldots M-1$ is

$$\frac{\partial}{\partial a_n} \sum_k (f(\mathbf{a},x_k) - y_k) \cdot \frac{\partial}{\partial a_m} f(\mathbf{a},x_k)$$
$$= \sum_k \left[ (f(\mathbf{a},x_k) - y_k) \cdot \frac{\partial^2}{\partial a_n \partial a_m} f(\mathbf{a},x_k) + \frac{\partial}{\partial a_n} f(\mathbf{a},x_k) \cdot \frac{\partial}{\partial a_m} f(\mathbf{a},x_k) \right].$$

The first term is generally discarded because it vanishes as the residual error vanishes anyway. The matrix of the partial derivatives is called the Jacobian matrix, and is written as follows:

$$J_{k,a} = \frac{\partial}{\partial a_n} f(\mathbf{a},x_k).$$

The residual vector is written as

$$r_k = (f(\mathbf{a},x_k) - y_k),$$

and Newton's method calculation of $\Delta\mathbf{a}$ is written in matrix-vector form as

$$\Delta\mathbf{a} = \left(\mathbf{J}^T \cdot \mathbf{J}\right)^{-1} \cdot \mathbf{J}^T \cdot \mathbf{r}.$$

Therefore, for model extraction, one is provided with (presumably measured) data in $\mathbf{y}$ corresponding to known values in $\mathbf{x}$ and a guess at the variables in the model $\mathbf{a}$. Then, on each iteration, a vector $\Delta\mathbf{a}$ is calculated and subtracted from $\mathbf{a}$ until convergence is achieved.

A variation on this method is to use a weighted mean-squared error as the fitting criteria. This allows more or less importance to be given to any set of data. For each $k$, a weight is provided as $w_k$. The weights are typically normalized such that $\sum w_k = 1$, for easier interpretation of the weighted variance. A matrix $\mathbf{W}$ is formed with the weights on the diagonal, and the formula for $\Delta\mathbf{a}$ is written as

$$\Delta\mathbf{a} = \left(\mathbf{J}^T \cdot \mathbf{W} \cdot \mathbf{J}\right)^{-1} \cdot \mathbf{J}^T \cdot \mathbf{W} \cdot \mathbf{r}. \tag{16.4}$$

The weighted variance is expressed as

$$\sigma^2 = \frac{1}{K} \cdot \mathbf{r}^T \cdot \mathbf{W} \cdot \mathbf{r}.$$

An example of Newton's method in multiple variables for model extraction is not provided because it is not used in its raw form. Instead, the more dependable Levenberg–Marquardt algorithm provided in §16.3 is used.

## 16.3 The Levenberg–Marquardt Algorithm

One potential problem with Newton's method is that, despite its rapid convergence, it can also diverge in certain circumstances. When it is used for the calculation of certain functions within a computer, the algorithm is tested for convergence within the reduced range for all possible seed values, usually at the boundary of each seed. Thus, its use is very different from that of model parameter extraction in that, properly designed, it is predictable. When used for model fitting, the data provided are completely uncontrolled.

When Newton's method is used for model fitting, it is always possible to test the convergence. In other words, the variance before and after each $\Delta a$ is calculated can be measured and, in instances of increasing variance, the iteration step can be discarded to prevent the calculation from diverging. Even better would be simply to scale $\Delta a$. The Levenberg–Marquardt algorithm [73] accomplishes this in an elegant manner.

In Levenberg–Marquardt, a matrix is created such that

$$\mathbf{H} = \mathbf{J}^T \cdot \mathbf{W} \cdot \mathbf{J}.$$

Prior to inversion, this matrix is adjusted by adding a value $\lambda$ multiplied by the diagonal values of $\mathbf{H}$:

$$\mathbf{H} = \mathbf{H} + \lambda \cdot \mathrm{diag}(\mathbf{H}) .$$

Thus, on each iteration, $\mathbf{\Delta a}$ is calculated as

$$\Delta \mathbf{a} = \left(\mathbf{H} + \lambda \cdot \mathrm{diag}(\mathbf{H})\right)^{-1} \cdot \mathbf{J}^T \cdot \mathbf{W} \cdot \mathbf{r}. \tag{16.5}$$

On each iteration, a test is performed on the new variance to see if it is reduced. If it is, then $\lambda$ is scaled by dividing it by some factor. In other words, successful iterations decrease the value of $\lambda$. If the iteration is unsuccessful, then $\lambda$ is scaled by multiplying it by some factor. Therefore, unsuccessful iterations increase the value of $\lambda$. The rationale for this can be seen by examining how $\lambda$ affects things in the limit. In the limit where $\lambda$ is zero, (16.5) becomes Newton's method equation, (16.4). In the limit where $\lambda$ approaches $\infty$, the diagonal of $\mathbf{H}$ becomes very large and $\Delta\mathbf{a}$ becomes a smaller and smaller step in the direction of the gradient. Thus $\lambda$ steers the convergence behavior between the rapid convergence of Newton's method and the slow convergence of a gradient walk depending on the success of the algorithm. Iteration is generally stopped when $\lambda$ reaches either an extremely large or extremely small value.

## 16.4 Python Fitting Code

Listing 16.1 and Listing 16.2 show the remarkably compact code for performing fitting according to the Levenberg–Marquardt algorithm. The **LevMar** class is shown to derive from the class **CallBacker**, which handles callbacks during the fit operation. The optional

**Listing 16.1** Levenberg–Marquardt algorithm initialization, functions, and derivatives

```
class LevMar(CallBacker):
    def __init__(self,callback=None):
        self.m_lambda=1000
        self.m_lambdamin=1e-15
        self.m_lambdamax=1e9
        self.m_lambdaMultiplier = 2.
        self.m_epsilon = 1e-19
        self.ccm=FitConvergenceMgr()
        CallBacker.__init__(self,callback)
    def fF(self,a):
        raise SignalIntegrityExceptionFitter('fF must be overloaded')
    def fPartialFPartiala(self,a,m,Fa=None):
        aplusDeltaa = copy.copy(a)
        aplusDeltaa[m][0]=a[m][0]+self.m_epsilon
        if Fa is None:
            Fa = self.fF(a)
        dFa = ((matrix(self.fF(aplusDeltaa))-matrix(Fa))/self.m_epsilon).tolist()
        return dFa
    def fJ(self,a,Fa=None):
        if Fa is None:
            Fa=self.fF(a)
        M = len(a)
        R = len(Fa)
        J = zeros((R,M)).tolist()
        for m in range(M):
            pFpam=self.fPartialFPartiala(a,m,Fa)
            for r in range(R):
                J[r][m]=pFpam[r][0]
        return J
    def AdjustVariablesAfterIteration(self,a):
        return a
    def Initialize(self,a,y,w=None):
        self.m_a = copy.copy(a)
        self.m_y = copy.copy(y)
        if w is None:
            self.m_sumw=len(y)
            self.m_W=1.0
        else:
            self.m_W = Matrix.diag(w)
            self.m_sumw = 0
            for r in range(w.rows()):
                self.m_sumw = self.m_sumw + w[r][0]
        self.m_Fa = self.fF(self.m_a)
        self.m_r=(matrix(self.m_Fa)-matrix(self.m_y)).tolist()
        self.m_mse=math.sqrt((matrix(self.m_r).getH()*
            self.m_W*matrix(self.m_r)).tolist()[0][0].real/self.m_sumw)
        self.m_lambdaTracking = [self.m_lambda]
        self.m_mseTracking = [self.m_mse]
        self.m_J = None
        self.m_H = None
        self.m_D = None
        self.m_JHWr = None
        self.ccm=FitConvergenceMgr()
...
```

**Listing 16.2** Levenberg–Marquardt algorithm `Solve()` method

```
class LevMar(CallBacker):
...
    def Iterate(self):
        if self.m_Fa is None:
            self.m_Fa=self.fF(self.m_a)
        if self.m_r is None:
            self.m_r=(matrix(self.m_Fa)-matrix(self.m_y)).tolist()
        if self.m_J is None:
            self.m_J=self.fJ(self.m_a,self.m_Fa)
        if self.m_H is None:
            self.m_H=(matrix(self.m_J).getH()*self.m_W*
                      matrix(self.m_J)).tolist()
        if self.m_D is None:
            self.m_D=zeros((len(self.m_H),len(self.m_H[0]))).tolist()
            for r in range(len(self.m_D)):
                self.m_D[r][r]=self.m_H[r][r]
        if self.m_JHWr is None:
            self.m_JHWr = (matrix(self.m_J).getH()*
            self.m_W*matrix(self.m_r)).tolist()
        Deltaa=((matrix(self.m_H)+matrix(self.m_D)*
                self.m_lambda).getI()*matrix(self.m_JHWr)).tolist()
        newa=(matrix(self.m_a)-matrix(Deltaa)).tolist()
        newa=self.AdjustVariablesAfterIteration(newa)
        newFa = self.fF(newa)
        newr=(matrix(newFa)-matrix(self.m_y)).tolist()
        newmse=math.sqrt((matrix(newr).getH()*self.m_W*
            matrix(newr)).tolist()[0][0].real/self.m_sumw)
        if newmse < self.m_mse:
            self.m_mse = newmse
            self.m_a = newa
            self.m_lambda = max(self.m_lambda/
                self.m_lambdaMultiplier,self.m_lambdamin)
            self.m_Fa = newFa
            self.m_r = newr
            self.m_J = None
            self.m_H = None
            self.m_D = None
            self.m_JHWr = None
        else:
            self.m_lambda = min(self.m_lambda*
                self.m_lambdaMultiplier,self.m_lambdamax)
        self.ccm.IterationResults(self.m_mse, self.m_lambda)
        self.m_lambdaTracking.append(self.m_lambda)
        self.m_mseTracking.append(self.m_mse)
    def Solve(self):
        self.Iterate()
        self.m_lastMse=self.m_mse
        self.m_MseAcc=self.m_lastMse
        while self.ccm.Continue():
            self.CallBack(self.ccm._IterationsTaken)
            self.Iterate()
        return self
```

callback function is provided during initialization, shown on line 2 in Listing 16.1, where the aforementioned value of $\lambda$ is initialized along with the min, max, and multiplicative factor used on successful iterations. An epsilon value is initialized that is used for numerical derivatives.

To use the **LevMar** class, a derived class is created that contains at a minimum an `__init__()` function (that should call the `__init__()` function on **LevMar** and possibly the `Initialize()` function) and a function `fF()` that provides $f(a)$. Notice that the derived class's `fF()` function overloads the function in the **LevMar** class on line 10 of Listing 16.1, which raises an exception if not overloaded. The `Initialize()` function on line 32 of Listing 16.1 initializes the guess at the variables in $\mathbf{a}$, the target values $\mathbf{y}$, and an optional vector of weights $\mathbf{w}$, which initializes the weights matrix $\mathbf{W}$. The function `fF()` is called once to establish the starting residual error and starting mean-squared error. Note that the intermediate calculations of the function, the residuals, the Jacobian, etc., are stored in member variables to avoid duplicate calculation when possible.

After initialization, the model is fitted with a call to `Solve()`, on line 45 of Listing 16.2. This function calls `Iterate()` on line 3, calling the callback function, and testing convergence on each iteration. The convergence test is not shown, but uses the mean-squared error and the value of $\lambda$ along with changes in these values to determine when to stop the fit. The `Iterate()` function should be self-explanatory based on the discussion in previous sections but there are some noteworthy items. One can see that before calculating any value, a test is made for existence in the member variable, with the existence carefully controlled by writing it to `None` whenever it needs to be obsoleted, which occurs whenever an iteration succeeds. When an iteration succeeds, the mean-squared error, the new values, and the function evaluation and residual (which was calculated with the new values to determine success of the iteration) are retained, with the remaining member variables being deleted. Successful iterations multiply $\lambda$. When an iteration fails, $\lambda$ is reduced and all values are retained for the next iteration that will calculate only a new value of $(\mathbf{H} + \lambda \cdot \mathrm{diag}(\mathbf{H}))^{-1}$.

Note also that, on each iteration, `AdjustVariablesAfterIteration()` on line 30 of Listing 16.1 is called. This function as it stands makes no changes, but it allows for an overloaded function in a derived class to enforce constraints on the adjusted variables. Some examples of adjustments might be the enforcement of realness or of positive values.

If only `fF()` is overloaded in the derived class, the Jacobian function `fJ()` on line 19 of Listing 16.1 will calculate numerical derivatives using `fPartialFPartiala()` on line 12. In many cases, numerical derivatives work fine,[3] but if analytic derivatives are known or have been computed, then the Jacobian and/or partial derivatives function can be overloaded in the derived class, as will be seen in the transmission line model fitting example in §16.5.

## 16.5 Transmission Line Model Fitting

Given the s-parameters of a transmission line, either measured or simulated, the transmission line parameters can be extracted by fitting a model. The Levenberg–Marquardt

[3]The absolute accuracy of the derivative in Newton's method is not required as the derivative is an approximation of the derivative in the vicinity of the answer anyway and $\mathbf{\Delta a}$ is also proportional to the residual that vanishes as the correct answer is approached.

**Listing 16.3 RLGCFitter** class initialization and `fF()` method

```
class RLGCFitter(LevMar):
    def __init__(self,sp,guess,callback=None):
        self.m_sp=sp
        self.f=self.m_sp.m_f
        self.Z0=self.m_sp.m_Z0
        v=self.VectorizeSp(sp)
        LevMar.__init__(self,callback)
        LevMar.Initialize(self, [[g] for g in guess], v)
        self.ones=[1 for _ in self.f]
        self.dZdR=self.ones
        self.dZdRse=[math.sqrt(f) for f in self.f]
        self.p2f=[2.*math.pi*f for f in self.f]
        self.dZdL=[1j*p2f for p2f in self.p2f]
        self.zeros=[0 for _ in self.f]
        self.dZdG=self.zeros
        self.dZdC=self.zeros
        self.dZddf=self.zeros
        self.dYdR=self.zeros
        self.dYdRse=self.zeros
        self.dYdL=self.zeros
        self.dYdG=self.ones
    def fF(self,a):
        (R,L,G,C,Rse,df)=(a[0][0],a[1][0],a[2][0],a[3][0],a[4][0],a[5][0])
        self.Z=[R+Rse*math.sqrt(f)+1j*2.*math.pi*f*L for f in self.f]
        self.Y=[G+2.*math.pi*f*C*(1j+df) for f in self.f]
        self.gamma=[cmath.sqrt(z*y) for (z,y) in zip(self.Z,self.Y)]
        self.Zc=[cmath.sqrt(z/y) for (z,y) in zip(self.Z,self.Y)]
        self.rho=[(zc-self.Z0)/(zc+self.Z0) for zc in self.Zc]
        self.rho2=[r*r for r in self.rho]
        self.eg=[cmath.exp(-g) for g in self.gamma]
        self.e2g=[egx*egx for egx in self.eg]
        self.D=[1-r2*e2 for (e2,r2) in zip(self.e2g,self.rho2)]
        self.S11=[r*(1-e2)/d for (r,e2,d) in zip(self.rho,self.e2g,self.D)]
        self.S12=[(1-r2)*e/d for (r2,e,d) in zip(self.rho2,self.eg,self.D)]
        S=[[[s11,s12],[s12,s11]] for (s11,s12) in zip(self.S11,self.S12)]
        vS=self.VectorizeSp(S)
        return vS
...
```

algorithm is used to deal with the nonlinearity of the problem. All that is needed is the model function to fit to and the partial derivatives of the function with respect to each of the variables to be supplied. To deal with the complexity, the function is considered as a function of many functions and the chain rule employed in the computation of the partials.

A reasonable model for a transmission line is defined in (7.16), which is a symmetric and reciprocal model, meaning $s_{11} = s_{22}$ and $s_{12} = s_{21}$, and therefore there are two equations that completely define the transmission line:

$$S_{11} = \frac{\rho \cdot \left(1 - e^{-2\cdot\gamma}\right)}{1 - \rho^2 \cdot e^{-2\cdot\gamma}}, \qquad S_{12} = \frac{\left(1 - \rho^2\right) \cdot e^{-\gamma}}{1 - \rho^2 \cdot e^{-2\cdot\gamma}}.$$

To get $S_{11}$ and $S_{12}$ in terms of circuit parameters, understand that

$$\rho = \frac{Z_c - Z0}{Z_c + Z0}, \quad Z_c = \sqrt{\frac{Z}{Y}}, \quad \gamma = \sqrt{Z \cdot Y},$$
$$Z = R + R_{se} \cdot \sqrt{f} + j \cdot 2\pi \cdot f \cdot L, \quad Y = G + 2\pi \cdot f \cdot C \cdot (j + \tan\delta),$$

**Listing 16.4 RLGCFitter** class `fJ()` Jacobian method

```
class RLGCFitter(LevMar):
...
    def fJ(self,a,Fa=None):
        if self.m_Fa is None: self.m_Fa=self.fF(a)
        (R,L,G,C,Rse,df)=(a[0][0],a[1][0],a[2][0],a[3][0],a[4][0],a[5][0])
        dZ=[self.dZdR,self.dZdL,self.dZdG,self.dZdC,self.dZdRse,self.dZddf]
        dYdC=[p2f*(1j+df) for p2f in self.p2f]
        dYddf=[p2f*C for p2f in self.p2f]
        dY=[self.dYdR,self.dYdL,self.dYdG,dYdC,self.dYdRse,dYddf]
        dgamma=[[1./(2.*cmath.sqrt(z*y))*(dz*y+z*dy)
                for (z,y,dz,dy) in zip(self.Z,self.Y,dZ[i],dY[i])]
                    for i in range(6)]
        dZc=[[-1./2*(-dz*y+z*dy)/(y*y*cmath.sqrt(z/y))
                for (z,y,dz,dy) in zip(self.Z,self.Y,dZ[i],dY[i])]
                    for i in range(6)]
        drho=[[2.*dzc*self.Z0/((zc+self.Z0)*(zc+self.Z0))
                for (zc,dzc) in zip(self.Zc,dZc[i])] for i in range(6)]
        D2=[d*d for d in self.D]
        dS11=[[-2.*r*e2*(r2-1)/d2*dg+(1-e2)*(1+r2*e2)/d2*dr
                for (r,r2,e2,d2,dg,dr) in
                    zip(self.rho,self.rho2,self.e2g,D2,dgdx,drdx)]
                        for (dgdx,drdx) in zip(dgamma,drho)]
        dS12=[[e*(r2-1)*(1+r2*e2)/d2*dg-2.*r*e*(1-e2)/d2*dr
                for (r,r2,e,e2,d2,dg,dr) in
                    zip(self.rho,self.rho2,self.eg,self.e2g,D2,dgdx,drdx)]
                        for (dgdx,drdx) in zip(dgamma,drho)]
        dS=[[[[dS11[i][n],dS12[i][n]],[dS12[i][n],dS11[i][n]]]
                for n in range(len(self.f))] for i in range(6)]
        vdS=[self.VectorizeSp(ds) for ds in dS]
        return [[vdS[m][r][0] for m in range(len(a))] for r in range(len(Fa))]
...
```

where $R$ is the series resistance in $\Omega$, $R_{se}$ is the resistance due to the skin effect in $\Omega/\sqrt{\text{Hz}}$, $L$ is the inductance in H, $G$ is the shunt conductance in S, $C$ is the capacitance in F, and $\tan\delta$ is the loss tangent (or dissipation factor), which is unitless. These six variables define a transmission line.

At a given frequency $f$, there are functions for the s-parameters in $S_{11}$ and $S_{22}$ in terms of the variables and $f$:

$$
\begin{aligned}
&S_z(R, R_{se}, L, G, C, \tan\delta, f)\\
&= S_z(\rho(Z_c(Z(R, R_{se}, L, f), Y(G, C, \tan\delta, f)), Z0), \gamma(Z(R, R_{se}, L, f), Y(G, C, \tan\delta, f))).
\end{aligned}
$$

The code that executes the initialization and function calculation is shown in Listing 16.3, where the **RLGCFitter** class is shown. It derives from the general purpose fitter **LevMar**. The initialization calculates and computes a lot of the partial derivatives of the variables over frequency, which in many cases are constant, unity, or zero. The initialization function also calculates many other constants to avoid recalculation.

The function `fF()` overrides the function in the **LevMar** base class and computes the function by chaining together all of the aforementioned functions, making use of many of the precalculated constants and storing many of the intermediate results for help in calculating the Jacobian, which will be described shortly. The function `fF()` computes the s-parameters for each frequency and then vectorizes them. Thus, for each frequency, four s-parameters with two of them duplicated are all placed in one long vector.

To perform the fit, the Jacobian is needed, which is a function of the derivatives of the s-parameters with respect to each of the variables. For any variable denoted as $x$:

$$\frac{\partial}{\partial x}S_{11} = \frac{\partial}{\partial x}\frac{\rho \cdot \left(1-e^{-2\cdot\gamma}\right)}{1-\rho^2 \cdot e^{-2\cdot\gamma}} = \frac{\partial}{\partial \gamma}S_{11} \cdot \frac{\partial}{\partial x}\gamma + \frac{\partial}{\partial \rho}S_{11} \cdot \frac{\partial}{\partial x}\rho$$
$$= \frac{-2\cdot\rho\cdot e^{-2\cdot\gamma}\cdot\left(\rho^2-1\right)}{\left(1-\rho^2\cdot e^{-2\cdot\gamma}\right)^2} \cdot \frac{\partial}{\partial x}\gamma + \frac{\left(1-e^{-2\cdot\gamma}\right)\cdot\left(1+\rho^2\cdot e^{-2\cdot\gamma}\right)}{\left(1-\rho^2\cdot e^{-2\cdot\gamma}\right)^2} \cdot \frac{\partial}{\partial x}\rho$$

and

$$\frac{\partial}{\partial x}S_{12} = \frac{\partial}{\partial x}\frac{\left(1-\rho^2\right)\cdot e^{-\gamma}}{1-\rho^2 \cdot e^{-2\cdot\gamma}} = \frac{\partial}{\partial \gamma}S_{12} \cdot \frac{\partial}{\partial x}\gamma + \frac{\partial}{\partial \rho}S_{12} \cdot \frac{\partial}{\partial x}\rho$$
$$= \frac{e^{-\gamma}\cdot\left(\rho^2-1\right)\cdot\left(1+\rho^2\cdot e^{-2\cdot\gamma}\right)}{\left(1-\rho^2\cdot e^{-2\cdot\gamma}\right)^2} \cdot \frac{\partial}{\partial x}\gamma + \frac{-2\cdot\rho\cdot e^{-\gamma}\cdot\left(1-e^{-2\cdot\gamma}\right)}{\left(1-\rho^2\cdot e^{-2\cdot\gamma}\right)^2} \cdot \frac{\partial}{\partial x}\rho,$$

where

$$\frac{\partial}{\partial x}\rho = \frac{\partial}{\partial x}\frac{Z_c - Z0}{Z_c + Z0} = 2\cdot\frac{\partial}{\partial x}Z_c \cdot \frac{Z0}{\left(Z_c+Z0\right)^2},$$
$$\frac{\partial}{\partial x}Z_c = \frac{\partial}{\partial x}\sqrt{\frac{Z}{Y}} = -\frac{1}{2}\cdot\frac{Z\cdot\frac{\partial}{\partial x}Y - Y\cdot\frac{\partial}{\partial x}Z}{\sqrt{\frac{Z}{Y}}\cdot Y^2},$$
$$\frac{\partial}{\partial x}\gamma = \frac{\partial}{\partial x}\sqrt{Z\cdot Y} = \frac{1}{2\cdot\sqrt{Z\cdot Y}}\cdot\left(Z\cdot\frac{\partial}{\partial x}Y + Y\cdot\frac{\partial}{\partial x}Z\right).$$

Finally, the derivatives of $Z$ and $Y$ with respect to each of the variables can be listed:

$$\frac{\partial}{\partial R}Z = 1, \quad \frac{\partial}{\partial R_{se}}Z = \sqrt{f}, \quad \frac{\partial}{\partial L}Z = j\cdot 2\pi\cdot f,$$
$$\frac{\partial}{\partial G}Z = \frac{\partial}{\partial C}Z = \frac{\partial}{\partial \tan\delta}Z = 0, \quad \frac{\partial}{\partial C}Y = 2\pi\cdot f\cdot\left(j+\tan\delta\right), \quad \frac{\partial}{\partial G}Y = 1,$$
$$\frac{\partial}{\partial \tan\delta}Y = 2\pi\cdot f\cdot C, \quad \frac{\partial}{\partial R}Y = \frac{\partial}{\partial R_{se}}Y = \frac{\partial}{\partial L}Y = 0.$$

Thus, using the chain rule, there is the following general equation for the partial derivatives:

$$\frac{\partial}{\partial x}S_z = \frac{\partial}{\partial \rho}S_z \cdot \frac{\partial}{\partial Z_c}\rho \cdot \left(\frac{\partial}{\partial Z}Z_c \cdot \frac{\partial}{\partial x}Z + \frac{\partial}{\partial Y}Z_c \cdot \frac{\partial}{\partial x}Y\right) + \frac{\partial}{\partial \gamma}S_z \cdot \left(\frac{\partial}{\partial Z}\gamma \cdot \frac{\partial}{\partial x}Z + \frac{\partial}{\partial Y}\gamma \cdot \frac{\partial}{\partial x}Y\right),$$

where $S_z$ represents one of the equations $S_{11}$ or $S_{12}$ and $x$ represents one of the variables $R$, $R_{se}$, $L$, $G$, $C$, or $\tan\delta$.

For example,

$$\frac{\partial}{R}S_{11} = \frac{\partial}{\partial \rho}S_{11} \cdot \frac{\partial}{\partial Z_c}\rho \cdot \left(\frac{\partial}{\partial Z}Z_c \cdot \frac{\partial}{\partial R}Z + \frac{\partial}{\partial Y}Z_c \cdot \frac{\partial}{\partial R}Y\right)$$
$$+ \frac{\partial}{\partial \gamma}S_{11} \cdot \left(\frac{\partial}{\partial Z}\gamma \cdot \frac{\partial}{\partial R}Z + \frac{\partial}{\partial Y}\gamma \cdot \frac{\partial}{\partial R}Y\right),$$

and, because $Y$ is not a function of $R$,

$$\frac{\partial}{\partial R}S_{11} = \frac{\partial}{\partial \rho}S_{11} \cdot \frac{\partial}{\partial Z_c}\rho \cdot \frac{\partial}{\partial Z}Z_c \cdot \frac{\partial}{\partial R}Z + \frac{\partial}{\partial \gamma}S_{11} \cdot \frac{\partial}{\partial Z}\gamma \cdot \frac{\partial}{\partial R}Z,$$

which can be simplified as

$$\frac{\partial}{\partial R}S_{11} = \left(\frac{\partial}{\partial \rho}S_{11} \cdot \frac{\partial}{\partial Z_c}\rho \cdot \frac{\partial}{\partial Z}Z_c + \frac{\partial}{\partial \gamma}S_{11} \cdot \frac{\partial}{\partial Z}\gamma\right) \cdot \frac{\partial}{\partial R}Z.$$

This is not simplified further as it is calculated in steps as shown in the code that executes the computation of the Jacobian for the **RLGCFitter** class in Listing 16.4. The Jacobian is a matrix with the same number of rows as data elements, but the number of columns is equal to the number of variables, where each row $r$ in $\mathbf{J}$ is defined as

$$J_{r*} = \left( \frac{\partial}{\partial R}F \quad \frac{\partial}{\partial L}F \quad \frac{\partial}{\partial G}F \quad \frac{\partial}{\partial C}F \quad \frac{\partial}{\partial R_{se}}F \quad \frac{\partial}{\partial \tan\delta}F \right).$$

While the `fJ()` algorithm in Listing 16.4 does not claim to be the most optimal, it reuses a lot of the initialized constants and many of the intermediate calculations of `fF()` in Listing 16.3.

### 16.5.1 RLGC Fit Example

A Python function is provided in Listing 16.3(a) for fitting an RLGC transmission line model. Measured s-parameters of a transmission line are supplied as an argument to the function. Prior to fitting, a guess is determined. The elements of the guess consist of the delay through the line, determined by calculating the 50 % threshold crossing of the step response (2.25 ns), and the impedance determined from the calculation of the reflection coefficient in the middle of the line (≈0.01). The line is assumed to be well behaved. From the reflection coefficient, the characteristic impedance is estimated as

$$Z_c = Z0 \cdot \frac{1+\rho}{1-\rho} = 50.92\,\Omega,$$

and the guess at the inductance and capacitance is calculated according to[4]

$$L = T_d \cdot Z0 = 114.74\,\text{nH}, \quad C = T_d/Z0 = 44.26\,\text{pF}.$$

The guess contains only the inductance and capacitance, with the resistance, conductance, skin-effect resistance, and dissipation factor assumed to be zero.

Figure 16.3(b) and Figure 16.3(c) show plots of the mean-squared error and $\lambda$ calculated for each iteration. It is typical to supply zero mean-squared error as the target of the fit and allow convergence to be determined by the behavior of the mean-squared error and $\lambda$; $\lambda$ will often move to its min or max value at the convergence point, but here convergence is determined by either the mean-squared error or by $\lambda$ not changing a lot. The changes in

[4] $Z_c$ should be used instead of $Z0$ for the guess at inductance and capacitance when the characteristic impedance is not close to 50 Ω.

```
def TlineFit(sp):
    stepResponse=sp.FrequencyResponse(2,1).ImpulseResponse().Integral()
    threshold=(stepResponse[len(stepResponse)-1]+stepResponse[0])/2.0
    for k in range(len(stepResponse)):
        if stepResponse[k]>threshold: break
    dly=stepResponse.Times()[k]
    rho=sp.FrequencyResponse(1,1).ImpulseResponse().Integral(scale=False).Measure(dly)
    Z0=sp.m_Z0*(1.+rho)/(1.-rho)
    L=dly*Z0; C=dly/Z0; guess=[0.,L,0.,C,0.,0.]
    (R,L,G,C,Rse,df)=[r[0] for r in si.fit.RLGCFitter(sp,guess).Solve().Results()]
    return si.sp.dev.TLineTwoPortRLGC(sp.f(),R,Rse,L,G,C,df,sp.m_Z0)
```

(a) Python code

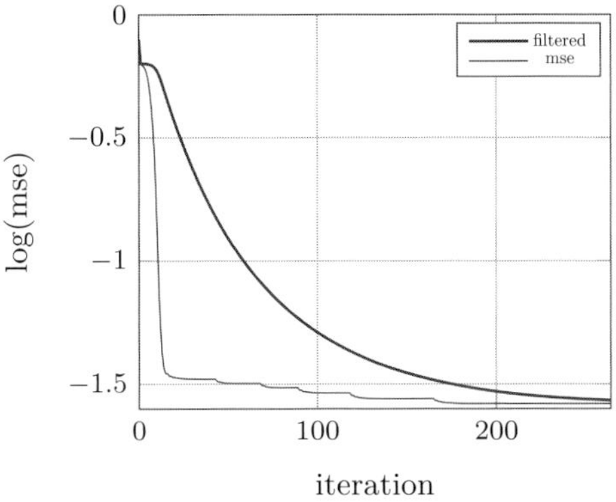


(b) Mean-squared error

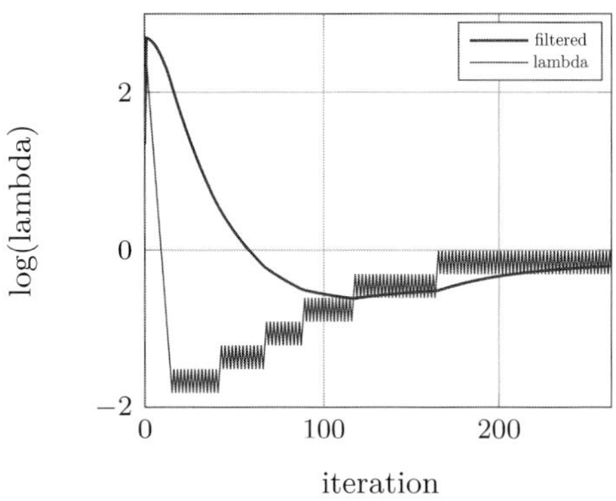


(c) $\lambda$

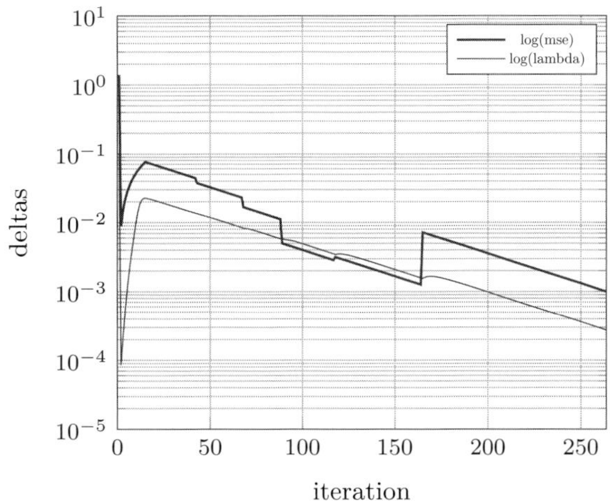


(d) Changes in mean-squared error and $\lambda$

**Figure 16.3** Fit convergence

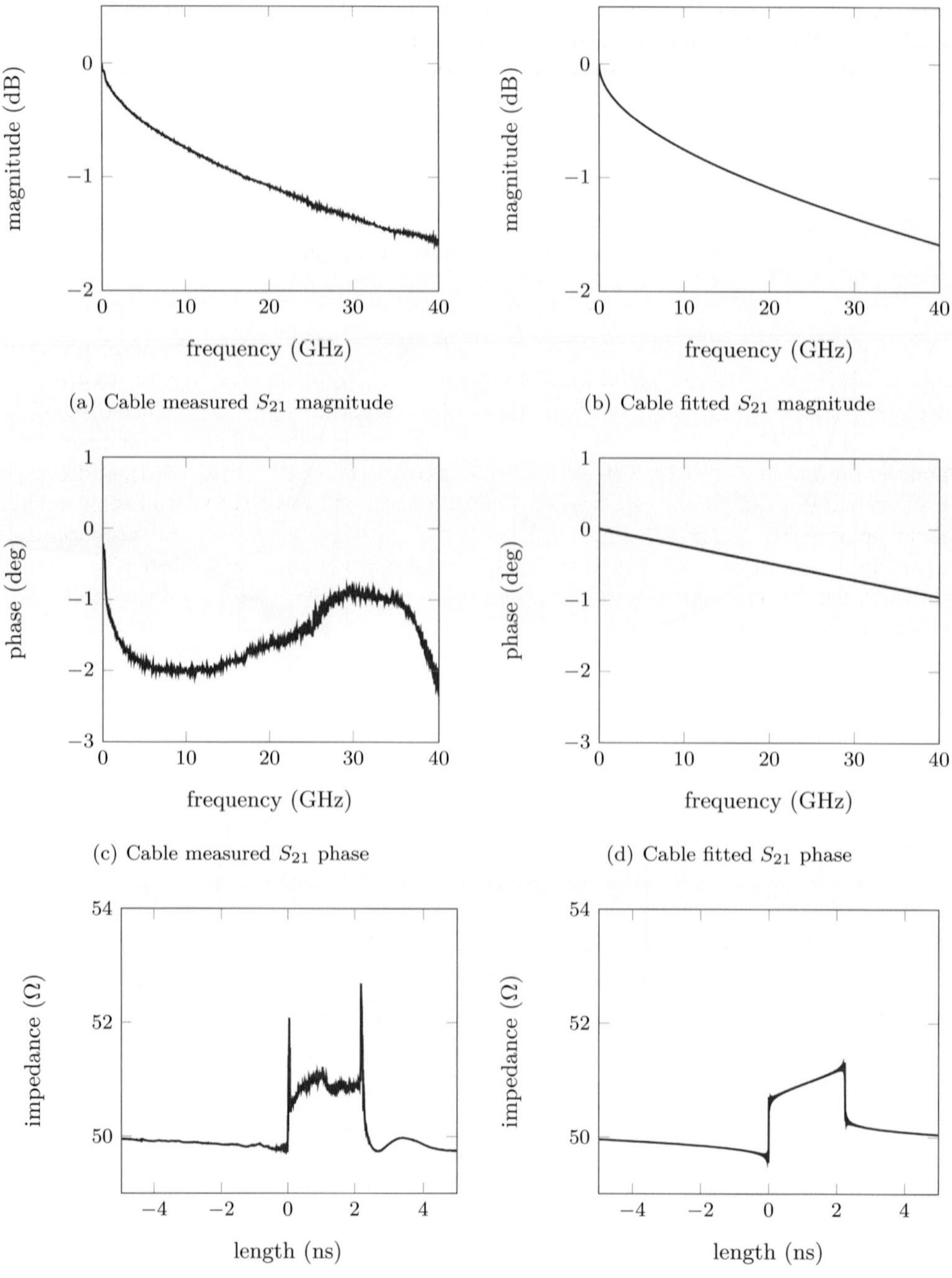


(a) Cable measured $S_{21}$ magnitude

(b) Cable fitted $S_{21}$ magnitude

(c) Cable measured $S_{21}$ phase

(d) Cable fitted $S_{21}$ phase

(e) Cable measured impedance profile

(f) Cable fitted impedance profile

**Figure 16.4** Cable nonlinear fit

mean-squared error and $\lambda$ are shown in Figure 16.3(d). The fit is terminated at iteration 264, when the filtered change in mean-squared error reaches $10^{-3}$.

The result of the fit is a transmission line with the following characteristics:

- resistance, $R = 1.045\,\Omega$;
- inductance, $L = 114.41\,\text{nH}$;
- conductance, $G = 260.57\,\mu\text{S}$;
- capacitance, $C = 44.13\,\text{pF}$;
- skin-effect resistance, $R_{se} = 48.56\,\mu\Omega/\sqrt{\text{Hz}}$;
- dissipation factor, $df = 0.00037$.

A comparison is provided in Figure 16.4. The original magnitude response, phase response (with 2.25 ns removed), and impedance are provided in Figures 16.4(a), 16.4(c), and 16.4(e), where a bend in the phase response is observed along with impedance discontinuities at the ends of the line due to the launch. The corresponding magnitude response, phase response, and impedance of the fitted line are shown in Figures 16.4(b), 16.4(d), and 16.4(f). The slight bend in the phase response is not accommodated by the model and some acausal behavior is observed in the impedance plot. Part of this may be due to the assumption of constant inductance, which is not completely accurate due to the skin effect [74]. It is up to the engineer to verify the accuracy of the measured data, the validity of the model, and whether the fit is good.

# Part IV

## *SignalIntegrity*

## Introduction

SOFTWARE TOOLS relieve engineers from dealing with complexity and speed up calculations. These tools are not meant to relieve engineers from understanding. One might be lucky, perhaps even for most of the time, in obtaining good results from simulations without understanding the underlying theory, but there are several dangers to this approach. It is important to develop proper expectations before attempting to simulate something for which the answer is unknown; otherwise, there is the danger of being fooled by the vagaries of simulation tools.

In principle, one should not only know the theory, but also understand the simulation tools themselves and the theory behind the simulation software itself. Unfortunately, much of the software for signal integrity is quite complex, meaning that someone spent a lot of money developing it – and they want their money back. This means that the software can be very expensive, and although the documentation might be included, the source code certainly will not be.

Ideally, one should be able to see the source code, get any problems fixed if they arise, change the software to their needs, and extend the functionality. Most readers would balk at the concept of modifying their simulation software, especially if they generally work with hardware rather than software. Regardless, this concept is what is behind the GNU project and the Free Software Foundation, and it is the general concept of open-source software. There are many people who participate in open-source software and many examples of great things that result from its use, but a list and discussion of these things would take up too much space.

In signal integrity, there are many tasks to be performed by software tools, and many of these tasks are relatively simple. Examples of these tasks have been presented in this book, such as the generation of the s-parameters of a network of connected devices, de-embedding, and linear simulation. Not only has all of the theory behind the execution of these tasks been presented in this book, but also the software has been provided. Finally, an open-source project exists with this software that is freely available to use, examine, modify, and fix, if necessary. Importantly, it can be understood, and understanding is the subject of this part of the book.

## The *SignalIntegrity* Project

*SignalIntegrity* is an open-source, Python based software project. At the time of writing, it is publicly hosted on GitHub,[1] a software hosting repository based on the popular open-source version control software *Git*. GitHub can be accessed at `https://github.com/` and Git can be found at `https://git-scm.com/`

The *SignalIntegrity* software, the theory and source code of which have been described in this book, can be found at:

```
https://github.com/TeledyneLeCroy/SignalIntegrity
```

GitHub is used for software development, and the intent is that the *SignalIntegrity* software will be added to by anyone interested in doing so.

The software is also hosted on the Python Packaging Index (PyPI) and can be installed on any system with Python installed and an internet connection by simply typing the following on the command line:

```
pip install SignalIntegrity
```

At PyPI, the software can be found at:

```
https://pypi.org/project/SignalIntegrity/
```

## License

*SignalIntegrity* is licensed under the GNU general public license (GPL), version 3. This is a license used by more than half of all free software packages. It is a "copyleft" license, meaning that the software is copyrighted, but it and any modifications to the software remain free.

While many of the algorithms presented in this book and in the software are patented, the intent is that these patents may be practiced, as long as they are practiced with this software. In the process of using this software, the license must be adhered to.

[1] GitHub® is a trademark of Microsoft Corporation.

The official text of the license is as follows:[2]

> Copyright (c) 2018 Teledyne LeCroy, Inc.
>
> All rights reserved worldwide.
>
> *SignalIntegrity* is free software: You can redistribute it and/or modify it under the terms of the GNU General Public License as published by the Free Software Foundation, either version 3 of the License, or any later version.
>
> This program is distributed in the hope that it will be useful, but WITHOUT ANY WARRANTY; without even the implied warranty of MERCHANTABILITY or FITNESS FOR A PARTICULAR PURPOSE. See the GNU General Public License for more details.
>
> You should have received a copy of the GNU General Public License along with this program. If not, see <`https://www.gnu.org/licenses/`>

The intent of the license is that the software is free, can be used and modified for any purpose, but any derivative works produced that are distributed must also adhere to this license; it must also be free.

## Project Organization

*SignalIntegrity* is broken into two separate and distinct parts, whose use depends on the problem solution:

- `SignalIntegrity.Lib` – the Python package containing functions and classes for scripted signal integrity solutions. This is covered in Chapter 17, **SignalIntegrity.Lib** Package.
- `SignalIntegrity.App` – the Python package containing *SignalIntegrityApp*, a GUI based Python application that allows for schematic entry and problem solving. *SignalIntegrityApp* is built separate from and on top of `SignalIntegrity.Lib` and is discussed in Chapter 18, *SignalIntegrityApp*.

The listings provided in this book and the functionality described are for version 1.1.6, released in November, 2019.

[2] It must be pointed out that the license for the software is subject to change. The license published with the software must be consulted and of course supersedes anything written here.

# 17

# SignalIntegrity.Lib Package

THE `SignalIntegrity.Lib` package is the underlying library in the *SignalIntegrity* software. Interestingly, it was begun about halfway through the writing of this book, when it became apparent that the theory being presented here would go unused unless it were elevated to some higher level of capability.

The `SignalIntegrity.Lib` package is extensively tested using the popular Python tool *unittest*. Unittest allows for test generation and provides a front end for running the tests, seeing which tests failed, etc. Currently there are about 1000 small tests that can test the entire software in about ten minutes. Using another popular Python tool called *coverage* enables the analysis of test code coverage, which is important in an interpreted language like Python. Unlike for compiled languages, one doesn't know for sure what will happen unless every line of code is executed. Coverage generates a .html report that shows about 98 % code coverage (the code that is not covered comprises untested small features). The goal is to eventually get 100 % code coverage either by removing the untested features or by adding tests. Unittest is run whenever changes are made to the software prior to checking code into version control. And mostly, as features are added, the tests are added right along with the features to test the operation as the feature is being developed. Many of the tests either write or test portions of this book in an effort to eliminate typographic errors. The reader will be the judge of the success of that effort.

A Python library is generally broken into a *package* hierarchy. This hierarchy can already be seen in that the package library is referred to as `SignalIntegrity.Lib`, the library level of *SignalIntegrity*. This package organization is described in §17.1. Finally, UML diagrams are provided for the common use cases of the software in §17.2.

## 17.1 Package Organization

The Python `SignalIntegrity.Lib` package is organized as Python *packages* and subpackages. These packages organize the software into *namespaces*. Namespaces in Python define how one accesses the function or object within the package.

The package and package namespaces are shown in Table 17.1. Most of the elements of these packages have been described theoretically throughout the rest of this book:

- `ChirpZTransform`: discussed in §12.5.2, The Chirp Z-Transform.
- `Conversions`: discussed in §3.4, S-Parameter Conversions.
- `Devices`: many have been discussed throughout the book in either Chapter 3, Scattering Parameters (for the passive devices) or Chapter 6, Sources (for the active devices).

**Table 17.1** `SignalIntegrity.Lib` package

| Namespace | Abbreviation | Contents |
|---|---|---|
| `ChirpZTransform` | `si.czt` | chirp z-transform |
| `Conversions` | `si.cvt` | network parameter and reference impedance conversion functions |
| `Devices` | `si.dev` | s-parameter device functions (single frequency or frequency independent) |
| `Fit` | `si.fit` | model extraction and fitting classes |
| `FrequencyDomain` | `si.fd` | frequency-domain classes |
| `Helpers` | `si.helper` | miscellaneous helper functions |
| `ImpedanceProfile` | `si.ip` | impedance profile |
| `Measurement` | `si.m` | measurement |
| `...Calibration` | `si.m.cal` | SOLT calibration |
| `...CalKit` | `si.m.calkit` | calibration kits and constants |
| `......Standards` | `si.m.calkit.std` | calibration standards |
| `...TDR` | `si.m.tdr` | TDR measurements |
| `Parsers` | `si.p` | netlist parser classes |
| `...Devices` | `si.p.dev` | netlist parser for devices |
| `Rat` | `si.rat` | rat function |
| `SParameters` | `si.sp` | s-parameters |
| `...Devices` | `si.sp.dev` | s-parameters of devices |
| `Splines` | `si.spl` | splines |
| `Subcircuits` | `si.sub` | subcircuits |
| `Symbolic` | `si.sy` | symbolic class |
| `SystemDescriptions` | `si.sd` | system descriptions |
| `Test` | `si.test` | test helpers |
| `TimeDomain` | `si.td` | time-domain classes |
| `...Filters` | `si.td.f` | filters and waveform processors |
| `...Waveform` | `si.td.wf` | waveform classes |
| `Wavelets` | `si.wl` | wavelets |

- `Fit`: discussed in Chapter 16, Model Extraction.
- `FrequencyDomain`: discussed in Chapter 12, Frequency Responses, Impulse Responses, and Convolution.
- `Helpers`: these are very generic and miscellaneous functions and classes.
- `ImpedanceProfile`: discussed in Chapter 14, The Impedance Profile.
- `Measurement`: discussed in Chapter 15, Measurement.
    - `Calibration`: discussed in §15.2, Calibration.
        - `CalKit`: discussed in §15.5, Calibration Standards.
            - `Standards`: discussed in §15.5, Calibration Standards.
    - `TDR`: discussed in §15.6, Time-Domain Reflectometry.
- `Parsers`: discussed in §8.4, The System Description Parser.
    - `Devices`: discussed in §8.5, Numeric Solutions.
- `Rat`: discussed in §12.5, Resampling.
- `SParameters`: discussed in §3.12, S-Parameter File Format.
    - `Devices`: discussed throughout Part I, Scattering Parameters.
- `Splines`: used for resampling; the details are not discussed here.
- `Subcircuits`: discussed in §8.6, Subcircuits.
- `Symbolic`: discussed in §8.3, Symbolics.
- `SystemDescriptions`: discussed extensively in Part II, Applications.
- `Test`: these are helper functions for unit testing.
- `TimeDomain`: discussed in Chapter 13, Waveforms and Filters.
    - `Filters`: discussed in Chapter 13, Waveforms and Filters.
    - `Waveform`: discussed in Chapter 13, Waveforms and Filters.
- `Wavelets`: discussed in §15.7.5, Wavelet Denoising.

The indented namespaces in Table 17.1 mean that they are referred to by their upper level namespace, a dot, and the indented namespace. Thus, where the top-level is `SignalIntegrity.Lib`, `SignalIntegrity.Lib.TimeDomain` refers to the `TimeDomain` package, and `SignalIntegrity.Lib.TimeDomain.Waveform` refers to the `Waveform` package. When the classes and functions are accessed from within Python, the following line is placed in the software:

```
import SignalIntegrity.Lib as si
```

This line says to make `si` the top level of the `SignalIntegrity.Lib` package. In Table 17.1, an abbreviation is seen for each package namespace. This is how it would be referred to in the software after the package has been imported. For example, to instantiate the class **SystemDescription** from within the `SignalIntegrity.Lib.SystemDescriptions` package, one would write `si.sd.SystemDescription()`.

There is complete online documentation for `SignalIntegrity.Lib` at the GitHub site. Here is an outline of where one would find commonly needed functions:

- To read an s-parameter file, one would instantiate `si.sp.SParameterFile()`. This is provided in Listing 3.15.
- S-parameters are instances of the class `si.sp.SParameters`. This is provided in Listing 3.16.
- S-parameters of devices that are frequency dependent can be found in the package `si.sp.dev`, which contains classes for various devices that are all derived from the `si.sp.SParameters` class. S-parameters of devices that are not frequency dependent are found in the package `si.dev`. These are functions that return a single matrix.
- All of the solutions that utilize the manual construction of system descriptions are found in the `si.sd` package. The class hierarchy for these solutions is described in §17.3.2. These produce results (transfer matrices or s-parameters) that are for a single frequency and are utilized by all of the parser solutions in the `si.p` package. These include:
  - `si.sd.SParametersNumeric()` shown in Listing 8.10;
  - `si.sd.SimulatorNumeric()` shown in Listing 9.5;
  - `si.sd.DeembedderNumeric()` shown in Listing 10.4;
  - `si.sd.VirtualProbeNumeric()` shown in Listing 11.4.
- All of the solutions that utilize a netlist are found in the `si.p` package. The class hierarchy for these solutions is described in §17.3.4. These include:
  - `si.p.SParametersNumericParser()` shown in Listing 8.11;
  - `si.p.SimulatorNumericParser()` shown in Listing 9.6;
  - `si.p.DeembedderNumericParser()` shown in Listing 10.5;
  - `si.p.VirtualProbeNumericParser()` shown in Listing 11.5.
- The transfer matrices produced in the simulation and virtual probing solutions are instances of the class `si.fd.TransferMatrices` (see Listing 9.7) found in the `si.fd` package, which contains all of the frequency-domain views of things.
- Waveforms are found in the package `si.td.wf`. All waveforms derive from the base **Waveform** class `si.td.wf.Waveform` (see Listing 13.4), and is used to read and write waveform files (see Listing 13.5).
- The `si.td.f` package contains the class `si.td.f.TransferMatricesProcessor` (see Figure 13.9(b)), which is supplied with an instance of `si.fd.TransferMatrices` and processes lists of instances of `si.td.wf.Waveform` to produce outputs that are also lists of instances of `si.td.wf.Waveform`. The interaction between waveforms, filters, transfer matrices, and transfer matrices processors is shown in the class hierarchy in §17.4.

This list so far covers most of the needs for processing outlined in Part II, Applications. For other common things:

- The `si.cvt` package contains all of the s-parameter conversion routines covered in Chapter 3.
- The `si.fit` package contains the model fitting classes covered in Chapter 16.

- The `si.ip` package contains the impedance profile classes covered in Chapter 14.
- The `si.m` package contains the ability to perform calibrated measurements covered in Chapter 15.

This covers the `SignalIntegrity.Lib` application programming interface (API) used to solve the application problems.

## 17.2 Universal Modeling Language Diagrams

Python is an object-oriented language, and the *SignalIntegrity* package is written in a manner to take advantage of this aspect. Therefore, much of the software is implemented using classes and objects. A class performs two functions:

- it defines a collection of data;
- it defines a connection between data held internally and member functions.

Usually, the definition is referred to as the *class* and an actual object of that type of class is referred to as an *instance of the class*. In this way, a filter class might define the storage of the internal filter taps and other things, along with methods for using these taps to perform filtering; a filter object would be an instance of this class. Many objects might be instances of one class. Sometimes the terminology is used loosely and an object is referred to as a class, but what is meant, strictly speaking, is that an object is an instance of that class.

In object-oriented programming, classes are made to work together, and, in order to understand the organization of object-oriented software, a type of diagram was created called a universal modeling language (UML) diagram. There are variations on how these diagrams are drawn. Here is how they are drawn in this text:

1. Classes are drawn as a box with at least the class name declared in it. If necessary, the box is broken into three sections, with the top section containing the class name, the next section containing the internal variables, and the final section containing the class methods. Class methods and variables are shown prefixed with a "+", and variables names are followed by a ":" with their type. One of the main tenets of object-oriented programming is *data hiding*. This means that often direct access to internal data and methods is restricted and hidden from users. This means that an external user typically does not need to worry about the internal workings. Unfortunately, the Python developers mostly tossed out this aspect, so data hiding is performed manually and is indicated usually by naming hidden variables and methods by prefixing the name with an underscore. These members are denoted as *private* with the generally available members declared as *public*. Often, only the public members will be shown.
2. Classes are generally shown inside a boxed region called the package or namespace. This corresponds to the package organization shown in Table 17.1.
3. Class relationships are depicted with certain lines and arrows connecting them. The definition of these lines and arrows are as follows:
   a) A solid line with an open arrow at the end indicates class *inheritance*. The class at the end of the arrow is called a base class and the object at the tail of the

arrow, a *derived* class. For brevity, one says that a derived class *is a* base class, as it inherits all of its functionality and data.

b) A solid line with a vee arrow at the end and a diamond arrow at the tail indicates containment. The class at the tail of the arrow where the diamond is located is called a *container* class. For brevity, one says that the container *has a* (instance of the) class pointed to. Sometimes, it has many such contained classes, and an indication might be made at the diamond showing how the class is contained. A "[]" is used to indicate a list or array of such classes, and a "[][]" to indicate a list of lists or a matrix of such classes.

c) A dashed line with a vee arrow at the end indicates class usage. The class at the tail of the arrow *uses* the class pointed at to do something.

d) (Non-standard) A dotted line with no arrows indicates that the classes are related in a special way. In the *SignalIntegrity* software, usually this means that the classes at each end are fundamentally representations of the same thing, but in a different abstract form. Usually classes related in this way will have member functions on each class to produce instances of the other class.

In the Python coding convention used, member variables will sometimes be preceded by "m_" and member properties will be preceded by "p". Member properties in Python are simply ways of accessing member variables through member access functions. Member functions utilize a naming convention called *CamelCase*. This means that the names are created by words joined without spaces but each word is capitalized. If a member function is not intended to be called externally (i.e. is private or protected in C++ language), then the name is preceded by the underscore "_".

## 17.3 *SignalIntegrity* Applications

Part II, Applications, covered the solution of the four main problems addressed in the book. These are:

1. s-parameter system solution;
2. de-embedding;
3. virtual probing;
4. simulation.

These solutions are accomplished in both numerical and symbolic ways. Symbolic solutions produce LaTeX code. All of these solutions built upon the concept of a system description, and all of the theory and class listings were shown, but because of the complicated class hierarchy it is helpful to see the relationship between these classes.[1]

### 17.3.1 System Descriptions

Figure 17.1 shows a UML diagram containing the classes discussed in Chapter 8. The classes in the diagram are shown inside large blocks that denote their location within a

[1] The seemingly over-complicated class hierarchy derives from the commonality in the solutions, and therefore code can be reused, but also from the need to fit class listings on a page. This serves the purpose of explaining these solutions in manageable chunks.

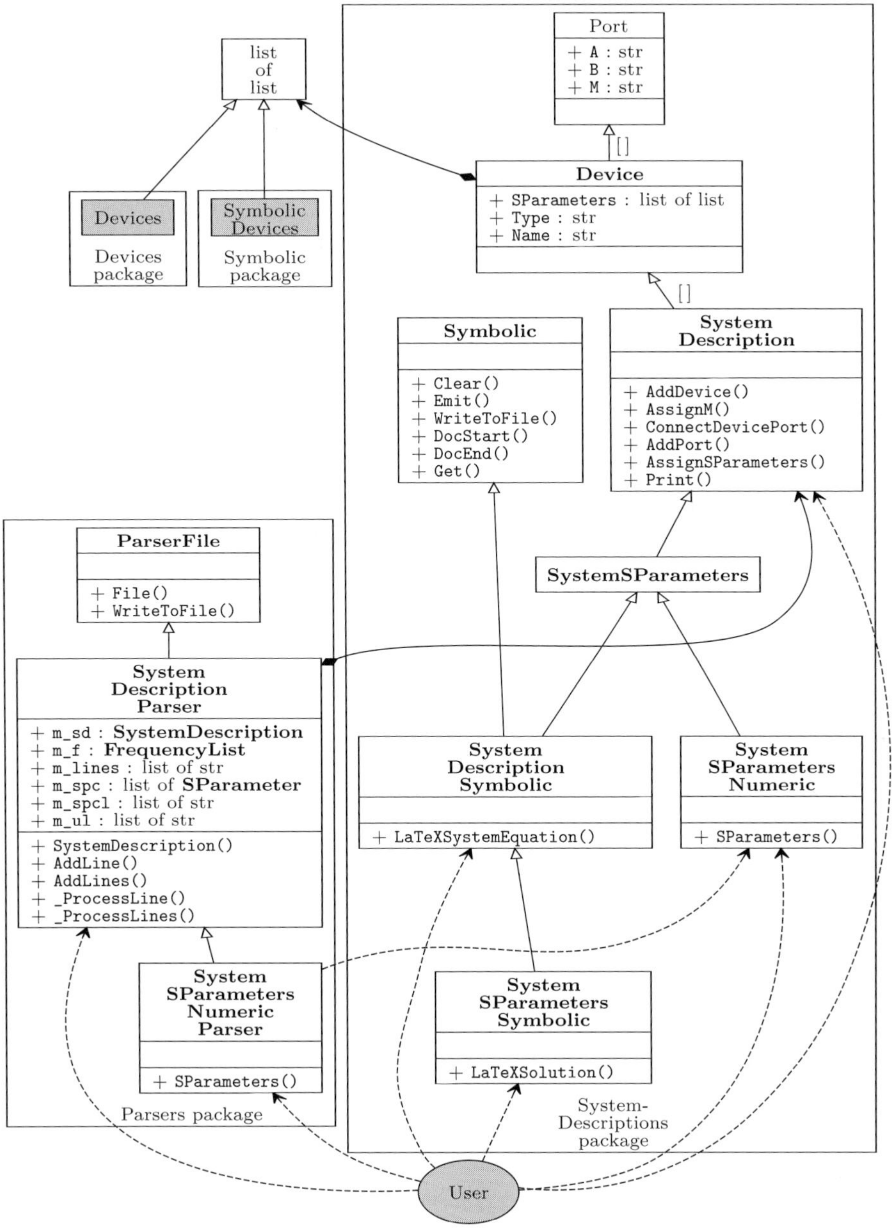


**Figure 17.1** System description and s-parameters solutions UML diagram

given *package.* Once the `SignalIntegrity.Lib` package is imported as `si` (at the top of all scripted examples[2]), the packages are accessed using dots and a package abbreviation. These packages and abbreviations are listed in Table 17.1. The two main packages shown are the `SystemDescriptions` package (accessed as `si.sd`) and the `Parsers` package (accessed as `si.p`). The user is in a bubble at the bottom and serves as a scripted user that instantiates and drives these classes. For various solution types, the user will instantiate the following classes:

1. For basic, manual system description construction, the **SystemDescription** class is instantiated and the system description is assembled using the member functions `AddDevice()`, `ConnectDevicePort()`, and `AddPort()`.
2. For manual symbolic system descriptions (the system equation only), the **SystemDescriptionSymbolic** class is instantiated and the system description is assembled using the member functions listed in item 1. The symbolic s-parameters are assigned through calls to `AssignSParameters()` with a list of list of LaTeX strings representing the s-parameter matrix, either generated by the user or by using devices in the `Symbolic` package (accessed as `si.sym`). The result is obtained by calling `LaTeXSystemEquation().Emit()`.
3. For manual system description construction for numeric s-parameters at a single frequency, the **SystemSParametersNumeric** class is instantiated and the system description is assembled using the member functions listed in item 1. The s-parameters are assigned through calls to `AssignSParameters()` with a list of list of complex numbers representing the s-parameter matrix either generated by the user, or by using devices in the `Devices` package (accessed as `si.dev`). The result is obtained by calling `SParameters()`.
4. For netlist generated construction of complete s-parameter solutions, the **SystemSParametersNumericParser** class is instantiated and the netlist is either read from a file with a call to `File()`, or the netlist lines are added with a call to `AddLines()` with a list of netlist lines. The s-parameter result is obtained by calling `SParameters()`.
5. For netlist generated construction of symbolic solutions, the **SystemDescriptionParser** class is instantiated and the netlist is either read from a file with a call to `File()`, or the netlist lines are added with a call to `AddLines()` with a list of netlist lines. In this case, the netlist should not contain any devices specified in device declarations, except for simple, frequency independent devices like ground, open, etc. The **SystemDescription** instance is obtained by calling `SystemDescription()` and supplied during construction to the **SystemSParametersSymbolic** class. The symbolic s-parameters are assigned manually prior to extracting the solution with a call to `LaTeXSolution().Emit()`.

### 17.3.2 Numerical System Description Solutions

Figure 17.2 shows the relationship between the classes in the `SystemDescriptions` package, with only those involved in numerical solutions shown. The user of these classes need only instantiate the classes pointed to by the user bubble at the bottom. These classes are:

[2]Sometimes this import is excluded in listings elsewhere, but is implied.

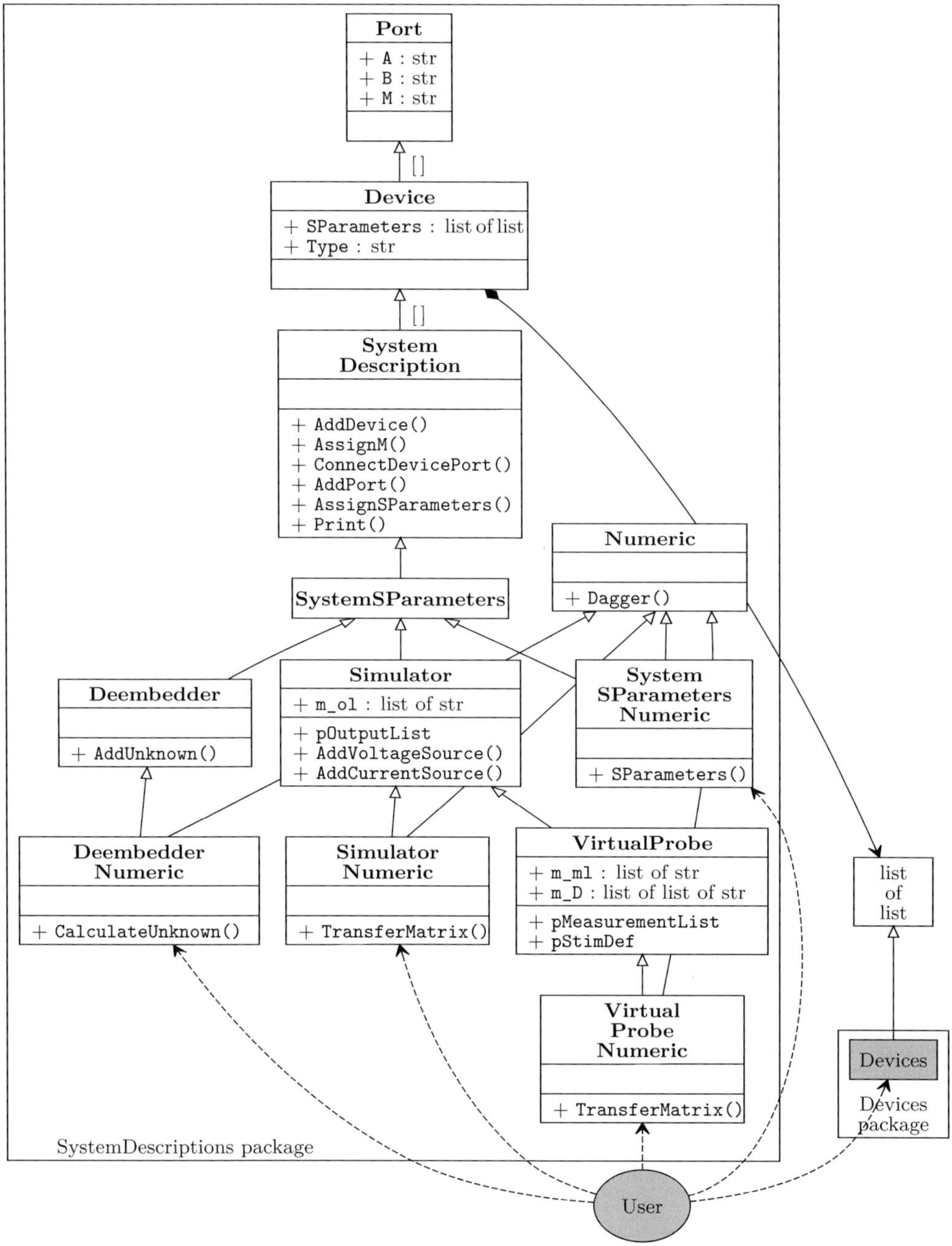


**Figure 17.2** System description numeric UML diagram

1. **SystemSParametersNumeric** – to solve systems of s-parameters.
2. **DeembedderNumeric** – to solve de-embedding problems.
3. **VirtualProbeNumeric** – to solve virtual probing problems.
4. **SimulatorNumeric** – to solve simulation problems.
5. **Devices** – these are actually functions in the `Devices` package that return lists of lists of complex numbers (i.e. a complex matrix) for various s-parameter devices.

Since all of the classes shown in Figure 17.2 inherit from the classes above them that are pointed to by the arrows, they all have the *is a* relationship, and therefore all of the derived classes at the bottom of the class hierarchy have all of the capability of the classes pointed to above them in the chain. For example, **DeembedderNumeric** is a **Deembedder** is a **SystemSParameters** is a **SystemDescription** is a list of **Device** is a list of **Port**. Therefore, looking downwards from the **SystemDescription** there is the **SystemSParameters** class from which everything else derives.

The **SystemDescription** class contains all of the functions to build a netlist. To solve a system of s-parameters, one instantiates the class **SystemSParametersNumeric**, uses the functions in **SystemDescription** to add devices to the system, add ports to the system, connect device ports together, and assign s-parameters to the devices,[3] and then calls the `SParameters()` method to solve for and retrieve the s-parameters. Prior to solving, the s-parameters of the devices added are supplied as lists of lists that are either computed by the user, or computed by calls to the functions in the `Devices` package that return these device matrices.

For de-embedding problems, the **Deembedder** class adds an additional method for adding an unknown device, and the **DeembedderNumeric** class adds a different calculation method to calculate the unknown device. To solve a de-embedding problem, one instantiates the class **DeembedderNumeric**, uses the functions in **SystemDescription** to add devices to the system, add ports to the system, connect device ports together, and assign s-parameters to the devices, and uses the added member in **Deembedder** to add unknown devices, and finally calls the `CalculateUnknown()` method to solve for and retrieve the s-parameters.

In the **Simulator** class there are methods for adding voltage and current sources to the system, along with a property[4] for accessing the output waveform list, held in `m_ol`. To solve a simulation problem, one instantiates the class **SimulatorNumeric**, uses the functions in **SystemDescription** to add devices to the system, connect device ports together, and assign s-parameters to the devices, and uses the added members in **Simulator** to add sources to the system. The list of device names and port numbers where output waveforms are to be produced is supplied to `pOutputList`, and finally a call to the `TransferMatrix()` method is made to obtain the transfer matrix. This transfer matrix is a list of list of complex numbers (a complex matrix) that, when multiplied by a list of complex numbers (a vector) representing the frequency content of the input waveforms listed in the order that the sources were added, produces a list of complex numbers (another vector) representing

[3]Here, assigning s-parameters means assigning a single s-parameter matrix at a single frequency.

[4]A property is just a set of methods for accessing an internal variable that make the code look like a simple variable access, as opposed to a method call. Externally, a property looks just like a variable.

the frequency content of the output waveforms listed in the order that they appear in the output list.

The **VirtualProbe** class derives from **Simulator** and adds two properties, `pMeasurementList` and `pStimDef`, for accessing the measurement list and the *stimdef* as explained in §11.6. To solve a virtual probing problem, one instantiates the class **VirtualProbeNumeric**, uses the functions in **SystemDescription** to add devices to the system, connect device ports together, and assign s-parameters to the devices, and uses the added properties in **Simulator** and **VirtualProbe** to set the measurement nodes, the output nodes, and the *stimdef*. Finally, a call to the `TransferMatrix()` method obtains the transfer matrix. The transfer matrix is a list of list of complex numbers (a complex matrix) that, if multiplied by a list of complex numbers (a vector) representing the frequency content of the waveforms listed in the order in the measurement list, produces a list of complex numbers (another vector) representing the frequency content of the output waveforms listed in the order that they appear in the output list.

The solution to a simulation or virtual probe problem is the transfer matrix. In order to obtain waveform solutions, the transfer matrices over many frequencies need to be converted to filters. This is described best in §17.4.

### 17.3.3 Symbolic System Description Solutions

Figure 17.3 shows the relationship between the classes in the `SystemDescriptions` package with only those involved in symbolic solutions shown. The actual classes instantiated by the user are the classes pointed to by the user bubble at the bottom. These classes are:

1. **SystemSParametersSymbolic** – to solve systems of s-parameters symbolically.
2. **DeembedderSymbolic** – to solve de-embedding problems symbolically.
3. **VirtualProbeSymbolic** – to solve virtual probing problems symbolically.
4. **SimulatorSymbolic** – to solve simulation problems symbolically.
5. **SymbolicDevices** – these are actually functions in the `Symbolic` package that return lists of lists of LaTeX understood strings for various s-parameter devices.

Much of Figure 17.3 is common with Figure 17.2. This is because:

- The methods used to construct the netlist and even assign the s-parameters are the same for symbolic and numeric solutions. S-parameter assignment does not actually care what type of data is in the list of list matrices.
- The class **SystemSParameters**, while not showing any public methods, provides private methods utilized for assembly of the node lists and weights matrices. These also do not care what type of data is in the s-parameter list of list matrices held in the **Device** class.
- The symbolic classes bypass the numeric solution classes in the class hierarchy and make use of the same helper functions in the **SystemSParameters**, **Deembedder**, **VirtualProbe**, and **Simulator** classes.

Therefore, the symbolic solutions are formed in much the same way as the numeric solutions, with the only exception being that the s-parameters supplied are list of list (matrix) of strings, usually LaTeX strings, that can be created from functions in the `Symbolic` package.

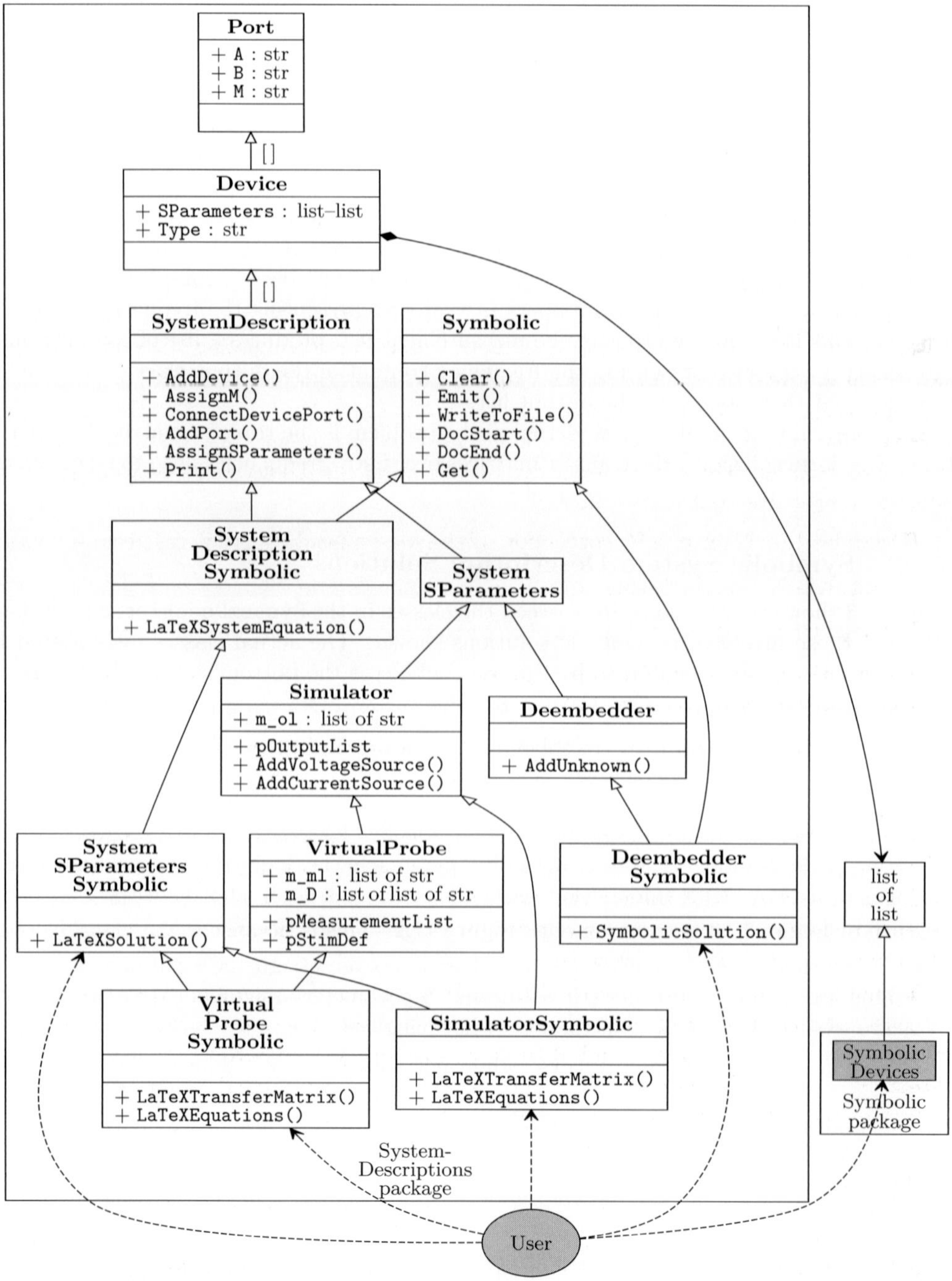


**Figure 17.3** System description symbolic UML diagram

The class structure makes use of multiple inheritance[5] and there is always a path from any of the instantiated solutions classes to the class **Symbolic**. This means that all of these classes inherit the capability of this class, which controls the LaTeX environment, the document settings, and the file output.

### 17.3.4 Parsers

The **SystemDescription** class and associated classes **SystemSParameters**, **Simulator**, **VirtualProbe**, and **Deembedder** are very useful for building the problem description for various problem solutions.

An extension of this to make it even easier to describe the problem is the `Parsers` package, which contains a number of analogous classes for constructing the same problem, except from a netlist as opposed to a scripted system description construction.

Figure 17.4 is a UML diagram showing the `Parsers` package and how it interacts with the `SystemDescriptions` package previously described, but now employing a netlist for system construction. There is a bubble at the bottom of the diagram representing the user. The user accesses the four classes directly:

1. **DeembedderNumericParser** – for solving de-embedding problems given a netlist.
2. **SimulatorNumericParser** – for solving simulation problems given a netlist.
3. **VirtualProbeNumericParser** – for solving virtual probing problems given a netlist.
4. **SystemSParametersNumericParser** – for solving for the s-parameters of circuits and systems given a netlist.

These numeric parser classes implement identically named methods as the numeric classes in the `SystemDescriptions` package. Recall that the numeric methods in the `SystemDescriptions` package compute either a list of list s-parameter matrix for de-embedding and system s-parameter solutions, or a list of list transfer matrix for simulation and virtual probing solutions. Here, the numeric parser classes return full instances of the class **SParameters**. A **TransferMatrices** instance along with a list of instances of the **Waveform** class representing input waveforms enables production of simulated or virtually probed output waveforms as described in §17.4.

The numeric parser classes derive from a base parser class for each solution, for the most part, and each of these classes implements only the functions `_ProcessLine()` and `_ProcessLines()`. Furthermore, all of these classes finally derive from a central base class **SystemDescriptionParser**, which does the heavy lifting in system description generation. The **SystemDescriptionParser** class has the methods `AddLine()` and `AddLines()`, which allow for the addition of lines of a netlist, along with the similar methods overloaded in the derived classes, `_ProcessLine()` and `_ProcessLines()`.

The user instantiates one of the four derived numeric parser classes depending on the type of problem being solved. Then, the user makes calls on `AddLines()` with a list of netlist lines describing the system. Alternatively, the user invokes the `File()` method in the base class **ParserFile** to read in the netlist. Finally, the user extracts the solution

[5] Multiple inheritance is often discouraged, as it can confuse things.

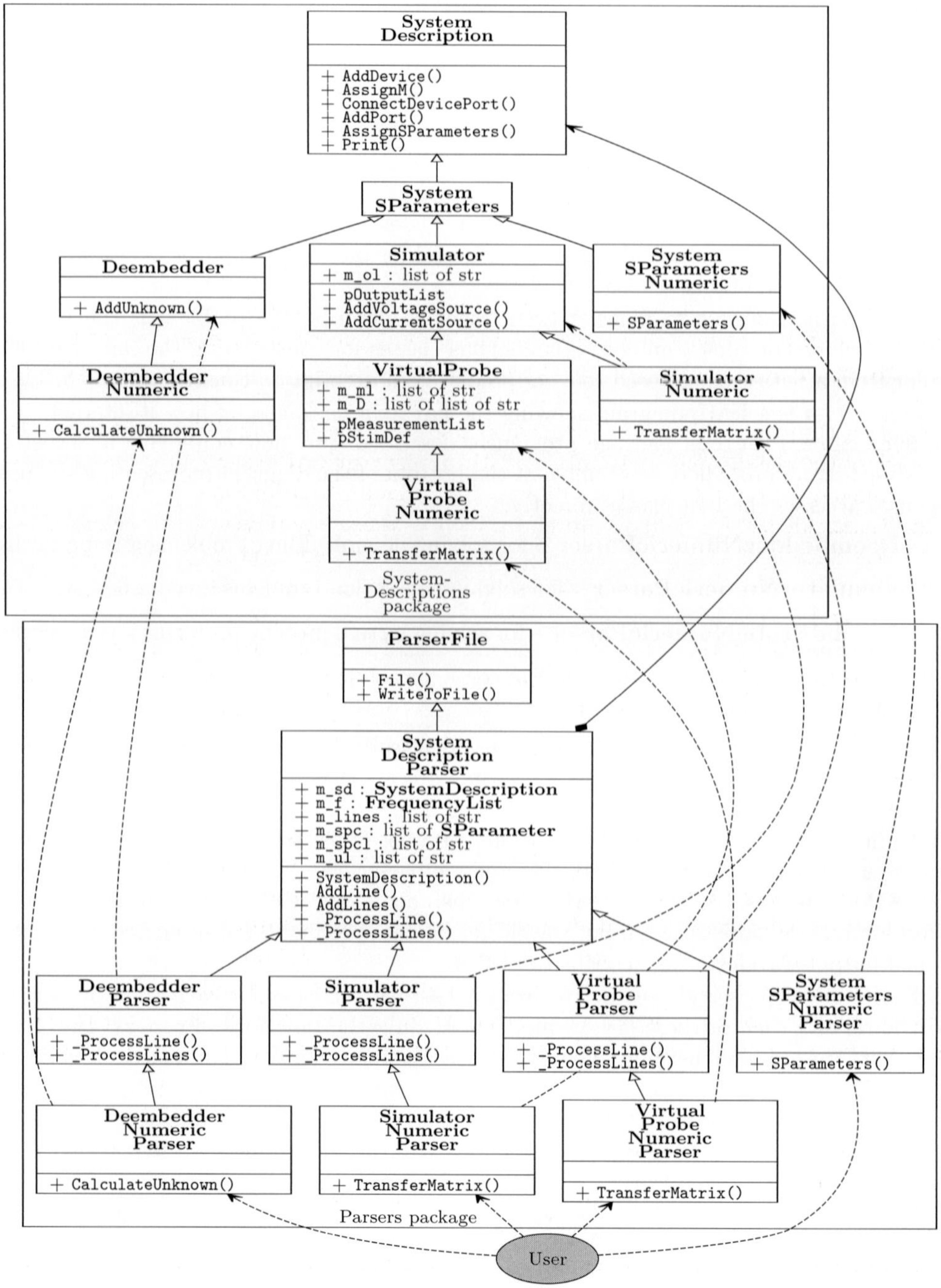


**Figure 17.4** Parsers UML diagram

to the problem in the form of either s-parameters or transfer matrices from the solution method.

In order to solve the problem, all of the solution methods call `_ProcessLines()` on the class they derive from. These classes will make calls to the base **SystemDescriptionParser** class to process lines and will process lines themselves. The **SystemDescriptionParser** class handles all of the netlist lines that are common to all of the solutions, like adding devices and connecting device ports together. It also parses lines which add named devices with parameters that are understood by the parser; the method is described in §17.3.5.

When the **SystemDescriptionParser** encounters a line `'device D1 2'`, it knows to add a two-port device named `'D1'` to the system, and it adds this device to its contained system description through a call to the `AddDevice()` method on **SystemDescription**. Similarly, when it encounters a line `'connect D2 1 D3 4'`, it knows to connect port 1 of the device named `'D2'` to port 4 of the device named `'D3'` in its contained system description through a call to the `ConnectDevicePort()` method on **SystemDescription**.

Each of the classes that derive from **SystemDescriptionParser** knows how to handle the commands specific to their solutions. If the **DeembedderParser** class encounters a line `'unknown D3 4'`, it knows to add a device whose s-parameters are unknown with four ports named `'D3'` to the system through a call to `AddUnknown()` on the associated **Deembedder** class. And if it encounters a line `'system file 4 device.s4p'`, it knows to read in and retain the s-parameters of the entire four-port system from the file specified.

The **SimulatorParser** class knows that to handle netlist lines `'voltagesource V 2'` or `'currentsource I 1'`, it needs to call `AddVoltageSource()` or `AddCurrentSource()` on the **Simulator** class with the device name and ports specified, and to handle the line `'output D1 3'` means to add the tuple `('D1',3)` to the `pOutputList` property on the **Simulator** class.

The **VirtualProbeParser** class handles other netlist lines not used in the other solutions. A line `'meas D1 3'` adds the tuple `('D1',3)` to the `pMeasurementList` property, and `'stimdef [[1.0], [-1.0], [1.0], [-1.0]]'` adds the list of list matrix to the `pStimDef` property of the **VirtualProbe** class. Note that since **VirtualProbe** is a **Simulator**, it also allows the addition of outputs to the `pOutputList` property.

It will be shown in §17.3.5 how the devices are parsed and added and how the numeric parser solutions are finally arrived at using the numeric classes in the `SystemDescriptions` package corresponding to the numeric parser classes in the `Parsers` package. Reference is again made to Figure 17.4 after dealing with the s-parameters of devices parsed from netlist lines.

### 17.3.5 Device Parser

Figure 17.5 shows a UML diagram showing how devices are added to the system using the `Parsers` package. This is the same diagram as Figure 17.4 with the addition of the `Devices` package within the `Parsers` package, and the removal of ancillary details in the `SystemDescriptions` package.

During the calculation of any of the numerical solutions used by method calls on the numeric parser classes, a call is made to `_ProcessLines()`, which parses the netlist. Many

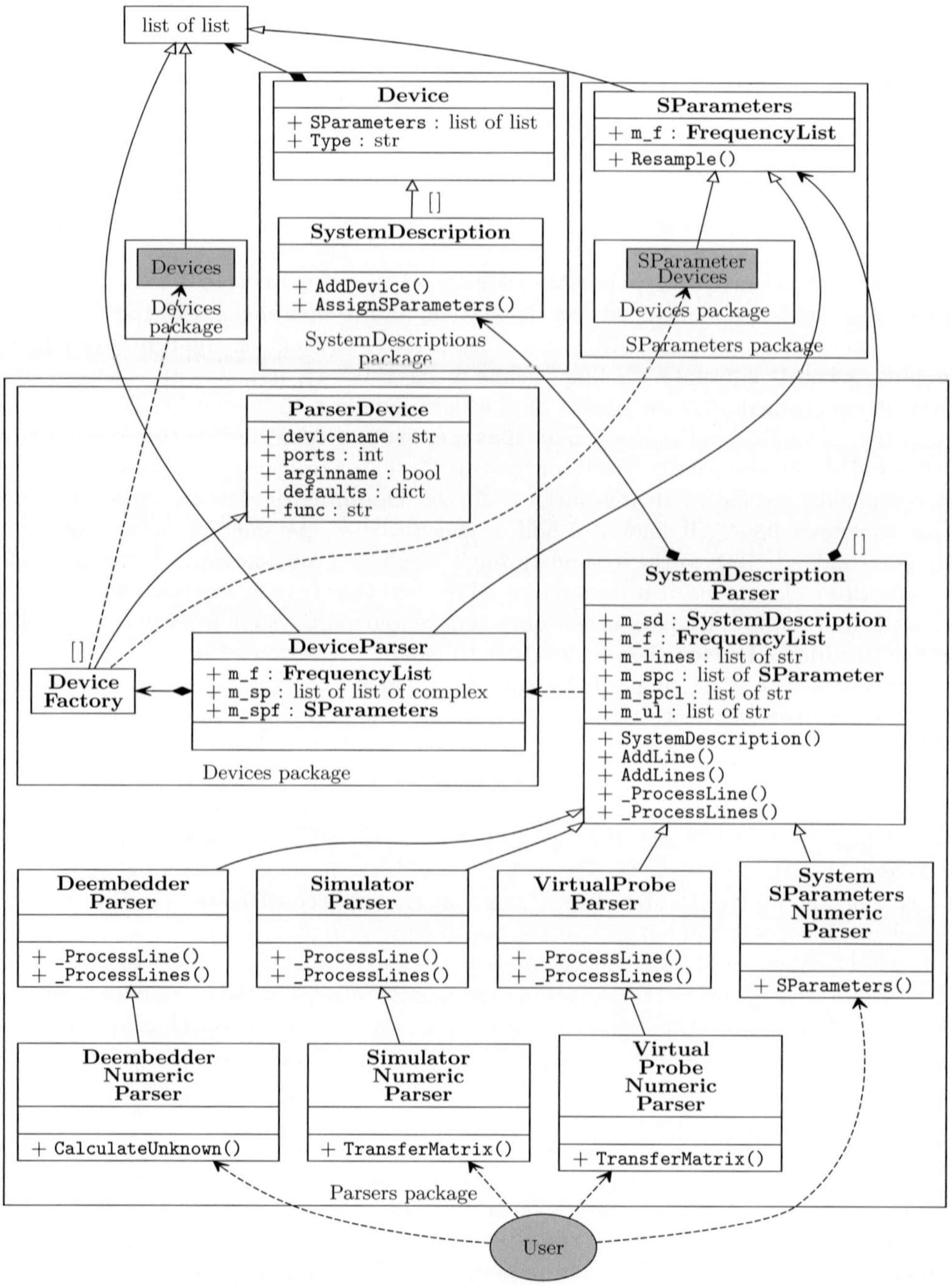


**Figure 17.5** Device parser UML diagram

**Table 17.2** List of devices and their settings in the **ParserDevice** class

| Name | Ports | Argument in name | Defaults | Frequency dependent | Device |
|---|---|---|---|---|---|
| `file` | any | `True` | `filename=None` | `True` | `sp.dev.SParameter-File(filename, 50.)` |
| `c` | 1 | `True` | `c=None df=0 esr=0 z0=50` | `True` | `sp.dev.TerminationC(f, c, z0, df, esr)` |
| `c` | 2 | `True` | `c=None df=0 esr=0 z0=50` | `True` | `sp.dev.SeriesC(f, c, z0, df, esr)` |
| `l` | 1 | `True` | `l=None` | `True` | `sp.dev.TerminationL(f, l, z0)` |
| `l` | 2 | `True` | `l=None` | `True` | `sp.dev.SeriesL(f, l, z0)` |
| `r` | 1 | `True` | `r=None` | `False` | `dev.TerminationZ(r)` |
| `r` | 2 | `True` | `r=None` | `False` | `dev.SeriesZ(r)` |
| `rse` | 2 | `True` | `r=None` | `True` | `sp.dev.SeriesRse(r)` |
| `shunt` | 2–4 | `True` | `r=None` | `False` | `dev.ShuntZ(ports,r)` |
| `m` | 4 | `True` | `m=None` | `True` | `sp.dev.Mutual(f, m)` |
| `ground` | 1 | `False` | | `False` | `dev.Ground()` |
| `open` | 1 | `False` | | `False` | `dev.Open()` |
| `thru` | 2 | `False` | | `False` | `dev.Thru()` |
| `directional-coupler` | 3–4 | `False` | | `False` | `dev.DirectionalCoupler(ports)` |
| `termination` | any | `False` | | `False` | `[[0]*ports]*ports` |
| `tee` | any | `False` | | `False` | `dev.Tee(ports)` |
| `mixedmode` | 4 | `True` | `'power'` | `False` | `dev.MixedModeConverter()` or `dev.MixedModeConverter-Voltage()` |
| `ideal-transformer` | 4 | `True` | `tr=1` | `False` | `dev.IdealTransformer(tr)` |
| `voltage-controlled-voltage-source` | 4 | `True` | `gain=None` | `False` | `dev.VoltageControlledVoltage-Source(gain)` |

**Table 17.2** List of devices and their settings in the **ParserDevice** class (continued)

| Name | Ports | Argument in name | Defaults | Frequency dependent | Device |
|---|---|---|---|---|---|
| `current-controlled-current-source` | 4 | `True` | `gain=None` | `False` | `dev.CurrentControlledCurrent-Source(gain)` |
| `current-controlled-voltage-source` | 4 | `True` | `gain=None` | `False` | `dev.CurrentControlledVoltage-Source(gain)` |
| `voltage-controlled-current-source` | 4 | `True` | `gain=None` | `False` | `dev.VoltageControlledCurrent-Source(gain)` |
| `voltage-amplifier` | 2–4 | `False` | `gain=None zo=0 zi=1e8 z0=50` | `False` | `dev.VoltageAmplifier(ports, gain, zi, zo)` |
| `current-amplifier` | 2–4 | `False` | `gain=None zo=1e8 zi=0 z0=50` | `False` | `dev.CurrentAmplifier(ports, gain, zi, zo)` |
| `trans-resistance-amplifier` | 2–4 | `False` | `gain=None zo=0 zi=0 z0=50` | `False` | `dev.Transresistance-Amplifier(ports, gain, zi, zo)` |
| `trans-conductance-amplifier` | 2–4 | `False` | `gain=None zo=1e8 zi=1e8 z0=50` | `False` | `dev.Transconductance-Amplifier(ports, gain, zi, zo)` |
| `opamp` | 3 | `False` | `zi=1e8 zd=1e8 zo=0 gain=1e8 z0=50` | `False` | `dev.OperationalAmplifier(zi, zd, zo, gain, z0)` |
| `tline` | 2,4 | `False` | `zc=50 td=0` | `True` | `sp.dev.TLineLossless(f, ports, zc, td)` |
| `telegrapher` | 2 | `False` | `r=0 rse=0 l=0 c=0 df=0 g=0 z0=50 sect=0` | `True` | `sp.dev.TLineTwoPortRLGC( f, r, rse, l, g, c, df, z0, sect)` |

**Table 17.2** List of devices and their settings in the **ParserDevice** class (continued)

| Name | Ports | Argument in name | Defaults | Frequency dependent | Device |
|---|---|---|---|---|---|
| telegrapher | 4 | False | rp=0 rsep=0 lp=0 cp=0 dfp=0 gp=0 rn=0 rsen=0 ln=0 cn=0 dfn=0 gn=0 lm=0 gm=0 cm=0 dfm=0 z0=50 sect=0 | True | sp.dev.TLineDifferential-RLGC(f, rp, rsep, lp, gp, cp, dfp, rn, rsen, ln, gn, cn, dfn, cm, dfm, gm, lm, z0, sect) |
| shortstd | 1 | False | od=0 oz0=50 ol=0 l0=0 l1=0.0 l2=0 l3=0 | True | m.calkit.std.ShortStandard(f, od, oz0, ol, l0, l1, l2, l3) |
| openstd | 1 | False | od=0 oz0=50 ol=0 c0=0 c1=0 c2=0 c3=0 | True | m.calkit.std.OpenStandard(f, od, oz0, ol, c0, c1, c2, c3) |
| loadstd | 1 | False | od=0 oz0=50 ol=0 tz0=50 | True | m.calkit.std.LoadStandard(f, od, oz0, ol, tz0) |
| thrustd | 2 | False | od=0 oz0=50 ol=0 | True | m.calkit.std.ThruStandard(f, od, oz0, ol) |

netlist lines were explained in §17.3.4. This section is concerned with how the parser handles netlist lines that begin with the token `'device'`. This token says that the netlist line contains instructions for adding a device to the netlist.

A netlist line takes the form `'device <name> <ports> <device name>'` followed by an argument or arguments in keyword pairs. The addition of a device is handled by the **SystemDescriptionParser** class during a call to `_ProcessLines()`. If the device token is encountered, the line is sent during the construction of a **DeviceParser** class. An instance of **DeviceParser**, after parsing, has either an instance of a list of list matrix in `m_sp`, or an instance of **SParameters** in `m_spf`, depending on the whether the created device is frequency dependent or not.

The **DeviceParser** has a static instance of **DeviceFactory**. A static instance is one that exists for all class instances. The **DeviceFactory** class is a list of instances of **Parser-Device**, which are definitions of each device recognized. The definitions are stored in the member variables of **ParserDevice**:

- `devicename`: the name of the device (the device name string in the netlist line).
- `ports`: an integer number of ports, or a string containing a dash or comma separated number of ports. A string `'2-4'` means two to four ports, and a string `'2,4'` means two or four ports.

- `arginname`: whether the argument (or first argument) requires a keyword pair or whether the device name implicitly defines the keyword. For example, a simple device like a capacitor has `arginname True` because there is only one argument and it is known to be the capacitance. A voltage amplifier has `arginname False` because it has three arguments, the gain, input impedance, and output impedance, and, in order to avoid confusion, all of these arguments are required to be preceded by the keyword corresponding to the argument in order for proper identification.
- `defaults`: a dictionary of default arguments specified by a keyword. These default arguments will be used for keyword variables when no keyword is supplied.
- `frequencydependent`: whether the device is frequency dependent.
- `func`: a string that, when evaluated, produces the device.

All devices have the list of frequencies and the number of ports available as an argument.

The list of devices defined in **ParserDevice** is provided in Table 17.2. The devices that are not frequency dependent are found in the `Devices` package, which contains functions that return a list of list matrix for all frequencies. All devices that are frequency dependent are found in the `SParameters.Devices` package, where devices that derive from **SParameters** are found. One exception to this is that the frequency dependent calibration kit standards are found in the `Measurement.Calibration.CalKit.Standards` package.

In Table 17.2, the device name is provided followed by the number of ports. If there is a range or selection of ports (as opposed to one number), then the number of ports is provided as an argument in the function. If `arginname` is `True`, the actual keyword for this argument is the first argument listed in the defaults. For example, a capacitor can be specified as device `'c 1 1e-12'` to specify a 1 pF capacitor. The understanding is that the default argument with keyword `'c'` is set to 1 pF, and the remaining arguments are set to their default. Typically, in this list, if an argument defaults to `None`, then it is unacceptable not to provide it.

After the device has been determined and evaluated, the s-parameters are either added to the member `m_spc`, if frequency dependent, or directly assigned to the device through a call to `AssignSParameters()` in **SystemDescription**, if not frequency dependent.

Frequency dependent devices also have the complete set of arguments for the device added to the member `m_spcl`. This is to avoid duplication of s-parameter calculations. Thus, in systems containing elements with exactly the same s-parameters, these are executed and stored once in the **SystemDescriptionParser** class.

After all of the lines have been processed in the netlist during the call to `_ProcessLines()` that was initiated by a call to one of the member functions on the numeric parser classes, the solution is formed by looping over all of the frequencies, assigning the s-parameter matrices corresponding to the frequency to the devices in the system through a call to `AssignSParameters()` in **SystemDescription**, and calling the numeric solution methods on the numeric classes found in the `SystemDescriptions` package, as shown in Figure 17.4. Remember that the numeric solutions in the numeric classes in the `SystemDescriptions` package provide a single list of list matrix (for one frequency). The numeric parsers automate the process of filling in the s-parameters for each frequency and computing the result for each frequency.

## 17.4 Waveform Processing

The organization of the classes involved in waveform processing is shown in Figure 17.6. Waveform processing involves affecting instances of the class **Waveform** defined in the `TimeDomain.Waveform` package with the instances of class **WaveformProcessor** defined in the `TimeDomain.Filters` package. As such, all waveforms being real-valued sequences with an associated time axis derive from the **Waveform** class. The **WaveformProcessor** class has a more complicated arrangement.

In Part II, Applications, there are two applications that involve waveform processing. These are simulation, described in Chapter 9 and virtual probing, described in Chapter 11. The solutions to this processing are instances of the class **TransferMatrices**, the theory of which is described in §13.5. The format of the data in the class **TransferMatrices** is similar to the format in **SParameters** in that there is a list of matrices and each matrix corresponds to a frequency. While s-parameters describe the port–port relationships, transfer matrices describe the input waveform to output waveform relationship. And transfer matrices don't need to be square like s-parameter matrices.

These lists of matrices can be easily converted into a matrix of instances of the class **FrequencyResponse** simply by reorganizing the data along with the frequency axis, where each element of the matrix describes the frequency response or transfer function of a filter that converts an input waveform into an output waveform. This is why in Figure 17.6 **TransferMatrices** *has a* **FrequencyResponse**,[6] and instances of **FrequencyResponse** are accessed through `FrequencyResponse()` and `FrequencyResponses()`. The class **FrequencyResponse**, along with the class **FrequencyContent**, derives from the class **FrequencyDomain** and inherits all of the functionality of frequency-domain elements that are complex-valued sequences with an associated frequency axis. **FrequencyContent** refers to frequency-domain versions of waveforms and **FrequencyResponse** refers to frequency responses associated with impulse responses and with filters. There is an association between the class **FrequencyResponse** and the class **ImpulseResponse**, and in fact, following the rules put forth in §12.4.2, there is a one-to-one correspondence between the two. In fact, there is a method on the **FrequencyResponse** class called `ImpulseResponse()`, and a method on the **ImpulseResponse** class called `FrequencyResponse()`. These methods convert one to the other.

Note that, on the **TransferMatrices** class, there is a method `ImpulseResponses()`. This method is simply a shortcut for obtaining the impulse responses corresponding to the frequency responses of the transfer matrices.

Instances of the class **ImpulseResponse** inherit all of the capabilities of the class **Waveform**, if one chooses to look at it that way, but mostly they are associated with the **FirFilter**. Impulse responses are easily converted into FIR filters with the values of the impulse response waveform forming the filter taps, and the waveform **TimeDescriptor** converted into a **FilterDescriptor** using the rules in §13.1.4. The method `FirFilter()` is used to obtain a **FirFilter** instance from an instance of **ImpulseResponse**.

[6] They are said to have frequency responses even though there is some conversion that takes place here, but the conversion is purely mechanical so this relationship is asserted.

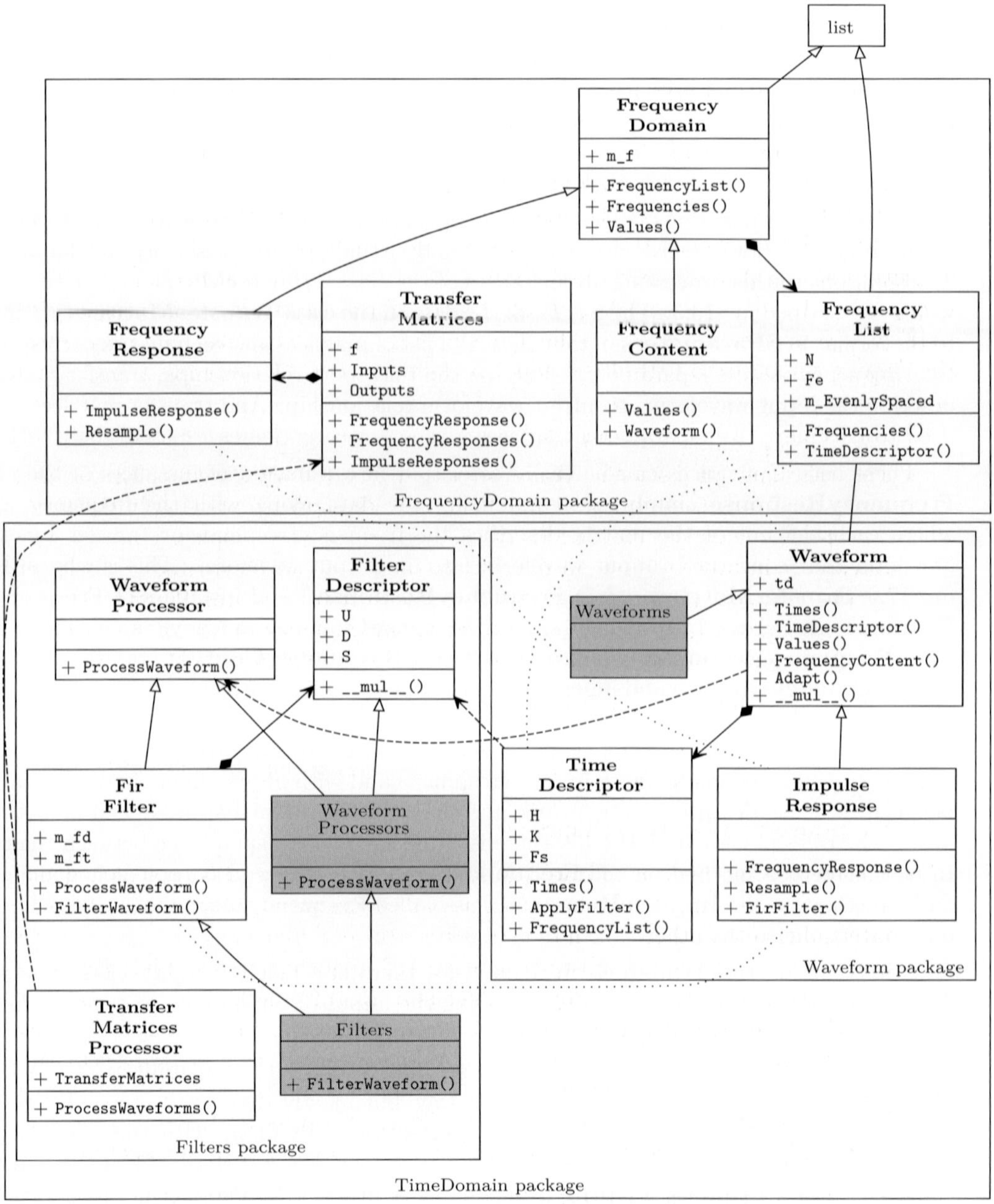


**Figure 17.6** Waveform processing UML diagram

The class **WaveformProcessor** is used to process waveforms. Waveform processors simply convert one waveform into another by processing them. They are recognized by the **Waveform** class, and multiplying an instance of **Waveform** by an instance of **WaveformProcessor** causes a call to `ProcessWaveform()` on **WaveformProcessor**. Thus, all waveform processors derive from the base **WaveformProcessor** class and overload the function `ProcessWaveform()` to perform the actual processing.

The **FirFilter** class derives from **WaveformProcessor** and provides two functions: `ProcessWaveform()` and `FilterWaveform()`. All FIR filters derive from **FirFilter** and can optionally overload the `FilterWaveform()` method. The `ProcessWaveform()` method in the base **FirFilter** class simply calls `FilterWaveform()`.

Thus, the act of processing a list of input waveforms with a transfer matrices processor to produce a list of output waveforms can be described as converting the transfer matrices to frequency responses, converting the frequency responses to impulse responses, converting the impulse responses to FIR filters, and processing the input waveforms through this array of filters.

In Figure 17.6, special relationships between the following classes are highlighted further:

- The relationships between **Waveform** and **FrequencyContent** classes and between the **ImpulseResponse** and **FrequencyResponse** are simply the time- and frequency-domain representations of each other and are completely interchangeable. Having an instance of one, an instance of the other can be created. This is enabled by the `FrequencyContent()` method on **Waveform** and the `Waveform()` method on **FrequencyContent**, and by the `FrequencyResponse()` method on **ImpulseResponse** and the `ImpulseResponse()` method on **FrequencyResponse**.

- The relationship between **ImpulseResponse** and **FirFilter** is such that they are simply different ways of viewing the same thing. The convolution of waveforms with impulse response (waveforms) is the same as filtering the waveform with a filter constructed from this impulse response. Thus, the method `FirFilter()` on **ImpulseResponse** provides the instance of **FirFilter** for this purpose.

- The relationship between **TimeDescriptor** and **FrequencyList** is such that the time axis for a waveform, the number of points and sample period of which define the sample rate and time duration, which in turn defines the end frequency and frequency resolution, which in turn define a list of evenly spaced frequency points. The converse of this is also true. This is enabled by the `FrequencyList()` method on **TimeDescriptor** and the `TimeDescriptor()` method on **FrequencyList**. **FrequencyList** is actually a base class for two classes of frequency lists: generic lists of frequencies and so-called evenly spaced frequency lists, which also contain the zero frequency point. A **TimeDescriptor** converts to an **EvenlySpacedFrequencyList** instance, and only an **EvenlySpacedFrequencyList** will convert back to a **TimeDescriptor**.

- The relationship between **TimeDescriptor** and **FilterDescriptor** is shown as a *uses* relationship, because when a waveform is filtered the filter descriptor is used in conjunction with the time descriptor to determine the time descriptor of the filtered waveform. This is performed through a call to `ApplyFilter()` on **TimeDescriptor** with an instance of **FilterDescriptor**.

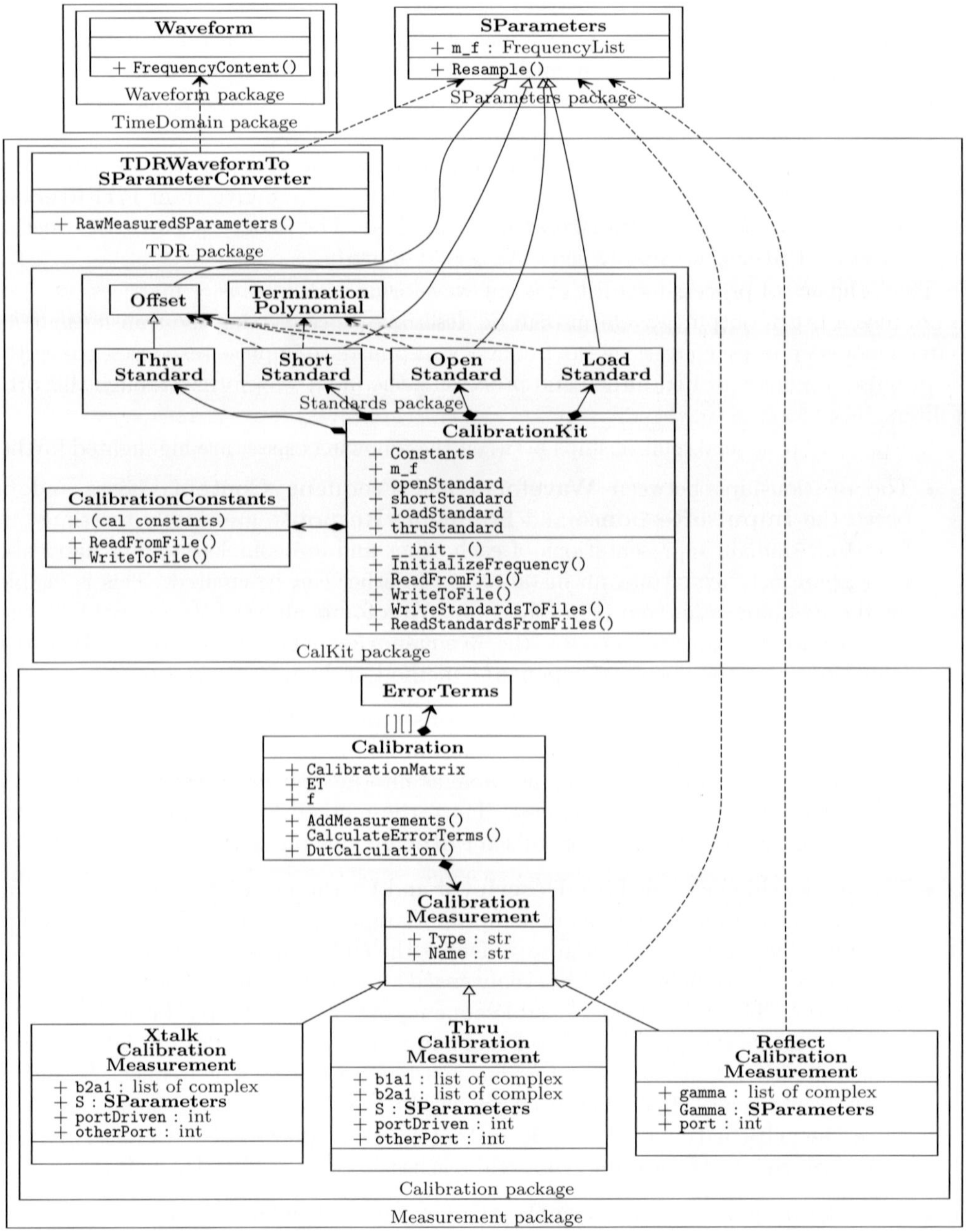


**Figure 17.7** Measurement UML diagram

Hopefully this section and Figure 17.6 make clear the class relationships involved in waveform processing with filters, and more specifically transfer matrices processing with lists of waveforms.

## 17.5 Measurement

The relationship between the classes in the `Measurement` package is shown in Figure 17.7. The method for computing s-parameters of a DUT from TDR waveform measurements involves the following:

1. A calibration kit is instantiated, usually by using the `ReadFromFile()` method on **CalibrationKit**. This instantiates a contained instance of **Calibration-Constants**. The frequencies for the DUT measurement are set through a call to `InitializeFrequency()`, which causes the calibration standards to be instantiated as instances of **ThruStandard**, **ShortStandard**, **OpenStandard**, and **Load-Standard** based on the constants in the **CalibrationConstants** instance.
2. Measurements of calibration standards are provided in the form of instances of **Waveform** to the `RawMeasuredSParameters()` method in **TDRWaveformTo-SParameterConverter** which calculates and returns raw measured s-parameters as instances of the class **SParameters**.
3. Depending on the calibration standards being measured, the classes **Thru-CalibrationMeasurement**, **ReflectCalibrationMeasurement**, and/or **Xtalk-CalibrationMeasurement** are instantiated with the raw measured s-parameters, along with the associated calibration standard read from the **CalibrationKit** instance.
4. A **Calibration** class is instantiated and the calibration measurements, which are all instances of **CalibrationMeasurement**, are added through calls on **Calibration** to `AddMeasurements()`.
5. After all of the measurements are added, an optional call to `CalculateErrorTerms()` is made on **Calibration**, which creates a list of list matrix of instances of the **Error-Terms** class; this is optional, because the **Calibration** object knows to calculate these if they are needed and have not already been calculated.
6. Finally, waveform measurements of the DUT are converted to raw measured s-parameters as instances of **SParameters** using the **TDRWaveformToSParameter-Converter** class and supplied to the instance of the **Calibration** class through a call to `DutCalculation()`, which calculates and returns an instance of **SParameters** containing the calibrated s-parameter measurement of the DUT.

Although the above steps are provided for TDR based s-parameter measurements, they can also be used for VNA measurements. In the case of VNA measurements, any of the steps involving measurements of waveforms and conversion using the **TDRWaveformTo-SParameterConverter** class instead measure the raw s-parameters directly.

# 18

# *SignalIntegrityApp*

*SignalIntegrityApp* IS BUILT ON TOP OF THE `SignalIntegrity.Lib` package. The `SignalIntegrity.Lib` package contains no links whatsoever to `SignalIntegrity.App` and can be operated alone. Because of this arrangement, mathematical and numerical operation of the software can be tested independent from the GUI. *SignalIntegrityApp* uses the standard Python GUI infrastructure, Tkinter, and makes calls to the `SignalIntegrity.Lib` package. The GUI remains untested to some extent, except by operating it. That being said, `SignalIntegrity.App` has a **SignalIntegrityAppHeadless** class that can be invoked to load projects and generate solutions. Using this headless class in testing covers about 50 % of the code in the loading of project files, generation of netlists, and handling of waveform and s-parameter results that come from the processing. The headless class offers no ability to edit the schematic and is mostly used for regression testing and for the construction of difficult simulation problems, automating the netlist generation and waveform processing.

The software organization for `SignalIntegrity.App` is not discussed in detail in this book, nor are the details of its operation. The operator manual for the *SignalIntegrityApp* software can be found at

`https://github.com/TeledyneLeCroy/SignalIntegrity/wiki/Documentation`

along with online up-to-date documentation on `SignalIntegrity.Lib`.

The underlying library is discussed in detail in Chapter 17 so that its organization is understood and it can be used effectively. Of course, the overall idea is that it is extended in an open-source software manner. Although this book contains all (or most) of the theory along with a discussion of much of the software and of the organization of the `SignalIntegrity.Lib` package, the line is drawn here, and the documentation on the website should be consulted for details about the software in its current state. In other words, this chapter will not be a user manual for the *SignalIntegrityApp* software. This chapter does discuss some of the underlying architecture of *SignalIntegrityApp* and provides a rather large example to demonstrate its usage.

## 18.1 Project File Format

The format of the project file used by *SignalIntegrityApp* is shown in Listing 18.1. Basically, the application has two main sections:

- `Drawing`,
- `CalculationProperties` (explained in §18.3).

**Listing 18.1** *SignalIntegrityApp* project file format

```
<Project>                                       % project properties
  <Drawing>                                     % drawing properties
    <DrawingProperties>                         % properties of the drawing canvas
      <Geometry>[string XxY+R+C]</Geometry>     % application window size and offset
      <Originy>[integer]</Originy>              % origin in schematic (x grid)
      <Originx>[integer]</Originx>              %
      <Height>[integer]</Height>                % height of canvas (in pixels)
      <Width>[integer]</Width>                  % width of canvas (in pixels)
      <Grid>[float]</Grid>                      % drawing grid (in pixels)
    </DrawingProperties>
    <Schematic>                                 % schematic properties
      <Wires>                                   % wires
        <Wire>                                  % wire
          <Vertices>                            % vertices in a wire
            <Vertex>                            % vertex
              <Coord>[tuple (x,y)]</Coord>      % coordinate of a vertex
            </Vertex>
            ...                                 % many vertices in a wire
          </Vertices>
        </Wire>
        ...                                     % many wires in a schematic
      </Wires>
      <Devices>                                 % devices
        <Device>                                % device
          <ClassName>[string]</ClassName>       % App class name of device
          <PartPicture>                         % picture of device
            <Index>[int]</Index>                % index of picture in table for part
            <Origin>[tuple (x,y)]</Origin>      % picture origin in schematic
            <MirroredHorizontally>[boolean]</MirroredHorizontally>  % horiz. mirrored
            <Orientation>[integer 0, 90, 180, or 270]</Orientation> % rotation
            <MirroredVertically>[boolean]</MirroredVertically>      % vert. mirrored
          </PartPicture>
          <PartProperties>                      % part properties
            <PartProperty>                      % number of ports property
              <Keyword>ports</Keyword>          % ports keyword
              <Value>[integer]</Value>          % number of ports in device
              <Visible>[boolean]</Visible>      % whether this is visible
              <KeywordVisible>[boolean]</KeywordVisible> % whether keyword is visible
            </PartProperty>
            <PartProperty>                      % reference designator property
              <Keyword>ref</Keyword>            % reference designator keyword
              <Value>[string]</Value>           % reference designator
              <Visible>[boolean]</Visible>      % whether this is visible
              <KeywordVisible>[boolean]</KeywordVisible> % whether keyword is visible
            </PartProperty>
            <PartProperty>                      % generic property
              <Keyword>[string]</Keyword>       % unique keyword for property
              <Value>[string]</Value>           % value of property
              <Visible>[boolean]</Visible>      % whether it is visible
              <KeywordVisible>[boolean]</KeywordVisible> % whether keyword is visible
            </PartProperty>
            ...                                 % many properties in device
          </PartProperties>
        </Device>
        ...                                     % many devices in schematic
      </Devices>
    </Schematic>
  </Drawing>
  <CalculationProperties>                         % calculation properties
    <UserSampleRate>[float]</UserSampleRate>      % user sample rate (S/s)
    <EndFrequency>[float]</EndFrequency>          % end frequency (Hz)
    <FrequencyPoints>[integer]</FrequencyPoints>  % number of frequency points
  </CalculationProperties>
</Project>
```

The `Drawing` is further broken down into:

- `DrawingProperties` – the geometry and grid of the canvas on which the schematic is drawn.
- `Schematic` – the circuit pertaining to the application being solved.

The `Schematic` is broken down into two categories:

- `Devices` – the circuit elements in the schematic.
- `Wires` – the interconnections of the circuit elements (with the understanding that circuit elements can be connected by abutting their pins without a wire).

Each element in `Wires` is a `Wire`, which is essentially a list of `Vertices`. The vertices can be grabbed and moved in the application, as can line segments (the line that virtually connects two vertices). If the virtual line segment connecting two vertices runs vertically or horizontally, and the segment intersects a pin or another vertex in another wire, for the sake of netlist generation the wire is connected.

In `Devices`, each `Device` has the following elements:

- `ClassName` – the name of the class defining the device within *SignalIntegrityApp*. This class knows how to produce the netlist line corresponding to the device and the part properties.
- `PartPicture` – the definition of how the part is drawn.
- `PartProperties` – a list of properties for the device. All devices have properties for the number of ports and a reference designator. Other properties depend on what the device is (e.g. a resistor has a resistance property). Note that there may be other part properties pertaining to a device held internally based on the device – this is only what is in the project file.

A `PartPicture` contains the following elements:

- `ClassName` – the name of the class that draws the device on the schematic.
- `Origin` – where the picture starts in the schematic (usually the upper left corner of an invisible bounding box).
- `Orientation` – the rotation of the picture.
- `MirroredHorizontally` – whether the part is flipped horizontally.
- `MirroredVertically` – whether the part is flipped vertically.

A `PartProperty` contains the following elements:

- `Keyword` – unique identifier of the property.
- `Value` – a string value of the property. If the property is something like resistance, then the string is converted to a float. In other words, the property keyword defines how the value is converted internally.
- `Visible` – whether the user can see the property value in the schematic picture.
- `KeywordVisible` – whether the keyword is shown in front of the property value in the schematic picture.

Note that the `PartProperty` might consist of other values internally based on the `Keyword`; these are only the things that are in the project file.

## 18.2 SignalIntegrityAppHeadless Application Programming Interface

Fortunately, except in rare instances, one never needs to look at the project file structure outlined in §18.1. As a project is edited with *SignalIntegrityApp*, this structure is modified automatically.

The ability to create a project file is very valuable in that it can describe complicated situations from which the netlist, which is the interface to the `SignalIntegrity.Lib` package, can be extracted. Fortunately, one does not need to choose between running the *SignalIntegrityApp* to provide solutions and scripting solutions. There is an intermediate solution involving drawing the schematic, setting the calculation properties using *SignalIntegrityApp*, saving this project file, and then opening that project from within a scripted solution. The API for this involves the use of the **SignalIntegrityAppHeadless** class.

This class is a stripped down version of *SignalIntegrityApp*. One need only instantiate `SignalIntegrity.App.SignalIntegrityAppHeadless()` as `app`, which has the following member functions available:

1. `app.OpenProjectFile()` – opens an existing project file.
2. `app.SaveProjectToFile()` – saves the project file.
3. `app.Drawing.schematic.NetList()` – obtains the netlist as a list of netlist lines.
4. `app.CalculateSParameters()` – calculates an s-parameter solution. The result is returned as a tuple (a kind of immutable list), where each element contains
   a) the resulting s-parameters;
   b) the recommended filename.
5. `app.Simulate()` – calculates a simulation solution. The result is returned as a tuple, where each element contains
   a) the source names;
   b) the output waveform names;
   c) the transfer matrices;
   d) the list of output waveforms.
6. `app.Deembed()` – calculates a de-embedding solution. The result is returned as a tuple, where each element contains
   a) the list of the names of the unknown devices;
   b) the s-parameters of the unknown devices (i.e. the de-embedded devices).
7. `app.VirtualProbe()` – calculates a virtual probing solution. The result is returned as a tuple, where each element contains
   a) the measure probe names;
   b) the output waveform names;
   c) the transfer matrices;
   d) the list of output waveforms.

**Table 18.1** Calculation properties

| Property | Type | Equation | Internal property name |
|---|---|---|---|
| User sample rate | base | $userSampleRate$ | `UserSampleRate` |
| End frequency | base | $endfrequency$ | `EndFrequency` |
| Number of frequency points | base | $frequencyPoints$ | `FrequencyPoints` |
| Base sample rate | derived | $baseSampleRate = 2 \cdot endFrequency$ | `BaseSampleRate` |
| Number of time points | derived | $timePoints = 2 \cdot frequencyPoints$ | `TimePoints` |
| Frequency resolution | derived | $frequencyResolution = \frac{endFrequency}{frequencyPoints}$ | `FrequencyResolution` |
| Time length of impulse response | derived | $impulseResponseLength = \frac{frequencyPoints}{endFrequency}$ | `ImpulseResponseLength` |
| User sample period | derived | $userSamplePeriod = \frac{1}{userSampleRate}$ | `UserSamplePeriod` |

Furthermore, one is not restricted to accepting the settings of the project file. The project file can be edited from within a scripted environment, but care must be taken. It is best explained through an example. If a project file is being used to calculate the s-parameters of a system, and one wants to set the number of frequency points to 1000 and the end frequency to 40 GHz, one need only type, after opening the project file, the following:

```
SignalIntegrity.App.Project['CalculationProperties.FrequencyPoints']=1000
SignalIntegrity.App.Project['CalculationProperties.EndFrequency']=40e9
```

This would be performed prior to the call to `CalculateSParameters()`. If the calculation properties are written into directly, however, §18.3 should be consulted.

## 18.3 Calculation Properties

The calculation properties internal to *SignalIntegrityApp* are shown in Table 18.1 and are universal for all of the applications. In other words, they apply equally to s-parameter calculation, simulation, virtual probing, and de-embedding. In s-parameter calculation and de-embedding, the end frequency and the number of points are obvious things to consider; however, these are purely frequency-domain criteria. The time-domain implications of these frequency-domain choices are not so obvious, but are linked, as shown in Table 18.1. In simulation and virtual probing applications, the time-domain implications are in the base sample rate and the impulse response length used for the filters calculated in the transfer matrices.

All of the properties are interrelated and derived from three base properties: the user sample rate, the end frequency, and the number of frequency points; these are the only properties stored in the project file. All other properties are based on these properties.

In *SignalIntegrityApp*, the number of frequency points specified equals the number of frequencies minus one. Said differently, the number of actual frequencies calculated is always the number shown plus one. So, for $N$ frequency points and an end frequency $Fe$, the actual frequency points are, for $n \in 0 \ldots N$,

$$f[n] = \frac{n}{N} \cdot Fe.$$

The impulse response is specified for $K = 2 \cdot N$ time points and a sample rate $Fs = 2 \cdot Fe$, where for $k \in 0 \ldots K-1$, the times are given by

$$t[k] = \left(k - \frac{K}{2}\right) \cdot \frac{1}{Fs}.$$

The user sample rate is the desired, final sample rate for output waveforms in simulation and virtual probing applications. The base sample rate property defines the sample rate used for processing the waveforms, and the results are upsampled (or downsampled) to the user sample rate.

In the application, when the user modifies any of these properties, the base properties on which the modified property depends are modified, and then all other derived properties are calculated using the following two rules:

1. The end frequency is modified only by directly editing it, or by changing the base sample rate or base sample period. It is not modified by making changes to any other properties.
2. The impulse response length is never modified by changing the end frequency, base sample rate, or base sample period, where the latter two dictate a change in end frequency. When the end frequency is modified, the number of frequency points (and therefore time points) are modified as well to hold the impulse response length and the corresponding frequency resolution constant.

These two rules are deemed to constitute proper behavior and match what a user is expecting. Said differently, the end frequency is a constant from the user's perspective, although it must be changed when changing the base sample rate or base sample period. If the user does not want that, the user sample rate should be modified. The impulse response length should be kept constant unless the user directly edits things that definitely affect it. While the end frequency and number of frequency points are held as base properties (because they are familiar to most users), the end frequency and impulse response length are the real base properties that dictate system behavior.

When a user modifies the project file from a script, they should ensure that the equalities in Table 18.1 hold. The application knows how to deal with the setting of any one of these parameters and applies the proper effect on the other properties.

When setting the calculation properties, one usually decides upon the end frequency and the impulse response length directly. The end frequency is clear – it is the end frequency of the s-parameter files used and is the highest frequency of interest in the system. This determines the base sample rate used for time-domain processing. Determining the impulse response length is a bit trickier: it is the length of the impulse response generated from the transfer matrices for processing, remembering that half the length is for positive time and half is for negative time. For example, an impulse response length of 50 ns dictates an impulse response that is 50 ns long, with 25 ns before zero time and 25 ns after. There is no sure-fire way to determine the right impulse response length except through an understanding of the system being simulated – and the transfer functions should always be viewed in the time domain during simulation.

Microwave (or other frequency-domain oriented) engineers might argue that frequency resolution requirements determine the number of frequency points to use and will talk about resonance, the quality factor, or other system aspects; these aspects all lead back to the impulse response length. If there is a sharp resonance in the system and one wants to know how many frequency points to measure or simulate, one simply looks at the length of the impulse response; when it dies down sufficiently, the impulse response length (or, more precisely, half of the length) is determined, and the reciprocal of twice that length in time is the correct, maximum frequency resolution to use.[1]

On a final note on this topic, the reader might wonder about the choice of half of the impulse response being before zero time and half after. After all, the system should be causal. But as was pointed out, especially in the discussion of virtual probing in Chapter 11, transfer matrices may be generated with non-causal filters when the measurement probing point is later in time than the output probing point. That aside, it is important to look at any unexpected non-causal behavior in the transfer matrices filters, and this would be impossible without considering the time before zero. Finally, as a result of resampling, upsampling, or truncation of s-parameters in frequency, some amount of Gibb's phenomenon [75] might be evident in the form of ringing; some amount of ringing might occur before time zero. Simply chopping this off would lead to bad errors in the simulation. So, while potentially somewhat wasteful, the simple decision was made to put exactly half of the impulse response before and half after zero time.

## 18.4 S-Parameter Viewing and Transfer Matrices

One of the basic tenets of s-parameters used for signal integrity is that the time-domain implications must be examined to ensure good results. For that reason, one is advised in all cases to examine the s-parameters of all of the elements in their circuit or system, especially the ones that were measured or produced by a simulator, and to examine all transfer matrices in simulation and virtual probing applications. The s-parameters of any

[1] The factor of two accounts for the assumption that half of the impulse response is before zero time and half after.

device can be viewed in the *SignalIntegrityApp*, including s-parameter files that are read in. Furthermore, when a time-domain simulation completes, the user has the option to view the transfer matrices, which are also shown using the s-parameter viewer. When s-parameter files are read in, they are shown in magnitude and phase in the frequency domain at the points provided. If the frequency points are evenly spaced and include DC, the time-domain impulse and step responses will also be shown. Otherwise, the user has the option (which should be taken) to resample the s-parameters onto an evenly spaced frequency spacing and to extrapolate the DC point, after which the time-domain views are shown.

All of these views contain important information:

1. The magnitude response and the phase response (if possible) should be checked for reality.
2. The impulse response should be zero prior to zero time and should settle before the time window ends, according to the impulse response length (see §18.3).
3. The step response should be zero prior to zero time and should reach a steady-state value before the time window ends, according to the impulse response length.
4. The step response shape should be sanity checked, as the shape of the step response carries the phase information, albeit in the time domain.

It is very important to understand that impulse response and step response length are directly related to the frequency resolution, which, given an end frequency for the s-parameters, depends also on the number of points.

While it may seem superfluous to check both the impulse and step response, one finds that certain bad effects are seen more easily with one or the other.

Also, the impedance profile can be viewed in lieu of the step response of the s-parameters on the diagonal.

The s-parameter viewer also provides other capabilities to manipulate s-parameters:

1. Passivity violations can be viewed and passivity enforced.
2. Causality violations can be viewed and causality enforced.
3. The impulse response length can be limited, or truncated, which can be used to clean up measurements and to minimize impulse response length requirements in waveform processing.
4. The reference impedance can be changed.
5. The DC point can be extrapolated and the s-parameters can be resampled.

While the VNA and some simulators will not calculate the DC point, it is advisable to supply s-parameters for simulation that are evenly spaced and include DC. Otherwise, the simulator will resample them behind the scenes. If DC point extrapolation and resampling are performed outside the simulator, one has the opportunity to view the time-domain responses and determine upfront whether there is a reasonable expectation for the s-parameters to simulate properly. If, having done this, the time-domain views are wrong, then the s-parameters will not simulate improperly.

As an example of frequency points calculation for s-parameters, consider a device that has an electrical length of 1 ns. It is reasonable to assume a positive impulse response length of 10 ns for a total impulse response length of 20 ns. This dictates a frequency resolution of

$1/20\,\text{ns} = 50\,\text{MHz}$. If the s-parameters are to 20 GHz, this is $20\,\text{GHz}/50\,\text{MHz} = 400$ points. It is reasonable, therefore, to supply 400 points from 50 MHz to 20 GHz in 50 MHz steps and let the resampling extract only the DC point. After all, the phase will be around $-18°$ at 50 MHz and around $-36°$ at 100 MHz, and any extrapolation algorithm will not have a problem with this. The trouble comes when VNA users try to measure too close to DC, where it has very poor dynamic range, which causes noisy measurements that cause extrapolation algorithms to fail. If the DC point extrapolation is unsuitable, then one can bump the frequency resolution to an extreme of, for example, 5 MHz, still providing evenly spaced points, then downsample the result if the impulse response length is seen to be excessive.

## 18.5 *SignalIntegrityApp* Equalization Example

In this final section of the book, the *SignalIntegrity* application is utilized in a manner to demonstrate its capability, both standing alone and in a combination of stand-alone and scripted environments.

Since signal integrity deals with the transfer of digital signals through channels, the example used will analyze the ability of a demo board to transmit serial data at a baud rate[2] of 5 GBd. While 5 GBd is not super-fast, the example will employ pulse amplitude modulation (PAM)-4 signaling, which has two bits for each symbol. The channel is shown in a simulation environment in Figure 18.1, where in Figure 18.1(a) the standard arrangement of voltage sources, terminations, and probes is shown, all of which require some discussion. There are four voltage sources, with two for the positive leg and two for the negative leg. In each leg, a source is provided for the differential and common modes. Since the negative leg is upside down, the common-mode source is also upside down and the differential-mode source is right side up. Thus, the sources labeled VCP and VCM provide the positive and negative leg common-mode voltages with the intent that they are driven with the same signal (i.e. only providing the common mode). The sources labeled VDP and VDM provide the differential-mode signals, and the intent is that they are also driven with the same signal, causing VDP and VDM to be equal and opposite and only providing differential-mode signals. There are tee terminations at the source and pi terminations at the receiver. Regarding the tee, it is easy to see that, when driven differentially, the source has an odd-mode termination of $50\,\Omega$ and therefore a differential-mode termination of $100\,\Omega$. The even-mode termination is $50 + 2 \cdot 20 = 90\,\Omega$ for a common-mode termination of $45\,\Omega$. The reason for this will be seen by examining the demo board. The pi termination is similar. The even-mode termination is $90\,\Omega$ and the common-mode termination is $90 \parallel 90 = 45\,\Omega$. The differential-mode termination is $(90 + 90) \parallel 225 = 100\,\Omega$ and the odd-mode termination is $50\,\Omega$. Remember, the differential mode is twice the odd mode and the common mode is half the even mode. All of these values can be obtained from Table 7.1.

As a sanity check, two projects are created that are shown in Figure 18.2. The mixed-mode tee and pi terminations are created as a project and are shown in Figure 18.2(a) and

[2] Baud (Bd) rate is the symbol rate where multiple bits may be encoded in each symbol. For non-return-to-zero (NRZ), the bit rate and baud rate are the same.

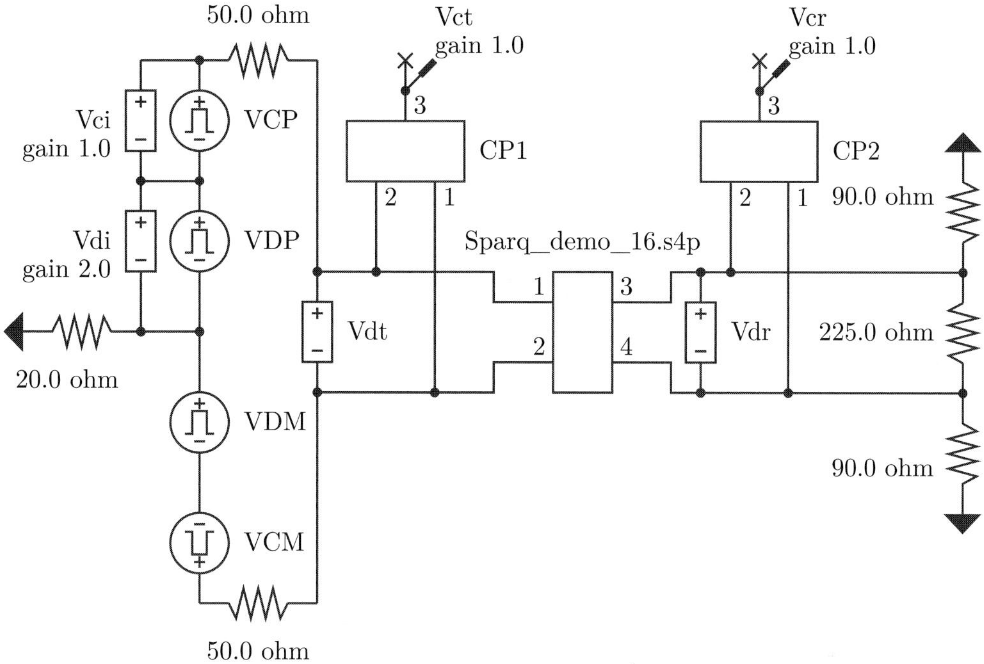


(a) Single-ended circuit

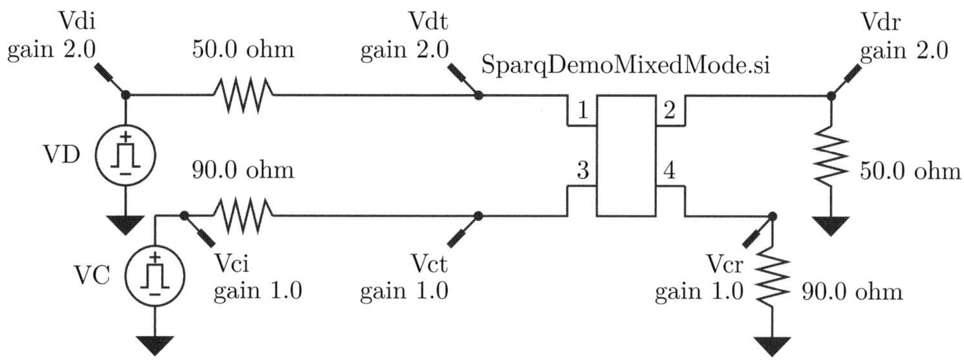


(b) Mixed-mode circuit

**Figure 18.1** Equivalent circuits

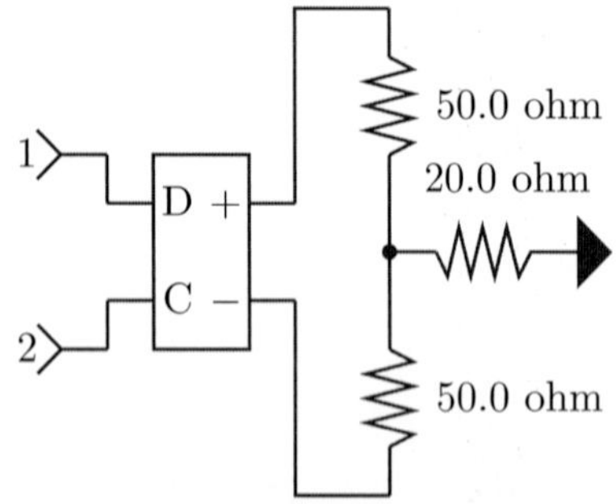


(a) Mixed-mode tee termination

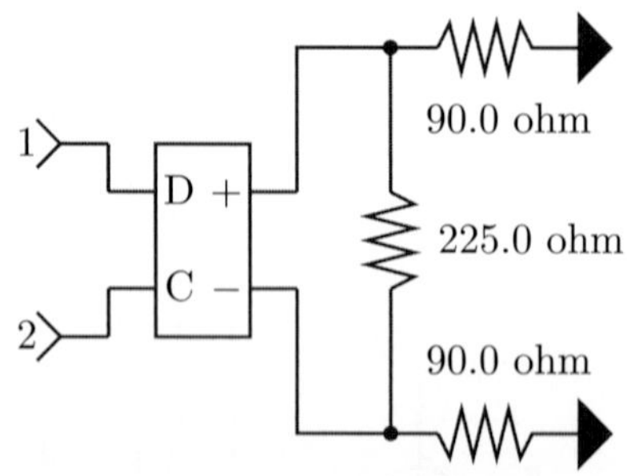


(b) Mixed-mode pi termination

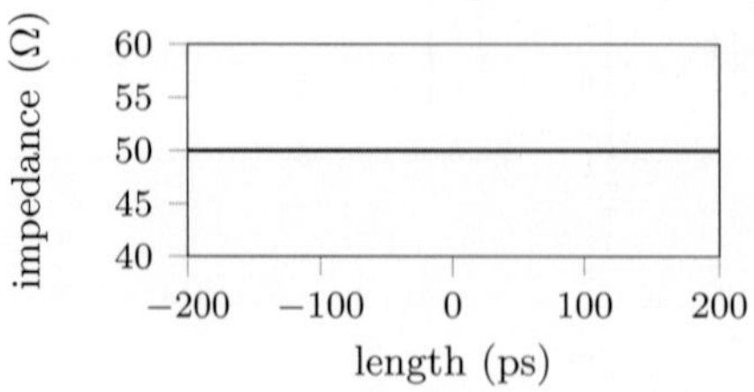


(c) Odd-mode SD1D1 impedance profile

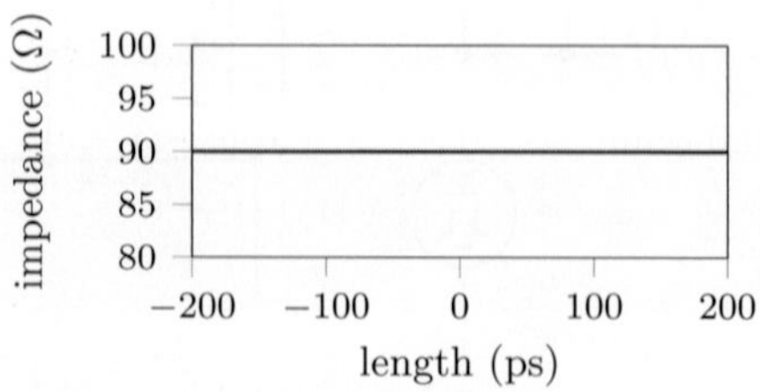


(d) Even-mode SC1C1 impedance profile (in 90 Ω reference impedance)

**Figure 18.2** Mixed-mode terminations

Figure 18.2(b). The s-parameters are generated and the impedance plots verified as 50 Ω odd-mode impedance in Figure 18.2(c) and 90 Ω even-mode impedance in Figure 18.2(d).

Returning to Figure 18.1, the differential probes are shown probing the various points in the circuit with the naming convention as Vxy, where x is either d or c for differential or common and y is either i, t, or r for input, transmitter, or receiver. The probe providing Vdi is doubled as the differential-mode voltage is twice the single-ended voltage of the source.

The *SignalIntegrity* application has a useful feature in that projects can be embedded as both waveform sources and as s-parameter elements. Here, CP1 and CP2 are projects that provide the s-parameters of a common-mode probe formed by combining two voltage amplifiers, as shown in Figure 18.3.

While the circuit in Figure 18.1(a) is useful for simulating the generated mixed-mode signals, the equivalent circuit in Figure 18.1(b) (which performs identically to that in Figure 18.1(a)) is much more useful. The reason that it is more useful is that the *SignalIntegrity* application utilizes the concept of transfer parameters, as discussed in §9.5. These transfer parameters are used not only to perform the simulation, but also to gain insight into the circuit performance. In simulations involving coupled lines, it is the mixed-mode transfer parameters that are of interest. Note here that the voltage sources VD and VC are really the odd and even modes, respectively, and the values match the values in VDP and VDM or

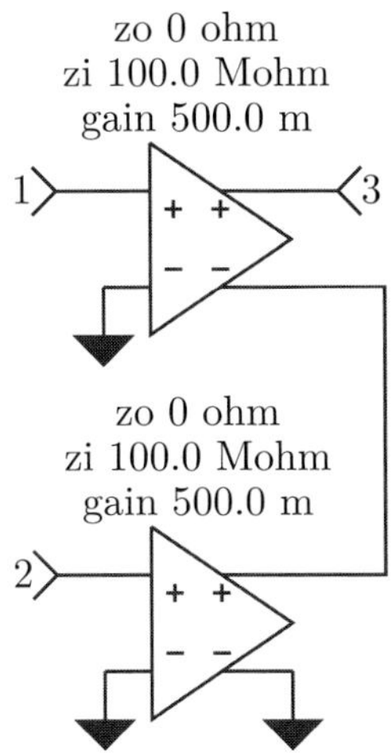


**Figure 18.3** Common-mode probe

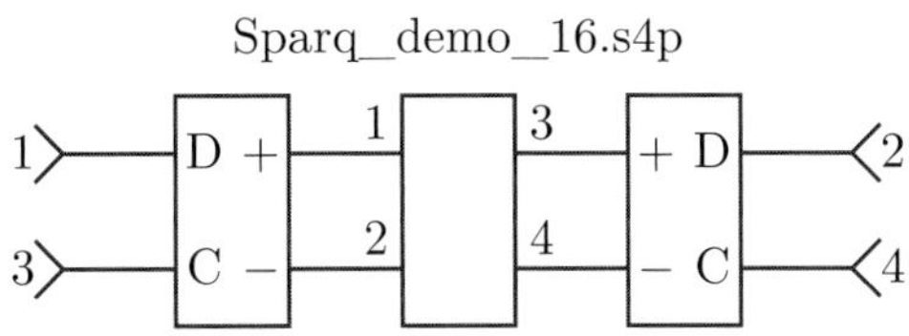


**Figure 18.4** Mixed-mode circuit

VCP and VCM. For the differential mode, though, all of the probe values must be multiplied by two, while the common-mode probe gains are unity. It is not appropriate to double the voltage of VD instead of the probe gains as the effect on the common mode due to the differential mode would be incorrectly scaled.

In the simulation shown in Figure 18.1(b), the mixed-mode s-parameters for the demo board are again provided by another project, the circuit for which is shown in Figure 18.4. The mixed-mode impedance s-parameters corresponding to Figure 18.4 are shown in Figure 18.5. The odd-mode impedance match is shown in magnitude in Figure 18.5(a) and in phase in Figure 18.5(b) (the way a microwave engineer would commonly view it). The impulse response is shown in Figure 18.5(c), and, most importantly, the impedance profile is shown in Figure 18.5(d). In Figure 18.5(d) it is seen that there is a connector discontinuity, a short length of line with 50 Ω impedance followed by about 1.4 ns of line with an impedance of 53 Ω, which rises, indicating some distributed series resistance, as mentioned in §14.6. By observing the construction of the demo board, one finds that there is a coaxial to microstrip transition onto the board, a portion of uncoupled line that joins to form the edge coupled line. The intent of the coupled line is for 50 Ω odd-mode (100 Ω differential-mode) impedance, but it is slightly off.

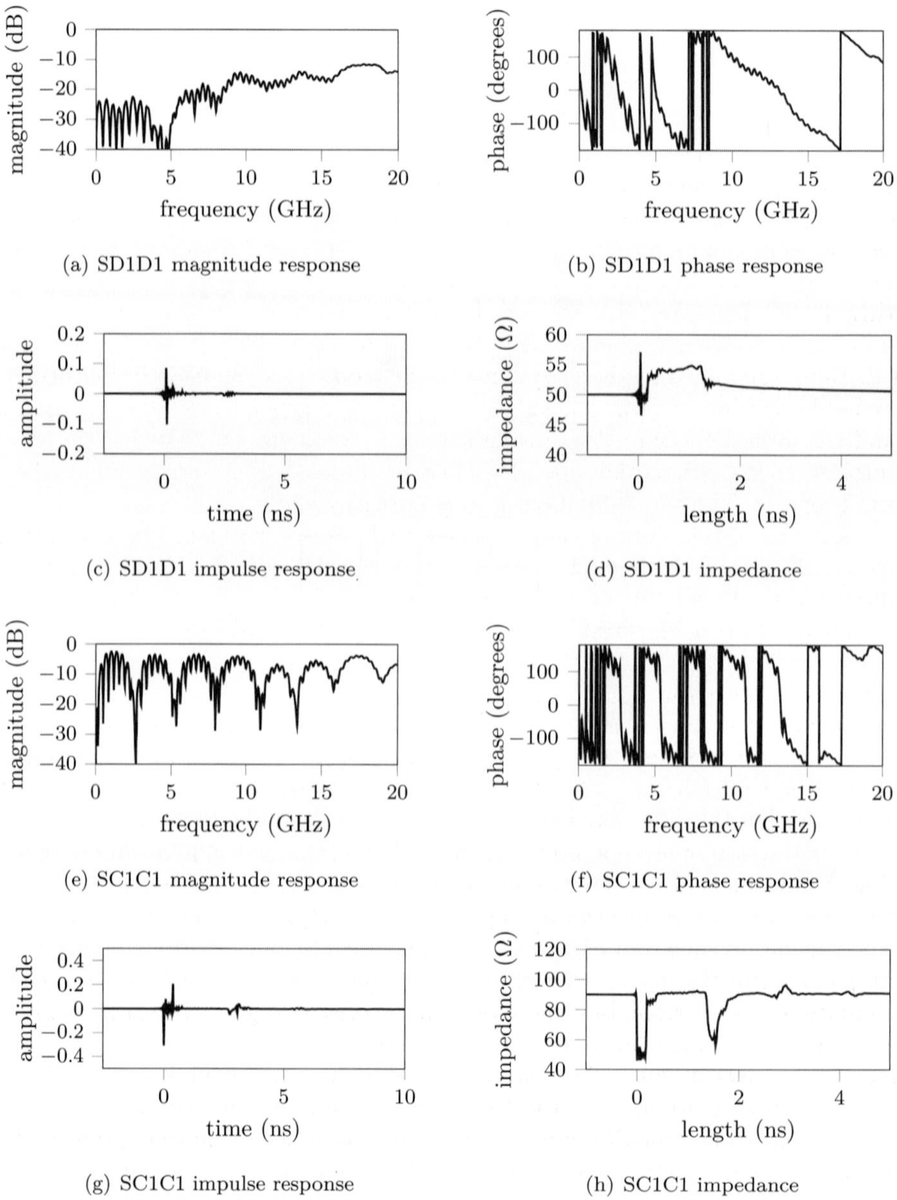


(a) SD1D1 magnitude response (b) SD1D1 phase response

(c) SD1D1 impulse response (d) SD1D1 impedance

(e) SC1C1 magnitude response (f) SC1C1 phase response

(g) SC1C1 impulse response (h) SC1C1 impedance

**Figure 18.5** Mixed-mode impedance s-parameters

The magnitude and phase of the even-mode match are shown in Figure 18.5(e) and Figure 18.5(f). All of the SC1C1 s-parameter plots are plotted in a reference impedance of 90 Ω, representing the even-mode termination shown in Figure 18.1(b). Here it is found that the even-mode match is not very good. The time-domain implications are shown in the SC1C1 impulse response shown in Figure 18.5(g). Again, the impedance profile plot shown in Figure 18.5(h) sheds the most light on what is happening. In the 90 Ω reference impedance, it is shown to drop to 50 Ω briefly, and then run for 90 Ω for about 1.4 ns, the even-mode characteristic impedance of the coupled portion of the line. The second dip is formed as the line becomes uncoupled again and the line ends. The bump around 3 ns is a secondary reflection that would not be shown if the peeling methods in §14.4 were utilized. However, the approximate methods discussed in §14.3 were used; these are more stable, but show more error in the face of severe mismatches.

Returning to Figure 18.1(b), some comments are required. In Figure 18.1(b), where mixed-mode s-parameters are concerned, the mixed-mode equivalent terminations must be provided. Here, the mixed-mode equivalents of the tee and pi source and receiver terminations are used according to §7.5. The odd-mode impedance terminates the differential-mode ports and the even-mode impedance terminates the common-mode ports. Also, the gain of the probes on the differential-mode leg must be doubled to show the differential-mode waveform properly. The common-mode probes are unity gain.

The main goal of this example is to see how well the system transmits a 5 GBd signal, so, to start, the mixed-mode thru s-parameters are examined, as shown in Figure 18.6. The differential-mode magnitude response is shown in Figure 18.6(a), where the response dips to around −5 dB at 2.5 GHz and to around −10 dB at 7.5 GHz. The implications of this are unclear for the moment regarding serial data transmission. The differential-mode phase response is shown in Figure 18.6(b), with 1.436 ns of delay removed, indicating the differential-mode electrical length of the channel. The differential-mode impulse and step responses are shown in Figure 18.6(c) and Figure 18.6(d), indicating that the channel can probably transmit a 5 Gb/s signal without too much distortion.

The common-mode thru response s-parameters are all shown in 90 Ω even-mode reference impedance. The magnitude response is shown in Figure 18.6(e), which looks very wiggly due to the large impedance discontinuities in the uncoupled regions of the line. The common-mode thru response phase is shown in Figure 18.6(f), with 1.586 ns of delay removed. This indicates that the common mode propagates more slowly than the differential mode. A different propagation velocity is always expected when lines are coupled. The common-mode impulse and step responses are shown in Figures 18.6(g) and 18.6(h), which show many reflections due to the impedance discontinuities.

While it is always useful to look at the mixed-mode s-parameters for a sanity check, a very important item to check is the length of the impulse and step responses. The s-parameter viewer in the *SignalIntegrity* application always shows the magnitude and phase response along with the impulse and step responses (or impedance, if desired for the diagonal s-parameters). These responses should be checked for causality problems and sufficient settling of the step response. Both impulse and step responses show settling and causality phenomena in slightly different ways, so both should be reviewed. Problems in this area are usually caused by insufficient number of points in the s-parameters, as outlined in §15.7.4. Here, we have 1000 points to 20 GHz for a spacing of 20 MHz, which leads to an impulse

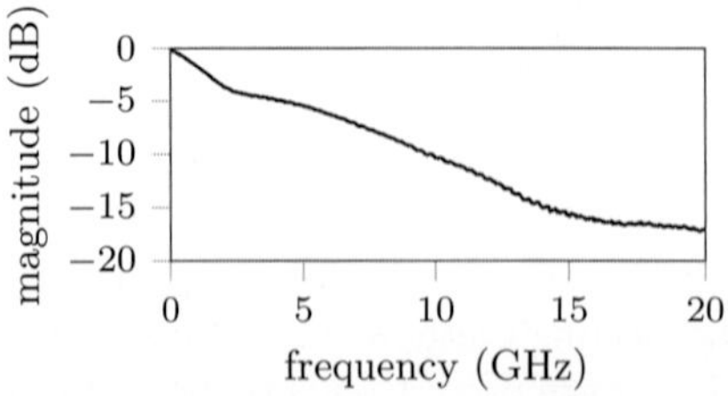


(a) SD2D1 magnitude response

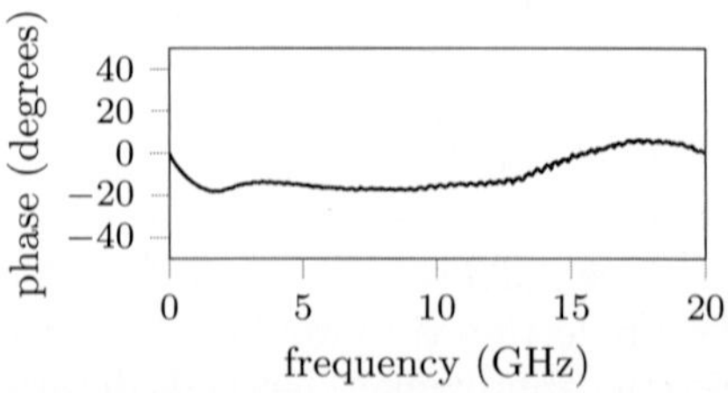


(b) SD2D1 phase response (−1.436 ns)

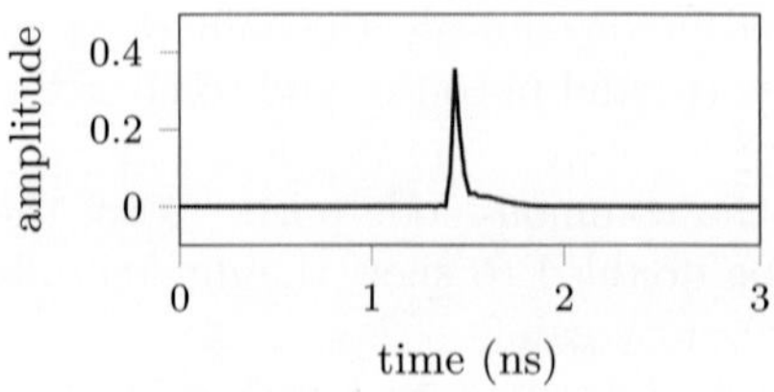


(c) SD2D1 impulse response

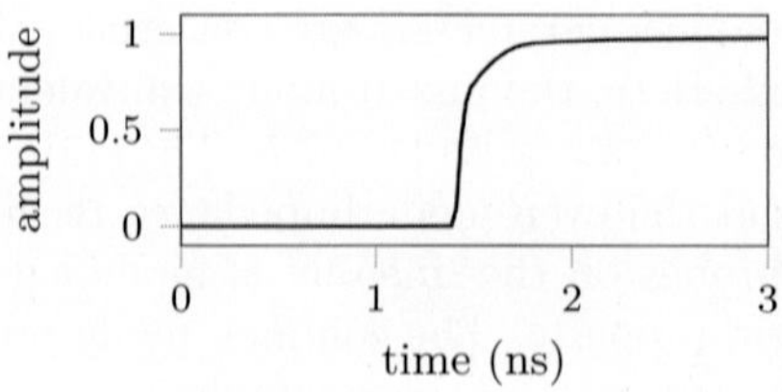


(d) SD2D1 step response

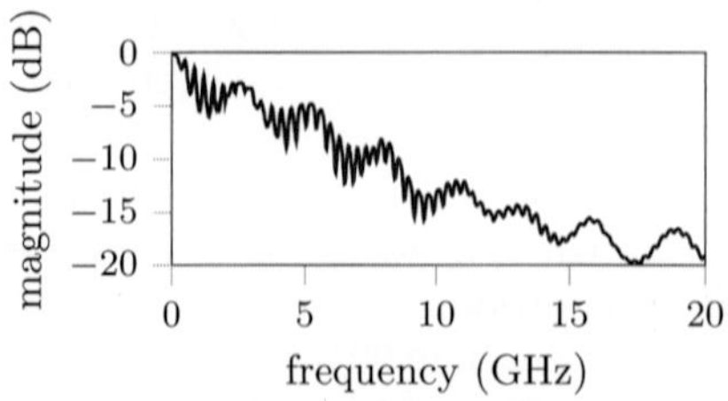


(e) SC2C1 magnitude response

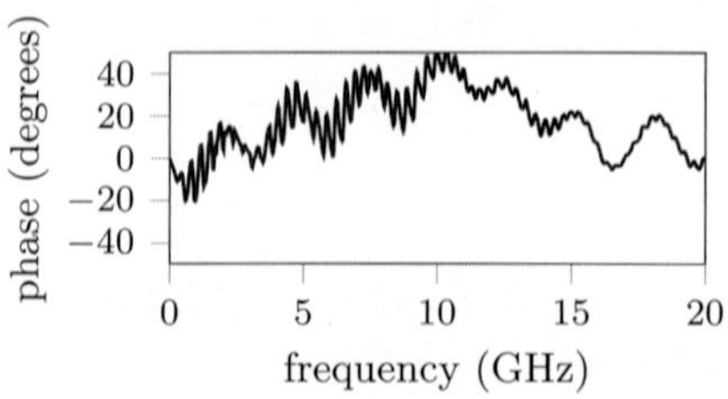


(f) SC2C1 phase response (−1.586 ns)

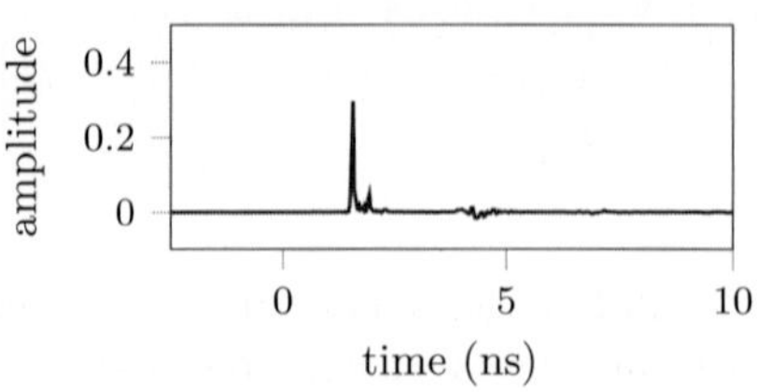


(g) SC2C1 impulse response

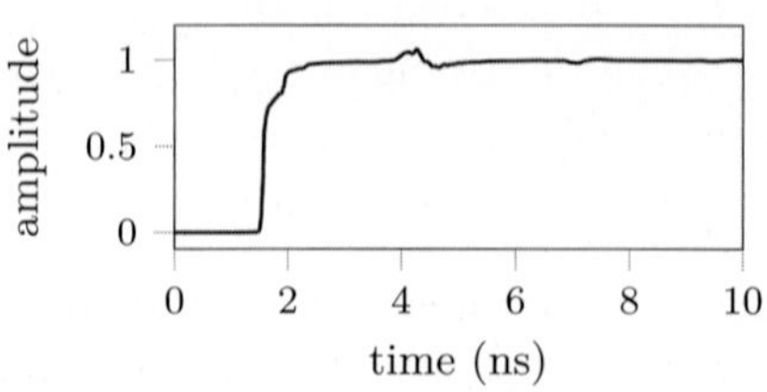


(h) SC2C1 step response

**Figure 18.6** Mixed-mode thru s-parameters

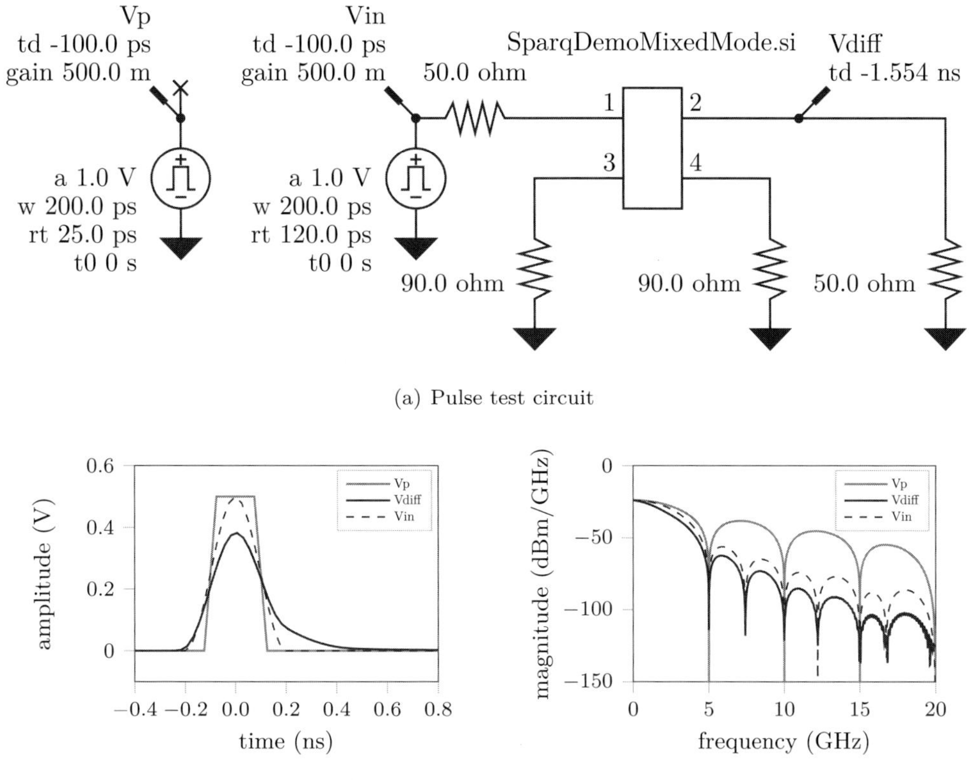


(a) Pulse test circuit

(b) Isolated pulses in time domain

(c) Isolated pulses spectral density

**Figure 18.7** Single-pulse test

response length of 50 ns (25 ns before and after zero time). When the impulse responses are examined, the mode conversion terms (not shown) show an impulse response length of about 20 ns, indicating that these are all correct. This 50 ns impulse response length must be accounted for in any simulation.

Continuing the analysis, the response to a 200 ps differential-mode pulse is simulated in Figure 18.7. A test circuit shown in Figure 18.7(a) injects a 200 ps differential-mode pulse into the demo board. The pulse has a risetime of 120 ps, which is the approximate risetime of the random bit pattern that will be injected later in the example. A probe labeled Vin probes the source with a gain of one-half, in order to match the size of the signal at the receiver. It is delayed by −100 ps in order to center the pulse at zero time. A probe labeled Vdiff is placed at the receiver. It is delayed by −1.554 ns to overlap with the input pulse. Note that there is an extra −100 ps compared with the delay of the channel. Finally, an input pulse with 25 ps risetime is placed driving an open and probed with a probe labeled

Vp, delayed −100 ps, and with a gain of one-half, for comparison. The small 25 ps risetime was added so that this fast pulse passes through the midpoint at exactly ±100 ps.

Some details must now be mentioned regarding the *SignalIntegrity* simulation application. Previously, it was ascertained that the impulse response length of the mixed-mode s-parameters of the demo board was 50 ns. Since this is the only device in the system, the simulation can be performed at 50 ns impulse response length, but, in preparation for more simulations with more interconnected devices, 100 ns impulse response length is used. It is important to understand that the impulse response length required for a simulation can theoretically be as high as the sum of the impulse response lengths of all components.[3]

Using 100 ns as the impulse response length means that the stimulus waveforms must account for the impulse response length, which will be removed from the final generated waveform. This means that if one wants to see 10 ns of output waveform, the source must apply a waveform that is at least 110 ns long. Here, the pulse generator starts at −55 ns and has a duration of 115 ns. The simulation will consume 50 ns from both the beginning and the end of the output waveform, leaving a waveform that starts at −5 ns and ends at +15 ns. Here, the simulation is run at 40 GS/s (as the s-parameters have an end frequency of 20 GHz). If the waveforms are upsampled, as seen later in this example, one extra sample would be consumed from each side of the waveform when linear interpolation is employed, and 64 samples are consumed from each side by default when sinc interpolation is employed, as indicated in §13.2.

The result of the pulse simulation is shown in the time domain in Figure 18.7(b) and in the frequency domain as spectral density in Figure 18.7(c). The time-domain simulation shows the sharp pulse, Vp, the spread input pulse, Vin, and the output of the channel, Vdiff. The spread in Vdiff indicates that there will be quite a bit of inter-symbol interference (ISI). The frequency-domain view in Figure 18.7(c) shows some more useful information. The Vp spectral content shows lobes with nulls at multiples of the 5 Gb/s bit rate (due to the 200 ps bit width) and an approximate drop in content of 10 dB/octave. The drop would be almost exactly 6 dB/octave if the pulse had zero risetime [76]. The Vin spectral content looks surprising in that, in addition to the nulls at multiples of the bit rate, there are additional nulls in between these expected nulls, at, for example, 7.5 GHz. This is a result of the raised cosine filter used to simulate the 120 ps risetime of the pulse. The spectral content of Vin is much lower than Vp, and the channel attenuates this further in Vdiff. Based on the limited risetime of the receiver, only the spectral content out to around 2.5 GHz is very interesting.

An equalizer is employed to improved the pulse. There are three ways in which equalizers are viewed in signal integrity:

1. As a means for boosting high frequencies (emphasis) or alternatively attenuating low frequencies (de-emphasis) to account for loss in the channel. This is the simplest, but least helpful, view of equalization in signal integrity, as ultimately signal integrity is about time-domain effects.

[3] This is for elements in series connections. The requirements can be even higher if feedback is employed, and depending on the impedances presented to the ports. To see what is meant by this, consider a two-port 100 pF shunt capacitance to ground and look at its step response in both 50 Ω and 500 Ω reference impedances.

```
def ZeroForcingEqualizer(project,waveform,bitrate,value,pre,taps):
    import SignalIntegrity.App as siapp; from numpy import matrix
    app=siapp.SignalIntegrityAppHeadless()
    app.OpenProjectFile(project)
    (_,outputWaveformLabels,_,outputWaveformList)=app.Simulate()
    pulsewf=outputWaveformList[outputWaveformLabels.index(waveform)]
    delay=pulsewf.td.TimeOfPoint(pulsewf.Values().index(max(pulsewf.Values())))
    print('delay: '+str(delay))
    H=pulsewf.td.H; ui=1./bitrate
    startTime=delay-(int((delay-H)/ui)-1)*ui
    endTime=delay+(int((pulsewf.Times()[-1]-delay)/ui-1)*ui)
    M=int((endTime-startTime)/ui+0.5); d=int((delay-startTime)/ui+0.5)
    x=[pulsewf.Measure(startTime+m*ui) for m in range(M)]
    X=[[0 if r-c < 0 else x[r-c] for c in range(taps)] for r in range(M)]
    r=[[value] if r == d+pre else [0] for r in range(len(X))]
    a=[v[0] for v in (matrix(X).getI()*matrix(r)).tolist()]
    print('results: '+str(a))
```

(a) Zero-forcing equalizer fitting function

```
delay:  5.86272511144e-12
results:  [-0.03763, 1.32098, -0.25341, 0.00600, -0.00863]
```

(b) Fit result

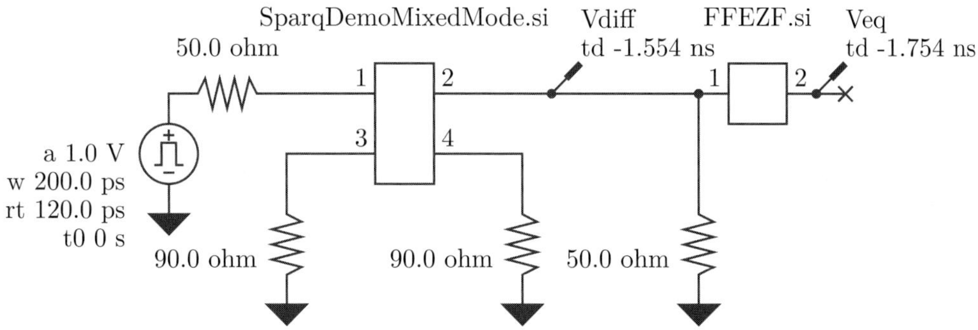


(c) Equalized pulse test circuit

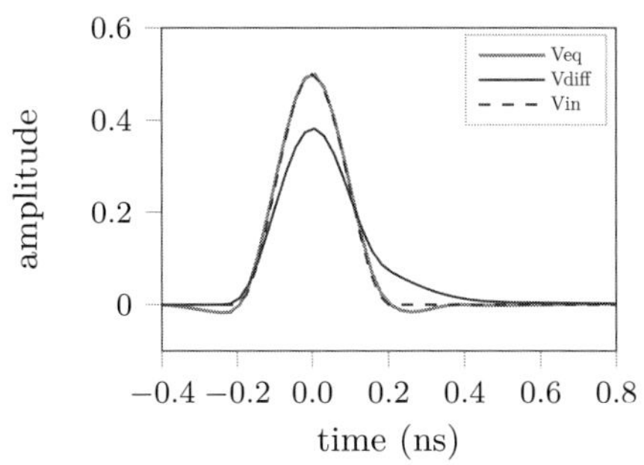


(d) Equalized pulses in time domain

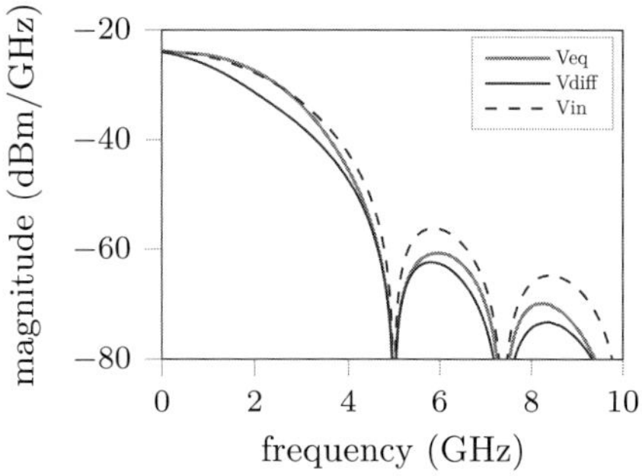


(e) Equalized pulses spectral density

**Figure 18.8** Zero-forcing equalizer

2. As a means for reducing ISI. The loss of the channel causes spreading of the pulse, which means that each transmitted symbol interferes with adjacent samples. Equalization can be used to narrow the pulse and to target the ISI that technically only occurs at exact multiples of the UI.

3. As a means for reducing the mean-squared error signal (which is the difference between the samples of the serial data waveform and the ideal samples).

As a first step, the ISI is reduced using a linear feed-forward equalizer (FFE) in the form shown later in Figure 18.12. The choice is made to place taps of the FFE exactly 200 ps apart and to sum weighted versions of the delayed signal seen at the output of the delay taps. Note that the chosen tap spacing is not a requirement of this type of equalizer, and often taps with less delay, like one-half of the UI, are employed.

Examining the Vdiff waveform in Figure 18.7(b), it seems reasonable to design an equalizer that removes the ISI caused by the pulse spreading. An equalizer, whose goal is to produce a value of 0.5 at time zero and a value of zero at all other times, can be created directly based on the pulse response. This type of equalizer design is called a *zero-forcing* equalizer, because its goal will be to force the pulse response to go through zero at all times other than zero. The Vdiff waveform is seen to be slightly lifted at −200 ps and at 600 ps, and therefore a five tap equalizer is created. In the equalizer, the output is taken from the tap that is delayed by 200 ps, creating one tap that is before the desired bit position (called the *cursor*). This means that there is one pre-cursor tap and three post-cursor taps.

A zero-forcing FFE equalizer example is provided in Figure 18.8. The tap values for the equalizer are produced by a small script shown in Figure 18.8(a), which shows the function `ZeroForcingEqualizer()`. This function takes arguments including: the name of the *SignalIntegrity* project to load (that generates the pulse response), the waveform in that project to equalize, the bit rate, the desired value at the cursor location, the number of pre-cursor taps, and the total number of taps. Although the project for the pulse response simulation was created in the *SignalIntegrity* GUI application, here it is loaded in a headless form accessible by external programs, and the simulation performed. The desired waveform Vdiff is extracted and this simulated waveform is used to determine the equalizer. The delay is calculated to the maximum point on the waveform determined to be the cursor location. Because the probe already had a delay of −1.554 ns, the delay is in addition to that delay. Ideally, it should be zero, in this case. In Figure 18.8(b), it is seen to be the small value of about 6 ps. Subsequently, the exact times are calculated such that one time is at the peak of the pulse and the others are at multiples of the UI. Then, these time points are sampled in the $M$ element list $\mathbf{x}$. The equation that solves for the taps is

$$\mathbf{X} \cdot \mathbf{a} = \mathbf{r},$$

where $\mathbf{a}$ is a vector containing the tap values, $\mathbf{X}$ is a preferably skinny matrix with the number of taps as the number of columns, containing $M$ rows, where each row contains delayed chunks of the samples in $\mathbf{x}$. The vector $\mathbf{r}$ contains the desired result of the multiplication $\mathbf{X} \cdot \mathbf{a}$, that is the desired value at one location and zero at all others. The result of this fit is shown in Figure 18.8(b) and is employed in an equalizer similar to that shown in Figure 18.12(a), but with five taps.[4]

The project containing the five tap equalizer is constructed in `FFEZF.si` and is included in the circuit shown in Figure 18.8(c). The simulation results are shown in the time domain in Figure 18.8(d) and in the frequency domain in Figure 18.8(e). Examination of the Veq waveform of Figure 18.8(d) shows that the zero-forcing objective was accomplished. At $-200$ ps, the lifting pulse is pulled down and at zero time the pulse reaches 0.5 V. The pulse is forced down to zero at exactly 200 ps. The zero forcing occurring at $\pm 200$ ps comes with some undershoot, and the equalizer lifts it back to zero again at 400 ps and back down again at 600 ps. Although this example does not illustrate any extreme behavior, these types of equalizers can be heavy handed in their approach as inter-sample behavior is completely ignored.

The frequency-domain behavior shown in Figure 18.8(e) can be confusing as it only serves to equalize the Veq signal to match Vin up to about 3 GHz. This is due to the Nyquist rate limitation of the equalizer; this will be discussed later in the example. There is some boost in the 6–9 GHz range, but this is clearly unimportant as Veq does not come close to matching Vin in this range, yet it is clear that this equalizer will perform well. As a note, the DC gain of the equalizer is the sum of the coefficients, which is very close to unity and indicated in Figure 18.8(e), where the spectral density at DC is largely unchanged.

The next step in the example is to examine the behavior with a PAM-4 signal. Therefore, a project is created as shown in Figure 18.9 containing the circuit shown in Figure 18.9(a). Here, a NRZ (also called PAM-2) generator is employed to generate a pseudo-random bit sequence (PRBS), NRZ waveform; PRBS waveforms are defined by a polynomial based on their length. The waveforms are pseudo random in nature in that they define a random waveform that repeats after about $2^P$ bits, where $P$ is the polynomial length [77]. PRBS generators are common in instrumentation, especially when tests are made with a bit error rate tester (BERT). Here, PRBS-15 repeats the pattern in approximately 6.5 µs.

In Figure 18.9(a), the PRBS waveform is buffered and delayed by transmission lines, and three amplifiers are used to combine the three waveforms. This produces a mostly random PRBS waveform, but tends to favor single-level transitions. It is not an optimal form of PAM-4 PRBS generation, but is serviceable. This construction is commonly used in high-speed optical communications testing, and often a three-way combiner is used to merge the waveforms. The PRBS waveform generated is shown in Figure 18.9(b).

Figure 18.10 provides a test using the generator in Figure 18.9, which is added as a waveform generating project in Figure 18.10(a) in place of the pulse generator. The resulting waveforms are shown in Figure 18.10(b), where a clear improvement is noted between Veq, the equalized waveform, and Vdiff, the unequalized waveform. Veq is seen to track Vin nearly exactly, except occasionally where it overshoots in between sample locations.

While the zero-forcing equalizer is a good equalizer in this case, it is worthwhile mentioning one more equalization objective – that of reducing the rms value of the error signal at the sample point locations, as mentioned previously. This method still uses the FFE as the equalizer, but uses a different objective in the determination of the tap weights, as provided

[4] The five tap equalizer is not shown, in the interest of space, but it is the same equalizer as provided in Figure 18.12(a) (with two additional post-cursor taps), and the tap values provided in Figure 18.8(b).

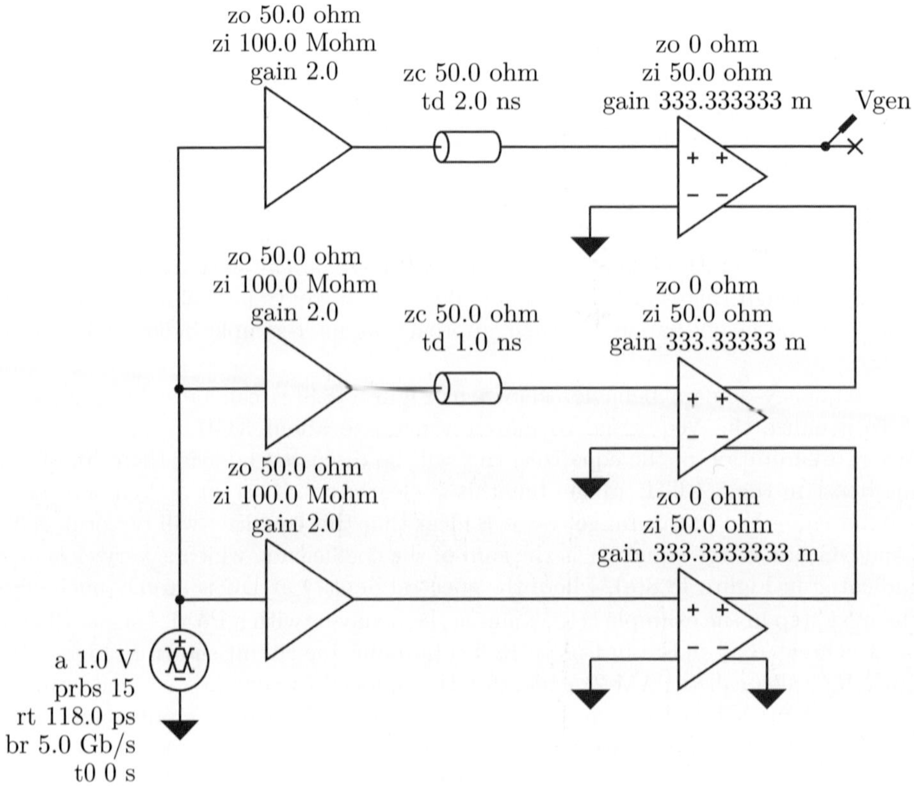


(a) Generation circuit

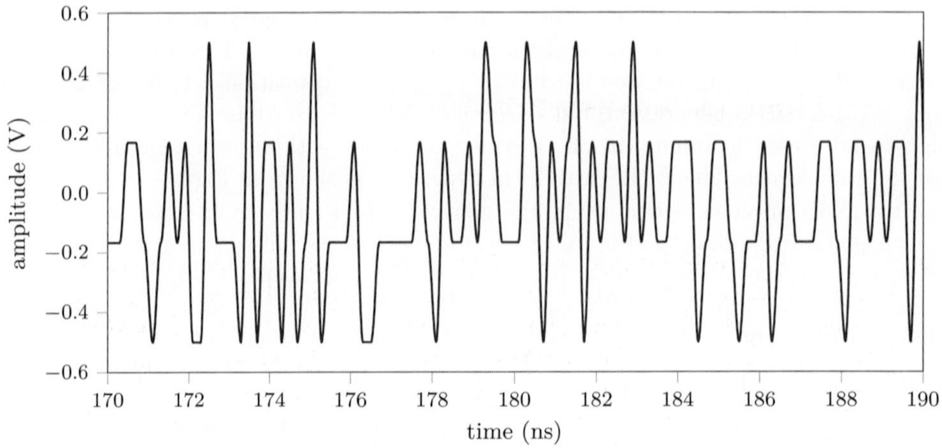


(b) PRBS waveform (zoomed)

**Figure 18.9** Pseudo-PRBS PAM-4 generator

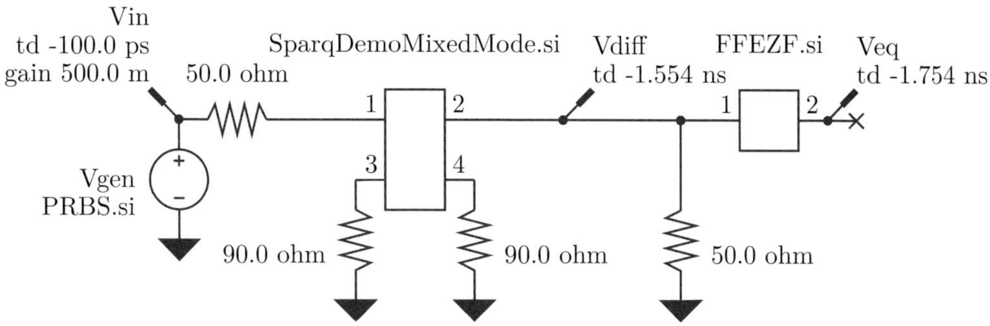


(a) PAM-4 analysis circuit

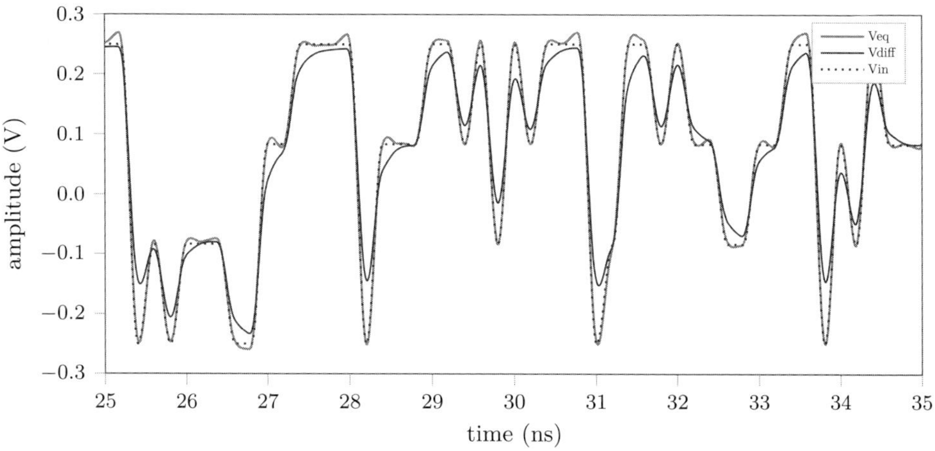


(b) PRBS waveforms

**Figure 18.10** PAM-4 equalization with zero-forcing equalizer

in Figure 18.11. Figure 18.11(a) shows a fitter derived from the **LevMar** class shown in Listing 16.1 and Listing 16.2 (used for model fitting in §16.5). Just as for the model fitting code in Listing 16.3, the **EqualizerFitter** class also derives from the **LevMar** class. To make a fitter that derives from the **LevMar** class, all that is needed is to provide the `fF()` method and preferably to override the `__init__()` method to transform the problem into the correct form during initialization. Here, the initialization method receives the sampled waveform, the correct levels, and the number of pre- and post-cursor taps. During initialization, the waveform is *decoded*, meaning that the closest of the levels provided is used as the correct value that the waveform should attain at the sample point. Thus, the system is set up to minimize the mean-squared error of the difference between $f(\mathbf{x}, \mathbf{a})$ and $\mathbf{y}$, where

```
class EqualizerFitter(si.fit.LevMar):
    def __init__(self,callback=None):
        si.fit.LevMar.__init__(self,callback)
    def Initialize(self,sampledWf,levels,pre,post):
        self.levels=levels; self.pre=pre; self.post=post
        a=[[0.] for _ in range(pre+post+1)]
        self.x=[[v] for v in sampledWf.Values()]
        y=[[m] for m in self.Decode(self.x)[pre:len(self.x)-post]]
        si.fit.LevMar.Initialize(self,a,y)
        self.m_epsilon=0.0000001
    def Decode(self,x):
        return [self.levels[min(list(zip([abs(v[0]-d)
            for d in self.levels],range(len(self.levels)))))[1]] for v in x]
    def fF(self,a):
        return [[sum([a[i][0]*self.x[k-i+self.pre][0]
            for i in range(self.pre+self.post+1)])]
                for k in range(self.pre,len(self.x)-self.post)]
    def AdjustVariablesAfterIteration(self,a):
        self.y=[[v] for v in self.Decode(self.x)[self.pre:len(self.x)-self.post]]
        return si.fit.LevMar.AdjustVariablesAfterIteration(self,a)
    def Results(self):
        return self.m_a
```

(a) **EqualizerFitter** class

```
def EqualizerFit(project,waveform,delay,bitrate):
    import SignalIntegrity.App as siapp
    import SignalIntegrity.Lib as si
    app=siapp.SignalIntegrityAppHeadless()
    app.OpenProjectFile(project)
    (_,outputWaveformLabels,_,outputWaveformList)=app.Simulate()
    prbswf=outputWaveformList[outputWaveformLabels.index(waveform)]
    H=prbswf.td.H; ui=1./bitrate
    dH=int(H/ui)*ui-delay+ui; lastTime=prbswf.Times()[-1]; dK=int((lastTime-ui-dH)/ui)
    decwftd=si.td.wf.TimeDescriptor(dH,dK,bitrate)
    decwf=si.td.wf.Waveform(decwftd,[prbswf.Measure(t) for t in decwftd.Times()])
    os.chdir(os.path.dirname(os.path.realpath(__file__)))
    fitter=EqualizerFitter()
    fitter.Initialize(decwf,[-0.25,-0.1667/2.,0.1667/2.,0.25],1,1)
    fitter.Solve()
    print(fitter.Results())
```

(b) Equalizer fit

```
[[-0.015044760441426711], [1.2365319540287978], [-0.19142847360919185]]
```

(c) Result

**Figure 18.11** Equalizer fitting Python code

the former is the equalized waveform and the latter consists of the correct levels obtained from the decoded waveform.

An important feature that has been added is that of *blind equalization*. Lucky [30] determined that, under most circumstances, it is not important that the waveform be decoded entirely properly at the beginning of the equalization process. In other words, the decoding of the waveform can be incorrect and the equalization can still succeed. That being said, one must not depend on the initial decoding forever, but should adapt the idea of what the correct levels look like as the fit progresses. Although a mathematical proof is not provided here, it absolutely works in practice. This is because of the fact that the equalizer does not set the bit, but only provides frequency response changes that emphasize or de-emphasize certain frequencies. Therefore, even in a closed-eye scenario, where many of the bits are decoded incorrectly, the fit serves to open the eye as it progresses, causing more and more bits to be decoded properly until the fit converges. Another way to look at this is that, when a bit is decoded incorrectly, it is usually on the edge, let's say halfway between two levels. In this case, the absolute error is the same, half a bit level, no matter how it is decoded, and the remedy is the same: to amplify the higher frequencies of rapid bit transitions, causing these bits to be decoded properly. Whichever way one looks at it, it is astounding that the equalizer values can be determined without knowing initially what the correct values are. The blind adaption feature is provided by overloading the method `AdjustVariablesAfterIteration()`. This method is usually used to constrain the fitted values during an iteration. An example might be to enforce DC gain being unity during the fit. Here, it is used to decode the waveform and declare the new objective $\mathbf{y}$ after each iteration. A wrongly decoded waveform would continuously adjust itself after each iteration step. A final note on this class is that there is an admission of laziness here as it is customary to overload the method `fJ()`, which calculates the Jacobian, as was seen for the model fitting in Listing 16.4. Here, a numerical approximation is employed despite the simplicity of the partial derivative calculations required.

As in the pulse simulation example, a function is provided that runs the simulation on the headless project created for the PRBS simulation project in Figure 18.10. This function is shown in Figure 18.11(b), which takes as arguments the project and waveform to be simulated, the amount of delay through the channel, and the bit rate. An improvement would be to provide the levels and the number of pre- and post-cursor taps as arguments, but they are hard coded here. Recognizing from the pulse simulation that the last two post-cursor taps are not required, they are eliminated and only one pre-cursor tap and one post-cursor tap are employed. The Vdiff waveform is sampled at locations containing the delay value at distances from it that are multiples of the reciprocal of the bit rate, which is 200 ps, and is provided to the **EqualizerFitter** class, and the equalizer tap values are solved for. The solution is shown in Figure 18.11(c).

The results shown in Figure 18.11(c) are utilized in Figure 18.12, with the circuit diagram in Figure 18.12(a) utilizing the tap weights calculated. The pre-cursor tap goes at the bottom where there is no delay, the cursor tap goes in the middle, and the post-cursor tap goes at the top where there is the most delay. The plots of the equalizer are all of $S_{21}$. In Figure 18.12(b), the equalizer is seen to supply 6–9 dB of gain.

The 6 dB of gain was initially confusing, but offers another educational moment regarding s-parameters. In Figure 18.12(a), the input impedance of the input buffer is high. It has

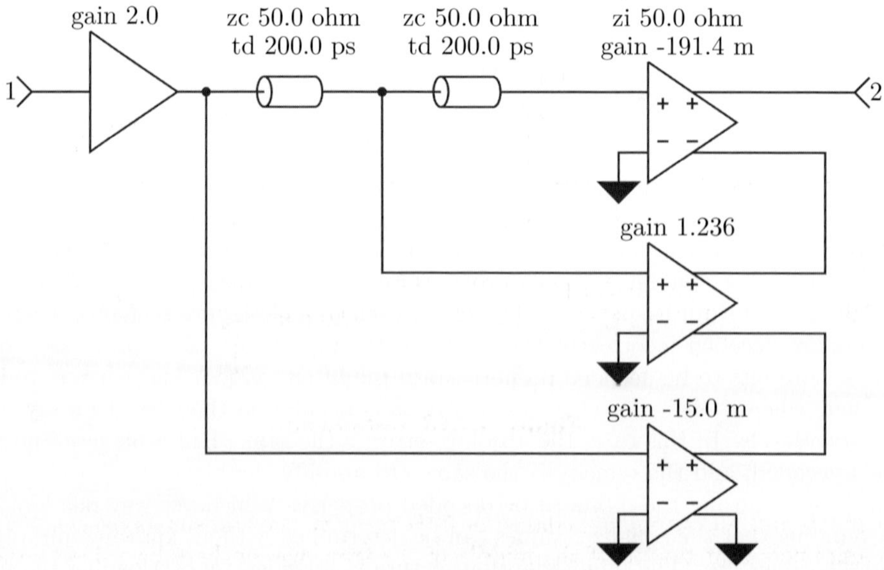


(a) Circuit

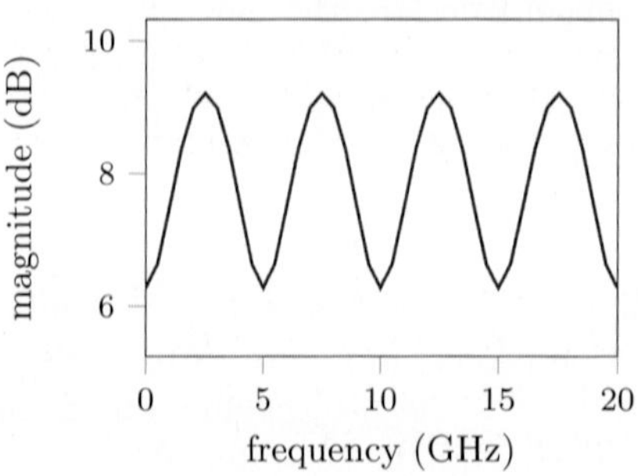


(b) Magnitude response

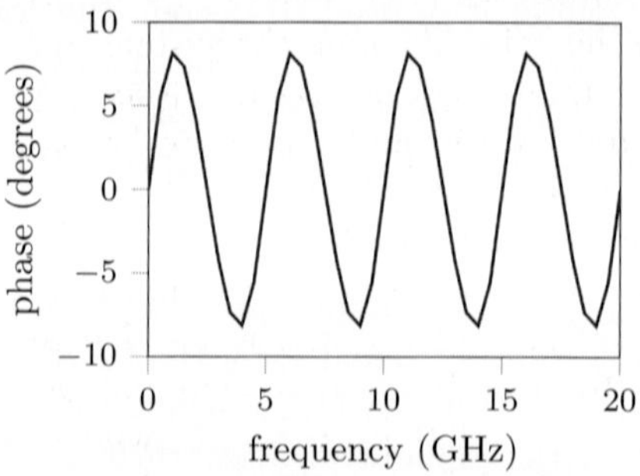


(c) Phase response (−200 ps)

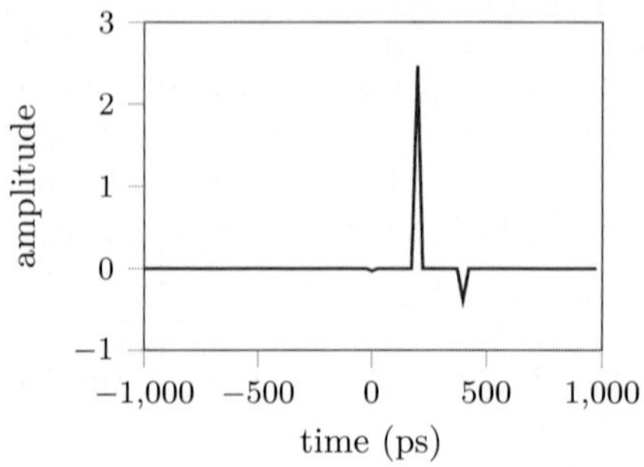


(d) Impulse response

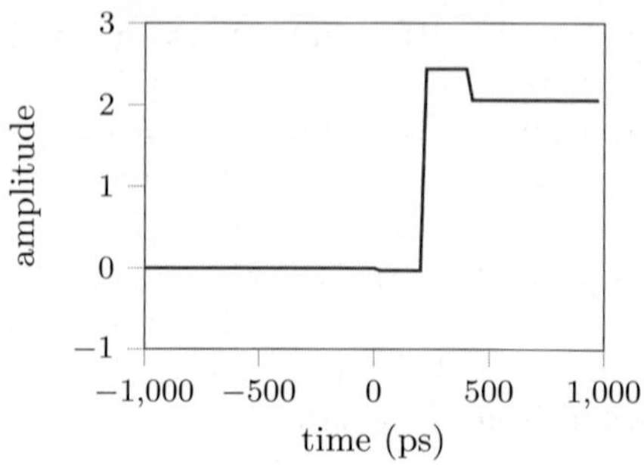


(e) Step response

**Figure 18.12** Feed-forward equalizer

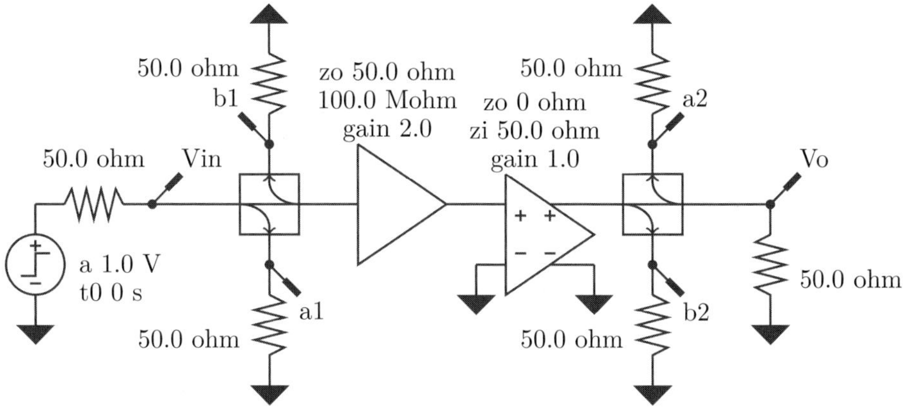


**Figure 18.13** Test circuit

a gain of 2.0 and an output impedance of 50 Ω to drive the transmission lines. While all of the amplifiers that tap off of the middle of the transmission lines have high impedance inputs, the top amplifier terminates the cascaded lines in 50 Ω, thus halving the size of waveform. The output amplifier has a low impedance output. This means that if 1 V is driven through a 50 Ω resistor, 1 V appears at an output 50 Ω resistor for unity gain! To resolve the gain dilemma, the system in Figure 18.13 is set up as a test; this is mostly the same circuit as Figure 18.12(a) without the extraneous equalization circuit items. Here, it is found that both Vo and Vin rise to 1 V at zero time. The circuit has been instrumented with directional couplers, as shown. The voltage reflected waveform at b2 is 1 V and the voltage incident waveform at a2 is zero, all as expected, but the voltage incident waveform at a1 rises to only 0.5 V, as does the voltage reflected waveform at b1. In other words, the voltage Vin is the sum of the incident and reflected voltage waveforms with the voltage formed equally from both, because the input impedance is high. Thus, $a_1 + b_1 = 0.5 + 0.5 = 1.0$ and $a_2 + b_2 = 0.0 + 1.0 = 1.0$, which means unity voltage gain. But the s-parameter $S_{21} = b_2/a_1 = 1.0/0.5 = 2.0$, and the problem is now resolved. It is important to remember that $S_{21}$ is a wave transfer function, not a voltage transfer function.

The voltage gain of the equalizer therefore ranges between 0.25 dB at DC to a maximum of 3.2 dB at 2.5 GHz, and repeats every 5 GHz. The repetition is due to the 200 ps tap delays,[5] making the effective sampling frequency of the equalizer 5 GHz and the Nyquist rate therefore 2.5 GHz. The effective sample rate can always be increased by making the tap delays smaller. The phase of the equalizer is shown in Figure 18.12(c) with 200 ps removed. The phase rises to about 8° at 1.25 GHz to zero at 2.5 GHz to about −8° at 3.75 GHz and again to zero at 5 GHz, where it repeats itself every 5 GHz. The rising phase at DC causes a seeming non-causality that leads to the name *feed forward*, as it seems that information about the bit is fed forward in the response. Remember, however, that 200 ps has been subtracted from the phase as the output comes delayed by one transmission line delay. A real equalizer would have a frequency response that cannot repeat endlessly, but, as seen

[5]This equalizer is also called a tapped delay line equalizer.

```
def EyePattern(project,waveform,delay,bitrate):
    import numpy as np
    import SignalIntegrity.App as siapp
    app=siapp.SignalIntegrityAppHeadless()
    app.OpenProjectFile(project)
    (_,outputWaveformLabels,_,outputWaveformList)=app.Simulate()
    prbswf=outputWaveformList[outputWaveformLabels.index(waveform)]
    ui=1./bitrate; times=prbswf.Times()
    timesInBit=[((t-delay)/3./ui-int((t-delay)/3./ui))*3.*ui for t in times]
    from PIL import Image
    R=400; C=600
    EyeWfCols=[int(t/3./ui*C) for t in timesInBit]
    EyeWfRows=[int((v+0.3)/0.6*R) for v in prbswf.Values()]
    bitmap=[[0 for c in range(C)] for _ in range(R)]
    for i in range(len(EyeWfRows)):
        bitmap[EyeWfRows[i]][EyeWfCols[i]]=bitmap[EyeWfRows[i]][EyeWfCols[i]]+1
    maxValue=(max([max(v) for v in bitmap]))
    bitmap=[[int((maxValue - float(bitmap[r][c]))/maxValue*255.0)
            for c in range(C)] for r in range(R)]
    img=Image.fromarray(np.squeeze(np.asarray(np.matrix(bitmap))).astype(np.uint8))
    img.save(waveform+'.png')
```

(a) Python code

(b) Vdiff

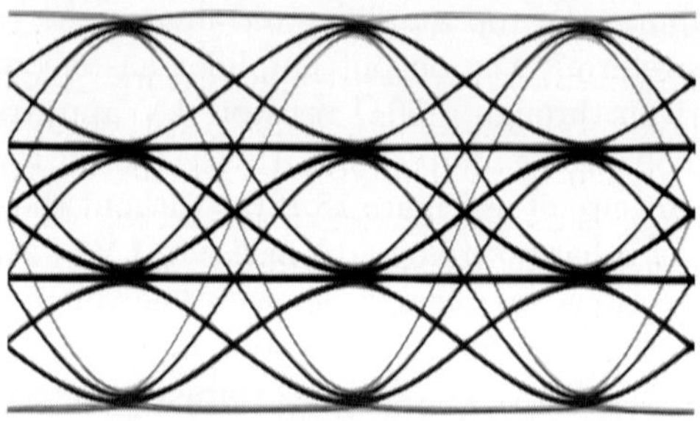

(c) Veq

(d) Veq (zero forcing)

**Figure 18.14** Eye diagrams

previously, the performance only seems to matter to 2.5 GHz anyway. That being said, these kinds of ideal equalizers are commonly implemented in software in oscilloscopes for standards testing.

The equalizer performance is seen clearly by looking at the impulse response in Figure 18.12(d) and especially clearly by looking at the step response in Figure 18.12(e) (remembering to divide it by half). Clearly, the equalizer emphasizes bits during transitions and does not emphasize bits when not transitioning.

While the waveforms shown in Figure 18.10(b) are interesting, the true figure of merit used in serial data testing is the eye diagram. An eye diagram is formed by overlaying portions of the waveforms, usually one or two UI wide with the sample point in the middle, which forms an eye shape, as in Figure 9.8(b). The best way to form the picture is to build a rasterized, or bitmap, image by making an array of pixels and counting the crossings made through each pixel. These counts can then be made into a bitmap by rendering this count as a particular color.

The eye diagrams are provided in Figure 18.14. A short script for producing an eye diagram is shown in Figure 18.14(a). The `EyePattern()` function shown takes the same arguments as the fitter, those being the project name, the waveform to use, the delay, and the bit rate. Prior to using this project, the output waveforms are generated upsampled to 2 TS/s, which makes each sample 0.5 ps. Since the bitmap is 600 pixels across, this makes it such that a waveform point never skips a bin horizontally. The 400 vertical pixels are chosen arbitrarily to give the picture a pleasing form factor. The times in each bit (actually three bits) are calculated, and, as the waveform traverses each pixel, the count in that pixel is incremented. The pixel counts are converted to a shade of gray and the Python package *pillow* is used to render this to a .png file for export. The bitmap pictures have their pixels scaled with a curve to improve contrast. In theory, this curve could have been applied by the script.

Figure 18.14(b) shows the result of applying the script to the Vdiff waveform. While no bit errors are seen in this plot, remember that this simulation only accounts for ISI effects. In a real system, there is also noise and jitter.[6] Both of these effects would probably close this eye, as there is very little margin to begin with. Figure 18.14(c) shows the result of applying the three tap equalizer shown in Figure 18.12. It is amazing that such a simple equalizer providing only 3 dB of gain has this effect on the eye. Finally, Figure 18.14(d) shows the result of applying the five tap zero-forcing equalizer as calculated previously. Note the pinched locations where the data is sampled, showing that all ISI has been removed at the sample times. Note also the slightly larger spreading of the waveform in between the sample points, as is characteristic of this type of equalizer design. While the second equalizer design can also have these effects, it tends to adapt better to other variations in response that can occur, and can also be used in conjunction with the clock recovered at the receiver to adapt to clock recovery and other deterministic jitter effects.[7]

[6]The analysis of jitter is actually an entire subject in itself and is beyond the scope of this book.

[7]Deterministic jitter is jitter that is correlated somehow with the pattern of bits being transmitted.

# Afterword

THIS BOOK WAS TYPESET by the author almost entirely in a free software environment; the development spanned four long-term support versions of Ubuntu Linux ending at 18.04 (see `https://ubuntu.com`).

The book was developed in LyX ending at version 2.3.2. LyX is a front-end for LaTeX and is a superior tool for such a large undertaking. LyX has many developers and was started by Matthias Ettrich (see `https://www.lyx.org`). It would have been impossible to produce this book without LyX.

The book was typeset using the LyX generated LaTeX code with the LuaTeX engine from the 2017 TeX live Debian distribution (see `https://tug.org/texlive`). LaTeX was created by Leslie Lamport and is derived from the TeX language created by Donald Knuth.

The fonts used are all from the Latin Modern family created by Boguslaw Jackowski and Janusz Marlan Nowacki (see `https://ctan.org/pkg/lm`).

The LaTeX document class is *memoir* written by Peter Wilson and Lars Madsen (see `https://ctan.org/pkg/memoir`).

The following LaTeX packages were used: tikz, cite, memhfixc, index, import, chapterfolder, acronym, showkeys, outliner, multicol, bbold, lettrine, alltt, lmodern, fontenc, pgfplots, bbm, float, mathtools, listings.

The index was created by MakeIndex written by Pehong Chen.

The references were created by BibTeX created by Oren Patashnik and Leslie Lamport.

There is not a single bitmap image in the entire document.

All of the schematics shown are in Ti*k*z, which is output directly from the *SignalIntegrity* application, but in the same format produced by TpX written by Alexander Tsyplakov (see `https://sourceforge.net/projects/tpx`). When necessary, the schematics were edited in TpX.

All of the plots were produced by Matplotlib created by John D. Hunter. Matplotlib is embedded in the *SignalIntegrity* application, which is capable of outputting PGF/Ti*k*z plots directly using matplotlib2tikz. Ti*k*z and PGFPlots were created by Till Tantau and Christian Feuersänger (see `https://ctan.org/pkg/pgfplots`).

Matplotlib2tikz was created by Nico Schlömer (see `https://github.com/nschloe/matplotlib2tikz`).

The eye patterns were rendered in the GNU Image Manipulation Program (GIMP) (see `https://www.gimp.org`) and converted to vector graphics using Inkscape (see `https://inkscape.org`).

The UML diagrams were produced by dot, which is part of the open-source Graph Visualization Software (Graphviz) (see `https://graphviz.org`), and edited in Inkscape.

The Python programming language (see `https://www.python.org`) was used to create the *SignalIntegrity* application, where many of the listings in that application are included.

A substantial number of tests of the book's math and software were made using Python, and many of the tests produced the LaTeX that was directly imported into the document.

I can't thank enough the developers of all the free software I used.

Much of the algebra in this book required the use of MathCad (see `https://www.ptc.com/en/products/mathcad`), originally from Mathsoft. MathCad 2000 was run on Microsoft Windows 2000 operated in a VirtualBox environment (see `https://www.virtualbox.org`) under Linux. TpX also operates on Windows 2000. Mathcad and Windows 2000 were the only non-free (and very old) software used.

## Errata

A book this large is bound to have errors. As errors are found, errata will be maintained online at:

```
https://github.com/TeledyneLeCroy/SignalIntegrity/wiki/
S-parameters-for-Signal-Integrity-Book
```

# Appendix A

## Terminology and Conventions

Within this book, the following conventions are maintained:

- $j$ is $\sqrt{-1}$ according to electrical engineering notation.
- To accommodate multiple character variables, a dot $(\cdot)$ is placed wherever multiplication takes place.
- $\lfloor x \rfloor$ is the floor of $x$, the next lower integer of $x$.
- $\lceil x \rceil$ is the ceiling of $x$, the next higher integer of $x$.
- $\lfloor x \rceil$ is the nearest integer to $x$.
- $\mathrm{Re}\,(x)$ is the real part of $x$.
- $\mathrm{Im}\,(x)$ is the imaginary part of $x$.
- $x^*$ is the complex conjugate of $x$.
- $(x * y)\,(t)$ is the function $x(t)$ convolved with the function $y(t)$.
- $\mathbf{x} * \mathbf{y}$ or $(x * y)\,[k]$ is the vector $\mathbf{x}$ convolved with the vector $\mathbf{y}$.
- $\overrightarrow{\mathbf{x} \cdot \mathbf{y}}$ means the Hadamard product, which is the element-wise product of two vectors. For vectors, $\overrightarrow{\mathbf{x} \cdot \mathbf{y}} = \mathrm{diag}\,(\mathbf{x}) \cdot \mathbf{y} = \mathrm{diag}\,(\mathbf{y}) \cdot \mathbf{x}$.
- $\arg\,(x)$ is the argument, or angle of $x$.
- $\approx$ means approximately equal to.
- $\vee$ is the logical OR operator.
- $\wedge$ is the logical AND operator.
- $\in$ means "is an element of."
- $i \in 0 \dots I-1$ means that the element $i$ ranges from 0 to $I-1$ in integer steps. In such ranges, the variable $k$ generally represents time steps, $n$ generally represents frequency steps, and $m$ is used for other things, like numbers of elements or sections.
- In matrices, $r$ is used to indicate a row and $c$ is used to indicate a column.
- $\|$ is used to calculate parallel circuit elements. For example, $R_1 \parallel R_2 = \left(\frac{1}{R_1} + \frac{1}{R_2}\right)^{-1}$.
- $\binom{n}{k}$ is the binomial coefficient defined as $\binom{n}{k} = \frac{n!}{k! \cdot (n-k)!}$.

## A.1 Matrices and Vectors

Within this book, the following conventions for matrices and vectors are maintained:

- Matrices are upper-case bold letters, such as $\mathbf{A}$.
- Vectors are lower-case bold letters, such as $\mathbf{a}$.
- Matrix elements are indicated as $A_{ij}$, where $i$ is the row and $j$ is the element.
- Vector elements are indicated by $a_i$, where $i$ is the row.
- Sometimes, when subscripts become confusing and programmatic methods are being shown, $A_{[i][j]}$ and $a_{[i]}$ are used to indicate access to the row and column of the matrix, or the row of the vector.
- $\mathbf{A}^T$ indicates the transpose of matrix $\mathbf{A}$.
- $\mathbf{A}^*$ indicates the complex conjugate of the matrix. All elements are conjugated.
- $\mathbf{A}^H$ indicates the Hermitian or conjugate transpose of matrix $\mathbf{A}$. All elements are conjugated and the matrix is transposed.
- $\mathbf{A_{ij}}$ indicates a block matrix formed by taking a submatrix of $\mathbf{A}$.
- $\left( \mathbf{A} \; \mathbf{B} \right)$ implies a single matrix formed by placing the elements $\mathbf{A}$ to the left of the elements of $\mathbf{B}$. This is referred to as an augmented matrix.
- $\left( \begin{smallmatrix} \mathbf{A} \\ \mathbf{B} \end{smallmatrix} \right)$ implies a single matrix formed by placing the elements of $\mathbf{A}$ above the elements of $\mathbf{B}$. This is referred to as stacking matrices.
- $|\mathbf{A}|$ means the determinant of $\mathbf{A}$.
- $\mathbf{A}_{i*}$ means a row vector formed by taking the $i$th row of $\mathbf{A}$.
- $\mathbf{A}_{*j}$ means a column vector formed by taking the $j$th column of $\mathbf{A}$.
- $\mathbf{0}$ is a zero matrix; a matrix filled with zeros.
- $\mathbf{1}$ is a matrix filled with ones.
- $\mathbb{I}$ is the identity matrix. This symbol is used to make it uniquely identifiable without being confused with other symbols and matrices.
- Usually the size of a matrix (the number of rows and columns) is obvious. In situations where the size of any matrix needs to be identified, its dimensions are written as $R \times C$, where $R$ is the number of rows and $C$ is the number of columns. A $P \times P$ square matrix showing its dimensions is written as $\underset{P}{\mathbf{A}}$. An $R \times C$ matrix showing its dimensions is written as $\underset{R \times C}{\mathbf{A}}$.
- $\operatorname{diag}(\mathbf{a})$ is a matrix with the elements of vector $\mathbf{a}$ placed along the diagonal and is zero elsewhere.
- $\operatorname{diag}(\mathbf{A})$ is a vector containing the diagonal elements of matrix $\mathbf{A}$.
- $\operatorname{vec}(\mathbf{A})$ is a vector formed by stacking each column of matrix $\mathbf{A}$.
- $\mathbf{A} \otimes \mathbf{B}$ denotes the Kronecker product of matrices $\mathbf{A}$ and $\mathbf{B}$.
- $\operatorname{index}(i, \mathbf{x})$ is the index of the occurrence of $i$ in the vector $\mathbf{x}$.

- $\mathbf{A}^{-1}$ is the inverse of the square matrix $\mathbf{A}$ such that $\mathbf{A}^{-1} \cdot \mathbf{A} = \mathbb{I}$.
- $\|\mathbf{A}\|_F$ is the Frobenius norm of a matrix $\mathbf{A}$ defined for an $R \times C$ matrix as

$$\|\mathbf{A}\|_F = \sqrt{\sum_{r=1}^{R} \cdot \sum_{c=1}^{C} |A_{r,c}|^2}.$$

- $\|\mathbf{a}\|_2$ is the Euclidian norm of a vector $\mathbf{a}$, defined as

$$\|\mathbf{a}\|_2 = \sqrt{\sum_{m=1}^{M} |a_m|^2} = \sqrt{\mathbf{a}^H \cdot \mathbf{a}}.$$

- $\|\mathbf{A}\|_2$ is the induced two-norm of a matrix $\mathbf{A}$, defined for an $R \times C$ matrix as

$$\|\mathbf{A}\|_2 = \max_{\left\|\underset{C}{\mathbf{x}}\right\|_2 = 1} \|\mathbf{A} \cdot \mathbf{x}\|_2 .$$

- The notation $\mathbf{A}^\dagger$ represents the pseudo-inverse of $\mathbf{A}$. If $\mathbf{A}$ is square, then $\mathbf{A}^\dagger$ is simply $\mathbf{A}^{-1}$. When $\mathbf{A}$ is not square, it is an inverse in a least-squares sense. See Appendix C, §C.2, for more information.

The pseudo-inverse is related to the SVD. More information about the pseudo-inverse and algorithms for solving linear equations and for dealing with rank deficient matrices can be found in [78].

## A.2 Normalization Factor and Reference Impedance

The reference impedance is denoted $Z0$ and $\sqrt{\mathcal{Z}0}$ is the normalization factor. Note the slight difference in symbols used.

In most common usage, the normalization factor and reference impedance are inter-related. For example, in traveling waves, $\sqrt{\mathcal{Z}0} \equiv \sqrt{Z0}$. To preserve the ability for this value to be chosen arbitrarily, as in the pseudo-waves definition, the $\sqrt{\mathcal{Z}0}$ symbol is used. It will not be optimized or combined with any variables $Z0$ in any equation to retain generality. See Chapter 2 and Chapter 5 for a complete discussion of this topic.

# Appendix B

# Telegrapher's Equations

THE TRANSMISSION LINE DERIVATION in Chapter 7 is based on the telegrapher's equations, set forth by Oliver Heaviside in the 1880s. In this appendix, it is shown how the telegrapher's equations are formed and how they relate to the ABCD transmission line element.

## B.1 Derivation of the Telegrapher's Equations

The RLGC section in Figure 7.1 can be used to derive the telegrapher's equations:[1]

$$\frac{v_1 - v_2}{Z} = i_1, \qquad \frac{i_1 + i_2}{Y} = v_2, \tag{B.1}$$

where

$$i_1 = i\left(x, s\right), \qquad i_2 = -i\left(x + \Delta x, s\right), \qquad v_2 = v\left(x, s\right), \qquad v_1 = v\left(x - \Delta x, s\right). \tag{B.2}$$

Approximations of the partial derivatives are as follows:

$$\frac{i\left(x + \Delta x, s\right) - i\left(x, s\right)}{\Delta x} \approx \frac{\partial}{\partial x} i\left(x, s\right), \qquad \frac{v\left(x, s\right) - v\left(x - \Delta x, s\right)}{\Delta x} \approx \frac{\partial}{\partial x} v\left(x, s\right),$$

which leads to

$$i\left(x + \Delta x, s\right) \approx i\left(x, s\right) + \frac{\partial}{\partial x} i\left(x, s\right) \cdot \Delta x, \qquad v\left(x - \Delta x, s\right) \approx v\left(x, s\right) - \frac{\partial}{\partial x} v\left(x, s\right) \cdot \Delta x.$$

Substituting these into (B.1):

$$\frac{i\left(x, s\right) - \frac{\partial}{\partial x} i\left(x, s\right) \cdot \Delta x - i\left(x, s\right)}{Y\left(s\right) \cdot \Delta x} = v\left(x, s\right), \qquad \frac{v\left(x, s\right) - \frac{\partial}{\partial x} v\left(x, s\right) \cdot \Delta x - v\left(x, s\right)}{Z\left(s\right) \cdot \Delta x} = i\left(x, s\right),$$

which leads to

$$\frac{\partial}{\partial x} i\left(x, s\right) = -v\left(x, s\right) \cdot Y\left(s\right), \qquad \frac{\partial}{\partial x} v\left(x, s\right) = -i\left(x, s\right) \cdot Z\left(s\right). \tag{B.3}$$

[1] Note carefully the sign of $i_2$ in Figure 7.1.

Using $Z(s) = R + s \cdot L$ and $Y(s) = G + s \cdot C$, and taking the inverse Laplace transform,

$$\frac{\partial}{\partial x} i\left(x,t\right) = -G \cdot v\left(x,t\right) - C \cdot \frac{\partial}{\partial t} v\left(x,t\right), \qquad \frac{\partial}{\partial x} v\left(x,t\right) = -R \cdot i\left(x,t\right) - L \cdot \frac{\partial}{\partial t} i\left(x,t\right). \tag{B.4}$$

The equations in (B.4) are the telegrapher's equations stated exactly, while the equations in (B.3) are the telegrapher's equations in the Laplace domain.

## B.2 Telegrapher's Equations Applied to ABCD Parameters

The ABCD parameter model in (7.2) implies that

$$\begin{pmatrix} v_1 \\ -i_1 \end{pmatrix} = \begin{pmatrix} A & B \\ C & D \end{pmatrix} \cdot \begin{pmatrix} v_2 \\ i_2 \end{pmatrix}. \tag{B.5}$$

Substituting (B.2) in (B.5) leads to

$$\begin{pmatrix} v\left(x,s\right) - \frac{\partial}{\partial x} v\left(x,s\right) \cdot \Delta x \\ -i\left(x,s\right) \end{pmatrix} = \begin{pmatrix} A & B \\ C & D \end{pmatrix} \cdot \begin{pmatrix} v\left(x,s\right) \\ -i\left(x,s\right) - \frac{\partial}{\partial x} i\left(x,s\right) \cdot \Delta x \end{pmatrix},$$

which is simplified as follows:

$$\begin{pmatrix} 1 & 0 \\ 0 & -1 \end{pmatrix} \cdot \begin{pmatrix} v\left(x,s\right) \\ i\left(x,s\right) \end{pmatrix} + \begin{pmatrix} -1 & 0 \\ 0 & 0 \end{pmatrix} \cdot \begin{pmatrix} \frac{\partial}{\partial x} v\left(x,s\right) \cdot \Delta x \\ \frac{\partial}{\partial x} i\left(x,s\right) \cdot \Delta x \end{pmatrix}$$
$$= \begin{pmatrix} A & B \\ C & D \end{pmatrix} \cdot \left[ \begin{pmatrix} 1 & 0 \\ 0 & -1 \end{pmatrix} \cdot \begin{pmatrix} v\left(x,s\right) \\ i\left(x,s\right) \end{pmatrix} + \begin{pmatrix} 0 & 0 \\ 0 & -1 \end{pmatrix} \cdot \begin{pmatrix} \frac{\partial}{\partial x} v\left(x,s\right) \\ \frac{\partial}{\partial x} i\left(x,s\right) \end{pmatrix} \right];$$

$$\left[ \begin{pmatrix} 1 & 0 \\ 0 & -1 \end{pmatrix} - \begin{pmatrix} A & B \\ C & D \end{pmatrix} \cdot \begin{pmatrix} 1 & 0 \\ 0 & -1 \end{pmatrix} \right] \cdot \begin{pmatrix} v\left(x,s\right) \\ i\left(x,s\right) \end{pmatrix}$$
$$= \left[ \begin{pmatrix} A & B \\ C & D \end{pmatrix} \cdot \begin{pmatrix} 0 & 0 \\ 0 & -1 \end{pmatrix} - \begin{pmatrix} -1 & 0 \\ 0 & 0 \end{pmatrix} \right] \cdot \begin{pmatrix} \frac{\partial}{\partial x} v\left(x,s\right) \cdot \Delta x \\ \frac{\partial}{\partial x} i\left(x,s\right) \cdot \Delta x \end{pmatrix};$$

$$\begin{pmatrix} \frac{\partial}{\partial x} v\left(x,s\right) \cdot \Delta x \\ \frac{\partial}{\partial x} i\left(x,s\right) \cdot \Delta x \end{pmatrix} = \left[ \begin{pmatrix} A & B \\ C & D \end{pmatrix} \cdot \begin{pmatrix} 0 & 0 \\ 0 & -1 \end{pmatrix} - \begin{pmatrix} -1 & 0 \\ 0 & 0 \end{pmatrix} \right]^{-1}$$
$$\cdot \left[ \begin{pmatrix} 1 & 0 \\ 0 & -1 \end{pmatrix} - \begin{pmatrix} A & B \\ C & D \end{pmatrix} \cdot \begin{pmatrix} 1 & 0 \\ 0 & -1 \end{pmatrix} \right] \cdot \begin{pmatrix} v\left(x,s\right) \\ i\left(x,s\right) \end{pmatrix};$$

$$\begin{pmatrix} \frac{\partial}{\partial x} v\left(x,s\right) \cdot \Delta x \\ \frac{\partial}{\partial x} i\left(x,s\right) \cdot \Delta x \end{pmatrix} = \begin{pmatrix} 1 & -B \\ 0 & -D \end{pmatrix}^{-1} \cdot \begin{pmatrix} 1-A & B \\ -C & D-1 \end{pmatrix} \cdot \begin{pmatrix} v\left(x,s\right) \\ i\left(x,s\right) \end{pmatrix};$$

and finally

$$\begin{pmatrix} \frac{\partial}{\partial x} v(x,s) \cdot \Delta x \\ \frac{\partial}{\partial x} i(x,s) \cdot \Delta x \end{pmatrix} = \frac{1}{D} \cdot \begin{pmatrix} D - (A \cdot D - B \cdot C) & B \\ C & 1 - D \end{pmatrix} \cdot \begin{pmatrix} v(x,s) \\ i(x,s) \end{pmatrix}. \tag{B.6}$$

The values of a differential ABCD element are implied by (7.3) as follows:

$$\begin{pmatrix} A & B \\ C & D \end{pmatrix} = \begin{pmatrix} 1 + Z(s) \cdot Y(s) \cdot \Delta x^2 & -Z(s) \cdot \Delta x \\ -Y(s) \cdot \Delta x & 1 \end{pmatrix}. \tag{B.7}$$

Substituting (B.7) into (B.6) results in

$$\begin{pmatrix} \frac{\partial}{\partial x} v(x,s) \\ \frac{\partial}{\partial x} i(x,s) \end{pmatrix} = \begin{pmatrix} 0 & -Z(s) \\ -Y(s) & 0 \end{pmatrix} \cdot \begin{pmatrix} v(x,s) \\ i(x,s) \end{pmatrix}. \tag{B.8}$$

Equation (B.8) is the same as (B.3), which leads to the telegrapher's equations shown in (B.4).

# Appendix C

# Matrix Algebra

VIRTUALLY ALL OF THE MATH in this book involves *linear algebra*. A full course on linear algebra is not necessary to deal with the topics in this book. All that is needed is some level of comfort dealing with systems of equations represented in matrix-vector form.

## C.1 Systems of Equations

Consider a situation where there are $U$ unknown variables in a system represented by, for $i \in 1 \ldots N$, $n_i$ and $K$ known variables in a system represented by, for $j \in 1 \ldots K$, $m_j$. The relationship between the unknown and known variables is such that each of the known variables is the sum of the products of a known weight $S_{ij}$ and the unknown variables. For each of the known variables, there exist the following equations:

$$\begin{aligned}
m_1 &= S_{11} \cdot n_1 + S_{12} \cdot n_2 + \ldots + S_{1U} \cdot n_U, \\
m_2 &= S_{21} \cdot n_1 + S_{22} \cdot n_2 + \ldots + S_{2U} \cdot n_U, \\
&\vdots \\
m_K &= S_{K1} \cdot n_1 + S_{K2} \cdot n_2 + \ldots + S_{KU} \cdot n_U.
\end{aligned}$$

Since there are $K$ such equations that are satisfied simultaneously, these are called simultaneous equations, or a system of equations. Within this text, the vector of knowns $\mathbf{m}$ are usually called stimuli, and the vector of unknowns $\mathbf{n}$ are called nodes, and $\mathbf{S}$ here is called the systems characteristics matrix. Each equation can be written as

$$m_j = \sum_{i=1}^{U} S_{ji} \cdot n_i.$$

Preferably, it is written in matrix-vector form as

$$\begin{pmatrix} S_{11} & S_{12} & \cdots & S_{1U} \\ S_{21} & S_{22} & \cdots & S_{2U} \\ \vdots & \vdots & \ddots & \vdots \\ S_{K1} & S_{K2} & \cdots & S_{KU} \end{pmatrix} \cdot \begin{pmatrix} n_1 \\ n_2 \\ \vdots \\ n_U \end{pmatrix} = \begin{pmatrix} m_1 \\ m_2 \\ \vdots \\ m_K \end{pmatrix},$$

or in a compact form as

$$\mathbf{S} \cdot \mathbf{n} = \mathbf{m}.$$

Since there is typically an equation per node, generally $K = U$ and $\mathbf{S}$ is $K \times U$. Given a stimulus vector $\mathbf{m}$, one solves for the node vector as

$$\mathbf{n} = \mathbf{S}^{-1} \cdot \mathbf{m}.$$

The notation $\mathbf{S}^{-1}$ means the inverse of $\mathbf{S}$ and is the matrix such that $\mathbf{S}^{-1} \cdot \mathbf{S} = \mathbb{I}$, the identity matrix, which is a matrix containing ones on the diagonal and zeros in the off-diagonal elements. In the case where $\mathbf{S}$ is square and invertible, there is one, unique solution.

Often, the stimulus is applied in a pattern for a number of cases $C$. In this situation, for each case $m \in 1 \ldots C$, the node and stimulus vector $\mathbf{n}$ and $\mathbf{m}$ each become a node and stimulus matrix $\mathbf{N}$ and $\mathbf{M}$. Matrix $\mathbf{N}$ is therefore $U \times C$ and $\mathbf{M}$ is $K \times C$:

$$\mathbf{S} \cdot \mathbf{N} = \mathbf{M}$$

and

$$\mathbf{N} = \mathbf{S}^{-1} \cdot \mathbf{M}.$$

This assumes $K = U$. This can be described as a system of equations, for $k \in 1 \ldots K$, $c \in 1 \ldots C$, as follows:

$$M_{k,c} = \sum_{u=1}^{U} S_{k,u} \cdot N_{u,c}, \tag{C.1}$$

or in matrix form as

$$\begin{pmatrix} S_{11} & S_{12} & \cdots & S_{1U} \\ S_{21} & S_{22} & \cdots & S_{2U} \\ \vdots & \vdots & \ddots & \vdots \\ S_{K1} & S_{K2} & \cdots & S_{KU} \end{pmatrix} \cdot \begin{pmatrix} N_{11} & N_{12} & \cdots & N_{1C} \\ N_{21} & N_{22} & \cdots & N_{2C} \\ \vdots & \vdots & \ddots & \vdots \\ N_{U1} & N_{U2} & \cdots & N_{UC} \end{pmatrix} = \begin{pmatrix} M_{11} & M_{12} & \cdots & M_{1C} \\ M_{21} & M_{22} & \cdots & M_{2C} \\ \vdots & \vdots & \ddots & \vdots \\ M_{K1} & M_{K2} & \cdots & M_{KC} \end{pmatrix}. \tag{C.2}$$

The rules for matrix multiplication have been demonstrated. Essentially, the sum of the products of each element in a row of $\mathbf{S}$ and a column of $\mathbf{N}$ forms an element in $\mathbf{M}$.

The equations in (C.1) and (C.2) are equivalent. The matrix $\mathbf{S}$ is a $K \times U$ element matrix, $\mathbf{N}$ is a $U \times C$ element matrix, and $\mathbf{M}$ is a $K \times C$ element matrix.

If these were scalars, the unknown $\mathbf{N}$ is solved for by dividing both sides by $\mathbf{S}$, but in matrix algebra one solves for the unknown matrix $\mathbf{N}$:

$$\mathbf{S}^{\dagger} \cdot \mathbf{S} \cdot \mathbf{N} = \mathbf{S}^{\dagger} \cdot \mathbf{M}, \tag{C.3}$$

where $\mathbf{S}^{\dagger}$ is defined such that, for $\mathbf{S}$ being an $R \times C$ element matrix,

$$\underset{\min(R,C)\times\min(R,C)}{\mathbb{I}} = \begin{pmatrix} 1 & 0 & \cdots & 0 \\ 0 & 1 & \cdots & 0 \\ \vdots & \vdots & \ddots & \vdots \\ 0 & 0 & \cdots & 1 \end{pmatrix} = \begin{cases} \mathbf{S}^{\dagger} \cdot \mathbf{S} & \text{if } R \geq C, \\ \mathbf{S} \cdot \mathbf{S}^{\dagger} & \text{if } R \leq C. \end{cases}$$

The identity matrix $\mathbb{I}$ is defined as a square matrix such that, when it is appropriately dimensioned and multiplied from the left of a vector, or from the right or left of a matrix, it leaves the vector or matrix unchanged; it is the matrix equivalent of unity. Since $\mathbf{S}^\dagger$ is defined such that $\mathbf{S}^\dagger \cdot \mathbf{S} = \mathbb{I}$ if $K \geq U$, and since multiplication by $\mathbb{I}$ doesn't change anything, the final solution for the unknown vector $N$ in (C.3) is

$$\mathbf{N} = \mathbf{S}^\dagger \cdot \mathbf{M}.$$

The elements of $\mathbf{N}$ are therefore a sum of products of the elements in the rows of $\mathbf{S}^\dagger$ and the columns of $\mathbf{M}$. If $\mathbf{Sd} = \mathbf{S}^\dagger$, then

$$\begin{pmatrix} N_{11} & N_{12} & \cdots & N_{1C} \\ N_{21} & N_{22} & \cdots & N_{2C} \\ \vdots & \vdots & \ddots & \vdots \\ N_{U1} & N_{U2} & \cdots & N_{UC} \end{pmatrix} = \begin{pmatrix} Sd_{11} & Sd_{12} & \cdots & Sd_{1K} \\ Sd_{21} & Sd_{22} & \cdots & Sd_{2K} \\ \vdots & \vdots & \ddots & \vdots \\ Sd_{U1} & Sd_{U2} & \cdots & Sd_{UK} \end{pmatrix} \cdot \begin{pmatrix} M_{11} & M_{12} & \cdots & M_{1C} \\ M_{21} & M_{22} & \cdots & M_{2C} \\ \vdots & \vdots & \ddots & \vdots \\ M_{K1} & M_{K2} & \cdots & M_{KC} \end{pmatrix}$$

and

$$N_{u,c} = \sum_{k=1}^{K} Sd_{u,k} \cdot n_{k,c}.$$

## C.2 The Moore–Penrose Pseudo-Inverse

If $\mathbf{S}$ is square (i.e. the number of rows and columns are the same), then $\mathbf{S}^\dagger = \mathbf{S}^{-1}$. In other words, given the system

$$\underset{U\times U}{\mathbf{S}} \cdot \underset{U\times C}{\mathbf{N}} = \underset{U\times C}{\mathbf{M}},$$

the solution is

$$\underset{U\times C}{\mathbf{N}} = \underbrace{\mathbf{S}^{-1}}_{U\times U} \cdot \underset{U\times C}{\mathbf{M}}, \tag{C.4}$$

where again, $\mathbf{S}^{-1} \cdot \mathbf{S} = \mathbf{S} \cdot \mathbf{S}^{-1} = \mathbb{I}$.

However, given the overconstrained case, $K > U$,

$$\underset{K\times U}{\mathbf{S}} \cdot \underset{U\times C}{\mathbf{N}} = \underset{K\times C}{\mathbf{M}}. \tag{C.5}$$

The matrix $\mathbf{S}$ is on the left and is tall and skinny. Both sides can be multiplied from the left by the Hermitian of $\mathbf{S}$ as follows:

$$\underbrace{\mathbf{S}^H}_{U\times K} \cdot \underset{K\times U}{\mathbf{S}} \cdot \underset{U\times C}{\mathbf{N}} = \underbrace{\mathbf{S}^H \cdot \mathbf{S}}_{U\times U} \cdot \underset{U\times C}{\mathbf{N}} = \underbrace{\mathbf{S}^H}_{U\times K} \cdot \underset{K\times C}{\mathbf{M}},$$

and solved as

$$\underset{U\times C}{\mathbf{N}} = \underbrace{\left(\mathbf{S}^H \cdot \mathbf{S}\right)^{-1}}_{U\times U} \cdot \underbrace{\mathbf{S}^H}_{U\times K} \cdot \underset{K\times C}{\mathbf{N}} = \underbrace{\left(\mathbf{S}^H \cdot \mathbf{S}\right)^{-1} \cdot \mathbf{S}^H}_{U\times K} \cdot \underset{K\times C}{\mathbf{M}} = \underbrace{\mathbf{S}^\dagger}_{U\times K} \cdot \underset{K\times C}{\mathbf{M}}. \tag{C.6}$$

Transposing all of the matrices in (C.5), and understanding that $(\mathbf{A} \cdot \mathbf{B})^H = \mathbf{B}^H \cdot \mathbf{A}^H$,

$$\underset{C\times U}{\mathbf{N}} \cdot \underset{U\times K}{\mathbf{S}} = \underset{C\times K}{\mathbf{M}}.$$

Now the matrix $\mathbf{S}$ is on the right and is short and fat. Both sides are multiplied from the right by the Hermitian of $\mathbf{S}$ as follows:

$$\underset{C\times U}{\mathbf{N}} \cdot \underset{U\times K}{\mathbf{S}} \cdot \underbrace{\mathbf{S}^H}_{K\times U} = \underset{C\times U}{\mathbf{N}} \cdot \underbrace{\mathbf{S}\cdot\mathbf{S}^H}_{U\times U} = \underset{C\times K}{\mathbf{M}} \cdot \underbrace{\mathbf{S}^H}_{K\times U},$$

and solved as

$$\underset{C\times U}{\mathbf{N}} = \underset{C\times K}{\mathbf{M}} \cdot \underbrace{\mathbf{S}^H}_{K\times U} \cdot \underbrace{\left(\mathbf{S}\cdot\mathbf{S}^H\right)^{-1}}_{U\times U} = \underset{C\times K}{\mathbf{M}} \cdot \underbrace{\left[\mathbf{S}^H\cdot\left(\mathbf{S}\cdot\mathbf{S}^H\right)^{-1}\right]}_{K\times U} = \underset{C\times K}{\mathbf{M}} \cdot \underbrace{\mathbf{S}^\dagger}_{K\times U}. \tag{C.7}$$

Thus, the three solutions for the three special cases in (C.4), (C.6), and (C.7) provide for a common definition of the † symbol as representing the Moore–Penrose pseudo-inverse [79], which is defined, for the purpose of this book, based on the dimensions of $\mathbf{S}$ as follows:[1]

$$\left(\underset{R\times C}{\mathbf{S}}\right)^\dagger = \underset{C\times R}{\mathbf{S}^\dagger} = \begin{cases} \mathbf{S}^{-1} & \text{if } R=C \text{ and } \mathbf{S}^{-1} \text{ exists,} \\ \left(\mathbf{S}^H\cdot\mathbf{S}\right)^{-1}\cdot\mathbf{S}^H & \text{if } R>C \text{ and } \left(\mathbf{S}^H\cdot\mathbf{S}\right)^{-1} \text{ exists,} \\ \mathbf{S}^H\cdot\left(\mathbf{S}\cdot\mathbf{S}^H\right)^{-1} & \text{if } R<C \text{ and } \left(\mathbf{S}\cdot\mathbf{S}^H\right)^{-1} \text{ exists.} \end{cases}$$

Note that the Moore–Penrose pseudo-inverse assumes that if the matrix is tall and skinny, it is on the left in an overconstrained equation and if it is short and fat, it is on the right in an overconstrained equation. Python's inverse in the linear algebra package provides the Moore–Penrose pseudo-inverse.

## C.3 Overconstrained Solutions

In this section, we examine the solutions found in §C.2 to understand the significance of the pseudo-inverse. If $\mathbf{S}$ is square (i.e. the number of rows and columns are the same), then

$$\underset{U\times U}{\mathbf{S}} \cdot \underset{U\times C}{\mathbf{N}} = \underset{U\times C}{\mathbf{M}}$$

and the solution is

$$\underset{U\times C}{\mathbf{N}} = \underbrace{\mathbf{S}^{-1}}_{U\times U} \cdot \underset{U\times C}{\mathbf{M}}.$$

In this particular case, if $\mathbf{S}^{-1}$ exists, a unique, exact solution for $\mathbf{N}$ can be found such that $\mathbf{S}\cdot\mathbf{N}=\mathbf{M}$. However, if given the overconstrained case, $K>U$:

$$\underset{K\times U}{\mathbf{S}} \cdot \underset{U\times C}{\mathbf{N}} = \underset{K\times C}{\mathbf{M}}.$$

In this case, it might not be possible to find such a solution for $\mathbf{N}$. Instead only a solution such that $\mathbf{S}^H\cdot\mathbf{S}\cdot\mathbf{N}=\mathbf{S}^H\cdot\mathbf{M}$ is found, but without justification for its significance.

To understand what this solution implies, consider that, for any given solution $\mathbf{N}$, there is a residual error matrix $\mathbf{R}=\mathbf{S}\cdot\mathbf{N}-\mathbf{M}$; a column of this residual error matrix pertains to a particular case $c\in 1\ldots C$:

$$R_{k,c} = \left(\mathbf{S}\cdot\mathbf{N}-\mathbf{M}\right)_{k,c} = \left(\mathbf{S}\cdot\mathbf{N}\right)_{k,c} - M_{k,c},$$

[1] If none of the conditions are met, then [79] should be referred to for the exact definition.

and the variance of the residual error for a given case is given by

$$\sigma_c^2 = \sum_{k=1}^{K} R_{k,c}^2 = \sum_{k=1}^{K} \left[ (\mathbf{S} \cdot \mathbf{N})_{k,c} - M_{k,c} \right]^2 = \sum_{k=1}^{K} \left[ \sum_{u=1}^{U} S_{k,u} \cdot N_{u,c} - M_{k,c} \right]^2.$$

A solution is desired for $\mathbf{N}_{*c}$ such that $\sigma_c^2$ is minimized. This is called a LMSE solution.[2] The condition for minimizing $\sigma_c^2$ is met when, for any $n \in 1 \ldots U$,

$$\frac{\partial}{\partial N_{n,c}} \sigma_c^2 = 0 = \frac{\partial}{\partial N_{n,c}} \sum_{k=1}^{K} \left[ \sum_{u=1}^{U} S_{k,u} \cdot N_{u,c} - M_{k,c} \right]^2,$$

meaning

$$\frac{\partial}{\partial N_{n,c}} \sum_{k=1}^{K} \left[ \sum_{u=1}^{U} S_{k,u} \cdot N_{u,c} - M_{k,c} \right]^2 = 2 \cdot \sum_{k=1}^{K} \left[ \sum_{u=1}^{U} S_{k,u} \cdot N_{u,c} - M_{k,c} \right] \cdot \frac{\partial}{\partial N_{n,c}} \sum_{k=1}^{K} \left[ \sum_{u=1}^{U} S_{k,u} \cdot N_{u,c} - M_{k,c} \right] = 0$$

or

$$\sum_{k=1}^{K} \left[ \sum_{u=1}^{U} S_{k,u} \cdot S_{k,n}^* \cdot N_{u,c} - M_{k,c} \cdot S_{k,n}^* \right] = 0.$$

While this is hard to see, this is the same as saying

$$\mathbf{S}^H \cdot \mathbf{S} \cdot \mathbf{N} - \mathbf{S}^H \cdot \mathbf{M} = 0$$

or

$$\mathbf{S}^H \cdot \mathbf{S} \cdot \mathbf{N} = \mathbf{S}^H \cdot \mathbf{M}.$$

Finally,

$$\mathbf{N} = \mathbf{S}^\dagger \cdot \mathbf{M} = \left( \mathbf{S}^H \cdot \mathbf{S} \right)^{-1} \cdot \mathbf{S}^H \cdot \mathbf{M}.$$

Thus, the Moore–Penrose pseudo-inverse offers the LMSE solution.

## C.4 Symbolic Block Matrix Inversion

In many areas of this book, symbolic solutions are provided for linear sets of equations involving matrices and, in particular, block matrices. The handling of block matrices in symbolic solutions requires some care that might not be obvious to readers unfamiliar with linear algebra and its applications.

[2]Technically, the sum of the squares is being minimized, but this minimizes the variance and the rms error as well.

For example, a system of equations might look like this:

$$\begin{pmatrix} \mathbf{A} & \mathbf{B} \\ \mathbf{C} & \mathbf{D} \end{pmatrix} \cdot \begin{pmatrix} \mathbf{X} \\ \mathbf{Y} \end{pmatrix} = \begin{pmatrix} \mathbf{W} \\ \mathbf{Z} \end{pmatrix}. \tag{C.8}$$

In this case, the solution for $\left(\begin{smallmatrix} \mathbf{X} \\ \mathbf{Y} \end{smallmatrix}\right)$ is expressed as

$$\begin{pmatrix} \mathbf{X} \\ \mathbf{Y} \end{pmatrix} = \begin{pmatrix} \mathbf{A} & \mathbf{B} \\ \mathbf{C} & \mathbf{D} \end{pmatrix}^{-1} \cdot \begin{pmatrix} \mathbf{W} \\ \mathbf{Z} \end{pmatrix}. \tag{C.9}$$

This is a perfectly good expression of the solution and is complete for the calculation if numerical techniques are utilized similar to those techniques that would be used to invert the matrix. In other words, the numeric solution would be to fill in the values of $\mathbf{A}$, $\mathbf{B}$, $\mathbf{C}$, and $\mathbf{D}$ and calculate the inverse of the matrix.

In symbolic solutions, it is often desired to have a better, reduced, answer that provides further insight. To this end, the inverse of the matrix is solved symbolically . This involves finding linear operations to perform on the matrix $\left(\begin{smallmatrix} \mathbf{A} & \mathbf{B} \\ \mathbf{C} & \mathbf{D} \end{smallmatrix}\right)$ to turn it into the identity matrix. This set of linear operations would be combined to form the matrix inverse.

If this is unclear, consider that there are three matrices that can be applied to the left of $\left(\begin{smallmatrix} \mathbf{A} & \mathbf{B} \\ \mathbf{C} & \mathbf{D} \end{smallmatrix}\right)$ to convert it to the identity matrix. In mathematical terms, there are three operator matrices $\mathbf{O}_3$, $\mathbf{O}_2$, and $\mathbf{O}_1$ such that

$$\mathbf{O}_3 \cdot \mathbf{O}_2 \cdot \mathbf{O}_1 \cdot \begin{pmatrix} \mathbf{A} & \mathbf{B} \\ \mathbf{C} & \mathbf{D} \end{pmatrix} = \begin{pmatrix} \mathbb{I} & \mathbf{0} \\ \mathbf{0} & \mathbb{I} \end{pmatrix}.$$

Here, it is clear that

$$\begin{pmatrix} \mathbf{A} & \mathbf{B} \\ \mathbf{C} & \mathbf{D} \end{pmatrix}^{-1} = \mathbf{O}_3 \cdot \mathbf{O}_2 \cdot \mathbf{O}_1.$$

A set of steps that could be performed would be as follows: First, select element $\mathbf{A}$ as the first pivot element. This means that all of the remaining elements in the first column are set to zero by subtracting multiples of the row containing the pivot element from the rows other than those containing the pivot element. In this case, this means retaining row 1 and subtracting $\mathbf{C} \cdot \mathbf{A}^{-1}$ times row 1 from row 2. The operation that performs this is as follows:

$$\begin{pmatrix} \mathbf{1} & \mathbf{0} \\ -\mathbf{C} \cdot \mathbf{A}^{-1} & \mathbf{1} \end{pmatrix} \cdot \begin{pmatrix} \mathbf{A} & \mathbf{B} \\ \mathbf{C} & \mathbf{D} \end{pmatrix} = \begin{pmatrix} \mathbf{A} & \mathbf{B} \\ \mathbf{0} & \mathbf{D} - \mathbf{C} \cdot \mathbf{A}^{-1}\mathbf{B} \end{pmatrix}.$$

The second step would be to select $\mathbf{D} - \mathbf{C} \cdot \mathbf{A}^{-1} \cdot \mathbf{B}$ as the final pivot element and to try to set all of the remaining elements in the second column to zero by retaining the second row and forming the first row by subtracting $\mathbf{B} \cdot \left(\mathbf{D} - \mathbf{C} \cdot \mathbf{A}^{-1} \cdot \mathbf{B}\right)^{-1}$ times the second row from the first row. The operation that performs this is as follows:

$$\begin{pmatrix} \mathbf{1} & -\mathbf{B} \cdot (\mathbf{D} - \mathbf{C} \cdot \mathbf{A}^{-1} \cdot \mathbf{B})^{-1} \\ \mathbf{0} & \mathbf{1} \end{pmatrix} \cdot \begin{pmatrix} \mathbf{A} & \mathbf{B} \\ \mathbf{0} & \mathbf{D} - \mathbf{C} \cdot \mathbf{A}^{-1} \cdot \mathbf{B} \end{pmatrix} = \begin{pmatrix} \mathbf{A} & \mathbf{0} \\ \mathbf{0} & \mathbf{D} - \mathbf{C} \cdot \mathbf{A}^{-1} \cdot \mathbf{B} \end{pmatrix}.$$

As the final step, the first row is multiplied by $\mathbf{A}^{-1}$ and the second row by $\left(\mathbf{D}-\mathbf{C}\cdot\mathbf{A}^{-1}\cdot\mathbf{B}\right)^{-1}$ to form the identity matrix:

$$\begin{pmatrix} \mathbf{A}^{-1} & \mathbf{0} \\ \mathbf{0} & \left(\mathbf{D}-\mathbf{C}\cdot\mathbf{A}^{-1}\cdot\mathbf{B}\right)^{-1} \end{pmatrix} \cdot \begin{pmatrix} \mathbf{A} & \mathbf{0} \\ \mathbf{0} & \mathbf{D}-\mathbf{C}\cdot\mathbf{A}^{-1}\cdot\mathbf{B} \end{pmatrix} = \begin{pmatrix} \mathbb{I} & \mathbf{0} \\ \mathbf{0} & \mathbb{I} \end{pmatrix}.$$

Thus,

$$\begin{pmatrix} \mathbf{A}^{-1} & \mathbf{0} \\ \mathbf{0} & \left(\mathbf{D}-\mathbf{C}\cdot\mathbf{A}^{-1}\cdot\mathbf{B}\right)^{-1} \end{pmatrix} \cdot \begin{pmatrix} \mathbf{1} & -\mathbf{B}\cdot\left(\mathbf{D}-\mathbf{C}\cdot\mathbf{A}^{-1}\cdot\mathbf{B}\right)^{-1} \\ \mathbf{0} & \mathbf{1} \end{pmatrix} \cdot \begin{pmatrix} \mathbf{1} & \mathbf{0} \\ -\mathbf{C}\cdot\mathbf{A}^{-1} & \mathbf{1} \end{pmatrix} \cdot \begin{pmatrix} \mathbf{A} & \mathbf{B} \\ \mathbf{C} & \mathbf{D} \end{pmatrix} = \begin{pmatrix} \mathbb{I} & \mathbf{0} \\ \mathbf{0} & \mathbb{I} \end{pmatrix},$$

and therefore

$$\begin{aligned} \begin{pmatrix} \mathbf{A} & \mathbf{B} \\ \mathbf{C} & \mathbf{D} \end{pmatrix}^{-1} &= \begin{pmatrix} \mathbf{A}^{-1} & \mathbf{0} \\ \mathbf{0} & \left(\mathbf{D}-\mathbf{C}\cdot\mathbf{A}^{-1}\mathbf{B}\right)^{-1} \end{pmatrix} \cdot \begin{pmatrix} \mathbf{1} & -\mathbf{B}\cdot\left(\mathbf{D}-\mathbf{C}\cdot\mathbf{A}^{-1}\mathbf{B}\right)^{-1} \\ \mathbf{0} & \mathbf{1} \end{pmatrix} \cdot \begin{pmatrix} \mathbf{1} & \mathbf{0} \\ -\mathbf{C}\cdot\mathbf{A}^{-1} & \mathbf{1} \end{pmatrix} \\ &= \begin{pmatrix} \mathbf{A}^{-1}+\mathbf{A}^{-1}\cdot\mathbf{B}\cdot\left(\mathbf{D}-\mathbf{C}\cdot\mathbf{A}^{-1}\cdot\mathbf{B}\right)^{-1}\cdot\mathbf{C}\cdot\mathbf{A}^{-1} & -\mathbf{A}^{-1}\mathbf{B}\cdot\left(\mathbf{D}-\mathbf{C}\cdot\mathbf{A}^{-1}\cdot\mathbf{B}\right)^{-1} \\ -\left(\mathbf{D}-\mathbf{C}\cdot\mathbf{A}^{-1}\cdot\mathbf{B}\right)^{-1}\cdot\mathbf{C}\cdot\mathbf{A}^{-1} & \left(\mathbf{D}-\mathbf{C}\cdot\mathbf{A}^{-1}\cdot\mathbf{B}\right)^{-1} \end{pmatrix}. \end{aligned} \tag{C.10}$$

Up to this point, the only extra care that has been taken is the maintenance of the internal matrix elements as matrix elements themselves during the matrix inversion process, but a more important point to realize is that the answer obtained, while being a possible answer, might not be the answer at all. The way to understand this is to imagine that the original matrix $\left(\begin{smallmatrix} \mathbf{A} & \mathbf{B} \\ \mathbf{C} & \mathbf{D} \end{smallmatrix}\right)$ is invertible, but that the internal matrix $\mathbf{A}=\mathbf{0}$ is the zero matrix. One can see that, if that were the case, the answer provided cannot be calculated because it involves the calculation of $\mathbf{A}^{-1}$. It has been assumed that $\mathbf{A}$ is a suitable pivot element in the first step *independent of what the actual value might be.* It turns out that it is possible to alter the selection of the pivot element easily by performing either row or column permutations. For a $2\times 2$ matrix, this means that the matrix can be left as it is (as in the first solution), the rows could be swapped, the columns could be swapped, or both rows and columns could be swapped, to give four possible combinations. The row permutation is considered first. Equation (C.8) can be rewritten as,

$$\begin{pmatrix} \mathbf{0} & \mathbb{I} \\ \mathbb{I} & \mathbf{0} \end{pmatrix} \cdot \begin{pmatrix} \mathbf{A} & \mathbf{B} \\ \mathbf{C} & \mathbf{D} \end{pmatrix} \cdot \begin{pmatrix} \mathbf{X} \\ \mathbf{Y} \end{pmatrix} = \begin{pmatrix} \mathbf{0} & \mathbb{I} \\ \mathbb{I} & \mathbf{0} \end{pmatrix} \cdot \begin{pmatrix} \mathbf{W} \\ \mathbf{Z} \end{pmatrix},$$

which allows (C.9) to be rewritten as

$$\begin{pmatrix} \mathbf{X} \\ \mathbf{Y} \end{pmatrix} = \left[\begin{pmatrix} \mathbf{0} & \mathbb{I} \\ \mathbb{I} & \mathbf{0} \end{pmatrix} \cdot \begin{pmatrix} \mathbf{A} & \mathbf{B} \\ \mathbf{C} & \mathbf{D} \end{pmatrix}\right]^{-1} \cdot \begin{pmatrix} \mathbf{0} & \mathbb{I} \\ \mathbb{I} & \mathbf{0} \end{pmatrix} \cdot \begin{pmatrix} \mathbf{W} \\ \mathbf{Z} \end{pmatrix}.$$

This means that the inverse can be described as

$$\begin{pmatrix} \mathbf{A} & \mathbf{B} \\ \mathbf{C} & \mathbf{D} \end{pmatrix}^{-1} = \left[\begin{pmatrix} \mathbf{0} & \mathbb{I} \\ \mathbb{I} & \mathbf{0} \end{pmatrix} \cdot \begin{pmatrix} \mathbf{A} & \mathbf{B} \\ \mathbf{C} & \mathbf{D} \end{pmatrix}\right]^{-1} \cdot \begin{pmatrix} \mathbf{0} & \mathbb{I} \\ \mathbb{I} & \mathbf{0} \end{pmatrix} = \begin{pmatrix} \mathbf{C} & \mathbf{D} \\ \mathbf{A} & \mathbf{B} \end{pmatrix}^{-1} \cdot \begin{pmatrix} \mathbf{0} & \mathbb{I} \\ \mathbb{I} & \mathbf{0} \end{pmatrix}.$$

Fortunately, this means the result in (C.10) can be reused by simply replacing the values of $\mathbf{A}$, $\mathbf{B}$, $\mathbf{C}$, and $\mathbf{D}$ in (C.10) with $\mathbf{C}$, $\mathbf{D}$, $\mathbf{A}$, and $\mathbf{B}$, respectively, and applying the column permutation on the right (i.e. swapping the columns in the result after replacement). This provides another form of the matrix inverse:

$$
\begin{aligned}
\begin{pmatrix} \mathbf{A} & \mathbf{B} \\ \mathbf{C} & \mathbf{D} \end{pmatrix}^{-1} &= \begin{pmatrix} \mathbf{C}^{-1}+\mathbf{C}^{-1}\cdot\mathbf{D}\cdot\left(\mathbf{B}-\mathbf{A}\cdot\mathbf{C}^{-1}\cdot\mathbf{D}\right)^{-1}\cdot\mathbf{A}\cdot\mathbf{C}^{-1} & -\mathbf{C}^{-1}\mathbf{D}\cdot\left(\mathbf{B}-\mathbf{A}\cdot\mathbf{C}^{-1}\cdot\mathbf{D}\right)^{-1} \\ -\left(\mathbf{B}-\mathbf{A}\cdot\mathbf{C}^{-1}\cdot\mathbf{D}\right)^{-1}\cdot\mathbf{A}\cdot\mathbf{C}^{-1} & \left(\mathbf{B}-\mathbf{A}\cdot\mathbf{C}^{-1}\cdot\mathbf{D}\right)^{-1} \end{pmatrix} \cdot \begin{pmatrix} \mathbf{0} & \mathbb{I} \\ \mathbb{I} & \mathbf{0} \end{pmatrix} \\
&= \begin{pmatrix} -\mathbf{C}^{-1}\cdot\mathbf{D}\cdot\left(\mathbf{B}-\mathbf{A}\cdot\mathbf{C}^{-1}\cdot\mathbf{D}\right)^{-1} & \mathbf{C}^{-1}+\mathbf{C}^{-1}\cdot\mathbf{D}\cdot\left(\mathbf{B}-\mathbf{A}\cdot\mathbf{C}^{-1}\cdot\mathbf{D}\right)^{-1}\cdot\mathbf{A}\cdot\mathbf{C}^{-1} \\ \left(\mathbf{B}-\mathbf{A}\cdot\mathbf{C}^{-1}\cdot\mathbf{D}\right)^{-1} & -\left(\mathbf{B}-\mathbf{A}\cdot\mathbf{C}^{-1}\cdot\mathbf{D}\right)^{-1}\cdot\mathbf{A}\cdot\mathbf{C}^{-1} \end{pmatrix}. \qquad \text{(C.11)}
\end{aligned}
$$

Equation (C.11) now provides a solution for the matrix inverse which does not involve inversion of $\mathbf{A}$, but which now involves inversion of $\mathbf{C}$. Unfortunately, it is possible that both $\mathbf{A}$ and $\mathbf{C}$ are zero, in which case a solution has not yet been found.

Similar to the row permutation scheme, a column permutation (i.e. the columns are swapped) can be performed. Equation (C.8) can again be rewritten as follows:

$$
\begin{pmatrix} \mathbf{A} & \mathbf{B} \\ \mathbf{C} & \mathbf{D} \end{pmatrix} \cdot \begin{pmatrix} \mathbf{0} & \mathbb{I} \\ \mathbb{I} & \mathbf{0} \end{pmatrix} \cdot \begin{pmatrix} \mathbf{0} & \mathbb{I} \\ \mathbb{I} & \mathbf{0} \end{pmatrix} \cdot \begin{pmatrix} \mathbf{X} \\ \mathbf{Y} \end{pmatrix} = \begin{pmatrix} \mathbf{W} \\ \mathbf{Z} \end{pmatrix},
$$

which allows (C.9) to be written as

$$
\begin{pmatrix} \mathbf{X} \\ \mathbf{Y} \end{pmatrix} = \begin{pmatrix} \mathbf{0} & \mathbb{I} \\ \mathbb{I} & \mathbf{0} \end{pmatrix} \cdot \left[ \begin{pmatrix} \mathbf{A} & \mathbf{B} \\ \mathbf{C} & \mathbf{D} \end{pmatrix} \cdot \begin{pmatrix} \mathbf{0} & \mathbb{I} \\ \mathbb{I} & \mathbf{0} \end{pmatrix} \right]^{-1} \cdot \begin{pmatrix} \mathbf{W} \\ \mathbf{Z} \end{pmatrix}.
$$

This means that the inverse can be described as

$$
\begin{pmatrix} \mathbf{A} & \mathbf{B} \\ \mathbf{C} & \mathbf{D} \end{pmatrix}^{-1} = \begin{pmatrix} \mathbf{0} & \mathbb{I} \\ \mathbb{I} & \mathbf{0} \end{pmatrix} \cdot \left[ \begin{pmatrix} \mathbf{A} & \mathbf{B} \\ \mathbf{C} & \mathbf{D} \end{pmatrix} \cdot \begin{pmatrix} \mathbf{0} & \mathbb{I} \\ \mathbb{I} & \mathbf{0} \end{pmatrix} \right]^{-1} = \begin{pmatrix} \mathbf{0} & \mathbb{I} \\ \mathbb{I} & \mathbf{0} \end{pmatrix} \cdot \begin{pmatrix} \mathbf{B} & \mathbf{A} \\ \mathbf{D} & \mathbf{C} \end{pmatrix}^{-1}.
$$

Again, the result in (C.10) can be reused by simply replacing the values of $\mathbf{A}$, $\mathbf{B}$, $\mathbf{C}$, and $\mathbf{D}$ in (C.10) with $\mathbf{B}$, $\mathbf{A}$, $\mathbf{D}$, and $\mathbf{C}$, respectively, and applying the row permutation on the left (i.e. swapping the rows in the result after replacement). This provides another form of the matrix inverse:

$$
\begin{aligned}
\begin{pmatrix} \mathbf{A} & \mathbf{B} \\ \mathbf{C} & \mathbf{D} \end{pmatrix}^{-1} &= \begin{pmatrix} \mathbf{0} & \mathbb{I} \\ \mathbb{I} & \mathbf{0} \end{pmatrix} \cdot \begin{pmatrix} \mathbf{B}^{-1}+\mathbf{B}^{-1}\cdot\mathbf{A}\cdot\left(\mathbf{C}-\mathbf{D}\cdot\mathbf{B}^{-1}\cdot\mathbf{A}\right)^{-1}\cdot\mathbf{D}\cdot\mathbf{B}^{-1} & -\mathbf{B}^{-1}\cdot\mathbf{A}\cdot\left(\mathbf{C}-\mathbf{D}\cdot\mathbf{B}^{-1}\cdot\mathbf{A}\right)^{-1} \\ -\left(\mathbf{C}-\mathbf{D}\cdot\mathbf{B}^{-1}\cdot\mathbf{A}\right)^{-1}\cdot\mathbf{D}\cdot\mathbf{B}^{-1} & \left(\mathbf{C}-\mathbf{D}\cdot\mathbf{B}^{-1}\cdot\mathbf{A}\right)^{-1} \end{pmatrix} \\
&= \begin{pmatrix} -\left(\mathbf{C}-\mathbf{D}\cdot\mathbf{B}^{-1}\cdot\mathbf{A}\right)^{-1}\cdot\mathbf{D}\cdot\mathbf{B}^{-1} & \left(\mathbf{C}-\mathbf{D}\cdot\mathbf{B}^{-1}\cdot\mathbf{A}\right)^{-1} \\ \mathbf{B}^{-1}+\mathbf{B}^{-1}\cdot\mathbf{A}\cdot\left(\mathbf{C}-\mathbf{D}\cdot\mathbf{B}^{-1}\cdot\mathbf{A}\right)^{-1}\cdot\mathbf{D}\cdot\mathbf{B}^{-1} & -\mathbf{B}^{-1}\cdot\mathbf{A}\cdot\left(\mathbf{C}-\mathbf{D}\cdot\mathbf{B}^{-1}\cdot\mathbf{A}\right)^{-1} \end{pmatrix}.
\end{aligned}
$$

Finally, both a row and column permutation can be performed. Equation (C.8) can again be rewritten as follows:

$$
\begin{pmatrix} \mathbf{0} & \mathbb{I} \\ \mathbb{I} & \mathbf{0} \end{pmatrix} \cdot \begin{pmatrix} \mathbf{A} & \mathbf{B} \\ \mathbf{C} & \mathbf{D} \end{pmatrix} \cdot \begin{pmatrix} \mathbf{0} & \mathbb{I} \\ \mathbb{I} & \mathbf{0} \end{pmatrix} \cdot \begin{pmatrix} \mathbf{0} & \mathbb{I} \\ \mathbb{I} & \mathbf{0} \end{pmatrix} \cdot \begin{pmatrix} \mathbf{X} \\ \mathbf{Y} \end{pmatrix} = \begin{pmatrix} \mathbf{0} & \mathbb{I} \\ \mathbb{I} & \mathbf{0} \end{pmatrix} \cdot \begin{pmatrix} \mathbf{W} \\ \mathbf{Z} \end{pmatrix},
$$

which allows (C.9) to be rewritten as

$$
\begin{pmatrix} \mathbf{X} \\ \mathbf{Y} \end{pmatrix} = \begin{pmatrix} \mathbf{0} & \mathbb{I} \\ \mathbb{I} & \mathbf{0} \end{pmatrix} \cdot \left[ \begin{pmatrix} \mathbf{0} & \mathbb{I} \\ \mathbb{I} & \mathbf{0} \end{pmatrix} \cdot \begin{pmatrix} \mathbf{A} & \mathbf{B} \\ \mathbf{C} & \mathbf{D} \end{pmatrix} \cdot \begin{pmatrix} \mathbf{0} & \mathbb{I} \\ \mathbb{I} & \mathbf{0} \end{pmatrix} \right]^{-1} \cdot \begin{pmatrix} \mathbf{0} & \mathbb{I} \\ \mathbb{I} & \mathbf{0} \end{pmatrix} \cdot \begin{pmatrix} \mathbf{W} \\ \mathbf{Z} \end{pmatrix}.
$$

This means that the inverse can be described as:

$$\begin{aligned}\begin{pmatrix}\mathbf{A} & \mathbf{B}\\ \mathbf{C} & \mathbf{D}\end{pmatrix}^{-1} &= \begin{pmatrix}\mathbf{0} & \mathbb{I}\\ \mathbb{I} & \mathbf{0}\end{pmatrix}\cdot\left[\begin{pmatrix}\mathbf{0} & \mathbb{I}\\ \mathbb{I} & \mathbf{0}\end{pmatrix}\cdot\begin{pmatrix}\mathbf{A} & \mathbf{B}\\ \mathbf{C} & \mathbf{D}\end{pmatrix}\cdot\begin{pmatrix}\mathbf{0} & \mathbb{I}\\ \mathbb{I} & \mathbf{0}\end{pmatrix}\right]^{-1}\cdot\begin{pmatrix}\mathbf{0} & \mathbb{I}\\ \mathbb{I} & \mathbf{0}\end{pmatrix}\\ &= \begin{pmatrix}\mathbf{0} & \mathbb{I}\\ \mathbb{I} & \mathbf{0}\end{pmatrix}\cdot\begin{pmatrix}\mathbf{D} & \mathbf{C}\\ \mathbf{B} & \mathbf{A}\end{pmatrix}^{-1}\cdot\begin{pmatrix}\mathbf{0} & \mathbb{I}\\ \mathbb{I} & \mathbf{0}\end{pmatrix}.\end{aligned}$$

Again, the result in (C.10) can be reused by simply replacing the values of **A**, **B**, **C**, and **D** in (C.10) with **D**, **C**, **B**, and **A**, respectively, and applying the row and column permutations (i.e. swapping both rows and columns in the result after replacement). This provides yet another form of the matrix inverse:

$$\begin{aligned}\begin{pmatrix}\mathbf{A} & \mathbf{B}\\ \mathbf{C} & \mathbf{D}\end{pmatrix}^{-1} &= \left(\begin{smallmatrix}\mathbf{0} & \mathbb{I}\\ \mathbb{I} & \mathbf{0}\end{smallmatrix}\right)\cdot\begin{pmatrix}\mathbf{D}^{-1}+\mathbf{D}^{-1}\cdot\mathbf{C}\cdot\left(\mathbf{A}-\mathbf{B}\cdot\mathbf{D}^{-1}\cdot\mathbf{C}\right)^{-1}\cdot\mathbf{B}\cdot\mathbf{D}^{-1} & -\mathbf{D}^{-1}\cdot\mathbf{C}\left(\mathbf{A}-\mathbf{B}\cdot\mathbf{D}^{-1}\cdot\mathbf{C}\right)^{-1}\\ -\left(\mathbf{A}-\mathbf{B}\cdot\mathbf{D}^{-1}\cdot\mathbf{C}\right)^{-1}\cdot\mathbf{B}\cdot\mathbf{D}^{-1} & \left(\mathbf{A}-\mathbf{B}\cdot\mathbf{D}^{-1}\cdot\mathbf{C}\right)^{-1}\end{pmatrix}\cdot\left(\begin{smallmatrix}\mathbf{0} & \mathbb{I}\\ \mathbb{I} & \mathbf{0}\end{smallmatrix}\right)\\ &= \begin{pmatrix}\left(\mathbf{A}-\mathbf{B}\cdot\mathbf{D}^{-1}\cdot\mathbf{C}\right)^{-1} & -\left(\mathbf{A}-\mathbf{B}\cdot\mathbf{D}^{-1}\mathbf{C}\right)^{-1}\cdot\mathbf{B}\cdot\mathbf{D}^{-1}\\ -\mathbf{D}^{-1}\cdot\mathbf{C}\cdot\left(\mathbf{A}-\mathbf{B}\cdot\mathbf{D}^{-1}\cdot\mathbf{C}\right)^{-1} & \mathbf{D}^{-1}+\mathbf{D}^{-1}\cdot\mathbf{C}\cdot\left(\mathbf{A}-\mathbf{B}\cdot\mathbf{D}^{-1}\cdot\mathbf{C}\right)^{-1}\cdot\mathbf{B}\cdot\mathbf{D}^{-1}\end{pmatrix}.\end{aligned}$$

This implies that the equation for the matrix inverse is produced by choosing one of four possible equations for each element in the matrix:

$$\begin{aligned}&\begin{pmatrix}\mathbf{A} & \mathbf{B}\\ \mathbf{C} & \mathbf{D}\end{pmatrix}^{-1}\\ &= \begin{pmatrix}\begin{cases}\mathbf{A}^{-1}+\mathbf{A}^{-1}\cdot\mathbf{B}\cdot\left(\mathbf{D}-\mathbf{C}\cdot\mathbf{A}^{-1}\cdot\mathbf{B}\right)^{-1}\cdot\mathbf{C}\cdot\mathbf{A}^{-1} & \\ -\mathbf{C}^{-1}\cdot\mathbf{D}\cdot\left(\mathbf{B}-\mathbf{A}\cdot\mathbf{C}^{-1}\cdot\mathbf{D}\right)^{-1} & r\\ -\left(\mathbf{C}-\mathbf{D}\cdot\mathbf{B}^{-1}\cdot\mathbf{A}\right)^{-1}\cdot\mathbf{D}\cdot\mathbf{B}^{-1} & c\\ \left(\mathbf{A}-\mathbf{B}\cdot\mathbf{D}^{-1}\cdot\mathbf{C}\right)^{-1} & rc\end{cases} & \begin{cases}-\mathbf{A}^{-1}\cdot\mathbf{B}\cdot\left(\mathbf{D}-\mathbf{C}\cdot\mathbf{A}^{-1}\cdot\mathbf{B}\right)^{-1} & \\ \mathbf{C}^{-1}+\mathbf{C}^{-1}\cdot\mathbf{D}\cdot\left(\mathbf{B}-\mathbf{A}\cdot\mathbf{C}^{-1}\cdot\mathbf{D}\right)^{-1}\cdot\mathbf{A}\cdot\mathbf{C}^{-1} & r\\ \left(\mathbf{C}-\mathbf{D}\cdot\mathbf{B}^{-1}\cdot\mathbf{A}\right)^{-1} & c\\ -\left(\mathbf{A}-\mathbf{B}\cdot\mathbf{D}^{-1}\cdot\mathbf{C}\right)^{-1}\cdot\mathbf{B}\cdot\mathbf{D}^{-1} & rc\end{cases}\\ \begin{cases}-\left(\mathbf{D}-\mathbf{C}\cdot\mathbf{A}^{-1}\cdot\mathbf{B}\right)^{-1}\cdot\mathbf{C}\cdot\mathbf{A}^{-1} & \\ \left(\mathbf{B}-\mathbf{A}\cdot\mathbf{C}^{-1}\cdot\mathbf{D}\right)^{-1} & r\\ \mathbf{B}^{-1}+\mathbf{B}^{-1}\cdot\mathbf{A}\cdot\left(\mathbf{C}-\mathbf{D}\cdot\mathbf{B}^{-1}\cdot\mathbf{A}\right)^{-1}\cdot\mathbf{D}\cdot\mathbf{B}^{-1} & c\\ -\mathbf{D}^{-1}\cdot\mathbf{C}\cdot\left(\mathbf{A}-\mathbf{B}\cdot\mathbf{D}^{-1}\cdot\mathbf{C}\right)^{-1} & rc\end{cases} & \begin{cases}\left(\mathbf{D}-\mathbf{C}\cdot\mathbf{A}^{-1}\cdot\mathbf{B}\right)^{-1} & \\ -\left(\mathbf{B}-\mathbf{A}\cdot\mathbf{C}^{-1}\cdot\mathbf{D}\right)^{-1}\cdot\mathbf{A}\cdot\mathbf{C}^{-1} & r\\ -\mathbf{B}^{-1}\cdot\mathbf{A}\cdot\left(\mathbf{C}-\mathbf{D}\cdot\mathbf{B}^{-1}\cdot\mathbf{A}\right)^{-1} & c\\ \mathbf{D}^{-1}+\mathbf{D}^{-1}\cdot\mathbf{C}\cdot\left(\mathbf{A}-\mathbf{B}\cdot\mathbf{D}^{-1}\cdot\mathbf{C}\right)^{-1}\cdot\mathbf{B}\cdot\mathbf{D}^{-1} & rc\end{cases}\end{pmatrix}.\end{aligned} \tag{C.12}$$

In (C.12), each of the four cases for each element is shown with no code or the codes $r$, $c$, or $rc$ to represent row, column, or row and column permutation. While any of the cases can be chosen for each element depending on what is suitable for the symbolic solution, generally, the first, second, third, or fourth equation for each element should be chosen depending on whether **A**, **C**, **B**, or **D** is invertible, respectively. This is because if the matrix inverse exists, and at least one matrix is invertible, then the solution can be shown to exist despite the existence of other matrix inverses within the equations.

It should be noted that everything described here is for symbolic matrix inversion and should not generally be used to supplant a full numerical inversion of the matrix in the case of numerical solutions unless one knows for sure the possibilities for the block matrix values and the invertibility of the matrices. Generally, the full numerical matrix inverse is superior in that it makes use of all values of the matrix when determining pivot elements. Here, four possible solutions have been shown for the inversion of a $2 \times 2$ block matrix; there are actually many more possible solutions for the inverse depending on the order of the internal block matrices. In addition, there are cases where none of these four solutions exist and the matrix is actually invertible! For example, if the internal block matrices are all $2 \times 2$ (meaning that the flattened form of the matrix is actually $16 \times 16$), then using partial pivoting (i.e. row permutations only), there are actually $4! = 24$ possibilities for the solutions; for full pivoting (i.e. row and column permutations), there are $4^2 \cdot 3^2 \cdot 2^2 = 576$ possibilities. Only four have been shown (retaining the original block matrix form) and none of these four might be possible. A case in point is the following block matrix:

$$\begin{pmatrix} \mathbf{A} & \mathbf{B} \\ \mathbf{C} & \mathbf{D} \end{pmatrix} = \begin{pmatrix} \begin{pmatrix} 1 & 0 \\ 0 & 0 \end{pmatrix} & \begin{pmatrix} 0 & 0 \\ 1 & 0 \end{pmatrix} \\ \begin{pmatrix} 0 & 1 \\ 0 & 0 \end{pmatrix} & \begin{pmatrix} 0 & 0 \\ 0 & 1 \end{pmatrix} \end{pmatrix}.$$

In this matrix, none of the block matrices is invertible, but the matrix inverse does exist.

## C.5 Swapping Order of Matrix Multiplication

During some of the derivations, the combination $(\mathbb{I} - \mathbf{A}) \cdot (\mathbb{I} + \mathbf{A})^{-1}$ comes about, yet, when checking against other formulas, one finds this combination reversed in order as $(\mathbb{I} + \mathbf{A})^{-1} \cdot (\mathbb{I} - \mathbf{A})$. Surprisingly, they are equal; a quick algebraic proof is given as follows:

$$\begin{aligned}
(\mathbb{I} - \mathbf{A}) \cdot (\mathbb{I} + \mathbf{A})^{-1} &= (\mathbb{I} + \mathbf{A})^{-1} - \mathbf{A} \cdot (\mathbb{I} + \mathbf{A})^{-1} \\
&= (\mathbb{I} + \mathbf{A})^{-1} \cdot \left[\mathbb{I} - (\mathbb{I} + \mathbf{A}) \cdot \mathbf{A} \cdot (\mathbb{I} + \mathbf{A})^{-1}\right] \\
&= (\mathbb{I} + \mathbf{A})^{-1} \cdot \left[\mathbb{I} - (\mathbf{A} + \mathbf{A} \cdot \mathbf{A}) \cdot (\mathbb{I} + \mathbf{A})^{-1}\right] \\
&= (\mathbb{I} + \mathbf{A})^{-1} \cdot \left[\mathbb{I} - \mathbf{A} \cdot (\mathbb{I} + \mathbf{A}) \cdot (\mathbb{I} + \mathbf{A})^{-1}\right] \\
(\mathbb{I} - \mathbf{A}) \cdot (\mathbb{I} + \mathbf{A})^{-1} &= (\mathbb{I} + \mathbf{A})^{-1} \cdot (\mathbb{I} - \mathbf{A}) .
\end{aligned} \tag{C.13}$$

# Appendix D

# Symbolic Device Solutions

## D.1 Four-Port Voltage Amplifier

```
import SignalIntegrity.Lib as si
sdp=si.p.SystemDescriptionParser()
sdp.AddLines(['device DV 4','device ZI 2','device ZO 2','connect ZI 1 DV 2',
    'connect ZI 2 DV 1','connect ZO 1 DV 4','port 1 ZI 1 2 ZI 2 3 ZO 2 4 DV 3'])
ssps=si.sd.SystemSParametersSymbolic(sdp.SystemDescription(),size='small')
ssps.AssignSParameters('DV',si.sy.VoltageControlledVoltageSource('\\alpha'))
ssps.AssignSParameters('ZI',si.sy.SeriesZ('Z_i'))
ssps.AssignSParameters('ZO',si.sy.SeriesZ('Z_o'))
ssps.LaTeXSolution(size='big').Emit()
```

(a) Python code

$$\mathbf{Wi} = \left[ \mathbb{I} - \begin{pmatrix} 0 & 0 & 0 & 0 & 0 & 0 & 0 & 0 & 1 & 0 \\ 0 & 0 & 0 & 0 & 0 & 0 & 1 & 0 & 0 & 0 \\ 0 & 0 & 0 & 0 & 0 & 0 & \alpha & 0 & -\alpha & 0 \\ 0 & 0 & 0 & 0 & 0 & 0 & 0 & \frac{Z_i}{Z_i+2\cdot Z0} & 0 & \frac{2\cdot Z0}{Z_i+2\cdot Z0} \\ 0 & 0 & 0 & 0 & 0 & 0 & 0 & \frac{2\cdot Z0}{Z_i+2\cdot Z0} & 0 & \frac{Z_i}{Z_i+2\cdot Z0} \\ 0 & 0 & \frac{Z_o}{Z_o+2\cdot Z0} & 0 & 0 & 0 & 0 & 0 & 0 & 0 \\ 0 & -\frac{1}{3} & 0 & \frac{2}{3} & 0 & 0 & 0 & 0 & 0 & 0 \\ 0 & \frac{2}{3} & 0 & -\frac{1}{3} & 0 & 0 & 0 & 0 & 0 & 0 \\ -\frac{1}{3} & 0 & 0 & 0 & \frac{2}{3} & 0 & 0 & 0 & 0 & 0 \\ \frac{2}{3} & 0 & 0 & 0 & -\frac{1}{3} & 0 & 0 & 0 & 0 & 0 \end{pmatrix} \right]^{-1}$$

$$\mathbf{S} = \begin{pmatrix} -\frac{1}{3} & 0 & 0 & 0 \\ 0 & -\frac{1}{3} & 0 & 0 \\ 0 & 0 & \frac{Z_o}{Z_o+2\cdot Z0} & 0 \\ 0 & 0 & 0 & 0 \end{pmatrix} + \begin{pmatrix} 0 & \frac{2}{3} & 0 & \frac{2}{3} & 0 & 0 & 0 & 0 & 0 & 0 \\ \frac{2}{3} & 0 & 0 & 0 & \frac{2}{3} & 0 & 0 & 0 & 0 & 0 \\ 0 & 0 & \frac{2\cdot Z0}{Z_o+2\cdot Z0} & 0 & 0 & 0 & 0 & 0 & 0 & 0 \\ 0 & 0 & 0 & 0 & 0 & 1 & -\alpha & 0 & \alpha & 0 \end{pmatrix} \cdot \mathbf{Wi} \cdot \begin{pmatrix} 0 & 0 & 0 & 0 \\ 0 & 0 & 0 & 0 \\ 0 & 0 & 0 & 1 \\ 0 & 0 & 0 & 0 \\ 0 & 0 & 0 & 0 \\ 0 & 0 & \frac{2\cdot Z0}{Z_o+2\cdot Z0} & 0 \\ \frac{2}{3} & 0 & 0 & 0 \\ \frac{2}{3} & 0 & 0 & 0 \\ 0 & \frac{2}{3} & 0 & 0 \\ 0 & \frac{2}{3} & 0 & 0 \end{pmatrix}$$

(b) LaTeX processed equations

**Figure D.1** Symbolic four-port voltage amplifier solution

Figure D.1 shows the Python code (Figure D.1(a)) and the symbolic solution (Figure D.1(b)) for the four-port voltage amplifier provided in Figure 6.15.

```
import SignalIntegrity.Lib as si
sdp=si.p.SystemDescriptionParser()
sdp.AddLines(['device DV 4',
    'port 1 DV 1 2 DV 3 3 DV 2',
    'connect DV 2 DV 4'])
ssps=si.sd.SystemSParametersSymbolic(sdp.SystemDescription(),size='small')
ssps.AssignSParameters('DV',si.sy.VoltageAmplifier(4,'\\alpha','Z_i','Z_o'))
ssps.LaTeXSolution(size='biggest').Emit()
```

(a) Python code

$$\mathbf{W_{ba}} = \begin{pmatrix} \frac{Z_i}{Z_i+2\cdot Z0} & 0 & 0 \\ \frac{2\cdot Z_i\cdot Z0\cdot\alpha}{(Z_i+2\cdot Z0)\cdot(Z_o+2\cdot Z0)} & \frac{Z_o}{Z_o+2\cdot Z0} & 0 \\ 0 & 0 & -\frac{1}{3} \end{pmatrix}$$

$$\mathbf{W_{bx}} = \begin{pmatrix} 0 & 0 & \frac{2\cdot Z0}{Z_i+2\cdot Z0} & 0 \\ 0 & 0 & -\frac{2\cdot Z_i\cdot Z0\cdot\alpha}{(Z_i+2\cdot Z0)\cdot(Z_o+2\cdot Z0)} & \frac{2\cdot Z0}{Z_o+2\cdot Z0} \\ \frac{2}{3} & \frac{2}{3} & 0 & 0 \end{pmatrix}$$

$$\mathbf{W_{xa}} = \begin{pmatrix} \frac{2\cdot Z0}{Z_i+2\cdot Z0} & 0 & 0 \\ -\frac{2\cdot Z_i\cdot Z0\cdot\alpha}{(Z_i+2\cdot Z0)\cdot(Z_o+2\cdot Z0)} & \frac{2\cdot Z0}{Z_o+2\cdot Z0} & 0 \\ 0 & 0 & \frac{2}{3} \\ 0 & 0 & \frac{2}{3} \end{pmatrix}$$

$$\mathbf{W_{xx}} = \begin{pmatrix} 0 & 0 & \frac{Z_i}{Z_i+2\cdot Z0} & 0 \\ 0 & 0 & \frac{2\cdot Z_i\cdot Z0\cdot\alpha}{(Z_i+2\cdot Z0)\cdot(Z_o+2\cdot Z0)} & \frac{Z_o}{Z_o+2\cdot Z0} \\ -\frac{1}{3} & \frac{2}{3} & 0 & 0 \\ \frac{2}{3} & -\frac{1}{3} & 0 & 0 \end{pmatrix}$$

$$\mathbf{S} = \mathbf{W_{ba}} + \mathbf{W_{bx}} \cdot [\mathbb{I} - \mathbf{W_{xx}}]^{-1} \cdot \mathbf{W_{xa}}$$

(b) LaTeX processed equations

**Figure D.2** Symbolic three-port voltage amplifier solution

## D.2 Three-Port Voltage Amplifier

Figure D.2 shows the Python code (Figure D.2(a)) and the symbolic solution (Figure D.2(b)) for the three-port voltage amplifier provided in Figure 6.16.

```
import SignalIntegrity.Lib as si
sdp=si.p.SystemDescriptionParser()
sdp.AddLines(['device DV 4','device G 1 ground','port 1 DV 1 2 DV 3',
    'connect DV 2 G 1','connect DV 4 G 1'])
ssps=si.sd.SystemSParametersSymbolic(sdp.SystemDescription(),size='small')
DV=si.sy.VoltageAmplifier(4,'\\alpha','Z_i','Z_o')
ssps.AssignSParameters('DV',DV)
ssps.LaTeXSolution(size='biggest').Emit()
```

(a) Python code

$$\mathbf{W_{ba}} = \begin{pmatrix} \frac{Z_i}{Z_i+2\cdot Z0} & 0 \\ \frac{2\cdot Z_i\cdot Z0\cdot\alpha}{(Z_i+2\cdot Z0)\cdot(Z_o+2\cdot Z0)} & \frac{Z_o}{Z_o+2\cdot Z0} \end{pmatrix}$$

$$\mathbf{W_{bx}} = \begin{pmatrix} 0 & 0 & 0 & \frac{2\cdot Z0}{Z_i+2\cdot Z0} & 0 & 0 \\ 0 & 0 & 0 & -\frac{2\cdot Z_i\cdot Z0\cdot\alpha}{(Z_i+2\cdot Z0)\cdot(Z_o+2\cdot Z0)} & 0 & \frac{2\cdot Z0}{Z_o+2\cdot Z0} \end{pmatrix}$$

$$\mathbf{W_{xa}} = \begin{pmatrix} \frac{2\cdot Z0}{Z_i+2\cdot Z0} & 0 \\ -\frac{2\cdot Z_i\cdot Z0\cdot\alpha}{(Z_i+2\cdot Z0)\cdot(Z_o+2\cdot Z0)} & \frac{2\cdot Z0}{Z_o+2\cdot Z0} \\ 0 & 0 \\ 0 & 0 \\ 0 & 0 \\ 0 & 0 \end{pmatrix}$$

$$\mathbf{W_{xx}} = \begin{pmatrix} 0 & 0 & 0 & \frac{Z_i}{Z_i+2\cdot Z0} & 0 & 0 \\ 0 & 0 & 0 & \frac{2\cdot Z_i\cdot Z0\cdot\alpha}{(Z_i+2\cdot Z0)\cdot(Z_o+2\cdot Z0)} & 0 & \frac{Z_o}{Z_o+2\cdot Z0} \\ 0 & 0 & 0 & 0 & -1 & 0 \\ -\frac{1}{3} & \frac{2}{3} & \frac{2}{3} & 0 & 0 & 0 \\ \frac{2}{3} & \frac{2}{3} & -\frac{1}{3} & 0 & 0 & 0 \\ \frac{2}{3} & -\frac{1}{3} & \frac{2}{3} & 0 & 0 & 0 \end{pmatrix}$$

$$\mathbf{S} = \mathbf{W_{ba}} + \mathbf{W_{bx}} \cdot \left[\mathbb{I} - \mathbf{W_{xx}}\right]^{-1} \cdot \mathbf{W_{xa}}$$

(b) LaTeX processed equations

**Figure D.3** Symbolic two-port voltage amplifier solution

## D.3 Two-Port Voltage Amplifier

Figure D.3 shows the Python code (Figure D.3(a)) and the symbolic solution (Figure D.3(b)) for the two-port voltage amplifier provided in Figure 6.17.

```
import SignalIntegrity.Lib as si
sdp=si.p.SystemDescriptionParser()
sdp.AddLines(['device DC 4','device ZI 2','device ZO 2',
    'port 1 ZI 1 2 DC 2 3 DC 4 4 DC 3',
    'connect ZI 2 DC 1','connect ZO 1 DC 4','connect ZO 2 DC 3'])
ssps=si.sd.SystemSParametersSymbolic(sdp.SystemDescription(),size='small')
ssps.AssignSParameters('DC',si.sy.CurrentControlledCurrentSource('\\beta'))
ssps.AssignSParameters('ZI',si.sy.SeriesZ('Z_i'))
ssps.AssignSParameters('ZO',si.sy.SeriesZ('Z_o'))
ssps.LaTeXSolution(size='big').Emit()
```

(a) Python code

$$\mathbf{Wi} = \left[\mathbb{I} - \begin{pmatrix} 0 & 0 & 0 & 0 & 0 & 0 & 0 & 0 & 0 & 0 \\ 0 & 0 & 0 & -\beta & 0 & 0 & 0 & 0 & 1 & 0 \\ 0 & 0 & 0 & \beta & 0 & 0 & 1 & 0 & 0 & 0 \\ \frac{Z_i}{Z_i+2\cdot Z0} & 0 & 0 & 0 & 0 & 0 & 0 & 0 & 0 & 0 \\ 0 & 0 & 0 & 0 & 0 & 0 & 0 & \frac{Z_o}{Z_o+2\cdot Z0} & 0 & \frac{2\cdot Z0}{Z_o+2\cdot Z0} \\ 0 & 0 & 0 & 0 & 0 & 0 & 0 & \frac{2\cdot Z0}{Z_o+2\cdot Z0} & 0 & \frac{Z_o}{Z_o+2\cdot Z0} \\ 0 & 0 & -\frac{1}{3} & 0 & \frac{2}{3} & 0 & 0 & 0 & 0 & 0 \\ 0 & 0 & \frac{2}{3} & 0 & -\frac{1}{3} & 0 & 0 & 0 & 0 & 0 \\ 0 & -\frac{1}{3} & 0 & 0 & 0 & \frac{2}{3} & 0 & 0 & 0 & 0 \\ 0 & \frac{2}{3} & 0 & 0 & 0 & -\frac{1}{3} & 0 & 0 & 0 & 0 \end{pmatrix}\right]^{-1}$$

$$\mathbf{S} = \begin{pmatrix} \frac{Z_i}{Z_i+2\cdot Z0} & 0 & 0 & 0 \\ 0 & 0 & 0 & 0 \\ 0 & 0 & -\frac{1}{3} & 0 \\ 0 & 0 & 0 & -\frac{1}{3} \end{pmatrix} + \begin{pmatrix} \frac{2\cdot Z0}{Z_i+2\cdot Z0} & 0 & 0 & 0 & 0 & 0 & 0 & 0 & 0 & 0 \\ 0 & 0 & 0 & 1 & 0 & 0 & 0 & 0 & 0 & 0 \\ 0 & 0 & \frac{2}{3} & 0 & \frac{2}{3} & 0 & 0 & 0 & 0 & 0 \\ 0 & \frac{2}{3} & 0 & 0 & 0 & \frac{2}{3} & 0 & 0 & 0 & 0 \end{pmatrix} \cdot \mathbf{Wi} \cdot \begin{pmatrix} 0 & 1 & 0 & 0 \\ 0 & \beta & 0 & 0 \\ 0 & -\beta & 0 & 0 \\ \frac{2\cdot Z0}{Z_i+2\cdot Z0} & 0 & 0 & 0 \\ 0 & 0 & 0 & 0 \\ 0 & 0 & 0 & 0 \\ 0 & 0 & \frac{2}{3} & 0 \\ 0 & 0 & \frac{2}{3} & 0 \\ 0 & 0 & 0 & \frac{2}{3} \\ 0 & 0 & 0 & \frac{2}{3} \end{pmatrix}$$

(b) LaTeX processed equations

**Figure D.4** Symbolic four-port current amplifier solution

## D.4 Four-Port Current Amplifier

Figure D.4 shows the Python code (Figure D.4(a)) and the symbolic solution (Figure D.4(b)) for the four-port current amplifier provided in Figure 6.18.

```
import SignalIntegrity.Lib as si
sdp=si.p.SystemDescriptionParser()
sdp.AddLines(['device D 4',
    'port 1 D 1 2 D 3 3 D 2',
    'connect D 2 D 4'])
ssps=si.sd.SystemSParametersSymbolic(sdp.SystemDescription(),size='small')
ssps.AssignSParameters('D',si.sy.CurrentAmplifier(4,'\\beta','Z_i','Z_o'))
ssps.LaTeXSolution(size='biggest').Emit()
```

(a) Python code

$$\mathbf{W_{ba}} = \begin{pmatrix} \frac{Z_i}{Z_i+2\cdot Z0} & 0 & 0 \\ \frac{2\cdot Z_o\cdot Z0\cdot\beta}{(Z_i+2\cdot Z0)\cdot(Z_o+2\cdot Z0)} & \frac{Z_o}{Z_o+2\cdot Z0} & 0 \\ 0 & 0 & -\frac{1}{3} \end{pmatrix}$$

$$\mathbf{W_{bx}} = \begin{pmatrix} 0 & 0 & \frac{2\cdot Z0}{Z_i+2\cdot Z0} & 0 \\ 0 & 0 & -\frac{2\cdot Z_o\cdot Z0\cdot\beta}{(Z_i+2\cdot Z0)\cdot(Z_o+2\cdot Z0)} & \frac{2\cdot Z0}{Z_o+2\cdot Z0} \\ \frac{2}{3} & \frac{2}{3} & 0 & 0 \end{pmatrix}$$

$$\mathbf{W_{xa}} = \begin{pmatrix} \frac{2\cdot Z0}{Z_i+2\cdot Z0} & 0 & 0 \\ -\frac{2\cdot Z_o\cdot Z0\cdot\beta}{(Z_i+2\cdot Z0)\cdot(Z_o+2\cdot Z0)} & \frac{2\cdot Z0}{Z_o+2\cdot Z0} & 0 \\ 0 & 0 & \frac{2}{3} \\ 0 & 0 & \frac{2}{3} \end{pmatrix}$$

$$\mathbf{W_{xx}} = \begin{pmatrix} 0 & 0 & \frac{Z_i}{Z_i+2\cdot Z0} & 0 \\ 0 & 0 & \frac{2\cdot Z_o\cdot Z0\cdot\beta}{(Z_i+2\cdot Z0)\cdot(Z_o+2\cdot Z0)} & \frac{Z_o}{Z_o+2\cdot Z0} \\ -\frac{1}{3} & \frac{2}{3} & 0 & 0 \\ \frac{2}{3} & -\frac{1}{3} & 0 & 0 \end{pmatrix}$$

$$\mathbf{S} = \mathbf{W_{ba}} + \mathbf{W_{bx}} \cdot \left[\mathbb{I} - \mathbf{W_{xx}}\right]^{-1} \cdot \mathbf{W_{xa}}$$

(b) LaTeX processed equations

**Figure D.5** Symbolic three-port current amplifier solution

## D.5 Three-Port Current Amplifier

Figure D.5 shows the Python code (Figure D.5(a)) and the symbolic solution (Figure D.5(b)) for the three-port current amplifier provided in Figure 6.19.

```
import SignalIntegrity.Lib as si
sdp=si.p.SystemDescriptionParser()
sdp.AddLines(['device D 4','device G1 1 ground','device G2 1 ground',
    'port 1 D 1 2 D 3',
    'connect D 2 G1 1','connect D 4 G2 1'])
ssps=si.sd.SystemSParametersSymbolic(sdp.SystemDescription(),size='small')
D=si.sy.CurrentAmplifier(4,'\\beta','Z_i','Z_o')
ssps.AssignSParameters('D',D)
ssps.LaTeXSolution(size='biggest').Emit()
```

(a) Python code

$$\mathbf{W_{ba}} = \begin{pmatrix} \frac{Z_i}{Z_i+2\cdot Z0} & 0 \\ \frac{2\cdot Z_o\cdot Z0\cdot\beta}{(Z_i+2\cdot Z0)\cdot(Z_o+2\cdot Z0)} & \frac{Z_o}{Z_o+2\cdot Z0} \end{pmatrix}$$

$$\mathbf{W_{bx}} = \begin{pmatrix} 0 & 0 & \frac{2\cdot Z0}{Z_i+2\cdot Z0} & 0 \\ 0 & 0 & -\frac{2\cdot Z_o\cdot Z0\cdot\beta}{(Z_i+2\cdot Z0)\cdot(Z_o+2\cdot Z0)} & \frac{2\cdot Z0}{Z_o+2\cdot Z0} \end{pmatrix}$$

$$\mathbf{W_{xa}} = \begin{pmatrix} \frac{2\cdot Z0}{Z_i+2\cdot Z0} & 0 \\ -\frac{2\cdot Z_o\cdot Z0\cdot\beta}{(Z_i+2\cdot Z0)\cdot(Z_o+2\cdot Z0)} & \frac{2\cdot Z0}{Z_o+2\cdot Z0} \\ 0 & 0 \\ 0 & 0 \end{pmatrix}$$

$$\mathbf{W_{xx}} = \begin{pmatrix} 0 & 0 & \frac{Z_i}{Z_i+2\cdot Z0} & 0 \\ 0 & 0 & \frac{2\cdot Z_o\cdot Z0\cdot\beta}{(Z_i+2\cdot Z0)\cdot(Z_o+2\cdot Z0)} & \frac{Z_o}{Z_o+2\cdot Z0} \\ -1 & 0 & 0 & 0 \\ 0 & -1 & 0 & 0 \end{pmatrix}$$

$$\mathbf{S} = \mathbf{W_{ba}} + \mathbf{W_{bx}} \cdot [\mathbb{I} - \mathbf{W_{xx}}]^{-1} \cdot \mathbf{W_{xa}}$$

(b) LaTeX processed equations

**Figure D.6** Symbolic two-port current amplifier solution

## D.6 Two-Port Current Amplifier

Figure D.6 shows the Python code (Figure D.6(a)) and the symbolic solution (Figure D.6(b)) for the two-port current amplifier provided in Figure 6.20.

```
import SignalIntegrity.Lib as si
sdp=si.p.SystemDescriptionParser()
sdp.AddLines(['device D 4','device ZI 4','device ZO 4',
    'port 1 ZI 1 2 ZI 2 3 ZO 1 4 ZO 2',
    'connect ZI 3 D 2','connect ZI 4 D 1','connect ZO 3 D 4','connect ZO 4 D 3'])
ssps=si.sd.SystemSParametersSymbolic(sdp.SystemDescription(),size='small')
ssps.AssignSParameters('D',si.sy.VoltageControlledCurrentSource('\\delta'))
ssps.AssignSParameters('ZI',si.sy.ShuntZ(4,'Z_i'))
ssps.AssignSParameters('ZO',si.sy.ShuntZ(4,'Z_o'))
ssps.LaTeXSolution(size='biggest').Emit()
```

(a) Python code

$$\mathbf{W_{ba}} = \begin{pmatrix} \frac{-Z0}{2\cdot(Z_i+Z0)} & \frac{Z0}{2\cdot(Z_i+Z0)} & 0 & 0 \\ \frac{Z0}{2\cdot(Z_i+Z0)} & \frac{-Z0}{2\cdot(Z_i+Z0)} & 0 & 0 \\ 0 & 0 & \frac{-Z0}{2\cdot(Z_o+Z0)} & \frac{Z0}{2\cdot(Z_o+Z0)} \\ 0 & 0 & \frac{Z0}{2\cdot(Z_o+Z0)} & \frac{-Z0}{2\cdot(Z_o+Z0)} \end{pmatrix}$$

$$\mathbf{W_{bx}} = \begin{pmatrix} \frac{Z0}{2\cdot(Z_i+Z0)} & \frac{2\cdot Z_i+Z0}{2\cdot(Z_i+Z0)} & 0 & 0 & 0 & 0 & 0 & 0 \\ \frac{2\cdot Z_i+Z0}{2\cdot(Z_i+Z0)} & \frac{Z0}{2\cdot(Z_i+Z0)} & 0 & 0 & 0 & 0 & 0 & 0 \\ 0 & 0 & \frac{Z0}{2\cdot(Z_o+Z0)} & \frac{2\cdot Z_o+Z0}{2\cdot(Z_o+Z0)} & 0 & 0 & 0 & 0 \\ 0 & 0 & \frac{2\cdot Z_o+Z0}{2\cdot(Z_o+Z0)} & \frac{Z0}{2\cdot(Z_o+Z0)} & 0 & 0 & 0 & 0 \end{pmatrix}$$

$$\mathbf{W_{xa}} = \begin{pmatrix} 0 & 0 & 0 & 0 \\ 0 & 0 & 0 & 0 \\ 0 & 0 & 0 & 0 \\ 0 & 0 & 0 & 0 \\ \frac{2\cdot Z_i+Z0}{2\cdot(Z_i+Z0)} & \frac{Z0}{2\cdot(Z_i+Z0)} & 0 & 0 \\ \frac{Z0}{2\cdot(Z_i+Z0)} & \frac{2\cdot Z_i+Z0}{2\cdot(Z_i+Z0)} & 0 & 0 \\ 0 & 0 & \frac{2\cdot Z_o+Z0}{2\cdot(Z_o+Z0)} & \frac{Z0}{2\cdot(Z_o+Z0)} \\ 0 & 0 & \frac{Z0}{2\cdot(Z_o+Z0)} & \frac{2\cdot Z_o+Z0}{2\cdot(Z_o+Z0)} \end{pmatrix}$$

$$\mathbf{W_{xx}} = \begin{pmatrix} 0 & 0 & 0 & 0 & 0 & 1 & 0 & 0 \\ 0 & 0 & 0 & 0 & 1 & 0 & 0 & 0 \\ 0 & 0 & 0 & 0 & -2\cdot\delta\cdot Z0 & 2\cdot\delta\cdot Z0 & 0 & 1 \\ 0 & 0 & 0 & 0 & 2\cdot\delta\cdot Z0 & -2\cdot\delta\cdot Z0 & 1 & 0 \\ \frac{Z0}{2\cdot(Z_i+Z0)} & \frac{-Z0}{2\cdot(Z_i+Z0)} & 0 & 0 & 0 & 0 & 0 & 0 \\ \frac{-Z0}{2\cdot(Z_i+Z0)} & \frac{Z0}{2\cdot(Z_i+Z0)} & 0 & 0 & 0 & 0 & 0 & 0 \\ 0 & 0 & \frac{Z0}{2\cdot(Z_o+Z0)} & \frac{-Z0}{2\cdot(Z_o+Z0)} & 0 & 0 & 0 & 0 \\ 0 & 0 & \frac{-Z0}{2\cdot(Z_o+Z0)} & \frac{Z0}{2\cdot(Z_o+Z0)} & 0 & 0 & 0 & 0 \end{pmatrix}$$

$$\mathbf{S} = \mathbf{W_{ba}} + \mathbf{W_{bx}} \cdot \left[\mathbb{I} - \mathbf{W_{xx}}\right]^{-1} \cdot \mathbf{W_{xa}}$$

(b) LaTeX processed equations

**Figure D.7** Symbolic four-port transconductance amplifier solution

## D.7 Four-Port Transconductance Amplifier

Figure D.7 shows the Python code (Figure D.7(a)) and the symbolic solution (Figure D.7(b)) for the four-port transconductance amplifier provided in Figure 6.21.

```
import SignalIntegrity.Lib as si
sdp=si.p.SystemDescriptionParser()
sdp.AddLines(['device D 4',
    'port 1 D 1 2 D 3 3 D 2',
    'connect D 2 D 4'])
ssps=si.sd.SystemSParametersSymbolic(sdp.SystemDescription(),size='small')
ssps.AssignSParameters('D',si.sy.TransconductanceAmplifier(4,'\\delta','Z_i','Z_o'))
ssps.LaTeXSolution(size='biggest').Emit()
```

(a) Python code

$$\mathbf{W_{ba}} = \left(\begin{array}{ccc} \frac{Z_i}{Z_i+2\cdot Z0} & 0 & 0 \\ \frac{2\cdot Z_i\cdot Z_o\cdot Z0\cdot\delta}{(Z_i+2\cdot Z0)\cdot(Z_o+2\cdot Z0)} & \frac{Z_o}{Z_o+2\cdot Z0} & 0 \\ 0 & 0 & -\frac{1}{3} \end{array}\right)$$

$$\mathbf{W_{bx}} = \left(\begin{array}{cccc} 0 & 0 & \frac{2\cdot Z0}{Z_i+2\cdot Z0} & 0 \\ 0 & 0 & -\frac{2\cdot Z_i\cdot Z_o\cdot Z0\cdot\delta}{(Z_i+2\cdot Z0)\cdot(Z_o+2\cdot Z0)} & \frac{2\cdot Z0}{Z_o+2\cdot Z0} \\ \frac{2}{3} & \frac{2}{3} & 0 & 0 \end{array}\right)$$

$$\mathbf{W_{xa}} = \left(\begin{array}{ccc} \frac{2\cdot Z0}{Z_i+2\cdot Z0} & 0 & 0 \\ -\frac{2\cdot Z_i\cdot Z_o\cdot Z0\cdot\delta}{(Z_i+2\cdot Z0)\cdot(Z_o+2\cdot Z0)} & \frac{2\cdot Z0}{Z_o+2\cdot Z0} & 0 \\ 0 & 0 & \frac{2}{3} \\ 0 & 0 & \frac{2}{3} \end{array}\right)$$

$$\mathbf{W_{xx}} = \left(\begin{array}{cccc} 0 & 0 & \frac{Z_i}{Z_i+2\cdot Z0} & 0 \\ 0 & 0 & \frac{2\cdot Z_i\cdot Z_o\cdot Z0\cdot\delta}{(Z_i+2\cdot Z0)\cdot(Z_o+2\cdot Z0)} & \frac{Z_o}{Z_o+2\cdot Z0} \\ -\frac{1}{3} & \frac{2}{3} & 0 & 0 \\ \frac{2}{3} & -\frac{1}{3} & 0 & 0 \end{array}\right)$$

$$\mathbf{S} = \mathbf{W_{ba}} + \mathbf{W_{bx}} \cdot \left[\mathbb{I} - \mathbf{W_{xx}}\right]^{-1} \cdot \mathbf{W_{xa}}$$

(b) LaTeX processed equations

**Figure D.8** Symbolic three-port transconductance amplifier solution

## D.8 Three-Port Transconductance Amplifier

Figure D.8 shows the Python code (Figure D.8(a)) and the symbolic solution (Figure D.8(b)) for the three-port transconductance amplifier provided in Figure 6.22.

```python
import SignalIntegrity.Lib as si
sdp=si.p.SystemDescriptionParser()
sdp.AddLines(['device D 4','device G1 1 ground','device G2 1 ground',
    'port 1 D 1 2 D 3',
    'connect D 2 G1 1','connect D 4 G2 1'])
ssps=si.sd.SystemSParametersSymbolic(sdp.SystemDescription(),size='small')
D=si.sy.TransconductanceAmplifier(4,'\\delta','Z_i','Z_o')
ssps.AssignSParameters('D',D)
ssps.LaTeXSolution(size='biggest').Emit()
```

(a) Python code

$$\mathbf{W_{ba}} = \begin{pmatrix} \frac{Z_i}{Z_i+2\cdot Z0} & 0 \\ \frac{2\cdot Z_i\cdot Z_o\cdot Z0\cdot\delta}{(Z_i+2\cdot Z0)\cdot(Z_o+2\cdot Z0)} & \frac{Z_o}{Z_o+2\cdot Z0} \end{pmatrix}$$

$$\mathbf{W_{bx}} = \begin{pmatrix} 0 & 0 & \frac{2\cdot Z0}{Z_i+2\cdot Z0} & 0 \\ 0 & 0 & -\frac{2\cdot Z_i\cdot Z_o\cdot Z0\cdot\delta}{(Z_i+2\cdot Z0)\cdot(Z_o+2\cdot Z0)} & \frac{2\cdot Z0}{Z_o+2\cdot Z0} \end{pmatrix}$$

$$\mathbf{W_{xa}} = \begin{pmatrix} \frac{2\cdot Z0}{Z_i+2\cdot Z0} & 0 \\ -\frac{2\cdot Z_i\cdot Z_o\cdot Z0\cdot\delta}{(Z_i+2\cdot Z0)\cdot(Z_o+2\cdot Z0)} & \frac{2\cdot Z0}{Z_o+2\cdot Z0} \\ 0 & 0 \\ 0 & 0 \end{pmatrix}$$

$$\mathbf{W_{xx}} = \begin{pmatrix} 0 & 0 & \frac{Z_i}{Z_i+2\cdot Z0} & 0 \\ 0 & 0 & \frac{2\cdot Z_i\cdot Z_o\cdot Z0\cdot\delta}{(Z_i+2\cdot Z0)\cdot(Z_o+2\cdot Z0)} & \frac{Z_o}{Z_o+2\cdot Z0} \\ -1 & 0 & 0 & 0 \\ 0 & -1 & 0 & 0 \end{pmatrix}$$

$$\mathbf{S} = \mathbf{W_{ba}} + \mathbf{W_{bx}}\cdot\left[\mathbb{I} - \mathbf{W_{xx}}\right]^{-1}\cdot\mathbf{W_{xa}}$$

(b) LaTeX processed equations

**Figure D.9** Symbolic two-port transconductance amplifier solution

## D.9 Two-Port Transconductance Amplifier

Figure D.9 shows the Python code (Figure D.9(a)) and the symbolic solution (Figure D.9(b)) for the two-port transconductance amplifier provided in Figure 6.23.

```
import SignalIntegrity.Lib as si
sdp=si.p.SystemDescriptionParser()
sdp.AddLines(['device D 4','device ZI 2','device ZO 2',
    'port 1 ZI 1 2 D 2 3 ZO 2 4 D 3',
    'connect ZI 2 D 1','connect ZO 1 D 4'])
ssps=si.sd.SystemSParametersSymbolic(sdp.SystemDescription(),size='small')
ssps.AssignSParameters('D',si.sy.CurrentControlledVoltageSource('\\gamma'))
ssps.AssignSParameters('ZI',si.sy.SeriesZ('Z_i'))
ssps.AssignSParameters('ZO',si.sy.SeriesZ('Z_o'))
ssps.LaTeXSolution(size='big').Emit()
```

(a) Python code

$$\mathbf{Wi} = \left[\mathbb{I} - \begin{pmatrix} 0 & 0 & 0 & 0 \\ 0 & 0 & \frac{\gamma}{2\cdot Z0} & 0 \\ \frac{Z_i}{Z_i+2\cdot Z0} & 0 & 0 & 0 \\ 0 & \frac{Z_o}{Z_o+2\cdot Z0} & 0 & 0 \end{pmatrix}\right]^{-1}$$

$$\mathbf{S} = \begin{pmatrix} \frac{Z_i}{Z_i+2\cdot Z0} & 0 & 0 & 0 \\ 0 & 0 & 0 & 0 \\ 0 & 0 & \frac{Z_o}{Z_o+2\cdot Z0} & 0 \\ 0 & \frac{\gamma}{2\cdot Z0} & 0 & 0 \end{pmatrix} + \begin{pmatrix} \frac{2\cdot Z0}{Z_i+2\cdot Z0} & 0 & 0 & 0 \\ 0 & 0 & 1 & 0 \\ 0 & \frac{2\cdot Z0}{Z_o+2\cdot Z0} & 0 & 0 \\ 0 & 0 & -\frac{\gamma}{2\cdot Z0} & 1 \end{pmatrix} \cdot \mathbf{Wi} \cdot \begin{pmatrix} 0 & 1 & 0 & 0 \\ 0 & -\frac{\gamma}{2\cdot Z0} & 0 & 1 \\ \frac{2\cdot Z0}{Z_i+2\cdot Z0} & 0 & 0 & 0 \\ 0 & 0 & \frac{2\cdot Z0}{Z_o+2\cdot Z0} & 0 \end{pmatrix}$$

(b) LaTeX processed equations

**Figure D.10** Symbolic four-port transresistance amplifier solution

## D.10 Four-Port Transresistance Amplifier

Figure D.10 shows the Python code (Figure D.10(a)) and the symbolic solution (Figure D.10(b)) for the four-port transresistance amplifier provided in Figure 6.24.

```
import SignalIntegrity.Lib as si
sdp=si.p.SystemDescriptionParser()
sdp.AddLines(['device D 4',
    'port 1 D 1 2 D 3 3 D 2',
    'connect D 2 D 4'])
ssps=si.sd.SystemSParametersSymbolic(sdp.SystemDescription(),size='small')
ssps.AssignSParameters('D',si.sy.TransresistanceAmplifier(4,'\\gamma','Z_i','Z_o'))
ssps.LaTeXSolution(size='biggest').Emit()
```

(a) Python code

$$\mathbf{W_{ba}} = \begin{pmatrix} \frac{Z_i}{Z_i+2\cdot Z0} & 0 & 0 \\ \frac{2\cdot\gamma\cdot Z0}{(Z_o+2\cdot Z0)\cdot(Z_i+2\cdot Z0)} & \frac{Z_o}{Z_o+2\cdot Z0} & 0 \\ 0 & 0 & -\frac{1}{3} \end{pmatrix}$$

$$\mathbf{W_{bx}} = \begin{pmatrix} 0 & 0 & \frac{2\cdot Z0}{Z_i+2\cdot Z0} & 0 \\ 0 & 0 & -\frac{2\cdot\gamma\cdot Z0}{(Z_o+2\cdot Z0)\cdot(Z_i+2\cdot Z0)} & \frac{2\cdot Z0}{Z_o+2\cdot Z0} \\ \frac{2}{3} & \frac{2}{3} & 0 & 0 \end{pmatrix}$$

$$\mathbf{W_{xa}} = \begin{pmatrix} \frac{2\cdot Z0}{Z_i+2\cdot Z0} & 0 & 0 \\ -\frac{2\cdot\gamma\cdot Z0}{(Z_o+2\cdot Z0)\cdot(Z_i+2\cdot Z0)} & \frac{2\cdot Z0}{Z_o+2\cdot Z0} & 0 \\ 0 & 0 & \frac{2}{3} \\ 0 & 0 & \frac{2}{3} \end{pmatrix}$$

$$\mathbf{W_{xx}} = \begin{pmatrix} 0 & 0 & \frac{Z_i}{Z_i+2\cdot Z0} & 0 \\ 0 & 0 & \frac{2\cdot\gamma\cdot Z0}{(Z_o+2\cdot Z0)\cdot(Z_i+2\cdot Z0)} & \frac{Z_o}{Z_o+2\cdot Z0} \\ -\frac{1}{3} & \frac{2}{3} & 0 & 0 \\ \frac{2}{3} & -\frac{1}{3} & 0 & 0 \end{pmatrix}$$

$$\mathbf{S} = \mathbf{W_{ba}} + \mathbf{W_{bx}} \cdot \left[\mathbb{I} - \mathbf{W_{xx}}\right]^{-1} \cdot \mathbf{W_{xa}}$$

(b) LaTeX processed equations

**Figure D.11** Symbolic three-port transresistance amplifier solution

## D.11 Three-Port Transresistance Amplifier

Figure D.11 shows the Python code (Figure D.11(a)) and the symbolic solution (Figure D.11(b)) for the three-port transresistance amplifier provided in Figure 6.25.

```
import SignalIntegrity.Lib as si
sdp=si.p.SystemDescriptionParser()
sdp.AddLines(['device D 4','device G1 1 ground','device G2 1 ground',
    'port 1 D 1 2 D 3',
    'connect D 2 G1 1','connect D 4 G2 1'])
ssps=si.sd.SystemSParametersSymbolic(sdp.SystemDescription(),size='small')
D=si.sy.TransresistanceAmplifier(4,'\\gamma','Z_i','Z_o')
ssps.AssignSParameters('D',D)
ssps.LaTeXSolution(size='biggest').Emit()
```

(a) Python code

$$\mathbf{W_{ba}} = \begin{pmatrix} \frac{Z_i}{Z_i+2\cdot Z0} & 0 \\ \frac{2\cdot\gamma\cdot Z0}{(Z_o+2\cdot Z0)\cdot(Z_i+2\cdot Z0)} & \frac{Z_o}{Z_o+2\cdot Z0} \end{pmatrix}$$

$$\mathbf{W_{bx}} = \begin{pmatrix} 0 & 0 & \frac{2\cdot Z0}{Z_i+2\cdot Z0} & 0 \\ 0 & 0 & -\frac{2\cdot\gamma\cdot Z0}{(Z_o+2\cdot Z0)\cdot(Z_i+2\cdot Z0)} & \frac{2\cdot Z0}{Z_o+2\cdot Z0} \end{pmatrix}$$

$$\mathbf{W_{xa}} = \begin{pmatrix} \frac{2\cdot Z0}{Z_i+2\cdot Z0} & 0 \\ -\frac{2\cdot\gamma\cdot Z0}{(Z_o+2\cdot Z0)\cdot(Z_i+2\cdot Z0)} & \frac{2\cdot Z0}{Z_o+2\cdot Z0} \\ 0 & 0 \\ 0 & 0 \end{pmatrix}$$

$$\mathbf{W_{xx}} = \begin{pmatrix} 0 & 0 & \frac{Z_i}{Z_i+2\cdot Z0} & 0 \\ 0 & 0 & \frac{2\cdot\gamma\cdot Z0}{(Z_o+2\cdot Z0)\cdot(Z_i+2\cdot Z0)} & \frac{Z_o}{Z_o+2\cdot Z0} \\ -1 & 0 & 0 & 0 \\ 0 & -1 & 0 & 0 \end{pmatrix}$$

$$\mathbf{S} = \mathbf{W_{ba}} + \mathbf{W_{bx}} \cdot \left[\mathbb{I} - \mathbf{W_{xx}}\right]^{-1} \cdot \mathbf{W_{xa}}$$

(b) LaTeX processed equations

**Figure D.12** Symbolic two-port transresistance amplifier solution

## D.12 Two-Port Transresistance Amplifier

Figure D.12 shows the Python code (Figure D.12(a)) and the symbolic solution (Figure D.12(b)) for the two-port transresistance amplifier provided in Figure 6.26.

## D.13 Two-Port Voltage Feedback Amplifier

The two-port voltage amplifier with voltage series feedback is shown in Figure D.13(a). The Python code (Figure D.13(b)) is utilized to generate the symbolic solution, shown in Figure D.13(c).

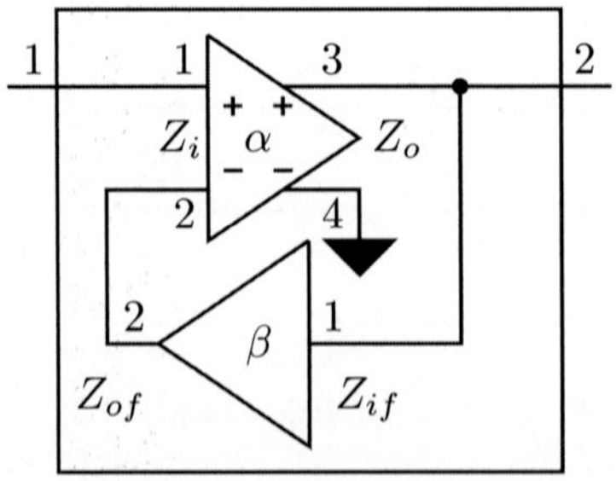


(a) Circuit representation

```python
import SignalIntegrity.Lib as si
sdp=si.p.SystemDescriptionParser()
sdp.AddLines(['device D 4','device F 2','device G 1 ground','port 1 D 1 2 D 3',
    'connect D 2 F 2','connect D 3 F 1','connect D 4 G 1'])
ssps=si.sd.SystemSParametersSymbolic(sdp.SystemDescription(),size='small')
ssps.AssignSParameters('D',si.sy.VoltageAmplifier(4,'\\alpha','Z_i','Z_o'))
ssps.AssignSParameters('F',si.sy.VoltageAmplifier(2,'\\beta','Z_{if}','Z_{of}'))
ssps.LaTeXSolution(size='biggest').Emit()
```

(b) Python code

$$\mathbf{W_{ba}} = \begin{pmatrix} \frac{Z_i}{Z_i+2\cdot Z0} & 0 \\ 0 & -\frac{1}{3} \end{pmatrix}$$

$$\mathbf{W_{bx}} = \begin{pmatrix} 0 & 0 & 0 & 0 & \frac{2\cdot Z0}{Z_i+2\cdot Z0} & 0 & 0 & 0 \\ 0 & \frac{2}{3} & 0 & \frac{2}{3} & 0 & 0 & 0 & 0 \end{pmatrix}$$

$$\mathbf{W_{xa}} = \begin{pmatrix} \frac{2\cdot Z0}{Z_i+2\cdot Z0} & 0 \\ \frac{2\cdot Z_i\cdot Z0\cdot\alpha}{(Z_i+2\cdot Z0)\cdot(Z_o+2\cdot Z0)} & 0 \\ -\frac{2\cdot Z_i\cdot Z0\cdot\alpha}{(Z_i+2\cdot Z0)\cdot(Z_o+2\cdot Z0)} & 0 \\ 0 & 0 \\ 0 & 0 \\ 0 & 0 \\ 0 & \frac{2}{3} \\ 0 & \frac{2}{3} \end{pmatrix}$$

$$\mathbf{W_{xx}} = \begin{pmatrix} 0 & 0 & 0 & 0 & \frac{Z_i}{Z_i+2\cdot Z0} & 0 & 0 & 0 \\ 0 & 0 & 0 & 0 & -\frac{2\cdot Z_i\cdot Z0\cdot\alpha}{(Z_i+2\cdot Z0)\cdot(Z_o+2\cdot Z0)} & \frac{2\cdot Z0}{Z_o+2\cdot Z0} & \frac{Z_o}{Z_o+2\cdot Z0} & 0 \\ 0 & 0 & 0 & 0 & \frac{2\cdot Z_i\cdot Z0\cdot\alpha}{(Z_i+2\cdot Z0)\cdot(Z_o+2\cdot Z0)} & \frac{Z_o}{Z_o+2\cdot Z0} & \frac{2\cdot Z0}{Z_o+2\cdot Z0} & 0 \\ 0 & 0 & 0 & 0 & 0 & 0 & 0 & \frac{Z_{if}-Z0}{Z_{if}+Z0} \\ \frac{Z_{of}-Z0}{Z_{of}+Z0} & 0 & 0 & 0 & 0 & 0 & 0 & \frac{2\cdot\beta\cdot Z_{if}\cdot Z0}{(Z_{if}+Z0)\cdot(Z_{of}+Z0)} \\ 0 & 0 & -1 & 0 & 0 & 0 & 0 & 0 \\ 0 & -\frac{1}{3} & 0 & \frac{2}{3} & 0 & 0 & 0 & 0 \\ 0 & \frac{2}{3} & 0 & -\frac{1}{3} & 0 & 0 & 0 & 0 \end{pmatrix}$$

$$\mathbf{S} = \mathbf{W_{ba}} + \mathbf{W_{bx}} \cdot \left[\mathbb{I} - \mathbf{W_{xx}}\right]^{-1} \cdot \mathbf{W_{xa}}$$

(c) LaTeX processed equations

**Figure D.13** Symbolic two-port voltage amplifier with voltage-series feedback

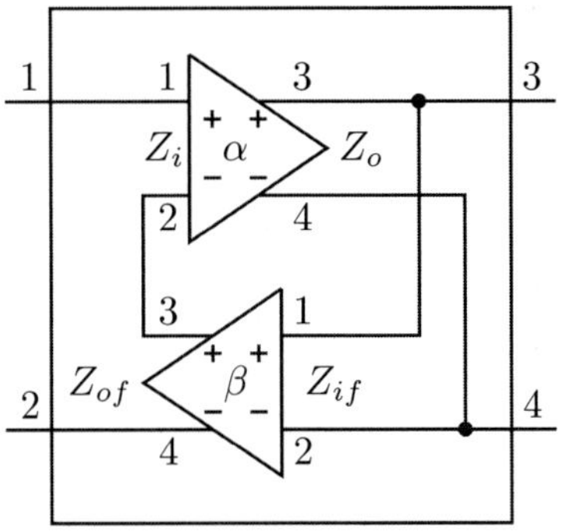


(a) Circuit representation

```
import SignalIntegrity.Lib as si
sdp=si.p.SystemDescriptionParser()
sdp.AddLines(['device D 4','device F 4','connect D 2 F 3',
    'connect D 3 F 1','connect D 4 F 2','port 1 D 1 2 F 4 3 D 3 4 F 2'])
ssps=si.sd.SystemSParametersSymbolic(sdp.SystemDescription(),size='small')
DSp=ssps[ssps.IndexOfDevice('D')].SParameters
DSp[1][1]=DSp[0][0]; DSp[1][0]=DSp[0][1]
DSp[0][2]=0; DSp[0][3]=0; DSp[1][2]=0; DSp[1][3]=0
DSp[2][1]='-'+DSp[2][0]; DSp[3][0]='-'+DSp[2][0]
DSp[3][1]=DSp[2][0]; DSp[3][2]=DSp[2][3]; DSp[3][3]=DSp[2][2]
DS=si.sy.VoltageAmplifier(4,'\\alpha','Z_i','Z_o')
ssps._AddEq('D_{11}={\\scriptstyle '+DS[0][0]+'}')
ssps._AddEq('D_{12}={\\scriptstyle '+DS[0][1]+'}')
ssps._AddEq('D_{31}={\\scriptstyle '+DS[2][0]+'}')
ssps._AddEq('D_{33}={\\scriptstyle '+DS[2][2]+'}')
ssps._AddEq('D_{34}={\\scriptstyle '+DS[2][3]+'}')
DSp=ssps[ssps.IndexOfDevice('F')].SParameters
DSp[1][1]=DSp[0][0]; DSp[1][0]=DSp[0][1]
DSp[0][2]=0; DSp[0][3]=0; DSp[1][2]=0; DSp[1][3]=0
DSp[2][1]='-'+DSp[2][0]; DSp[3][0]='-'+DSp[2][0]
DSp[3][1]=DSp[2][0]; DSp[3][2]=DSp[2][3]; DSp[3][3]=DSp[2][2]
FS=si.sy.VoltageAmplifier(4,'\\beta','Z_{if}','Z_{of}')
ssps._AddEq('F_{11}={\\scriptstyle '+FS[0][0]+'}')
ssps._AddEq('F_{12}={\\scriptstyle '+FS[0][1]+'}')
ssps._AddEq('F_{31}={\\scriptstyle '+FS[2][0]+'}')
ssps._AddEq('F_{33}={\\scriptstyle '+FS[2][2]+'}')
ssps._AddEq('F_{34}={\\scriptstyle '+FS[2][3]+'}')
ssps.LaTeXSolution(size='biggest').Emit()
```

(b) Python code

**Figure D.14** Symbolic four-port voltage amplifier with voltage-series feedback

## D.14 Four-Port Voltage Feedback Amplifier

The four-port voltage amplifier with voltage series feedback is shown in Figure D.14(a). The Python code (Listing D.14(b)) is utilized to generate the symbolic solution, shown in Figure D.14(c).

$$D_{11} = \frac{Z_i}{Z_i+2\cdot Z0}$$

$$D_{12} = \frac{2\cdot Z0}{Z_i+2\cdot Z0}$$

$$D_{31} = \frac{2\cdot Z_i\cdot Z0\cdot \alpha}{(Z_i+2\cdot Z0)\cdot(Z_o+2\cdot Z0)}$$

$$D_{33} = \frac{Z_o}{Z_o+2\cdot Z0}$$

$$D_{34} = \frac{2\cdot Z0}{Z_o+2\cdot Z0}$$

$$F_{11} = \frac{Z_{if}}{Z_{if}+2\cdot Z0}$$

$$F_{12} = \frac{2\cdot Z0}{Z_{if}+2\cdot Z0}$$

$$F_{31} = \frac{2\cdot Z_{if}\cdot Z0\cdot \beta}{(Z_{if}+2\cdot Z0)\cdot(Z_{of}+2\cdot Z0)}$$

$$F_{33} = \frac{Z_{of}}{Z_{of}+2\cdot Z0}$$

$$F_{34} = \frac{2\cdot Z0}{Z_{of}+2\cdot Z0}$$

$$\mathbf{W_{ba}} = \begin{pmatrix} D_{11} & 0 & 0 & 0 \\ 0 & F_{33} & 0 & 0 \\ 0 & 0 & -\frac{1}{3} & 0 \\ 0 & 0 & 0 & -\frac{1}{3} \end{pmatrix}$$

$$\mathbf{W_{bx}} = \begin{pmatrix} 0 & 0 & 0 & 0 & 0 & D_{12} & 0 & 0 & 0 & 0 \\ F_{34} & 0 & 0 & 0 & 0 & 0 & -F_{31} & 0 & 0 & F_{31} \\ 0 & \frac{2}{3} & 0 & \frac{2}{3} & 0 & 0 & 0 & 0 & 0 & 0 \\ 0 & 0 & \frac{2}{3} & 0 & \frac{2}{3} & 0 & 0 & 0 & 0 & 0 \end{pmatrix}$$

$$\mathbf{W_{xa}} = \begin{pmatrix} D_{12} & 0 & 0 & 0 \\ D_{31} & 0 & 0 & 0 \\ -D_{31} & 0 & 0 & 0 \\ 0 & 0 & 0 & 0 \\ 0 & 0 & 0 & 0 \\ 0 & F_{34} & 0 & 0 \\ 0 & 0 & \frac{2}{3} & 0 \\ 0 & 0 & \frac{2}{3} & 0 \\ 0 & 0 & 0 & \frac{2}{3} \\ 0 & 0 & 0 & \frac{2}{3} \end{pmatrix}$$

$$\mathbf{W_{xx}} = \begin{pmatrix} 0 & 0 & 0 & 0 & 0 & D_{11} & 0 & 0 & 0 & 0 \\ 0 & 0 & 0 & 0 & 0 & -D_{31} & 0 & D_{33} & D_{34} & 0 \\ 0 & 0 & 0 & 0 & 0 & D_{31} & 0 & D_{34} & D_{33} & 0 \\ 0 & 0 & 0 & 0 & 0 & 0 & F_{11} & 0 & 0 & F_{12} \\ 0 & 0 & 0 & 0 & 0 & 0 & F_{12} & 0 & 0 & F_{11} \\ F_{33} & 0 & 0 & 0 & 0 & 0 & F_{31} & 0 & 0 & -F_{31} \\ 0 & \frac{2}{3} & 0 & -\frac{1}{3} & 0 & 0 & 0 & 0 & 0 & 0 \\ 0 & -\frac{1}{3} & 0 & \frac{2}{3} & 0 & 0 & 0 & 0 & 0 & 0 \\ 0 & 0 & -\frac{1}{3} & 0 & \frac{2}{3} & 0 & 0 & 0 & 0 & 0 \\ 0 & 0 & \frac{2}{3} & 0 & -\frac{1}{3} & 0 & 0 & 0 & 0 & 0 \end{pmatrix}$$

$$\mathbf{S} = \mathbf{W_{ba}} + \mathbf{W_{bx}} \cdot [\mathbb{I} - \mathbf{W_{xx}}]^{-1} \cdot \mathbf{W_{xa}}$$

(c) LaTeX processed equations

**Figure D.14** Symbolic four-port voltage amplifier with voltage-series feedback (continued)

```python
import SignalIntegrity.Lib as si
sdp=si.p.SystemDescriptionParser()
sdp.AddLines(['device A 4','device ZI1 2','device ZI2 2',
    'device G 1 ground','port 1 ZI1 1 2 ZI2 1 3 A 4',
    'connect A 3 G 1','connect ZI1 2 A 1','connect ZI2 2 A 2'])
ssps=si.sd.SystemSParametersSymbolic(sdp.SystemDescription(),size='small')
ssps.AssignSParameters('ZI1',si.sy.ShuntZ(2,'Zi'))
ssps.AssignSParameters('ZI2',si.sy.ShuntZ(2,'Zi'))
ssps.AssignSParameters('A',si.sy.VoltageAmplifier(4, 'G', 'Zd', 'Zo'))
ssps.LaTeXSolution(size='biggest').Emit()
```

(a) Python code

$$\mathbf{W_{ba}} = \begin{pmatrix} \frac{-Z0}{2\cdot Zi+Z0} & 0 & 0 \\ 0 & \frac{-Z0}{2\cdot Zi+Z0} & 0 \\ 0 & 0 & \frac{Zo}{Zo+2\cdot Z0} \end{pmatrix}$$

$$\mathbf{W_{bx}} = \begin{pmatrix} \frac{2\cdot Zi}{2\cdot Zi+Z0} & 0 & 0 & 0 & 0 & 0 \\ 0 & \frac{2\cdot Zi}{2\cdot Zi+Z0} & 0 & 0 & 0 & 0 \\ 0 & 0 & 0 & -\frac{2\cdot Zd\cdot Z0\cdot G}{(Zd+2\cdot Z0)\cdot(Zo+2\cdot Z0)} & \frac{2\cdot Zd\cdot Z0\cdot G}{(Zd+2\cdot Z0)\cdot(Zo+2\cdot Z0)} & \frac{2\cdot Z0}{Zo+2\cdot Z0} \end{pmatrix}$$

$$\mathbf{W_{xa}} = \begin{pmatrix} 0 & 0 & 0 \\ 0 & 0 & 0 \\ 0 & 0 & \frac{2\cdot Z0}{Zo+2\cdot Z0} \\ \frac{2\cdot Zi}{2\cdot Zi+Z0} & 0 & 0 \\ 0 & \frac{2\cdot Zi}{2\cdot Zi+Z0} & 0 \\ 0 & 0 & 0 \end{pmatrix}$$

$$\mathbf{W_{xx}} = \begin{pmatrix} 0 & 0 & 0 & \frac{Zd}{Zd+2\cdot Z0} & \frac{2\cdot Z0}{Zd+2\cdot Z0} & 0 \\ 0 & 0 & 0 & \frac{2\cdot Z0}{Zd+2\cdot Z0} & \frac{Zd}{Zd+2\cdot Z0} & 0 \\ 0 & 0 & 0 & \frac{2\cdot Zd\cdot Z0\cdot G}{(Zd+2\cdot Z0)\cdot(Zo+2\cdot Z0)} & -\frac{2\cdot Zd\cdot Z0\cdot G}{(Zd+2\cdot Z0)\cdot(Zo+2\cdot Z0)} & \frac{Zo}{Zo+2\cdot Z0} \\ \frac{-Z0}{2\cdot Zi+Z0} & 0 & 0 & 0 & 0 & 0 \\ 0 & \frac{-Z0}{2\cdot Zi+Z0} & 0 & 0 & 0 & 0 \\ 0 & 0 & -1 & 0 & 0 & 0 \end{pmatrix}$$

$$\mathbf{S} = \mathbf{W_{ba}} + \mathbf{W_{bx}} \cdot \left[\mathbb{I} - \mathbf{W_{xx}}\right]^{-1} \cdot \mathbf{W_{xa}}$$

(b) LaTeX processed equations

**Figure D.15** Symbolic operational amplifier solution

## D.15 Operational Amplifier

Figure D.15 shows the Python code (Figure D.15(a)) and the symbolic solution (Figure D.15(b)) for the operational amplifier provided in Figure 6.27.

```
import SignalIntegrity.Lib as si
sdp=si.p.SystemDescriptionParser()
sdp.AddLines(['device DV 4','device DC 4','port 1 DC 4 2 DC 3 3 DC 1 4 DV 3',
    'connect DC 2 DV 4','connect DC 4 DV 2','connect DC 3 DV 1'])
ssps=si.sd.SystemSParametersSymbolic(sdp.SystemDescription(),size='small')
ssps.AssignSParameters('DV',si.sy.VoltageControlledVoltageSource('a'))
ssps.AssignSParameters('DC',si.sy.CurrentControlledCurrentSource('a'))
ssps.LaTeXSolution(size='big').Emit()
```

(a) Python code

$$\mathbf{Wi} = \left[ \mathbb{I} - \begin{pmatrix} 0 & 0 & 0 & 0 & 0 & 0 & 0 & 0 & 0 & 1 \\ 0 & 0 & 0 & 0 & 0 & 0 & 0 & 1 & 0 & 0 \\ 0 & 0 & 0 & 0 & 0 & 0 & 0 & a & 0 & -a \\ 0 & 0 & 0 & 0 & 0 & 0 & 0 & 0 & 0 & 0 \\ 0 & 0 & a & 0 & 0 & 0 & 0 & 0 & 1 & 0 \\ 0 & 0 & -a & 0 & 0 & 0 & 1 & 0 & 0 & 0 \\ 0 & \frac{2}{3} & 0 & 0 & 0 & -\frac{1}{3} & 0 & 0 & 0 & 0 \\ 0 & -\frac{1}{3} & 0 & 0 & 0 & \frac{2}{3} & 0 & 0 & 0 & 0 \\ \frac{2}{3} & 0 & 0 & 0 & -\frac{1}{3} & 0 & 0 & 0 & 0 & 0 \\ -\frac{1}{3} & 0 & 0 & 0 & \frac{2}{3} & 0 & 0 & 0 & 0 & 0 \end{pmatrix} \right]^{-1}$$

$$\mathbf{S} = \begin{pmatrix} -\frac{1}{3} & 0 & 0 & 0 \\ 0 & -\frac{1}{3} & 0 & 0 \\ 0 & 0 & 0 & 0 \\ 0 & 0 & 0 & 0 \end{pmatrix} + \begin{pmatrix} 0 & \frac{2}{3} & 0 & 0 & 0 & \frac{2}{3} & 0 & 0 & 0 & 0 \\ \frac{2}{3} & 0 & 0 & 0 & \frac{2}{3} & 0 & 0 & 0 & 0 & 0 \\ 0 & 0 & 1 & 0 & 0 & 0 & 0 & 0 & 0 & 0 \\ 0 & 0 & 0 & 1 & 0 & 0 & 0 & -a & 0 & a \end{pmatrix} \cdot \mathbf{Wi} \cdot \begin{pmatrix} 0 & 0 & 0 & 0 \\ 0 & 0 & 0 & 0 \\ 0 & 0 & 0 & 1 \\ 0 & 0 & 1 & 0 \\ 0 & 0 & -a & 0 \\ 0 & 0 & a & 0 \\ \frac{2}{3} & 0 & 0 & 0 \\ \frac{2}{3} & 0 & 0 & 0 \\ 0 & \frac{2}{3} & 0 & 0 \\ 0 & \frac{2}{3} & 0 & 0 \end{pmatrix}$$

(b) LaTeX processed equations

**Figure D.16** Symbolic ideal transformer solution

## D.16 Ideal Transformer

Figure D.16 shows the Python code (Figure D.16(a)) and the symbolic solution (Figure D.16(b)) for the ideal transformer provided in Figure 6.29.

## D.17 Transistor

Figure D.17 shows the NPN transistor schematically in Figure D.17(a). Python code (Figure D.17(b)) is utilized to generate the symbolic solution, shown in Figure D.17(c) and Figure D.17(d). This is the solution of the more complex form of the transistor provided in Figure 6.28.

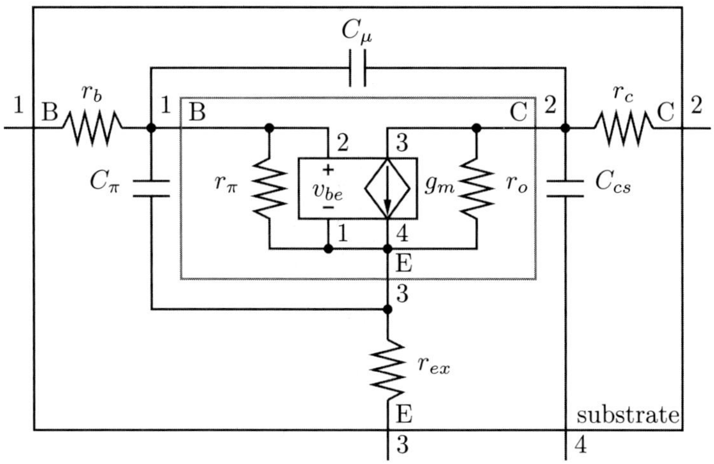


(a) Circuit representation

```
import SignalIntegrity.Lib as si
sdp=si.p.SystemDescriptionParser()
sdp.AddLines(['device T 3','device rb 2','device Cms 4','device Cps 4','device rx 2',
              'device rc 2','device Ccs 3',
              'port 1 rb 1 2 rc 2 3 rx 2 4 Ccs 3',
              'connect rb 2 Cms 1','connect Cms 3 Cps 1','connect Cps 3 T 1',
              'connect T 3 Cps 4','connect Cps 2 rx 1','connect T 2 Ccs 1',
              'connect Ccs 2 Cms 4','connect Cms 2 rc 1'])
ssps=si.sd.SystemSParametersSymbolic(sdp.SystemDescription(),size='small')
symbolic=si.sd.Symbolic(size='small')
rb=si.sy.SeriesZ('r_b')
symbolic._AddEq('\\mathbf{rb}='+ssps._LaTeXMatrix(rb))
rc=si.sy.SeriesZ('r_c')
symbolic._AddEq('\\mathbf{rc}='+ssps._LaTeXMatrix(rc))
rx=si.sy.SeriesZ('r_ex')
symbolic._AddEq('\\mathbf{rx}='+ssps._LaTeXMatrix(rx))
Cms=si.sy.ShuntZ(4,'\\frac{1}{C_{\\mu}\\cdot s}')
symbolic._AddEq('\\mathbf{Cms}='+ssps._LaTeXMatrix(Cms))
Ccs=si.sy.ShuntZ(3,'\\frac{1}{C_{cs}\\cdot s}')
symbolic._AddEq('\\mathbf{Ccs}='+ssps._LaTeXMatrix(Ccs))
Cps=si.sy.ShuntZ(4,'\\frac{1}{C_{\\pi}\\cdot s}')
symbolic._AddEq('\\mathbf{Cps}='+ssps._LaTeXMatrix(Cps))
T=si.sy.TransconductanceAmplifierThreePortWithoutDenom('-g_m','r_{\\pi}','r_o')
D=si.sy.TransconductanceAmplifierThreePortDenom('-g_m','r_{\\pi}','r_o')
symbolic._AddEq('\\mathbf{T}='+'\\frac{1}{'+D+'}')
symbolic._AddEq('\\cdot '+ssps._LaTeXMatrix(T))
symbolic.Emit()
ssps.LaTeXSolution(size='biggest').Emit()
```

(b) Python code

**Figure D.17** Symbolic transistor solution

$$\mathbf{rb} = \begin{pmatrix} \frac{r_b}{r_b+2\cdot Z0} & \frac{2\cdot Z0}{r_b+2\cdot Z0} \\ \frac{2\cdot Z0}{r_b+2\cdot Z0} & \frac{r_b}{r_b+2\cdot Z0} \end{pmatrix}$$

$$\mathbf{rc} = \begin{pmatrix} \frac{r_c}{r_c+2\cdot Z0} & \frac{2\cdot Z0}{r_c+2\cdot Z0} \\ \frac{2\cdot Z0}{r_c+2\cdot Z0} & \frac{r_c}{r_c+2\cdot Z0} \end{pmatrix}$$

$$\mathbf{rx} = \begin{pmatrix} \frac{r_{ex}}{r_{ex}+2\cdot Z0} & \frac{2\cdot Z0}{r_{ex}+2\cdot Z0} \\ \frac{2\cdot Z0}{r_{ex}+2\cdot Z0} & \frac{r_{ex}}{r_{ex}+2\cdot Z0} \end{pmatrix}$$

$$\mathbf{Cms} = \begin{pmatrix} \frac{-Z0}{2\cdot(\frac{1}{C_\mu\cdot s}+Z0)} & \frac{Z0}{2\cdot(\frac{1}{C_\mu\cdot s}+Z0)} & \frac{2\cdot\frac{1}{C_\mu\cdot s}+Z0}{2\cdot(\frac{1}{C_\mu\cdot s}+Z0)} & \frac{Z0}{2\cdot(\frac{1}{C_\mu\cdot s}+Z0)} \\ \frac{Z0}{2\cdot(\frac{1}{C_\mu\cdot s}+Z0)} & \frac{-Z0}{2\cdot(\frac{1}{C_\mu\cdot s}+Z0)} & \frac{Z0}{2\cdot(\frac{1}{C_\mu\cdot s}+Z0)} & \frac{2\cdot\frac{1}{C_\mu\cdot s}+Z0}{2\cdot(\frac{1}{C_\mu\cdot s}+Z0)} \\ \frac{2\cdot\frac{1}{C_\mu\cdot s}+Z0}{2\cdot(\frac{1}{C_\mu\cdot s}+Z0)} & \frac{Z0}{2\cdot(\frac{1}{C_\mu\cdot s}+Z0)} & \frac{-Z0}{2\cdot(\frac{1}{C_\mu\cdot s}+Z0)} & \frac{Z0}{2\cdot(\frac{1}{C_\mu\cdot s}+Z0)} \\ \frac{Z0}{2\cdot(\frac{1}{C_\mu\cdot s}+Z0)} & \frac{2\cdot\frac{1}{C_\mu\cdot s}+Z0}{2\cdot(\frac{1}{C_\mu\cdot s}+Z0)} & \frac{Z0}{2\cdot(\frac{1}{C_\mu\cdot s}+Z0)} & \frac{-Z0}{2\cdot(\frac{1}{C_\mu\cdot s}+Z0)} \end{pmatrix}$$

$$\mathbf{Ccs} = \begin{pmatrix} \frac{-Z0}{2\cdot\frac{1}{C_{cs}\cdot s}+3\cdot Z0} & \frac{2\cdot\frac{1}{C_{cs}\cdot s}+2\cdot Z0}{2\cdot\frac{1}{C_{cs}\cdot s}+3\cdot Z0} & \frac{2\cdot Z0}{2\cdot\frac{1}{C_{cs}\cdot s}+3\cdot Z0} \\ \frac{2\cdot\frac{1}{C_{cs}\cdot s}+2\cdot Z0}{2\cdot\frac{1}{C_{cs}\cdot s}+3\cdot Z0} & \frac{-Z0}{2\cdot\frac{1}{C_{cs}\cdot s}+3\cdot Z0} & \frac{2\cdot Z0}{2\cdot\frac{1}{C_{cs}\cdot s}+3\cdot Z0} \\ \frac{2\cdot Z0}{2\cdot\frac{1}{C_{cs}\cdot s}+3\cdot Z0} & \frac{2\cdot Z0}{2\cdot\frac{1}{C_{cs}\cdot s}+3\cdot Z0} & \frac{2\cdot\frac{1}{C_{cs}\cdot s}-Z0}{2\cdot\frac{1}{C_{cs}\cdot s}+3\cdot Z0} \end{pmatrix}$$

$$\mathbf{Cps} = \begin{pmatrix} \frac{-Z0}{2\cdot(\frac{1}{C_\pi\cdot s}+Z0)} & \frac{Z0}{2\cdot(\frac{1}{C_\pi\cdot s}+Z0)} & \frac{2\cdot\frac{1}{C_\pi\cdot s}+Z0}{2\cdot(\frac{1}{C_\pi\cdot s}+Z0)} & \frac{Z0}{2\cdot(\frac{1}{C_\pi\cdot s}+Z0)} \\ \frac{Z0}{2\cdot(\frac{1}{C_\pi\cdot s}+Z0)} & \frac{-Z0}{2\cdot(\frac{1}{C_\pi\cdot s}+Z0)} & \frac{Z0}{2\cdot(\frac{1}{C_\pi\cdot s}+Z0)} & \frac{2\cdot\frac{1}{C_\pi\cdot s}+Z0}{2\cdot(\frac{1}{C_\pi\cdot s}+Z0)} \\ \frac{2\cdot\frac{1}{C_\pi\cdot s}+Z0}{2\cdot(\frac{1}{C_\pi\cdot s}+Z0)} & \frac{Z0}{2\cdot(\frac{1}{C_\pi\cdot s}+Z0)} & \frac{-Z0}{2\cdot(\frac{1}{C_\pi\cdot s}+Z0)} & \frac{Z0}{2\cdot(\frac{1}{C_\pi\cdot s}+Z0)} \\ \frac{Z0}{2\cdot(\frac{1}{C_\pi\cdot s}+Z0)} & \frac{2\cdot\frac{1}{C_\pi\cdot s}+Z0}{2\cdot(\frac{1}{C_\pi\cdot s}+Z0)} & \frac{Z0}{2\cdot(\frac{1}{C_\pi\cdot s}+Z0)} & \frac{-Z0}{2\cdot(\frac{1}{C_\pi\cdot s}+Z0)} \end{pmatrix}$$

$$\mathbf{T} = \frac{1}{3\cdot Z0^2+(2\cdot r_o+2\cdot r_\pi+g_m\cdot r_\pi\cdot r_o)\cdot Z0+r_o\cdot r_\pi}$$

$$\cdot \begin{pmatrix} r_o\cdot r_\pi+Z0\cdot(2\cdot r_\pi+g_m\cdot r_\pi\cdot r_o)-Z0^2 & 2\cdot Z0^2 & 2\cdot Z0^2+2\cdot r_o\cdot Z0 \\ 2\cdot Z0^2-2\cdot g_m\cdot r_\pi\cdot r_o\cdot Z0 & r_o\cdot r_\pi+Z0\cdot(2\cdot r_o+g_m\cdot r_\pi\cdot r_o)-Z0^2 & 2\cdot Z0^2+Z0\cdot(2\cdot r_\pi+2\cdot g_m\cdot r_\pi\cdot r_o) \\ 2\cdot Z0^2+Z0\cdot(2\cdot r_o+2\cdot g_m\cdot r_\pi\cdot r_o) & 2\cdot Z0^2+2\cdot r_\pi\cdot Z0 & r_o\cdot r_\pi-Z0^2-g_m\cdot r_\pi\cdot r_o\cdot Z0 \end{pmatrix}$$

(c) LaTeX processed equations

**Figure D.17** Symbolic transistor solution (continued)

$$\mathbf{W_{ba}} = \begin{pmatrix} rb_{11} & 0 & 0 & 0 \\ 0 & rc_{22} & 0 & 0 \\ 0 & 0 & rx_{22} & 0 \\ 0 & 0 & 0 & Ccs_{33} \end{pmatrix}$$

$$\mathbf{W_{bx}} = \begin{pmatrix} 0 & 0 & 0 & 0 & rb_{12} & 0 & 0 & 0 & 0 & 0 & 0 & 0 & 0 & 0 & 0 & 0 \\ 0 & 0 & 0 & 0 & 0 & rc_{21} & 0 & 0 & 0 & 0 & 0 & 0 & 0 & 0 & 0 & 0 \\ 0 & 0 & 0 & 0 & 0 & 0 & 0 & 0 & 0 & rx_{21} & 0 & 0 & 0 & 0 & 0 & 0 \\ 0 & Ccs_{31} & 0 & 0 & 0 & 0 & 0 & Ccs_{32} & 0 & 0 & 0 & 0 & 0 & 0 & 0 & 0 \end{pmatrix}$$

$$\mathbf{W_{xa}} = \begin{pmatrix} 0 & 0 & 0 & 0 \\ 0 & 0 & 0 & 0 \\ 0 & 0 & 0 & 0 \\ rb_{21} & 0 & 0 & 0 \\ 0 & 0 & 0 & 0 \\ 0 & 0 & 0 & 0 \\ 0 & 0 & 0 & 0 \\ 0 & 0 & 0 & 0 \\ 0 & 0 & 0 & 0 \\ 0 & 0 & 0 & 0 \\ 0 & 0 & 0 & 0 \\ 0 & 0 & 0 & 0 \\ 0 & 0 & rx_{12} & 0 \\ 0 & rc_{12} & 0 & 0 \\ 0 & 0 & 0 & Ccs_{13} \\ 0 & 0 & 0 & Ccs_{23} \end{pmatrix}$$

$$\mathbf{W_{xx}} =$$

$$\begin{pmatrix} 0 & 0 & 0 & 0 & 0 & 0 & 0 & 0 & 0 & 0 & T_{11} & T_{13} & 0 & 0 & T_{12} & 0 \\ 0 & 0 & 0 & 0 & 0 & 0 & 0 & 0 & 0 & 0 & T_{21} & T_{23} & 0 & 0 & T_{22} & 0 \\ 0 & 0 & 0 & 0 & 0 & 0 & 0 & 0 & 0 & 0 & T_{31} & T_{33} & 0 & 0 & T_{32} & 0 \\ 0 & 0 & 0 & 0 & rb_{22} & 0 & 0 & 0 & 0 & 0 & 0 & 0 & 0 & 0 & 0 & 0 \\ 0 & 0 & 0 & Cms_{11} & 0 & 0 & 0 & 0 & Cms_{13} & 0 & 0 & 0 & 0 & Cms_{12} & 0 & Cms_{14} \\ 0 & 0 & 0 & Cms_{21} & 0 & 0 & 0 & 0 & Cms_{23} & 0 & 0 & 0 & 0 & Cms_{22} & 0 & Cms_{24} \\ 0 & 0 & 0 & Cms_{31} & 0 & 0 & 0 & 0 & Cms_{33} & 0 & 0 & 0 & 0 & Cms_{32} & 0 & Cms_{34} \\ 0 & 0 & 0 & Cms_{41} & 0 & 0 & 0 & 0 & Cms_{43} & 0 & 0 & 0 & 0 & Cms_{42} & 0 & Cms_{44} \\ Cps_{13} & 0 & Cps_{14} & 0 & 0 & 0 & Cps_{11} & 0 & 0 & 0 & 0 & 0 & Cps_{12} & 0 & 0 & 0 \\ Cps_{23} & 0 & Cps_{24} & 0 & 0 & 0 & Cps_{21} & 0 & 0 & 0 & 0 & 0 & Cps_{22} & 0 & 0 & 0 \\ Cps_{33} & 0 & Cps_{34} & 0 & 0 & 0 & Cps_{31} & 0 & 0 & 0 & 0 & 0 & Cps_{32} & 0 & 0 & 0 \\ Cps_{43} & 0 & Cps_{44} & 0 & 0 & 0 & Cps_{41} & 0 & 0 & 0 & 0 & 0 & Cps_{42} & 0 & 0 & 0 \\ 0 & 0 & 0 & 0 & 0 & 0 & 0 & 0 & 0 & rx_{11} & 0 & 0 & 0 & 0 & 0 & 0 \\ 0 & 0 & 0 & 0 & 0 & rc_{11} & 0 & 0 & 0 & 0 & 0 & 0 & 0 & 0 & 0 & 0 \\ 0 & Ccs_{11} & 0 & 0 & 0 & 0 & 0 & Ccs_{12} & 0 & 0 & 0 & 0 & 0 & 0 & 0 & 0 \\ 0 & Ccs_{21} & 0 & 0 & 0 & 0 & 0 & Ccs_{22} & 0 & 0 & 0 & 0 & 0 & 0 & 0 & 0 \end{pmatrix}$$

$$\mathbf{S} = \mathbf{W_{ba}} + \mathbf{W_{bx}} \cdot [\mathbb{I} - \mathbf{W_{xx}}]^{-1} \cdot \mathbf{W_{xa}}$$

(d) LaTeX processed equations (concluded)

**Figure D.17** Symbolic transistor solution (continued)

```
import SignalIntegrity.Lib as si
sdp=si.p.SystemDescriptionParser().AddLines([
    'device MM 4 mixedmode','device R1 2','device R2 2','device R3 1','connect MM 1 R1 2',
    'connect R1 1 R2 2 R3 1','connect MM 2 R2 1','port 1 MM 3 2 MM 4',])
ssps=si.sd.SystemSParametersSymbolic(sdp.SystemDescription(),size='small')
ssps.AssignSParameters('R1',si.sy.SeriesZ('Z_1'))
ssps.AssignSParameters('R2',si.sy.SeriesZ('Z_2'))
ssps.AssignSParameters('R3',si.sy.ShuntZ(1,'Z_3'))
ssps.LaTeXSolution(size='big')
ssps.Emit()
```

(a) Python code

$$\mathbf{Wi} = \left[ \mathbb{I} - \begin{pmatrix} 0 & 0 & 0 & 0 & 0 & 0 & 0 & 0 & 0 & 0 \\ 0 & 0 & 0 & 0 & 0 & 0 & 0 & 0 & 0 & 0 \\ \frac{2\cdot Z0}{Z_1+2\cdot Z0} & 0 & 0 & 0 & 0 & 0 & 0 & 0 & \frac{Z_1}{Z_1+2\cdot Z0} & 0 \\ \frac{Z_1}{Z_1+2\cdot Z0} & 0 & 0 & 0 & 0 & 0 & 0 & 0 & \frac{2\cdot Z0}{Z_1+2\cdot Z0} & 0 \\ 0 & \frac{Z_2}{Z_2+2\cdot Z0} & 0 & 0 & 0 & 0 & 0 & \frac{2\cdot Z0}{Z_2+2\cdot Z0} & 0 & 0 \\ 0 & \frac{2\cdot Z0}{Z_2+2\cdot Z0} & 0 & 0 & 0 & 0 & 0 & \frac{Z_2}{Z_2+2\cdot Z0} & 0 & 0 \\ 0 & 0 & 0 & 0 & 0 & 0 & 0 & 0 & 0 & \frac{Z_3-Z0}{Z_3+Z0} \\ 0 & 0 & \frac{2}{3} & 0 & 0 & -\frac{1}{3} & \frac{2}{3} & 0 & 0 & 0 \\ 0 & 0 & -\frac{1}{3} & 0 & 0 & \frac{2}{3} & \frac{2}{3} & 0 & 0 & 0 \\ 0 & 0 & \frac{2}{3} & 0 & 0 & \frac{2}{3} & -\frac{1}{3} & 0 & 0 & 0 \end{pmatrix} \right]^{-1}$$

$$\mathbf{S} = \begin{pmatrix} 0 & 0 & 0 & \frac{1}{\sqrt{2}} & -\frac{1}{\sqrt{2}} & 0 & 0 & 0 & 0 & 0 \\ 0 & 0 & 0 & \frac{1}{\sqrt{2}} & \frac{1}{\sqrt{2}} & 0 & 0 & 0 & 0 & 0 \end{pmatrix} \cdot \mathbf{Wi} \cdot \begin{pmatrix} \frac{1}{\sqrt{2}} & \frac{1}{\sqrt{2}} \\ -\frac{1}{\sqrt{2}} & \frac{1}{\sqrt{2}} \\ 0 & 0 \\ 0 & 0 \\ 0 & 0 \\ 0 & 0 \\ 0 & 0 \\ 0 & 0 \\ 0 & 0 \\ 0 & 0 \end{pmatrix}$$

(b) LaTeX processed equations

**Figure D.18** Symbolic mixed-mode tee termination

## D.18 Tee Termination

Figure D.18 shows the Python code (Figure D.18(a)) and the symbolic solution (Figure D.18(b)) for the tee termination provided in Figure 7.16.

```
import SignalIntegrity.Lib as si
sdp=si.p.SystemDescriptionParser().AddLines([
    'device MM 4 mixedmode','device Z3 2','device Z1 1','device Z2 1',
    'connect MM 1 Z3 2 Z1 1','connect MM 2 Z3 1 Z2 1','port 1 MM 3 2 MM 4'])
ssps=si.sd.SystemSParametersSymbolic(sdp.SystemDescription(),size='small')
ssps.AssignSParameters('Z3',si.sy.SeriesZ('Z_3'))
ssps.AssignSParameters('Z1',si.sy.ShuntZ(1,'Z_1'))
ssps.AssignSParameters('Z2',si.sy.ShuntZ(1,'Z_2'))
ssps.LaTeXSolution(size='big')
ssps.Emit()
```

(a) Python code

$$\mathbf{Wi} = \left[\mathbb{I} - \begin{pmatrix} 0&0&0&0&0&0&0&0&0&0&0&0\\ 0&0&0&0&0&0&0&0&0&0&0&0\\ 0&0&0&0&0&0&\frac{2\cdot Z0}{Z_3+2\cdot Z0}&0&0&\frac{Z_3}{Z_3+2\cdot Z0}&0&0\\ 0&0&0&0&0&0&\frac{Z_3}{Z_3+2\cdot Z0}&0&0&\frac{2\cdot Z0}{Z_3+2\cdot Z0}&0&0\\ 0&0&0&0&0&0&0&0&\frac{Z_1-Z0}{Z_1+Z0}&0&0&0\\ 0&0&0&0&0&0&0&0&0&0&0&\frac{Z_2-Z0}{Z_2+Z0}\\ \frac{2}{3}&0&0&-\frac{1}{3}&\frac{2}{3}&0&0&0&0&0&0&0\\ -\frac{1}{3}&0&0&\frac{2}{3}&\frac{2}{3}&0&0&0&0&0&0&0\\ \frac{2}{3}&0&0&\frac{2}{3}&-\frac{1}{3}&0&0&0&0&0&0&0\\ 0&\frac{2}{3}&-\frac{1}{3}&0&0&\frac{2}{3}&0&0&0&0&0&0\\ 0&-\frac{1}{3}&\frac{2}{3}&0&0&\frac{2}{3}&0&0&0&0&0&0\\ 0&\frac{2}{3}&\frac{2}{3}&0&0&-\frac{1}{3}&0&0&0&0&0&0 \end{pmatrix}\right]^{-1}$$

$$\mathbf{S} = \begin{pmatrix} 0&0&0&0&0&0&0&\frac{1}{\sqrt{2}}&0&0&-\frac{1}{\sqrt{2}}&0\\ 0&0&0&0&0&0&0&\frac{1}{\sqrt{2}}&0&0&\frac{1}{\sqrt{2}}&0 \end{pmatrix} \cdot \mathbf{Wi} \cdot \begin{pmatrix} \frac{1}{\sqrt{2}}&\frac{1}{\sqrt{2}}\\ -\frac{1}{\sqrt{2}}&\frac{1}{\sqrt{2}}\\ 0&0\\ 0&0\\ 0&0\\ 0&0\\ 0&0\\ 0&0\\ 0&0\\ 0&0\\ 0&0\\ 0&0 \end{pmatrix}$$

(b) LaTeX processed equations

**Figure D.19** Symbolic mixed-mode pi termination

## D.19 Pi Termination

Figure D.19 shows the Python code (Figure D.19(a)) and the symbolic solution (Figure D.19(b)) for the pi termination provided in Figure 7.17.

```
import SignalIntegrity.Lib as si
sdp=si.p.SystemDescriptionParser().AddLines([
    'device MM 4 mixedmode','device S 2','connect S 1 MM 1',
    'connect MM 2 S 2','port 1 MM 3 2 MM 4'])
ssps=si.sd.SystemSParametersSymbolic(sdp.SystemDescription(),size='small')
ssps.LaTeXSolution().Emit()
```

(a) Python code

$$\mathbf{S} = \begin{pmatrix} 0 & 0 & \frac{1}{\sqrt{2}} & -\frac{1}{\sqrt{2}} \\ 0 & 0 & \frac{1}{\sqrt{2}} & \frac{1}{\sqrt{2}} \end{pmatrix} \cdot \left[ \mathbb{I} - \begin{pmatrix} 0 & 0 & 0 & 0 \\ 0 & 0 & 0 & 0 \\ S_{11} & S_{12} & 0 & 0 \\ S_{21} & S_{22} & 0 & 0 \end{pmatrix} \right]^{-1} \cdot \begin{pmatrix} \frac{1}{\sqrt{2}} & \frac{1}{\sqrt{2}} \\ -\frac{1}{\sqrt{2}} & \frac{1}{\sqrt{2}} \\ 0 & 0 \\ 0 & 0 \end{pmatrix}$$

(b) LaTeX processed equations

**Figure D.20** Symbolic mixed-mode termination

## D.20 Mixed-Mode Termination

Figure D.20 shows the Python code (Figure D.20(a)) and the symbolic solution (Figure D.20(b)) for the mixed-mode termination provided in Figure 7.15.

# References


[1] H. Johnson and M. Graham, *High-Speed Digital Design: A Handbook of Black Magic*, 1st edn., Prentice Hall Modern Semiconductor Design. Upper Saddle River, NJ: Prentice Hall, 1993.

[2] H. Johnson and M. Graham, *High-Speed Signal Propagation: Advanced Black Magic*, Prentice Hall Modern Semiconductor Design. Upper Saddle River, NJ: Prentice Hall PTR, 2003.

[3] S. Ghosh, *Network Theory: Analysis and Synthesis.* New Delhi: Prentice Hall of India, 2005, p. 353.

[4] G. Matthaei, L. Young, and E. M. T. Jones, *Microwave Filters, Impedance-Matching Networks, and Coupling Structures*, 1st edn. New York, NY: McGraw-Hill, 1964, p. 26.

[5] K. Kurokawa, "Power waves and the scattering matrix," *IEEE Trans. Microw. Theory Tech.*, vol. MTT-13, pp. 194–202, Mar. 1965.

[6] R. W. Anderson, "S-parameter techniques," Hewlett-Packard, Application Note 95-1, Feb. 1967.

[7] R. Marks and D. Williams, "A general waveguide circuit theory," *J. Res. Natl. Inst. Stand. Technol.*, vol. 97, no. 5, pp. 553–562, 1992.

[8] J. W. Nilsson, *Electric Circuits*, 2nd edn. Reading, MA: Addison-Wesley, 1984.

[9] EIA/IBIS Open Forum, "Touchstone 1.1 file format specification," 2002, version 1.1, Draft. [Online]. Available: http://www.vhdl.org/pub/ibis/connector/touchstone_spec11.pdf

[10] TechAmerica, "Touchstone 2.0 file format specification," Apr. 2009, version 2.0. [Online]. Available: http://www.eda.org/ibis/touchstone_ver2.0/touchstone_ver2_0.pdf

[11] S. J. Mason, "Feedback theory – further properties of signal flow graphs," *Proc. IRE*, pp. 920–926, Jul. 1956.

[12] "S-parameter design," Hewlett-Packard Corporation, Application Note 154, Mar. 1990.

[13] R. B. Northrop, *Signals and Systems Analysis in Biomedical Engineering.* Boca Raton, FL: CRC Press, 2003, ch. B, pp. 373–377.

[14] M. L. Edwards, *S-Parameters, Signal Flow Graphs, and Other Matrix Representations*, Sep. 2001, ch. 3. [Online]. Available: https://web.archive.org/web/20040325002849/http://www.apl.jhu.edu/Classes/Notes/Penn/EE774/Chap_03r.pdf

[15] D. A. Smolyansky and S. D. Corey, "Characterization of differential interconnects from time-domain reflectometry measurements," *Microw. J.*, vol. 43, pp. 68–80, Mar. 2000.

[16] P. J. Pupalaikis, "Validation methods for s-parameter measurement based models of differential transmission lines," in *Proc. DesignCon.* Santa Clara, CA: IEC, Feb. 4–7, 2008, pp. 1482–1586.

[17] D. E. Bockelman and W. R. Eisenstadt, "Combined differential and common-mode scattering parameters: theory and simulation," *IEEE Trans. Microw. Theory Techn.*, vol. 43, pp. 1530–1539, Jul. 1995.

[18] A. Ferrero and M. Pirola, "Generalized mixed-mode s-parameters," *IEEE Trans. Microw. Theory Tech.*, vol. 54, pp. 458–463, Feb. 2006.

[19] S. Smith, S. Sedig, and V. Balasubramanian, "Theory and measurement of unbalanced differential-mode transmission lines," in *Proc. DesignCon.* Santa Clara, CA: IEC, Feb. 6–9, 2006, pp. 1219–1243.

[20] H. Johnson and M. Graham, *High-Speed Digital Design: A Handbook of Black Magic*, 1st edn., Prentice Hall Modern Semiconductor Design. Upper Saddle River, NJ: Prentice Hall, 1993, ch. 1, p. 8.

[21] S. Agili, V. Balasubramanian, and A. Morales, "De-embedding techniques in signal integrity: a comparison study," in *Proc. Conf. Information Sciences and Systems.* Baltimore, MD: Johns Hopkins University, Feb. 2005, pp. 1–6.

[22] J. S. Martens, "Methods for embedding and de-embedding balanced networks," U.S. Patent 6 665 628, Jan. 15, 2002.

[23] G. Antonini, A. Scogna, and A. Orlandi, "S-parameters characterization of through, blind, and buried via holes," *IEEE Trans. Mobile Comput.*, vol. 2, pp. 174–184, 2003.

[24] J. Song, F. Ling, G. Flynn, W. Blood, and E. Demircan, "A de-embedding technique for interconnects," in *IEEE 10th Topical Meeting on Electrical Performance of Electronic Packaging (Cat. No. 01TH8565)*, Cambridge, MA, Oct. 2001, pp. 129–132.

[25] P. J. Pupalaikis and K. Doshi, "Method for de-embedding device measurements," U.S. Patent 8 566 058, Oct. 22, 2013.

[26] P. J. Pupalaikis, L. W. Jacobs, and S. M. Sekel, "Virtual probing," U.S. Patent 7 660 685, Feb. 9, 2010.

[27] P. J. Pupalaikis, L. W. Jacobs, and S. M. Sekel, "Virtual probing," U.S. Patent 8 170 820, May 1, 2012.

[28] H. A. Affel, "Adjustable equalizer," U.S. Patent 1 511 013, Jul. 8, 1920.

[29] R. L. Easton, "Undersea cable systems – a survey or explanation to an unknown lady in Philadelphia," *IEEE Commun. Soc. Newsletter*, c. 1970.

[30] R. W. Lucky, "Techniques for adaptive equalization of digital communication systems," *Bell Syst. Tech. J.*, vol. 73, pp. 1349–1387, Sep. 1966.

[31] B. Casper, P. J. Pupalaikis, and J. Zerbe, "Techforum on serial equalization," in *Proc. DesignCon.* Santa Clara, CA: IEC, Feb. 2007.

[32] J. Cavazos, "Oscilloscopes evolve from center instrument to the only instrument on your bench," *Electron. Des.*, May 2013.

[33] P. J. Pupalaikis, "Advanced tools for high speed serial data measurements," in *Proc. DesignCon.* Santa Clara, CA: IEC, Feb. 2007.

[34] J. Kenney and P. J. Pupalaikis, "Timing measurement problems and solutions in source terminated memory systems with inaccessible probing points," in *Proc. DesignCon.* Santa Clara, CA: IEC, Feb. 2010.

[35] *Wavelink Series High Bandwidth Differential Probes (13-25 GHz) User's Manual*, Teledyne LeCroy, Chestnut Ridge, NY.

[36] D. Maliniak, "Oscilloscope probes influence measurements," *Electron. Des.*, May 2014.

[37] M. T. Jong, *Methods of Discrete Signal and System Analysis.* New York, NY: McGraw-Hill, 1982.

[38] H. Johnson, "See beyond the edge," *EDN*, Oct. 2005.

[39] "TDR techniques for characterization techniques for characterization and modeling of electronic packaging," TDA Systems, Application Note, 2001.

[40] O. Kreidler, "Signal integrity analysis of gigabit interconnects," Tektronix, Inc., Application Note, Jun. 2007.

[41] D. Smolyansky, "TDR and s-parameter measurements: how much performance do you need?" *EDN*, Sep. 2007.

[42] A. Blankman, "De-embedding Gigaprobes using time domain gating with the LeCroy SPARQ," Teledyne LeCroy/Cascade Microtech, Application Note, 2011.

[43] P. J. Pupalaikis, "Method for de-embedding in network analysis," U.S. Patent 9 194 930, Nov. 24, 2015.

[44] I. Novak, Y. Li, E. Kunz, "Determining PCB trace impedance by TDR: Challenges and possible solutions," in *Proc. DesignCon.* Santa Clara, CA: UBM, Feb. 2013.

[45] A. Ferrero and U. Pisani, "Two-port network analyzer calibration using an unknown thru," *IEEE Microw. Guided Wave Lett.*, vol. 2, pp. 505–507, Jan. 1993.

[46] D. C. DeGroot, J. A. Jargon, and K. L. Reed, "Equivalent circuit models for coaxial OSLT standards," in *Proc. 54th ARFTG Conf.* Atlanta, GA: Georgia Institute of Technology, Dec. 1999.

[47] P. Wittwer and P. J. Pupalaikis, "A general closed-form solution to multi-port scattering parameter calculations," in *Proc. 72nd ARFTG Microwave Measurement Symp.*, Portland, OR, Dec. 2008, pp. 137–143.

[48] J. A. Jargon, R. B. Marks, and D. K. Rytting, "Robust SOLT and alternative calibrations for four-sampler vector network analyzers," *IEEE Trans. Microw. Theory Tech.*, vol. 47, no. 10, pp. 2008–2013, Oct. 1999.

[49] R. B. Marks, "Formulations of the basic vector network analyzer error model including switch-terms," in *Proc. 50th ARFTG Conf. Dig.*, vol. 32, Dec. 1997, pp. 115–126.

[50] L. Hayden, "VNA error model conversion for N-port calibration comparison," in *Proc. 69th ARFTG Conf.*, Honolulu, HI, Jun. 2007, pp. 1–10.

[51] J. V. Butler, D. Rytting, M. F. Iskander, R. Pollard, and M. Vanden Bossche, "16-term error model and calibration procedure for on wafer network analysis measurements (MMICs)," in *Proc. IEEE MTT-S Int. Microw. Symp. Dig.*, Jul. 1991, pp. 1125–1127.

[52] V. Teppati and A. Ferrero, "On-wafer calibration algorithm for partially leaky multiport vector network analyzers," *IEEE Trans. Microw. Theory Tech.*, vol. 53, no. 11, pp. 3665–3671, Nov. 2005.

[53] A. Mostkwoski and M. Stark, "Complex numbers," in *Introduction to Higher Algebra*, A. Mostkowski and M. Stark, Eds. Oxford, UK: Pergamon, 1964, pp. 57–103.

[54] J. R. Andrews, B. A. Bell, and E. E. Baldwin, "Reference flat pulse generator," National Bureau of Standards, Boulder, CO. National Engineering Lab, Report Number NBS-TN-1067, Oct. 1983.

[55] A. Agoston and J. E. Carlson, "Fast transition flat pulse generator," U.S. Patent 4 758 736, Jul. 19, 1988.

[56] A. Agoston, J. B. Rettig, S. P. Kaveckis, J. E. Carlson, and A. E. Finkbeiner, "Dual channel time domain reflectometer," U.S. Patent 4 755 742, Jul. 05, 1988.

[57] P. J. Pupalaikis and K. Doshi, "TDR-based s-parameters," in *Modern RF and Microwave Measurement Techniques*, V. Teppati, A. Ferrero, and M. Sayed, Eds., Cambridge RF and Microwave Engineering. Cambridge, UK: Cambridge University Press, 2013, ch. 11, pp. 279–306.

[58] P. J. Pupalaikis, K. Doshi, and A. Sureka, "Time-domain reflectometry step to s-parameter conversion," U.S. Patent 8 706 433, Apr. 22, 2014.

[59] P. J. Pupalaikis, K. Doshi, and A. Sureka, "Time-domain reflectometry step to s-parameter conversion," U.S. Patent 10 396 907, Aug. 27, 2019.

[60] R. Miller, "Waveform translator for dc to 75 GHz oscillography," U.S. Patent 6 242 899, Jun. 05, 2001.

[61] S. Ems, S. Kreymerman, and P. J. Pupalaikis, "Time-domain reflectometry in a coherent interleaved sampling timebase," U.S. Patent 8 390 300, Mar. 5, 2013.

[62] W. L. Gans and N. S. Nahman, "Continuous and discrete Fourier transform of step-like waveforms," *IEEE Trans. Instrum. Meas.*, vol. IM-31, pp. 97–101, Jun. 1982.

[63] A. M. Nicolson, "Forming the fast Fourier transform of a step response in time-domain metrology," *Electron. Lett.*, vol. 9, pp. 317–318, Jul. 1973.

[64] P. Triverio, S. Grivet-Talocia, M. S. Nakhla, F. G. Canavero, and R. Achar, "Stability, causality, and passivity in electrical interconnect models," *IEEE Trans. Adv. Packag.*, vol. 30, no. 4, pp. 795–808, Nov. 2007.

[65] C. D. Meyer, *Matrix Analysis and Applied Linear Algebra.* Philadelphia, PA: Society for Industrial and Applied Mathematics, 2000.

[66] K. Doshi, A. Sureka, and P. J. Pupalaikis, "Fast and optimal algorithms for enforcing reciprocity, passivity and causality in s-parameters," in *Proc. DesignCon.* Santa Clara, CA: UBM, Feb. 2012.

[67] J. D. Irwin and R. M. Nelms, *Basic Engineering Circuit Analysis*, 9th edn. Hoboken, NJ: Wiley Publishing, 2008.

[68] I. Daubechies, *Ten Lectures on Wavelets.* Philadelphia, PA: Society for Industrial and Applied Mathematics, 1992. [Online]. Available: https://epubs.siam.org/doi/abs/10.1137/1.9781611970104

[69] A. Haar, "Zur theorie der orthogonalen funktionensysteme," *Math. Ann.*, no. 69, pp. 331–371, 1910.

[70] P. J. Pupalaikis, "Wavelet denoising for TDR dynamic range improvement," in *Proc. DesignCon.* Santa Clara, CA: IEC, Feb. 2011.

[71] P. J. Pupalaikis, A. Sureka, and K. Doshi, "Wavelet denoising for TDR dynamic range improvement," U.S. Patent 8 843 335, Sep. 23, 2014.

[72] K. Levenberg, "A method for the solution of certain non-linear problems in least squares," *Quart. Appl. Math.*, vol. 2, pp. 164–168, 1944.

[73] D. Marquardt, "An algorithm for least-squares estimation of nonlinear parameters," *SIAM J. Appl. Math.*, vol. 11, pp. 431–441, 1963.

[74] H. Johnson and M. Graham, *High-Speed Digital Design: A Handbook of Black Magic*, 1st edn., Prentice Hall Modern Semiconductor Design. Upper Saddle River, NJ: Prentice Hall, 1993, ch. 2, pp. 67–79.

[75] E. Hewitt and R. E. Hewitt, "The Gibbs–Wilbraham phenomenon: an episode in Fourier analysis," *Arch. Hist. Exact Sci.*, vol. 21, no. 2, pp. 129–160, Jun. 1979.

[76] P. J. Pupalaikis and E. Yudin, "Eye patterns in scopes," in *Proc. DesignCon.* Santa Clara, CA: IEC, Feb. 2005.

[77] P. Alfke, "Efficient shift registers, LFSR counters, and long pseudo-random sequence generators," Xilinx, Application Note. [Online]. Available: http://www.xilinx.com/support/documentation/application_notes/xapp052.pdf

[78] C. Meyer, *Matrix Analysis and Applied Linear Algebra.* Philadelphia, PA: Society for Industrial and Applied Mathematics, 2001.

[79] C. D. Meyer, *Matrix Analysis and Applied Linear Algebra.* Philadelphia, PA: Society for Industrial and Applied Mathematics, 2000.

# Index